Lecture Notes in Electrical E

Volume 500

** **Indexing: The books of this series are submitted to ISI Proceedings, EI-Compendex, SCOPUS, MetaPress, Springerlink** **

Lecture Notes in Electrical Engineering (LNEE) is a book series which reports the latest research and developments in Electrical Engineering, namely:

- Communication, Networks, and Information Theory
- Computer Engineering
- Signal, Image, Speech and Information Processing
- Circuits and Systems
- Bioengineering
- Engineering

The audience for the books in LNEE consists of advanced level students, researchers, and industry professionals working at the forefront of their fields. Much like Springer's other Lecture Notes series, LNEE will be distributed through Springer's print and electronic publishing channels.

For general information about this series, comments or suggestions, please use the contact address under "service for this series".

To submit a proposal or request further information, please contact the appropriate Springer Publishing Editors:

Asia:

China, *Jessie Guo, Assistant Editor* (jessie.guo@springer.com) (Engineering)

India, *Swati Meherishi, Senior Editor* (swati.meherishi@springer.com) (Engineering)

Japan, *Takeyuki Yonezawa, Editorial Director* (takeyuki.yonezawa@springer.com) (Physical Sciences & Engineering)

South Korea, *Smith (Ahram) Chae, Associate Editor* (smith.chae@springer.com) (Physical Sciences & Engineering)

Southeast Asia, *Ramesh Premnath, Editor* (ramesh.premnath@springer.com) (Electrical Engineering)

South Asia, *Aninda Bose, Editor* (aninda.bose@springer.com) (Electrical Engineering)

Europe:

Leontina Di Cecco, Editor (Leontina.dicecco@springer.com)
(Applied Sciences and Engineering; Bio-Inspired Robotics, Medical Robotics, Bioengineering; Computational Methods & Models in Science, Medicine and Technology; Soft Computing; Philosophy of Modern Science and Technologies; Mechanical Engineering; Ocean and Naval Engineering; Water Management & Technology)

Christoph Baumann (christoph.baumann@springer.com)
(Heat and Mass Transfer, Signal Processing and Telecommunications, and Solid and Fluid Mechanics, and Engineering Materials)

North America:

Michael Luby, Editor (michael.luby@springer.com) (Mechanics; Materials)

More information about this series at http://www.springer.com/series/7818

Amit Kumar · Stefan Mozar
Editors

ICCCE 2018

Proceedings of the International Conference on Communications and Cyber Physical Engineering 2018

Editors
Amit Kumar
BioAxis DNA Research Centre Pvt Ltd.
Hyderabad, Andhra Pradesh
India

Stefan Mozar
Dynexsys
Sydney, NSW
Australia

ISSN 1876-1100 ISSN 1876-1119 (electronic)
Lecture Notes in Electrical Engineering
ISBN 978-981-13-4361-2 ISBN 978-981-13-0212-1 (eBook)
https://doi.org/10.1007/978-981-13-0212-1

This Springer imprint is published by the registered company Springer Nature Singapore Pte Ltd.
The registered company address is: 152 Beach Road, #21-01/04 Gateway East, Singapore 189721, Singapore

ICCCPE-2018 Conference Committee

Chief Patron

Shri. Ch. Malla Reddy, MP

Patron

Shri. Ch. Gopal Reddy

Co-patron

Shri Ch. Srisailam Reddy
Shri Ch. Bhopal Reddy

Conference Chair

Dr. Amit Kumar, IEEE Strategic Development and Environmental Assessment Committee

Technical Program Chair

Prof. Keeley Crockett, Manchester Metropolitan University, UK, IEEE CIS Women in Engineering

Program Chair

Prof. Atul Negi, Central University of Hyderabad, India, IEEE CIS

Publication Chair

Dr. Shaik Fahimuddin, AITS Rajampet
Smita Sinha, Weather Channel, Atlanta, USA

Finance Chair

Dr. Siddapuram Arvind, CMR Institute of Technology Hyderabad
Anukul Mishra, BDRC Hyderabad

IOT Track

Dr. S. K. Sinha, Indian Institute of Science (IISC), Bangalore

Convenor

Dr. Vinit Kumar Gunjan, CMR Institute of Technology Hyderabad

Organizing Co-chair

Dr. M. Janga Reddy, Principal, CMRIT

Co-convener

Dr. D. Baswaraj

Publication Co-chair

Dr. Vijender Kr. Solanki, CMRIT

Technical Program Committee

Dr. Stefan G. Mozar, CEO Dynexsys Sydney, Australia, IEEE Fellow
Prof. Jan Hasse, Helmut Schmidt University of the Federal Armed Forces Hamburg, Germany
Prof. Ponnuthurai Nagaratnam Suganthan, Nanyang Technological University, Singapore, IEEE CIS
Parijat Parimal, Townsville, Queensland, Australia
Prof. Yaji Sripada, University of Aberdeen, UK
Dr. Mehdi Khamassi, Computational Neuroscience and Neurorobotics Paris, France
Dr. Vitoantonio Bevilacqua, Polytechnic of Bari, Italy
Dr. Maode Ma, Nanyang Technological University, Singapore
Dr. Samuel Gan, A*STAR, Singapore
Mrs. Madhumita Chakravarti, RCI, DRDO, India
Shishir Rajan, JD Power Asia Pacific, Singapore
Amitava Ghosh, Gruppo Industriale Tosoni (Transport Division), Verona, Italy
Dr. Preeti Bajaj, GHRCE, IEEE India Council
Dr. Allam Appa Rao, Former VC, JNTU Kakinada
Dr. Syed Musthak Ahmed, SREC, Warangal
Varun Kumar, Teledata Security System and Building Automation, Milan, Italy
Dr. Nataraj, SJBIT, Bangalore
Dr. Naresh Babu M., SVEC, Tirupati

Prof. Eduard Babulak, Fort Hays State University, USA
Dr. Satheesh P., MVGR, Vizianagaram
Dr. Saragur Srinidhi, ACCS Bangalore, Cleveland State University, USA
Dr. Vinit Kumar Gunjan, IEEE Computer Society
Dr. K. V. Mahendra Prasanth, SJBIT, Bangalore
Dr. Krishna, SJBIT, Bangalore
Jatin Jogi, Cognizant, Atlanta, USA
Dr. Kallam Suresh, Associate Professor, Annamacharya Institute of Technology and Sciences, Rajampet, Andhra Pradesh
Maruthi Pathapati, Vidcentum, IEEE Hyderabad Section
Dr. M. Rajasekhara Babu, SCOPE, VIT University, Vellore
Dr. Thinagaran Perumal, Universiti Putra Malaysia, Malaysia
Dr. Arige Subramanyam, Professor in CSE and Dean Administration, AITS Rajampet

College Committee

Registration Committee
Dr. Naga Rama Devi
Mrs. Y. Sucharitha
Mr. N. Vijay Kumar
Mrs. T. N. Chitti
Mrs. S. Sridevi
Mr. G. Praveen Kumar

Reception Committee
Mrs. B. Sunitha Devi
Dr. K. Kenanya Kumar
Ms. S. Sarika
Mrs. Haripriyapatra
Mrs. B. Vasavi
Mr. C. H. Mahender Redddy

Transport and Accommodation Committee
Mr. A. Balaram
Mr. K. Morarjee
Mr. K. Kishore Kumar
Mr. G. Krishna Lava Kumar
Mr. S. Asif
Mr. U. Veeresh
Mr. R. Akhilesh Reddy
Mrs. Y. Prathima

Web site
Mr. T. Nagaraju

Photography and Media
Mr. P. Pavan Kumar
Mr. M. D. Ahmed Ali
Mr. K. Srinivasa Rao

Food Committee
Mr. V. Ramulu
Mr. H. V. Ramana Rao
Mrs. P. Rupa
Mrs V. Alphonsa
Mrs. B. Shirisha

Session Conduction Committee
Dr. D. Baswaraj
Dr. Vijender Kumar Solanki
Dr. Puja. S. Prasad

Video Conferencing Committee
Mr. Sri Rama Lakshmi Reddy
Mrs. G. Swetha
Mr. Rony Preetham
Mr. M. Mahipal Reddy
Mrs. B. Annapoorna

Overall Conference Committee Coordinator
Dr. M. Janga Reddy
Dr. Vinit Kumar Gunjan

Preface

The 2018 International Conference on Communications and Cyber Physical Engineering brought together scientists, researchers, educationalists, and developers to share new ideas and establish cooperation and provided a platform for the presentation of their original work and a means to exchange ideas, information, techniques, and applications in cutting-edge areas of the engineering sciences. This volume of proceedings from the conference provides an opportunity for readers to engage with a selection of refereed papers that were presented during the conference.

Participants were encouraged to present papers on all topics listed in the call for papers, which enable interdisciplinary discussions of new ideas and the latest research and development. Also, prominent foreign and local experts delivered keynote speeches and plenary lectures at the conference.

A conference would not be possible without the contribution of many people. First and foremost, we would like to thank the authors for contributing and presenting their latest work at the conference. Without their contribution, this conference would not have been possible. The editors would like to thank members of the Advisory Committee, Technical Program Committee, and Organizing Committee for their support and their suggestions during the preparation and organization of this conference.

The editors would also like to thank the eminent professors and academicians for accepting to give invited talks and chair the sessions, and for sparing valuable time in spite of their busy schedules.

Hyderabad, India — Amit Kumar
Sydney, Australia — Stefan Mozar

Contents

Moving Object Recognition and Detection Using Background Subtraction

Loveleen Kaur and Usha Mittal

Abstract Motion detection and object recognition algorithms are a significant research area in computer vision and involve building blocks of numerous high-level methods in video scrutiny. In this paper, a methodology to identify a moving object with the use of a motion-based segmentation algorithm, i.e. background subtraction, is explained. First, take a video as an input and to extract the foreground from the background apply a Gaussian mixture model. Then apply morphological operations to enhance the quality of the video because during capture the quality of a video is degraded due to environmental conditions and other factors. Along with this, a Kalman filter is used to detect and recognize the object. Finally, vehicle counting is complete. This method produces a better result for object recognition and detection.

Keywords Motion segmentation · Background subtraction · GMM
Morphological operations · Kalman filter

1 Introduction

Moving object recognition and detection is the initial phase in the analysis of video [1]. This is a fundamental technique for surveillance applications, for direction of independent vehicles, for effective video compression, for smart tracking of moving objects, remote sensing, image processing, robotics, and medical imaging. The tracking of moving object is the primary step in the recognition of objects. The main aim of detection is to extract moving objects that are of interest from video successions with a background, which can be static or dynamic [2]. Recognizing the moving objects in respect to the entire image is its major function. In this paper,

L. Kaur (✉) · U. Mittal
Lovely Professional University, Phagwara, Punjab, India
e-mail: loveleenkaur1993@gmail.com

U. Mittal
e-mail: usha.20339@lpu.co.in

A. Kumar and S. Mozar (eds.), *ICCCE 2018*,
Lecture Notes in Electrical Engineering 500,
https://doi.org/10.1007/978-981-13-0212-1_1

there are different techniques available for moving object detection from video sequences, i.e. background subtraction, frame differencing, and optical flow. The rest of the paper is ordered as follows: Section 2 defines the techniques of motion segmentation and their comparison. Section 3 defines the methodology that is used to detect the moving object. Section 4 defines the results of existing techniques and the proposed technique, which shows that the existing algorithm treats two objects as single objects that are near to each other. However, the proposed method works in an efficient manner. It predicts two objects separately according to the distance between them.

2 Motion Segmentation

In this section, motion segmentation is the method of segmenting digital image into numerous parts and arrangements of pixels usually called super-pels. The objective of this method is to segment the video/image into different parts and the modification of an image into approximate extent, i.e. more imperative and not easy to separate. Motion segmentation breaks the video into different segments. This is moving towards picking a form for every pixel in an image to such a degree that pixels with a comparative name share certain behaviors. The following are techniques of motion segmentation.

2.1 Frame Differencing

Frame differencing technique depends on frame distinction that attempts to recognize movement regions by making use of the different successive frames in a video. This technique is very flexible in a static environment. In this manner, it is great at giving initial coarse movement regions. From the outcomes, it is a simple strategy for distinguishing moving objects in a static environment with the use of some threshold value. This threshold value restrains the noise. However, if the background is not static, this technique is very sensitive to any movement and it is hard to separate true and false movement. As a result, this method can be used to recognize the possible object-moving zone in order for the optical flow calculation to distinguish real object movement [3].

The frame differencing strategy uses a couple of frames based on a series of time images to deduct and acquire frames of different images. This method is the same as background subtraction and after image subtraction it gives moving target data through some threshold value. This procedure is easy to execute and computationally complex. However, it fails to detect whole pixels of moving objects [3].

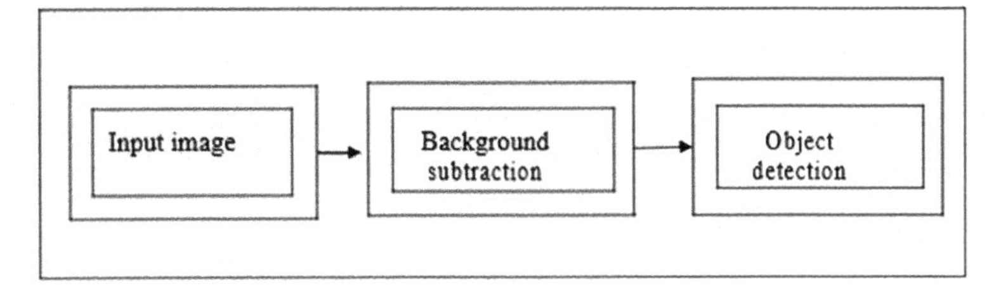

Fig. 1 Background subtraction model

2.2 *Background Subtraction*

This technique is designed to segment motion from still images. An algorithm will recognize movement zones by withdrawing the present picture pixel-by-pixel from the moving background picture that is made by averaging pictures after an initialization period. This methodology is fundamental and easy to recognize. It exactly removes the characteristics of objective data, notwithstanding the fact that it is sensitive to the modification of the outside atmosphere, so it is important that the background is recognized. The moving object is depicted utilizing a background subtraction method [4].

For the identification of movement, two images ideally of a similar size are taken from video. One image is introduced as the background image, in which the moving item is not present and the second image is the present image. Every image has two parts, one is the foreground and the other is the background demonstration. The foreground model is the prototypical model in which the movement of an object is available, while the background model defines the movement of item that is not present. Initialization of the image is a procedure that introduces the background image. In the video, the number of frames with respect to time, one of these frames is introduced as the background image that contains background information. Consequently, the initialization of the background is a fundamental preprocessing procedure for the location of movement [4] (Fig. 1).

2.3 *Optical Flow*

The optical flow method uses the field of optical flow for moving objects over characteristics of flow distribution to see moving regions in an image. This method also uses a clustering process based on flow distribution characteristics [5]. The use of this technique provides complete information about the movement of object. This system can perceive motion in video progressions even from a moving camera and moving background.

However, this method is difficult to implement, sensitive to noise, and a large number of calculations are required. It also uses vector characteristics to target the motion that changed according to time information in order to track the motion region from the image or video. The optical flow [6] procedures are computationally complex and cannot be used without supporting hardware.

3 Proposed Methodology

This part discusses the strategy and how it identifies the movement of objects in video and CCTV footage etc. There are different methodologies for the identification of the movement of objects, but here a background subtraction strategy is used. The following are the main steps to be performed (Fig. 2).

3.1 Gaussian Mixture Model (GMM) for Background Subtraction

This method is a unique extension of a Gaussian probability function. The GMM [7] make any background uneven of the distribution of density, so is consistently used in areas of image processing that give good results. Use of the GMM includes

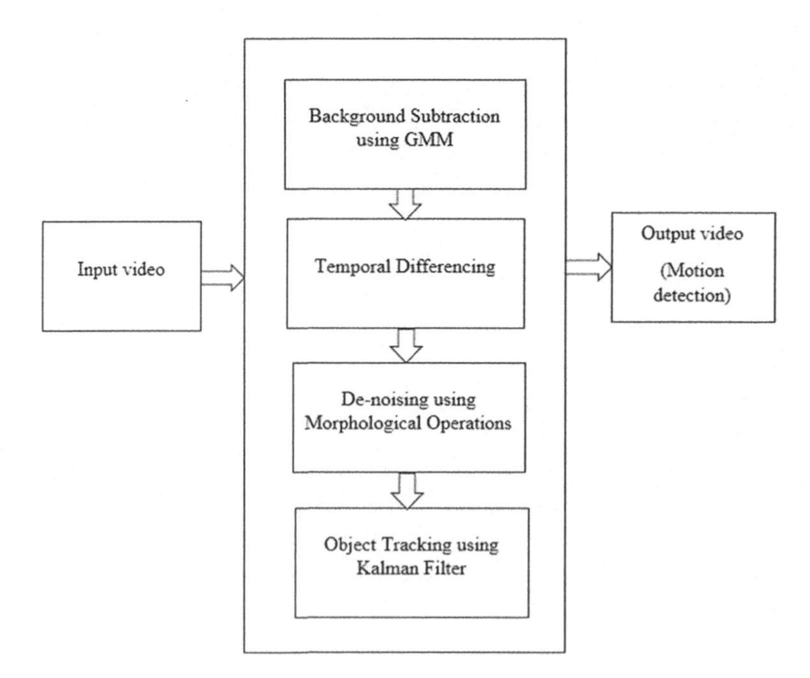

Fig. 2 Steps of the proposed model

the mixture of Gaussian probability density functions. Each Gaussian probability density function has its own specific mean, standard deviation, and the weights can be interpreted by the use of frequency, before the value of the maximum weight is found. The high frequency of occurrence, and then find the weight on the Gaussian probability density function [7]. For this technique, the possibility of a color at a given pixel is given by:

$$P(I_{s,t}) = \sum_{i=1}^{k} \omega_{i,s,t} N\left(\mu_{i,s,t}, \Sigma_{i,s,t}\right) \tag{1}$$

where N is the *i*th Gaussian model. Note that for computational purposes, as suggested by Stauffer and Grimson, the covariance matrix $\Sigma_{i,s,t}$ can be assumed to be diagonal, $\Sigma = \sigma^2$ Id. In their method, parameters of the matched component (i.e. the nearest Gaussian for which Is, t is within 2.5 standard deviations of its mean) are updated as follows:

$$\omega_{i,s,t} = (1 - \alpha)\omega_{i,s,t-1} + \alpha \tag{2}$$

$$\mu_{i,s,t} = (1 - \rho) \cdot \mu_{i,s,t-1} + \rho \cdot I_{s.t} \tag{3}$$

$$\sigma^2_{i,s,t} = (1 - \rho) \cdot \sigma^2_{i,s,t-1} + \rho \cdot d_2\left(I_{s,t}, \mu_{i,s,t}\right) \tag{4}$$

where α is a user-defined learning rate, ρ is a second learning rate defined as $\rho = \alpha \cdot N(\mu_{(i,s,t)}, \Sigma_{(i,s,t)})$ and d_2 is the distance well-defined in the above equation. Parameters μ and σ of matchless distributions remain the same while their weight is reduced [7].

3.2 *Temporal Differencing (Frame Differencing)*

The technique of frame differencing depends on frame distinction that attempts to recognize movement regions by making use of the different successive frames in a video. This technique is very flexible in a static environment. In this manner, it is very good at identifying initial coarse movement regions. From the outcomes, it is seen to be a simple strategy for distinguishing moving objects in a static environment with the use of some threshold value. This threshold value restrains the noise. However, if the background is not static, this technique is very sensitive to any movement and it is hard to separate true and false movement. As a result, this method can be used to recognize the possible object-moving zone in order for the optical flow calculation to distinguish real object movement [3].

The frame differencing strategy uses a couple of frames based on series of time image to deduct and acquires frames from different images. This method is the same as background subtraction and after image subtraction it provides moving target data through some threshold value. This procedure is easy to execute and is computationally complex. However, it fails to detect whole pixels of moving objects [3].

3.3 Morphological Operations

These operations are used to remove noise. Ordinarily utilized operations are erosion, closing, dilation, opening, diminishing, thickening, and skeletalization and so on [8]. The following is a description of the operations:

Dilation: Dilation is used on binary images; however, some forms work on gray scale [8]. The essential impact of this operator on binary images is to enhance the borders of regions of foreground pixels bit by bit, i.e. commonly white pixels. In this way, holes inside those regions become smaller, while areas of foreground pixels grow in size.

Erosion: Erosion is an operator that is also used on binary images. However, some adaptations work on gray scale [8]. The fundamental impact of the operator is to dissolve or erode the borders of regions of pixels in the foreground, i.e. white pixels. In this way, region of foreground pixels shrink in size while gaps inside those areas become larger.

Opening: This is a combinational operation of erosion and dilation. The union set operation is also used to find the points of the opened image [9]. This operation makes smooth the rough draft of an object, clears the narrow bridges, and eliminates minor extensions present in the object.

Closing: This is also a combinational operation of erosion and dilation [9]. It is different from the opening operation in the sense of order of occurrence of erosion and dilation operations. This operation smoothes the areas of shapes, and when all is said and done, mixes narrow breaks and thin holes. Therefore, it dispenses with little openings and fills holes in the objects limits.

Opening by reconstruction: The opening by reconstruction operator restores the original shapes of the objects that remain after erosion. When the erosion operator is applied, it eliminates small objects and subsequently the dilation operator re-establishes the objects' shapes that remain in the morphological opening [9]. On the other hand, this restoration's accuracy depends on the similarity between the shapes and the structuring element. This operator performs the recreation of an image from another image, i.e. a marker image. It applies a series of restrictive dilations on the second input image called the marker image utilizing the first input image, i.e. the original image as a conditional image. If the conditional dilation is essential, so the third input, i.e. the structuring element, is important.

Closing by reconstruction: This is an important morphological operation. It performs image reconstruction from another image, i.e. the marker image [9]. This operator applies a series of conditional dilations on the second input image, i.e. the marker image, utilizing the original image as the conditional image, i.e. the first input. If the conditional dilation operator is vital, a third input image, a structuring

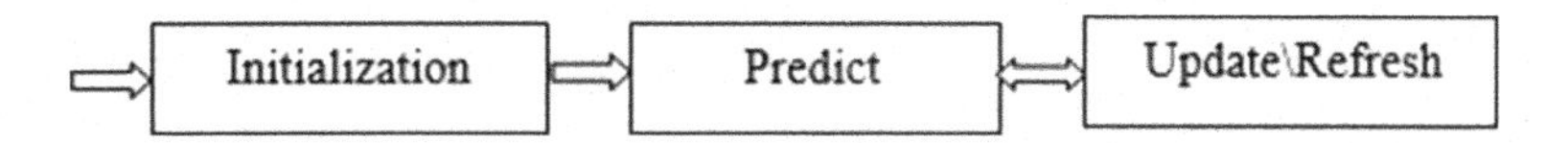

Fig. 3 The operation of a Kalman filter

component is compulsory. It is exceptionally valuable for removing features with the use of markers on binary images, and to remove dark features smaller than the structuring elements and produces a similar image.

3.4 Kalman Filter for Object Tracking

This is generally called Linear Quadratic Estimation (LQE). A Kalman filter is a technique that uses a measurement series with respect to time. It contains noise and errors and produces an approximation of unknown variability that is likely to be more correct than those according to a single measurement alone. Specifically, the Kalman filter [10] works repetitively on streams of noisy input data to make a quantifiable estimation. Fundamental applications are guiding vehicles, control of navigation systems, particularly in aircraft and transport. This technique works in a two-stage process. One part is predicting, and the other is refreshing (Fig. 3).

This predicted state is also called a priori estimation. Although it is an estimation method of the state at the present time step, it excludes perception data from the present time step. In the refresh stage, the current a priori prediction is joined with current estimations data to refine the state. This enhanced estimate is termed the posterior state estimate. The Kalman filter is an arrangement of scientific conditions that gives a productive computational recursive intended to assess the condition of a procedure in a few perspectives [10].

4 Research Outcomes

The simulation results are presented as follows and show object detection and tracking of moving objects under different Matlab RGB frames. The performance is good and the results are shown below. "Figure 4" is the input frame, "Fig. 5" is the output achieved after GMM [11] is implemented. "Figure 6" is the output achieved after applying morphological operations and "Fig. 7" gives the output after the Kalman filter [10] is applied and object get tracked from kalman filter from Kalman filter. Finally, "Fig. 8" shows the predicted object after applying different operations. The following screenshot shows the comparison between the existing technique and the proposed method. The existing technique shows the difference in result that predicts two objects as a one object that is near to other object.

Video 1

Existing Technique

Fig. 4 Input frame 101

Fig. 5 Output after GMM (noisy frame)

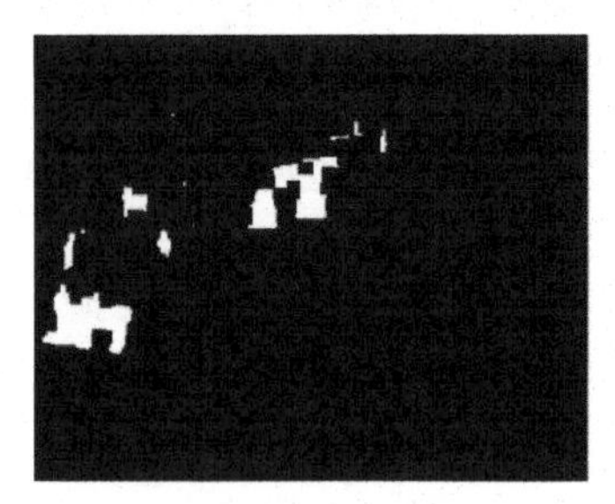

Fig. 6 Output after morphological operations

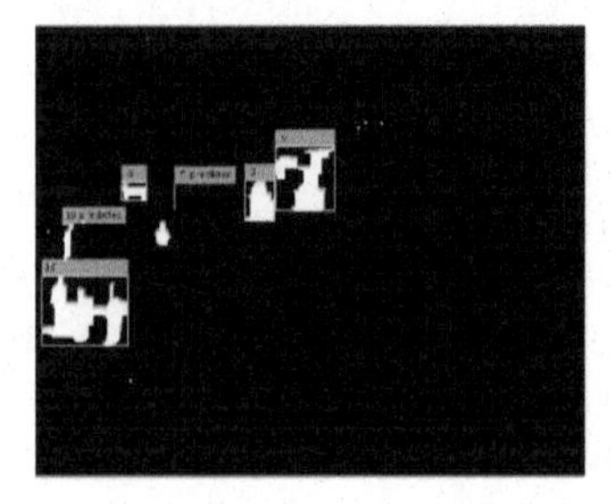

Fig. 7 Output after tracking using the Kalman filter

Fig. 8 Output showing the movement of an object in RGB

Table 1 Comparison of Motion Segmentation Techniques

Methods	Accuracy	Complexity	Description
Background subtraction	Moderate	Moderate	• Cannot manage multimodal background • Require low memory
Optical flow	Moderate	High	• Large computations required • Provides complete information regarding object movement
Frame differencing	High	Low to moderate	• Best for static background detection
Proposed method	High	Moderate	• Predicts each object separately from the group

Comparison of Motion Segmentation Techniques

See (Table 1).

5 Conclusion

Moving object tracking is assessed for different surveillance and vision investigations. Segmentation techniques additionally having the capacity and additionally having the capacity to separate extra data, for example, transient differencing that permits perceiving the moving objects; background subtraction takes into consideration enhanced object identification and along the following lines. The extraction of two frames of input shows the effects of movement of object in detection part. This will provide good data on the movement of the object. This background subtraction technique will be utilized to recognize both background and foreground. This calculation is quick and uncomplicated, ready to recognize moving object and this is accurate method. This technique is extremely good to detect objects from video observation applications. The primary stage is the division of the object utilizing a Gaussian mixture model that gives better comprehension of grouped objects. At that point, the second stage is employed to enhance the quality of video by applying morphological operations to the yield after GMM for a better outcome. Denoising is utilized to improve the nature of video. In the following stage vehicle detection and tracking is done using a Kalman filter algorithm, and produces excellent outcomes. These algorithms can also be extended to evening movement in the future with good results.

Acknowledgements I would like to express my supreme appreciation to Ms. Usha Mittal for her unceasing help with the paper.

References

1. Gonzalez R, Woods R (2008) Digital image processing. Prentice Hall, US
2. Zhou Y, Zhang J (2010) A video semantic object extraction method based on motion feature and visual attention. In: 2010 IEEE international conference on intelligent computing and intelligent systems, pp 1–5
3. Shaikh S, Saeed K, Chaki N (2014) Moving object detection approaches, challenges and object tracking. In: Moving object detection using background subtraction. SpringerBriefs in Computer Science. Springer, Cham. https://doi.org/10.1007/978-3-319-07386-6_2
4. Dinesh P (2014) Moving object detection using background subtraction. In: Special issue on IEEE sponsored international conference on intelligent systems and control (ISCO'15), pp 5–15
5. Hirai J, Yamaguchi T, Harada H (2009) Extraction of moving object based on fast optical flow estimation. In: ICROS-SICE international joint conference 2009, pp 2691–2695
6. Parikh MC, Maradia KG (2015) Moving object segmentation in a video sequence using optical flow and motion histogram technique. Int J Comput Appl 116(16):1–7
7. Trinayani K, Sirisha B (2015) Moving vehicle detection and tracking using GMM and Kalman filter on highway traffic. Int J Eng Technol Manag Appl Sci 3(5):309–315
8. Mittal U, Anand S (2013) Effect of morphological filters on medical image segmentation using improved watershed segmentation. Int J Comput Sci Eng Technol 4(6):631–638
9. Srisha R, Khan AM (2013) Morphological operations for image processing: understanding and its applications. In: NCVSComs-13, pp 17–19
10. Chauhan AK, Krishan P (2013) Moving object tracking using gaussian mixture model and optical flow. Int J Adv Res Comput Sci Softw Eng 3(4):243–246
11. Zhao T, Nevatia R (2004) Tracking multiple humans in complex situations. IEEE Trans Pattern Anal Mach Intell 26(9):1208–1221
12. Zhou DZD, Zhang HZH (2005) Modified GMM background modeling and optical flow for detection of moving objects. In: 2005 IEEE international conference on systems, man and cybernetics, vol 3, pp 2224–2229. https://doi.org/10.1109/ICSMC.2005.1571479

Fog Computing: Overview, Architecture, Security Issues and Applications

Kishore Dasari and Mounika Rayaprolu

Abstract There is a famous saying that goes "Necessity is the mother of invention". In today's globalized world people are getting stuck with many problems like data management, time management, and security and privacy concerns etc. There are traditional methods like cloud computing, cloudlet, and mobile management techniques to sort out the processing, storing, and executing of the data. But with the passage of time, the world is exploring new areas and these traditional methods are on the wane in terms of data handling. In this paper we discuss the technology that helps in data management, time management and security issues. We also addresses some real time scenarios.

Keywords Cloud computing · IoT · Fog computing · Edge computing

1 Introduction

Computing is a term that can be defined as a unique process which utilizes computer technology for computing a particular task or goal. Cloud computing, social computing, grid computing, and parallel computing all come under the umbrella of computing. It is accurate to say that computing computes the data.

In the late 1990s when the internet was not widely used, technocrats used a technology called cloud computing to compute data. Cloud computing is a technology whereby remote servers are hosted on the internet to store, manage and process data.

In the early twentieth century the internet saw an immense growth and has totally changed the world. Internet use began to grow rapidly. With the huge usage of the internet, the concept of the "Internet of Things" (IoT) came into existence.

K. Dasari (✉)
Faculty of Computer Science Engineering, PPDCET, Vijayawada, India
e-mail: Kishoredasari99@gmail.com

M. Rayaprolu
PPDCET, Vijayawada, India

A. Kumar and S. Mozar (eds.), *ICCCE 2018*,
Lecture Notes in Electrical Engineering 500,
https://doi.org/10.1007/978-981-13-0212-1_2

The IoT is generating a huge amount of data and it is apt to say that a huge volume of varied high-velocity data is being generated by the IOT. So, analyzing and processing these three "v"s (volume, variety, and velocity) of data using cloud computing is creating latency. To minimize these drawbacks, technocrats have come forward with a technology called fog computing.

2 Overview of Fog Computing

In this digital and technological world, the birth of any technology happens by considering three "w"s (what, why, and when). In this section we focus on what fog computing is, why use fog computing, and when to consider using fog computing.

2.1 What Is Fog Computing?

Fog computing, also called edge computing, is an extension of cloud computing. Fog computing acts as a bridge between cloud computing and the IoT.

Fog computing is a term coined by Cisco that analyzes the IoT data at the network edge where it is generated [1]. A large amount of heterogeneous data is analyzed, stored and processed at the current position. Fog computing sends only the historical data to the cloud for storage and analysis rather than sending a stream of data.

2.2 Why Fog Computing?

Handling the high volume, variety and velocity of IoT data exposes a few cons. They can be:

Unreliable network: Sharing of devices is the main policy in cloud computing. In that process the network plays a key role. A network can be defined as a group of interconnecting devices. So in such scenarios ICT devices like laptops and mobiles are not connected to a specific device. Instead they are interconnected among each other which results in congestion, causing unreliability in the network.

Latency: To act on data in cloud computing, the data collected from devices is sent to the cloud server where the data is analyzed and processed, and is then sent back to the device. This creates a time delay.

Lack of mobility support: The main motto of the IoT is to interconnect all the devices through the internet and provide a platform for communication. However, cloud computing is not competent enough to support the dynamic processing of data. So it is clear that cloud computing is not suitable for mobile processing.

Location awareness: Location awareness refers to a process of identifying the location using devices. As cloud computing is static, it does not support the collection of data from a diverse geographical range.

Security issues: Security is one of the major issues in today's world. There are many security parameters like authentication and encryption etc. In cloud computing, in order to process data it should be transferred from the devices to the cloud server. In this process the data may be lost in the middle or it may be stolen by unauthorized users.

So, to overcome all these drawbacks a novel model called fog computing came into being.

2.3 When to Consider Fog Computing?

When data has to be collected at the extreme edge: In the IoT, data is collected from end devices. Consider a scenario of a smart home. A smart home consists of sensors such as temperature, humidity, etc. When the number of members increases at home, the temperature sensor collects the data and makes changes accordingly.

Millions of heterogeneous data across large geographic area gets generated: Heterogeneous data can be also defined as distinct data. As fog computing is associated with the IoT we get data from several different resources. For instance, consider a smart city. To perpetuate a smart city we need traffic management, weather forecasting, water management, electricity management, hospitality, transportation, food facility, health care, etc. Collecting and processing such diverse data in cloud computing decreases efficiency and increases latency.

3 Characteristics of Fog Computing

The characteristics of fog computing are displayed below:

Low latency: Fog computing collects the data from the device and acts on the data where it is generated. This reduces the time gap.

Widespread geographical distribution: The fog nodes are capable of collecting data from a diverse geographical range.

Mobility: Fog computing furnishes the distributed infrastructure. With this, all the fog devices spread at the network edge. This creates mobility in data collecting and data processing.

Predominant role of wireless access: As fog computing is designed to act on IoT data, it is sufficient to say that the accessing of information happens through the internet thus supporting wireless access [2].

Strong presence of streaming and real time applications: Fog computing is now used widely in virtual and augmented reality, and artificial intelligence in order to handle the vast amount of data.

Heterogeneity: Fog computing supports the IoT, so it communicates with different kinds of devices. It is not confined to homogeneous data as it doesn't stick to a single device.

4 Working of Fog Computing

Fog computing extends the cloud to be nearer to things that evolve and act on the IoT data with the help of a device called a fog node. Fog nodes can be industrial controllers, switches, routers, and embedded servers.

Developer's port IoT applications for fog nodes at the network edge ingest data from IoT devices. Then the fog IoT application directs heterogeneous data to the appropriate place for analysis.

This can be briefly explained as: The most time sensitive data is analyzed on the fog node nearer to whatever is producing the data. Data that can wait seconds or minutes for action is moved along to an aggregation node for analysis and action. Less time-sensitive data is sent to the cloud for historical and big data analytics.

Fog nodes:

Receives information from IoT devices using any protocol (wireless protocol or wired protocol) [3]

Runs IoT embedded applications for real time control and analysis with millisecond response time

Provide transient storage and send periodic data to the cloud.

Cloud platform:

Receives and interfaces data summaries from many fog nodes

Can send new applications to fog nodes.

5 Architecture of Fog Computing

The architecture of fog computing, which is also known as open fog architecture, is a three-layer architecture (Fig. 1).

The first layer comprises embedded sensors that act as fog nodes and collect the data from end devices.

The second layer possesses multiserver edge nodes and performs fog data services such as data reduction, control response, and data visualization.

The third layer is the core layer that sends historical data to the cloud.

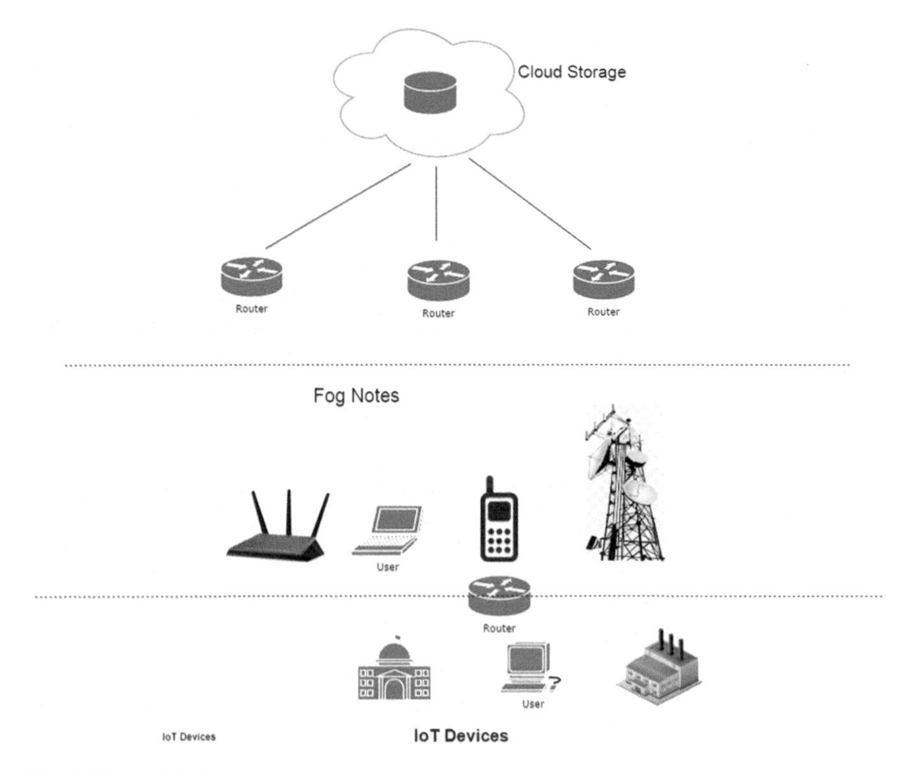

Fig. 1 Fog architecture

6 Security Issues and Mechanisms

In today's global world security is the major concern for any individual. In this section we discuss some of the security concerns, their case studies and mechanisms for respective issues in fog computing [4].

The major security issues can be divided into:

6.1 *Data Issues*

Data is a precious piece of fact. Data threats in the fog affect the data and cause insecurity for the information present in the server.

Data breach: A data breach is an issue where the confidential information of organization is appropriate unauthorized users. From statistics it has been shown that from 2005 to June 2015, the number of data threats was 6,284.

Mechanism: In order to prevent this data breach in the fog layer we use a decoy technique. In this decoy technique we use decoy data, i.e. replicated data to confuse the attackers. This decoy technique is similar to a honeypot.

Data loss: Data loss occurs by data deletion, data corruption, or a fault in the data storage. Statistics have shown that in 2013, around 44% of data servers have been attacked by brute force method thus leading to data loss.

Mechanism: To avoid this problem we use a data recovery technique. In this we use a server called a data backup server. The original information is stored at a main server and from there streams of information are stored at a data backup server.

6.2 Network Issues

Network is the key factor in fog computing. Providing security to the network is the essential goal. Basic network issues can be:

Account hijacking: This is the process where the attacker tries to hack the account in order to steal the identity of the user.

Mechanism: A combination of decoy and data recovery techniques gives rise to the solution to account hijacking.

Denial of service (DOS): A DOS is a process in which the communication of source and destination is prevented. A man in the middle attack (MITM) is an example of a DOS.

Mechanism: To prevent a MITM attack we use an encryption technique. In this we provide a strong encryption between the client and the server. In this, the server authenticates the client's request by processing a digital certificate and then a connection is established.

7 Applications

Web optimization: Researchers from Cisco are employing fog computing to enhance the performance of websites. Traditionally, for every HTTP request, the web page makes a round for content, style sheets, redirections, scripts, and images. Fog nodes can help in retrieving, integrating, and executing them simultaneously. This minimizes latency.

Smart meters: With the expansion of the smart grid, a vast amount of data is collected, processed, and transmitted from smart meters with the help of a Data Integration Unit. The data integration process takes long time because of the low bandwidth capacity of hardware. This can be prevented using fog computing. Initially, a fog-based router is attached with smart meters that assemble the data reading of all sub-meters with a predefined time. All values are then moved to a second fog platform for data reduction [5].

Intelligent food traceability: Food is one of the basic needs for any living being. As a human it is our responsibility to preserve the quality of food. Fog computing uses a solution called food traceability management. The quality of a food item is predicted by a cyber physical stream (CPS) that makes decisions. This quality information is sent to the fog network, where the entire supply chain is traceable.

Smart agriculture: The IoT is strong enough to render information such as crop yields, rainfall, pest infestation, and soil using sensors. Fog computing collects this data from sensors using fog nodes and the data is analyzed periodically.

Augmented brain computer interaction: Communication is the main theme of the IoT. Fog computing is a technology that supports the IoT. So fog computing is creating a platform to communicate on. Considering this objective, fog computing has introduced a real time brain detection system using multi-tier fog architecture.

8 Future Work

Mobile phones have become an essential part of every human life. It is not an exaggeration to say human survival is almost not possible or is at least a bit tough without a mobile. And today's world is racing to introduce novel approaches in mobile communication. Currently, we are using the fourth generation of mobiles (4G). However, the intensive use of mobiles has bought a massive growth in the consumption of mobile data. This is paving way for the fifth generation of mobiles (5G). Fog computing is helping in delivering this 5G approach with better service quality.

9 Conclusions

Time and tide waits for no man. With the passing of time everything in the world changes. And this applies to the internet. A rapid growth in the internet has occurred and this has led to the Internet of Things. To manage this IoT a technology called fog computing has been developed.

In this paper we have gone through the evolution of the IoT, the drawbacks of cloud computing, an overview of fog computing, the architecture and working of fog computing, fog computing security issues and their mechanisms, and fog computing applications. By realizing the nature and functions the fog computing is the best technology to cope with the IoT data and to thus deliver a quality service to customers.

References

1. Stojmenovic I et al (2016) An overview of fog computing and its security issues. Concurr Comput Pract Exp 28(10):2991–3005
2. Gia TN et al (2015) Fog computing in healthcare internet of things: a case study on ECG feature extraction. In: 2015 IEEE international conference on computer and information technology; ubiquitous computing and communications; dependable, autonomic and secure computing; pervasive intelligence and computing (CIT/IUCC/DASC/PICOM). IEEE
3. Hassan MA et al (2015) Help your mobile applications with fog computing. 2015 12th annual IEEE international conference on sensing, communication, and networking-workshops (SECON Workshops). IEEE
4. Mahmud R, Kotagiri R, Buyya R (2018) Fog computing: a taxonomy, survey and future directions. In: Internet of everything. Springer, Singapore, pp 103–130
5. Modi C et al (2013) A survey on security issues and solutions at different layers of cloud computing. J Supercomput 63(2):561–592

Proposal of Linear Specific Functions for R-L-C as Fundamental Elements in Terms of Considered Specific Electric Constants

Vineet Kumar

Abstract In this paper electric physical quantities which can be characterized as being of either of the scalar or phasor type, are discussed. The scalar physicalities of resistance, inductance and capacitance are regarded as being of a fundamental type as they provide the initial basis of circuit designing, while the phasor physicalities are voltage and current etc. These fundamental elements may also be regarded as conventional in comparison to those of semiconductors devices, which are introduced later. In taking these scalar elements into consideration, the linearized specific functions are consider in terms of the considered specific electric constants possible from the ground of already existing electromagnetic constants of the wave propagating through the free space, namely intrinsic impedance, permeability, and permittivity.

Keywords Conventional elements • Farad multiplier • Henry multiplier
Linear specific function • Ohmic multiplier • Phasor physicality
Specific electric constants • Scalar physicality

1 Introduction

In branch of electrical, one can say that all of the different physicality's in concern to same can be resolved in the two categories, namely scalar and phasor, which are in direct analogy to the scalar and vector quantities of physics. Here, scalar physicalities represent the complex function with a non-existing imaginary part, i.e. having only length and zero inclination (or no direction) in any frame, whereas, phasor physicality represent the complex function having both length and inclination. Besides the phasor representing itself as a vector, the only differences are that:

V. Kumar (✉)
Department of Electrical Engineering, Kurukshetra University,
Kurukshetra 136119, India
e-mail: vineet05k@gmail.com

A. Kumar and S. Mozar (eds.), *ICCCE 2018*,
Lecture Notes in Electrical Engineering 500,
https://doi.org/10.1007/978-981-13-0212-1_3

1. Over the base of arithmetic operations other than addition '+' and subtraction '−', the rest are not acceptable.
2. Unlike vectors, phasors are available on the background of complex type. Examples of electrical phasor quantities are voltage, current, impedance, power, etc., while electrical scalar quantities are resistance (R), inductance (L), capacitance (C) etc [1].

These $R, L \& C$ scalar electrical physicalities are regarded as fundamental elements, which are the only option at beginning of all circuit designs of all, based on the fact that these fundamental elements are also regard as being of the conventional type. After the discovery of the diode around 1906, the era of mixed circuit designing with the uses of both conventional and non-conventional elements came into play, where the non-conventional electric elements are regard as semiconductor devices, such as diodes, transistors, etc., which to some extent possess the property of fundamental elements as well.

First, after the introduction in Sect. 1, this paper proposes specific electric constants for all of the conventional electric elements, i.e. $R, L \& C$, in terms of electromagnetic constants of free space in Sect. 2. In Sect. 3, after the proposition of specific constants of such the linear specific functions for the conventional elements of same also find out. Finally, the paper presents our conclusions in Sect. 4.

2 Proposition of Specific Electric Constants for Conventional Elements in Terms of Electromagnetic Constants

Charge, the intrinsic property of a particle, is responsible for the electric forces of attraction and repulsion in the presence of other particles with similar properties, but a dissimilar nature, which in the case of zero valued velocity function imposing on it, i.e. in a stationary state, it determines the electrostatic field denoted by $\vec{E}(x, y, z)$ [2, 3]. In case of a non-zero constant valued velocity over the function imposing on the charged particle, it determines the magnetostatic field denoted by $\vec{H}(x, y, z)$. Next, with the time over function imposing on it, these two fields further modify to become $\vec{E}(x, y, z, t)$ & $\vec{H}(x, y, z, t)$. In case of static or time-invariant conditions, these two fields are independent of each other, but in case of dynamic or time variant condition these two fields are dependent on each other.

Besides field vectors $\vec{E}$ & $\vec{H}$, there are other field vectors denoted by $\vec{D}$ & $\vec{B}$ respectively in correspondent of it relating by $\vec{D} = \epsilon\vec{E}$ and $\vec{B} = \mu\vec{H}$ for linear, isotropic and homogeneous media. The other fundamental relation omitted in regard of this is $\vec{J} = \sigma\vec{E} + \rho\vec{v}$, where $\vec{J}$ & $\vec{v}$ are the current density and the velocity vector respectively. Taking these three constitutive relations into account, we conclude that the plane wave propagating through a medium needs three constants in general, which are regarded as a constants of electromagnetic to describe the same of it.

The constants $\mu\,\&\,\epsilon$ for any medium represent the ability of the medium to store energy within it for the respective field vectors of the electromagnetic wave through which it propagates. Taking these two energy-storing elements into account in the case of a free space condition, we have three constants in particular, which are collectively regarded as specific electric constants essential to describe the linear specific functions of the conventional elements, as described in the following section.

2.1 Specific Resistance Constant

In case of a plane wave propagating through free space, the modification of the electromagnetic constants $\sigma = 0, \mu = \mu_o\,\&\,\epsilon = \epsilon_o$ takes place, which determines the intrinsic impedance of free space as $\eta_o = \sqrt{\mu_o/\epsilon_o} = 120\pi\,\Omega \cong 377\,\Omega$. Taking this free space electromagnetic intrinsic impedance into consideration as a reference, the electric resistance of a material can be defined in general as a linear function which may provide the platform for a detailed description of the same as,

$$R_{\eta_o} = \eta_o \tag{1}$$

The term R_{η_o} of Eq. (1) may be termed the specific resistance constant.

2.2 Specific Inductance Constant

The term μ_o as employed previously to define η_o is another electromagnetic constant known as the permeability of free space and having value $4\pi \times 10^{-7}\,\mathrm{T}\,\frac{\mathrm{m}^2}{\mathrm{A}}\,(or\,\frac{Henry}{meter})$, which came into existence with the magnetic field vector. Over the spatial distance of 1 m this free space permeability constant is modified and becomes,

$$L_{\mu_o} = (1\,\mathrm{m}) \times \mu_o \tag{2}$$

The term L_{μ_o} of Eq. (2) may be called the specific inductance constant, which represents the inductance of free space. In the case of a dielectric medium ground, it also has the same constant, but like that of permeability it is also modified to become $L_\mu = L_{\mu_o} L_{\mu_r}$ responsible for the proportion of energy stored in the medium, such that the higher the value of it for given condition, the greater the storing of energy.

2.3 Specific Capacitance Constant

Like μ_o, there was another term omitted, ϵ_o, employed previously to define the term η_o. This is the third electromagnetic constant known as the permittivity of free space and having a value of $8.85419 \times 10^{-12} \frac{\mathrm{C}^2}{\mathrm{Nm}} \left(or \frac{Farad}{meter}\right)$, which came into the existence with the electric field vector. Over the spatial distance of 1 m this free space permittivity constant is modified to become,

$$C_{\in_o} = (1\,\mathrm{m}) \times \epsilon_o \tag{3}$$

The term $C_{\in_o}$ of Eq. (3) is called the specific capacitance constant and represents the capacitance of free space. In the case of dielectric medium permittivity modified and becomes $C_{\in} = C_{\in_o} C_{\in_r}$ responsible for the proportion of energy stored in the medium, such that the higher the value of it for given condition, the greater the storing of energy. Now, these specific electric constants R_{η_o}, L_{μ_o} & $C_{\in_o}$ in free space condition, apart from the material space specific electric constants R_η, L_μ & $C_\in$ with $R_\eta = R_{\eta_o}$ are going to be employed to find the functions of conventional elements.

3 Linear Specific Electric Functions for Conventional Elements

The consideration of the dimensionless positive quantity *X*, as independent variable 1 for the term $\omega\sqrt{LC}$ in paper [4] further yields to the relation of the form $X(f, l) = \frac{2\pi}{c} \cdot fl$ in the case of uniform distribution of power line parameters, over its entire length irrespective of any prior assumption made on the ground level for energy storing conventional elements (i.e. inductance and capacitance). Here, *l* is the length of the powerline and *f* is the operating frequency. Based on the consideration of multipliers for each the elements, for instance *y* & *z*, such that $(y, z) \geq 0$ with respect to the specific electric constant of particulars, i.e. L_{μ_o} & $C_{\in_o}$, it gives the same relation but with the condition that $y = z = \frac{l}{1\,\mathrm{m}}$ must hold.

Taking these two conventional energy storing elements as a reference, one can consider the dimensionless multiplier for electric resistance as well, let it be *x*, which with respect to the resistance specific constant such that with a suitable selection of *x* it gives the corresponding resistance for it. Representing the conventional electric elements in terms of a multiplier along with suitable electromagnetic constants is the main objective of this section, which can generally be employed in the similar cases discussed earlier.

3.1 Specific Resistance Function

The electric resistance of a material is the property by virtue of which it opposes the flow current through it. In between voltage and current, mathematically it is responsible for the holding of the linear relation between them via an operator of proportional type, which is given by $V = RI$ and is responsible for the losses of energy in the form of heat, regarded as being of the real (or active) type.

Now, the electrical resistance of an element for any network, and of any value, can be considered as a linear function with respect to a particular constant, that is, the resistance electric constant which is equal to the intrinsic impedance of an electromagnetic wave in a vacuum, and is given by,

$$R = R(x) = xR_{\eta_o} \tag{4}$$

Equation (4) may be termed as the specific resistance function or, simply, the resistance function, where the dimensionless multiplier 'x' introduced may be called the Ohmic multiplier.

3.2 Specific Inductance Function

The electric inductance of a material is the property by virtue of which it opposes any change in the magnitude or direction of current through it. In between voltage and current, it is mathematically responsible for the holding of the linear relation between them via a differential type operator, which is given by $V = L\frac{d}{dt}I$ or $V = (LD)I$ and conversely for current and voltage it holds between them via an operator of integral type, which is given by $I = 1/L\int V\,dt$ or $I = \left(\frac{1}{LD}\right)V$. This element is responsible for the power of lagging VAR associated with it, corresponding to magnetic energy, and which does not dissipate any energy but only stores it. It is regarded as being of the imaginary (or reactive) type. Now, the inductance of an element for any network, and being of any value, can be considered as a linear function with respect to a particular constant, that is, the inductance electric constant which is equal to the permeability of an electromagnetic wave propagating in a vacuum over 1 m, and is given by,

$$L = L(y) = y \cdot L_{\mu_o} \tag{5}$$

Equation (5) may be called the specific inductance function or, simply, the inductance function, where the dimensionless multiplier 'y' introduced may be called the Henry multiplier.

3.3 Specific Capacitance Function

The electric capacitance of a material is the property by virtue of which it shows the capability of storing electric charge within it. In between voltage and current, mathematically it is responsible for the hold of the linear relation between them via an operator of the integral type, which is given by $V = 1/C \int I\, dt$ or $V = \left(\frac{1}{CD}\right)I$ and conversely for current and voltage it holds the relation between them via an operator of the differential type, which is given by $I = C\frac{d}{dt}V$ or $I = (CD)V$. This element is responsible for the power of leading *VAR* associated with it, and corresponding to electric energy, which does not dissipate any energy but only stores it, and is regarded as being of the imaginary (or reactive) type. Now, the capacitance of an element for any network, and of any value, can be considered as a linear function of a constant, that is, the capacitance electric constant which is equal to the permittivity of an electromagnetic wave propagating in vacuum over a 1 m, and is given by,

$$C = C(z) = zC_{\in_o} \tag{6}$$

Equation (6) may be called the specific capacitance function, where the scalar multiplier 'z' introduced may be called the Farad multiplier.

4 Conclusion

As stated earlier, the main objective of this paper is to find the linear specific functions for conventional electric elements in terms of new specific electric constants, which are being the left out task for an assumption of paper [4]. Here, the linear specific functions of such are given by the series of Eqs. from (4) to (6). Whereas, on the other hand the new specific electric constants on the ground of already available electromagnetic constants in support of such are given by the range of Eqs. from (1) to (3). This finding of linear specific functions for conventional elements not only fulfills an assumption of paper [4], but may prove to be a better explanation of the energized powerline by helping in the formulation of new complex functions for the different physicalities by using the same in collective, so consider here.

Acknowledgements The author thanks all those with whom fruitful discussions on the topic took place, and also the referees as well for their direct and indirect support which helped to improve the paper considerably.

References

1. Steinmetz CP (1897) Theory and calculation of alternating current phenomena. The W. J. Johnston Company, 253 Broadway, New York
2. Elliott RS (1981) Electromagnetic theory: a simplified representation. IEEE Trans Educ E-24 (4):294–296
3. Elliott RS (1979) Some useful analogies in the teaching of electromagnetic theory. IEEE Trans Educ E-22(1):7–10
4. Kumar V (2017) Open voltage of 2–port network power line not always longer than close voltage. In: IEEE Xplore, 30 Oct 2017, pp 494–502. https://doi.org/10.1109/icstm.2017.8089210

Efficient Video Delivery Over a Software-Defined Network

R. Thenmozhi and B. Amudha

Abstract This paper proposes a framework called SDN Streamer for an OpenFlow controller in order to provide QoS support for scalable video streaming over an OpenFlow network. OpenFlow is a protocol that decouples control and forwarding layers of routing. Abstracting control from the forwarding plane lets administrators dynamically adjust network-wide traffic flow and keeps the network agile. Software-Defined Networking (SDN) aims to improve the reliability of multimedia streaming while reducing utilization of server resources. It optimizes video delivery using a Scalable Video Coding (SVC) algorithm that sends layers of different quality via discrete paths. Dynamic rerouting capability is ensured using a Lagrange Relaxation-based Aggregate Cost (LARAC) algorithm. Unlike Dijikstra's algorithm, the LARAC algorithm does not calculate least "hop-counts" to find the optimal path but calculates the optimal path based on "link statistics." The SDN Streamer can guarantee seamless video delivery with little or no video artifacts experienced by the end-users. This project makes use of HTML5 for browser display. The SDN server allows everyone to access the media files irrespective of their platform, device or browser. Performance analysis shows there is significant improvement on the video's overall PSNR under network congestion.

Keywords Software-defined networking (SDN) • Scalable video coding (SVC) • Lagrange relaxation-based aggregate cost (LARAC) Peak signal-to-noise ratio (PSNR)

R. Thenmozhi (✉)
Department of Information Technology, Valliammai Engineering College, Chennai, India
e-mail: thenusoma@gmail.com

B. Amudha
Department of Computer Science and Engineering, SRM University, Chennai, India

A. Kumar and S. Mozar (eds.), *ICCCE 2018*,
Lecture Notes in Electrical Engineering 500,
https://doi.org/10.1007/978-981-13-0212-1_4

1 Introduction

The key technique of SDN is OpenFlow, which provides additional configuration options in the data plane. The principle of communication between the data plane and control plane is allowed by the OpenFlow design [1] and also permits the complete network to be controlled through an Application Programming Interface (API) and distributed artificial intelligence in the network [2]. The networking foundation of Software-Defined Networking (SDN) is an emerging network architecture where network control is decoupled from the forwarding plane and is directly programmable. SDN is defined by two characteristics, namely decoupling of the control and data planes, and programmability on the control plane. Software-Defined Networking is a promising Internet architecture for delivering multimedia with end-to-end quality of service (QoS) [3, 4]. The reliability of multimedia streaming is improved while the utilization of server resources in the SDN is reduced. This optimized video delivery was accomplished by using video over a SDN as described by Owens II and Durresi [5].

This enables multicasting the multimedia video files from one system to another connected over an OpenFlow-enabled distributed network. It enables optimal dynamic management of network resources and on-demand QoS provisioning. OpenQoS introduced by Gilmez et al. [6] is the open flow controller that is used for delivering multimedia data with end to end QoS. HTML5 is used to stream the video without delay and packet loss by programming routing mechanisms to the centralized SDN controller.

HTML5 uses adaptive streaming which segments the video into small chunks. It means that the video is encoded at multiple bitrates and resolutions and adapts to larger or smaller chunks automatically as network conditions change. The aim is to construct an OpenFlow-enabled network to study the data transmission over a SDN and to monitor the packet delay and loss performance of multimedia streaming over the networks to facilitate timely delivery of data.

Video QoS is improved by implementing the LARAC algorithm and the Scalable Video Coding algorithm. The QoS is achieved by conveying the best available delivery node based on the network conditions and dynamically changing routing paths between network routers, so that QoS parameters, such as bandwidth, packet loss, jitter, delay, and throughput can be guaranteed. When the delivery node is dynamically changed, the available paths from the assigned node experience congestions. The first assigned delivery node sends a redirection message to the SDN controller via an OpenFlow request message so that the video is delivered through another available delivery node that can provide the client with higher networking performance. The routing path is selected based on network link characteristics such as available bandwidth and link utilization. Link utilization is identified by the LARAC algorithm which is used to ensure whether the link is congested or non-congested.

A Floodlight open source SDN controller developed by Shahid et al. [7], implemented Dijkstra's Algorithm to determine the critical path for the shortest distance between any source and destination in the entire available network. A multimedia traffic model network has been developed by Golaup and Aghvami [8] which disassociates the traffic developed by the multimedia component into a number of basic fundamental components and tries to arrive at an optimal traffic model for each component. They also developed a Graphics User Interface for representing the traffic modeling of the multimedia routing aspects.

A modular framework was developed by Bustos-Jiménez et al. [9] for deriving a relation between Qualtiy of Experience and QoS for multimedia transmissions and is commonly known as Boxing Experience and was developed through open source software. Amiri et al. [10] proposed a SDN controller to transfer huge data in cloud computing without traffic congestion which reduces end to end delay. Based on the end to end delay the controller disperses and distributes the game traffic load through different network paths. Kassler et al. [11] developed a software system for arriving at QoE paths for multimedia services. The system uses Openflow to arrive at various paths for the network components.

The excessive traffic load in today's internet era is addressed by Noghani and OğuzSunay [12]. They arrived at a SDN-based framework for both medium and heavy network loads aimed at delivering high quality video through this technique. Dobrijevic et al. [13] experimented with an ant colony optimization (ACO) approach for flow routing in SDN environments which utilized QoE for multimedia services. Tranoris et al. [14] developed SDN with OpenFlow for controlling computer network protocols for applications involving video-conferencing, multiplayer gaming, etc.

The concept of a scalable video encoder is to split single-stream video in a multi-stream flow, often referred to as layers. A scalable video coding (SVC) encoder encodes the input video sequence into complementary layers. The layered structure of scalable video content can be distinguished as the combination of a base layer and several additional enhancement layers. The base layer corresponds to the lowest supported video performance, whereas the enhancement layers allow for the refinement of the base layer. A receiver in a slow-bandwidth network would receive only the base layer, hence producing a video. On the other hand, the second receiver in a network with higher bandwidth can process and combine both layers, yielding a video with full frame rate and eventually a smoother video. The addition of enhancement layers improves the resolution of the decoded video sample.

1.1 Application of the Framework

- Live streaming using HTML5 with minimised delay in performance.
- The networking components that make up massive data centre platforms can all be managed from the SDN controller.
- Creation of websites for performing video streaming without flash support.
- Live conferencing with reduction in jitter.

- Webinars using a higher level of SVC algorithm.
- HTML5 video with flash fullback for HTTP Streaming of data.
- Top companies use it to implement service chaining.

2 Related Work

The OpenFlow [15] protocol removes the disadvantage of the brutality of static protocols, and opens the possibility of fast improvement and led the research community to investigate new paradigms. This paper presents an analysis of five policies such as route management, route discovery, traffic analysis, call admission and topology management.

Valdivieso Caraguay et al. [16] describe the SDN architecture and analyze the opportunities to provide new multimedia services. Moreover, a SDN framework is also presented to provide QoS for different multimedia services. The framework uses OpenFlow, network virtualization and establishes functional boxes and interfaces to test different routing algorithms. Then, the modules of "network performance" and "QoS Routing Algorithm" are implemented to demonstrate the effectiveness of the framework. The experiments with video streaming information show a quality optimization (PSNR, SSIM, MOS) in comparison with the best effort engine. Xu et al. [3] contemplated and proposed OpenQoS, which is a novel OpenFlow controller design for multimedia, delivering end-to-end Quality of Service (QoS) support.

The fast failure healing mechanism [17] is added to the video streamer making it capable of recovering from connection failure using a different path. The control plane in the SDN performs forwarding of packets and also aids in links alteration at data level dynamically. It also provides a programmable network and can find isolated networks by means of control plane. Nunes et al. [18] has carried out a detailed survey study on SDNs and their impact on past, present and future management of traffic networks in internet protocols involving multimedia using OpenFlow techniques.

Foukas et al. [1] describe SDNs as a relatively new paradigm of a programmable network which change the way that networks are designed and managed by introducing an abstraction that decouples the control from the data plane. In this technique, the controller, which is a software control program, is responsible for overall control over the network. The task of decision making is performed by the controller, whereas the hardware is only responsible for forwarding packets to the intended destination as per the controller's instructions, typically a set of packet-handling rules.

3 Existing System

Organizations are increasingly confronted with the limitations that accompany the hardware-centric approach. Network functionality is mainly implemented in a dedicated appliance. A "dedicated appliance" refers to one or multiple switches, routers and/or application delivery controllers. Most functionality within this appliance is implemented in dedicated hardware. A traditional network configuration is time consuming and prone to error. A network administrator has to manually configure the multiple devices in a traditional network.

A multi-vendor support environment requires a high level of knowledge and expertise. Existing system lack in the provision of flexibility to the user and the network administrators. A traditional system's components or switches are proprietary switches and it is difficult to configure them in a multi-tenant environment. These problems are addressed by the introduction of a SDN.

3.1 Drawbacks of the Existing System

Increased traffic congestion, lack of dynamic bandwidth allocation, greater call/session/request dropping probability, limited QoS and QoE, increased overheads and costs at the network layer, lack of reliability and robustness, scalability, and managing of data between various resources is tedious.

4 Proposed Work

The proposed system as shown in Fig. 1 is to decouple the control plane and the data plane. The data plane implements a set of forwarding operations. Other proposed tasks are: creating an open interface for communication between layers, OpenFlow-enabled switches (laptops) for streaming the multimedia file, streaming the multimedia file over HTTP, guaranteeing route management, route calculation, call admission, traffic policing, and flow management.

The advantage of the proposed system are: monitoring of network resource and end-to-end QoS support, application-layer aware QoS, differential services, virtualization, packet-type discrimination, QoS routing, video optimization, cost reduction, rapid deployment of new services, moving away from proprietary hardware, better visualization of the network.

A software-defined networking (SDN) architecture (or SDN architecture) defines how a networking and computing system can be built using a combination of open,

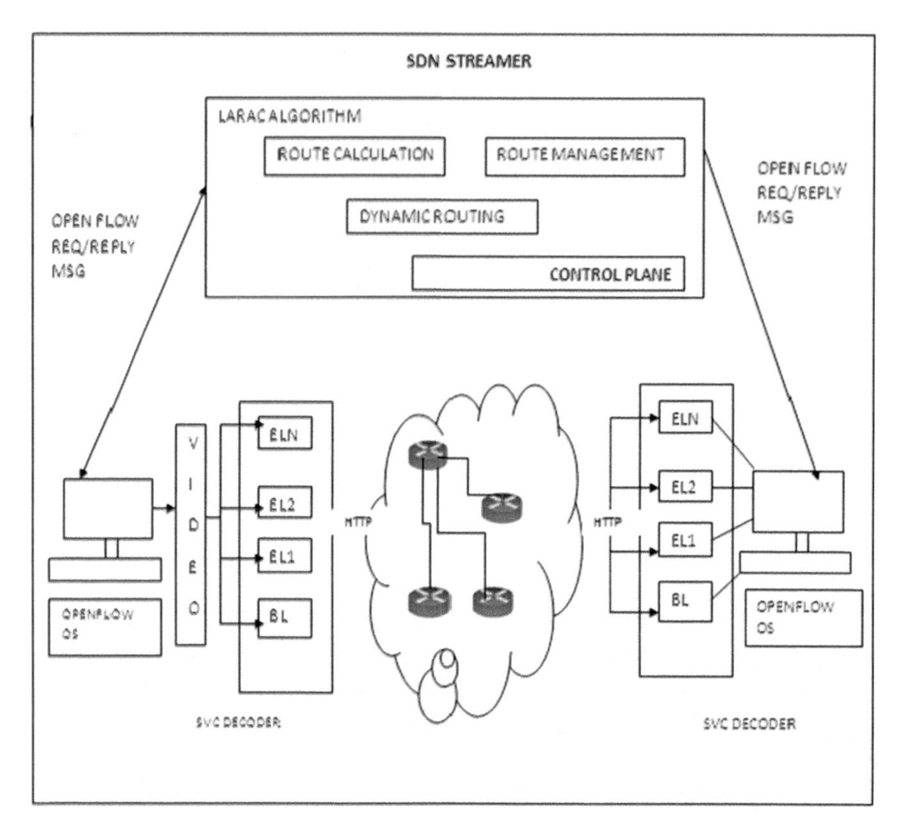

Fig. 1 Architecture diagram

software-based technologies and commodity networking hardware that separate the control plane and the data layer of the networking stack.

SDN architectures generally have three components or groups of functionality:

SDN Applications
SDN Applications are programs that communicate behaviors and needed resources with the SDN controller via application programming interfaces (APIs). In addition, the applications can build an abstracted view of the network by collecting information from the controller for decision-making purposes. These applications could include networking management, analytics, or business applications used to run large data centers. For example, an analytics application might be built to recognize suspicious network activity for security purposes.

SDN Controller
The SDN controller is a logical entity that receives instructions or requirements from the SDN application layer and relays them to the networking components. The controller also extracts information about the network from the hardware devices

and communicates back to the SDN applications with an abstract view of the network, including statistics and events relating to what is happening.

SDN Networking Devices
The SDN networking devices control the forwarding and data processing capabilities of the network. This includes forwarding and processing of the data path.

The SDN streamer is comprised of four modules. These modules are responsible for efficient transmission of multimedia files over the network.

- Creation of network
- Routing algorithm implementation
- Setup to stream video
- OpenQoS for delay and jitter reduction.

5 SDN Streamer

5.1 Creation of Network

In this module a sample network is created in a mininet simulator as depicted in Fig. 2. The torus network consists of switches at each corner. In this module, the controllers which are residing along the data plane are decoupled into a centralized SDN switch controller. Here 127.0.0.1 (localhost) acts as the SDN controller. All of the individual switches are then added to the controller and are globally managed by the controller. As such, the centralized controller can have access to all the switches connected to it, and the switches are OpenFlow enabled. So, this whole network resembles a SDN, which results in reliable transmission of data packets. And then, the torus network is tested with multimedia streaming functions. Intent is added with the source and destination.

The network shows the optimal path between the source and destination. The same can be implemented in real network conditions. In this module, ONOS has to be initialized. A cubical network with 4 × 4 switches is constructed. Using mininet, 64 switches are interlinked forming a visual pattern. Hosts are pinged with the switches. The command for creating a torus network in mininet is: sudomn – topo = torus 4, 4–controller remote.

5.2 Routing Algorithm Implementation

The SDN controller is a logical entity that monitors the OpenFlow enabled switches by sending OpenFlow FEATURE_REQUEST and FEATURE_REPLY messages. In this module, the various possible links between all the switches are analyzed. Using the LARAC algorithm, the path link which is the shortest (minimum hops) is

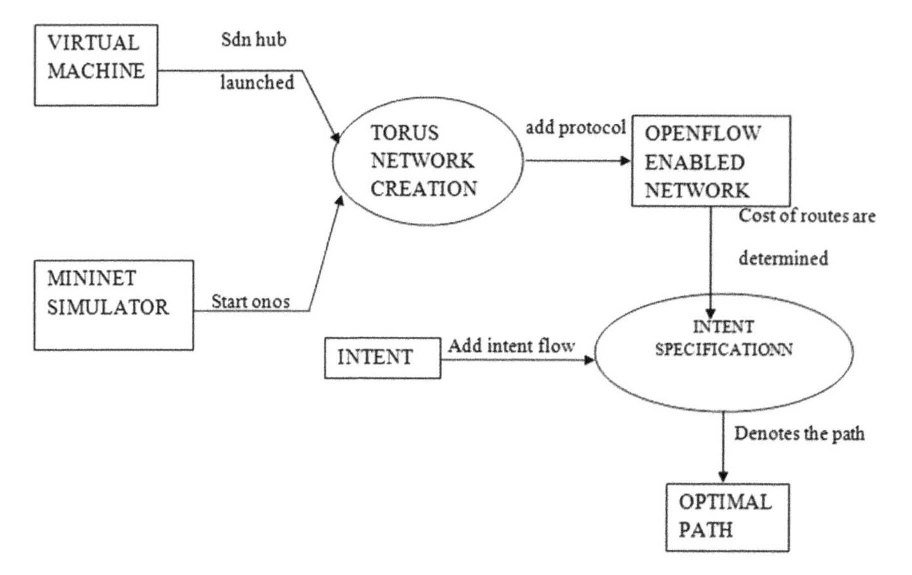

Fig. 2 Diagram of the creation of a network

calculated. This path is concluded to be the most efficient link for the transfer of data packets between the switches. The centralized controller maintains a flow table that contains all the possible links between the switches.

At the end of this module, an intent between the hosts is created. The controller has a buffer that stores the link statistics. The controller determines the optimal path specified in the flow table. If the current path is congested, then it finds the next optimal path. The congested path is one in which 75% (or more) of the bandwidth is already utilized. Using this LARAC algorithm, the centralized controller provides the optimal path to the switches which is free of congestion. This optimal path is free of traffic and facilitates easy transmission of data packets to the switches.

5.3 Setup to Stream Video

In this module, the multimedia data is segmented into various layers where each layer represents a different quality as shown in Fig. 3. These are all performed using the SVC algorithm, and then the layers are combined at the receiver end. The ways to perform this methodology are given below:

Web server creation is performed by registering with Amazon Web Service. EC2 instance is launched from a preferred AMI (Amazon Machine Image). Configuring security groups is used to add IP addresses which form a network to make use of the instance. A scalable video encoder is used to split the single-stream video in a multi-stream flow. The base layer corresponds to the lowest supported video

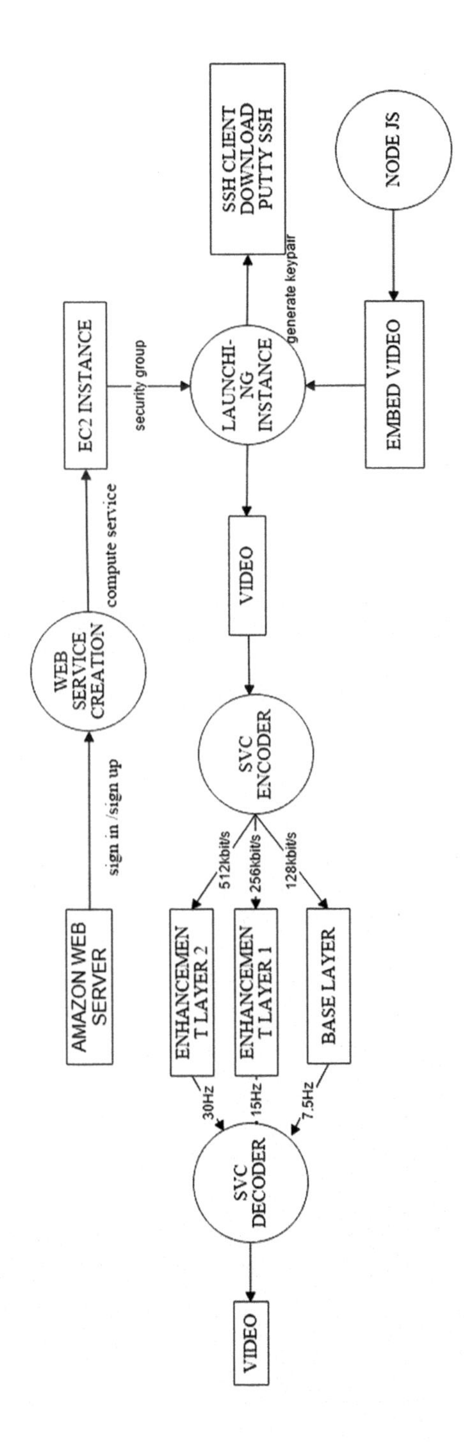

Fig. 3 Setup for streaming video

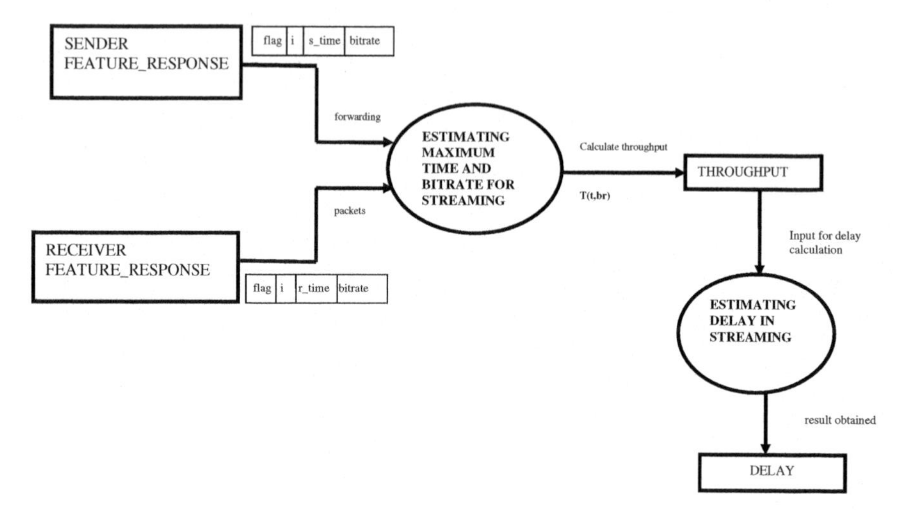

Fig. 4 Open QoS for delay and jitter reduction

performance, whereas the enhancement layers allow for the refinement of the base layer. The addition of enhancement layers improves the resolution of the decoded video sample.

5.4 *OPEN QoS for Delay and Jitter Reduction*

This module is mainly focused on the calculation of delay and jitter reduction and is represented in Fig. 4. Based on the calculation of delay and jitter, the traditional network and the software defined network are compared. The client decides to request one base and two enhancement layers and sends an HTTP GET message to the corresponding ports of the server at the same time. Hence, the client throughput is maximized since it does not have to wait for downloading packets of a video layer to request the next layer packets for a segment.

6 Conclusion

In this paper we proposed a framework for a video streamer to optimally stream video by streamlining a SDN with enhanced services with dynamic routing, route management, and route calculation. While comparing it to the traditional network, we added functionalities to the SDN by incorporating SVC and LARAC algorithms to minimize the delay and jitter in the network.

References

1. Foukas X, Marina MK, Kontovasilis K (2015) Software defined networking concepts. The University of Edinburgh & NCSR "Demokritos". http://homepages.inf.ed.ac.uk/mmarina/papers/sdn-chapter.pdf
2. Jimenez JM, Romero O, Rego A, Dilendra A, Lloret J (2015) Study of multimedia delivery over software defined networks. Netw Protoc Algorithms 7(4). Universidad Politécnica de alencia, ISSN: 1943-3581
3. Xu C, Chen B, Qian H (2015) Quality of service guaranteed resource management dynamically in software defined network. J Commun 10(11)
4. Wilczewski G (2015) Utilization of the software-defined networking approach in a model of a 3DTV service. J Telecommun Inf Technol
5. Owens II H, Durresi A (2015) Video over software-defined networking (VSDN). Comput Netw 1–16. Accepted 15 Sept 2015
6. Gilmez HE, Dane ST, Bagci KT, Tekalp AM (2012) OpenQoS: an OpenFlow controller design for multimedia delivery with end-to-end quality of service over software-defined networks. In: Signal and information processing association annual summit and conference (APSIPA ASC), 3–6 Dec 2012. Asia-Pacific and Published by IEEE Explore
7. Shahid A, Fiaidhi J, Mohammed S (2016) Implementing innovative routing using software defined networking (SDN). Int J Multimedia Ubiquitous Eng 11(2):159–172
8. Golaup A, Aghvami H (2006) A multimedia traffic modeling framework for simulation-based performance evaluation studies. Comput Netw 50:2071–2087. Available online 7 Nov 2005
9. Bustos-Jiménez J, Alonso R, Faúndez C, Méric H, Boxing experience: measuring QoS and QoE of multimedia streaming using NS3, LXC and VLC. In: 8th IEEE workshop on network measurements, WNM 2014, Edmonton, Canada
10. Amiri M, Al Osman H, Shirmohammadi S, Abdallah M (2015) An SDN controller for delay and jitter reduction in cloud gaming. ResearchGate
11. Kassler A, Skorin-Kapov L, Dobrijevic O, Matijasevic M, Dely P (2012) Towards QoE-driven multimedia service negotiation and path optimization with software defined networking. In: 2012 20th international conference on software, telecommunications and computer networks (SoftCOM), 11–13 Sept 2012. Split, published by IEEE Explorer, pp 1–5
12. Noghani KA, OğuzSunay M (2014) Streaming multicast video over software-defined networks. IEEE
13. Dobrijevic O, Santl M, Matijasevic M (2015) Ant colony optimization for QoE-centric flow routing in software-defined networks. ISSN: 978-3-901882-77-7. IFIP
14. Tranoris C, Denazis S, Mouratidis N, Dowling P, Tynan J (2013) Integrating OpenFlow in IMS networks and enabling for future internet research and experimentation. In: Galis A, Gavras A (eds) FIA 2013. LNCS, vol 7858, pp 77–88
15. Sharma S, Staessens D, Colle D, Pickavet M, Piet D (2012) A demonstration of fast failure recovery in software defined networking. In: Institute for computer sciences, social informatics and telenetworking, vol 44. Publisher Springer, Berlin, Heidelberg, pp 411–414
16. Valdivieso Caraguay AL, Barona López LI, GarcíaVillalba LJ (2014) SDN: evolution and opportunities in the development IoT applications. Int J Distrib Sens Netw. Article ID: 735142, 10 pp. http://dx.doi.org/10.1155/2014/735142
17. Xia W, Wen Y, Heng Foh C, Niyato D, Xie H (2015) A survey on software-defined networking. IEEE Commun Surv Tutor 17(1) (First Quarter 2015)
18. Nunes BAA, Mendonca M, Nguyen X-N, Obraczka K, Turletti T (2014) A survey of software-defined networking: past, present, and future of programmable networks. IEEE Commun Soc Inst Electron Electron Eng 16(3):1617–1634

A Viewpoint: Discrimination Between Two Equivalent Statements of Kirchhoff's Current Law from the Ground of Precedenceness

Vineet Kumar

Abstract In this paper, on the ground of precedenceness, the two equivalent statements in regard to the basic law of electrical from Kirchhoff's current are discriminated. Here, this viewpoint of statements discrimination for the law of same does not means to regard that there is a differences in between, but to regard that the statement of one out of the two exist due to the existence of other. In addition, the current regulation function is discussed, which is always limited to the range of 0 to 1 as it determines the ratio of totality at each node for either sides of the branches collected by the converger and diverger respectively. In the case of a condition without unity, it determines that the law of conservation of charge does not hold.

Keywords Algebraic statement · Converger · Current regulation function Diverger · Equality statement · Precedence ground

1 Introduction

Two or more statements in concern to the action of same irrespective of any other conditional elements as additional, then, other than being equivalent of it they may show the case of precedenceness, which reflects the discrimination in between. However, this discrimination of statements in concern to same does not mean to have any difference but this only shows the case that the existence of one is due to the existence of other. Based on such here in case of electrical as well, the law as a base Kirchhoff's current law (KCL) with statement of two, named here as equality statement and algebraic statement rather than equivalent can also be discriminate form the ground of precedenceess. Where the equality statement for KCL concludes to be at first place on the ground of precedenceness, while the algebraic statement at the second place, which means that the algebraic statement exist if the equality

V. Kumar (✉)
Department of Electrical Engineering, Kurukshetra University, Kurukshetra 136119, India
e-mail: vineet05k@gmail.com

A. Kumar and S. Mozar (eds.), *ICCCE 2018*,
Lecture Notes in Electrical Engineering 500,
https://doi.org/10.1007/978-981-13-0212-1_5

statement exist rather than equivalent of it. Now, the equivalent statements of two for law of same (i.e. KCL) [1–3] are given as,

1. *Equality Statement for Kirchhoff's Current Law*: At any node of electrical circuit, the sum of the incoming currents is equal to the sum of the outgoing currents.
2. *Algebraic Statement for Kirchhoff's Current Law*: The algebraic sum of all the currents meeting at a node is equal to zero.

The structure of this paper is as follows. It begins with the introduction in Sect. 1, then the equality statement for KCL is obtained in Sect. 2 and this is further taken into consideration for the finding of another algebraic statement for KCL in Sect. 3. Finally, the paper ends with the conclusion in Sect. 4. The set of equations shown here in a step-by-step process, as require, from Sects. 2 to 3 satisfy the objective of such.

2 Finding of the Equality Statement for KCL

Based on the conservation principle for any flowing system, here in the case of charge, the sum of the incoming charges and the sum of outgoing charges along the way of different to and from the space volume, over an interval of Δt, must be equal. If the sum of the incoming and outgoing charges over the time interval Δt are denoted by $Q^i|\Delta t$ and $Q^o|\Delta t$ respectively, then, with respect to the node n of the electrical circuit along the l and m different ways, it is given as,

$$Q^i|\Delta t = \underbrace{\Delta Q_1^i|\Delta t + \Delta Q_2^i|\Delta t + \cdots + \Delta Q_l^i|\Delta t}_{I} = \sum_{k=1}^{l} \Delta Q_k^i|\Delta t \tag{1}$$

$$Q^o|\Delta t = \underbrace{\Delta Q_1^o|\Delta t + \Delta Q_2^o|\Delta t + \cdots + \Delta Q_m^o|\Delta t}_{II} = \sum_{k=1}^{m} \Delta Q_k^o|\Delta t \tag{2}$$

Taking part I and II of Eqs. (1) and (2) respectively over the divisor Δt in accordance with the equality condition at same node, it gives Eq. (3) as,

$$\frac{\Delta Q_1^i|\Delta t}{\Delta t} + \frac{\Delta Q_2^i|\Delta t}{\Delta t} + \cdots + \frac{\Delta Q_l^i|\Delta t}{\Delta t} = \frac{\Delta Q_1^o|\Delta t}{\Delta t} + \frac{\Delta Q_2^o|\Delta t}{\Delta t} + \cdots + \frac{\Delta Q_m^o|\Delta t}{\Delta t} \tag{3}$$

On using the relation of $\Delta Q_k|\Delta t = (en_k)v_k A_k \Delta t$ in Eq. (3) it gives $\sum_{k=1}^{l} n_k^i v_k^i A_k^i = \sum_{k=1}^{m} n_k^o v_k^o A_k^o$, so that in case of identical charge concentration, i.e. $n_1^i = \cdots = n_l^i = n_1^o = \cdots n_m^o = n$, it reduces to become $\sum_{k=1}^{l} v_k^i A_k^i = \sum_{k=1}^{m} v_k^o A_k^o$, which resembles the equation of continuity of hydrodynamics. Where v_k^i is the acquired velocity and A_k^i is the area of cross-section through which charges flow. As current

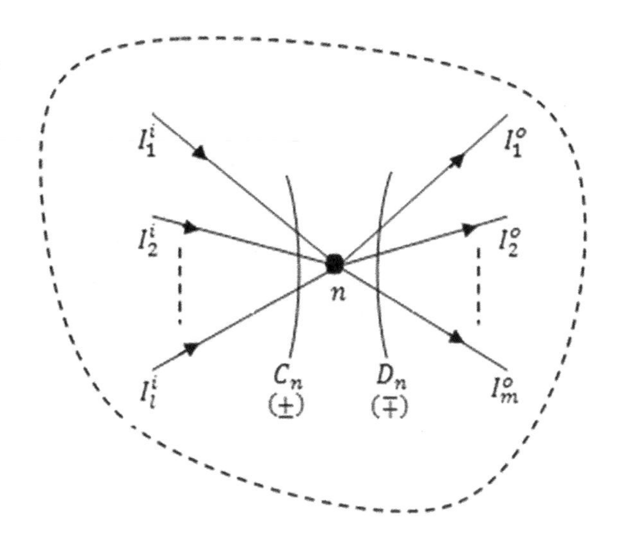

Fig. 1 Representation of the branches meeting at node n of electrical circuit carrying currents $I_1^i, I_2^i, \ldots, I_l^i$ as incoming and $I_1^o, I_2^o, \ldots, I_m^o$ as outgoing, such that the nature of convention so employ for converger C_n and diverger D_n must be opposite either as $(\pm)$ or $(\mp)$

is defined as the rate of flow of charge, therefore in accordance with Fig. 1, Eq. (3) can be further modified to become,

$$I_1^i + I_2^i + \cdots I_l^i = I_1^o + I_2^o + \cdots I_m^o \tag{4}$$

Keeping Eq. (4) in consideration, the current regulation function at a node n can simply be obtained by $R_n = \left(I_1^o + I_2^o + \cdots I_m^o\right) / \left(I_1^i + I_2^i + \cdots I_l^i\right)$, which is the ratio of the sum of outgoing currents to the sum of incoming currents. With respect to the dynamic term $n_k v_k A_k$, this current regulation function is also represented by $R_n = \left(\sum_{k=1}^{m} n_k^o v_k^o A_k^o\right) / \left(\sum_{k=1}^{l} n_k^i v_k^i A_k^i\right)$ which in the case of the ideal condition holds equality to 1, but in case of the practical condition it satisfy the less than unity value to it. This less-than-unity value of current regulation function at any node indicates that the law of conservation of charge is not obeyed.

3 Finding of the Algebraic Statement for KCL

Based on simple arithmetical operations like addition '+' and subtraction '−', for the current carried by branches of different but connected to the node of same, all the terms of either hand side, i.e. of *L.H.S.* or of *R.H.S.*, of Eq. (4) can be taken from one to the other, in two ways. The taking of all the terms of either side way over equality as one of the elements of sign conventional space from to another are followed as,

Case *(i): All terms taken from R.H.S to L.H.S.*

$$I_1^i + I_2^i + \cdots I_l^i - I_1^o - I_2^o - \cdots - I_m^o = 0 \tag{5}$$

or equivalently,

$$I_1^i + I_2^i + \cdots I_l^i + (-I_1^o) + (-I_2^o) + \cdots + (-I_m^o) = 0) \equiv \left(\sum_{k=1}^{l+m} (\pm I_k) = 0\right) \tag{5.1}$$

Case *(ii): All terms taken from L.H.S to R.H.S.*

$$(-I_1^i) + (-I_2^i) + \cdots + (-I_l^i) + I_1^o + I_2^o + \cdots + I_m^o = 0) \equiv \left(\sum_{k=1}^{l+m} (\pm I_k) = 0\right) \tag{6}$$

Equations (5), (5.1) and (6) as obtained here over the two elements of sign conventional space, which is nothing but the arithmetical operations; regard the algebraic statement of Kirchhoff's current law for lumped element model on any domain. Taking of these three equations and getting back to the Eq. (4) in reverse is a matter of simple task but for that one need to have the realization of such conservation principle which itself first provide by the Eq. (4) as a base following from Eqs. (1) to (3). On the basis of above from Eqs. (1) to (6), the two equivalent statements of KCL are discriminated from the ground of precedenceness, where the equality statement for KCL concludes to be at first place whiles the algebraic statement at the second place. Next, these two-side way options of taking the entire terms of base Eq. (4) indicate that the nature of both converger and diverger is independent of the elements of sign conventional space. Where, converger is the collecting of all current carrying branches that regard the current directions heading toward the node while diverger is the collecting of all branches that regard the current directions heading away from the node.

4 Conclusion

As stated, the main objective of this paper is to satisfy the realization that the two equivalent statements concerning KCL on the ground of precedenceness does not stand at same place. To these ends, Eq. (4) is first obtained in satisfaction with the law of conservation of charge, which when moulded in accordance with the frame of sign convention it yields Eqs. (5) and (6) respectively to meet the objective.

Acknowledgements The author would like to thank the referees for their direct and indirect support which helped to improve the paper considerably.

References

1. Steinmetz CP (1897) Theory and calculation of alternating current phenomena. The W. J. Johnston Company, 253 Broadway, New York
2. Elliott RS (1979) Some useful analogies in the teaching of electromagnetic theory. IEEE Trans Educ E-22(1):7–10
3. Chakrabarti A (2008) Circuit theory analysis and synthesis, 5th revised edn

A DES-Based Mechanism to Secure Personal Data on the Internet of Things

Pragya Chandi, Atul Sharma, Amandeep Chhabra and Piyush Gupta

Abstract The Internet of Things (IoT) helps users in their day to day activities such that they can communicate with each other through sensors very easily and in less time. Communication through intermediate sensor nodes may violate security because it may harm confidential user data or information by either modifying it or not forwarding to it. To deal with such problems, a number of cryptographic secure mechanisms are available that provide symmetric and asymmetric keys to secure data and each mechanism has its own pros and cons. In this paper, a secure DES-based mechanism is proposed in which a DES algorithm is used to transfer the data between users through sensor nodes without no loss of security.

Keywords IoT (internet of things) · Sensors · Security · Cryptography and DES (data encryption standard)

1 Introduction

The internet is most essential piece of the IoT. It began as a feature of DARPA (Defense Advanced Research Projects Agency) in 1962, and then gained ground with ARPANET in 1969. In the 1980s, specialist business organizations begin supporting the work of ARPANET, enabling it to advance towards the current internet.

P. Chandi (✉) · A. Sharma · A. Chhabra · P. Gupta
Department of Computer Science and Engineering, University Institute of Engineering & Technology (UIET), Kurukshetra University, Kurukshetra, Haryana, India
e-mail: pragyachandi@gmail.com

A. Sharma
e-mail: atulsharma2204@gmail.com

A. Chhabra
e-mail: aman6sep85@gmail.com

P. Gupta
e-mail: piyushgpt.er@gmail.com

A. Kumar and S. Mozar (eds.), *ICCCE 2018*,
Lecture Notes in Electrical Engineering 500,
https://doi.org/10.1007/978-981-13-0212-1_6

Kevin Ashton, the executive director of Auto-ID Labs at MIT, was the first to present the idea of the IoT, while giving an introduction for Procter and Gamble. Kevin Ashton presumed that radio frequency identification (RFID) was one of the earlier conditions for the IoT. The articulation "Auto-ID" is used for any expansive class of identification innovations utilized as a part of industry in order to mechanize as well as increase proficiency and reduce errors. These advances are biometrics, sensors, scanner tags, brilliant cards, and voice acknowledgment. Be that as it may, since 2003 the fundamental Auto-ID innovation has been RFID. The middle begin working from earliest reference point on electronic tag could be put on every single protest on the planet, enabling each to be exceptionally distinguished, followed and furthermore be conceivably controlled. we expected to put all put away information in memory and RFID is shabby then again littlest chip of Silicon is costly. The Internet was the main place to begin, and from that point the "Internet of Objects" or the "Internet of Things" turned into an unmistakable reference.

Another essential segment in building up a practical IoT was IPV6's exceptional insightful conclusion to expand address space. IoT is striking in light of the fact that a protest that can describe itself carefully progresses toward becoming an option that is more prominent than when the question existed without anyone else's input.

2 Realizing the Concept

One of the main examples of an IoT is from 1980s, and was a Coca Cola machine, at Carnegie Melon University. Software engineers interfaced the refrigerated machine with the internet to verify whether there was a drink accessible, and that it was cold, before making the excursion. By the year 2013, the IoT had formed into to a framework utilizing different advances, extending from the internet to remote correspondence and from small-scale electromechanical frameworks (MEMS) to installed frameworks. The conventional fields of robotization (counting the mechanization of structures and homes), remote sensor systems, GPS, and control frameworks etc., all help the IoT. Some of the general and key characteristics identified during the research study are as follows:

- Interconnectivity: [1] With reference to the IoT, anything can be interconnected with the worldwide data and correspondence framework. Digital physical frameworks (CPSs), likewise called the IoT [2] have broadened to new levels. For instance, the quantity of associated gadgets on Earth as of now surpasses the quantity of individuals. By 2020, there will, all things considered, be around seven associated gadgets for each individual. Things-related services: The IoT is an internet of three things:
  (1) People to people, (2) People to machine/things, (3) Things/machine to things/machine.
  Collaborating through the internet the earth of an assortment of things/objects that through remote and wired associations and exceptional tending to plans can

connect with each other and coordinate with different things/objects to make new applications/administrations and achieve shared objectives.
- Heterogeneity: based on various equipment stages and systems every one of the gadgets in the IoT are heterogeneous. They can connect with different gadgets or administration stages through various systems.
- Dynamic changes: The position, area and condition of gadgets change progressively, e.g., resting and awakening, associated or potentially disengaged, and in addition the setting of gadgets includes area and speed. The quantity of gadgets can change enormously. Objects can speak with each other and with the client.
- Enormous scale: The quantity of gadgets that should be overseen and that speak with each other will be no less than a request of greatness bigger than the gadgets associated with the present Internet.
- Safety: Although the IoT will increase profits, we should not disregard security. As both the makers and beneficiaries of the IoT, we should plan for wellbeing.

3 Architecture of the IoT

Coding Layer: The coding layer is the substructure of the IoT and offers identification to the every last protest of intrigue. In this layer, each protest is relegated an interesting ID which makes it simple to distinguish the objects [3] (Fig. 1).

Perception Layer: This is additionally called the gadget layer of the IoT and gives a physical importance to each question. It has information sensors in various structures like RFID labels, IR sensors or other sensor systems [4] which are utilized to detect the temperature, stickiness, speed and area and so forth of the objects. This layer gathers all the helpful data of the objects from the sensor gadgets

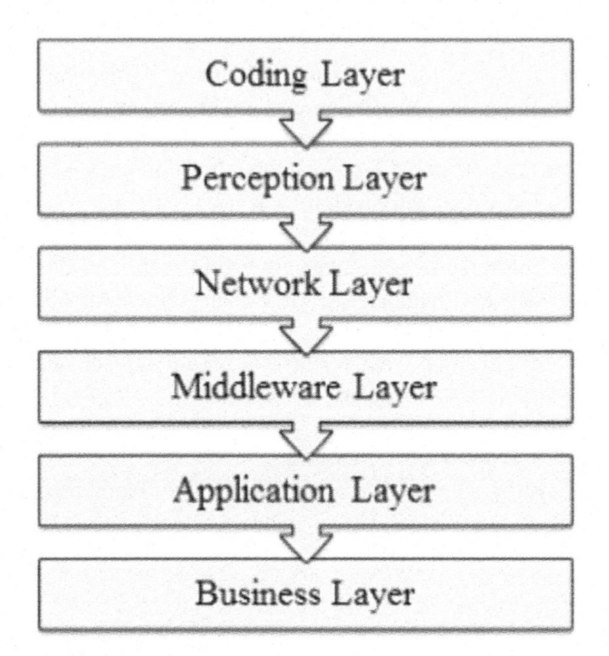

Fig. 1 Six-layered architecture of the IoT

and connects with them and believers the data into advanced signs which is then passed onto the following layer i.e. Network Layer.

Network Layer: The point of this layer is to get the data as advanced signs from the past layer and transmit it to the handling frameworks in the middleware layer through transmission mediums like Wi-Fi, Bluetooth, WiMaX, Zigbee, GSM, 3G and so on, with protocols like IPv4, IPv6, MQTT, DDS and so forth [5].

Middleware Layer: This layer forms the data obtained from the sensor gadgets. It incorporates advancements like cloud processing. Utilizing some intelligent processing equipment, the data is handled and a completely robotized move is made in light of the prepared consequences of the data.

Application Layer: This layer fulfills the uses of the IoT for any sorts of industry, in view of the handled information. This layer is extremely useful in the huge-scale improvement of the IoT and is arranged [6] in light of the fact that applications advance the improvement of the IoT.

Business Layer: This layer controls the applications and administrations of the IoT and furthermore is in charge of all the research identified with the IoT. It creates distinctive plans of action for successful business systems.

4 Need for Security in the IoT

To enhance security, an IoT gadget that should be straightforwardly available over the internet ought to be fragmented into its own system and have organize get to limited. The system portion should then be checked to distinguish potential atypical movement, and move ought to be made if there is an issue. Essentially, if your ice chest or TV has an internet association, at that point it turns into an IoT gadget. In 2016, the Mirai botnet propelled one of the greatest DDoS assaults ever recorded. More than 1 terabyte per second overflowed the system. This assault was so unique that it was the first to be completed with IoT gadgets. A rundown of potential difficulties:

- Security: Increased robotization and digitization produce new security concerns.
- Enterprise: Security issues could present dangers.
- Consumer Privacy: Potential of security breaches.
- Data: Lots of information will be produced, both for huge information and individual information.
- Storage Management: Industry needs to make sense of what to do with the information in a financially savvy way.
- Server Technologies: More interest in servers will be important.
- Data Center Network: WAN connections are streamlined for human interface applications, and the IoT is required to drastically change designs by transmitting information consequently.

A. Cryptography Process:

Cryptography is a concept of two main different process on the basis of two different keys. Encryption: This is a process that is used to convert plain text into secure encrypted text, i.e. cipher text. Decryption: This is a process that is used to convert encrypted text into plain text. These are two main important parts of cryptography and they are differentiated by the number of keys they have. I. Private key/symmetric key: This possess only has one key. II. Public key/asymmetric key: This key possesses two different keys meaning we encrypt our information using one key while decrypting our information using another key. So, here we can see how encryption and decryption are performed with the help of two different keys. First, the plain text will be delivered from the sender side, and with the help of a public key it will encrypt and convert into cipher text and then the cipher text will be decrypted with the help of another key, i.e. a private key, and this key will decrypt that cipher text into plain text and the recipient will get the original plain text.

5 Proposed Work

Data encryption standard algorithm (DES): Another symmetric key encryption is DES. It is a 64-bit piece figure which implies that it encodes information 64 bits at once. It is for the most part in light of figure called fiestal piece figure. It utilizes a similar key and works on encryption and for unscrambling [7]. Encryption of a piece of the message happens in 16 states or adjusts. From the info key, sixteen 64-bit keys are created, one for each round. In each cycle, eight supposed S-boxes are utilized. These S-encloses are settled the determination of the standard. Utilizing the S-boxes, gatherings of six bits are mapped to gatherings of four bits. The substance of these S-boxes has been dictated by the United States' National Security Agency (NSA). The S-boxes have all the earmarks of being haphazardly filled, however this isn't the situation. As of late it has been found that these S-boxes, first used in the 1970s, are safe against an assault called differential cryptanalysis, which was first seen in the 1990s.

The piece of the message is isolated into two parts. The correct half is extended from 32 to 48 bits utilizing another settled table. The outcome is joined with the subkey for that round utilizing the XOR operation. Utilizing the S-boxes, the 48 coming about bits are then changed again to 32 bits, which are accordingly permutated again utilizing yet another settled table. This is then completely rearranged and the right half is presently joined with the left half utilizing the XOR operation. In the following round, this blend is utilized as the new left half.

The figure ought to ideally make this procedure a bit clearer. In the figure, the left and right parts are signified by L0 and R0, and in consequent adjusts as L1, R1, L2, R2 etc. The capacity f is in charge of the considerable number of mappings portrayed previously.

A. Proposed Algorithm:

1. Start
2. Input text to be encrypted.

3. Divide this data into rows and columns.
4. Generate secret key.
5. Divide this key into 16 blocks.
6. Apply f-function to each block and perform XOR operation on it.
7. After processing all blocks, a secure encrypted key has been generated.
8. Use this key to encrypt the data.
9. Reverse these steps to decrypt the encrypted data.
10. End

In the proposed algorithm, 1st input data value then divide this data into two blocks of 8-bits each. Now the EOR operation is performed on these blocks with some f-function. An f-function is used to generate random values which will be helpful for generating a strongly encrypted key.

6 Results

To implement the proposed mechanism, MATLAB is used. MATLAB is a language which is based on matrix calculations and it is also known as a fourth generation programming language. It provides an easy to use environment where problems can be expressed in the form of a matrix or uses the numerical notations (Figs. 2, 3 and 4).

Figure 3 illustrates the encryption and decryption processes of the proposed mechanism. First give some input data and then the proposed algorithm converts

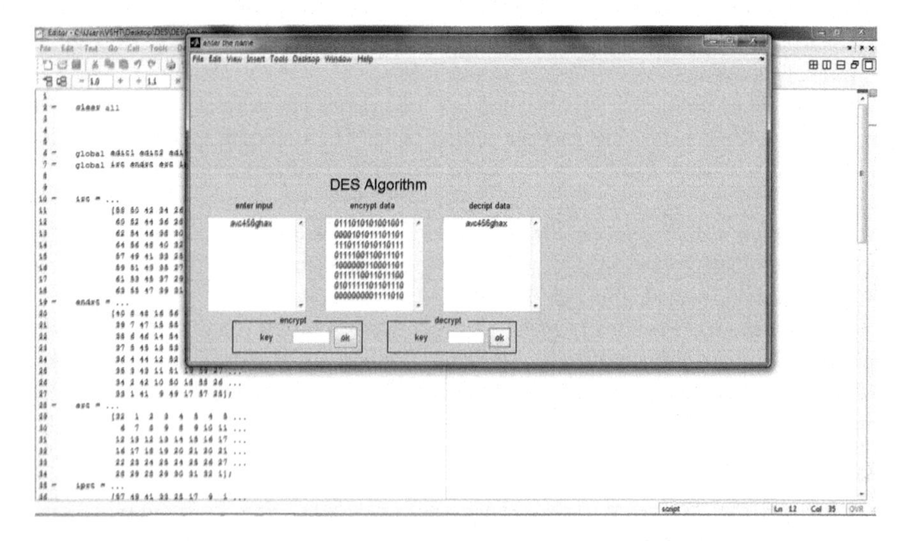

Fig. 2 DES-based secure data transmission

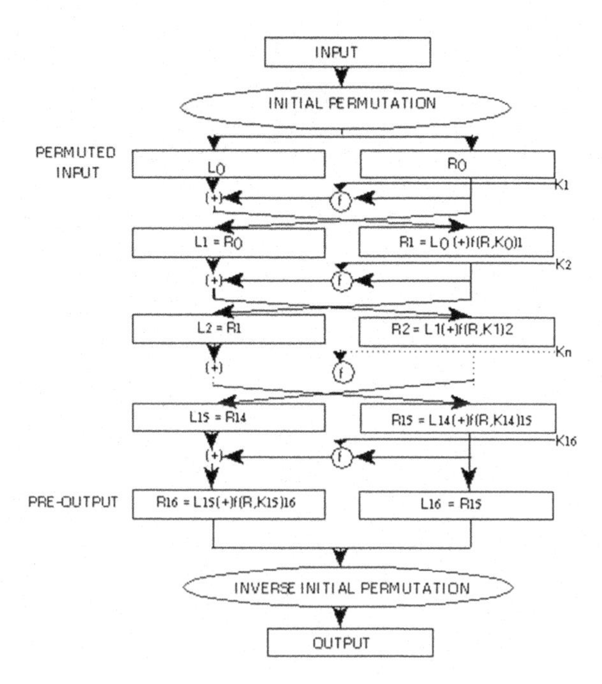

Fig. 3 Proposed mechanism

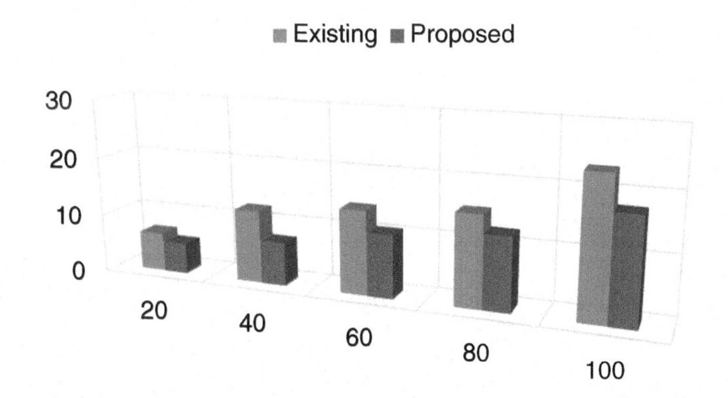

Fig. 4 Comparison of existing and proposed work

this data into cipher text and this cipher text is transmitted towards the destination where the destination user enters the secure key and decrypts the original message, as shown in Fig. 4.

7 Conclusion

In the IoT, secure transmission of personal data is a challenging task due to intermediate sensors nodes. In this paper, a DES algorithm is used to encrypt the message and this encrypted message is transferred from one user to another user in a secure manner. The DES algorithm provides security as well as taking less time to perform encryption and decryption processes compared to other cryptography algorithms. MATLAB is used to simulate the proposed mechanism. The simulation results shows that the proposed mechanism takes less time to perform encryption and decryption. In future, we will continue to work on it, try to use other security algorithms, and also try to enhance the existing DES algorithm.

References

1. Patel KK, Patel SM (2016). Internet of things-IOT: definition, characteristics, architecture, enabling technologies, application & future challenges
2. Ford S (2014) The internet's next big idea: connecting people, information, and things
3. Cheng S, Cai Z, Li J (2015) Curve query processing in wireless sensor networks. IEEE Trans Veh Technol 64(11):5198–5209
4. Huang B, Yu J, Yu D, Ma C (2014) SINR based maximum link scheduling with uniform power in wireless sensor networks. KSII Trans Internet Inf Syst 8(11)
5. Hu C, Li H, Huo Y, Xiang T, Liao X (2016) Secure and efficient data communication protocol for wireless body area networks. IEEE Trans Multi-Scale Comput Syst 2(2):94–107
6. "Nest thermostat". https://nest.com/thermostat/meet-nest-thermostat/
7. Vermesan O, Friess P (eds) (2013) Internet of things: converging technologies for smart environments and integrated ecosystems. River Publishers
8. Evans D (2011) The internet of things: how the next evolution of the internet is changing everything. White paper, Cisco Internet Business Solutions Group (IBSG)
9. Greenough J (2015) The 'internet of things' will be the world's most massive device market and save companies billions of dollars. Available via Business Insider
10. Vermesan O, Friess P (eds) (2014) Internet of things-from research and innovation to market deployment, vol 29. River Publishers, Aalborg
11. Jie Y, Pei JY, Jun L, Yun G, Wei X (2013) Smart home system based on IoT technologies. In: 2013 fifth international conference on computational and information sciences (ICCIS). IEEE, pp 1789–1791
12. Wang M, Zhang G, Zhang C, Zhang J, Li C (2013) An IoT-based appliance control system for smart homes. In: 2013 fourth international conference on intelligent control and information processing (ICICIP). IEEE, pp 744–747
13. Song T, Li R, Mei B, Yu J, Xing X, Cheng X (2017) A privacy preserving communication protocol for IoT applications in smart homes. IEEE Internet Things J
14. Hu C, Cheng X, Zhang F, Wu D, Liao X, Chen D (2013) OPFKA: secure and efficient ordered-physiological-feature-based key agreement for wireless body area networks. In: 2013 proceedings IEEE, INFOCOM. IEEE, pp 2274–2282
15. Naglic M, Souvent A (2013) Concept of smart home and smart grids integration. In: 2013 4th international youth conference on energy (IYCE). IEEE, pp 1–5
16. Fogli D, Lanzilotti R, Piccinno A, Tosi P (2016) AmI@Home: a game-based collaborative system for smart home configuration. In: Proceedings of the international working conference on advanced visual interfaces. ACM, pp 308–309

17. Li H, He Y, Sun L, Cheng X, Yu J (2016) Side-channel information leakage of encrypted video stream in video surveillance systems. In: 2016-the 35th annual IEEE international conference on computer communications, IEEE INFOCOM. IEEE, pp 1–9
18. Poslad S (2011) Ubiquitous computing: smart devices, environments and interactions. Wiley
19. Pantelopoulos A, Bourbakis NG (2010) A survey on wearable sensor-based systems for health monitoring and prognosis. IEEE Trans Syst Man, Cybern Part C Appl Rev 40(1):1–12
20. Sutherland IE (1968) A head-mounted three dimensional display. In: Proceedings of the fall joint computer conference, part I, 9–11 December 1968. ACM, pp 757–764
21. Miner CS, Chan DM, Campbell C (2001) Digital jewelry: wearable technology for everyday life. In: CHI'01 extended abstracts on human factors in computing systems. ACM, pp 45–46
22. Billinghurst M, Starner T (1999) Wearable devices: new ways to manage information. Computer 32(1):57–64
23. Spangler WS, Kreulen JT, Chen Y, Proctor L, Alba A, Lelescu A, Behal A (2010) A smarter process for sensing the information space. IBM J Res Dev 54(4):1–13
24. Klein C, Kaefer G (2008) From smart homes to smart cities: opportunities and challenges from an industrial perspective. In: International conference on next generation wired/wireless networking. Springer, Berlin, Heidelberg, pp 260–260
25. Yovanof GS, Hazapis GN (2009) An architectural framework and enabling wireless technologies for digital cities & intelligent urban environments. Wirel Pers Commun 49 (3):445–463
26. Schonwalder J (2010) Internet of things: 802.15.4, 6LoWPAN, RPL, COAP. http://www.utwente.nl/ewi/dacs/Colloquium/archive/2010/slides/2010-utwente-6lowpan-rpl-coap.pdf. Last accessed, 10 April 2014
27. Gershenfeld N, Krikorian R, Cohen D (2004) Internet of things. Scientific American

A Reputation-Based Mechanism to Detect Selfish Nodes in DTNs

Rakhi Sharma and D. V. Gupta

Abstract A delay tolerant network (DTN) is a complete wireless network. In a DTN there is no base station as it is in the case of existing wireless networks. Nodes may behave selfishly to transmit a message to save their own resources, such as energy. The cooperation requires detecting routes and transmitting the packets for other nodes, even though it consumes network bandwidth, buffer, and energy. A selfish node is a node that may be unwilling to cooperate to transfer packets. Such a node wants to preserve its own energy while using the services of others and consuming their resources. Many approaches have been used in the literature to implement the concept of non-cooperation in a simulated environment. However, none of them is capable of reflecting real cases and thus, the implementation of non-cooperative behavior needs improvement. In this paper, we focus on malicious and selfish node behavior, and we present a new classification and comparison between existing methods and algorithms to implement selfish nodes. Finally, we propose a new algorithm to implement selfish nodes in a DTN environment.

Keywords Delay tolerant network (DTN) · Selfish node · Reputation
Watchdog and cooperation

1 Introduction

A delay tolerant network (DTN) is a networking architecture [1] composed of nodes that cooperate with each other to forward packets, with associated connectivity variables delays, high error rates, and intermittent connectivity [2, 3]. DTNs follow a store, carry and forward mechanism, this mechanism consist of storing the data in a

R. Sharma (✉)
JMIT, Radaur, India
e-mail: rakhisharma2k7@gmail.com

D. V. Gupta
College of Engineering Roorkee, Roorkee, India
e-mail: dvgupta.rke@gmail.com

A. Kumar and S. Mozar (eds.), *ICCCE 2018*,
Lecture Notes in Electrical Engineering 500,
https://doi.org/10.1007/978-981-13-0212-1_7

node's buffer and forwarding it to the next available node. The DTN architecture implements store-and-forward message switching by adding a new protocol layer called the "bundle layer" on top of heterogeneous region specific lower layers. The bundle layer ties together the region specific lower layers so that application programs can communicate across multiple regions [4] (Fig. 1).

There are a number of nodes present in the environment as some of them are selfish nodes [5]. A selfish node is a node that doesn't want to cooperate in the transmission of packets in order to save its own energy [5, 6]. Such nodes in the network decrease the network performance. In this paper, we classify selfish behavior from different aspects, and the reasons for non-cooperation of nodes, such as malicious attacks, buffer limitations and energy constraints etc. We explicitly consider why a particular node does not cooperate as it generally should, and are mainly interested in the resulting less cooperative behavior [7].

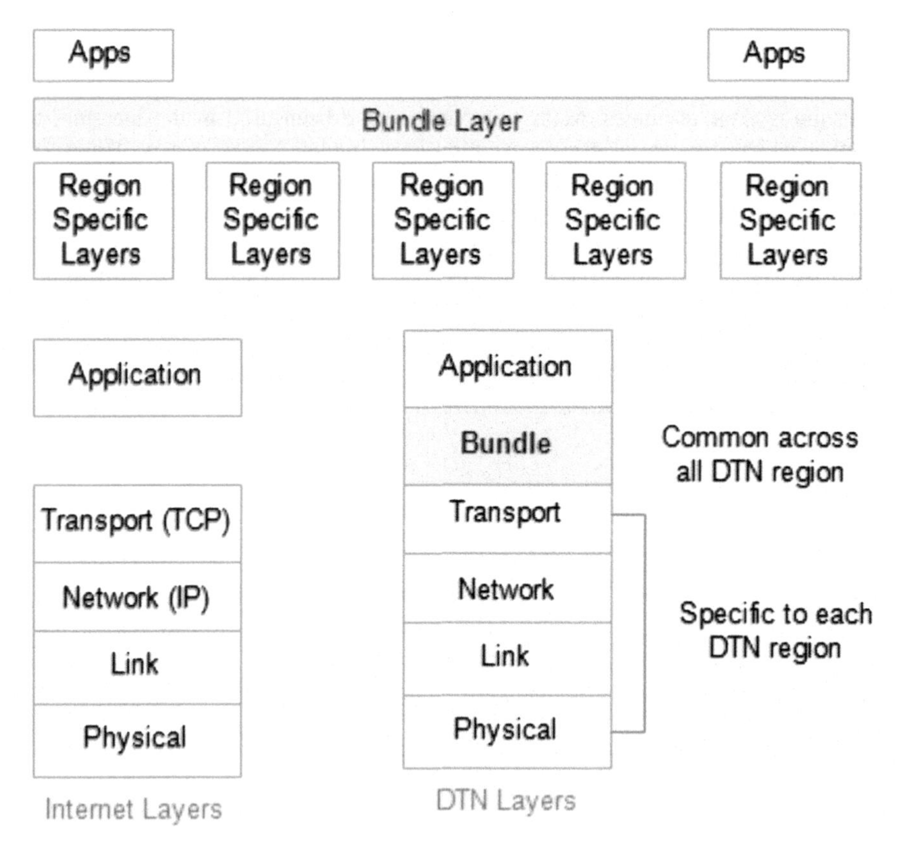

Fig. 1 DTN-architecture [4]

A. Causes of non-cooperation

A node might be unwilling to take part in a DTN directing convention, since they would prefer not to spend assets, for example, power and cushion on sending bundles of others, [8] as a few or every one of them attempt to augment their faculty benefits [9]. A node might be ready but not be able to collaborate due to resources [10], the gadgets are typically asset constrained, as powerlessness to manage the cost of the vitality (e.g. fueled by non-rechargeable batteries), support restrictions, which may make their clients be selfish.

Nodes can be selfish because of the current of security saving conventions as indicated by their protection targets (personality protection, area protection, message and substance security, and additionally connections security), nodes are stressed of unveiling their private versatility information [11, 12].

2 Related Work

Asuquo et al. [13] planned a collaborative trust management scheme (CTMS) which depended on the Bayesian identification guard dog way to deal with identifying selfish and malicious conduct in DTN nodes.

Chen et al. [14] recognized and redress the selfish and malicious conduct of nodes and upgrade the participation among nodes. They proposed a novel technique that recognizes selfish and malicious nodes rapidly and effortlessly by creating a negative reaction to the DTN.

Mariyam Benazir and Umarani [15] introduced a proficient motivating incentive compatible routing protocol (ICRP) with numerous duplicates for two bounce DTNs in view of the algorithmic diversion hypothesis. It takes both the experience likelihood and transmission fetched into thought to manage the mischievous activities of selfish nodes. They built up a mark conspire in light of bilinear guide to keep the malicious nodes from altering.

Cai et al. [16] proposed a provenance-based trust system, in particular PROVEST, that intends to accomplish exact shared trust appraisal and amplify the conveyance of correct messages obtained by goal nodes while limiting message deferral and correspondence fetched under asset-obliged network situations. Provenance alludes to the historical backdrop of responsibility for esteemed protest or data.

Cho and Chen [17] Proposed a refined approach to compute the selfish level of the nodes and looked at the subject utilizing cooperative capacity.

Table 1 Comparison of various trust management techniques [18–21]

Techniques	Security	Drop rate	Overhead	Transmission delay
Reputation-based	High	Low	High	High
Barter-based	Low	High	Low	Low
Credit-based	Low	Low	High	High
Watchdog based	High	Low	High	Low

3 Comparative Analysis

In this section is a comparison between different existing techniques, such as reputation-based, credit-based, barter-based and watchdog with different performance metrics, such as security, drop rate, packet overhead ratio and time delay. The analysis shows that reputation and watchdog techniques are much better than the other techniques [18] (Table 1).

4 Proposed Mechanism

In this paper we propose a mechanism in which selfish nodes will be detected in a DTN. In the proposed mechanism, a TA, i.e. a trusted authority, is used to monitor the traffic of the network and record data transmission information of each node. This data transmission contains information about the number of messages sent by a node, the number of messages received by a node and the number of messages dropped by a node. On the basis of this information, the TA assigns some reputation value to all nodes and if a node has a reputation value lower than some Th, i.e. threshold value, then that node may be treated as a selfish node, otherwise the node is normal (Fig. 2).

Proposed Algorithm

1. Start
2. Set nodes in network
3. Set TA in network
4. Now TA checks routing information of each node
5. A reputation value is assigned to each node based on past history
6. If node reputation < 0.5
   Then
7. Node is a selfish node
   Else
8. Node is a normal node
9. End

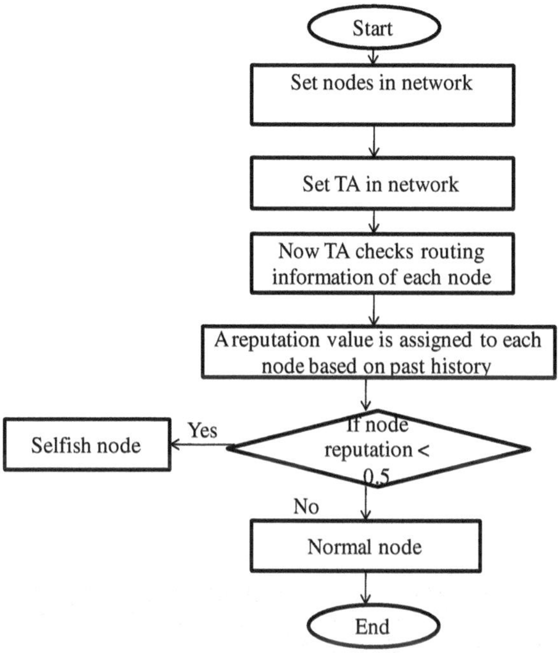

Fig. 2 Flow chart of the proposed mechanism

Start
Set nodes in network
Set TA in network
Now TA checks routing information of each node
A reputation value is assigned to each node based on past history
If node reputation < 0.5
Yes
Selfish node
No
Normal node
End

5 Conclusion

DTNs are extremely sensible, making it impossible for uncooperative practices to occur. The speculation of remote devices will soon hand DTNs over a standout amongst the most critical association techniques to the Internet. Be that as it may, the current of selfish nodes in the earth diminishes the execution of DTNs, so for a superior assessment of a calculation of recognition selfish nodes, we ought to have reenactment mirror the truth of uncooperative nodes.

In this paper, we initially grouped the diverse reasons of non-collaboration, and the selfish conduct in DTNs. We dissected and analyzed the diverse existing systems for executing selfish conduct, utilizing irregular numbers, a rate of selfish nodes, and a few parameters to depict misbehavior of nodes. Last, we proposed a calculation to actualize the non-participation in DTNs, our reason is to propose another calculation that mirror the truth of selfish nodes, taking in thought the reasons of the unwillingness of nodes to coordinate.

In future, it is planned to keep taking a shot at it and would propose a superior plan to actualize a wide range of misbehavior nodes in DTN, and implement on ONE simulator.

References

1. Jagadale PS (2014) A probabilistic misbehavior detection scheme in DTN survey. Int J Innov Res Comput Commun Eng 2(11):6784–6789
2. Magaia N, Rogerio Pereira P Correia MP (2013) Selfish and malicious behavior in delay-tolerant networks. In: Future networks mobile summit, pp 1–10
3. Benamar N, Singh KD, Benamar M, Ouadghiri DE, Bonnin J-M (2014) Routing protocols in vehicular delay tolerant networks: a comprehensive survey computer. Communication 48:141–158
4. Krug S, Schellenberg S (2015) Impact of traffic and mobility patterns on network performance in disaster scenarios. In: Proceedings of 10th ACM MobiCom work challenged networks, pp 9–12
5. Gamit V, Patel H (2014) Evaluation of DTN routing protocols. Int J Eng Sci Res Technol 3(2):1–5
6. Dias JA, Rodrigues JJ, Shu L, Ullah S (2014) Performance evaluation of a cooperative reputation system for vehicular delay-tolerant networks. EURASIP J Wirel Commun Netw 1:1–5
7. Benamar N, Benamar M, Ahnana S, Saiyari FZ, El Ouadghiri MD, Bonnin JM (2013) Are VDTN routing protocols suitable for data collection in smart cities: a performance assessment. J Theory Appl Inf Technol 58(3):589–600
8. Benamar N, Singh KD, Benamar M, El Ouadghiri D, Bonnin JM (2014) Routing protocols in vehicular delay tolerant networks: a comprehensive survey. Comput Commun 48:141–158
9. Doddamani ML, Shanwad V (2014) Delay tolerant network. Int J Sci Technol 2:50–52
10. Loudari SE, Benamar M, Benamar N, Habbal A (2015) The impact of energy consumption on the performance of DTN routing protocols. In: Fourth international conference internet applications protocols services, pp 147–154
11. Miao J, Hasan O, Ben Mokhtar S, Brunie L, Yim K (2012) An investigation on the unwillingness of nodes to participate in mobile delay tolerant network routing. Int J Inf Manag 1–11
12. Benamar M, Benamar N, El Ouadghiri D (2015) The effect of cooperation of nodes on VDTN routing protocols. In: International conference on wireless networks and mobile communications (WINCOM), pp 1–7
13. Asuquo P, Cruickshank H, Anyigor Ogah CP, Lei A, Sun Z (2016) A collaborative trust management scheme for emergency communication using delay tolerant networks, pp 1–6
14. Chen K, Shen H, Yan L (2015) Multicent: a multifunctional incentive scheme adaptive to diverse performance objectives for DTN routing. IEEE Trans Parallel Distrib Syst 26(6): 1643–1653
15. Mariyam Benazir SA, Umarani V (2016) Detection of selfish & malicious behavior using dtn-chord monitoring in mobile networks. In: International conference on information communication and embedded system (ICICES), pp 1–5
16. Cai Y, Fan Y, Wen D (2016) An incentive-compatible routing protocol for two-hop delay tolerant networks. IEEE Trans Veh Technol 1–11
17. Cho J-H, Chen I-R (2016) PROVEST: provenance-based trust model for delay tolerant networks. IEEE Trans Depend Secur Comput 1–15
18. Sharma A, Singh D, Sharma P, Dhawan S (2015) Selfish nodes detection in delay tolerant networks. future trends. In: Computer and knowledge management ABLAZE, pp 407–410
19. Sharma A (2014) A credit based routing mechanism to contrast selfish nodes in delay tolerant networks. In: International conference on parallel, distributed and grid computing, pp 295–300

20. Jiang Q, Men C, Yu H, Cheng X (2015) A secure credit-based incentive scheme for opportunistic networks. In: 7th international conference on intelligent human-machine systems and cybernetics, vol 1, pp 87–91
21. Zhu H, Du S, Gao Z, Dong M Cao Z (2014) A probabilistic misbehavior detection scheme toward efficient trust establishment in delay-tolerant networks. IEEE Trans Parallel Distrib Syst 25(1):22–32
22. Subramaniyan S, Johnson W, Subramaniyan K (2014) A distributed framework for detecting selfish nodes in MANET using record- and trust-based detection (RTBD) technique. EURASIP J Wirel Commun Netw 205:210

Improved Target Detection in Doppler Tolerant Radar Using a Modified Hex Coding Technique

Majid Alotaibi

Abstract In every corner of the globe, nations want to improve the monitoring mechanism of the country, so that no one can enter their territory in an unwanted manner easily. Well-known equipment, called Radar, is commonly used for monitoring. However, only a small amount of work is done to monitor multiple moving targets in the presence of Doppler. This important issue diverts the attention of the research community away from working on this platform. In the present literature, the merit factor (MF) is improved by increasing the amplitude of the main lobe. However, these particular approaches did not attach more importance to the effects of noise side peaks of fast moving targets. The drawback of noise peaks masks slow-moving targets and cannot be clearly seen by the radar receiver. As a result it reduces the performance of the Doppler radar system. In this paper, an approach is presented which not only improves multiple moving target detection, but also reduces the energy of code generation. This approach is simple and effective in detecting multiple moving targets at the desired Doppler. The presented technique is called Improved Target Detection in Doppler Tolerant Radar Using a Modified Hex Coding Technique. MATLAB is used to formalize the results by simulation.

Keywords Doppler tolerant radar code · Hex code · Multiple moving targets Matlab

1 Introduction

The monitoring of day-to-day activity by a country's surveillance system is an important factor in observing various activities. Radar is the only equipment to monitor such activities in the country. However, the state of art of the work mainly focuses towards the development of immobile object recognition. To achieve the

M. Alotaibi (✉)
Department of Computer Engineering, College of Computer and Information Systems, Umm Al-Qura University, Makkah, Saudi Arabia
e-mail: mmgethami@uqu.edu.sa

A. Kumar and S. Mozar (eds.), *ICCCE 2018*,
Lecture Notes in Electrical Engineering 500,
https://doi.org/10.1007/978-981-13-0212-1_8

goal of target detection probability, several approaches have been presented to improve the merit factor (MF) of the received echo by means of the auto correlation technique. However, these approaches result in noise side peaks which restrict the technique to use in detecting stationary targets and makes it less useful for finding multiple small moving objects as the noise side peaks of the auto-correlated signal as the side peaks of noise in the auto-correlated signal acquire the echoes or noise from many small moving targets. To enhance the current approaches, a variety of constraints such as attitude, altitude, and Range discovery were analyzed. Also for multiple moving target discovery processes, several approaches are being proposed to augment the discovery probability of multiple and moving targets, which requires an array of Doppler filter bank. Current radar for moving and multiple target recognition employs the numerous matchless radiating (k) aerial arrays to acquire an intelligent and sharp autocorrelation reply, and thus the object finding probability is improved. Deviation of the acknowledged constraints is similar to the phase and processing of the acknowledged signal that can execute different operations such as tracking and finding. This result of tracking and finding targets shows the level of autonomy of the transmitted signal, therefore the transmitted signal cannot shift while the acknowledged signal could be shifted more than once, and can be represented by 'p' for simplicity. This scheme is able to broadcast unreliable signals from 'k' matchless aerials and the received signals are jointly processed subsequent to the acknowledged signal by 'p' matchless receiving aerials which results in the enhancement of the accuracy of detection of moving and multiple targets. However, emission by multiple moving aerials results in the need for enormous power, moreover the side noise peaks are more because of acknowledged echoes from the moving and multiple targets. Thus slow affecting targets are masked by these side noise peaks and also the range of the radar is affected by this method. Consequently, power consumption and range presentation is lost with the enhanced probability, and relatively it is not up to the mark. In this paper, multiple moving target detection is upgraded in terms of range and Doppler by using different windowing techniques. The main objective of this paper is to reduce the amplitude of side noise spikes and to increase the amplitude of the main lobe. To achieve this goal we are using windowing techniques to reduce range noise side spikes and make the detection of moving targets much easier. The identified targets can be shown on a Doppler vs. delay plot which arises from the ambiguity function. The key role of this paper is to present a comprehensive detection of moving targets in the presence of Doppler at different ranges. The presented approach is very simple but very affective for multiple moving target detection and also minimizes the transmission power by sending a simple digital code which discords one major portion of the detection process called the range gates. The rest of the paper is organized as follows. In Sect. 2, a literature survey is presented. The proposed approach is discussed in Sect. 3, and the conclusion is given in Sect. 4.

2 Literature Survey

Rafiuddin and Bhangdia [1] presented an approach in which the authors use p1 and p3 series of poly-phase codes along with hyperbolic frequency modulation (HFM). The presented approach enhanced the merit value of the received echo. However, the presented approach increases the delay therefore it cannot fulfill the purpose of moving and multiple target detection. Lewis and Kretschmer Jr. [2] develop an approach in which they proved that in place of poly-phase codes, bi-phase codes can be used to enhance the synchronization of the primary surveillance radar (PSR) by shrinking the bits of the broadcast signal and in that way security can be improved. Also at the same time, the poly-phase codes (i.e. P1 and P2) can be suitably created using a linear frequency modulated waveform technique (LFMWT) on step evaluation. This approach also improves the transmission capacity of the receiver. Lewis and Kretschmer Jr. [3] presented another method using P3 and P4 codes generated by the use of linear frequency modulation waveform (LFMW) to give improved target detection probability when compared with P1 and P2.

Kretschmer Jr. and Lewis [4] proposed an another approach using a set of codes called P3 and P4 codes to enhance the signal-to-noise ratio and they also demonstrated that such codes are more capable of getting a better response in terms of target detection probability when compared to other codes of the poly-phase family. But in the presence of Doppler these codes showed a very poor response of probability of target detection. Lewis [5] proposed a technique, known as the sliding window technique (SWT), to reduce the noise peaks which are caused due to the range-time noise spikes produced. However the presented approach is inadequate to decrease the noise spikes up to a certain level, and as a consequence has finite appliances in Doppler tolerant radars. Kretschmer Jr. and Welch [6] offered a technique in which they used autocorrelation of poly-phase codes to remove the noise elements that are present with the signal. But the presented approach fails to locate high velocity targets in the occurrence of Doppler, because autocorrelation of poly-phase codes creates noise spikes at close to zero Doppler. As a result, this approach is unsuitable for moving and multiple target discoveries. This particular approach also begins with the use of an amplitude weighting function (AWF) utilizing poly-phase codes to reduce noise spikes on the receiver side. However, there is an extra power loss in the method and merely an inspection on correlating the sending and the receiving power at source and destination respectively is made.

Sahoo and Panda [7] proposed a compaction window approach to decrease the effect of the noise peaks in Doppler tolerant radars. However, the presented approach increases delay and thus fails to create a larger window or enhance the capacity of windows to recognize the moving and multiple targets exactly. Singh et al. [8] proposed coding technique to enhance the size of the window in which they used Hex coding to enhance the probability of moving and multiple target detection. Though due to huge mathematical complexity it devours extra power and boosts delay, therefore it is valid to distinguish slow moving targets only. Singh et al. [9] proposed a method called the matrix coding technique (MCT), where no doubt the number of windows are greater in number in comparison with the existing approaches to obtain an obvious image of the present position of the moving target. But this approach is restricted to find immobile and sluggish targets only, because the duration of the calculated code vector is less and this reduces the merit factor (MF) of the auto-correlated signal and results in side noise peaks approximately around zero Doppler.

In this paper, a technique called Improved Target Detection in Doppler Tolerant Radar Using a Modified Hex Coding Technique is proposed. This technique improves the probability of target detection by creating multiple numbers of windows with respect to the desired Doppler. It also improves the merit factor of the auto-correlated signal and reduces the power consumed by the received echo.

3 Proposed Approach

In the present approach, equal weighted binary hex codes from 0 to 15 are considered, which are divisible by 3 (such as 3, 6, 9 and 12) and can be represented in the binary system as 0011, 0110, 1001 and 1100 given as

$$H_c = \prod_{k=1}^{j} Pk \tag{1}$$

where H_c is an equal weighted hex code, P = 3 and $1 \leq j \leq 4$.

The concatenation binary series of H_c can be represented as below

0 0 1 1 0 1 1 0 1 0 0 1 1 1 0 0

A matrix N × N can be obtained by taking the above series as the first row and column of the matrix. The other elements of the matrix can be developed by using ex-or operation shown in the equations

$$R_{22} = R_{12} \oplus R_{21} \tag{2}$$

$$R_{23} = R_{21} \oplus R_{13} \tag{3}$$

. . . .

. . . .

$$R_{2n} = R_{21} \oplus R_{1n} \tag{4}$$

Generalizing the above, we get,

$$R_{n(n-1)} = R_{n1} \oplus R_{1(n-1)} \quad (5)$$

and

$$R_{nn} = R_{n1} \oplus R_{1n} \quad (6)$$

where R is the radar matrix.

Matlab finds its use in image processing as it is feasible and holds good for testing of the algorithm as it is a growing database with in-built libraries. In this approach, matlab is used to simulate the results by transmitting the matrix blocks and detecting the moving targets masked in the noise at the desired Doppler. Figure 1 shows the Doppler frequency v/s normalized amplitude when a binary matrix of equal weighted hex code (see Table 1) is transmitted. From the figure we observe two clear windows from 8 to 12 kHz and from 14 to 40 kHz where we can easily detect the target as the amplitude of the noise peaks is much lower than the threshold limit, i.e. 0.2 (as per the literature).

Quadratic residues are widely used in acoustics, graph theory, cryptography, etc. Quadratic residues are used to get a clear window for detecting the moving targets which are masked in the side lobes. A quadratic residue of 15 is taken as it is close to 16 (the total number of bits in the presented approach) and odd values provide a greater number of changes than even values.

$$Q_r(15) = \{1, 4, 6, 9, 10\}$$

where Q_r is the quadratic residue.

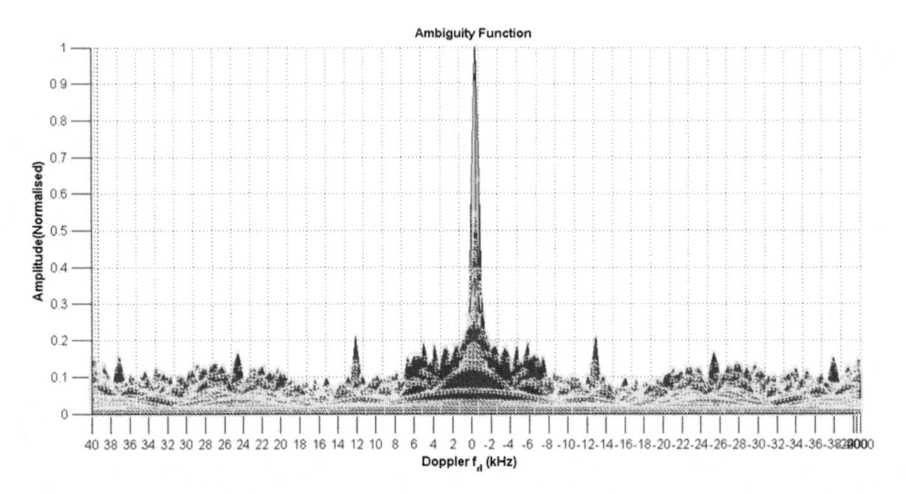

Fig. 1 Ambiguity function of the table

Table 1 Binary matrix of equal weighted hex code

0	0	1	1	0	1	1	0	1	0	0	1	1	1	0	0
0	0	1	1	0	1	1	0	1	0	0	1	1	1	0	0
1	1	0	0	1	0	0	1	0	1	1	0	0	0	1	1
1	1	0	0	1	0	0	1	0	1	1	0	0	0	1	1
0	0	1	1	0	1	1	0	1	0	0	1	1	1	0	0
1	1	0	0	1	0	0	1	0	1	1	0	0	0	1	1
1	1	0	0	1	0	0	1	0	1	1	0	0	0	1	1
0	0	1	1	0	1	1	0	1	0	0	1	1	1	0	0
1	1	0	0	1	0	0	1	0	1	1	0	0	0	1	1
0	0	1	1	0	1	1	0	1	0	0	1	1	1	0	0
0	0	1	1	0	1	1	0	1	0	0	1	1	1	0	0
1	1	0	0	1	0	0	1	0	1	1	0	0	0	1	1
1	1	0	0	1	0	0	1	0	1	1	0	0	0	1	1
1	1	0	0	1	0	0	1	0	1	1	0	0	0	1	1
0	0	1	1	0	1	1	0	1	0	0	1	1	1	0	0
0	0	1	1	0	1	1	0	1	0	0	1	1	1	0	0

Table 2 New code with zeros and ones changed

Equal weighted hex code word	0	0	1	1	0	1	1	0	1	0	0	1	1	1	0	0
Quadratic residue of 15	Q1 ↓			Q2 ↓		Q3 ↓			Q4 ↓	Q5 ↓						
New code with zeros and ones changed (C_{01})	1	0	1	0	0	0	1	0	0	1	0	1	1	1	0	0

Where C_{01} is the code generated after the zeros and ones change

Consider the positions of $Q_r(15)$ in Eq. (1), 16 bits are generated by complementing the binary digits present at positions 1, 4, 6, 9 and 10 as depicted in Table 2.

A matrix of 16 × 16 is obtained (shown in Table 3) by taking C_{01} as first row and column of the matrix and rest of the elements in the matrix are generated in the same manner as Eqs. (2)–(6) and Table 1.

Figure 2 shows the normalized amplitude v/s Doppler frequency graph by transmitting binary matrix of equal weighted hex code with ones and zeros changed (Table 3) which has two clear windows, at 7 kHz to 11 kHz and 14 kHz to 34 kHz, respectively.

Similarly, we can generate C_1 and C_0 codes (shown in Tables 4 and 6) by changing only ones and only zeros in the binary code of Eq. (1) and developing their respective matrix as given in Tables 5 and 7.

Table 3 Binary matrix of equal weighted hex code with ones and zeros changed

1	0	1	0	0	0	1	0	0	1	0	1	1	1	0	0
0	0	1	0	0	0	1	0	0	1	0	1	1	1	0	0
1	1	0	0	0	0	0	1	1	0	1	0	0	0	1	1
0	0	1	0	0	0	1	0	0	1	0	1	1	1	0	0
0	0	1	0	0	0	1	0	0	1	0	1	1	1	0	0
0	0	1	0	0	0	1	0	0	1	0	1	1	1	0	0
1	1	0	1	1	1	0	1	1	0	1	0	0	0	1	1
0	0	1	0	0	0	1	0	0	1	0	1	1	1	0	0
0	0	1	0	0	0	1	0	0	1	0	1	1	1	0	0
1	1	0	1	1	1	0	1	1	0	1	0	0	0	1	1
0	0	1	0	0	0	1	0	0	1	0	1	1	1	0	0
1	1	0	1	1	1	0	1	1	0	1	0	0	0	1	1
1	1	0	1	1	1	0	1	1	0	1	0	0	0	1	1
1	1	0	1	1	1	0	1	1	0	1	0	0	0	1	1
0	0	1	0	0	0	1	0	0	1	0	1	1	1	0	0
0	0	1	0	0	0	1	0	0	1	0	1	1	1	0	0

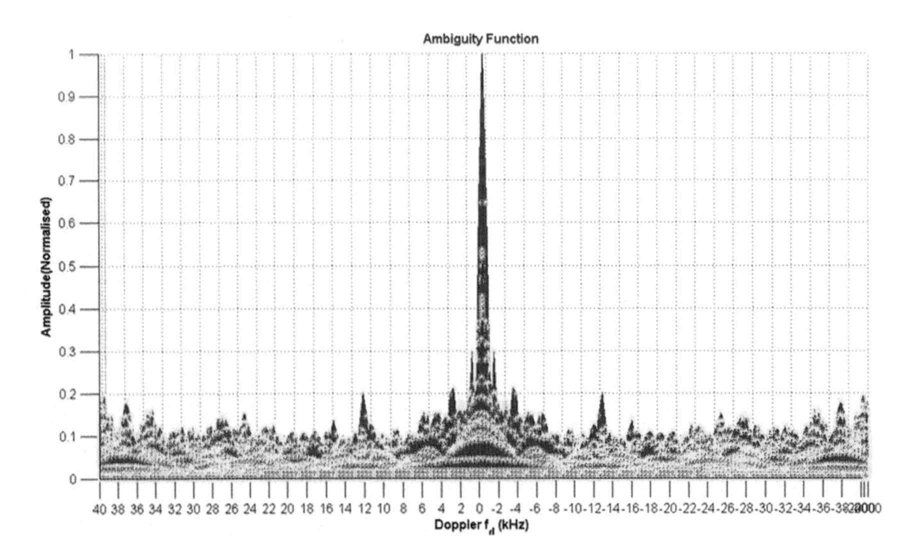

Fig. 2 Ambiguity function of Table 3

Figure 3 shows the ambiguity function received after transmitting a binary matrix of equal weighted hex code with ones changed (Table 5) to detect multiple moving targets. From Fig. 3 we can observe one small window from 7 to 11 kHz and a huge window from 14 to 34 kHz.

Table 4 New code with only ones changed

Equal weighted hex code word	0	0	1	1	0	1	1	0	1	0	0	1	1	1	0	0
Quadratic residue of 15	Q1 ↓			Q2 ↓		Q3 ↓			Q4 ↓	Q5 ↓						
New code word with only ones changed C_1	0	0	1	0	0	0	1	0	0	0	0	1	1	1	0	0

Where C_1 is the code generated after the ones have been changed

Table 5 Binary matrix of equal weighted hex code with ones changed

0	0	1	0	0	0	1	0	0	0	0	1	1	1	0	0
0	0	1	0	0	0	1	0	0	0	0	1	1	1	0	0
1	1	0	1	1	1	0	1	1	1	1	0	0	0	1	1
0	0	1	0	0	0	1	0	0	0	0	1	1	1	0	0
0	0	1	0	0	0	1	0	0	0	0	1	1	1	0	0
0	0	1	0	0	0	1	0	0	0	0	1	1	1	0	0
1	1	0	1	1	1	0	1	1	1	1	0	0	0	1	1
0	0	1	0	0	0	1	0	0	0	0	1	1	1	0	0
0	0	1	0	0	0	1	0	0	0	0	1	1	1	0	0
0	0	1	0	0	0	1	0	0	0	0	1	1	1	0	0
0	0	1	0	0	0	1	0	0	0	0	1	1	1	0	0
1	1	0	1	1	1	0	1	1	1	1	0	0	0	1	1
1	1	0	1	1	1	0	1	1	1	1	0	0	0	1	1
1	1	0	1	1	1	0	1	1	1	1	0	0	0	1	1
0	0	1	0	0	0	1	0	0	0	0	1	1	1	0	0
0	0	1	0	0	0	1	0	0	0	0	1	1	1	0	0

Table 6 New code with only zeros changed

Equal weighted hex code word	0	0	1	1	0	1	1	0	1	0	0	1	1	1	0	0
Quadratic residue of 15	Q1 ↓			Q2 ↓		Q3 ↓			Q4 ↓	Q5 ↓						
New code word with zeros changed only C_0	1	0	1	1	0	1	1	0	1	1	0	1	1	1	0	0

Where C_0 is the code generated after the zeros have been changed

The ambiguity function simulation result in Fig. 4 gives a clear windows from Doppler frequency 5 to 11 kHz and from 13 to 40 kHz.

Table 7 Binary matrix of equal weighted hex code with zeros changed

1	0	1	1	0	1	1	0	1	1	0	1	1	1	0	0
0	0	1	1	0	1	1	0	1	1	0	1	1	1	0	0
1	1	0	0	1	0	0	1	0	0	1	0	0	0	1	1
1	1	0	0	1	0	0	1	0	0	1	0	0	0	1	1
0	0	1	1	0	1	1	0	1	1	0	1	1	1	0	0
1	1	0	0	1	0	0	1	0	0	1	0	0	0	1	1
1	1	0	0	1	0	0	1	0	0	1	0	0	0	1	1
0	0	1	1	0	1	1	0	1	1	0	1	1	1	0	0
1	1	0	0	1	0	0	1	0	0	1	0	0	0	1	1
1	1	0	0	1	0	0	1	0	0	1	0	0	0	1	1
0	0	1	1	0	1	1	0	1	1	0	1	1	1	0	0
1	1	0	0	1	0	0	1	0	0	1	0	0	0	1	1
1	1	0	0	1	0	0	1	0	0	1	0	0	0	1	1
1	1	0	0	1	0	0	1	0	0	1	0	0	0	1	1
0	0	1	1	0	1	1	0	1	1	0	1	1	1	0	0
0	0	1	1	0	1	1	0	1	1	0	1	1	1	0	0

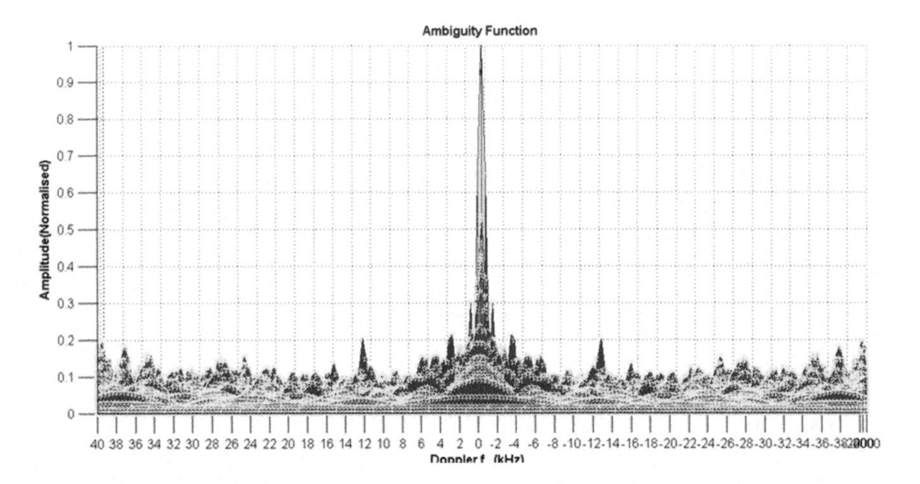

Fig. 3 Ambiguity function of Table 5

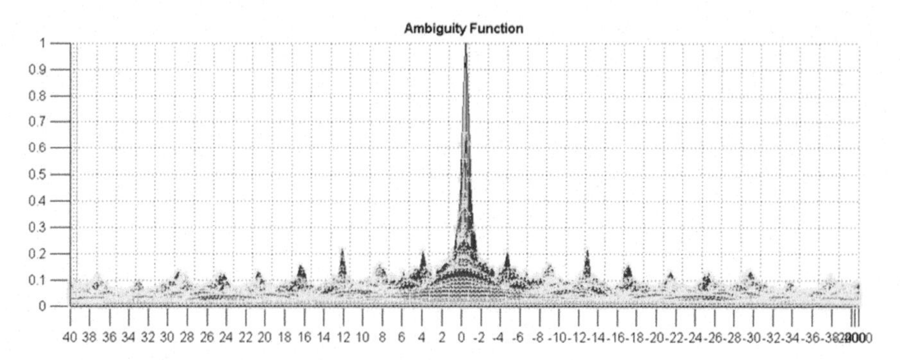

Fig. 4 Ambiguity function of Table 7

4 Conclusion

In this paper, a simple binary matrix coding approach is presented using a quadratic residue technique to detect multiple moving targets simultaneously. In this approach multiple clear windows are created with respect to the Doppler in order to obtain accurate information about multiple moving targets. This approach is more effective and simple. The approach is validated by simulation results obtained using MATLAB.

References

1. Rafiuddin SSA, Bhangdia VK (2013) Empirical analysis on Doppler tolerant radar codes. Int J Sci Eng Res 4(5):1579–1582
2. Lewis BL, Kretschmer FF Jr (1981) A new class of polyphase pulse compression codes and techniques. IEEE Trans Aerosp Electron Syst AES-17(3):364–372
3. Lewis BL, Kretschmer FF Jr (1982) Linear frequency modulation derived polyphase pulse compression codes. IEEE Trans Aerosp Electron Syst AES-18(5):637–641
4. Kretschmer FF Jr, Lewis BL (1983) Doppler properties of polyphase pulse compression waveforms. IEEE Trans Aerosp Electron Syst 19(4):521–531
5. Lewis BL (1993) Range-time-side lobe reduction technique for FM-derived polyphase PC codes. IEEE Trans Aerosp Electron Syst AES-29(3):834–840
6. Kretschmer FF Jr, Welch LR (2000) Side lobe reduction techniques for polyphase pulse compression codes. In: IEEE international radar conference, May 2000. pp 416–421
7. Sahoo AK, Panda G (2011) Doppler tolerant convolution windows for radar pulse compression. Int J Electron Commun Eng 4(1):145–152
8. Singh RK, Elizabath Rani D, Ahmad SJ (2016) RSBHCWT: re-sampling binary hex code windowing technique to enhance target detect. Indian J Sci Technol 9(47):1–5
9. Singh RK, Elizabath Rani D, Ahmad SJ (2017) HQECMT: hex quadratic residue Ex-OR coded matrix technique to improve target detection in Doppler tolerant radar. Int J Sci Res (PONTE) 73(1):21–28

Enhanced Packet Loss Calculation in Wireless Sensor Networks

Saud S. Alotaibi

Abstract Wireless Sensor Networks (WSNs) are autonomous and structure-less dynamic networks which consist of spatially distributed sensor nodes to support real-time applications. However, due to limited resource availability these networks face certain challenges. Many researchers address bandwidth and delay using different approaches to increase the quality of service (QoS). Almost all researchers address loss calculation using sliding window flow control protocol, which may not always give an optimum solution. So, accurate loss calculation is necessary to increase the packet delivery ratio (PDR) which in turn increases QoS. In this paper, a mathematical model is proposed to enhance the loss calculation in WSNs using Poisson theory.

Keywords WSN · Poisson random process · Active nodes · Link capacity PDR

1 Introduction

A wireless sensor network (WSN) is a network consisting of huge independent sensor gadgets, distributed in space to track environmental conditions, such as light, temperature, and humidity. They are cost effective, easily deployed and used for multimedia applications but are incorporated with storage, battery and bandwidth limitations. QoS is very important in WSN to transfer the data from source to destination node with minimal interruption and loss. It may be difficult to manage the delay and sequential order of the packets. If delay is present in the network links it should be minimal and the same delay should be present in every link of the network. Packet loss in the network is mostly due to the failure of the path chosen for communication, interruptions in packet transfer from source to destination, and jamming of intermediate nodes which act as routers. Packet loss can also occur if

S. S. Alotaibi (✉)
Department of Information Systems, College of Computer and Information Systems, Umm Al-Qura University, Makkah, Saudi Arabia
e-mail: ssotaibi@uqu.edu.sa

A. Kumar and S. Mozar (eds.), *ICCCE 2018*,
Lecture Notes in Electrical Engineering 500,
https://doi.org/10.1007/978-981-13-0212-1_9

the packets are not in the proper order in which they are sent due to communication collapse.

The order of packets can be conserved by using error correcting methods and route failure in the network can be avoided by using arrangements of nodes and making route decisions in advance. QoS should guarantee a large bandwidth, optimal delay from source to destination, and interruption-free routing. The dynamic nature of WSN, along with adjustable bandwidth, makes multimedia data transfer feasible over WSN. However, there are some issues that come together with it.

In the proposed approach a mathematical technique using the Poisson principle to calculate the data packets loss is presented, which improves the packet delivery ratio (PDR). The rest of the paper is organized as follows. Related work is in Sect. 2, Sect. 3 presents the proposed approach and simulation results, and the conclusion to the paper is in Sect. 4.

2 Related Work

The issues and challenges facing WSN are addressed in several ways by the research community. The model proposed by Avrachenkov and Antipolis [1] discussed the fact that size of the buffer required for the routers is comparably small. However, it uses general transmission speed and delay in the network link. This model may fail to evaluate the loss of packets and link utilization. Wei et al. [2] proposed a model to minimize packet loss in high speed networks and discussed various difficulties with the current TCP approach. The approach presented by Sarker and Johansson [3] gives a minimal outcome for loss of packets and delay analysis using an LTE (long-term evolution) system, which may not hold good for multimedia transmission and the time required to examine the path behavior in the network is a delayed process.

Katabi et al. [4] proposed an approach for high speed routers to control internet congestion and make the performance of TCP stable with an increase in the bandwidth-delay product. Kelly [5] presented a lossless and minimum delay protocol technique for TCP-IP networks. However, it may fail to increase the utilization of link in the network. The model presented by Zanella et al. [6] used a Markov chain and developed an analytic model to improve the performance of the network using TCP Westwood (TCPW). Xu et al. [7] presented a new scheme for control of congestion that mitigates the round trip time (RTT) injustice using *additive increase* and *binary search increase* policies to control window size. Leith et al. [8] propose a H-TCP to control overcrowding which is feasible for deployment in networks with greater speed and distance.

The model presented by Li et al. [9] gives a cross layer technique to reduce the loss in the network. This can be done by assigning the paths, considering the resource authorization. However, this approach fails when the packets to be transferred are greater compared to the routers in the network. Wang et al. [10]

proposed an approach based on TCP-FIT, an AIMD method to control the loss of packets from source to destination. However, it cannot give the exact size of the window as expected. Chen et al. [11] proposed a protocol based on CARM (congestion aware routing protocol) to maximize the QoS by managing data packet loss. This approach may not be appropriate when there are fewer routers and if the router node is not present in the network. The approach presented by Kaur and Singh [12] controls the packet loss by adjusting the measurements of the WSN, power and dynamic nodes. However, this may not be valid when node mobility is increased and due to consistent monitoring of the data packets from source to destination, more energy is required. In the presented approach, the packet loss is calculated accurately by developing a mathematical technique based on Poisson's ratio, which is simple to implement.

3 Proposed Approach

To accomplish exact transmission of multimedia data packets, loss must be lie within acceptable limits. So the estimation of exact packet loss is essential to focus on for multimedia transmission in WSN. To judge the window size, loss of packet information is the most important constraint. In the present literature, the size of the window is improved if constructive acknowledgement is being received; otherwise the size of the window is decreased. The size of the window is not as per the requirement of the network as it is pre-defined. This results in the poor usage of resources. In this approach, active sensor nodes are marked as a Poisson random process and calculate the loss of packet, busy and idle periods at the front side of the router. This loss of packet assessment is helpful for explicit requirement of the buffer size which results in an improved QoS. The investigational results reveal the utility in manipulating the optimal control in order to get a better QoS for multimedia data transmission in WSN.

Mathematical Modeling

Consider a WSN comprising a number of wireless sensor nodes extending over a physical area. Figure 1 represents the input station 'S' that cooperates with the router nodes and transmits the data packets towards the other router node till the message reaches destination node 'D'.

Each wireless sensor node (station) communicates with any other wireless sensor node over duplex channels either directly or indirectly. Two wireless sensor stations can directly communicate when both fall within in the coverage range of each other. Indirect transmission can also takes place between any two wireless sensor stations which are at a distance. For any successful transmission session, a route must be recognized prior to the beginning of the transmission session between any two wireless sensor stations through some intermediate wireless sensor stations called routers. So, congestion may happen at some of these router stations for some other

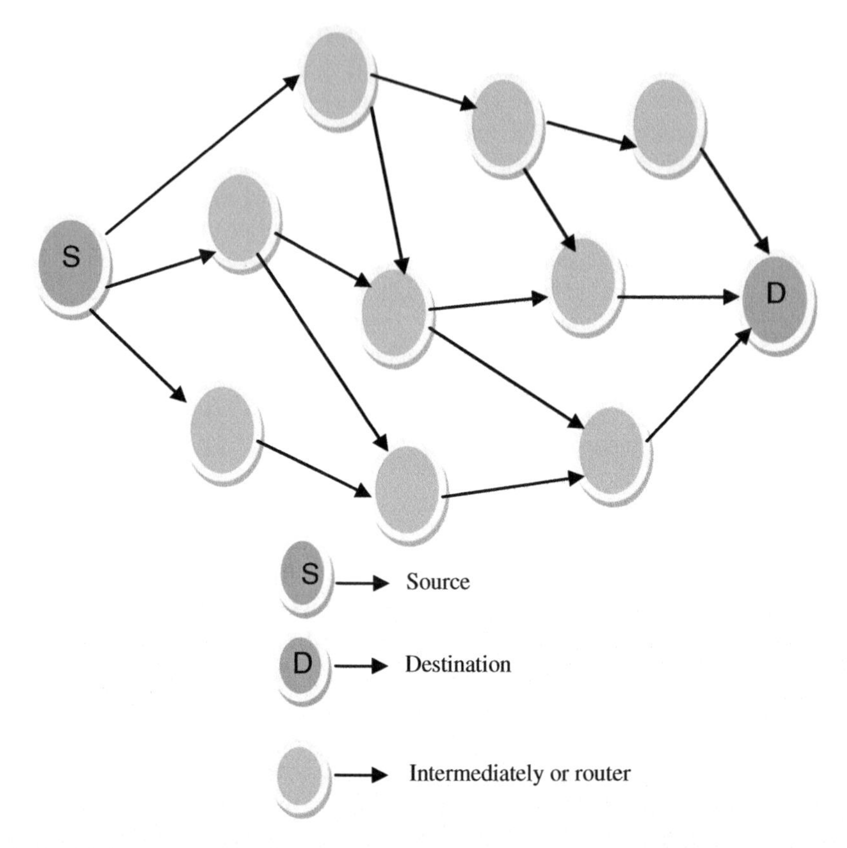

Fig. 1 Wireless sensor network

trans-receiver pairs. We consider the following assumptions for the purpose of simplicity

- Duplex channels are used to transfer data packets between two stations (sensor nodes).
- A node can be sleeping or active. A sleeping node (station) is one which cannot send data packets.
- A time period T is fixed through which an active sensor node (station) finishes the data packet communication and the same is being received at the intermediate sensor node (i.e. a router station). A sensor node is restricted to sending only one data packet on its own in a time period T.
- An intermediate sensor node (i.e. a router) can transmit q number of data packets in the time period T where $q > 1$.
- An intermediate senor node acts as router for number of Trans- Receiver pairs. The router collects the data packets from all these active sensor nodes and transfers them to the next sensor node via channels in the subsequent time

period T at a fixed rate of q data packets *per time period* T. So the service rate of one data packet is T/q per second.

- Enough bandwidth is set aside for multi-hop routing prior to the data packet starting to move forward.
- An intermediate sensor node (i.e. a router station) has a buffer of size q packets long in order to store the incoming data packets coming from the active sensor nodes with the up-series router.
- Loss can happen at the intermediate sensor node (router), if more data packets enter at the node in a given amount of time period T and the data packets go down, hence these data packets are vanished.
- To estimate the loss of packets due to congestion on the router, let the amount of active sensor nodes within the network be a random variable with a mean value μ per time period and $\mu \leq q$.
- The loss of packet information can be helpful in proposing an optimum approach for controlling the loss by increasing the buffers at intermediate sensor nodes (routers). This approach increases the bandwidth of the router channels and reduces the incoming data rate at the intermediate node (router).

Let the data packets coming from the active sensor nodes to the intermediate node (router) in one service time T/q form a Poisson random process with an average value γ of data packets, as

$$\gamma = \frac{\mu}{q} \tag{1}$$

So at this point, two cases take place:

1. Active input nodes η_{act} (stations) in which no congestion occurs at the intermediate sensor node (router), where $\eta_{act} \leq q$
2. In active input stations in which congestion occurs due to overflow of data packets at the intermediate node (router), when $\eta_{act} > q$ thus packet loss occurs so extra data packets will be dropped, if no additional buffer facility is provided (i.e. a secondary buffer).

Probability that $\eta_{act} \leq q$ is

$$\mathrm{P}(\eta_{act} \leq q) = \mathrm{P} = \sum_{j=0}^{q} \frac{e^{-\gamma}\gamma^{j}}{jI} \tag{2}$$

If g is the total number of packets generated in the time τ/α, then

$$g = \sum_{j=0}^{\infty} P(\eta_{act} = i)i = \sum_{j=0}^{q} \frac{e^{-\gamma}\gamma^{j}}{jI} \tag{3}$$

The probability that $q > \eta_{act}$ is

$$P(\eta_{act} > q) = 1 - P\ (\eta_{act} \leq \mathbf{q}) = \sum_{j-q+1}^{\infty} \frac{e^{-\gamma}\gamma^{j}}{jI} \quad (4)$$

Suppose $\mathbf{P_D}$ represents the packets lost at the intermediate sensor node (router) 0 due to more packet flow towards this station. Then $\mathbf{P_D}$ can be calculated as

$$P_D = \sum_{j-q+1}^{\infty} \frac{e^{-\gamma}\gamma^{j}}{jI}(j - q) \quad (5)$$

The packets lost 'L' at the front side of the intermediate node (router) can be calculated as

$$L = \frac{P_D}{g} \quad (6)$$

The idle time of the router, i.e. T idle can be calculated as

$$I_{time} = \sum_{j=0}^{q-1} \frac{e^{-\gamma}\gamma^{j}}{jI} \times \left(\frac{q-j}{q}\right) \quad (7)$$

The busy time (B_{time}) of the router can be calculated simply as

$$B_{Time} = 1 - I_{time} \quad (8)$$

The link utilization % U_T is given by

$$\%\,U_T = B_{Time} \times 100 \quad (9)$$

The above equations can be useful to improve the QoS of multimedia packet transmission. Table 1 shows the results when q = 2.

It can be seen from the Table 1 that when number of active input stations are increased, the loss of packets increased also. This can be controlled by increasing the output capacity of the outgoing link from the router. However, as the medium remains constant, so it is difficult to change the output capacity of the router link. Since multimedia is a loss-tolerant application, accurate estimation of packet loss is very important. Therefore, the estimation of delay and loss must be set aside, which calculates the exact control to the received multimedia data traffic of the sensor router. This can minimize complications in multimedia data transmission, such as delay, jitter, and jamming to the highest degree possible. So, in the presented approach, the data traffic is continuously monitored at various values of '*q*'.

This particular approach therefore helps in calculating the received multimedia data traffic and exploitation of the channel thereby promoting an enhanced QoS for

Table 1 Experimental results for g, P_D, L, I_{time}, B_{time} and % U_T for q = 2, T = 5 ms

(μ)	(g)	(P_D)	(L)	% L	I_{time}	B_{time}	(% U_T)
1	0.5	0.0163	0.0326	3.26	0.758	0.242	24.2
2	1.0	0.1036	0.1036	10.36	0.552	0.458	45.8
3	1.5	0.2809	0.187	18.7	0.39	0.61	61.0
4	2.0	0.5414	0.2707	27.07	0.2707	0.7293	72.93
5	2.5	0.8694	0.3477	34.77	0.1847	0.8153	81.53
6	3.0	1.249	0.4163	41.63	0.1245	0.8755	87.55
8	4.0	2.11	0.527	52.7	0.055	0.945	94.5
10	5.0	3.048	0.6096	60.96	0.0236	0.9764	97.64
12	6.0	4.019	0.6698	66.98	$9.85 * 10^{-3}$	0.9901	99.01
15	7.5	5.505	0.7334	73.34	$2.71 * 10^{-3}$	0.9972	99.72
16	8.0	6.003	0.7503	75.03	$1.68 * 10^{-3}$	0.9983	99.83
20	10.0	8.00	0.80	80	$2.72 * 10^{-4}$	0.9997	99.97
25	12.5	10.500	0.84	84	$2.69 * 10^{-5}$	0.9999	99.99

where μ: Active nodes; **g**: Packets generated; $\mathbf{P_D}$: Packets dropped; **L**: Fraction of packets lost $\mathbf{U_T}$: % Utilization; $\mathbf{I_{time}}$: Idle time; $\mathbf{B_{time}}$: Busy time

multimedia transmission. The loss can be controlled by optimizing the sensor node buffer at the application layer, which acts here as a router between any trans-receiver pair.

Figure 2 shows the variation of active nodes v/s packets generated, when q = 2 and T = 5 ms. From Fig. 2 it can be seen that the number of packets generated increases linearly with an increase in the active nodes in the network. Figure 3 depicts the exponential increase in the packet loss count with increments in the active nodes, which reduces the efficiency of the network. Therefore, to reduce the packet drop one needs to increase the output link capacity (bandwidth) as the bandwidth is limited. Sliding window flow control protocol may be the optimum solution when the size of the window is small. However, if the window size is large and loss is due to the packet number being very near to the window size (e.g., if the

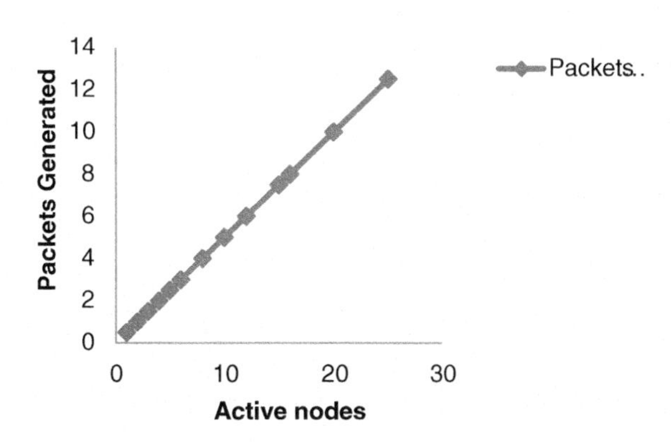

Fig. 2 Active nodes versus packets generated

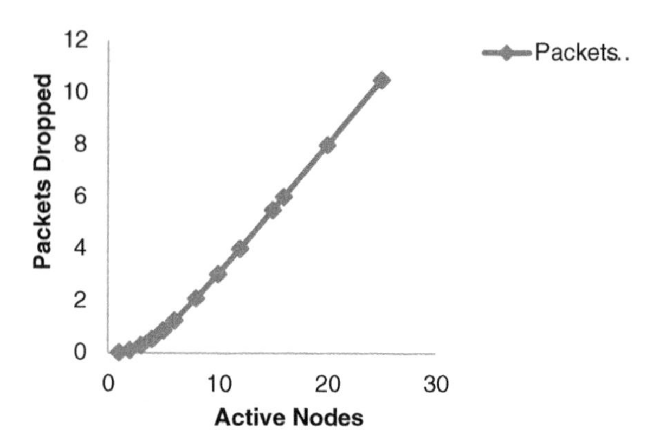

Fig. 3 Active nodes versus packets dropped

size of the window is 64 and the loss occurs at 63), the utilization of the link is very poor. To improve the utility of the link one needs to get the exact information about the loss of packets. The mathematical model present in this paper helps to calculate the accurate packet loss in WSNs

4 Conclusion and Future Scope

The presented mechanism of loss calculation is at the top of windowing techniques used to enhance the packets lost at the router. The presented approach is also simple and effective for WSNs. In this paper, a well-known technique called Poisson's distribution of probability is used to calculate the packet loss with greater accuracy compared to conventional methods, in order to enhance the PDR. The future scope of this paper is the development of a model which can control the loss of packets in the network at the desired level.

References

1. Avrachenkov K, Antipolis S (2005) Optimal choice of the buffer size in the internet routers. In: Proceedings of the 44th IEEE conference on decision and control, and the European control conference 2005, Seville, Spain
2. Wei D, Jin XC, Low SH, Hegde S (2006) FAST TCP: motivation, architecture, algorithms, performance. IEEE/ACM Trans Netw 14(6)
3. Sarker Z, Johansson I (2012) Improving the interactive real time video communication with network provided congestion notification. In: IAB/IRTF workshop on congestion control for interactive real-time communication, Newyork
4. Katabi D, Handley M, Rohrs C (2002) Congestion control for high bandwith-delay product networks. In: Proceedings of ACM SIGCOMM 2002
5. Kelly T (2002) On engineering a stable and scalable TCP variant. Cambridge University Engineering, Department Technical Report, CUED/FINFENG/TR 435

6. Zanella A, Procissi G, Gerla M, Sanadidi MY (2001) TCP westwood: analytic model performance evaluation. In: Proceedings of IEEE Globecom
7. Xu L, Harfoush K, Rhee L (2004) Binary increase congestion control for fast long-distance networks. In: Proceedings of IEEE INFOCOM
8. Leith D, Shorten R, Li Y (2005) H-TCP: a framework for congestion control in high-speed and long-distance networks. HI Technical Report. http://www.hamilton.ie/net/htcp/
9. Li C, Wang J, Li M (2017) An efficient cross-layer optimization algorithm for data transmission in wireless sensor networks. Int J Wireless Inf Netw 1–8
10. Wang J, Wen J, Zhang J, Han Y (2011) TCP-FIT: an improved TCP congestion control algorithm and its performance. ISSN: 978-1-4244-9921-2/11/©2011. IEEE
11. Chen H, Jones M, Jayalath ADS (2012) Congestion-aware routing protocol for mobile Ad hoc networks. ISSN: 1-4244-0264-6/07©2012. IEEE
12. Kaur P, Singh R (2013) A systematic approach for congestion control in wireless Ad hoc network using Opnet. Int J Adv Res Comput Commun Eng 2(3)

Enhanced Security of MANETs Against Black Hole Attacks Using AS Technique

Ishrath Unissa and Syed Jalal Ahmad

Abstract A mobile ad hoc network (MANET) is an autonomous structureless arrangement of mobile nodes to figure a momentary network. Communication between any two nodes is possible directly if the two nodes belong to the same sensing range; otherwise communication can be achieved by means of the nodes which are present between source and destination. As the network nodes are mobile, any node can enter or leave the network at any particular time interval. Thus, whichever node is present in between the source and destination can perform as a router or the host node in the arranged network. Therefore this poses security challenges to MANETs. This paper presents a solution to black hole attacks. The presented method is easy to use and efficient in detecting black hole attacks. The presented approach is validated by the use of network simulator 2 (NS2).

Keywords MANET · NS2 · Black hole · Security · Authentication
Hop count

1 Introduction

A mobile ad hoc network (MANET) is an arrangement of mobile nodes to form a network and doesn't require any infrastructure for its deployment. When communication occurs between any two nodes directly, it is called single hop communication, or else communication can be achieved by means of intermediate nodes called multiple hop communication. This type of transfer of information is also called indirect communication. As the nodes are mobile, they can enter or leave the network at any moment without any information being given to the other network

I. Unissa (✉)
Mahatma Gandhi Institute of Technology, Hyderabad, India
e-mail: ishrathunnisa94@gmail.com

S. J. Ahmad
GNITC, Hyderabad, India
e-mail: jalal0000@yahoo.com

A. Kumar and S. Mozar (eds.), *ICCCE 2018*,
Lecture Notes in Electrical Engineering 500,
https://doi.org/10.1007/978-981-13-0212-1_10

nodes. This particular character gives birth to security issues in the network. A MANET also does not provide any guarantee that the route from source to destination is free from attacker nodes. Due to the dynamic topology of the network various attacks have been noticed, such as spoofing attacks which happens when an assailant attempts to corrupt the node which is present in the path in which packet transmission takes place [1]. A Sybil attack [2], is a type of attack where the assailant not only symbolizes the network node but in addition it takes for granted the individuality of numerous nodes and accordingly does not succeed in locating the protocol redundancy number. Of all assaults, one of the most famous is famous is a black hole attack. This particular attack is produced by a malevolent node (a malicious node) transferring an extremely rapid response with maximum receiver sequence number representing the shortest route. So this type of attacker node can simply alter the data. To deter such attacks, the research group devoted a lot of effort towards studying them. A black hole assault is mitigated in the exiting approaches either by distributing keys among the nodes or by considering the node energies [3, 4]. However, both approaches may not be the optimum security methods, because due to dynamic topology, distribution of the key is not possible. Malicious nodes can also enter the network with different energies, and can, therefore, corrupt the node and access the data easily.

In this paper we are using an ASCII security technique (AST) to provide security against black hole attacks in MANETs called "*Enhanced Security of MANETs against Black Hole Attack Using AS Technique*" and using an IPV4 header to represent the security information to identify the black hole attack. In this approach, two different securities are being provided with respect to hop count (i.e. an odd parity of security type is used at even hops and an even parity of security type is used at odd hops, (user defined)) in order to improve security of MANETs against black hole attack. The presented security technique not only validates the active nodes in the network between end users, but also saves energy as well as reducing the processing time needed to validate the node, as the proposed approach reduces complexity by means of using a simple code vector.

The body of the paper is represented as follows, in Sect. 2 we present the related work, Sect. 3 illustrates the proposed approach and node-matching process, Sect. 4 presents the simulation results, and we conclude our paper in Sect. 5.

2 Related Work

Mirchiardi and Molva [5] addressed a method to detect the misbehavior and response of a mobile node in an ad hoc network. However, the response of this method is poor when collisions take place in the path during transmission of data packets. This approach is also not perfect as a result of the smaller amount of transmission power. Sanjeev and Manpreet [3] presented an approach in which the authors used a two-hop model to authenticate network nodes, which provides a protected communication between source and destination. However, the presented method necessitates more power and needs more processing time, so it cannot be used to secure the system as the delay involved is more. Hu et al. [6] addressed a protected method by the use of an on-demand routing protocol, in which the authors try to increase the network life span and also secure the control messages between end users. This system may not able to authenticate if the attacker node enters the network between the end users. So, the black hole can enter the network easily and corrupt routing information.

Sharma et al. [7] give a way to authenticate the active nodes within the network by adjusting the acknowledgement time of the transmitted node to obtain the repeat request from other neighboring nodes. In this approach they believe that the acknowledgement time is accurately identical to half of the route reply (RREP). However, this assumption is not correct when the network has multi hops and the two available and optimum routes from sender to receiver have sufficient time variation to accept requests (due to huge queuing delays and the propagation time of path 1 when compared to path 2). Deng et al. [8] proposed a protocol to rout the packets from source to destination, in which each node at the intermediate level needs to send an acknowledgement message to the transmitted node. This particular approach fails when the packet drop ratio is increasing, which in turn increases the delay between end users. Chanderkant [4] presented a model in which he tries to secure the MANET by identifying the attacker nodes based on energy parameters. However, this particular approach fails when a number of black hole attacker nodes penetrates the network with dissimilar energies. Lu et al. [9] addressed a routing protocol based on a SAODV (secure Ad hoc on-demand distance vector) to avoid black hole attack in MANETs. However, the presented approach addresses only a few of the security limitations of AODV, therefore it cannot remove the black hole completely from the network. Deswal and Sing [10] presented an improved version of the SAODV protocol by assigning a code word to the entire routing nodes. However, this approach is not valid when a new active node is entering the network, and cannot take part in it as the node does not have the code word assigned to it. As a result this reduces the network performance. Kukreja et al. [11] presented a security model against malicious node attack in MANETs by taking into consideration the power as a major parameter to sense the malicious node within the network. However, this approach fails when the attacker nodes are heterogeneous and take part in the networks. Adnan et al. [12] presented an approach to secure the MANET based on energy parameters. However, energy parameters alone may not be adequate to recognize the malicious node, as the attacker nodes can take part with dissimilar energies in the network with respect to time.

In this paper, we proposed an approach which enhances network performance by providing security at each hop based on the hop count.

3 Proposed Approach

ASCII Code is a well-known code of digital communication used to represent text files. Here, ASCII code is primarily used to generate bits of the code vector, as this code is almost readily available in a digital system. We are also getting 7 bits directly from each alphabet that is used to represent part of a security code vector to save more energy in comparison with the existing approaches. As a result, such types of coding techniques can secure the network and make the system intelligent.

In this paper, we are using an ASCII security technique (AST) to provide security against black hole attack in MANETs. An IPV4 packet header is used to store the security information. The final 8-bit security code of the approach can be obtained using following steps:

1. Consider ASCII string "SECURITY" where each character is represented in binary code of length 7. See Table 1.
2. Append one even/odd parity bit to the above binary code to obtain 8 bits for each character.
3. Extract 4 hamming bits from 8 bits which $2^{(z)}$ where z = 0,1,2,3 ... of 2 locations.
4. Perform x-or operation among 4 hamming bits to get a single bit for each character.

The ASCII character 'S' has the decimal equivalent 83 and the binary representation of 83 is 1010011. In a similar manner, the rest of the binary codes are generated for the entire string as shown in Table 1.

Appending even and odd parities:
In this method of security, even and odd parity bits are being added on the row directions (see Tables 2 and 3) to improve the security of the system, as if we provide a constant security at all hops one can easily attack by continuously monitoring the system. To detect the malicious nodes in the network, two different security code vectors have been used to enhance the trust of the node.

Table 1 The 7-bit binary code of ASCII characters for AS alphabets

AS letters	ASCII equivalent of AS letters represented in binary
S	1 0 1 0 0 1 1
E	1 0 0 0 1 0 1
C	1 0 0 0 0 1 1
U	1 0 1 0 1 0 1
R	1 0 1 0 0 1 0
I	1 0 0 1 0 0 1
T	1 0 1 0 1 0 0
Y	1 0 1 1 0 0 1

Table 2 Appending even parity check bit

AS letters	ASCII code binary form	Even parity (EP)	Complete security code
S	1 0 1 0 0 1 1	0	1 0 1 0 0 1 1 0
E	1 0 0 0 1 0 1	1	1 0 0 0 1 0 1 1
C	1 0 0 0 0 1 1	1	1 0 0 0 0 1 1 1
U	1 0 1 0 1 0 1	0	1 0 1 0 1 0 1 0
R	1 0 1 0 0 1 0	1	1 0 1 0 0 1 0 1
I	1 0 0 1 0 0 1	1	1 0 0 1 0 0 1 1
T	1 0 1 0 1 0 0	1	1 0 1 0 1 0 0 1
Y	1 0 1 1 0 0 1	0	1 0 1 1 0 0 1 0

Table 3 Appending odd parity check bit

AS letters	ASCII code	Odd parity (OP)	Complete security code
S	1 0 1 0 0 1 1	1	1 0 1 0 0 1 1 1
E	1 0 0 0 1 0 1	0	1 0 0 0 1 0 1 0
C	1 0 0 0 0 1 1	0	1 0 0 0 0 1 1 0
U	1 0 1 0 1 0 1	1	1 0 1 0 1 0 1 1
R	1 0 1 0 0 1 0	0	1 0 1 0 0 1 0 0
I	1 0 0 1 0 0 1	0	1 0 0 1 0 0 1 0
T	1 0 1 0 1 0 0	0	1 0 1 0 1 0 0 0
Y	1 0 1 1 0 0 1	1	1 0 1 1 0 0 1 1

From Table 2, the 8 bits for the character 'S' after appending the even parity bit is 10100110, in the same way the 8 bits are calculated for the characters 'E', 'C', 'U', 'R', 'I', 'T', 'Y' by appending even parity bits.

In the next step we consider the hamming bits out of the 8-bit code given by

$$H_b = (2^n)\ \text{bit positions of the code} \quad (1)$$

where H_b are the hamming bits and n = 0, 1, 2, 3, …

Here the hamming bits positions are $\{H_{b1}, H_{b2}, H_{b3}, H_{b4}\} = \{1, 2, 4, 8\}$ positions of code.

Therefore, the hamming bits for character 'S' (1 0 1 0 0 1 1 0) = 1 0 0 0

Hamming bits for character 'E' (1 0 0 0 1 0 1 1) = 1 0 0 1
Hamming bits for character 'C' (1 0 0 0 0 1 1 1) = 1 0 0 1
Hamming bits for character 'U' (1 0 1 0 1 0 1 0) = 1 0 0 0
Hamming bits for character 'R' (1 0 1 0 0 1 0 1) = 1 0 0 1
Hamming bits for character 'I' (1 0 0 1 0 0 1 1) = 1 0 1 1
Hamming bits for character 'T' (1 0 1 0 1 0 0 1) = 1 0 0 1
Hamming bits for character 'Y' (1 0 1 1 0 0 1 0) = 1 0 1 0

The last step of this approach is to perform an x-or operation among the hamming bits to develop a single bit for each character of the string "SECURITY" as given below

$$F_{sb}(\mathrm{c}) = H_{b1} \oplus H_{b2} \oplus H_{b3} \oplus H_{b4} \tag{2}$$

where $F_{sb}(\mathrm{c})$ is the final security bit and 'c' represents characters of the string.

So we get the final security bit (after an x-or operation) for the character 'S' as F_{sb}('S') = 1, F_{sb}('E') = 0, F_{sb}('C') = 0, F_{sb}('U') = 1, F_{sb}('R') = 0, F_{sb}('I') = 1, F_{sb}('T') = 0 and F_{sb}('Y') = 0.

The complete security bit generated for the string is represented as

$$F_{sc}(S_{ep}) = \{\text{`S', `E', `C', `U', `R', `I', `T', `Y'}\} = \{1\,0\,0\,1\,0\,1\,0\,0\} \tag{3}$$

where $F_{sc}(S_{ep})$ is the final security code for the string using even parity.

Table 3 shows the appending of odd parity check bits, where the same steps are followed as for $F_{sc}(S_{ep})$ to get $F_{sb}(S_{op})$ and $F_{sc}(S_{op})$ is the final security code of the string using odd parity

$$F_{sc}(S_{op}) = \{0\,1\,1\,0\,1\,0\,1\,1\} = \overline{F_{sb}(S_{ep})} \tag{4}$$

The source node initially sends the data packets to the next anchoring node along with the given security code. In this approach we are using $F_{sc}(S_{ep})$ at odd hops and $F_{sc}(S_{op})$ at even hops (user defined). If the final security code vector with respect to the hop count matches the transmitted code vector, then it will hand over the data block to the anchoring node. This code block-matching process will continue till the data reaches the destination. After matching the code block we also check the packet delivery ratio to enhance the trust.

$$\text{PDR} = P.R/P.T \tag{5}$$

where PDR is the packet delivery ratio, $P.R$ is the number of packets received, and $P.T$ is the number of packets transmitted.

Node Matching Process

Consider Fig. 1, in which a black hole wants to enter and take part with the active nodes. Initially, the source node transmits towards its neighboring nodes. However, those neighboring nodes can also be black hole attacker nodes. However, during the matching process, only active nodes can access the data. This is because a black hole attack is not be able to synchronize with the source node within a specified amount of time (i.e. TTL: time to live), due to the unavailability of the resulting security code of hop 1 to match with the source node. Thus it cannot take part

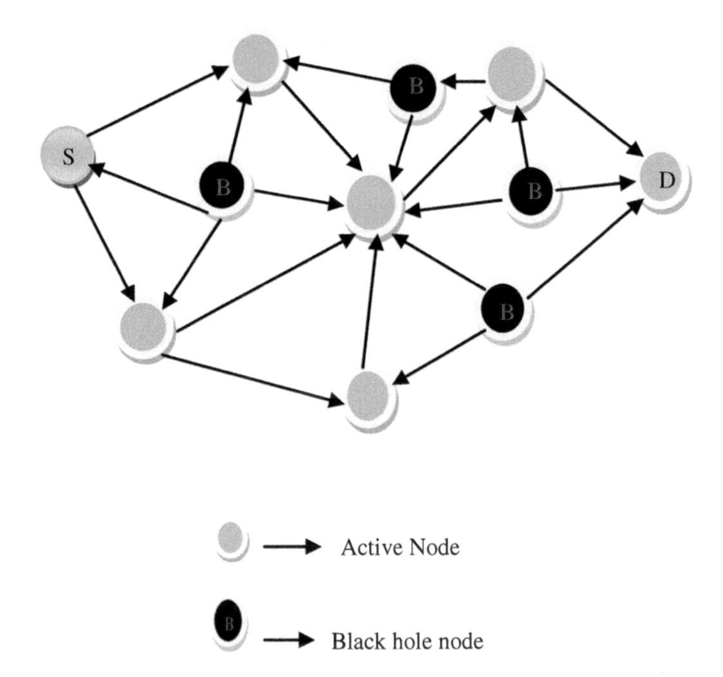

Fig. 1 A 4-hop WSN with black hole attacker nodes

within the network. This node-matching process will continue at each and every intermediately node till the destination node is reached. In this way, a black hole node can be easily judged and removed from the network.

4 Simulation Results

In this section we are presenting simulated results using NS2, and comparing them with existing approaches, such as AODV and Chandrakant approaches.

Figure 2 shows the variation of packet delivery ratio (PDR) versus simulation time (ST) of the source destination pair. From the figure it has been observed that if the source destination pair are very far from each other, there will be a greater chance that the maximum number of black holes can enter the network. However, our approach still produces a greater PDR in comparison with the other two approaches. This indicates that our approach does not allow a black hole into the network. Initially the PDR of the Chanderkant approach is higher than our approach, the reason behind this is that, initially, fewer black holes will interact with the network with the same energy, so cannot be allowed to take part in the network. Thus there is a greater packet delivery ratio (PDR) in comparison with our approach.

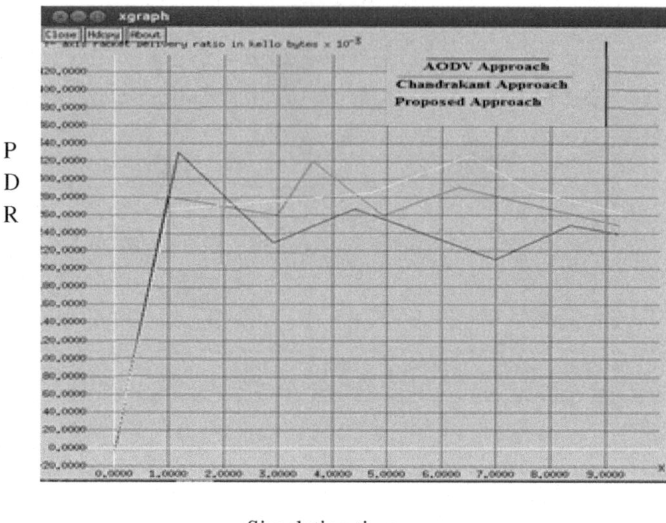

Fig. 2 Packet delivery ratio versus simulation time

However, over time a greater number of black holes will come with different energies and interact with the network, so black holes cannot be judged by the Chanderkant approach, and therefore black holes can easily enter the network and degrade the performance of the network. Our approach still shows a higher PDR than the other two approaches, so energy may not always be an appropriate parameter by which to identify the black hole. Table 4 shows the simulation parameters used during the testing process. To maintain the routing table we use a location aware and energy efficient routing protocol (LAEERP) [13].

Table 4 Simulation parameters

Network parameters	Values
Time for simulation	60 s
No. of nodes	2–100
Link layer	Logical link (LL)
Medium access control	802.11
Queue type	Drop tail
Type of antenna	Omni antenna
Protocol for routing	LAEERP
Type of traffic	Video
Network area	1500 m × 1500 m

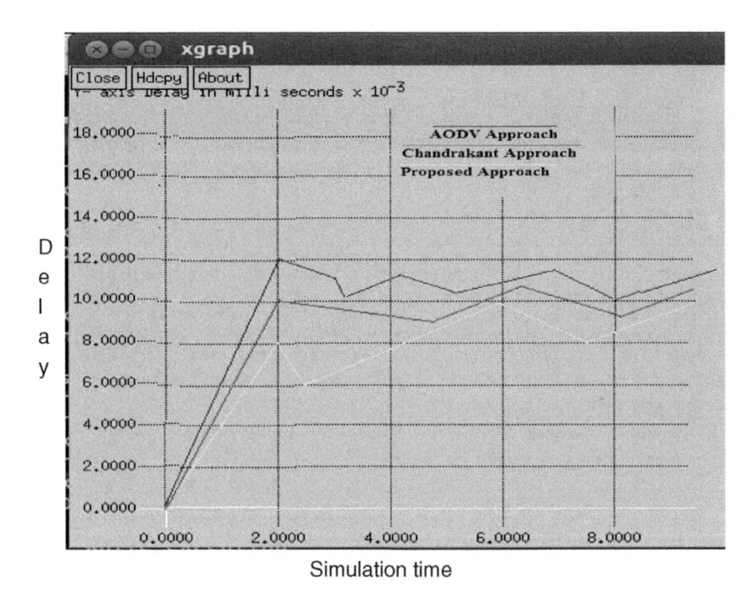

Fig. 3 Delay versus simulation time

Figure 3 represents the variation in delay with respect to simulation time. It shows that the delay of existing approaches (i.e. AODV and Chanderkant) is greater than with our approach. This is because when the number of hops increases, the existing approaches take more time to calculate the energy of the node. Moreover, the simulation results reveal that if router nodes (intermediate nodes) are busy with further source-destination pairs for communication, our approach still increases the PDR.

5 Conclusion

In this paper, a security model has been proposed called ASCII security technique. In this approach, two different security codes can be given (namely even and odd) with respect to odd and even hops respectively. Our approach is simpler and very effective in improving the trust between nodes within the network. It also provides complete security against black hole attacks in the network using fewer overheads and less energy resulting in an increased lifetime of the network. We compared our approach with both AODV and Chanderkant approaches, and the results demonstrated the applicability of our approach.

References

1. Karlof C, Wagner D (2003) Secure routing in wireless sensor networks: attacks and counter measures. AdHoc Netw J Spec Issue Sens Netw Appl Protoc 1(2–3):293–315
2. Douceur JR (2002) The Sybil attack. In: Druschel P, Kaashoek MF, Rowstron A (eds) IPTPS 2002, vol 2429. LNCS. Springer, Heidelberg, pp 251–260
3. Sanjeev R, Manpreet S (2011) Performance analysis of malicious node aware routing for MANET using two hop authentication. Int J Comput Appl 25(3):17–24
4. Chandrakant N (2013) Self protecting nodes for secured data transmission in energy efficient MANETs. Int J Adv Res Comput Sci Softw Eng 3(6):673–675
5. Michiardi P, Molva R (2002) CORE: a collaborative reputation mechanism to enforce node cooperation in mobile Adhoc networks. In: Proceedings of the IFIP TC6/TC11 sixth joint working conference on communications and multimedia security: advanced communications and multimedia security, Portorosz, Slovenia, pp 107–121
6. Hu YC, Perrig A, Johnson DB (2005) Aridane: a secure on-demand routing protocol for Adhoc networks. Wireless Netw 11:21–38
7. Sharma VC, Gupta A, Dimri V (2013) Detection of black hole attack in MANET under AODV routing protocol. Int J Adv Res Comput Sci Softw Eng 3(6):438–443
8. Deng H, Li W, Agrawal DP (2002) Routing security in wireless Adhoc networks. IEEE Commun Mag 40(10):70–75
9. Lu S, Li L, Lem KY, Jia L (2009) SAODV: a MANET routing protocol that can withstand black hole attack. In: Proceedings of the 2009 international conference on computational intelligence and security (CIS 2009), vol 2, Beijng, China, pp 421–425
10. Deswal S, Sing S (2010) Implementation of routing security aspects in AODV. Int J Comput Theory Eng 2(1):135–138
11. Kukreja D, Dhurandher SK, Reddy BVR (2017) Power aware malicious nodes detection for securing MANETs against packet forwarding misbehavior attack. J Ambient Intell Humaniz Comput 1–16
12. Adnan A, Abubakar K, Channa MI, Haseeb K, Khan AW (2016) A trust aware routing protocol for energy constraint wireless sensor networks. Telecommun Syst 61(1):123–140
13. Ahmad SJ, Reddy VSK, Damodaram A, Krishna PR (2013) Location aware and energy efficient routing protocol for long distance MANETs. Int J Netw Virtual Organ (IJNVO) 13 (4):327–350 Inderscience

Design of a Smart Water-Saving Irrigation System for Agriculture Based on a Wireless Sensor Network for Better Crop Yield

Meeradevi, M. A. Supreetha, Monica R. Mundada and J. N. Pooja

Abstract Precision agriculture is a decision-support system that helps farmers to make better decisions in the management of their farms, thus increasing returns while preserving resources. An automated irrigation system facilitates continuous and efficient irrigation under conditions of water and labor scarcity. Overwatering of crops causes nutrients to flow off the land surface and this can lead to lower crop yields. This wireless technology helps farmers to address the problem of overwatering and underwatering their crops. Currently, automation is one of the more important aspects affecting human life. It not only provides comfort but also reduces energy, increases efficiency and saves time. The proposed system uses wireless technology to irrigate crops in need of water. Water requirement varies depending on the type of crop, for example, paddyfields needs more water while crops like ragi needs less water. The proposed system irrigates based on the water requirements of particular crops in particular areas and the system as designed also provides smart irrigation technology at a low cost, usable by Indian farmers. Temperature, humidity, moisture of the land, and the water level in the tank will be measured and sent to the user via GSM communication. The water pump is automatically operated through the messages and Android application. Data is stored in the cloud for analysis. The proposed moisture-sensing method has the ability to be incorporated into an automated drip irrigation scheme and perform automated, precision agriculture in conjunction with decentralized water control.

Keywords Irrigation · Wireless sensor networks · Global system for mobile communications (GSM) · Android application · Cloud · Sensors

Meeradevi (✉) · M. A. Supreetha · M. R. Mundada · J. N. Pooja
M S Ramaiah Institute of Technology, Bangalore, India
e-mail: meera_ak@msrit.edu

A. Kumar and S. Mozar (eds.), *ICCCE 2018*,
Lecture Notes in Electrical Engineering 500,
https://doi.org/10.1007/978-981-13-0212-1_11

1 Introduction

Irrigation makes available the right amount of water to crops by analyzing soil properties. Current irrigation systems requires the farmer's presence in the field to manually irrigate the crops. Automating the system allows the remote monitoring of crops and the efficient use of resources thereby saving the time and energy of the farmer [1]. Automated drip irrigation systems are a smart way to monitor crops along with soil parameters in order to increase crop yield. Farmers can use a wireless network to access real-time information on the current condition of their fields and the location of their equipment. Farmers use 3G and 4G network on their smartphones or tablets to access real-time information on crops on their farms remotely. Plant growth is affected if the soil is completely waterlogged. To prevent such inaccuracies and flaws in watering, the proposed irrigation system is automatic and based on the growth stage of plants. It uses a wireless sensor network and a microcontroller. Sensors scattered across the field communicate the collected information through a network in order for it to reach the base station, where the necessary computations are performed and actions are generated. The collected information is then sent to the cloud infrastructure. The real-time monitoring of the environmental parameters is carried out and the user is informed of the precise conditions of the crops in the field via an Android application [2]. Flow of information between the sensor's nodes occurs through the tree-based network formed by the nodes. This tree-based communication minimizes energy consumption and makes the network more scalable. The user can then select the growth stage of crop, which is one of the specifications that determine the amount of water to be supplied to the crops. The sensed information pertaining to the crops is obtained via the GSM module. The GSM module provides flexibility to supervise and manage the performance of the irrigation systems remotely. The proposed system also shows the status of the motor, wetness of land, temperature, and humidity parameters. It uploads the sensed data periodically to the local server whenever there is a change in any of the sensed parameters. All the data is continuously monitored and updated on the cloud for the purpose of analyzing the sensed parameters.

2 System Design

The proposed scheme uses a wireless sensor network (WSN) as a backbone. WSN is basically a network created among the sensor nodes so that communication between the nodes is possible in order to perform the desired task [3]. The proposed system uses the principles of the internet of things (IoT) in order to transmit information through the internet. The IoT allows devices to be controlled remotely or sense various factors using sensors. The user interface is provided through an Android app. The information collected by the nodes propagates in the form of a tree in order to reach the base station as shown in Fig. 1. Arduino UNO transfers

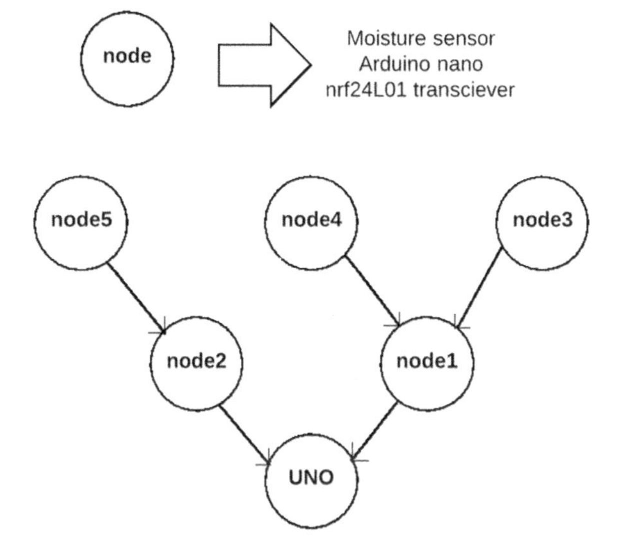

Fig. 1 Information flow in the network

the collected data serially to a nodeMCU which is connected to the internet and uploads the collected information to the cloud. The nodeMCU compares the collected data with the threshold value and initiates the actions to be performed. Once the data is uploaded to the cloud it can be accessed through an Android app. The app provides an interface for both automatic as well as manual operations and information from the app is transmitted to the nodeMCU through the cloud as show in Fig. 2. Here, all hardware actions such as opening of solenoid valves as well as relays are carried out by the nodeMCU.

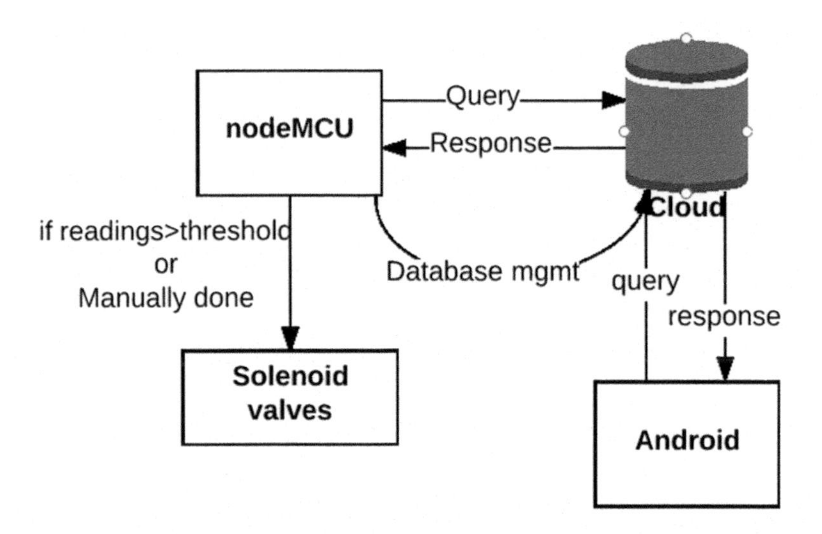

Fig. 2 Information exchange between user and nodeMCU through cloud

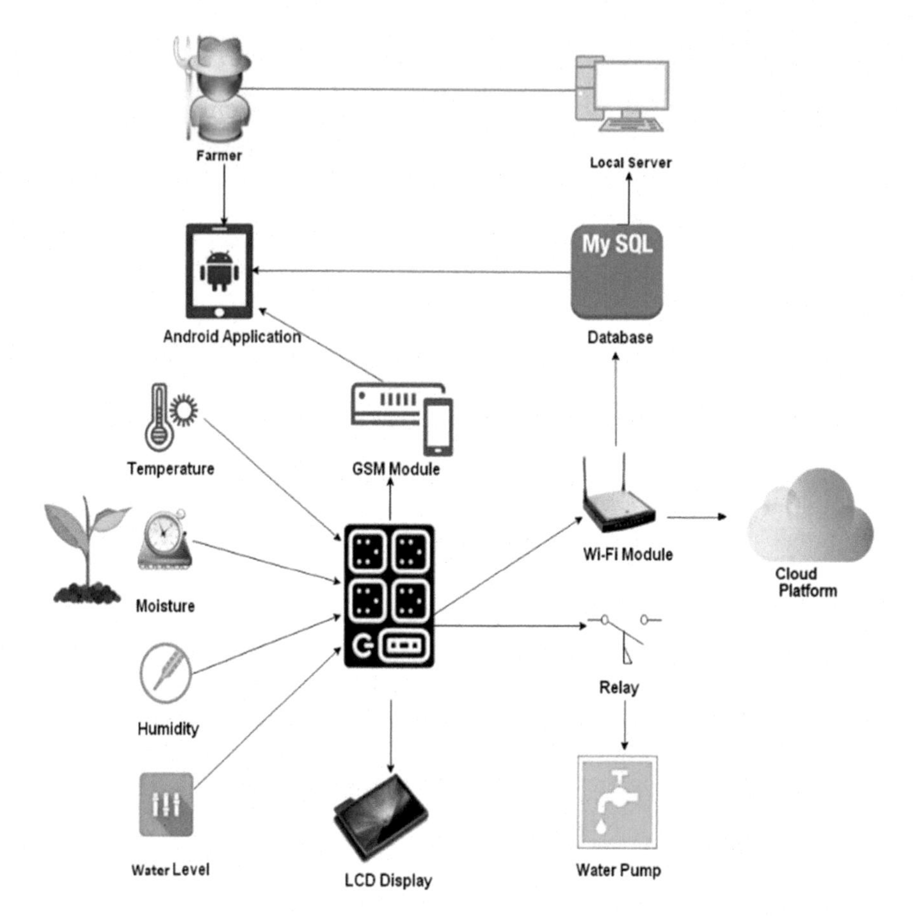

Fig. 3 Overall system architecture

The system architecture is as shown in Fig. 3. The heart of this system is the nodeMCU to which all the sensors (namely temperature, humidity, soil moisture, water level of tank) and the water pump are connected (Fig. 4). It is also connected to the GSM module for communication between the farmer and the hardware system. The wi-fi module is used for sending the data from the sensors to the cloud. For the user to interact with the system, an Android application has been developed. The data sent to the application is also stored in the local server.

Fig. 4 Moisture sensor in the soil

2.1 Flowchart

The flowchart as shown in Fig. 5 represents the entire system's operation. First the sensor data is collected and converted to digital format using ADC (analog to digital conversion). Each environmental parameter (temperature, humidity, soil moisture and water level of tank) are then compared with their respective threshold values and appropriate messages are displayed on the LCD screen [4]. Soil moisture plays a major role in irrigation and when the soil moisture value is more than its pre-defined threshold value the motor is automatically turned "OFF" else the motor is switched "ON" and a message is sent to the user alerting them of the motor status. When the user receives the message on his Android phone, the Android application reads the message automatically. If in the message, the motor status is "ON" then an SMS is automatically sent in reply to the message received after a certain time delay specified by the user based on the stage of plant growth. The details sent in the message are decoded and uploaded to the database on a local server which is in the same network that the phone connected to. When the reply message is received by the system, the motor status is checked. If the motor is "ON" then it is switched "OFF" else if it is "OFF" then no action is performed. The sensor data that is collected from the sensors is also periodically uploaded to the cloud platform for analysis. The data is uploaded once every 15 min. Different stages of crop growth need different a quantity of water, which can be handled using this system which is automated. In this proposed system the threshold value can be set based on the stage of crop growth.

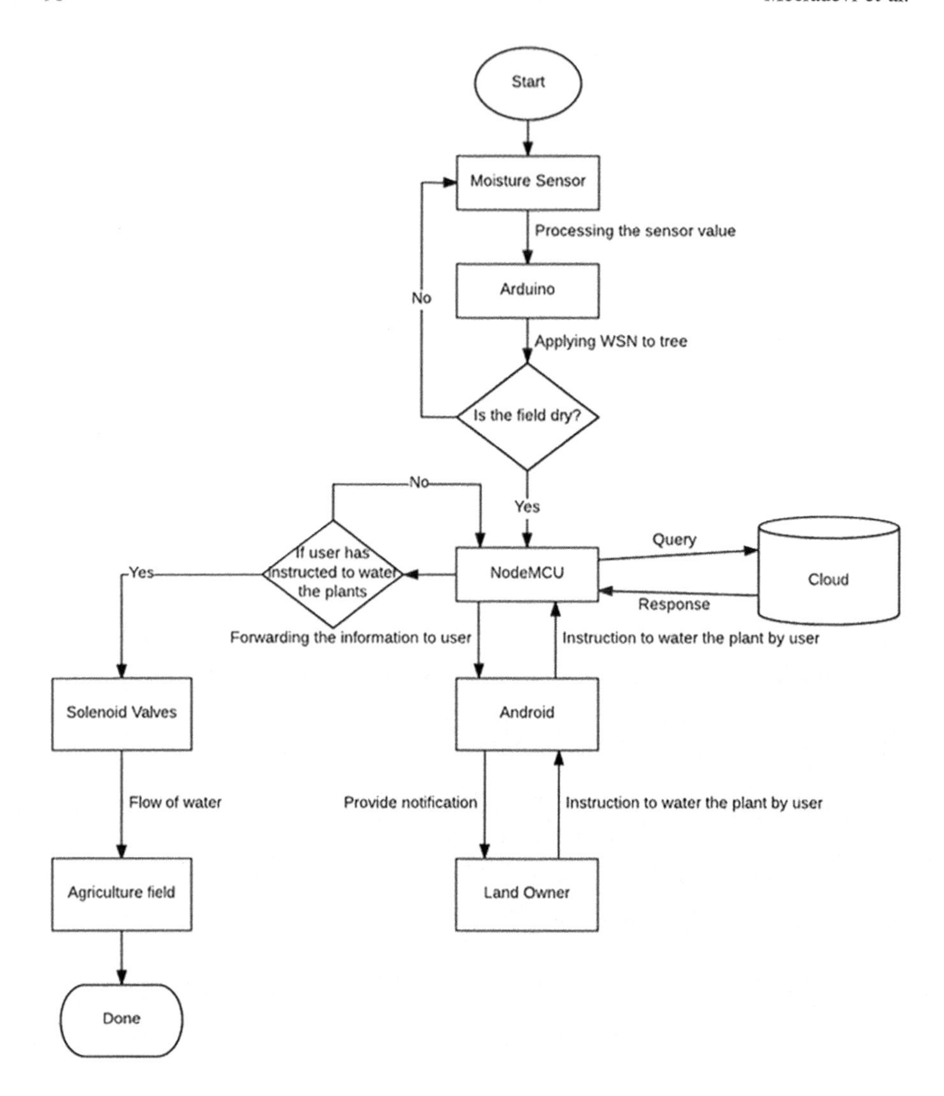

Fig. 5 System flowchart

3 Algorithms

3.1 Node Deployment Algorithm

Step 1: Assigning node ids for all nodes in the sensor network, base station being 00.

Step 2: Apart from the base station each node is given information regarding its parent node, so that a particular node can only send or get the data through the parent node.

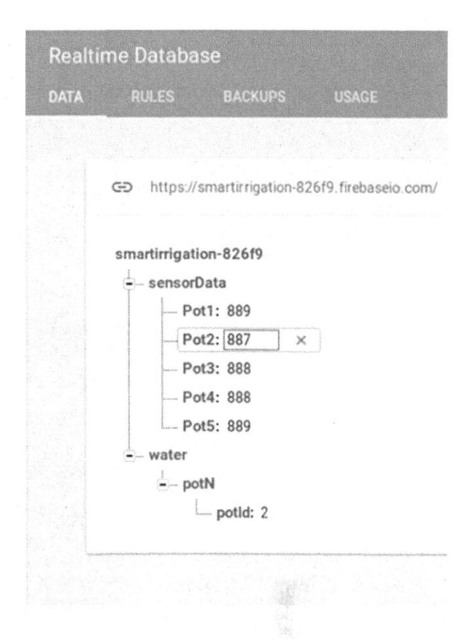

Fig. 6 Real-time values in firebase

Step 3: Once a tree network is formed, propagation of information is started. After a few iterations a network is formed in which every node will know the active nodes list.

Step 4: Once the information reaches the base node, serially communicate it to the nodeMCU.

Step 5: Perform required computations and simultaneously upload the collected information to the cloud.

Step 6: Query the database for the user's choice (Fig. 6).

Step 7: Based on the user's choice perform activation of solenoid valves, i.e. either automatic or manual (Fig. 7).

Step 8: Repeat step 3.

3.2 Algorithm for Irrigation

Step 1: Splash screen activity to the main screen.
Step 2: Main screen has two buttons.

- Pot readings.
- Water pots.

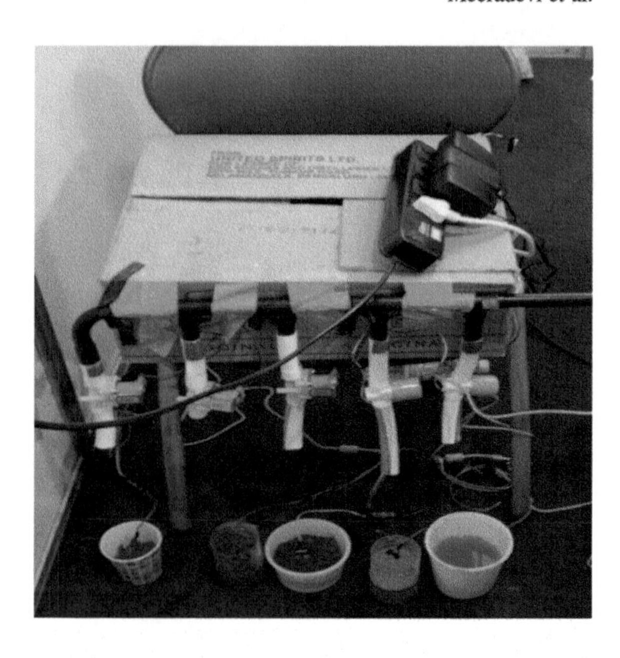

Fig. 7 Solenoid valves operating in the field

Step 2a: On a click of the pot readings button go to pot readings activity.

- Select a pot for individual moisture readings and the moisture values of selected nodes will be displayed.
- Open serial monitor to view moisture readings of all nodes.

Step 2b: On click water pots button go to water pots activity.

- Get dryness percentage of all nodes.
- Manual or automatic toggle button.
- On a click of manual make the drop-down content visible so that the user can manually choose which pot to water.
- If Automatic drop-down contents will be hidden.
- Display the recent pot watered.
- Notify the user regarding which pot is being watered and also notify once completed.
- If no node is selected or is being watered, then display a text view that no pot is selected.

4 Results

The Android app used by the end user queries the database informing it of details about the water content in the soil. The user can also manually water the crops in the field from remote location as shown in Fig. 8.

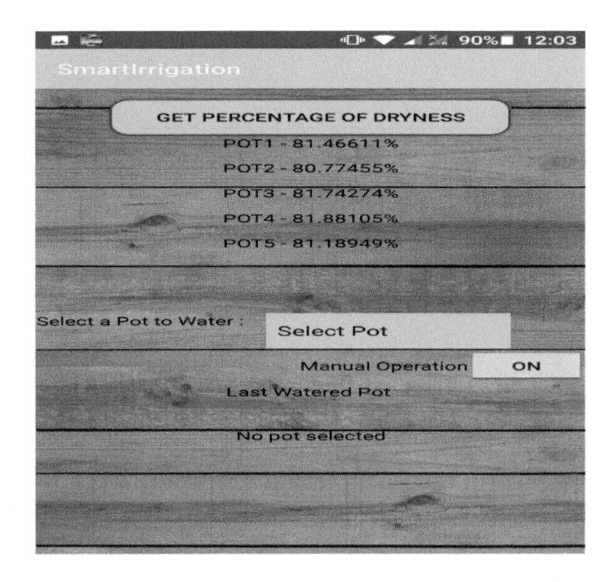

Fig. 8 Android application for user

4.1 Cloud Analysis

The cloud platform under use is ThingSpeak which allows uploading of data every 15 s once. The data was collected on 25th may 2017. The x-axis shows time and the y-axis shows the type of data sensed. The data from the three sensors, i.e. temperature, humidity, and soil moisture is uploaded once every 15 min to continuously monitor the parameters. This monitoring will help in analysing the effect of environmental parameters on plant growth [5]. The ThingSpeak platform produces dynamic graphs for each parameter uploaded and also allows the user to create other variations and graphs for different parameters. Given below are the graphs for temperature, humidity, and soil moisture along with the combined graph of all the parameters together (Figs. 9, 10, 11 and 12).

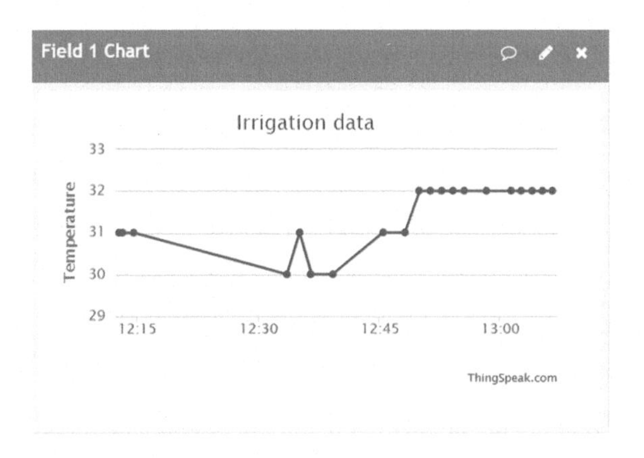

Fig. 9 Graph of temperature data

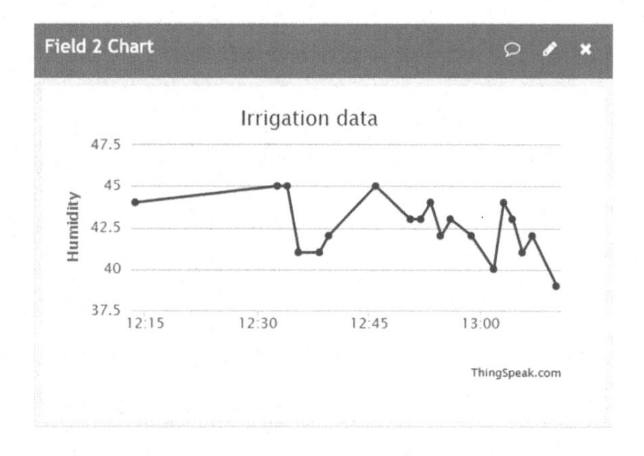

Fig. 10 Graph of humidity data

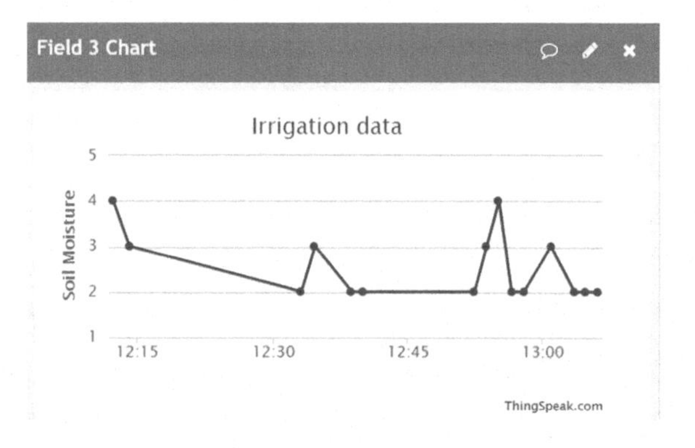

Fig. 11 Graph of soil moisture data

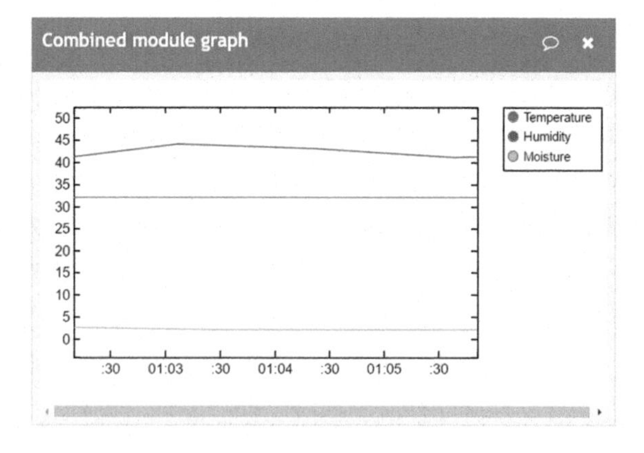

Fig. 12 Combined module graph

5 Benefits of Using Wireless Technology in Irrigation Management

1. Wireless technology has proven to be beneficial to agriculture farmer for irrigation management and water conservation by reducing the labour intensity of farming.
2. Creating pathways for more precise information about growing condition.
3. Provide farmers about real-time information.
4. Precision control of irrigation.
5. Impact of water availability.
6. Increased yield and quality.
7. Reduction in nutrient leaching.

6 Scope for Future Work

The system can be deployed in agricultural fields for automating the present irrigation system. The system allows the user to be away from the field and still get to know about the status of the field parameters and the motor. Since the Android application allows for automatic message transmission, it reduces the burden on the farmer of switching the motor on and off.

In future, the automated agriculture system could be made more dynamic. By making the system dynamic (whenever a node enters the system it finds its own path in the network without any assignment), the system will become more reliable and appealing. More sensors could be used to get a more accurate result. In future many data mining technique could also be used along with wireless sensor networks to get more accurate and faster results and data analytics algorithm prediction can be done on how to increase the crop yield based on actual data which is in cloud thereby helping the farmer to increase yields.

7 Conclusion

Farmers can remotely monitor and control irrigation decisions with the help of a user interface delivered through an Android application. The application also monitors the water level using the water level sensor in the tank for automatic irrigation. The amount of water required varies based on the growth stage of the crop and the type of crop. This prevents overwatering of the crops, which usually occurs due to human error in attempting to adjust the moisture levels. The system stores data in a database as well as in the cloud for future analysis. Thus the system provides a better backup of harvested data. The project uses sensors, such as a moisture sensor to retrieve the moisture values and compute the percentage of dryness in the field. The Android application provides a user interface for both manual operations as well as automatic operations performed in the field for irrigating crops. The WSN uses a tree-based structure for the transmission of information.

References

1. Hade AH, Sengupta MK (2014) Automatic control of drip irrigation system & monitoring of soil by wireless. IOSR J Agric Vet Sci (IOSR-JAVS) 7(4):57–61. E-ISSN: 2319-2380, 978-ISSN: 2319-2372, Ver. III. www.iosrjournals.org
2. Madli R, Hebbar S, Heddoori V (2016) Intelligent irrigation control system using wireless sensors and android application. Int J Comput Electr Autom Control Inf Eng 10
3. Balajibhanu B, Raghava Rao K, Ramesh JVN, Hussain MA (2014) Agriculture field monitoring and analysis using wireless sensor networks for improving crop production. ISSN: 978-1-4799-3156-9/14/$31.00 ©2014. IEEE
4. Shekhawat AS, Ahmed S, Kumar SG (2016) Intelligation: an IoT-based framework for smarter irrigation. In: National conference on product design
5. Villa-Medina JF, Nieto-Garibay A (2013) Automated irrigation system using a wireless sensor network and GPRS module. IEEE Trans Instrum Meas 63:166–176

SVM—A Way to Measure the Trust Ability of a Cloud Service Based on Rank

Sharmistha Dey, Vijender Kumar Solanki and Santanu Kumar Sen

Abstract Trust management is one of the most serious and demanding issues facing by cloud computing. In spite of having some surprising qualities, such as virtualization potential, highly optimized storage capacity, multi-tenancy features, and 24-7 service availability, cloud technology still faces security and authenticity issues, which have created an obstacle for adapting cloud computing as a widespread technology. Threat is a qualitative factor rather than its quantitative approach. Trust is also a quality factor, which can be more useful when trust can be established and proven quantitatively. Previously, cloud users had to show blind faith towards cloud service providers and vendors. Today, the importance of cloud auditing has increased. This paper focuses on the establishment of trust and measuring the quality of service for SaaS cloud service model, by using some measurement indices and with the help of some known parameters.

Keywords Alpha reliability · Audit trail · Confidentiality · Denial of services Non-return value of security investment · Trust ability · Zombie

1 Introduction

Trust management in cloud services is a high priority today. With the enhancement of dependency on virtual service infrastructure like the cloud, vulnerability increases. So it has become inevitable to evaluate the service provided by cloud vendors.

S. Dey (✉)
Maulana Abul Kalam Azad University of Technology, Kolkata, India
e-mail: papri.dey@gmail.com

V. K. Solanki
CMR Institute of Technology (Autonomous), Hyderabad, India
e-mail: spesinfo@yahoo.com

S. K. Sen
Guru Nanak Institute of Technology, Kolkata, India
e-mail: profsantanu.sen@gmail.com

A. Kumar and S. Mozar (eds.), *ICCCE 2018*,
Lecture Notes in Electrical Engineering 500,
https://doi.org/10.1007/978-981-13-0212-1_12

Cloud service models are of three basic types: IaaS (infrastructure as a service), PaaS (platform as a service) and SaaS (software as a service). Several deployment models are also present in the cloud, such as private cloud, public cloud, hybrid cloud, mobile cloud, etc. Risk factors vary depending on the service model employed.

There are several attacks enlisted below which is very essential in case of cloud.

(i) ***Cloud Malware Injection Attack***—With this attack, the client introduces an account in a distributed environment and the provider generates an image of the client's virtual system in the image repository system of the cloud. In the case of a cloud malware-injection attack, the intruders make attempts to inject malicious service or code using a script, which appears as one of the legitimate services running in the cloud. If the invader is successful in his attempt, then the service will suffer from eavesdropping. This attack is the foremost example of exploiting the service-to-cloud attack surface [14].

(ii) ***Malicious Insider***—This is known as one of the biggest security attacks in the cloud. This type of malware is found in emails and web applications. They are launched usually via VBscript or javascript.

(iii) ***Cross-Site Scripting Attack***—A cross-site scripting attack (or XSS attack) is an application level security threat where an attacker injects malicious codes into a link which appears to be from a faithful source. When a victim clicks on the URL, the embedded programming is automatically submitted as a part of the client's request and it is executed on the client's computer, which allows the attacker to take information without the user being aware. So, in spite of going to the original server address, the link will be directed to the malevolent site. XSS attacks have a significant impact on cloud computing.

(iv) ***Insecure API***—Over the last three years, attackers have tried hard to target the digital keys that are used to protect internet infrastructure. This attack was started by a Iranian hacker when he first broke a registry COMODO and hence broke the secure socket layer. The unknown attackers use unsafe APIs to steal significant information on security token of RSA algorithm, which is a device that generates one-time keys to strengthen online security.

(v) ***Denial of Services attack or Distributed Denial of Services Attack***—Denial of Service Attack(Dos) is a network level passive attack, very common attack for cloud. A modified version of DoS is distributed DoS (DDoS), which is even more serious. It is a special type of DoS attack where numerous compromised systems are used to make a zombie network, which is usually then infected with a trojan horse used to attack the server by continuously sending signals to it, creating a denial of service (DoS) using a divided compromised network in different layers. Victims of a DDoS attack consist of both the end targeted system and all systems maliciously used and prohibited by the hacker in the distributed attack [1–3].

2 Background Study

Many authors have worked on trust management. Huang and Nicol (2013), in "Trust mechanisms for cloud computing" illustrated a reputation-based trust mechanism and self-assessment in the case of trust management in the cloud [3]. This paper focuses on the semantics of trust in a policy-based trust mechanism in the cloud. According to the authors, reputation-based trust or SLA verification-based trust, may be a good service, but there may be some additional factors which influence users' ratings of a service provider and the ratings may be biased. The authors have proposed a framework but the focus of the paper was not primarily on computing trust or establishing trust mathematically.

In another paper "Developing Secure Cloud Storage System by Integrating Trust and Cryptographic Algorithms with Role based Access Control," the authors Bhise and Phursule [3] discussed role-based access control in the case of trust management in the cloud [4]. Their work shows the mathematical computation of trust establishment but only when the user is authorized by the owner or administrator. The authors have used familiar cryptographic algorithms like AES for encrypting and decrypting data and RSA for decrypting keys, in order to provide role-based access to the cloud. However, the size of cipher text as well as the key is constant [4].

Blomqvist [4] in his article, "The many faces of trust," discussed the idea of how trust is approached and defined in various disciplines. This work may be considered as the basis of the concept of trust management in case of the cloud [5]. Though it is a work on management, the need for trust management was made clear.

Another study, "Research on Trust Management Strategies in Cloud Computing Environment" by Li et al. [5], focused on strategies related to trust management in the cloud and the authors proposed a fuzzy comprehensive-based algorithm for establishing trust. They have provided a trust-based cloud transaction framework. A trust evaluation model was also proposed to establish trust quantitatively. Though they have asked for high trust accuracy in their model, no result was provided in their work relating to the actual cloud platform [6].

Chiregi and Navimipour [6], showed the impact of topological metrics on cloud service identification, in their work entitled "Trusted services identification in the cloud environment using the topological metrics" [7]. The paper evaluates reputation value and identifies trusted services in the cloud environment on the basis of three parameters, namely accessibility, dependability and ability. The topological metric approach has provided a quantitative and formalized approach to trust establishment. Quantitative measurement of reputation evaluation is a strong point of this work. Using a MATLAB simulation, the proposal has been shown to be successful, but the authors have shown trusted service having a direct relationship to reputation, which may not always be the case. They have not performed formal verification of trust evaluation, which is not only very challenging but also essential in cloud security auditing.

The "NIST Cloud Computing Standard Roadmap V 10.0, NIST", white paper by Hogan et al. [8] gives a clear vision of cloud architecture, several cloud services, and various roles in the cloud environment. This makes it easy to understand the various roles and hence their dependability and involvement [9]. This work is a roadmap of my proposal and have given my work an extra dimension.

In another conference paper entitled as, "SMI Cloud: A Framework for Comparing and Ranking Cloud Services" [10], authored by Garg et al. [7], a framework called SMI Cloud was introduced, and has provided the foundations for this paper, following the same concept of ranking cloud services based of some known and measured parameters [10].

As none of the proposed works focuses on a matrix-based approach, which may be easy to formalize and understand, this proposed model focuses on a matrix-based approach to trust establishment. The following section elaborates the concept of the four dimensions of a matrix-based trust evaluation system in the cloud environment.

3 An Analytical Study of Several Security Frameworks and Proposal of a Security and Vulnerability Matrix (SVM)

Since security has become a straightforward issue for in cloud, establishment of trust has become important in terms of selecting a suitable cloud service.

There are several frameworks and models providing support to measure cloud security and some of them rank cloud services, which in turn helps users to make decisions related to service selection [13].

Almost all existing models have covered the important QoS aspects for cloud service selection: availability, reliability, security, response time, and usability of the service.

The SMICloud framework [10], proposed in 2011 by Garg et al. in their paper "SMICloud: A Framework for Comparing and Ranking Cloud Services," compared several service providers based on user requirements, depending upon their rank. Their service measurement index (SMI) contains two types of key process index: Qualitative and Quantitative, based on the ISO standard.

This framework helps cloud users to choose the most suitable service provider and initiate service level agreements (SLAs) but this framework only uses some specific challenges to measure the quality of cloud services, on the basis of which the service has been ranked. This does not addresses broader issues like error percentage in the SLA or data retrieval capacity.

Another framework, COBIT [15], is a business framework for the management of enterprise information security, proposed by ISACA on 2012, highlights audit and control aspects in its first and second versions and governance and management of enterprise IT in version 5.0. COBIT introduces asset management and it separates governance from management. COBIT has also addressed the issue of meeting stakeholders' requirements like the SMI model. However, the main approach of the formation of the COBIT 5.0 framework is an holistic approach, which means the full participation of a person in selecting a service provider, which is quite impractical due to a lack of knowledge and awareness in some cloud users.

Since none of the models cover those aspects in a generalized and user friendly way, a new framework, called a security and vulnerability matrix (SVM), is proposed in this paper. The main objective of the proposed work is to choose the most suitable service provider on basis of user requirements, using a ranking of trustworthiness of a cloud service provider.

The proposed model can be used for all types of cloud service models. The proposed matrices are as follows [1, 10]:

1. **Confidentiality Matrix(C)**—To measure the confidentiality level kept by the provider
2. **Integrity Matrix(I)**—To measure service integrity
3. **Availability Matrix(A)**—To measure service availability in terms of uptime and downtime
4. **Reliability Matrix (RM)**—To determine the reliability of service and how free it is from security threats

The parameters for formulating the **confidentiality matrix** are as follows:

1. **Confidentiality Matrix** (Fig. 1)

   Here, TA = Tangible Asset.

$$\begin{aligned}&\textit{Lost TA Index}\\&= \textit{ROTA (Return of Total Tangible Asset) Company/Loss of TA of company}\\&\quad \textit{due to attacks against confidentiality}\end{aligned} \tag{1}$$

Here, a lost asset should be measured to determine its impact and it can be quantified by measuring the parameter termed return on total tangible asset (ROTA). TA index is measurable in terms of ROTA [2].

Where ROTA = return on total tangible asset. This is a ratio which measures a company's earnings before interest and taxes (EBIT) against its total net assets. To measure ROTA we have to do the following,

$$\begin{aligned}\textit{ROTA} &= \textit{EBIT/Total Net Asset, where EBIT}\\&= \textit{Net Income} + \textit{Interest Expense} + \textit{Taxes}\end{aligned} \tag{2}$$

$$\textit{Asset Loss Matrix (ALM)} = \begin{pmatrix} \textit{Lost TA Index} & \textit{Intangible Asset Loss} \\ \textit{ROSI Index} & \textit{Audit trail acceptability} \end{pmatrix} = \begin{pmatrix} wf1 & wf2 \\ wf3 & wf4 \end{pmatrix}$$

Fig. 1 Confidentiality matrix

By measuring this return, we can measure the risk to security also and again a rank can be issued to the service providers based on higher return record.

A cloud service security audit is mandatory for all service providers, and in order to check ***audit trail acceptability***, the audit trail is usually performed by a cloud auditor. This parameter may be determined by the following points: for how long have security records been kept, has dedicated storage been used or not, and how does the cloud service provider protect the audit trail from tampering etc. [2, 11].

$$\begin{aligned} &\boldsymbol{Return\ on\ Security\ Investment\ (ROSI)} \\ &= \boldsymbol{Risk\ Exposure\ x\%\ of\ Risk\ Mitigated - Solution} \frac{\boldsymbol{Cost}}{\boldsymbol{Solution\ Cost}} \end{aligned} \tag{3}$$

2. **Integrity matrix** (Fig. 2)

Error quotient in SLAs: SLAs or service level agreements are very significant for any cloud service. It is the agreement between the cloud service provider and a cloud tenant. Therefore, it is a parameter which measures the error in the response rate of service level agreements. How much a provider is usually able to cover the aspects mentioned in the SLA comes under this parameter and rank can be determined on this basis [9].

Data retrieval capacity relates to the integrity of the service. It is also essential for cloud service providers.

Information integrity check index can be measured by measuring standard deviations for those the integrity has been damaged and then make an indexing based on their percentage of deviation.

The **availability matrix** can be measured as [1, 8, 10, 12]:

$$\textbf{Availability} = \textbf{Total uptime} - \textbf{Total downtime/Total Uptime required} \tag{4}$$

The total uptime and downtime should be recorded and on the basis of probability of service availability, the providers will be given a rank, which will help users to decide on service selection when the criteria is service availability.

Reliability is the degree of measuring stability and consistency of the service provided by the service providers. Poor reliability degrades the accuracy of measurement and reduces capability to detect changes in measurements in experimental studies [9].

The reliability matrix has been formulated as in Fig. 3.

$$\begin{pmatrix} \textit{Error quotient in SLA} & \textit{Data retrieval capacity} \\ \textit{Information integrity check index} & \textit{Physical security hazards} \end{pmatrix} = \begin{pmatrix} wf1 & wf2 \\ wf3 & wf4 \end{pmatrix}$$

Fig. 2 Integrity matrix

Reliability Matrix

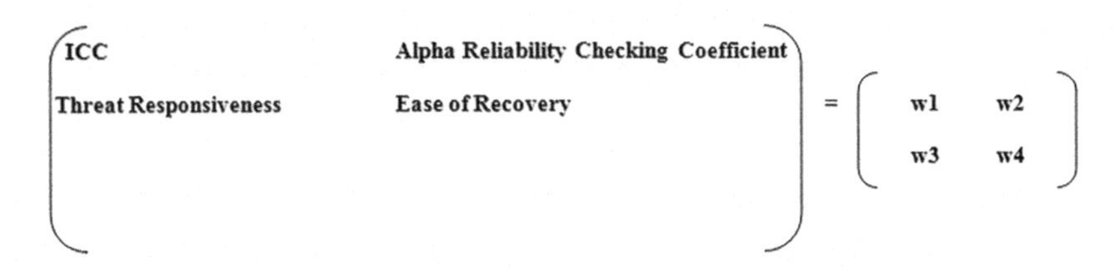

Fig. 3 Reliability matrix

ICC (intra class correlation coefficient) measures the reliability of the rating itself. Suppose the vendor has rated a cloud service. This parameter will crosscheck the rank that was already possessed by the service provider and in case of any discrepancy, it will be changed.

Alpha reliability checking coefficient is obtained by assuming each item represents a retest of a single item and it can be determined using the following formula:

$$r_{\propto} = \frac{k}{(k-1)} x \left(1 - \frac{\Sigma \sigma_i^2}{\sigma^2}\right) \tag{5}$$

Here, r is the alpha coefficient and k is the number of items.

Threat responsiveness of a system is a measure of response time towards a vulnerability in the system. If it takes a low time to response to a new threat, the value increases and may be assumed as 2, with the nominal value being 1.

Ease of recovery is related to the recovery capacity of a service provider. It can be measured in terms of time to recover from a failure.

It is essential to judge a qualitative parameter like trustworthiness in a quantitative manner and as this proposal formulates the parameters in a matrix format, unlike others, it is easier to determine the rank of the matrices and with the help of this to establish trust in the service providers. This can also help in the making of decisions for service selection based on this rank.

$$\boldsymbol{Trust} = \boldsymbol{Average}(\boldsymbol{Rank}(\boldsymbol{C}), \boldsymbol{Rank}(\boldsymbol{I}), \boldsymbol{Rank}(\boldsymbol{A}), \boldsymbol{Rank}(\boldsymbol{RM})) \tag{6}$$

The determination of trustworthiness of a service provider and allocating them a rank on the basis of that, makes the selection process easy and unbiased. The independence of the work is increased by the proposal written in this paper, and understandability also increases [1, 6, 11].

4 Conclusion and Future Work

Evaluation of trust in a quantitative manner has become essential for cloud services and determining trust worthiness using the defined matrices allows for cloud service selection on the basis of trust, where it is not only dependent on the reputation of the vendor, brand, or user rating. As a result, decision making is unbiased. The confidentiality, reliability or availability of a service can be easily measured but this proposal is needed to be implemented in a real scenario as previously it has only been based on a few specific parameters. An exploration of further parameters for future inclusion is ongoing.

References

1. Purohit GN, Jaiswal M, Pandey S (2012) Challenges involved in implementation of ERP on demand solution: cloud computing. Int J Comput Sci Issues 9(4):481–489
2. Sen S, Dey S, Roy D (2014) Design of quantifiable real-life security matrix for cloud computing. Int J Eng Sci Res Technol (IJESRT, India) 3(5):19–28
3. Bhise A, Phursule R (2017) Developing secure cloud storage system by integrating trust and cryptographic algorithms with role based access control. Int J Comput Appl, Found Comput Sci (New York, USA) 168(10):18–23. https://doi.org/10.1007/978-981-10-1678-3_13
4. Blomqvist K (1997) The many faces of trust, Scandanavian journal of management. Sci Direct (Netherland) 3(13):271–286. https://doi.org/10.1016/s0956-5221(97)84644-1
5. Li W et al (2012) Research on trust management strategies in cloud computing environment. J Comput Inf Syst (Zhongshan University, China) 4(4):1757–1767. https://doi.org/10.1016/j.kijoms.2016.06.002
6. Chiregi M, Navimipour N (2016) Trusted services identification in the cloud environment using the topological metrics. Karbala Int J Mod Sci (Elsevier, Karbala, Iraq) 2(4):203–210. https://doi.org/10.1016/j.kijoms.2016.06.002
7. Garg S, Versteeg S, Buyya R (2011) SMICloud: a framework for comparing and ranking cloud services. In: 4th IEEE international conference on utility and cloud computing, Washington DC, USA, December 2011, pp 210–218
8. Hogan M et al NIST cloud computing standard roadmap V 10.0, NIST whitepaper 500-291, July 2011, Dept of Commerce, USA
9. Mathur Mohit (2012) KLSI. Cloud computing black book, 1st edn. Wiley Publication, India
10. Tirodkar S et al (2014) Improved 3-dimensional security in cloud computing. Int J Comput Trends Technol (Seventh Sense Research Group, India) 5(9):242–247. https://doi.org/10.14445/22312803/ijctt-v9p145
11. Dey S, Sen S Four dimensional security and vulnerability matrix for cloud (4-SVM). In: International conference of research in intelligent and computing in engineering, March 2017. Gopeswar, India, pp 165–169. https://doi.org/10.15439/2017r41
12. Chou T (2013) Security threats on cloud computing vulnerabilities. Int J Comput Sci Inf Technol (AIRCC, Chennai, India) 5(3):79–88. https://doi.org/10.5121/ijcsit.2013.5306
13. Singh U, Joshi C, Gaud N (2016) Information security assessment by quantifying risk level of network vulnerabilities. Int J Comput Appl; Found Comput Sci (New York, USA) 156(2):37–44. https://doi.org/10.5120/ijca20169123

14. Huang J, Nicol DM (2013) Trust mechanisms for cloud computing. J Cloud Comput 2(1): 1–14
15. Garsoux M (2013) COBIT5: ISACAs new framework for IT Governance, Risk, Security and Auditing. Whitepaper published by ISACA

An Optimized Five-Layer Model with Rainfall Effects for Wireless Propagation in Forests

Mohammed Saleh H. Al Salameh

Abstract This paper presents a new propagation model for evaluating the fading of wireless communication signals in forests. The model considers rainfall and snowfall effects, and allows for the estimation of attenuation at varying frequencies in the VHF/UHF bands that are used by cognitive radios. The structure of the vegetation environment is represented here by five material layers, namely soil, scrubs and small plants under the trees, trunks of trees, foliage of trees, and free space. The model parameters are optimized using the least squares technique. The resulting model is verified by comparison with measured data where acceptable agreement is observed. The average rain rate $R_{0.01\%}$ that will probably be exceeded for at most 0.01% of the year is computed using real measured data in Jordan. $R_{0.01\%}$ is found to be 22.9 mm/h which agrees with the ITU recommended value of 22 mm/h.

Keywords Least squares · Rain · Forest · Propagation · Wireless Measurements

1 Introduction

There is a growing interest in establishing communications in forest environments. This includes battlefield communications, fire and rescue services, ambulance and emergency services, security, police, and private mobile radio systems.

On leave from Jordan University of Science and Technology, Irbid 22110, Jordan, salameh@just.edu.jo.

M. S. H. Al Salameh (✉)
Department of Electrical Engineering, American University of Madaba,
Kings Highway, Madaba, Jordan
e-mail: salameh@just.edu.jo

A. Kumar and S. Mozar (eds.), *ICCCE 2018*,
Lecture Notes in Electrical Engineering 500,
https://doi.org/10.1007/978-981-13-0212-1_13

Furthermore, cognitive radio automatically captures the best available spectrum for the best quality communications. This implies operation at varying frequencies. In that regard, the USA Federal Communications Commission (FCC) allows the operation of unlicensed radio systems in the VHF/UHF television broadcast frequency bands when the spectrum is not in use by other licensed services [1]. These VHF/UHF bands are suitable for cognitive radio operation. Thus, for reliable communications, it is important to model the forest communication channel behavior at different frequencies.

Path loss modeling in forest was investigated using a dissipative dielectric slab model for frequencies of 1–100 MHz [2]. The knife edge diffraction model, with two knife-edges, was used to characterize the terrain effects of the forest [3]. Characteristic curves were derived from measurements along road sections in large forests [4]. An empirical model was obtained from measurements in the rain forests of India [5]. A model was introduced which treated the trees as a statistically homogeneous random half-space medium of discrete, lossy scatterers at 11.2 GHz [6]. Attenuation associated with lossy trees and buildings was modeled by the uniform theory of diffraction for satellite mobile communications [7]. The effect of wind and rain on continuous wave fading in a tropical forest was approximated by Rician distribution function [8]. Propagation loss in a tropical forest was analyzed by a proposed empirical model at 240 and 700 MHz [9] for near ground communications. Comparison of propagation models in forest environments of Nigeria revealed that the direct ray model augmented by a suitable vegetation loss model is more accurate than the other investigated models [10]. A review was conducted on propagation in rain [11]. The empirical foliage loss models didn't always show accurate results [10]. Alternatively, the empirical path loss models are more attractive. Moreover, to the author's knowledge, there is limited research work on weather-induced effects related to the propagation of radiowaves in forest environments.

Wave attenuation due to snowfall depends on its liquid water content. Dry snow consists of ice and air which indicates that attenuation due to dry snowfall can be ignored [12]. Wet snow, in contrast, contains ice, air, and liquid water. Thus attenuation of wet snowfall is comparable to rain showers with big raindrops [13]. Based on this, rain attenuation will be considered in this paper, and the results also apply to wet snowfall weather conditions.

This paper presents a new optimized empirical path loss model augmented by rainfall losses, for estimating the attenuation of wireless signals in a forest at varying frequencies in the VHF/UHF bands. The model is optimized using the least squares method.

2 Theoretical Analysis

The path loss PL is expressed by [14]:

$$PL = -L_t + G_t - L + G_r - L_r \tag{1}$$

where L_t and L_r are the transmitter and receiver feeder losses respectively, L is the signal attenuation due to wave propagation, and G_t and G_r are the gains of the transmitting and receiving antennas, respectively. Propagation loss (L) in a forest environment can be written as:

$$L = L_0 + L_G + L_V + L_R \tag{2}$$

L_O is the free space propagation loss, L_G is ground effects loss, L_V is vegetation effects loss, and L_R is loss caused by weather conditions such as rainfall and snowfall.

Combining Eqs. (1) and (2) yields the total path loss in a forest area,

$$PL = -L_t + G_t - L_0 - L_G - L_V - L_R + G_r - L_r \tag{3}$$

The forest propagation scenario is represented in this paper by five layers as described in Fig. 1.

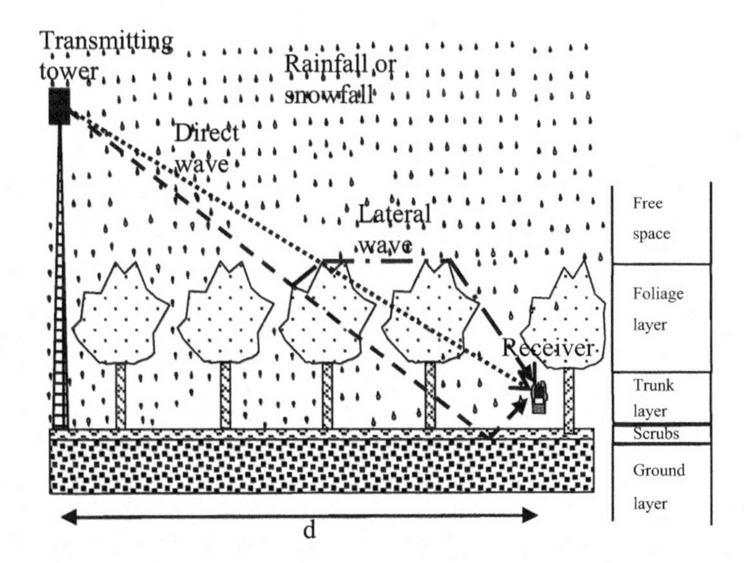

Fig. 1 Five-layer model for propagation in forest areas: ground, scrub, and small plants, tree trunks, foliage, and free space

3 Evaluation of Rain Attenuation

The specific rain attenuation γ in dB/km is [9]:

$$\gamma = aR^b \tag{4}$$

where R is the rain rate in mm/hr, and the parameters a and b can be found from [15]. In case the rain attenuation rate γ doesn't vary with distance r_r, rain loss is:

$$L_R = \gamma(r)r_r \tag{5}$$

For reliable radio communication systems, the rain rate value $R_{0.01\%}$ is considered, which indicates a rain rate that will probably be exceeded for at most 0.01% of the year, i.e.

$$\Delta t = 0.01\% \times (365\ \text{days/year}) \times (24\ \text{h/day}) = 0.876\ \text{h}$$

Accordingly, the reliability of the communication system will be 99.99%. The following model is based on the Rice-Holmberg model [16] where the value of $R_{0.01\%}$ is computed from the average annual rainfall R_Y by:

$$R_{0.01\%} = \frac{1}{0.03}\left[\ln(0.03\frac{R_Y\beta}{\Delta t})\right] = \frac{1}{0.03}\ln(\frac{R_Y}{105.5}) \tag{6}$$

β is the ratio of convectional rainfall to total rainfall accumulation. The value β = 27.7% for Jordan is estimated from data given in [17].

Equation (6) is used in this paper to calculate $R_{0.01\%}$ values for different areas in Jordan from measured annual rainfalls, as shown in Table 1. $R_{0.01\%}$ = 22.9 mm/h is obtained which agrees with the ITU recommended value of 22 mm/h [18].

4 Model Optimization

The proposed model in this paper is an experimental propagation path loss L_{Forest} model for forests [20]:

$$L_{Forest} = K + A\ \log(d) + Bd \tag{7}$$

where K, A, and B are parameters to be determined based on the measured data, and d is the distance in kilometers between the transmitter and receiver. The least squares method will be used to find the optimum values of parameters K, A, and B. Accordingly, it is necessary to minimize the sum of the squared errors (SE) between the measured data P_{mi} and the prediction model data P_{ri} at the measured data points i for a total of N data samples:

Table 1 Calculated $R_{0.01\%}$ values using measured rainfall in different areas of Jordan [19]

Area	Average annual rainfall (mm)	$R_{0.01\%}$ in mm/h
Amman	505.3	52.2
Marka	268.2	31.1
Madaba	324.4	37.4
Swaileh	475.6	50.2
Salt	514.4	52.8
Irbid	459	49.0
Samma	415.8	45.7
Ras Muneef	580.5	56.8
Zarqa	124.3	5.5
Dhulail	138.5	9.1
Ghabawi	84.8	0
Zizia	158.1	13.5
Mafraq	152.1	12.2
Safawi	71	0
Rwaished	79.7	0
Azraq	58.5	0
Baqura	389.3	43.5
Wadi El-Rayyan	296.4	34.4
Dair Alla	281	32.6
Ghor Safi	72.6	0
Tafileh	245.1	28.1
Shoubak	269.1	31.2
Wadi Mousa	176.2	17.1
El-Rabba	337.3	38.7
Ma'an	40.9	0
Qatraneh	96.8	0
El Jafer	32	0
Aqaba	26.9	0
Average $R_{0.01\%}$ for Jordan: 22.9 mm		

$$SE = \sum_{i=1}^{N} (P_{mi} - P_{ri})^2 \tag{8}$$

Substituting Eq. (7) into (8) yields:

$$SE = \sum_{i=1}^{N} (P_{mi} - [K + Alog(d_i) + Bd_i])^2 \tag{9}$$

To minimize errors, partial derivatives of SE should vanish:

$$\frac{\partial(SE)}{\partial K} = 0, \ \frac{\partial(SE)}{\partial A} = 0, \ \frac{\partial(SE)}{\partial B} = 0 \tag{10}$$

This gives the following three equations:

$$NK + \left(\sum_{i=1}^{N} log(d_i)\right)A + \left(\sum_{i=1}^{N} d_i\right)B = \sum_{i=1}^{N} P_{mi} \tag{11}$$

$$\left(\sum_{i=1}^{N} \log(d_i)\right)K + \left(\sum_{i=1}^{N} \{\log(d_i)\}^2\right)A + \left(\sum_{i=1}^{N} \mathrm{d_i}\log(d_i)\right)B = \sum_{i=1}^{N} \mathrm{P_{mi}}\log(d_i) \tag{12}$$

$$\left(\sum_{i=1}^{N} d_i\right)K + \left(\sum_{i=1}^{N} \mathrm{d_i}\log(d_i)\right)A + \left(\sum_{i=1}^{N} \{d_i\}^2\right)B = \sum_{i=1}^{N} \mathrm{P_{mi}}d_i \tag{13}$$

Substituting the measured data [21] into Eqs. (11–13), and solving these equations by Gauss elimination technique or iteration methods produces the following optimized parameter values: $K = -97.79$, $A = 32.33$ and $B = 0.1832$. Introducing these values into Eq. (7), the proposed optimized empirical model for medium forests with scrub and small plants covering the ground is expressed as:

$$L_{Forest} = -97.79 - 32.33 log(d) + 0.1832d \tag{14}$$

In order to take in the frequency dependence, the forest path loss is computed for different frequencies using the lateral wave ITU-R model integrated with a suitable propagation model [22]. Inserting a frequency dependence correction factor in addition to the rain loss L_R, (14) becomes:

$$L_{Forest} = -97.79 - 32.33\, log(d) + 0.1832d - 36\, log(f/92.1) + a\, R^b r_r \tag{15}$$

5 Results

The optimized model equations presented in the previous section were used to evaluate the path loss in the forest. The measured data [21] are in excellent agreement with the optimized model predictions. The root mean square error (RMSE) of the optimized model is only 4.5 dB with reference to the measured data.

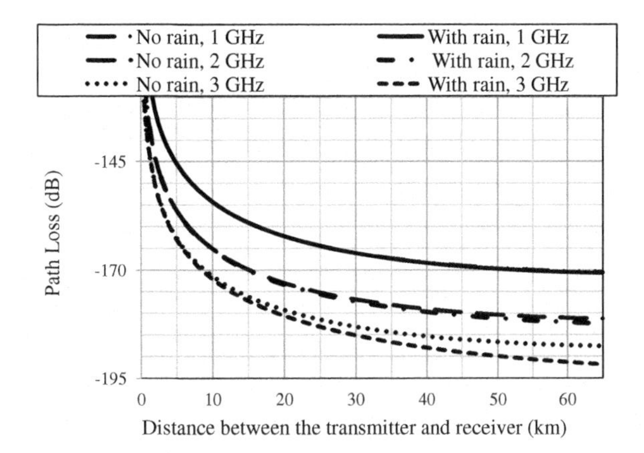

Fig. 2 Path losses with and without rainfall in forest areas, according to the new model

The path loss versus distance in a forest environment is shown in Fig. 2 for different frequencies and for the cases of rainfall and no rainfall. The curves are similar in the cases of rain and no rain at 1 GHz in the UHF band, even when a high rain rate of 228.8 mm/h is used. In fact, this is the average rain rate in Nigeria [23]. At 2 GHz, the rain effect is noticeable, and at 3 GHz the rain effect is clear especially at long distances from the transmitter. The path loss increases with frequency, rainfall rate, and distance from the transmitter as expected. Microsoft Excel computer programs were utilized in this paper in order to numerically evaluate the new model's predicted results.

6 Conclusions

This paper presents a new five-layer model for propagation in forest with scrub and small plants covering the soil under the trees. The model presented in this paper considers forest environment including vegetation losses, rainfall, and snowfall losses, in addition to the wave propagation losses. The model is optimized using the least squares method and allows for varying frequencies. Rainfall and snowfall are considered. The results computed by the new model agree well with the measurements where the root mean square error (RMSE) is only 4.5 dB. The rainfall effect is significant when the wave frequency is higher than 1 GHz. The rainfall rate exceeded for 0.01% of the time $R_{0.01\%}$ is found to be 22.9 mm/h for Jordan, which agrees with the ITU recommended value of 22 mm/h.

Acknowledgements The author would like to thank Jordan University of Science & Technology and the American University of Madaba for their continuous support.

References

1. FCC Report, Unlicensed operation in the TV broadcast bands, ET Docket No. 04-186, Nov 2008. https://apps.fcc.gov/edocs_public/attachmatch/FCC-08-260A1.pdf. Accessed Oct 2017
2. Tamir T (1967) On radio-wave propagation in forest environments. IEEE Trans Antennas Propag 15(6):806–817
3. Meeks ML (1983) VHF propagation over hilly, forested terrain. IEEE Trans Antennas Propag 31(3):483–489
4. Low K (1988) UHF measurement of seasonal field-strength variations in forests. IEEE Trans Veh Technol 37(3):121–124
5. Tewari RK, Swarup S, Roy MN (1990) Radio wave propagation through rain forests of India. IEEE Trans Antennas Propag 38(4):433–449
6. Al-Nuaimi MO, Hammoudeh AM (April 1994) Measurements and predictions of attenuation and scatter of microwave signals by trees. In: IEE proceedings—microwaves, antennas & propagation, vol 141, no 2, pp 70–76
7. Al Salameh MSH, Qasaymeh MM (2004) Effects of buildings and trees on satellite mobile communications. Int J Electron 91:611–623
8. Song Meng Y, Hui Lee Y, Chong Ng B (2009) The effects of tropical weather on radio-wave propagation over foliage channel. IEEE Trans Veh Technol 58(8):4023–4030
9. Song Meng Y, Hui Lee Y, Chong Ng B (2009) Empirical near ground path loss modeling in a forest at VHF and UHF bands. IEEE Trans Antennas Propag 57(5):1461–1468
10. Al Salameh MSH Vegetation attenuation combined with propagation models versus path loss measurements in forest areas. In: World symposium on web application and networking—international conference on network technologies and communication systems, 22–24 Mar, 2014. Hammamet, Tunisia, pp 120–124
11. Okamura S, Oguchi T (2010) Electromagnetic wave propagation in rain and polarization effects. In: Proceedings of the Japan academy, series B, vol 86
12. Sugiyama S, Enomoto H, Fujita S, Fukui K, Nakazawa F, Holmlund P (2010) Dielectric permittivity of snow measured along the route traversed in the Japanese-Swedish Antarctic expedition 2007/08. Ann Glaciol 51(55):9–15
13. Tamošiunaite M, Tamošiuniene M, Gruodis A, Tamošiunas S (2010) Prediction of electromagnetic wave attenuation due to water in atmosphere. 1. Attenuation due to rain. Int J Innov Info-technol Sci Bus Educ 2(9):3–10
14. Saunder SR, Aragon-Zavala A (2007) Antennas and propagation for wireless communication systems, 2nd edn. Wiley, Chichester, England
15. ITU-R recommendation P.838-3 (03/05) approved in 2005-03-08: Specific attenuation model for rain for use in prediction methods, International Telecommunication Union 2005. https://www.itu.int/rec/R-REC-P.838-3-200503-I/en
16. Ippolito LJ Jr (1986) Radiowave propagation in satellite communications. Springer, Heidelberg, Germany
17. Final Feasibility Study Report (12 147 RP 04), "Red sea—dead sea water conveyance study program," Appendix A—climate change study, April 2011. http://www.waj.gov.jo/sites/en-us/Documents/RSDS%20Project/vol1/Appendix%20A%20-%20Climate%20Change.pdf. Accessed Sept 2017
18. ITU-R recommendation P.837. approved in 2012: Characteristics of precipitation for propagation modelling. https://www.itu.int/rec/R-REC-P.837/en. Accessed Sept 2017
19. Jordanian Meteorological Department. http://jometeo.gov.jo/. Accessed Sept 2017
20. Meng YS, Lee YH, Ng BC (2009) Study of propagation loss prediction in forest environment. Prog Electromagn Res B 17:117–133
21. Michael AO (2012) Standardization of attenuation formula for radio waves propagation through free space (LOS) communication links. Sci J Phys 2012, article id sjp-281

22. Al Salameh MSH (2014) Lateral ITU-R foliage and maximum attenuation models combined with relevant propagation models for forest at the VHF and UHF bands. Int J Netw Commun 1(1):16–24. http://www.ijnngt.org/jr2vl1.php?page=2
23. The World Bank, Climate change knowledge portal. http://sdwebx.worldbank.org/climateportal/index.cfm?page=country_historical_climate&ThisCCode=NGA. Accessed Oct 2017

Leak Detection Methods—A Technical Review

R. Ramadevi, J. Jaiganesh and N. R. Krishnamoorthy

Abstract For safe transmission of various fluids or gases leakage detection in pipelines is very important. The leak of hazardous/dangerous fluids and gases can cause loss of property and lives (e.g., the Bhopal gas tragedy). Hence review of various available technologies should be necessary in order to identify a technology which provides an easy, adaptable, flexible, inexpensive, and efficient approach for real-time distributed data acquisition and monitoring. Based on review one can able to know that which technology has a very low false alarm rate and cost effective one etc. In this paper the performance and ability of the different systems is compared in terms of their leak detection capability.

Keywords Leak detection · Review methods · Pipeline · False alarm

1 Introduction

Pipelines are commonly used to deliver petroleum products, natural gas, liquid hydrocarbons, and water to consumers and industry for various applications. The movement of chemical products from place to place (e.g., natural gas, crude oil, and many other chemicals) is commonly carried out through a pipeline network. While transporting these products hundreds of miles, the pipes pass through various regions which include highly populated areas. It is essential to take measures and exercise care in those regions when chemicals are being transported. There have been many leakage accidents around the world, causing great losses of lives and properties.

R. Ramadevi (✉) · J. Jaiganesh · N. R. Krishnamoorthy
Sathyabama Institute of Science and Technology, Jeppiaar Nagar, Rajiv Gandhi Road, Chennai 600119, India
e-mail: ramadevi.eni@sathyabama.ac.in

J. Jaiganesh
e-mail: jaiganesh.eni@sathyabama.ac.in

N. R. Krishnamoorthy
e-mail: krishnamoorthy.eni@sathyabama.ac.in

A. Kumar and S. Mozar (eds.), *ICCCE 2018*,
Lecture Notes in Electrical Engineering 500,
https://doi.org/10.1007/978-981-13-0212-1_14

These types of accidents may occur in chemical industries, manufacturing industries, ships, or in any regions where pipelines are used. The reasons could be welding defects, corrosion, or erosion of external and internal walls in pipelines.

Pipeline degradation may also occurs because of stresses caused by changes in pressure and the deformation of the pipeline caused by soil dislocations, leading to the formation of micro-gaps and wear. When toxic chemicals are transported, the properties of those chemicals as well as suitable environmental conditions must be kept in mind in order to avoid any chemical reactions. So, it is necessary to study the advantages and disadvantages of existing leak detection methodologies.

In this work, seven important parameters are considered when comparing the performance of various methods. They are leak sensitivity, location estimate capability, operational change, availability, false alarm rate, maintenance requirement, and cost and power consumption. Out of these, the major parameter in almost all the methodologies faces problem is the false alarm rate.

A false alarm is highly undesirable for the following reasons.

- They generate additional work for the monitoring user.
- They reduce the confidence level of the user.
- A real leakage may be overlooked due to false alarms.

Close to 1,000 gas leakage incidents have occurred. Since LPG contains a propane and butane mixture which is highly inflammable and must be prevented to avoid any explosion. Concerned with environment protection and the costs of cleaning up oil spillages, more and more oil and gas production and transport companies are using pipeline leak detection systems on their main pipelines.

2 Leakage Detection Methodologies

Leakage detection methodologies are broadly categorized into three systems.

(a) **Hardware-based system**

These are systems that use hardware, special sensing devices for gas leak detection. As there are various types of sensors and instruments available it can be further subclassified as: acoustic [1], optical-based sensors, soil inspection [2], ultrasonic flow meters, and vapor sampling [3, 4].

(b) **Biological-based system**

This type of system does not use any sensing devices, instead it uses experienced personal to inspect the pipeline beds using either visual inspection or handheld instruments for measuring gas flow, or dogs trained to smell the leak [1]. In this system the pipeline is inspected for leak at regular interval of time among odor or sound and on hyper spectral imaging with advanced satellite (by [5–7]).

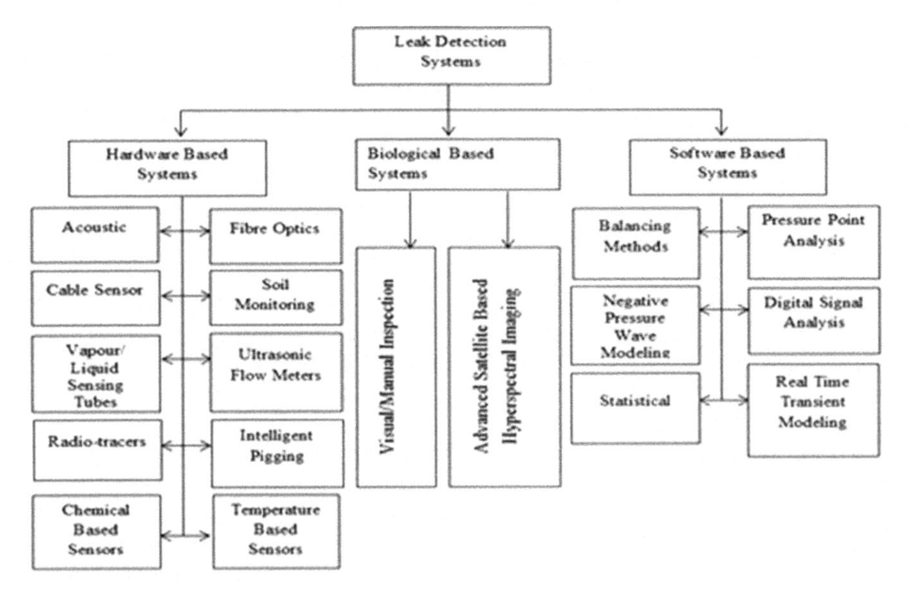

Fig. 1 Categorization of leak detection methods

(c) **Software-based system**

This type of method use different kinds of computer software package. The software implements different kinds of algorithms to monitor the condition of process parameters, such as pressure, temperature, flow rate, or other pipeline parameters. The software system depends on various techniques, namely pressure-based system-acoustic/negative pressure wave, pressure point analysis, real-time transient modeling by using a dynamic model-based system, statistical analysis and digital signal processing, flow/pressure change detection and mass/volume balance [8, 9]. Figure 1 presents major methods of leak detection techniques.

2.1 Hardware-Based System

2.1.1 Acoustic Method

The gas which is getting released at the leak point produces an acoustic signal as it flows through the pipe. This signal is used for leak detection and to record noise present inside the pipeline. Continuous monitoring can be attained by placing acoustic sensors outside the pipeline as shown in Fig. 2, which are placed at the desired distance (in meters) apart [10]. The gap between two acoustic sensors plays

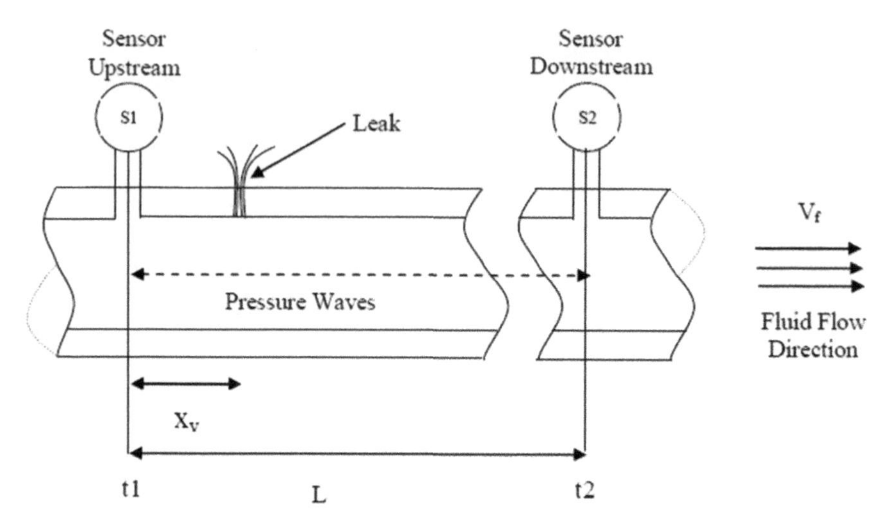

Fig. 2 Acoustic methods

a major role in determining sensitivity of the method. If the sensors are separated by a large distance, it will ultimately increase the risk of an undetected leakage, whereas placing them too close to each other will increase the cost [11].

When leakage occurs a noise signal is generated since fluid is moving out of the leak. The wave of this noise signal propagates the properties of fluid flowing through the pipeline and then the acoustic detector detects the corresponding wave and the leak [1, 12]. The problem with leak detection in longer pipelines is that it requires a large number of sensors and consequently increases the cost and is difficult to maintain, making it impractical also. Unwanted noise signals from the surroundings can be added to the original signal leading to difficulty in minute leakage detection.

2.1.2 Optical Methods

Optical methods are subdivided into two parts, namely active and passive [6, 13]. The active method uses a radiation source for scanning the area, whereas in passive methods there is no need for a radiation source because it depends on the radiation generated by the gas only. The active method illuminates the area above the pipeline bed by using a radiation source. The techniques for active monitoring technology include tunable diode laser absorption spectroscopy [14], laser-induced fluorescence [15], and coherent anti-raman spectroscopy (CARS) [16].

Active methods

The amounts of radiation which are absorbed or reflected by natural gas molecules is analyzed and if significant change or variation in absorbed and scattered light is

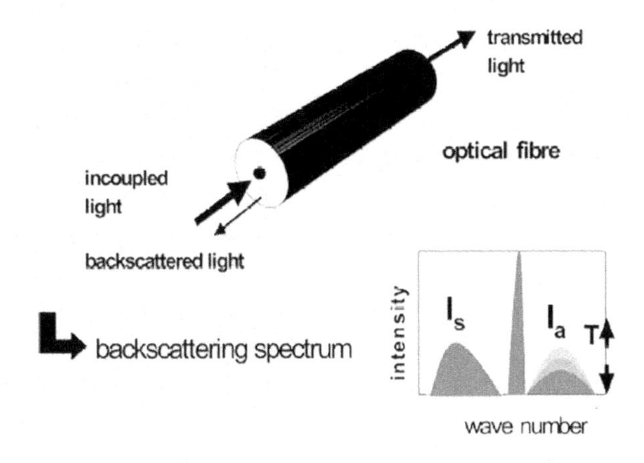

Fig. 3 Optical fiber use in backscatter imaging

detected above a pipeline bed, then a leak exists. There are different active methods for the optical detection of leaks, such as the LIDAR (light detection and ranging) method, diode laser absorption, millimeter wave radar systems, and backscatter.

Millimeter wave radar systems

In this method, the radar signature of the gas pipeline is generated. A gas like methane is lighter than air and the difference in density can produce a specific radar signature so as to detect a leak, but the major disadvantage is that it is highly expensive [17].

Backscatter imaging

This technique is also expensive, and for illuminating the scene a carbon dioxide laser is used. An infrared camera is used to capture the scattering signature, and the image revealed by the camera shows the location of leak on the pipe as shown in Fig. 3 [18, 19].

Passive methods

In the presence of hydrocarbons, the optical properties of fiber optics are affected thus providing another way of detecting gas leaks. Fiber optic sensing provides details of gas concentration and leak locations. Generally, lasers and optical detectors are used to record transmission characteristics.

Thermal imaging

To detect leaks, Weil [20] uses the difference in temperature of the leaked gas and the surrounding environment. This method is appropriate from ground and aerial vehicles, and is also successfully implemented on autonomous robots. Figure 4 shows the thermal image of a leaking pipeline. Thermal images are expensive. The major drawback is if the escaping gas has a similar temperature to that of the surrounding environment, then the leak cannot be detected.

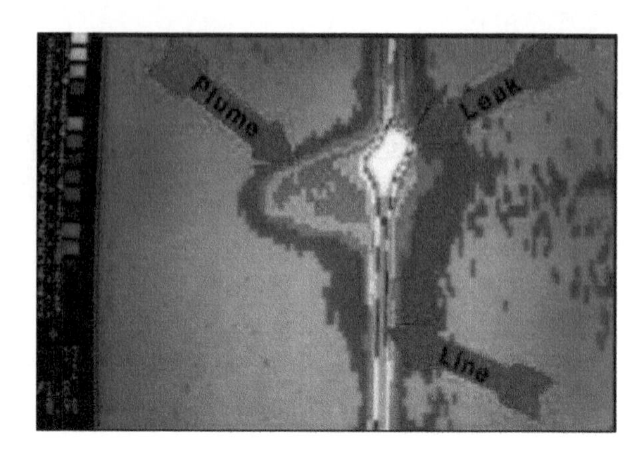

Fig. 4 Detection of a leak by thermal imaging. https://www.propublica.org/article/pipelines

Multi-spectral or multi-wavelength imaging technique

This method can be used in absorption or emission mode. Emission mode can lead to detection of leakages if the temperature of the gas escaping is much higher than the surrounding air. In absorption mode, absorption of background radiation is recorded at multiple wavelengths to generate a map of the gas concentration. The advantage is that leak detection takes place even if there is no significant difference between the escaping gas and the surrounding environment. It has a much lower possibility of generating a false alarm giving it added value. A major disadvantage of this method is that imaging sensors are highly expensive.

Gas filter correlation radiometry (GFCR)

Tolton [21], make use of a sample of the target gas as a spectral filter, where incoming radiation splits into two different directions when it passes through the narrow band pass filter. One of the cells is filled with the gas of interest (called the correlation cell) and the other one is empty. A spectral filter comprised of the correlation cell is used to remove the energy from the incoming beam at wavelengths corresponding to the absorption lines of the gas. Radiant fluxes from the two paths are measured using IR detectors and on the basis of the result it is decided if a gas leak is present. This method can detect leaks from an altitude of 300 m.

2.1.3 Soil Monitoring

Soil monitoring involves injecting the gas in the pipeline with an amount of tracer compound [22]. The tracer can be chemical or a non-hazardous or highly volatile gas, which will leave the pipe in exactly the same place as the leak (if a leak occurs). To monitor the surface above the pipeline, instrumentation is used to detect a leak by moving devices along it [23] or through probes installed in the soil close to the pipeline. Samples are collected and analyzed using a gas chromatograph [24]. Advantages of this method are a reduced false alarm rate and high sensitivity.

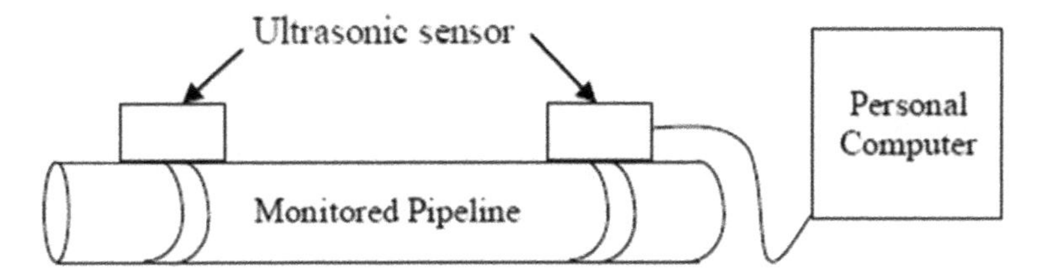

Fig. 5 Ultrasonic flow meter method. https://en.wikipedia.org/wiki/Leak_detection.4.1

A disadvantage is the high cost factor because a trace of the chemicals should be added continuously to the pipe during the detection process.

2.1.4 Ultrasonic Flow Meters

These systems were designed by Controlotron [25, 26] and later taken up by Siemens Industry Automation division [27]. In this system it is considered that the pipeline consists of a series of segments. Every segment is surrounded by two site stations which consist of a clamp-on flow meter, a temperature transducer, and a processing unit as shown in Fig. 5. All site stations measure or compute various parameters like volumetric flow rates, gas and ambient air temperature, sonic propagation velocity, and site diagnostic conditions. A master station collects the entire information obtained on or from various site stations. The computation process of the volume balance is done by the master station comparing the values obtained through site stations. The variation in the gas volume at the inlet and outlet of each pipeline segment provides necessary leak information. A small span of integration periods are used to show large leaks very quickly, while a long span of integration periods is needed to detect smaller leaks [28, 29].

This method provides accurate results but the major disadvantage is retrofitting to buried pipelines would be difficult.

2.2 Biological-Based System

In this process of detection, trained dogs are used because of their high sensitivity towards smell. The sensitivity based on various target/defect compounds, is in the range of 10 parts-per-billion (ppb) to 500 parts-per-trillion (ppt), in laboratory conditions [30]. A soap bubble screening method is also used for precisely locating smaller leaks [31, 6]. In this method, the operator sprays soap solution on different components of the pipeline, parts of the pipeline, or suspicious surfaces of the pipe. Usually it is preferable to apply this solution at the valves and piping joints because these areas are the places most prone to gas leaks. This method is rapid and the cost is low. Therefore it is helpful for routine inspection procedures. This method has the advantage that it requires no special equipment and results in the immediate localization of the leak upon detection, an advantage over the other techniques. The main disadvantage is the frequency of inspections determines the detection time, which is usually very small. The accuracy of detecting a leak greatly depends on the observation, experience, and scrupulousness of the employed personnel.

2.3 *Software-Based System*

A method based on software depends on information gained about flow, pressure, and temperature at certain regions of the pipeline. The performance efficiency and ability of a software-based leak detection system is determined by analyzing a series of factors of the existing methodologies. The necessary things to be kept in mind to evaluate the performance of leak detection systems are [32]: ability of the estimation of leak position, the speed of detection and the accuracy in determining leak size. The summarization of the very important parameters provided by each detection technique includes these criteria. Various abbreviations are used in the table below: yes (Y), no (N) for detection, slow (S), medium (M), fast (F) for detecting speed, and low (L) and high (H) for cost of the technique. A dash shows the inapplicability of the particular feature.

Table 1 provides a comparison of various leak detection techniques on the basis of power, size, location, response, and false alarms.

Table 2 shows information about assorted parameters, namely cost, monitoring speed, and easy usage, etc.

2.3.1 Mass/Volume Balance

The basic principle of this method is mass conservation between input and output. A change in the input and output gas mass or volume can be used for the determination of the leak [33, 7]. The amount of gas leaving a section/portion of pipeline is being removed from the amount of gas entering this section/portion and if the difference in the volume is above a certain predetermined limit, a leak alarm will be generated by the medium. The mass/volume can easily be computed using the readings collected by monitoring of some of the frequently used process parameters: flow, pressure, and temperature, along with various other parameters.

Leak in the pipe and its detection depend on calculating the change of inlet flow and outlet flow measurements in the pipeline. The meter accuracy and its tolerance is responsible for the sensitivity of the mass/balance method. The efficiency of this method mostly relies on the leak size, rapidity of measuring the balance, along with calculation by the system and the accuracy of the measuring instruments/devices being used. The installation of the system is easier when compared to other existing methods because it depends on instrumentation which is readily available. The operator can easily understand, learn and use it in an improved way, hence reducing

Table 1 Comparison of various leak detection techniques

Method	False alarm	Leak size	Location	Smallest leak	Response time
Acoustic emission	1 false alarm/ year	Nominal flow medium	±30 m	10% of pipeline diameter, 1-3%	15 s to 1 min
Fiber optic sensing	No	Large, medium and small leaks	1 m	50 ml/min	30 s to 5 min
vapor sensing	No	Large, medium and small leaks	0.5% of monitored area	100 l/hr	2–24 h
Ultrasonic flow meters	No	Nominal flow small	100 m range for 100 km pipeline	0.15%	Near real time
Volume balance	Many	Indicated by difference in flow	–	Greater than 5% of flow	Bigger leak-faster response
Reflected wave	Many	Related to size of propagated wave	Difficult to locate if near measuring section	10%	Fast
Pressure analysis	Many	Very small	Depends on position of transducers	50 ml/min	Delayed response
GLR	Very Less	Indicated by mass flow variation	Almost entire length	10%	Fast

errors. A further advantage of this system is its comparatively low cost. The usage of balancing techniques is restricted to leakage detection during varying flow or shut in and slack line conditions. It takes a longer time to detect small leakages. For example, a 1% leak needs approximately 40–60 min to detect [34]. This method is not favorable when it comes to locating the leak and another drawback is that unless thresholds are adapted, it generates frequent false alarms during transient states.

2.3.2 Pressure Change

In this process, pressure sensors have an important role and are mostly installed at the extremes of pipelines. Initially, a predefined limit is set when the steady state occurs and if the pressure falls below this limit (as shown in Fig. 6) then a leak exists. The usage of low pass filter is an advanced and improved technique for use with long pipelines, and is done with respect to the occurrence of pressure disturbances.

Table 2 Comparisons based on other parameters

	Visual inspection	Soap screening	Aconstic	LIDAR	Diode laser absorption	Millimeter radar	Backscatter imaging	Broadband absorption	Fiber optic cable	Thermal imaging
Cost	L	L	H	H	M	H	H	L	H	H
Detection speed	S	S	F	M	M	M	M	M	F	M
East retrofiting	–	–	Y	–	–	–	–	–	N	–
Easy usage	Y	Y	Y	Y	Y	Y	Y	N	Y	Y
Leak localization	Y	Y	Y	Y	Y	Y	Y	Y	Y	Y
Leak size estimation	Y	Y	Y	Y	Y	Y	Y	Y	Y	Y
	Multi-Spectral imaging	Gas filter correlation radiometry	Soil monitoring	Vapor sampling	Ulrasonic flow meters	Mass/ Volume balance	Negative pressure valve	Pressure point analysis	Statistical	Digital signal processing
Cost	H	M	H	H	H	L	L	L	H	H
Detection speed	M	M	S	F	F	F	F	F	F	F
East retrofiting	–	–	N	N	N	Y	Y	Y	Y	N
Easy usage	Y	Y	Y	Y	Y	Y	Y	Y	Y	Y
Leak localization	Y	Y	Y	Y	Y	N	Y	N	Y	Y
Leak size estimation	Y	Y	N	Y	N	Y	Y	N	Y	N

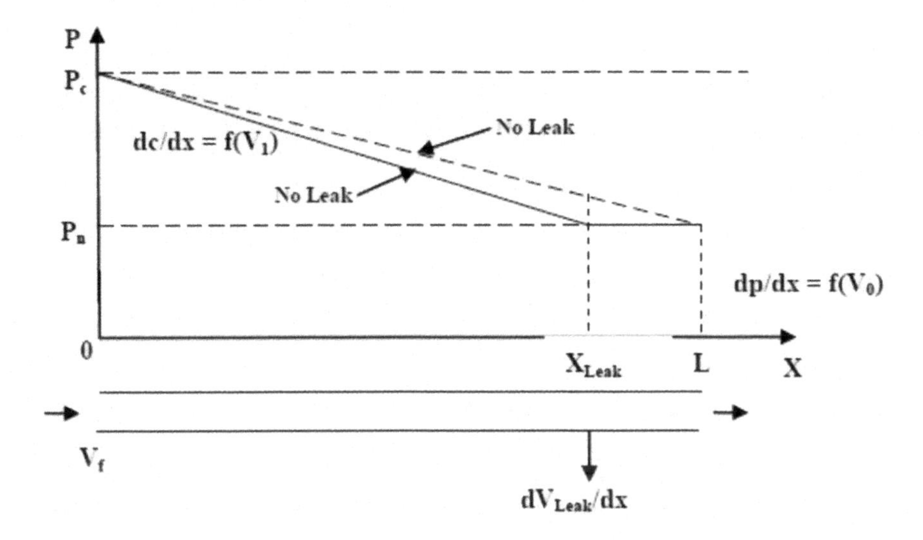

Fig. 6 Showing the variation in pressure

2.3.3 Change in Flow

In this method the operator uses a predefined figure like a reference figure, which is used as a model for possibilities in the change of flow. The leak detection here is assumed to take place when in a specific time period the rate of change of the flow observed is higher than a predefined figure.

2.3.4 Negative Pressure Wave

It is known that the spot where pressure drops or where there is a sudden variation in pressure leads to increase in the leakage probability, which generates the wave of pressure upstream and downstream. This generated wave is known as a negative pressure wave and the readings are collected by using pressure sensors which are placed at extremes of the pipe [35].

For determination of the leakage, the leakage algorithm collects the reading (information) from the pressure transducer placed on the pipeline. Different methods, including a support vector machine [36], are used for the same purpose. The time difference between the moments at which the two pressure transducers ends, senses the negative pressure wave and is used to identify the leak location. If the leak is near to one end of the pipe, then the corresponding transducer will be the first to receive the pulse and the amount of time required to receive the pulse at the other end is used to detect the leak location with a good degree of precision. Negative pressure wave-based leak detection systems, such as ATMOS Wave [37, 38], can estimate the size of the leak.

Another approach of detecting leaks by means of pressure waves is by manually or intentionally generating the transient pressure waves. This is done by closing and opening valves at intervals of time [39, 40]. The presence of a leak will partially reflect these pressure waves and allow for the detection and location of the leak. A disadvantage of using pressure waves is that it become impractical to detect leaks in long-distance pipeline.

2.3.5 Pressure Point Analysis

Pressure point analysis is a fast leak detection software technique based on the principle that in the presence of a leak, the pressure drops or changes will occur inside the pipeline [41]. This technique is made efficient by continuous measurements of the pressure at various locations on the pipeline. The presence of a leak can be detected by statistical analysis of the measured values and by comparing the measured mean value of pressure with the threshold set point value. If the measurement is below the threshold value, then leakage is detected, otherwise there is no leakage. The patent [41] of this leakage detection technique is with EFA Technologies Inc. which offers PPA™ as part of their LEAKNET™ leak detection system along with MassPack™, PPA™ which has been proven to work in different environment condition (high and low temperature, pressure) [42] and leak rates below 0.1% of flow but it is not a dependable technique during transient flow.

2.3.6 Statistical Analysis

An easier method of detecting gas leakages, without the need to design a mathematical model is by using a statistical analysis technique. The corresponding analysis is done on various measuring parameters like flow and pressure at different locations along the entire pipeline bed. The system will generate a leak alarm only if it detects a pattern consisting of a relative change in flow and pressure parameters [43].

The thresholds for leakage are set after a tuning period of the system during which the parameter is placed under different operating conditions in the absence of a leak. To reduce the false alarm rate, the tuning process is done for a long period of time [44]. During the tuning period, the initial data will be affected in the presence of a leak and the system behavior will be considered as normal due to which the leak would not be detected unless and until it grows large enough in size to go beyond the threshold limits given during the tuning process.

A leakage of 0.5% was detected [44] but even smaller leaks can be detected. Instruments being used should have a high resolution and be accurate and precise. Statistical analysis can also be used for determining the leak location and position. The main advantage of this technique is its flexibility of use, being adaptive and robust to different pipeline configurations. The main disadvantages of using this method are the difficulty in estimating leak volume and considerably high costs.

2.3.7 Digital Signal Processing (DSP)

In this method, leaks can be detected by measuring flow rate, pressure, and temperature parameters obtained by using digital signal processing [45]. The response of a known flow change is measured during the setup phase. Measurement of parameters is used together with DSP to identify changes in the system response. DSP allows the leak response to be recognized from noisy data. In the beginning, this technique was provided only for liquid pipelines [46] but later it was even considered for gas pipelines. There is no requirement for a mathematical model for the pipeline; its main motive is to extract leak information from noisy data. Similar to a statistical approach, if during the set-up phase a leak is already present in system, it would never be detected until its size grows. Disadvantages are its high expense, implementation difficulty.

3 Conclusions

A review of the various leak detection techniques has been presented in this paper. Comparison of different leak detection techniques based on various features such as cost, false alarm rate, approximate leak location capability etc., has been provided. Of all the techniques, the optical fiber method is the most effective in all aspects except for cost and maintenance factors. Acoustic methods provided reasonable detection sensitivity but under low surrounding noise it became incapable. Hence, an ultrasonic flow meter is used for surrounding noise. Biological methods like the surveying of pipelines depend greatly on the experience and meticulousness of the employed personnel. So, it cannot be used frequently, and as a result software methods were introduced. Software methods helped to continuously monitor real-time leak detection, providing better accuracy on the position of leaks and the size of leaks.

References

1. Hough JE (1988) Leak testing of pipelines uses pressure and acoustic velocity. Oil Gas J (United States) 86:47: n. pag. Print
2. Tracer Research Corporation (2003) Patent product described in the website of Tracer Research Corporation. www.tracerresearch.com
3. Bose JR, Olson MK (1993) TAPS's leak detection seeks greater precision. Oil Gas J, April Issue, pp 43–47
4. Turner NC (1991) Hardware and software techniques for pipeline integrity and leak detection monitoring. In: Proceedings of offshore Europe, vol 91. Aberdeen, Scotland
5. Carlson BN (1993) Selection and use of pipeline leak detection methods for liability management into Journal of Applied Engineering (JOAE), 2(2), February-2014 (Volume-II, Issue-II) 32 the 21st century. In: Pipeline infrastructure II, Proceedings of the international conference of the American Society of Chemical Engineers, ASCE

6. Murvay P-S, Silea I (2012) A survey on gas leak detection and localization techniques. J Loss Prev Process Ind
7. Scott SL, Barrufet MA (2003) Worldwide assessment leak detection capabilities for single and multiphase pipelines, project report prepared for The Minerals Management Service. Offshore Technology Research Centre, Texas
8. Griebenow G, Mears M (1988) Leak detection implementation: modelling and tuning methods. Am Soc Mech Eng Pet Div 19:9–18
9. Liou JCP, Tian J (1994) Leak detection: a transient flow simulation approach. Am Soc Mech Eng Pet Div 60:51–58
10. Brodetsky I, Savic M (1993) Leak monitoring system for gas pipelines. In: IEEE international conference on acoustics, speech, and signal processing, ICASSP-93, vol 3. IEEE, pp 17–20
11. Loth J, Morris G, Palmer G (2003) Technology assessment of on-line acoustic monitoring for leaks/infringements in underground natural gas transmission lines. Technical Report, West Virginia University, USA
12. Klein WR (1993) Acoustic leak detection. Am Soc Mech Eng Pet Div (Publication) PD, vol 55, pp 57–61
13. Reichardt TA, Einfeld W, Kulp TJ (1999) Review of remote detection for natural gas transmission pipeline leaks. Report prepared for NETL, Sandia National Laboratories, Albuquerque, NM
14. Hanson RK, Varghese PL, Schoenung SN, Falcone FK (1980) Absorption spectroscopy of combustion gases using a tunable IR diode laser. In: Laser probes for combustion chemistry, ACS symposium series, vol 134, pp 413–426
15. Crosley DR, Smith GP (1983) Laser induced fluorescence spectroscopy for combustion diagnostics. Opt Eng 22:545–553
16. Eckbreth AC, Boncyzk PA, Verdieck JF (1979) Combustion diagnostics by laser, Raman and fluorescence techniques. Prog Energy Combust Sci 5:253–322
17. Gopalsami N, Raptis AC (2001) Millimeter-wave radar sensing of airborne chemicals. IEEE Trans Microw Theory Tech 49:646–653
18. Kulp TJ, Kennedy R, Delong M, Garvis D (1993) The development and testing of a backscatter absorption gas imaging system capable of imaging at a range of 300 m. Appl Laser Radar Technol; In: Proceedings of society of photo-optical instumentation engineering, vol 1936, pp 204–212
19. Kasai N, Tsuchiya C, Fukuda T, Sekine K, Sano T, Takehana T (2011) Propane gas leak detection by infrared absorption using carbon infrared emitter and infrared camera. NDT E Int 44(1):57–60
20. Weil G (1993) Non contract, remote sensing of buried water pipeline leaks using infrared thermography. ASCE, New York, NY, USA, pp 404–407
21. Tolton T, Boyd A (2004) Concept for a gas-filter correlation radiometer to remotely sense the atmospheric carbon dioxide column from space. J Atmos Ocean Technol 21:837. https://doi.org/10.1175/1520-0426(2004)021%3c0837%3AACFAGC%3e2.0.CO%3B2
22. Lowry W, Dunn S, Walsh R, Merewether D, Rao D (2000) Method and system to locate leaks in subsurface containment structures using tracer gases, US Patent 6,035,701
23. Praxair Technology Inc (2007) Seeper trace leak detection for in-situ gas storage, sequestration and EOR sites
24. Thompson G, Golding R (1993) Pipeline leak detection using volatile tracers. Leak Detect Underg Storage Tanks 1161:131–138
25. Controlotron Corporation (2005) Controlotron non-intrusive ultrasonic flowmeters. Brochure
26. Controlotron Corporation (2006) Controlotron system 1010 family of non-intrusive ultrasonic flowmeters. Brochure
27. Siemens Industry Inc Clamp-on leak detection solution for enhanced pipeline management. Brochure, Aug 2011
28. Bloom D (2004) Non-intrusive system detects leaks using mass measurement. Pipeline Gas J 231(7):20–21

29. Siemens Industry Inc SITRANS F: Ultrasonic flowmeters FUS-LDS leak detection system operating instructions, Sept 2011
30. Johnston J (1999) Canine detection capabilities: operational implications of recent R & D findings. Institute for Biological Detection Systems, Auburn University
31. Liu AE (2008) Overview: pipeline accounting and leak detection by mass balance, theory and hardware implementation
32. Stafford M, Williams N (1996) Pipeline leak detection study. Technical Report, Bechtel Limited
33. Parry B, Mactaggart R, Toerper C (1992) Compensated volume balance leak detection on a batched LPG pipeline. In: Proceedings of the international conference on offshore mechanics and arctic engineering. American Society of Mechanical Engineers, pp 501–501
34. Doorhy J (2011) Real-time pipeline leak detection and location using volume balancing. Pipeline Gas J 238(2):65–66
35. Silva R, Buiatti C, Cruz S, Pereira J (1996) Pressure wave behaviour and leak detection in pipelines. Comput Chem Eng 20:S491–S496
36. Chen H, Ye H, Chen L, Su H (2004) Application of support vector machine learning to leak detection and location in pipelines. In: Proceedings of the 21st IEEE instrumentation and measurement technology conference, 2004. IMTC 04, vol 3. IEEE, pp 2273–2277
37. de Joode AS, Hoffman A (2011) Pipeline leak detection and theft detection using rarefaction waves. In: 6th pipeline technology conference
38. Twomey M (2011) A complimentary combination. World Pipelines, pp 85–88
39. Elaoud S, Hadj-Taleb L, Hadj-Taleb E (2010) Leak detection of hydrogen-natural gas mixtures in pipes using the characteristics method of specified time intervals. J Loss Prev Process Ind 23(5):637–645
40. Mpesha W, Gassman S, Chaudhry M (2001) Leak detection in pipes by frequency response method. J Hydraul Eng 127:134–147
41. Farmer E, et al (1989) A new approach to pipe line leak detection. Pipe Line Ind (USA) 70(6): 23–27
42. Scott S, Barrufet M (2003) Worldwide assessment of industry leak detection capabilities for single & multiphase pipelines. Technical Report, Department of Petroleum Engineering, Texas A&M University
43. Zhang X (1993) Statistical leak detection in gas and liquid pipelines. Pipes Pipelines Int 38(4): 26–29
44. Zhang J, Di Mauro E (1998) Implementing a reliable leak detection system on a crude oil pipeline. In: Advances in pipeline technology, Dubai, UAE
45. USDT (2007) Leak detection technology study for pipes act. Technical Report, U.S. Department of Transportation
46. Golby J, Woodward T (1999) Find that leak [digital signal processing approach]. IEE Rev 45(5):219–221
47. Mandal PC (2014) Gas leak detection in pipelines & repairing system of titas gas. J Appl Eng (JOAE), 2(2), pp 23–34
48. Sivathanu Y, Narayanan V (2003) Natural gas leak detection in pipelines, Technology Status Report

Text Message Classification Using Supervised Machine Learning Algorithms

Suresh Merugu, M. Chandra Shekhar Reddy, Ekansh Goyal and Lakshay Piplani

Abstract In recent years, as the popularity of mobile phone devices has increased, the short message service (SMS) has grown into a multi-billion dollar industry. At the same time, a reduction in the cost of messaging services has resulted in the growth of unsolicited messages, known as spam, one of the major problems that not only causes financial damage to organizations but is also very annoying for those who receive them. **Findings:** Thus, the increasing volume of such unsolicited messages has generated the need to classify and block them. Although humans have the cognitive ability to readily identify a message as spam, doing so remains an uphill task for computers. **Objectives:** This is where machine learning comes in handy by offering a data-driven and statistical method for designing algorithms that can help computer systems identify an SMS as a desirable message (HAM) or as junk (SPAM). But the lack of real databases for SMS spam, limited features and the informal language of the body of the text are probable factors that may have caused existing SMS filtering algorithms to underperform when classifying text messages. **Methods/Statistical Analysis:** In this paper, a corpus of real SMS texts made available by the University of California, Irvine (UCI) Machine Learning Repository has been leveraged and a weighting method based on the ability of individual words (present in the corpus) to point towards different target classes (HAM or SPAM) has been applied to classify new SMSs as SPAM and HAM. Additionally, different supervised machine learning algorithms such as support vector machine, k-nearest neighbours, and random forest have been compared on the basis of their performance in the classification of SMSs. **Applications/Improvements:** The results of this comparison are shown at the end of the paper along with the desktop application for the same which helps in classification of SPAM and HAM. This is also developed and executed in python.

S. Merugu (✉) · M. C. S. Reddy
Research and Development Centre, CMR College of Engineering & Technology, Hyderabad 501401, Telangana, India
e-mail: msuresh@cmrcet.org

E. Goyal · L. Piplani
Department of Computer Science and Engineering, Maharaja Agrasen Institute of Technology, New Delhi, India

A. Kumar and S. Mozar (eds.), *ICCCE 2018*,
Lecture Notes in Electrical Engineering 500,
https://doi.org/10.1007/978-981-13-0212-1_15

Keywords Text message classification · SPAM · Machine learning algorithms

1 Introduction

The mobile phone market has experienced substantial growth in recent years and people are becoming increasingly reliant on cell phones. As a result, the short message service (SMS), or text messaging, has become one of the most widely used communication channels [1]. Today, this service is almost free, or is offered at negligible prices [2]. It is the most widely used data application with an estimated 4.77 billion active users in 2017; an increase of about 3.4% over last year's estimate. However, with the exponential rise in the popularity of this medium of communication, it is now being increasingly misused to distribute spam messages. According to Korea Information Security (KISA), SMS has become a very popular means of circulating spam. According to one study, cell phone users in the United States received over a billion spam messages, whereas those in China received 8.27 spam SMS per week [3].

Spam, received through any communication channel, whether through e-mail or as a text message, is undesirable and a nuisance. Robust mechanisms exist that are able to tackle spam received through e-mail. Hence, development of methods that are able to classify incoming text messages as spam or ham, and possibly inform receivers of the arrival of spam, is very much needed. Text classification is the process of identifying the most suitable target class to which the message belongs based on the body of the message [4]. Construction of effective classification algorithms is one of the most strenuous yet rewarding tasks in the realm of machine learning [5].

In this paper, the goal is to apply different machine learning algorithms in an attempt to solve the SMS spam classification problem, compare their performance to gain insight and further explore the problem, and design an application based on one of these algorithms that can filter SMS spam with a high degree of accuracy. This paper used a dataset of 5,574 text messages available at the UCI Machine Learning repository [6, 7]. The corpus consists of actual text messages exchanged between people. It contains a collection of 425 SMS spam messages, a subset of 3,375 SMS randomly chosen non-spam (ham) messages of the NUS SMS Corpus (NSC), a list of 450 SMS non-spam messages collected from Caroline Tag's PhD Thesis, and the SMS Spam Corpus v.0.1 Big (1,002 SMS non-spam and 322 spam messages, all publicly available). The dataset is a large text file in which each line starts with the label of the message, followed by the text message string. Actually, this corpus serves as the central dataset that the server of our application uses to formulate its classification model [8]. In addition to this central dataset, each client maintains its own personalized dataset, which is populated as the client flags a received text message as incorrectly classified. Declaration of a message as incorrectly classified by a client causes the message and the correct label to be added to the client's personal dataset. Once the server and client classification models are developed, they can individually calculate the probability that a text message is spam. A linear combination of these probabilities is finally used to display an incoming text message as spam or ham at the client. Classification techniques such as naïve Bayes, SVM, and other methods are compared on the basis of their performance on the central corpus and the most suitable algorithm is chosen to develop the classification models [9].

This paper is organized as follows. In Sect. 2, it describes previous work that attempts to classify text messages as spam or ham. In Sect. 3, proposed and tested application work is described. In Sect. 4, the performances of various algorithms is compared. Finally, Sect. 5 describes the conclusion and future potential of this work.

2 Background Study

There have been numerous studies on active learning for text classification using machine learning techniques. The popular techniques for text classifications are naïve Bayes classifier, support vector machine, k-nearest neighbor, neural networks and decision trees. In spite of the fact that so much been proposed so far, the existing text classification methods are far from being infallible and there is a lot of room for improvement.

While [2, 7, 10] have attempted to propose sound solutions to the problem of identifying spam messages by employing various machine learning algorithms and various feature representations (such as bag of words (BoWs), BoWs augmented by statistical features, such as the proportion of upper case letters or punctuation in the text, orthogonal sparse word bigrams, character bigrams and trigrams) [11]. Another study [12] proposed a system of SMS classification using naïve Bayes classifier and an a priori algorithm using the same dataset available at the UCI repository.

A further study [10] proposed a term frequency-inverse document frequency (TF-IDF) approach to convert text into numerical features. In this method, terms with a high frequency of occurrence in a given document but lower frequency of occurrence across all documents present in the corpus are considered to have a very good ability to differentiate between target classes (spam and ham in this case). Though this approach places more emphasis on the ability to differentiate between classes, it ignores the fact that the term that frequently appears in the documents belonging to the same class, can be more representative of the characteristic of that class. The weight is associated with each word by applying the TF-IDF technique and the word which has a high frequency in those documents [13].

3 Methodology

This section describe the steps performed to create the proposed application.

a. ***Preparing the dataset: separating labels and text messages***

Each line of the dataset used starts with the label of the message (spam or ham) followed by the actual text. Hence, the first step is to create two new files, one for the label and the other for the actual text of the corresponding message.

In summary:

- Two new files are created using the original dataset; one contains the text string and the other contains the corresponding label (0 for HAM and 1 for SPAM)
- Each message is processed to remove punctuation and any extra white spaces. Each word in a message is converted to lowercase.

b. ***Splitting the dataset into training and test sets***

The example messages are split into training (90%) and testing (10%) sets. This is done by loading the strings of the text messages and the corresponding labels from the two files into an array-like structure, and then randomly selecting strings and corresponding labels to form the training and test sets.

c. ***Removal of high frequency tokens***

Stop words of a language are those that have a high frequency of occurrence in a given text of that particular language. Hence, before creating TF-IDF vectors, English stop words are removed because, due to the high frequency of occurrence of these words, they will not contribute much to the differentiating ability of the classification model. For the same reason, token appearing is more than 50% of the example messages.

d. ***Feature extraction using TF-IDF***

The example text messages are strings of variable length. In order to feed this data into a classification model, each string needs to be converted into a fixed-size numerical feature vector. This is accomplished through a popular method called term frequency-inverse document frequency (TFIDF). TF-IDF uses a BoW model; it generates a list of all unique words, or tokens, appearing across all text strings present in the dataset. Each text string is then converted into a numerical feature vector, where each feature corresponds to the frequency of occurrence of a token present in the BoW. The frequency is calculated as shown below:

$$\mathrm{TF-IDF}_{i,j} = \mathrm{TF}_{i,j} * \log(\mathrm{N}/\mathrm{DF}_i)$$

where,

$\mathrm{TF\text{-}IDF}_{i,j}$ Term frequency-inverse document frequency of word ‘i’ in document ‘j’.

$\mathrm{TF}_{i,j}$ Total occurrences of word ‘i’ in document ‘j’.

DF_i Total number of documents containing word ‘i’.
N Total number of documents.

It is interesting to note that IDF is a normalizing term; tokens that appear in a large number of documents have a low TF-IDF value, which represents their ineffectiveness in differentiating between target classes. Hence, the steps followed here as:

(1) Using the text messages in the training set, the TF-IDF model is set up. This is done to ensure that feature vectors in both the training and test sets have the same attributes.
(2) The BoW for this model consists of all unique tokens appearing in the example messages in the training set. These tokens will serve as attributes for feature vectors of both the training and test sets.
(3) We configure the model to neglect all tokens that are English stop words or that appear in more than 50% of the documents.
(4) Finally, numerical feature vectors are obtained according to the TF-IDF model.
(5) These feature vectors are very lengthy and a number of features in each vector have a value of 0. Hence, we use univariate feature selection to retain only the top 10% of the best features. This greatly reduces the size of the vectors, making the ensuing computation easier.
(6) Along with the corresponding numerical labels (0 for HAM and 1 for SPAM), these vectors are used to train and subsequently test the classification models.

e. ***The server***

(1) First, the classification model (naive Bayes/support vector machine/k nearest neighbors/random forest) is trained using the prepared training dataset, and tested using the test dataset. The test dataset accuracy, and the time taken for training and testing is calculated.
(2) Repeat steps 3–4 until the server is closed.
(3) If there is a connection request from a new client:

(a) Accept the request
(b) Broadcast to all existing participants that a new participant has joined

(4) If a message is received from a client.

(a) If the client requests correction of the label of a text message:

i. Increment the number of such requests received.
ii. If the number of such requests exceeds the specified threshold, then:

- Process the message by removing punctuation and extra white spaces.
- Add the processed message and the corresponding label (0 for HAM, and 1 for SPAM).

(b) else:

i. Using the trained model, calculate the probability *probServer* of the message being spam.

ii. Broadcast the message to all the participants along with the probability calculated above.

f. ***The client***

(1) Request to connect to the server.
(2) If the connection is not successful, exit.
(3) Repeat steps 4–5 until client is closed.
(4) Using the client's own personal dataset, train a classification model. The model is created only if the client's dataset contains examples of both ham and spam, and at least a specified total number of examples, using the same feature extraction procedure described above. The feature selection procedure is skipped as the size of the client's personal dataset is relatively small and losing features might lead to a loss of substantial information.
(5) If a message is received from the server:

(a) Calculate the probability *probClient* of the message being spam using the client's personal classification model. The final probability of the message being spam is calculated as

$$P = 0.6*(probClient) + 0.4*(probServer)$$

The message is spam if $P \geq 0.5$.

(b) Output the message along with the label.
(c) If the user says the message is incorrectly classified:

i. Send the message and the correct label back to the server.
ii. Add the message and the correct label to the client's own dataset. To prevent overcrowding by a single message, the message is added only if it occupies less than a specified fraction of the total length of the dataset.
iii. Correct the display label of the message.

(6) If the user enters a text to be sent, send the message to the server.

4 Results and Discussion

This whole process of executing the experimental dataset used Intel Core^{TM} i7 and a machine with 4 GB ram. The whole system is implemented using Python language and the UCI data repository is used for training the system. But for training, the whole test suite is not used. Only a certain percentage of test suites is used, with the rest used for prediction. In this case, 90% was used for training with the other 10% used for testing. This prevents the decision surface from over-fitting into the training dataset.

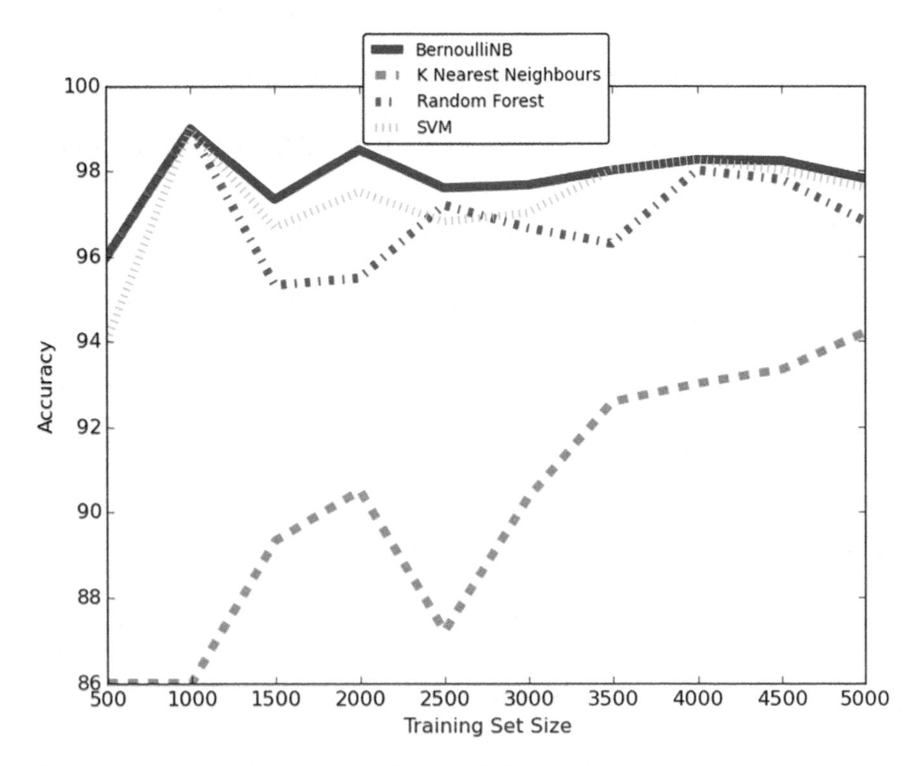

Fig. 1 Learning curve for various algorithms applied to the dataset

The proposed application was tested using several algorithms, such as naïve Bayes, support vector machine, random forest, and k-nearest neighbor. For each of these techniques, the training set size was varied from 500 to 5,000, against which corresponding accuracies were plotted as shown in Fig. 1.

K-nearest neighbor produced the worst accuracies of all and random forest, support vector machine, and naïve Bayes performed the best, with each one performing better than the former, with an accuracy of close to 98%.

From the plot it can also be seen that support vector machine works as well as naïve Bayes at some points—or even better. So the question may arise as to why it isn't possible to use support vector machine? This can be explained by analyzing the plot plotted between the training times required by these techniques for the varied data set sizes as shown in Fig. 2. It can be observed that naïve Bayes is considerably faster at training the model than support vector machine, and for greater sizes of dataset this becomes an important consideration. Hence, this paper selected the naïve Bayes technique as it is fast and gives an acceptable accuracy of 97.6%.

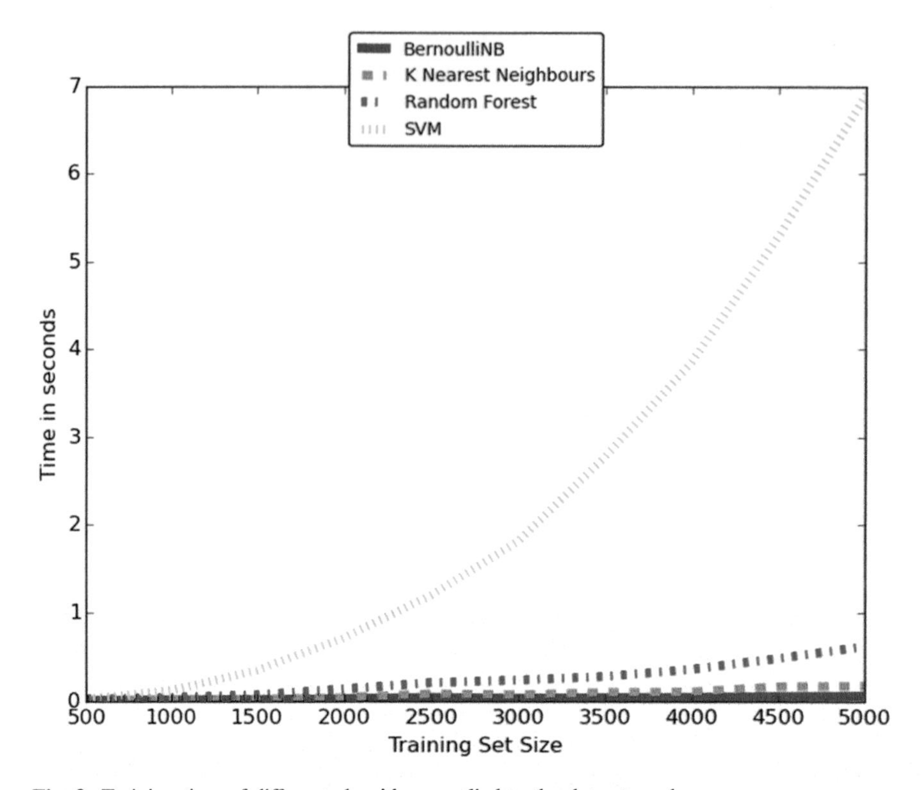

Fig. 2 Training time of different algorithms applied to the dataset used

The best machine learning technique for SMS classification in this study is the naïve Bayes method. Here the question may arise as to how we may be sure that the dataset used by anyone is optimal and that adding more data will not result in increased accuracy of any of the techniques? This can be explained by looking at the following plots in Fig. 3 between test error and training error against the data set size. It can be seen that training set error and test set error are close to each other. Therefore, there is no problem with high variance, and gathering more data may not result in much of an improvement in the performance of the learning algorithm, which proves that our dataset is optimal.

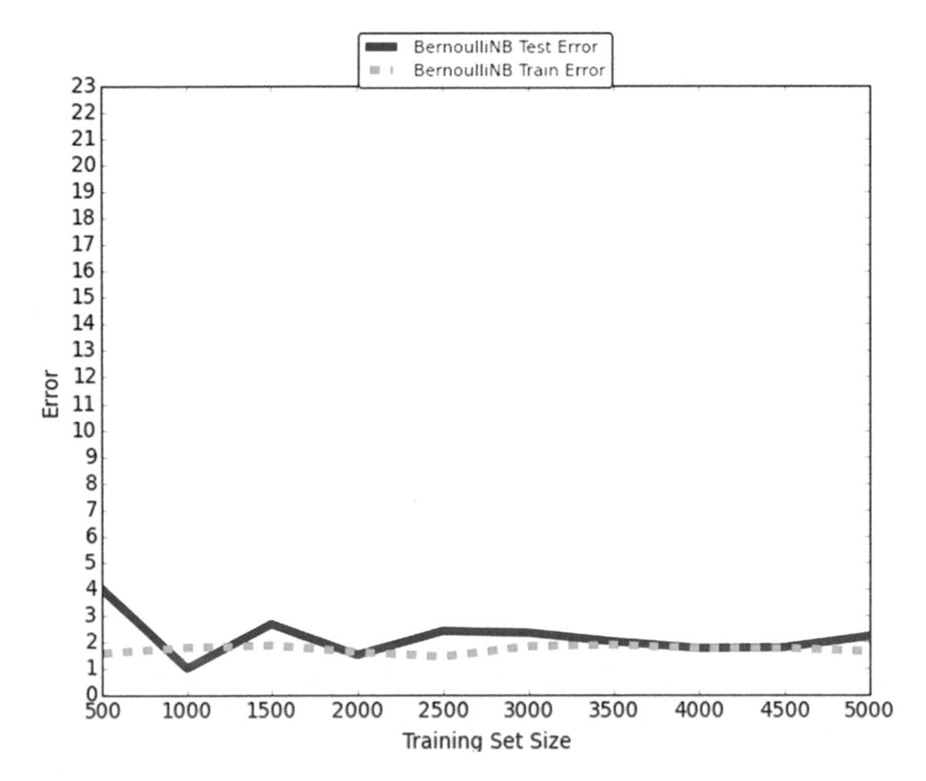

Fig. 3 Training and test error for Bernoulli NB

5 Conclusion

In this paper, the applicability of some of the most popular machine learning algorithms to the problem of SMS spam classification is reviewed and tested. The comparison of the performance of these algorithms on the dataset present in the UCI Machine Learning Repository is presented. The experimental observations shows very promising results with the long-used machine learning algorithm for text classification, naïve Bayes, proving to be the best of the lot for SMS spam classification in terms of both accuracy and training time. The desktop application for the same has been developed successfully.

References

1. Gandhi K, Pandit Rao R, Lahane VB (2015) A survey on OSN message filtering. Int J Comput Appl 113(17):19–22
2. Soundararajan K, Eranna U, Mehta S (2012) A neural technique for classification of intercepted e-mail communications with multilayer perceptron using BPA with LMS learning. Int J Adv Electr Electron Eng 1(13):141–150

3. Wang Q, Han X, Wang X (2009) Studying of classifying junk messages based on the data mining. In: Management and service science, Wuhan, China, pp 1–4
4. Hidalgo JMG, Almeida TA, Yamakami A On the validity of a new SMS spam collection. In: Proceedings of the 11th IEEE international conference on machine learning and applications (ICMLA'12), vol 2, Boca Raton, FL, USA, December 2012, pp 240–245
5. Suresh M, Jain K (2017) Subpixel level mapping of remotely sensed imagery to extract fractional abundances using colorimetry. E J Remote Sens Spat Sci. https://doi.org/10.1016/j.ejrs.2017.02.004
6. Aha D (1987) UCI machine learning repository: SMS spam collection data set. http://archive.ics.uci.edu/ml/datasets/SMS+Spam+Collection. Accessed on 21 Dec 2016
7. Almeida TA, Hidalgo JMG (2011) SMS spam collection (v. 1) (Online). http://www.dt.fee.unicamp.br/~tiago/smsspamcollection/. Accessed on 11 Nov 2016
8. Suresh M, Jain K (2015) Semantic driven automated image processing using the concept of colorimetry. In: Second international symposium on computer vision and the internet (VisionNet'15); Procedia Computer Science (Elsevier) 58:453–460. https://doi.org/10.1016/j.procs.2015.08.062
9. Suresh M, Jain K (2016) Colorimetry-based edge preservation approach for color image enhancement. J Appl Remote Sens (SPIE) 10(3):035011. https://doi.org/10.1117/1.jrs.10.035011
10. Al-Talib GA, Hassan HS (2013) A study on analysis of SMS classification using TF-IDF weighting. Int J Comput Netw Commun Secur 1(5):189–194
11. Caragea C, McNeese N, Jaiswal A et al Classifying text messages for the Haiti earthquake. In: Proceedings of the 8th international information systems for crisis response and management conference (ISCRAM'12), Lisbon, Portugal, May 2011
12. Ahmed, Guan D, Chung T (2014) SMS classification based on naïve Bayes classifier and Apriori algorithm frequent item set. Int J Mach Learn Comput 4(2):183–187
13. Almeida TA, Hidalgo JMG, Yamakami A Contributions to the study of SMS spam filtering: new collection and results. In: Proceedings of the 2011 ACM symposium on document engineering (DOCENG'11), Mountain View, CA, USA, pp 259–262, Sept 2011

Error Assessment of Fundamental Matrix Parameters

Bankim Chandra Yadav, Suresh Merugu and Kamal Jain

Abstract Stereo image matching comprises of establishing epipolar geometry based on fundamental matrix estimation. Accuracy of the epipoles is governed by the fundamental matrix. A stereo image pair may contain errors on a systematic and/or random basis which determine the accuracy of the fundamental matrix required for matching. The algorithm used to extract the image pair correspondence, and the method used to estimate the correct parameters of the matrix, controls its accuracy further. A performance analysis of widely adopted matrix estimators over the point pairs found by correspondence determiners is undertaken in this chapter. The methods are modified for the best combination of results, based on the properties of the resulting fundamental matrix. Permutations are analyzed over the possible paths for obtaining the matrix parameters with the expected characteristics, followed by error analysis. Amongst the estimators analyzed, RAndom SAmple Consensus (RANSAC) and M-estimator SAmple Consensus (MSAC) estimators were found to produce the best results over the features detected by the Harris–Stephens corner detector.

B. C. Yadav (✉) · K. Jain
Department of Civil Engineering, Indian Institute of Technology Roorkee, Uttarakhand, India
e-mail: bcyadav0808@outlook.com

K. Jain
e-mail: kjainfceiitr@gmail.com

S. Merugu
CMR College of Engineering and Technology, Hyderabad, India
e-mail: suresh1516@gmail.com

A. Kumar and S. Mozar (eds.), *ICCCE 2018*,
Lecture Notes in Electrical Engineering 500,
https://doi.org/10.1007/978-981-13-0212-1_16

1 Introduction

The fundamental matrix is a third-order square matrix which establishes a relation between the corresponding homogeneous coordinates of an image pair. It is a basic element in the works of photogrammetry, computer vision, and overlapping fields. The conventional technique of 3D picturization is by image pair acquisition using a stereo rig. The information related to the extrinsic parameters, camera calibration, and other metric information is captured in matrices viz. the essential matrix [5], camera calibration matrix, and fundamental matrix [6].

The fundamental matrix defines the relationship between corresponding point pairs in terms of the underlying projective relationship without requiring the external camera parameters. For the purpose of context, a brief idea of the fundamental matrix is given here without delving into its derivation.

1.1 Fundamental Matrix

"The Fundamental matrix is the algebraic representation of epipolar geometry," as per the original introduction given by [6]. Consider a stereo rig. Given two points $\vec{x_1}$ and $\vec{x_2}$ in some stereo images an epipolar line exists for $\vec{x_1}$ in the second image. This line obtained in the second image is the projection of the line passing from $\vec{x_1}$ and center of the first camera. Hence there exists a mapping between the corresponding points and lines as such:

$$\begin{array}{ccc} \vec{x_1} & \rightarrow & \vec{l_2} \\ \vec{x_2} & \rightarrow & \vec{l_1} \end{array}$$

Such mappings describing the projections between points and lines in a stereo pair are described by a matrix called the fundamental matrix.

2 Related Research Work

The concept of the essential matrix was introduced by [5]. It is the predecessor of the fundamental matrix and can be used for calibrated cameras only. Reference [6] introduced the fundamental matrix which does not require inner camera parameters unlike its precursor. Gaps exist in performance evaluation and stability analysis of the constituent parameters of the fundamental matrix. These date back two decades, constituing the influential works of [5–7, 9, 11, 12]. More recent works include [14], where the author tried to perform matrix estimation by the transformation of points in a projective space.

References [1, 10] analyzed the linear, iterative, and robust methods, and attempted to rectify tri-stereo images using the fundamental matrix. Reference [2] studied the possibility of motion segmentation using an affine fundamental matrix.

Reference [15] considered the normalization of the homographies before finding the matrix, moving on towards matrix estimation with horizontal and vertical feature lines [16]. Reference [4] considered rank-minimization of the fundamental matrix. The most recent works include [3, 13], proposing a reduction of constraints on the matrix estimation process and modification of the RANSAC method for estimation. RANSAC is a repetitive method for finding the optimum parameters from an observed dataset containing outliers. The presence and hence influence of the outliers is minimized.

It is evident that a multitude of tasks in vision-related fields depend on the accuracy of these mappings. Evaluation of our results is carried out depending on the correctness of the parameters defining the underlying epipolar geometry or projective relations. We attempt to perform stability analysis using a simplified approach while trying to minimize the errors in the degenerate solutions of the fundamental matrix. This represents the first step in subsequent works in photogrammetry and extended fields.

3 Correspondence Determiners

Three feature-extraction methods for establishing correspondence are used in this chapter. These methods are well known, have been used for two decades, and are fittingly researched upon:

1. Speeded up robust features (SURF).
2. Maximally stable extremal regions (MSER) [8].
3. Harris/Harris-Stephens corner detector.

4 Matrix Estimators

For choosing solvers of matrix estimations we have tried to include all the major prevalent robust methods, metioned in Sect. 2, pertaining to stereo vision. The basic algorithm for the estimators remains the same with necessary modifications made to additional parameters in the design of the algorithm. All the work is performed in Matlab®.[1] The following estimators have been used in our work:

1. Least trimmed squares.
2. M-estimator sample consensus.
3. Random sample consensus.
4. Least median of squares.
5. Normalized eight-point algorithm.[2]

5 Data Used

The data used for this work is chosen from an old publicly available stereo data set,[3] offering a variety of stereo images for analysis. Twelve stereo images, in pairs, were chosen belonging to different domains viz. satellite imageries, aerial imageries and image pairs of circuit boards, earthquake scenes, living room and vegetables. The permutations described in Sect. 6 are carried out for each image pair and the collective results given in Tables 1, 2, 3, 4 and 5.

6 Analysis and Results

While working with the correspondence determiners and matrix estimators mentioned in Sects. 3 and 4, an attempt was made to achieve the best obtainable results. Implementing modifications over the point-pair finders, the results did not seem to vary much and hence we show the analysis of the matrix estimators. In all there were a total of 129 permutations for the findings.

Table 1 Matrix estimator: normalized eight-point algorithm

Method	Analysis parameters		
	$\sum_{er}$	μ_{er}	σ_d
Harris	**0.203**[4]	**0.008**	**0.016**
MSER	−5.539	−0.163	0.947
SURF	−0.751	−0.007	0.048

[1]Version: 8.5.0.197613 (R2015a).

[2]Primarily developed for solution of the essential matrix.

[3]http://www.cs.cmu.edu/afs/cs/project/vision/vasc/idb/www/html_permanent/index.html.

Every method produced parameters of the fundamental matrix with variations up to its third decimal place for all nine parameters. The matrix is represented as: $\begin{bmatrix} f_{11} \, f_{12} \, f_{13} \\ \vdots \;\; \ddots \;\; \vdots \\ f_{31} \, f_{32} \, f_{33} \end{bmatrix}$ following the condition:

$$[x_2 \; 1] * \begin{bmatrix} f_{11} \, f_{12} \, f_{13} \\ \vdots \;\; \ddots \;\; \vdots \\ f_{31} \, f_{32} \, f_{33} \end{bmatrix} * \begin{bmatrix} x_1 \\ 1 \end{bmatrix} = 0 \tag{1}$$

where $[x_2 1]$ and $[x_1 1]$ represent the homologous coordinates of the point pairs, detected in the target and query image, respectively, and x_2 and x_1 are the components of the vectors with reference to Sect. 1.1. Deviations from the above properties are noted in the analysis and the results are also examined to find the nature of the errors, to help choose the best methods for the determination of the said matrix. The results are analyzed in terms of the following parameters:

N_t Number of random trials for obtaining inliers.
D_{ty} Distance type, that is, algebraic or Sampson.
D_{th} Distance threshold.
C_f Required confidence value.
I_p Inlier minimum percentage specified.
$\sum_{er}$ Absolute sum of the deviations from null.
μ_{er} Mean of the absolute deviations from null.
σ_d Standard deviation of the errors.

Opting for a particular feature detector and estimator allows a fundamental matrix to be obtained. Out of the eight parameters listed the last three parameters, viz. $\sum_{er}$, μ_{er}, and σ_d serve to provide the necessary information regarding the nature of the error in the fundamental matrices obtained using various methods.[4] The sum of the deviations $\sum_{er}$ and mean of the deviations μ_{er} are noted for all the features. For a random distribution of errors the mean is zero. It should be noted that these are iterative methods and do not serve to produce the same results in each and every iteration. Hence the method producing the minimum for both is chosen.

Furthermore, the standard deviation of the matrix parameters and the errors in each case were identified. The latter was chosen as the fitness criteria of the method since it produces closer values for error analysis. The method yielding the minimum values in the individual criteria of $\sum_{er}$, μ_{er}, and σ_d was chosen. Thereafter, the method producing a minimum in all criteria was chosen. These are emboldened

[4]Boldfaced values with minimum of $\sum_{er}$, μ_{er}, and σ_d are taken at once.

in Tables 1, 2, 3, 4 and 5. Among those in each category of Correspondence Determiners, i.e., Harris-Stephens, MSER and SURF, parameter combinations yielding minimum values were identified. Here the MSER detector provided an average performance in all cases of the matrix estimators.

Here, the minima of $\sum_{er}$, μ_{er}, and σ_d were obtained in the case of the Harris–Stephens detector with both the RANSAC estimator and MSAC estimator performing equally well. MSER feature detector—MSAC estimator performs next best, followed by Harris detector with LTS estimator. Hence, amongst the chosen detectors and estimators, the Harris detector seems to produce the best restults with RANSAC and MSAC estimators. Hence the fundamental matrices obtained from the cornerness metric, based on maximum gradient change, are found to be less error prone than those found from descriptor-dominant detectors.

Table 2 Matrix estimator: least median of squares

Method	Analysis parameters				
	D_{ty}	N_t	$\sum_{er}$	μ_{er}	σ_d
Harris	Sampson	500	**1.468**	**0.054**	**0.297**
		1000	−19.583	−0.725	3.561
		500	34.880	1.292	6.230
	Algebraic	500	34.880	1.292	6.230
		1000	1.614	0.060	0.161
		2000	2.885	0.107	0.218
MSER	Sampson	500	12.802	0.377	2.840
		1000	31.721	0.933	4.748
		500	6.810	0.200	1.214
	Algebraic	500	6.810	0.200	1.214
		1000	**0.365**	**0.011**	**0.897**
		2000	4.877	0.143	0.872
SURF	Sampson	500	49.485	0.471	5.137
		1000	77.900	0.742	7.387
		500	21.999	0.210	2.389
	Algebraic	500	21.999	0.210	2.389
		1000	**5.305**	**0.051**	**0.172**
		2000	6.256	0.060	0.299

Table 3 Matrix estimator: random sample consensus

Method	Analysis parameters						
	D_{ty}	D_{th}	C_f	n_t	$\sum_{er}$	μ_{er}	σ_d
Harris	Sampson	0.01	98	500	−11.314	−0.419	2.044
				1000	0.378	0.014	0.088
				2000	0.721	0.027	0.146
			99	500	14.119	0.523	2.703
				1000	−11.636	−0.431	2.133
				2000	−11.054	−0.409	2.032
	Algebraic	0.01	98	500	0.192	0.007	0.017
				1000	0.902	0.033	0.177
				2000	**0.203**	**0.008**	**0.016**
			99	500	0.902	0.033	0.177
				1000	0.434	0.016	0.041
				2000	0.434	0.016	0.041
MSER	Sampson	0.0001	98	500	−5.179	−0.152	0.563
				1000	−15.375	−0.452	1.644
				2000	79.774	2.346	6.240
			99	500	5.752	0.169	1.053
				1000	40.451	1.190	6.464
				2000	1.985	0.058	0.729
	Algebraic	0.0001	98	500	4.772	0.140	0.809
				1000	6.028	0.177	0.802
				2000	**0.757**	**0.022**	**1.037**
			99	500	4.924	0.145	0.836
				1000	6.154	0.181	0.689
				2000	4.202	0.124	0.826
SURF	Sampson	0.0001	98	500	0.911	0.009	0.096
				1000	**3.821**	**0.036**	**0.395**
				2000	9.372	0.089	0.335
			99	500	6.268	0.060	0.525
				1000	3.612	0.034	0.471
				2000	6.805	0.065	0.548
	Algebraic	0.0001	98	500	7.140	0.068	0.319
				1000	4.750	0.045	0.164
				2000	6.285	0.060	0.334
			99	500	3.629	0.035	0.507
				1000	9.081	0.086	0.254
				2000	6.403	0.061	0.299

Table 4 Matrix estimator: M-estimator sample consensus

Method	Analysis parameters						
	D_{ty}	D_{th}	C_f	n_t	$\sum_{er}$	μ_{er}	σ_d
Harris	Sampson	0.01	98	500	−3.320	−0.123	0.529
				1000	−23.682	−0.877	3.897
				2000	13.498	0.500	2.200
			99	500	−10.365	−0.384	1.853
				1000	3.693	0.137	0.273
				2000	−2.184	−0.081	0.399
	Algebraic	0.01	98	500	0.434	0.016	0.041
				1000	0.902	0.033	0.177
				2000	**0.203**	**0.008**	**0.016**
			99	500	0.222	0.008	0.017
				1000	0.902	0.033	0.177
				2000	0.434	0.016	0.041
MSER	Sampson	0.0001	98	500	5.129	0.151	2.274
				1000	13.633	0.401	2.615
				2000	−3.247	−0.096	1.731
			99	500	12.946	0.381	2.211
				1000	2.602	0.077	0.374
				2000	−30.006	−0.883	3.000
	Algebraic	0.0001	98	500	14.029	0.413	1.996
				1000	**−0.931**	**−0.027**	**1.034**
				2000	4.464	0.131	0.788
			99	500	0.489	0.014	1.021
				1000	2.023	0.059	0.833
				2000	2.106	0.062	0.713
SURF	Sampson	0.0001	98	500	3.849	0.037	0.256
				1000	13.839	0.132	0.649
				2000	33.413	0.318	3.189
			99	500	**3.582**	**0.034**	**0.321**
				1000	8.039	0.077	0.516
				2000	5.555	0.053	0.927
	Algebraic	0.0001	98	500	4.943	0.047	0.330
				1000	8.911	0.085	0.274
				2000	4.354	0.041	0.177
			99	500	9.830	0.094	0.340
				1000	9.505	0.091	0.310
				2000	3.663	0.035	0.322

Table 5 Matrix estimator: least trimmed squares

Method	Analysis parameters					
	D_{ty}	I_p	N_t	$\sum_{er}$	μ_{er}	σ_d
Harris	Sampson	50	500	0.705	0.026	0.239
			1000	**−3.258**	**−0.121**	**0.901**
			2000	−17.564	−0.651	3.061
		60	500	10.961	0.406	1.772
			1000	−30.589	−1.133	5.619
			2000	−10.365	−0.384	1.853
	Algebraic	50	500	3.142	0.116	0.236
			1000	3.021	0.112	0.232
			2000	2.858	0.106	0.231
		60	500	2.529	0.094	0.199
			1000	2.998	0.111	0.235
			2000	2.809	0.104	0.225
MSER	Sampson	50	500	7.655	0.225	1.662
			1000	−16.689	−0.491	2.130
			2000	4.538	0.133	0.938
		60	500	6.600	0.194	1.284
			1000	3.531	0.104	1.081
			500	13.074	0.385	3.002
	Algebraic	50	500	**2.253**	**0.066**	**0.702**
			1000	4.877	0.143	0.872
			2000	3.714	0.109	0.751
		60	500	14.693	0.432	1.663
			1000	−7.561	−0.222	1.163
			500	2.457	0.072	0.770
SURF	Sampson	50	500	−2.167	−0.021	0.406
			1000	−29.379	−0.280	2.759
			2000	5.666	0.054	0.620
		60	500	8.198	0.078	0.905
			1000	7.433	0.071	0.927
			500	−29.405	−0.280	3.290
	Algebraic	50	500	3.041	0.029	0.181
			1000	9.702	0.092	0.303
			2000	10.148	0.097	0.296
		60	500	1.567	0.015	0.123
			1000	**0.717**	**0.007**	**0.100**
			500	2.995	0.029	0.345

7 Summary and Future Scope

Here we performed an analysis of different methods, while introducing modifications to these methods, for obtaining the best combination of detectors and estimators for the computation of the fundamental matrix following ideal properties. From the specific detectors chosen, and major estimators covered, the Harris–Stephens detector with RANSAC and MSAC estimators tend to produce minimum errors in the parameters of the fundamental matrix.

From the point of view of analysis and algorithm development, a more rigorous analysis of the additional parameters discussed in Sect. 6 is required, while working on the design of the algorithms used for the analysis.

References

1. Armangué X, Salvi J (2003) Overall view regarding fundamental matrix estimation. Imag Vision Comput 21(2):205–220
2. Basah SN, Hoseinnezhad R, Bab-Hadiashar A (2014) Analysis of planar-motion segmentation using affine fundamental matrix. IET Comput Vis 8(6):658–669. http://digital-library.theiet.org/content/journals/10.1049/iet-cvi.2013.0224
3. Boudine B, Kramm S, Akkad NE, Bensrhair A, Saaidi A, Satori K (2016) A flexible technique based on fundamental matrix for camera self-calibration with variable intrinsic parameters from two views. J Vis Commun Imag Represent 39:40–50
4. Cheng Y, Lopez JA, Camps O, Sznaier M (2015) A convex optimization approach to robust fundamental matrix estimation. In: 2015 IEEE conference on computer vision and pattern recognition (CVPR), pp 2170–2178. IEEE. http://ieeexplore.ieee.org/document/7298829/
5. Longuet-Higgins HC (1981) A computer algorithm for reconstructing a scene from two projections. Nature 293(5828):133–135. http://www.nature.com/doifinder/10.1038/293133a0
6. Luong QT, Faugeras OD (1996) The fundamental matrix: theory, algorithms, and stability analysis. Int J Comput Vis 17(1):43–75. http://link.springer.com/10.1007/BF00127818
7. Luong QT, Deriche R, Faugeras O, Papadopoulo T (1993) On determining the fundamental matrix: analysis of different methods and experimental results
8. Nistér D, Stewénius, H (2008) Linear time maximally stable extremal regions. European conference on computer vision. Springer, Berlin, Heidelberg
9. Stehman SV, Czaplewski RL (1998) Design and analysis for thematic map accuracy assessment: fundamental principles. Remot Sens Environ 64(3):331–344
10. Sun C (2013) Trinocular stereo image rectification in closed-form only using fundamental matrices. In: 2013 IEEE international conference on image processing, pp 2212–2216. IEEE. http://ieeexplore.ieee.org/document/6738456/
11. Torr P, Murray D (1997) The development and comparison of robust methods for estimating the fundamental matrix. Int J Comput Vis 24(3):271–300. http://link.springer.com/10.1023/A:1007927408552
12. Torr P, Zisserman A, Maybank S (1998) Robust detection of degenerate configurations while estimating the fundamental matrix. Comput Vis Imag Understand 71(3):312–333
13. Wang L, Zhang Z, Liu Z (2016) Efficient image features selection and weighting for fundamental matrix estimation. IET Comput Vis 10(1):67–78. http://digital-library.theiet.org/content/journals/10.1049/iet-cvi.2014.0436
14. Zhang Z, Loop C (2001) Estimating the fundamental matrix by transforming image points in projective space
15. Zhou F, Zhong C, Zheng Q (2015) Method for fundamental matrix estimation combined with feature lines. Neurocomputing 160:300–307
16. Zhou Y, Kneip L, Li H (2015) A revisit of methods for determining the fundamental matrix with planes. In: 2015 International conference on digital image computing: techniques and applications (DICTA), pp 1–7. IEEE. http://ieeexplore.ieee.org/document/7371221/

A Two-Band Convolutional Neural Network for Satellite Image Classification

Anju Unnikrishnan, V. Sowmya and K. P. Soman

Abstract The advent of neural networks has led to the development of image classification algorithms that are applied to different fields. In order to recover the vital spatial factor parameters, for example, land cover and land utilization, image grouping is most important in remote sensing. Recently, benchmark classification accuracy was achieved using convolutional neural networks (CNNs) for land cover classification. The most well-known tool which indicates the presence of green vegetation from multispectral pictures is the Normalized Difference Vegetation Index (NDVI). This chaper utilizes the success of the NDVI for effective classification of a new satellite dataset, SAT-4, where the classes involved are types of vegetation. As NDVI calculations require only two bands of information, it takes advantage of both RED- and NIR-band information to classify different land cover. The number and size of filters affect the number of parameters in convolutional networks. Restricting the aggregate number of trainable parameters reduces the complexity of the function and accordingly decreases overfitting. The ConvNet Architecture with two band information, along with a reduced number of filters, was trained, and high-level features obtained from a tested model managed to classify different land cover classes in the dataset. The proposed architecture, results in the total reduction of trainable parameters, while retaining high accuracy, when compared with existing architecture, which uses four bands.

A. Unnikrishnan (✉) · V. Sowmya · K. P. Soman
Center for Computational Engineering and Networking (CEN),
Amrita School of Engineering Coimbatore, Amrita Vishwa Vidyapeetham, India
e-mail: anjuruk@gmail.com

V. Sowmya
e-mail: v_sowmya@cb.amrita.edu

K. P. Soman
e-mail: kp_soman@amrita.edu

A. Kumar and S. Mozar (eds.), *ICCCE 2018*,
Lecture Notes in Electrical Engineering 500,
https://doi.org/10.1007/978-981-13-0212-1_17

Keywords Image classification · SAT-4 · Normalized difference vegetation index · Convolutional networks · Trainable parameters

1 Introduction

Satellite images contain valuable and rich sources of interesting information. Captured by various imaging satellites, such imagery provides spatial, spectral, temporal, radiometric, and geometric resolution. For the proper use of satellite images, it is essential to analyze and interpret the spectral information of the data. Satellite imagery plays an important role in the analysis of various fields like the environment, agriculture, and forestry [1]. The reshaping rate of Earth is rapidly increasing, so the need for realization of temporal changes in land cover has arisen. In order to maintain standardization and to develop sustainable use of land systems, land cover regions must be classified and monitored [2]. Interpretation of land resources can be done by calculating the Normalized Difference Vegetation Index (NDVI), which indicates the level of greeness present in vegetation, and is associated directly with photosynthetic capacity [2–4]. The formula to calculate the NDVI is given in Eq. 1.

$$NDVI = \frac{NIR - RED}{NIR + RED} \tag{1}$$

where NIR and RED stands for the measured spectral reflectance in near-infrared and red (visible) regions respectively. The spectral reflectance of each band is computed by the ratio of reflected radiation to incoming radiation. In NIR wavelengths, reflected radiation is much higher than in the red wavelengths, which indicates the presence of dense vegetation. Using radiometric models which use various satellite parameters, measured reflectances can be transformed into digital numbers. By analyzing the magnitude of digital numbers present in the bands, interpretation of different types of vegetations is possible. A high value digital number in NIR and a low value in red indicates vegetation with high chlorophylian activity. Additionally, trees are identified with smaller values in NIR and low values in the red band.

In order to precisely classify different types of vegetation, RED and NIR band information is sufficient since the computation of NDVI is done using only these two bands. This motivated us to reduce the computational complexity of the existing deep neural architecture available for the SAT-4 dataset [4].

With the advent of deep neural network technologies, large amounts of labeled datasets can be trained properly [3]. Parallel processing and a non-linear relationship between inputs and outputs can be modeled in neural networks. Parameters in the model can be reduced by considering the shift invariant properties of images

[5]. Studies have shown that it is possible to automatically construct high-level features hierarchically using deep learning based classification methods [6]. Accurate classification of remotely sensed data can be done through certain deep architectures [7]. Compared to other image classification algorithms, convolutional neural networks (CNNs) require little pre-processing [8]. CNNs find applications in image recognition, computer vision, video analysis, natural language processing, drug discovery, checkers, semantic segmentation, and scene classification [9–11]. Programs that combine both sequential and parallel information can be learned in a more effective way by using the concept of recurrent neural networks (RNNs) [12]. The bifold contributions of this work include:

1. Training the architecture using information from two bands with the same hyperparameters as the existing architecture and comparison in terms of accuracy and total trainable parameters.
2. Training the architecture using information from two bands with a reduced number of filters in the convolution layers and comparison in terms of accuracy and total trainable parameters.

The results indicate a reduction in computation with high accuracy rates implying the potential capability of the advanced deep-learning framework.

The organization of the chapter is as follows: Sect. 2 describes the proposed architectures in detail, showing its variation from existing CNN architecture for land cover classification. Sect. 3 presents an overview of the overall experimental results and inferences. Sect. 4 concludes the chapter by considering future work.

2 ConvNet Architecture for Satellite Image Classification

ConvNet architecture is composed of three stages, and each stage consists of three distinct types of layers, such as a filter bank layer, a non-linearity layer, and a pooling layer [4]. The network architecture transforms the input image volume into an output which holds the class scores. Using a differentiable function, input 3D volume from each layer is transformed to output 3D volume. The network learns the parameters from the convolution and fully connected layers.

The architecture consists of 10 layers. The core building block, which does the computational heavy lifting, is the convolution layer, which stands as the first layer in the network. This is followed by a tangent layer in which output size remains unchanged. Next is a max-pooling layer, which narrows down the image size by the amount of pooling factor as well as protecting against overfitting. These three layers constitute the first stage in the network architecture. The next stage follows the same pattern (CONV-TANH-POOL). Reshape layers transform the given output volume

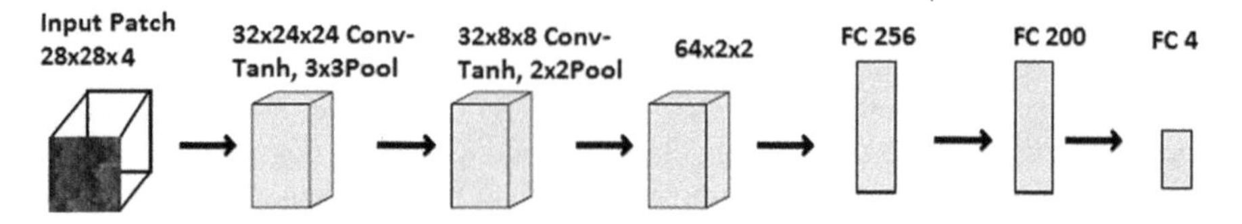

Fig. 1 An illustration of four-band architecture with input $28 \times 28 \times 4$, two convolutional layers, two fully connected layers, and a soft-max layer

into a 1D tensor. The next layer is a fully connected layer which transforms the inputs to some hidden units, followed by a non-linearity layer. The final layer is a linear one which results in an output volume of size which is equal to the number of classes. Here, a soft-max classifier is managed to classify the images into different classes. This existing architecture, using four bands, is shown in Fig. 1.

2.1 Experimental Procedure

The experiment is performed on the multispectral image dataset SAT-4 [3, 4]. The DeepSat dataset known as SAT-4, consists of images in the form of patches. These image patches are extracted from the National Agriculture Imagery Program (NAIP) dataset [4]. The SAT-4 dataset consists of a total of 500,000 images arranged in two different groups of patches. Of these, 400,000 image patches are used for training (four fifths of the entire dataset) with the remaining 100,000 used for testing (one fifth of the entire dataset). The dataset consists of four broad land-cover classes such as barren land, trees, and grasslands, with all other land cover grouped together in the fourth class. Each image consists of four bands—red, green, blue, and NIR. The size of each image patch is 28×28. The implementation is done using open source Torch library. Two architectures are proposed in this study.

1. Two-band ConvNet architecture for satellite image classification.
2. Two-band ConvNet architecture with a reduced number of filters for satellite image classification (modified two-band architecture).

2.1.1 Two-Band ConvNet Architecture for Satellite Image Classification

The architecture of two band differs from existing four band only in its first layer, where the dimension of the former is $28 \times 28 \times 2$. Input images are $28 \times 28 \times 2$, where 2 refers to the RED and NIR bands which are convolved with 32 filters, each of size 5×5, producing an output volume of size $32 \times 24 \times 24$. This is followed by a

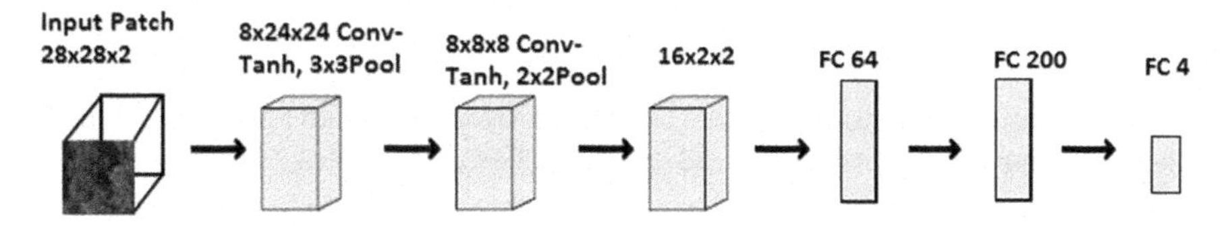

Fig. 2 A brief representation of ConvNet architecture using two-band input (proposed), two convolutional layers employed with reduced numbers of filters, two fully connected layers, and a four-way, soft-max layer

tanh activation function. The image is downsampled by a factor of three in the max-pooling layer, which produces an output volume of size $32 \times 8 \times 8$. Next, a second convolution layer with 64 filters, each of size 5×5, is convolved with $32 \times 8 \times 8$ to produce an output of $64 \times 4 \times 4$. Again the dimension is reduced by a factor of two resulting in an image of size $64 \times 2 \times 2$. The reshape layer converts the given volume of size $64 \times 2 \times 2$ to an output volume of size $(64 \times 2 \times 2) \times 1 = 256 \times 1$. This is followed by two, fully connected layers. In these layers, 256 inputs are mapped into 200 hidden units. Finally, a soft-max classifier is used to classify the image into one of four classes.

2.1.2 Two-Band ConvNet Architecture with a Reduced Number of Filters for Satellite Image Classification (Modified Two-Band Architecture)

This architecture is proposed in order to minimize the layer-wise computational complexity and is achieved by reducing the number of filters in each convolution layer.

The architecture is shown in Fig. 2. Input images are $28 \times 28 \times 2$, where 2 refers to the RED and NIR bands which are convolved with 8 filters of size 5×5 producing an output volume of size $8 \times 24 \times 24$. This is followed by a tanh activation function. The image is downsampled by a factor of three in the max-pooling layer, which produces an output volume of size $8 \times 8 \times 8$. The second convolution layer with 16 filters, each of size 5×5, is convolved with $8 \times 8 \times 8$ to produce an output of $16 \times 4 \times 4$. Again the dimension is reduced by a factor of two resulting in an image of size $16 \times 2 \times 2$. The reshape layer converts the given volume of size $16 \times 2 \times 2$ to an output volume of size $(16 \times 2 \times 2) \times 1 = 64 \times 1$. This is followed by two, fully connected layers. In these layers, 64 inputs are mapped into 200 hidden units. Finally, a four-way, soft-max classifier is used to classify the images.

Several regularization methods exists, which prevent overfitting. The simplest way is by limiting the number of parameters [9]. Reducing the number of filters results in a reduction of trainable parameters. It also results in minimum computational complexity for the weight matrix. While designing the architecture, layers near the input will tend to have fewer filters, with higher layers having more.

The formula for calculating the size of the weight matrix in each layer is given by $m_1 \times n_1 \times p_1 \times p_2$, where n_1 refers to the number of input channels in the first layer, m_1 is the number of output channels in the first layer, and p_1 and p_2 refer to the filter size. Trainable parameters in each layer are calculated using the formula given by $(F \times F \times D + 1)K$, where F refers to the filter size, D refers to the depth, and K refers to the number of filters in that layer.

3 Experimental Results and Discussion

The experimental results performed on the SAT-4 dataset (using only two bands) are presented and a comparative study with existing architecture (using four bands) is summarized in Table 1. The results are estimated on the basis of accuracy, precision, and number of trainable parameters. Accuracy and precision are calculated using the formula [4]:

$$Accuracy = \frac{TP + TN}{TP + FN + FP + TN} \tag{2}$$

$$Precision = \frac{TP}{TP + FP} \tag{3}$$

where TP is the quantity of effectively classified patches; TN is the quantity of the patches that do not have a place in a particular class and are not grouped correctly; FN is the quantity of patches that have a place in a particular class, however, have not been grouped accurately; and FP is the quantity of patches that do not have a place in a particular class and have been wrongly classified.

Comparing the proposed two-band architecture and existing four-band architecture using the same hyperparameters, shows a reduction of weight matrix computation in the first layer of the network and no change for rest of the layers. The computation of the weight matrix is explained below. The number of input channels in first layer for the four-band and two-band architecture is four and two, respectively. The number of output channels in the first layer (i.e., the number of filters) and filter size are 32 and 5×5, which remains the same for both. Hence:

Weight matrix size for four bands is $32 \times 4 \times 5 \times 5$.
Weight matrix size for two bands is $32 \times 2 \times 5 \times 5$.

Table 1 Comparison of classification accuracy and precision rates on the SAT-4 dataset after the application of existing and proposed architecture

Classes	Architecture					
	4 Band		Two band (proposed)		Modified two band (proposed)	
	Accuracy (%)	Precision (%)	Accuracy (%)	Precision (%)	Accuracy (%)	Precision (%)
Barren land	99.83	99.54	99.16	99.45	97.45	98.29
Trees	99.95	99.90	99.64	99.52	98.74	98.25
Grasslands	99.81	99.59	98.52	98.59	96.43	95.50
Other	99.96	99.95	99.86	99.88	99.42	99.50
Overall	99.86	99.75	99.46	99.36	98.01	97.88

Regarding the experiments performed on the two-band architecture, with the reduced number of filters, results show a reduction in size of the weight matrix over each layer in the network, since the number of filters were changed from 32 and 64 to 8 and 16, respectively. The computation of the size of the weight matrix and the trainable parameters are discussed in the following subsections.

3.1 Computation of Weight Matrix Size

1. The first convolution layer with 2 input channels and 8 output channels, with filter size 5×5, results in a weight matrix of size $8 \times 2 \times 5 \times 5$.
2. The second convolution layer with 8 input channels and 16 output channels, with filter size 5×5, results in a weight matrix of size $16 \times 8 \times 5 \times 5$.
3. The fully connected layer with 64 inputs and 200 hidden units, results in a weight matrix of size 200×64.
4. The fully connected layer with 200 inputs and 4 outputs, results in a weight matrix of size 4×200.

This can be compared with the four-band model which results in a weight matrix of size $32 \times 4 \times 5 \times 5$ in the first convolution layer, $64 \times 32 \times 5 \times 5$ in the second convolution layer, 200×256 in the first linear layer, and 4×200 in the second linear layer.

3.2 Trainable Parameters

As a result of reducing the number of filters, layer-wise trainable parameters are calculated (Table 2).

The proposed architectures result in the reduction of the total number of trainable parameters while retaining high accuracy.

Table 2 Comparison of trainable parameters in convolution and fully connected layers of the proposed architecture against the existing architecture

4 Band		Two band (proposed)		Modified two band (proposed)	
Layers	Parameters	Layers	Parameters	Layers	Parameters
Conv(4 ->32)	3232	Conv(2 ->32)	**1632**	Conv(2 ->8)	**408**
Conv(32 ->64)	51,264	Conv(32 ->64)	51,264	Conv(8 ->16)	**3216**
Linear(256 ->200)	51,400	Linear(256 ->200)	51,400	Linear(64 ->200)	**13,000**
Linear(200 ->4)	804	Linear(200 ->4)	804	Linear(200 ->4)	**804**

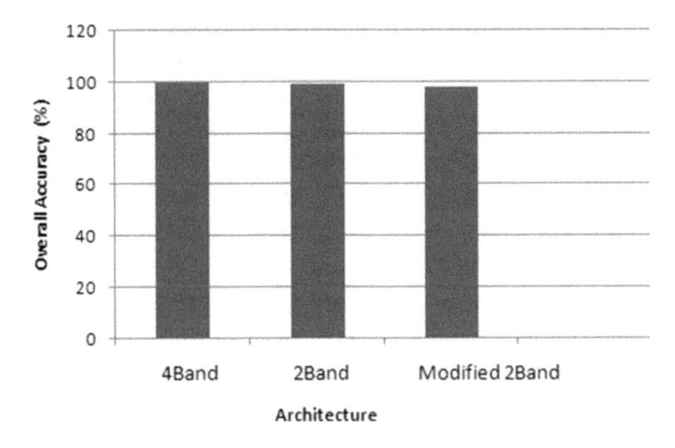

Fig. 3 Graph showing the overall performance accuracy of the proposed and existing architecture

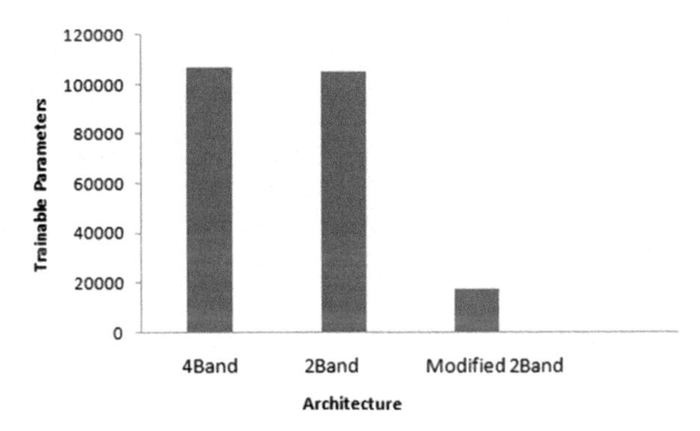

Fig. 4 Graph showing the variation in trainable parameters for proposed and existing architectures

The total number of trainable parameters using four-band architecture is 1,06,700. The total number of trainable parameters using two-band architecture is **1,05,100**. The total number of trainable parameters using the modified two-band architecture is **17,428**.

Graphs showing the response of existing and proposed architectures versus overall accuracy and trainable parameters are plotted in Figs. 3 and 4, respectively.

Reducing the number of filters limits the total trainable parameters in the proposed two-band ConvNet architectures for satellite image classification. Hence, without degrading the performance of the existing architecture, significant reduction in terms of trainable parameters has been achieved by the proposed architecture.

4 Conclusion

This chapter utilizes the ability of the NDVI to act as a stand-alone parameter for accurate classification of landcover, something which led to the results presented in this chapter, in which the existing four-band CNN was modified into a two-band version, with a reduced number of filters used for satellite image classification. The experimental results show that the benchmark accuracy for SAT-4 image classification can be achieved using the proposed architectures with fewer trainable parameters. For future work, further reduction of the number of trainable parameters can be analyzed, with the same being applied to the SAT-6 dataset.

References

1. Dixon KDM, Ajay A, Sowmya V, Soman KP (2016) Aerial and satellite image denoising using least square weighted regularization method. Indian J Sci Technol 9(30)
2. Jeevalakshmi D, Narayana Reddy S, Manikiam B (2016) Land cover classification based on NDVI using LANDSAT8 time series: a case study Tirupati region. In: proceedings of IEEE international conference on communication and signal processing (ICCSP), pp 1332–1335
3. Basu S, Ganguly S, Mukhopadhyay S, DiBiano R, Karki M, Nemani R (2015) Deepsat: a learning framework for satellite imagery. In: proceedings of 23rd SIGSPATIAL international conference on advances in geographic information systems, p 37
4. Papadomanolaki M, Vakalopoulou M, Zagoruyko S, Karantzalos K (2016) Benchmarking deep learning frameworks for the classification of very high resolution satellite multispectral data. ISPRS Ann Photogramm Remote Sens Spat Inf Sci 3(7):83–88
5. Marmanis D, Wegner JD, Galliani S, Schindler K, Datcu M, Stilla U (2016) Semantic segmentation of aerial images with an ensemble of CNSS. ISPRS Ann Photogramm Remote Sens Spat Inf Sci 3:473–480
6. Makantasis K, Karantzalos K, Doulamis A, Doulamis N (2015) Deep supervised learning for hyperspectral data classification through convolutional neural networks. In: IEEE international geoscience and remote sensing symposium (IGARSS), pp 4959–4962
7. Lu D, Weng Q (2007) A survey of image classification methods and techniques for improving classification performance. Int J Remote Sens 28(5):823–870
8. Krizhevsky A, Sutskever I, Hinton GE (2012) Imagenet classification with deep convolutional neural networks. In: Advances in neural information processing systems, pp 1097–1105
9. Kaiser P, Wegner JD, Lucchi A, Jaggi M, Hofmann T, Schindler K (2017) Learning aerial image segmentation from online maps. IEEE Trans Geosci Remote Sens 55(11):6054–6068
10. Sachin R, Sowmya V, Govind D, Soman KP (2017) Dependency of various color and intensity planes on CNN based image classification. In: International symposium on signal processing and intelligent recognition systems, pp 167–177
11. Vakalopoulou M, Karantzalos K, Komodakis N, Paragios N (2015) Building detection in very high resolution multispectral data with deep learning features. In: IEEE international geoscience and remote sensing symposium (IGARSS), pp 1873–1876
12. Schmidhuber J (2015) Deep learning in neural networks: an overview. Neural Netw 61:85–117

Dimensionally Reduced Features for Hyperspectral Image Classification Using Deep Learning

K. S. Charmisha, V. Sowmya and K. P. Soman

Abstract Hyperspectral images (HSIs) cover a wide range of spectral bands in the electromagnetic spectrum with a very finite interval, and with high spectral resolution of data. The main challenges encountered with HSIs are those associated with their large dimensions. To overcome these challenges we need a healthy classification technique, and we need to be able to extract required features. This chapter analyzes the effect of dimensionality reduction on vectorized convolution neural networks (VCNNs) for HSI classification. A VCNN is a recently introduced deep-learning architecture for HSI classification. To analyze the effect of dimensionality reduction (DR) on VCNN, the network is trained with dimensionally reduced hyperspectral data. The network is tuned in accordance with the learning rate and number of iterations. The effect of a VCNN is analyzed by computing overall accuracy, classification accuracy, and the total number of trainable parameters required before and after DR. The reduction technique used is dynamic mode decomposition (DMD), which is capable of selecting most informative bands using the concept of eigenvalues. Through this DR technique for HSI classification using a VCNN, comparable classification accuracy is obtained using the reduced feature dimension and a lesser number of VCNN trainable parameters.

Keywords Hyperspectral images · Dimensionality reduction · Convolution neural network · Dynamic mode decomposition · Trainable parameters Learning rate

K. S. Charmisha (✉) · V. Sowmya · K. P. Soman
Centre for Computational Engineering and Networking, Amrita School of Engineering, Amrita Vishwa Vidyapeetam,
Coimbatore, India
e-mail: charmisha99@gmail.com

V. Sowmya
e-mail: v_sowmya@cb.amrita.edu

K. P. Soman
e-mail: kp_soman@amrita.edu

A. Kumar and S. Mozar (eds.), *ICCCE 2018*,
Lecture Notes in Electrical Engineering 500,
https://doi.org/10.1007/978-981-13-0212-1_18

1 Introduction

Hyperspectral image (HSI) processing, also known as imaging spectroscopy, is new technology that is currently being investigated by researchers and scientists in order to detect and identify terrestrial vegetation, land use and land cover, minerals, other background materials, and man-made materials. HSI are in the form of cube where the X-Y plane contains spatial information and the Z plane contains spectral information. Each pixel in the image possesses a continuous spectrum, which helps in characterizing objects with high precision and detail for a given scene. In the field of remote sensing, HSI classification is one of the major areas of research [6]. The commonly used classifiers for HSI are support vector machines (SVM), minimum spanning forest (MSF), probability-based multinominal logistic regression (MLR), etc., [6].

In recent years, deep learning based methods have achieved benchmark results in many fields [2]. Convolution neural networks (CNN) give better classification rates on vision-related tasks [2]. The advantages of using CNN compared to conventional techniques are:

1. Real-time operations: special hardware devices are being manufactured and designed. Computations can be carried out in parallel.
2. Adaptive learning: based on the data given for training, an ability to learn is developed for performing tasks.
3. Self-organisation: during its learning time, it can create its own representation and organization of data.

CNNs are biologically inspired deep-learning models, which use a single neural network that is trained, end to end, from raw image pixel values to classifier outputs [2]. Hence, CNNs can be used for HSI classification.

The main challenges which are encountered with HSIs are those associated with their large dimensions. The bands of HSIs are highly correlated. Hence, we need a strong dimensionality reduction (DR) technique to remove highly correlated bands without feature loss. The advantage of DR is that it increases the ease of handling high dimensional data and reduces classification time.

The most commonly used reduction techniques for HSI are principal component analysis (PCA) [6], which is a standard reduction technique used to lower the dimension and singular-value decomposition (SVD) [8], which is also a reduction technique to reduce unwanted feature information. A new hyperspectral DR technique called dynamic mode decomposition (DMD) has recently been incorporated for HSI classification [7]. This method is used in static HSIs to find spectral variations.

In this chapter, the 3D hyperspectral data is converted to 2D, and the data is fed to the network as pixel vectors for classification, and is thus called a vectorized convolution neural network (VCNN). The main objective of this chapter is to analyze the effect of DR on a VCNN without any pre-processing. The effect of DR on a VCNN is analyzed by computing the total number of trainable parameters required to train the network before and after performing DR. So, we choose a combination of VCNN and DMD techniques inorder to obtain precise classification accuracy.

The organization of the chapter is as follows: Sect. 2 provides a background study of a VCNN and DMD; Sect. 3 gives the methodology or the overview of the overall flow of the algorithm; Sect. 4 analyzes the experimental results; and Sect. 5 provides the conclusion.

2 Background Theory of Vectorized Convolution Neural Networks and Dynamic Mode Decomposition

2.1 Vectorized Convolution Neural Network

The 3D hyperspectral data $m \times n \times b$, where m denotes scan lines, n denotes samples, and b denotes bands, containing hundreds of spectral bands, can be illustrated as 2D curves [5]. A CNN is a feed-forward neural network that is similar to an ordinary neural network. There may be any number of layers depending on the typical network but, as the number of filters and layers in a network increases, the computation time and complexity may also increase. So, proper care has to be taken while selecting a network. Here, the input to the network is given as the pixel vectors, and hence is called a VCNN [5].

The architecture is shown in [5]. This network consists of five layers in total [5]. The first layer is the input layer (L1), the second is the convolution layer (C1), and third is the max-pooling layer (P1). We have two fully connected layers (F1) and (L2), where L2 is the output layer. Each pixel sample can be represented as a 2D image whose height is equal to 1 [5]. Weights and biases are the trainable parameters. These are intialized to zero and then are automatically updated. The size of input layer L1 is $1 \times n_1$ where n_1 is the number of bands; the output of this is input to the next layer. The size of the convolution layer C1 is $20 \times (1 \times n_2)$ where n_2 is calculated by the formulae $n_2 = (n_1 - k_1/s) + 1$, where s represents the stride which is equal to 1 and k_1 is given as $k_1 = n_1/9$. Here, 20 filters are used and each kernal has a size of $1 \times k_1$. The number of trainable parameters between L1 and C1 is given by the computation $20 \times (k_1 + 1)$. the next layer is the max-pooling layer, which has $20 \times (1 \times n_3)$ number of nodes where n_3 is computed by formulae $n_3 = n_2/k_2$. Here, k_2 denotes the kernal size $(1 \times k_2)$. The fully connected layer F1 has $1 \times n_4$ number of nodes. The number of trainable parameters in this layer is calculated as $(20 \times n_3 + 1) \times n_4$. The last layer L2 is the output layer having n_5 number of nodes, with trainable parameters being $(n_4 + 1) \times n_5$, where n_5 indicates the number of classes. The total number of trainable parameters for a VCNN is calculated as:

$$20 \times (k_1 + 1) + (20 \times n_3 + 1) \times n_4 + (n_4 + 1) \times n_5 \tag{1}$$

The activation function used in this network is the RELU activation function and the softmax function is used to predict the class labels.

2.2 *Dynamic Mode Decomposition*

This concept is used in fluid dynamics. It attempts to extract the dynamic information from flow fields [9]. DMD is used to find the dynamcity of a non-linear system. Spectral decomposition of the map results in an eigenvalue and eigenvector representation (referred to as dynamic modes) [9]. Here, each band data is vectorized and appended as columns, creating a 2D matrix.

The process of DMD is illustrated in a paper [7] taken from the concept raised by [9]. The final equation shown below explains the DMD process. The computation of $\tilde{S}$ is given in [5]:

$$\tilde{S} = (V\Sigma^{-1})^{-1}S(V\Sigma^{-1}) = U^H X_2{}^m V\Sigma^{-1} \tag{2}$$

The final output matrix obtained from Eq. 2 is arranged in a descending fashion of information (from most informative bands to least informative bands). Hence, we can reduce the redundant information using this process.

3 Methodology

The objective of this work is to analyze the effect of DR on a VCNN. Before performing a DR technique, 3D HSI data is converted into a 2D matrix by vectorizing each band and appending them as columns of a matrix. Without any preprocessing, the raw data is split into training and testing sets: 80% for training and 20% for testing. The input data is normalized in the range $[-1, +1]$, so that entire data lies in the same scale, which makes any analysis or comparisons easier. In order to analyze the effect of DR techniques, we should first classify hyperspectral data using a VCNN without DR. Before classification, the network is tuned by varying hyper-parameters like learning rate and the number of iterations (Fig. 1).

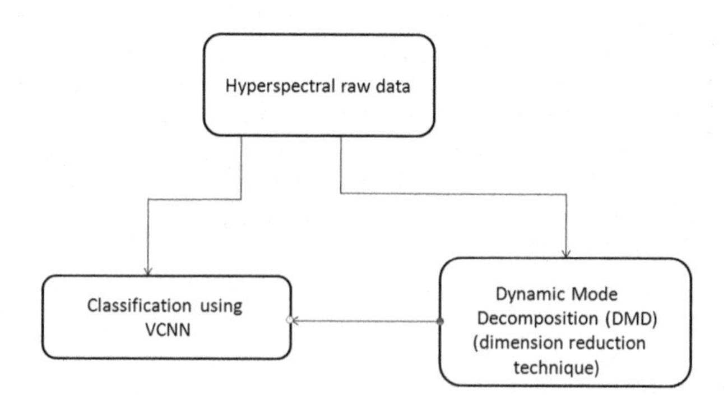

Fig. 1 Block diagram of the proposed HSI classification using a VCNN

After performing DR through DMD, 50% of the information is reduced. This effect is validated through VCNN classification. This process is extended to more than a 50% reduction of information, provided there is no information loss for feature extraction. The effect of DR on a VCNN is analyzed by computing the total number of trainable parameters of the network before and after DR. Evaluation of the classification results are undertaken through calculation of classification accuracies and visual interpretation.

4 Experimental Results

4.1 *Dataset Description*

The dataset used here is Indian Pines. Details of the dataset are as follows: The spatial configuration is 145×145 pixels with a spectral configuration of 220 bands. The wavelength is in the range 0.4–2.2 μm with spectral resolution of 10 nm and a spatial resolution of 20 m. Groundtruth has 16 classes in total but only 8 classes are considered for classification, chosen based on maximum number of pixels available, since the neural network requires a sufficient amount of data for training. The sensor used is the Airborne Visible/Infrared Imaging Spectrometer (AVIRIS). Table 1 shows the classes taken for classification based on samples. Here, 80% of the total pixels are chosen for training and 20% are chosen for testing.

4.2 *Results and Analysis*

The network is tuned by setting two core parameters—learning rate and number of iterations. The parameters are varied one at a time: when the learning rate is varied,

Table 1 The number of training and testing samples for Indian Pines

Classes choosen	Total pixels	Training pixels (80%)	Testing Pixels (20%)
Corn-nortill	1428	1142	286
Corn-mintill	830	664	166
Grass-pasture	483	386	97
Hey-windrowed	478	382	96
Soyabean-notill	972	777	195
Soyabean-mintill	2455	1964	491
Soyabean-clean	593	474	119
Woods	1265	1012	253
		6801	1703

the number of iterations are fixed, and vice versa. We vary the parameters until a high classification accuracy is obtained. First, the learning rate is fixed at a small number, as a lower learning rate always avoids overfitting. Then, we vary the number of iterations until a high classification accuracy is obtained. This process is repeated for both raw data, with 220 bands, and dimensionally reduced data, with 110 bands, i.e., one half of the total number of bands, and with 73 bands, i.e., two thirds of the total number of bands. From Fig. 2, we can see that, at 12,000 iterations we are able to get a high classification accuracy (CA) of 84.33% for raw data as compared to other iteration values. Hence, the iteration is fixed to **12,000**. For a reduced dimension with 110 bands we get around 83.62% CA for **10,000** iterations. Hence, here the iteration is fixed to 10,000. Similarly with 73 bands the **15,000** iteration produces a high CA of 83.67%. So, the iteration is fixed to 15,000.

The learning parameter value is varied until we get a high CA. Figure 3 shows the CA for the corresponding learning rate. From Fig. 3 we can see that among all learning rate values, a high CA of 84.33% is obtained for raw data (220 bands) with **0.07**. So, learning rate is set to 0.07. For a reduced dimension with 110 bands, we get around 83.62% CA for **0.08** iterations. Hence, here the learning rate is fixed to 0.08. Similarly, with 73 bands the learning rate,with a high CA of 83.67%, is produced for **0.08**. So, the learning rate is fixed to 0.08.

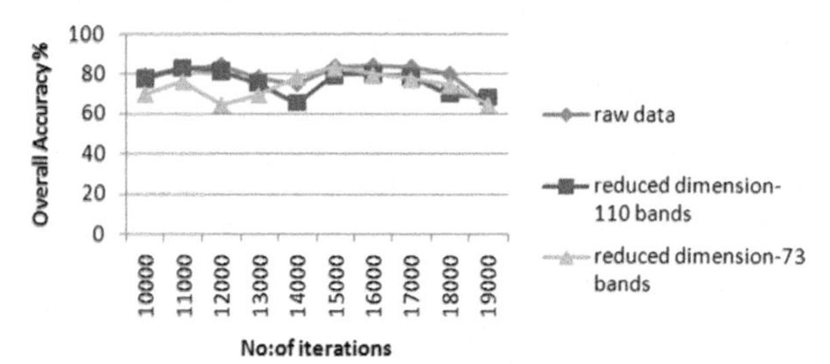

Fig. 2 Plot of overall accuracy versus number of iterations considered for a VCNN with and without DR

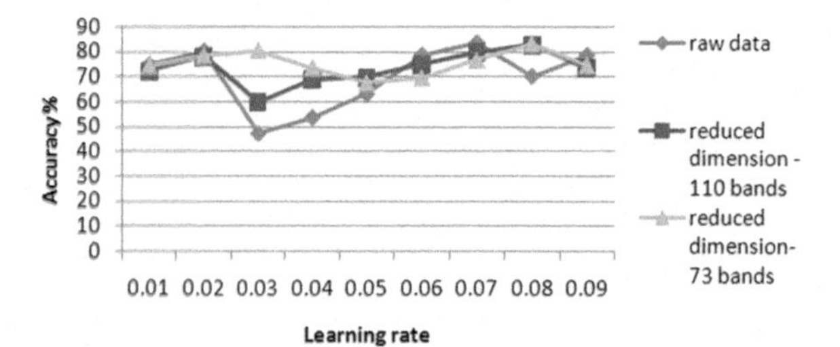

Fig. 3 Plot of overall accuracy versus learning rate considered for a VCNN with and without DR

1. Before DR, the layer parameters of the VCNN are as follows. $n_1 = 220$ and $n_5 = 8$ are the input and output channel sizes. $k_1 = 24$, $n_2 = 197$, $k_2 = 5$, and $n_3 = 39$, are the nodes in the fully connected layer $n_4 = 100$ [3]. The input dimension is 1×220 and the number of nodes in the convolution layer is $20 \times (1 \times 197)$. The trainable parameters between the input and convolution layer number 500. The max-pooling layer has $20 \times (1 \times 39)$ nodes, the dimension of the fully connected layer is 1×100. The trainable parameters of this layer are given as 78,100. The output layer contains 1×8 nodes where 8 is number of classes. The total number of trainable parameters is **79,408**, computed using Eq. 1.
2. After DR through DMD the number of bands are reduced from 220 to 110, i.e., 50% of the bands are reduced. The layer parameters of the VCNN are as follows. $n_1 = 110$ and $n_5 = 8$ are the input and output channel sizes. $k_1 = 12$, $n_2 = 99$, $k_2 = 3$, and $n_3 = 33$ are the nodes in he fully connected layer $n_4 = 100$. The input dimension is 1×110 and the number of nodes in the convolution layer is $20 \times (1 \times 99)$. The trainable parameters between the input and convolution layer are 260. The max-pooling layer has $20 \times (1 \times 33)$ nodes, the dimension of the fully connected layer is 1×100. The trainable parameters of this layer are given as 66,100. The output layer has 808 trainable parameters. The total number of trainable parameters is **67,168**.
3. On furthur reduction of bands from 110 to 73, i.e., with more than 50% of the bands being reduced from the original bands (220), the layer parameters of the VCNN are as follows. $n_1 = 73$ and $n_5 = 8$ are the input and output channel sizes. $k_1 = 8$, $n_2 = 66$, $k_2 = 3$, and $n_3 = 22$ are the nodes in the fully connected layer $n_4 = 100$. The input dimension is 1×73 and the number of nodes in the convolution layer is $20 \times (1 \times 66)$. The trainable parameters between the input and conolution layer number 180. The max-pooling layer has $20 \times (1 \times 22)$ nodes, the dimension of the fully connected layer is 1×100. The trainable parameters of this layer are given as 44,100. The output layer contains 1×8 nodes where 8 represents the number of classes. This layer has 808 trainable parameters. The total number of trainable parameters is **45,088**.

4.3 Classification Accuracy

Classification accuracy is recorded and measured before and after DR. Classwise and overall accuracy can be interpreted as shown:

$$\begin{aligned} &Classwise accuracy\,(CA) \\ &= \frac{Number of\ pixels\ which\ are\ correctly\ classified\ in\ each\ class}{Total\ number\ of\ pixels\ in\ each\ class} \times 100 \\ &Overall accuracy\,(OA) \\ &= \frac{Total\ number of\ pixels\ which\ are\ correctly\ classified}{Total\ number of\ pixels} \times 100 \end{aligned}$$

The classification accuracies are **84.63%**, **83.62%**, and **83.67%** for raw data, reduced dimension data with 220 bands, and reduced dimension data with 110 bands, respectively. The performance of the VCNN classifier can be evaluated by classwise accuracies, with Fig. 4 showing classwise accuracies for raw data and DR hyperspectral data.

The number of trainable parameters, and their corresponding accuracies, are graphically shown in Fig. 5. It can be seen that the total number of trainable parameters is reduced before and after DR. Though the trainable parameters are reduced, the classification accuracy obtained before and after DR is comparable. By selecting the most informative bands and discarding the redundant bands, the VCNN achieves almost the same classification accuracy with a reduced number of trainable parameters.

When this dimension reduction is further reduced to three quarters (55) of the total bands, we observe that the same classification accuracy is retained until two thirds of the total bands are reduced (73), whereas for a total band reduction of three quarters (55), we are able to reduce the trainable parameters significantly to 33,048, with a CA of around 80.32%, which shows that further reduction in the feature dimension leads to misclassification.

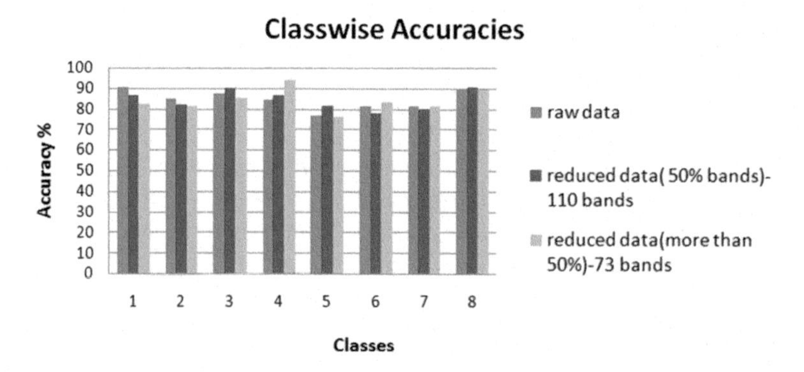

Fig. 4 Classwise accuracy before and after DR

Fig. 5 Comparison of trainable parameters and accuracies with and without DR

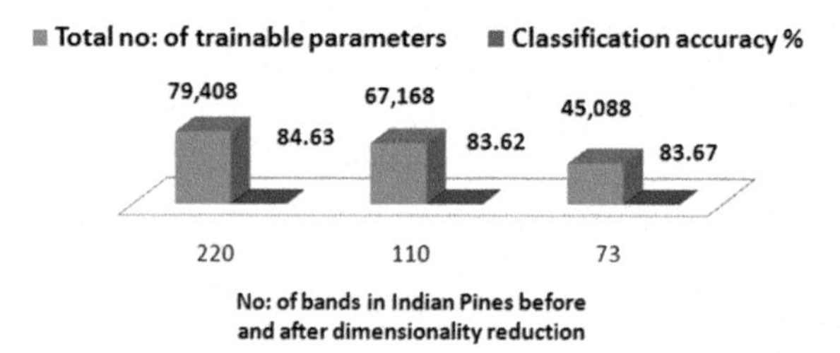

5 Conclusion

The effect of DR on a VCNN is analyzed in this chapter. Based on the accuracy of measurement parameters, the performance of this classifier is evaluated. From experimental results, it is evident that, though there is dimension reduction, a VCNN is able to achieve almost the same classification accuracy as that of HSI raw data. The total number of trainable parameters were also reduced, which led to easier handling of data for the VCNN. Also, the experimental results show that two thirds of the total number of available bands represents the maximum possible reduction in feature dimension, resulting in comparable classification accuracy.

References

1. Abadi M, Agarwal A, Barham P, Brevdo E, Chen Z, Citro C, Corrado GS et al (2016) Tensorflow: large-scale machine learning on heterogeneous distributed systems. arXiv: 603.04467
2. Aswathy C, Sowmya V, Soman KP (2015) ADMM based hyperspectral image classification improved by denoising using Legendre Fenchel transformation. Ind J Sci Technol 8(24):1
3. Chen, Y, Jiang H, Li C, Jia X, Ghamisi P (2016) Deep feature extraction and classification of hyperspectral images based on convolutional neural networks. IEEE Trans Geosci Remote Sens 54(10):6232–6251
4. Deepa Merlin Dixon K, Sowmya V, Soman KP (2017) Effect of denoising on vectorized convolutional neural network for hyperspectral image classification. In: International conference on Nextgen electronic technologies: silicon to software (ICNETS2). LNEE Springer proceedings
5. Hu W, Huang Y, Wei L, Zhang L, Li H (2015) Deep convolutional neural networks for hyperspectral image classification. J Sens 2015
6. Koonsanit K, Jaruskulchai C, Eiumnoh A (2012) Band selection for dimension reduction in hyper spectral image using integrated information gain and principal components analysis technique. Int J Mach Learn Comput 2(3):248
7. Megha P, Sowmya V, Soman KP (2017) Effect of dynamic mode decomposition based dimension reduction technique on hyperspectral image classification. In: International conference on Nextgen electronic technologies: silicon to software (ICNETS2). LNEE Springer proceedings
8. Sadek RA (2012) SVD based image processing applications: state of the art, contributions and research challenges. arXiv: 1211.7102
9. Schmid Peter J (2010) Dynamic mode decomposition of numerical and experimental data. J Fluid Mech 656:5–28

Asymptotic Symbol Error Rate Analysis of Weibull/Shadowed Composite Fading Channel

Puspraj Singh Chauhan and Sanjay Kumar Soni

Abstract In this work, we derive the asymptotic expressions of the average symbol error probability (SEP) of a wireless system over the Weibull-lognormal fading channel. First, we evaluate an approximation of the multipath distribution at the origin then the composite distribution is obtained by averaging the approximate multipath probability density function (PDF) with respect to shadowing. The result is further extended to include maximal ratio combining (MRC), equal gain combining (EGC), and selection combining (SC) PDF at the origin. The derived expressions of the composite PDF are further utilized to evaluate the average SEP for both coherent and non-coherent modulation schemes. The derived expressions have been corroborated with Monte-Carlo simulations.

Keywords Probability distribution function · Diversity · Symbol error probability

1 Introduction

A composite model is a class of mathematical model which includes both multipath and shadowing phenomena simultaneously and hence is a more realistic model. Among the available class of composite fading models, Weibull-lognormal (WLN) draws its significance from the fact that the Weibull distribution is known to characterize the multipath effects of an indoor and outdoor channel, based on its excellent matching with the measurements conducted in related environments [1–4]. The shadowing effect of the channel is best captured by the lognormal (LN) distribution [5]. Moreover, the LN distribution is shown to characterize a number of wireless applications such as an outdoor scenario, fading phenomenon in an indoor environment, radio channels affected by body worn devices, ultra wideband indoor channels, and

P. S. Chauhan · S. K. Soni (✉)
G B Pant Institute of Engineering and Technology, Pauri, UK, India
e-mail: puspraj.chauhan@gmail.com

S. K. Soni
e-mail: sanjoo.ksoni@gmail.com

A. Kumar and S. Mozar (eds.), *ICCCE 2018*,
Lecture Notes in Electrical Engineering 500,
https://doi.org/10.1007/978-981-13-0212-1_19

weak-to-moderate turbulence channels found in free-space optical communications channels [5–7].

In the performance analysis of a wireless system, the closed-form solution facilitates better interpretation of system behavior. Yet, sometimes the complexities of the expression defies the basic purpose of the system optimization [8]. This motivates us to go for an asymptotic analysis of system performance. In the literature, various work related to asymptotic behavior of a system has been carried out [9–11]. For example, in [9], asymptotic bit error rate (BER) analysis has been presented for maximal-ratio combining with transmit antenna selection in flat Nakagami-m fading channels. In [10], simplified expressions of the BER for the $\eta - \mu/Gamma$ composite fading channel in a high-power regime are derived. The asymptotic BER expressions for the $\alpha - \eta - \mu$ fading channel have been derived for both coherent and non-coherent modulation schemes [11]. To date, the asymptotic analysis over W-LN fading channel with diversity reception has not been reported in the open literature. Recently, authors of a current paper have reported asymptotic closed-form expressions of the average symbol error probability (SEP) with maximal ratio combining (MRC) diversity [12]. The common approach adopted to derive the asymptotic solutions of the average SEP over the composite fading channel is to first derive the composite distribution by averaging the multipath with respect to shadowing, approximate the distribution at the origin as suggested in [8], then deduct the average SEP. Generally, composite distribution following the previous concept may lead to a result having a summation term, and thus the solution may not be tractable as far as the derivation of the probability density function (PDF) of the MRC, equal gain combining (EGC), and selection combining (SC) output is concerned, and usually does not lead to the closed-form solution.

In this chapter, we obtain the asymptotic expressions for the average SEP with all three diversity schemes such as MRC, EGC, and SC. While deriving the asymptotic solutions we have followed the following approach. First, we evaluate an approximation of the multipath distribution at the origin then the composite distribution is obtained by averaging the approximate multipath PDF with respect to shadowing. The result is further extended to include MRC, EGC, and SC PDFs at the origin. These expression have been used to evaluate the closed-form solutions of the average SEP. Furthermore, we have compared the performance of MRC, EGC, and SC in the context of error probability over the composite fading channel.

2 System Model

The Weibull envelope "X" has the PDF given as follows [13]:

$$f_X(x) = \frac{cA}{\Omega^{c/2}} x^{c-1} \exp\left[-A\left(\frac{x^2}{\Omega}\right)^{c/2} \right] \tag{1}$$

where Ω is the average fading power $\Omega = \mathrm{E}[X^{\frac{c}{2}}]$, $\mathrm{A} = [\Gamma(1 + \frac{2}{c})]^{c/2}$ and $\Gamma(.)$ is the Gamma function. Here c is the multipath parameter and the channel condition improves as $c \to \infty$. As a special case, when $c = 1, 2$ the Weibull distribution reduces to the exponential and Rayleigh distributions, respectively. An LN random variable (RV) "Z" has the PDF [5]:

$$f_Z(z) = \frac{1}{z\sigma\sqrt{2\pi}} exp\left[-\left(\frac{\ln z - \mu}{\sqrt{2}\sigma} \right)^2 \right] : z > 0 \tag{2}$$

where σ and μ are the mean and standard deviation of ln(Z). The expected value of Z is $E[Z] = \Gamma = Z_{avg} = exp\ (\mu + \sigma^2/2)$. As such, and by using Taylor's series, the $f_X(x)$ given in (1) can be rewritten as:

$$f_X(x) = \frac{cA}{\Omega^{\frac{c}{2}}} x^{c-1} + \mathcal{O} \tag{3}$$

where $\mathcal{O}$ stands for higher order terms. First, substitute (3) and (2) into the definition of the composite distribution [14, Eq. (3)], then setting $t = (ln(z) - \mu)/\sqrt{2}\sigma$, employing the identity [15, Eq. (3.323.2^{10})], and finally following the conversion $\gamma = x^2\rho$, $\bar{\gamma} = \Omega\rho$ and $f_Y(\gamma) = f_X(\sqrt{\gamma/\rho})/2\sqrt{\gamma\rho}$, where $\rho = \frac{E_s}{N_0}$, E_s is the energy per symbol and N_0 is the one-sided power spectral density of the additive white Gaussian noise (AWGN) [13], the signal-to-noise ratio (SNR) distribution of the composite distribution around origin can be given as:

$$f_Y(\gamma) \approx \frac{cAe^{-\frac{\mu c}{2}} e^{\frac{\sigma^2 c^2}{8}}}{2\rho^{\frac{c}{2}}} \gamma^{\frac{c}{2}-1} \tag{4}$$

The simplified PDF of (4) does not contain any summation term, thus enabling us to derive the PDF of the diversity combiner output in a convenient way, which is presented next.

2.1 Maximal Ratio Combining Probability Density Function at the Origin

For MRC with L independent and identically distributed (i.i.d.) diversity branches, the instantaneous SNR of the combiner output is given by:

$$\gamma_{mrc} = \sum_{j=1}^{L} \gamma_j \tag{5}$$

where γ_j is the instantaneous SNR of the jth branch. Since the L WLN RVs are i.i.d., the moment-generating function (MGF) of γ_{mrc} is expressed as $M_{\gamma_{mrc}}(s) = \prod_{j=1}^{L} M_{\gamma_j}(s)$, where $M_{\gamma_j}(s)$ is the jth branch MGF and is deduced by taking the Laplace transform of (4) with the aid of [15, Eq. (3.381.4)]. Thus, assuming the average SNR of each branch to be same, i.e., $\rho_1 = \rho_2 = \ldots = \rho_L = \rho$, the MGF of γ_{mrc} can readily be shown as:

$$M_{\gamma_{mrc}}(s) \approx \left(\frac{cAe^{-\frac{\mu c}{2}} e^{\frac{\sigma^2 c^2}{8}} \Gamma(\frac{c}{2})}{2(s\rho)^{\frac{c}{2}}} \right)^L \tag{6}$$

The PDF of the RV Y_{mrc} is deduced by performing the inverse Laplace transform of (6) with the aid of [15, Eq. (3.381.4)], yields [12]:

$$f_{Y_{mrc}} \approx \frac{(\vartheta)^L (\Gamma(\frac{c}{2}))^L}{\rho^{\frac{Lc}{2}} \Gamma(\frac{Lc}{2})} \gamma^{\frac{Lc}{2}-1} \tag{7}$$

where $\vartheta = \dfrac{cAe^{-\frac{\mu c}{2}} e^{\frac{\sigma^2 c^2}{8}}}{2}$.

2.2 *Equal Gain Combining Probability Density Function at the Origin*

For L i.i.d. diversity branches, the instantaneous SNR of the EGC output is given as:

$$\gamma_{egc} = \left(\frac{1}{\sqrt{L}} \sum_{j=1}^{L} \sqrt{\gamma_j} \right)^2 \tag{8}$$

The above equation can be further be expressed by taking the square-root of both sides as:

$$x_{egc} = \sum_{j=1}^{L} \frac{x_j}{\sqrt{L}} \tag{9}$$

In a similar context to MRC, the MGF for EGC is expressed as $M_{x_{egc}} = \prod_{j=1}^{L} M_{x_j}$ $(s/\sqrt{L})$. Now, following a similar approach to MRC, and with the aid of [13], the SNR distribution around the origin is deduced as:

$$f_{Y_{egc}}(\gamma) \approx \frac{\alpha^L (\Gamma(c))^L (\sqrt{L})^{Lc}}{2\Gamma(Lc)\rho^{\frac{Lc}{2}}} \gamma^{\frac{Lc}{2}-1} \tag{10}$$

where $\alpha = Ace^{-\frac{\mu c}{2}} e^{\frac{\sigma^2 c^2}{8}}$.

2.3 Selection Combining Probability Density Function at the Origin

The simplest approach for combining the signals from the channel branches is the SC method. From the practical point of view, this algorithm has the easiest implementation. In this, the output or branch is picked which has the highest SNR which can be defined mathematically as $Y_{sc} = max(Y_j), j = 1, 2...L$. The PDF of the output SNR is defined as [16]:

$$f_{Y_{sc}} = L(F_Y(\gamma))^{L-1} f_Y(\gamma) \tag{11}$$

where $F_Y(\gamma)$ is the cumulative distribution function (CDF). The CDF can be obtained by substituting (4) in the definition $F_Y(\gamma_t h) = F(Y < \gamma_{th})$ [16] and after some straight forward mathematical simplification:

$$F_Y(\gamma_{th}) = \frac{\vartheta \gamma_{th}^{\frac{c}{2}}}{\frac{c}{2}\rho^{\frac{c}{2}}} \tag{12}$$

Further, substituting (4) and (12) into (11) results in the closed-form expression of the SC distribution:

$$f_{Y_{sc}}(\gamma) \approx \frac{L\vartheta^L}{(\frac{c}{2})^{L-1}\rho^{\frac{Lc}{2}}} \gamma^{\frac{Lc}{2}-1} \tag{13}$$

3 Average Symbol Error Probability Analysis

In this section, we analyse the performance of the composite fading channel over average SEP for both coherent and non-coherent modulation schemes. The general expression of the average SEP over a fading channel is obtained by taking an ensemble average of the instantaneous error probability over the fading distribution. The general expression of the average SEP over a fading channel is given by [16]:

$$\bar{P}_e = \int_0^\infty P_e(\gamma) f_Y(\gamma) d\gamma \tag{14}$$

where $P_e(\gamma)$ is the instantaneous symbol error rate (SER) of the modulation technique.

3.1 Coherent Average Symbol Error Probability

The generalized probability of error for coherent modulation schemes is given by [11, Eq. (17)]:

$$P_e(\gamma) = A_p erfc(\sqrt{B_p\gamma}) \tag{15}$$

where constants A_p and B_p, for different modulation techniques, are given in [11, Table I] for various constellation size. $erfc(.)$ is the complementary error function and is defined as $erfc(x) = \frac{2}{\sqrt{\pi}}\int_x^{\infty} exp(-t^2)dt$.

3.1.1 Average Symbol Error Probability for Maximal Ratio Combining

By substituting (7) and (15) in (14), letting $t = \sqrt{B_p\gamma}$, and using [17, Eq. (2.8.2.1)], the asymptotic average SEP can be obtained as:

$$\bar{P}_{e,coh}^{mrc,asy} \approx \frac{2A_p\vartheta^L(\Gamma(\frac{c}{2}))^L\Gamma(\frac{Lc+1}{2})}{cL\sqrt{\pi}(B_p\rho)^{\frac{cL}{2}}\Gamma(\frac{Lc}{2})} \tag{16}$$

The result of the asymptotic average SEP can also be expressed in terms of coding gain (G_c) and diversity gain (G_d), i.e., $P_e^{asym} \approx (G_c.\bar{\gamma})^{-G_d}$ [8, eq. (1)] as:

$$G_d = \frac{Lc}{2} \qquad G_c = \left(\frac{2A_p\vartheta^L(\Gamma(\frac{c}{2}))^L\Gamma(\frac{Lc+1}{2})}{cL\sqrt{\pi}B_p^{\frac{cL}{2}}\Gamma(\frac{Lc}{2})}\right)^{-\frac{2}{cL}} \tag{17}$$

3.1.2 Average Symbol Error Probability for Equal Gain Combining

By substituting (10) and (15) in (14), and following a similar procedure as defined above, it follows immediately that:

$$\bar{P}_{e,coh}^{egc,asy} \approx \frac{A_p\alpha^L(\Gamma(c))^L(\sqrt{L})^{Lc}\Gamma(\frac{Lc+1}{2})}{\sqrt{\pi}\Gamma(Lc+1)(\rho B_p)^{\frac{Lc}{2}}} \tag{18}$$

Diversity and coding gain are expressed as:

$$G_d = \frac{Lc}{2} \qquad G_c = \left(\frac{A_p\alpha^L(\Gamma(c))^L(\sqrt{L})^{Lc}\Gamma(\frac{Lc+1}{2})}{\Gamma(Lc+1)\sqrt{\pi}B_p^{\frac{cL}{2}}}\right)^{-\frac{2}{cL}} \tag{19}$$

3.1.3 Average Symbol Error Probability for Selection Combining

By substituting (13) and (15) in (14), and following a similar procedure as defined in Sect. 3.1.1, it follows immediately that:

$$\bar{P}_{e,coh}^{sc,asy} \approx \frac{2A_p\vartheta^L\Gamma(\frac{Lc+1}{2})}{c\sqrt{\pi}(\frac{c}{2})^{L-1}(\rho B_p)^{\frac{Lc}{2}}} \tag{20}$$

The values of diversity and coding gain are expressed as:

$$G_d = \frac{Lc}{2} \qquad G_c = \left(\frac{2A_p\vartheta^L\Gamma(\frac{Lc+1}{2})}{c\sqrt{\pi}(\frac{c}{2})^{L-1}B_p^{\frac{cL}{2}}}\right)^{-\frac{2}{cL}} \tag{21}$$

3.2 Non-coherent Average Symbol Error Probability

The instantaneous SEP for different non-coherent modulation schemes is given by [12, Eq. (18)]:

$$P_e(\gamma) = A_n exp(-B_n\gamma) \tag{22}$$

where the parameters A_n and B_n are defined in [12, Table 2].

3.2.1 Average Symbol Error Probability Maximal Ratio Combining

The asymptotic average SEP is derived by substituting (7) and (22) in (14), which with the aid of [15, Eq. (3.381.4)], yields:

$$\bar{P}_{e,non}^{mrc,asy} \approx \frac{A_n(\vartheta)^L(\Gamma(\frac{c}{2}))^L}{(B_n\rho)^{\frac{Lc}{2}}} \tag{23}$$

The diversity and coding gain are expressed as:

$$G_d = \frac{Lc}{2} \qquad G_c = \left(\frac{A_n(\vartheta)^L(\Gamma(\frac{c}{2}))^L}{B_n^{\frac{cL}{2}}}\right)^{-\frac{2}{cL}} \tag{24}$$

3.2.2 Average Symbol Error Probability Equal Gain Combining

The closed-form asymptotic solution to average SEP is derived by substituting (10) and (22) in (14), and repeating similar steps to those defined above:

$$\bar{P}_{e,non}^{egc,asy} \approx \frac{A_n\alpha^L(\Gamma(c))^L(\sqrt{L})^{Lc}\Gamma(\frac{Lc}{2})}{2\Gamma(Lc)(\rho B_n)^{\frac{Lc}{2}}} \tag{25}$$

Diversity and coding gain are expressed as:

$$G_d = \frac{Lc}{2} \qquad G_c = \left(\frac{A_n \alpha^L (\Gamma(c))^L (\sqrt{L})^{Lc} \Gamma(\frac{Lc}{2})}{2\Gamma(Lc) B_n^{\frac{cL}{2}}} \right)^{-\frac{2}{cL}} \tag{26}$$

3.2.3 Average Symbol Error Probability Selection Combining

The closed-form asymptotic solution is evaluated by substituting (13) and (22) in (14), and repeating similar steps to those defined in Sect. 3.2.1:

$$\bar{P}^{sc,asy}_{e,non} \approx \frac{L A_n \vartheta^L \Gamma(\frac{Lc}{2})}{(\frac{c}{2})^{L-1} (\rho B_n)^{\frac{Lc}{2}}} \tag{27}$$

The values of diversity and coding gain are expressed as:

$$G_d = \frac{Lc}{2} \qquad G_c = \left(\frac{L A_n \vartheta^L \Gamma(\frac{Lc}{2})}{(\frac{c}{2})^{L-1} B_n^{\frac{cL}{2}}} \right)^{-\frac{2}{cL}} \tag{28}$$

4 Numerical Analysis

In this section, the asymptotic behavior of the average SEP for the WLN fading channel has been presented graphically. The Monte-Carlo simulations are also included in all the figures to validate the accuracy of the derived expressions.

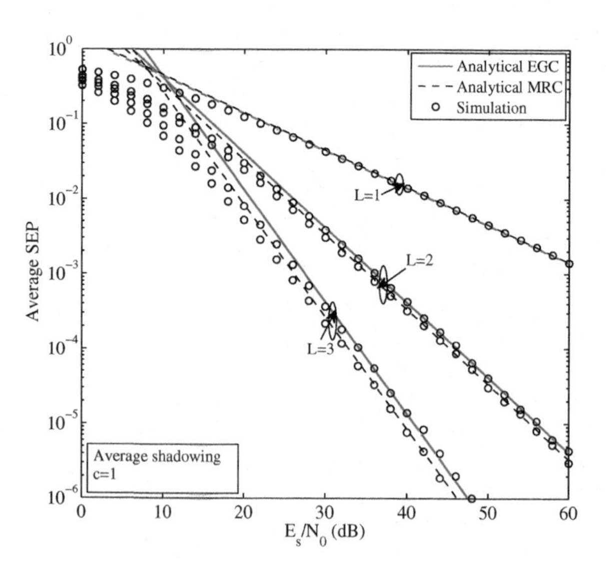

Fig. 1 Average SEP for MPAM with the MRC and EGC diversity scheme and constellation size $M = 4$

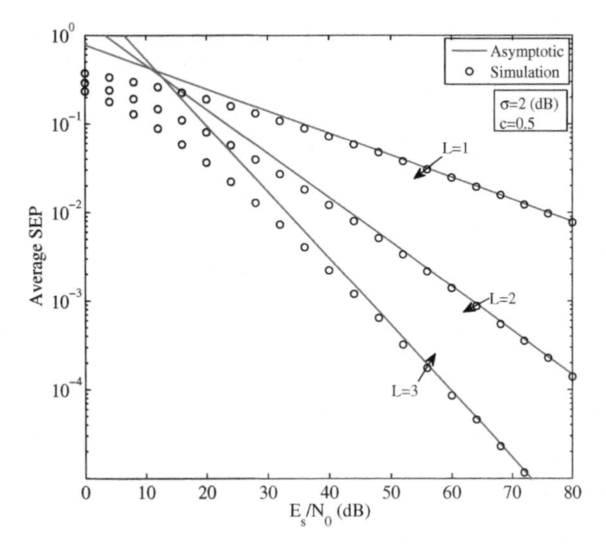

Fig. 2 Average SEP for BPSK with the SC diversity scheme

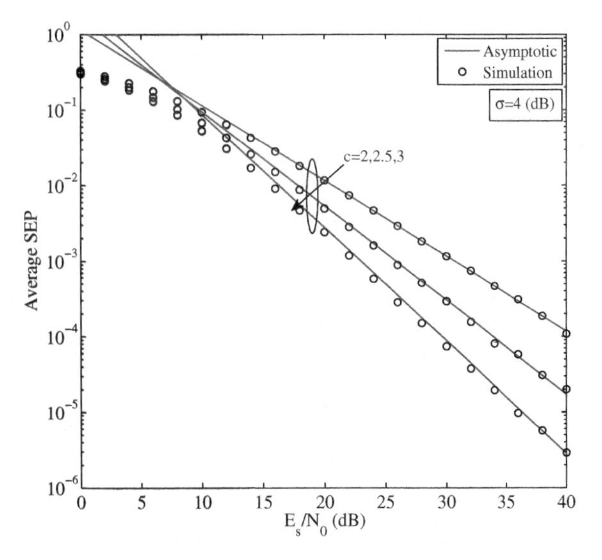

Fig. 3 Average SEP for non-coherent DBPSK with $c = 2, 2.5, 3$

In Fig. 1, asymptotic plots of the average SEP for coherent M-ary pulse amplitude modulation (MPAM), with MRC (16) and EGC (18) side by side, are presented against E_s/N_0. The parameters under consideration are $c = 1$, infrequent light shadowing [18, 19], constellation size $M = 4$, and diversity order $L = 1, 2, 3$. It is clear from the figure that the asymptotic plot converges at high SNR and coincides with Monte-Carlo simulations. It is also observed from the plot that MRC is superior to EGC for all the diversity schemes, and the separation increases with increase in diversity order. The average SEP for coherent binary phase shift keying (BPSK) versus E_s/N_0 is given in Fig. 2 with the SC diversity scheme. The Monte-Carlo simulations are also included and shown to coincide with the closed-form solution at high S-NR. In Fig. 3, the plot illustrates the non-coherent differential BPSK (DBPSK) (23) scheme versus E_s/N_0. As expected, it is revealed from the figure that increasing parameter c means that system performance improves.

5 Conclusion

The closed-from expressions of diversity PDF at the origin for the composite W-LN fading channel have been presented. The derived results were then extended to evaluate the asymptotic expressions of the average SEP for both coherent and non-coherent modulation schemes. It was shown that the asymptotic plot merges with Monte-Carlo simulations at high SNRs, verifying the accuracy of the derived expressions.

References

1. Sagias NC, Karagiannidis GK, Bithas PS, Mathiopouls PT (2005) On the correlated Weibull fading model and its applications. In: Ieee transactions on vehicular technology conference, pp 2149–2153
2. Cheng J, Tellambura C, Beaulieu NC (2004) Performance of digital linear modulations on weibull slow-fading channels. IEEE Trans Commun 52(8):1265–1268
3. Ibdah Y, Ding Y (2015) Mobile-to-mobile channel measurements at 1.85 GHz in suburban environments. IEEE Trans Commun 63(2):466–475
4. Bessate A, Bouanani FEL (2016) A very tight approximate results of MRC receivers over independent Weibull fading channels. Phys Commun 21:30–40
5. Khandelwal V (2014) Karmeshu: a new approximation for average symbol error probability over log-normal channels. IEEE Wirel Commun Lett 3(1):58–61
6. Navidpour SM, Uysal M, Kavehrad M (2007) BER performance of free-space optical transmission with spatial diversity. IEEE Trans Wirel Commun 6(8):2813–2819
7. Héliot F, Xiaoli C, Hoshyar R, Tafazolli R (2009) A tight closed-form approximation of the log-normal fading channel capacity. IEEE Trans Wirel Commun 8(6):2842–2847
8. Wang Z, Giannakis GB (2003) A Simple and General Parametrization Quantifying Performance in Fading Channels. IEEE Trans Commun 51(8):1389–1398
9. Chen Z, Chi Z, Li Y, Vucetic B (2009) Error performance of maximal-ratio combining with transmit antenna selection in flat Nakagami-m fading channels. IEEE Trans Wirel Commun 8(1):424–431
10. Zhang H, Matthaiou M, Tan Z, Wang H (2012) Performance Analysis of digital communication systems over composite $\eta\mu$/Gamma fading channels. IEEE Trans Veh Technol 61(7):3114–3124
11. Badarneh OS, Aloqlah MS (2016) Performance analysis of digital communication systems over $\alpha-\eta-\mu$ fading channels. IEEE Trans Veh Technol 65(10):7972–7982
12. Chauhan PS, Tiwari D, Soni SK (2017) New analytical expressions for the performance metrics of wireless communication system over weibull/lognormal composite fading. Int J Electron Commun (AEU) 82:397–405
13. Simon MK, Alouini M (2005) Digital communication over fading channels, (2nd ed.), New York, Wiley

14. Shanker PM (2004) Error rates in generalized shadowed fading channels. Wirel Person Commun 28:233–238
15. Gradshteyn IS, Ryzhik IM (2007) Table of integrals, series, and products. (7th ed.), Academic Press, California
16. Rana V, Chauhan PS, Soni SK, Bhatt M (2017) A new closed-form of ASEP and channel capacity with MRC and selection combining over Inverse Gaussian shadowing. Int J Electron Commun (AEU) 74:107–115
17. Prudnikov AP, Brychkov YA, Marichev OI (1986) Integrals and Series Volume 2: Special Functions, 1st edn. Gordon and Breach Science Publishers
18. Loo C (1985) A statistical model for a land mobile satellite link. IEEE Trans Veh Technol 34:122–127
19. Loo C (1990) Digital transmission through a land mobile satellite channel. IEEE Trans Commun 38:693–697

Vehicle Detection and Categorization for a Toll Charging System Based on TESSERACT OCR Using the IoT

A. Vijaya Krishna and Shaik Naseera

Abstract In India the main transport system is the road network. Government design different plans for transport system like national highways under development. The administration consents to arrangement with the privately owned businesses who manufacture the foundation for national highways for a definite time. The private agencies deduct the amount from the vehicles which are passed on that recently developed the highways. Vehicle detection is the crucial step in the toll collection management system. There are various ways of implementing a toll charging system including manual toll charging, RFID systems and barcodes. However, these techniques are error prone while charging the toll system. In this paper we propose a framework using Tesseract OCR and Raspberry Pi. If an input is passed to Raspberry Pi, then the Raspberry Pi detects and charges an amount for the vehicle by using the web server and its database. Finally an alert message is pushed to the vehicle owner's mobile number after deducting the amount from the user's account.

Keywords ETC · Sensors · Raspberry Pi · GSM · Open CV

1 Introduction

In India we find the chance to watch for the most part expansive of National thruways. Government designs different stages to finish the tasks under development. The administration consents to arrangement with the privately owned businesses who manufacture the foundation like streets for some period [1]. The contributed sum is collected from vehicles passing on the recently manufactured road. This gathered whole is called as toll imposes [1]. Individuals must choose

A. Vijaya Krishna (✉) · S. Naseera
VIT University, Vellore, Tamil Nadu, India
e-mail: Vijay.merits@gmail.com

S. Naseera
e-mail: naseerakareem@gmail.com

A. Kumar and S. Mozar (eds.), *ICCCE 2018*,
Lecture Notes in Electrical Engineering 500,
https://doi.org/10.1007/978-981-13-0212-1_20

between limited options to for paying the toll impose for utilizing the passage. PC vision is an essential field in the capture of high dimensional information from cameras in the toll system [1, 2].

The main steps taken as part of this procedure are obtaining, handling, and breaking down the picture and changing it into a number or a representative shape [2]. It is used to understand the scene electronically and the procedure is proportionate to the capacity of human vision. The numerical or emblematic data of a scene is chosen in light of the proper model developed with the help of protest geometry, material science, measurement, and learning hypothesis [2]. The scene in mind is converted into image(s) or video(s), including many pictures, using camera (s) concentrated from various areas on a scene [3]. Picture handling, picture investigation and machine vision are likewise firmly identified with PC vision. Picture handling and picture examination manage 2D pictures. In the preparation of a picture it is changed into another by applying a few operations, for example, differential upgrade, edge location, and geometrical changes [4, 5].

Manual toll collection is not suitably for collecting tolls as it exceptionally time consuming. This strategy causes quite a long waiting time at tollbooths as vehicles need to wait until their turn comes.

1.1 Other ETCs

An electronic toll collection system (ETC) is the best alternative to a manual toll collection system. They use diverse innovations to better aid toll collection. They primarily attempt to stay away from manual intercession at the tollbooth. Some use barcodes, RFID labels and so forth for recognition and these are extremely helpful for ongoing picture handling [6]. However, ETC systems suffer from problems of type classification and toll estimation. Vehicle grouping depends on parameters, such as the length of the vehicle, distance between two wheels of the vehicle, and the zone occupied by the vehicle in the picture. There is no exact parameter can be utilized for usage with relating to its imperatives and deducting toll charging is also a major problem in ETC systems.

2 Proposed System

The proposed framework depends on vehicle identification and uses Open CV libraries with an implanted Linux stage. With this model it is likewise conceivable to check the number of vehicles going through the toll corner (Fig. 1).

Raspberry Pi is a charge card measured single-board PC created in the UK. It is one of the mainstream of installed Linux-based advancement sheets [1]. Additionally it is used to check the data rundown of toll gathered vehicles. It will help the manager to check whether a toll charge is entered accurately or not.

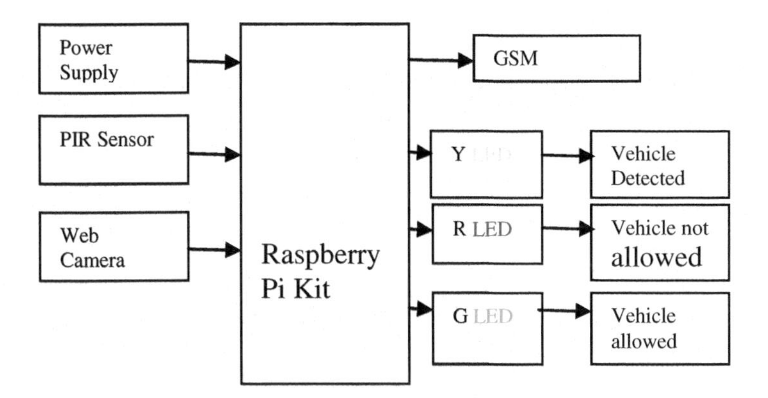

Fig. 1 Block diagram of the proposed system

In this proposed model we need to use high picture catching computerized camera to capture unmistakable pictures of vehicles. For viable reason, we have utilized after camera only for exhibit. Assist this data is passed to the Raspberry Pi which is having web server set up on it. At the point when Raspberry Pi comes to know the vehicle, at that point it get to the web server data and as indicated by the sort of the vehicle, fitting toll is charged.

3 System Design

Vehicles discovery must be pertinent to various natural conditions like light, brilliance, activity status evolving and so on. In the proposed framework, while performing tests we have made a constant situation. The vehicles proceeding along the road and a camera is placed near to the tollbooth area. This camera captures pictures of vehicles and sends them to the framework. These pictures are only the edges changed over from the video by the framework (Fig. 2).

An arrangement framework like the one proposed here can provide critical data for a specific outline situation.

3.1 Overview of Tesseract OCR

Vehicle number plate recognition is crucial in ETC systems. Character recognition is used to obtaining the registration numbers of vehicles [7]. In our model we uses Tesseract OCR to recognise vehicle registration numbers. The pipeline of the Tesseract OCR motor is given in Fig. 3. The initial step is adaptive thresholding, which converts the picture into a twofold form using Otsu's strategy [8]. The

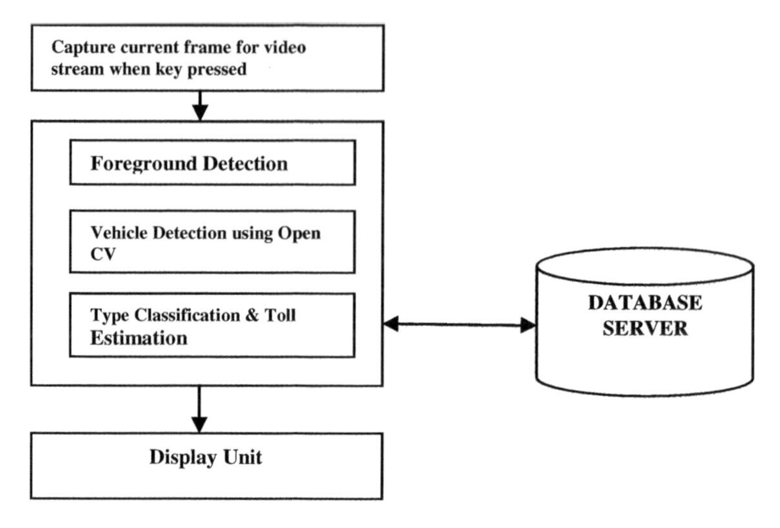

Fig. 2 System architecture of video subsystem design

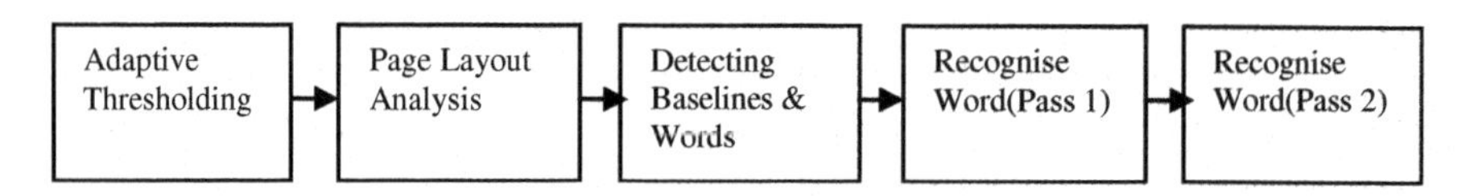

Fig. 3 Architecture of Tesseract OCR

subsequent stage is page design investigation, which is used to extricate the content squares.

In the following stage, the baselines of each line are distinguished and the content is partitioned into words using distinct spaces and fluffy spaces [9].

In the following stage, the character plots are removed from the words. Acknowledgment of the content is then begun as two-pass process. In the primary pass, word acknowledgment is completed using the static classifier.

3.1.1 Adaptive Thresholding

In Tesseract OCR, Otsu's technique [8] is employed to perform bunching-based picture thresholding. The pixels in the photo are spoken to in L dimension levels 0, 1, …, L where each esteem compares to a potential limit. In Otsu's strategy we look for the edge that limits the intra-class change, characterized as a weighted whole of the differences of the two classes:

$$\sigma_{\omega}^{2}(t) = \omega_0(t)\sigma_{0}^{2}(t) + \omega_1(t)\sigma_{1}^{2}(t)$$

Weights ω_0 and ω_1 are the probabilities of the two shading classes isolated by a limit t and square σ_{0}^{2} and σ_{1}^{2} are the differences of these two classes. Otsu demonstrates that limiting the intra-class change is the same as augmenting between class fluctuations [8]

$$\sigma_{b}^{2}(t) = \omega_0(t)\omega_1(t)[\mu_0(t) - \mu_1(t)]^2$$

$$\mu_0(t) = \sum_{i=0}^{t-1} \frac{p(i)}{\omega_0(t)}$$

$$\mu_1(t) = \sum_{i=1}^{L-1} \frac{p(i)}{\omega_1(t)}$$

Every one of these qualities can be calculated from the twofold grayscale input given. The ideal edge t* that augments is chosen by a consecutive search of the diverse estimations of t. To represent varieties inside the picture, local adaptive thresholding is performed in Tesseract, where Otsu's calculation is connected to little estimated rectangular divisions of the picture.

3.1.2 Page Layout Analysis

Page design examination, one of the initial steps of OCR, isolates a picture into ranges of content and non-content, and also partly multi-segments the content into segments [9]. The page design investigation in Tesseract depends on recognizing tab stops in a recorded picture.

3.1.3 Baseline Fitting and Word Detection

Tesseract utilizes an exceptionally novel calculation to find the lines of content on a page. The calculation performs well even within the sight of wrecked and coupled characters, spot clamor and page tilt [10].

Once the content lines have been discovered, the baselines are better fitted utilizing a quadratic spline and a minimum squares fit [10].

Tesseract discovers words by measuring holes in a restricted vertical range between the standard and mean line. Spaces that are near the limit at this stage are made fluffy, with the goal that an official choice can be made after word acknowledgment.

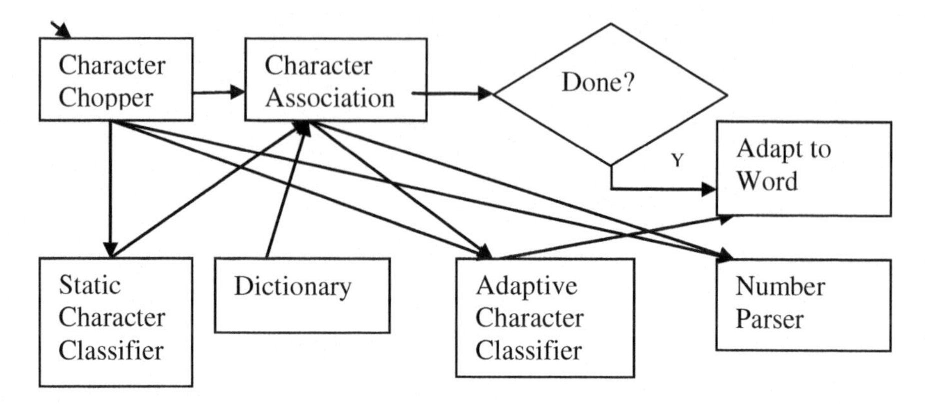

Fig. 4 Block diagram of the word recognizer

3.1.4 Word Recognition

In this part the distinguished words are portioned into characters. Tesseract tests the content lines to decide if they are of settled pitch (having a steady separation of words and characters) amid the word recognition step [9]. For a settled pitch content, Tesseract slashes the words into characters using the pitch. Whatever remains of the word acknowledgment step applies only to nonfixed-pitch content (Figs. 4 and 5).

The word recognizer first characterizes each blob, and shows the outcomes to a lexicon hunt to discover a word in the mix of classifier decisions for each blob in the word. If the word result is unsuitable, Tesseract hacks the blob with most exceedingly awful certainty from the character classifier [9]. Applicant slash focii are found from the inward vertices of a polygonal estimate of the framework.

After slashing, conceivable outcomes are depleted, the associator makes an A* (best first) pursuit of the division chart of conceivable mixes of the maximally cleaved blobs into hopeful characters [9]. At each progression in the best-first pursuit, any new blob blends are grouped, and the classifier comes about are given to the word reference once more.

The yield for a word is the character string present in the lexicon that had the best general separation-based rating.

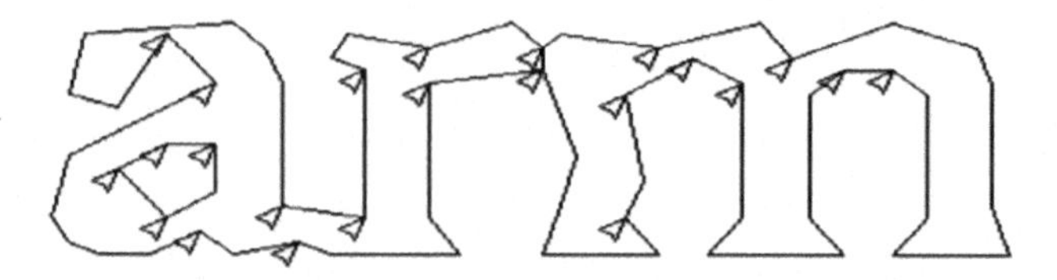

Fig. 5 Candidate chop points

4 Implementation

4.1 OpenCV—Open Computer Vision Library

OpenCV is an open source PC vision library. The library is composed in C and C++ and runs under Linux, Windows, and Mac OS X. The OpenCV library contains more than 500 capacities that traverse numerous regions of vision, including processing plant item review, therapeutic imaging, security, UI, camera adjustment, stereo vision, and mechanical technology.

4.2 Virtual Network Computing (VNC)

VNC is a graphical desktop sharing system. By proper authentication we can connect to the VNC. This creates a virtual Linux operating system on our operating system. With this we run our project if a vehicle is detected, and it displays the number of the vehicle.

5 Simulation Results

(see Figs. 6, 7, 8, 9 and 10).

Fig. 6 VNC viewer authentication

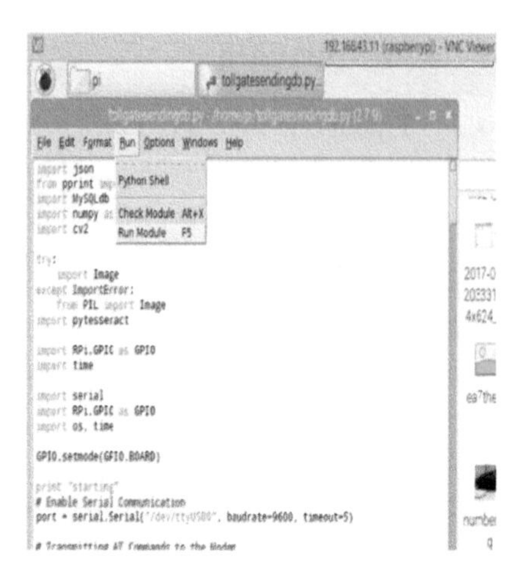

Fig. 7 Sample code of a toll system

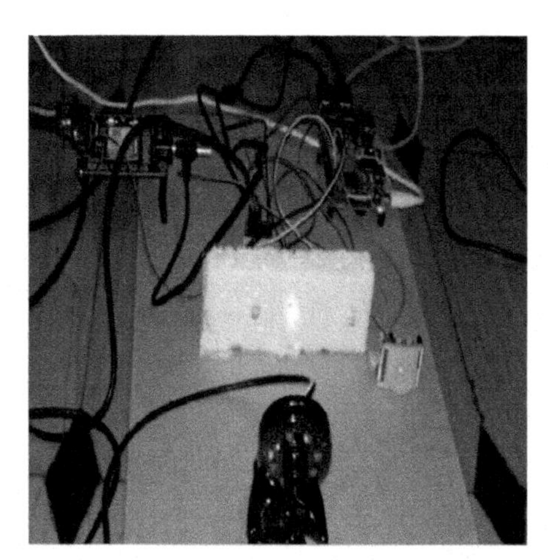

Fig. 8 Proposed toll system model with Raspberry Pi Kit

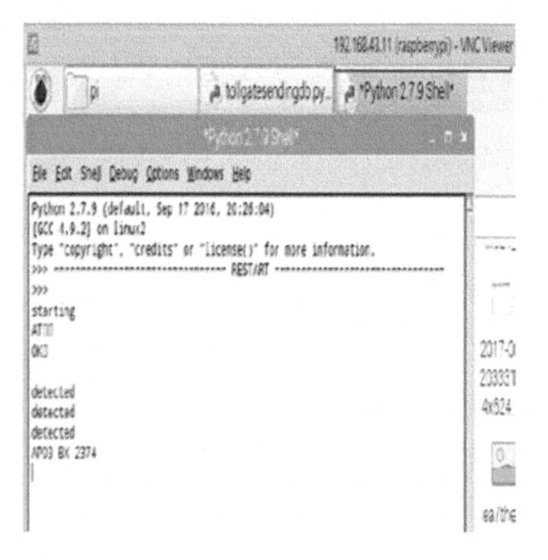

Fig. 9 Detection and recognition of vehicle number

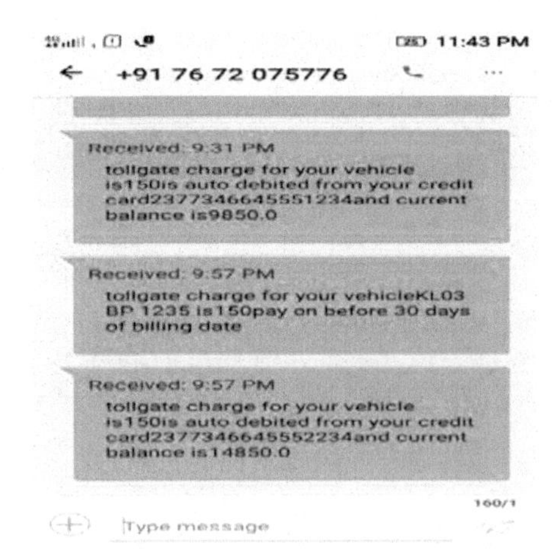

Fig. 10 Pushing a vehicle detection and toll charging message to a customer using the internet of things

6 Conclusion

The technique which is used for implementation is very efficient and more practical than any other methods of toll estimation. As for development, if an embedded Linux system is used, then processing speed will be fast. On the other hand, OpenCV plays a crucial role in vehicle detection. It has libraries which can be used for vehicle detection and one can also extend its use according to requirements. At the tollbooth, a major task of toll collection will be completed with less human effort. This idea gives very less expensive toll collection system concept. Also the system is transparent in terms of toll collection and provides a reliability that it can also work in adverse climatic condition.

References

1. Suryatali A, Dharmadhikari VB Computer vision based vehicle detection for toll collection system using embedded Linux. In: 2015 international conference on circuit, power and computing technologies (ICCPCT), 2015 March, pp 1–7. IEEE
2. Hsieh JW, Yu SH, Chen YS, Hu WF (2006) Automatic traffic surveillance system for vehicle tracking and classification. IEEE Trans Intell Transp Syst 7(2):175–187
3. Robert K (2009) Video-based traffic monitoring at day and night vehicle features detection tracking. In: ITSC '09. 12th international IEEE conference on intelligent transportation systems, 2009 October, pp 1–6. IEEE
4. Scott J, Pusateri MA, Cornish D (2009) Kalman filter based video background estimation. In: Applied imagery pattern recognition workshop (AIPRW), 2009, October IEEE, pp 1–7. IEEE
5. Rother C, Kolmogorov V, Blake A (2004) Grabcut: interactive foreground extraction using iterated graph cuts. ACM Trans Graph (TOG) 23(3):309–314. ACM
6. Khan AA, Yakzan AIE, Ali M (2011) Radio frequency identification (RFID) based toll collection system. In: 2011 third international conference on computational intelligence, communication systems and networks (CICSyN), 2011 July, pp 103–107. IEEE

7. Qadri MT, Asif M (2009) Automatic number plate recognition system for vehicle identification using optical character recognition. In: International conference on education technology and computer, 2009 April, ICETC '09, pp 335–338. IEEE
8. Merler M, Kender JR (2009) Semantic keyword extraction via adaptive text binarization of unstructured unsourced video. In: 2009 16th IEEE international conference on image processing (ICIP), 2009 November, pp 261–264. IEEE
9. Smith R (2007) An overview of the Tesseract OCR engine. In: Ninth international conference on document analysis and recognition, 2007 September, ICDAR 2007, vol 2, pp 629–633. IEEE
10. Smith R (1995) A simple and efficient skew detection algorithm via text row accumulation. In: Proceedings of the third international conference on document analysis and recognition, 1995 August, vol 2, pp 1145–1148. IEEE

Transmission Spectrum of a Typical Waveguide in Photonic Crystal with Tunable Width: Simulation and Analysis

Neeraj Sunil, V. Jayakrishnan, Harish Somanathan and Alok Kumar Jha

Abstract In this paper a typical waveguide in 2D photonic crystal of air holes in dielectric slab structure has been simulated to explore the possible transmission spectrum as shown later in various figures. The waveguide width is variable and correspondingly its transmission spectrum changes. This may be improved upon and used to design optical communication devices and photonic sensors. The algorithms used for simulation are finite difference time domain (FDTD), and plane wave expansion method (PWEM).

Keywords Photonic band gap · Waveguide · PBG · FDTD MEEP · PWEM · MPB

1 Introduction

Photonic crystal is a periodically modulated dielectric material. The periodic variation of the dielectric constant gives rise to a particular photonic band gap (PBG), and a unique photonic band structure [1, 2].

The band structure of the crystal depends upon various parameters, such as difference in refractive index and lattice structure etc. A basic hexagonal lattice of air holes in a dielectric slab is shown in Fig. 1.

N. Sunil · V. Jayakrishnan · H. Somanathan · A. K. Jha (✉)
Department of Electronics & Communication Engineering, Amrita School of Engineering, Amrita Vishwa Vidyapeetham, Bengaluru, India
e-mail: alok_jha@blr.amrita.edu

N. Sunil
e-mail: neeraj.sunil95@gmail.com

V. Jayakrishnan
e-mail: jkrishnan95v@gmail.com

H. Somanathan
e-mail: harisvirgonps@gmail.com

A. Kumar and S. Mozar (eds.), *ICCCE 2018*,
Lecture Notes in Electrical Engineering 500,
https://doi.org/10.1007/978-981-13-0212-1_21

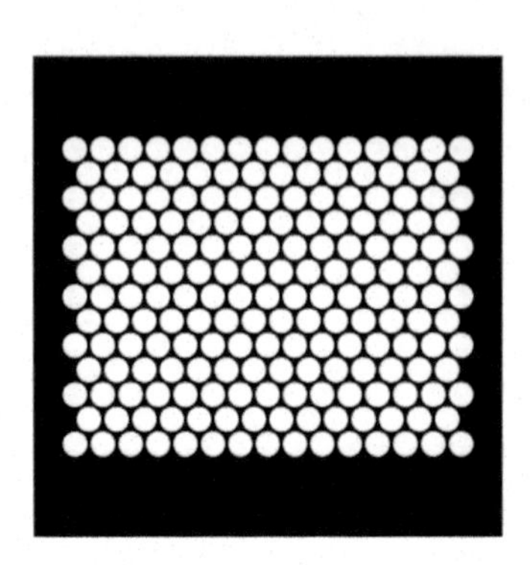

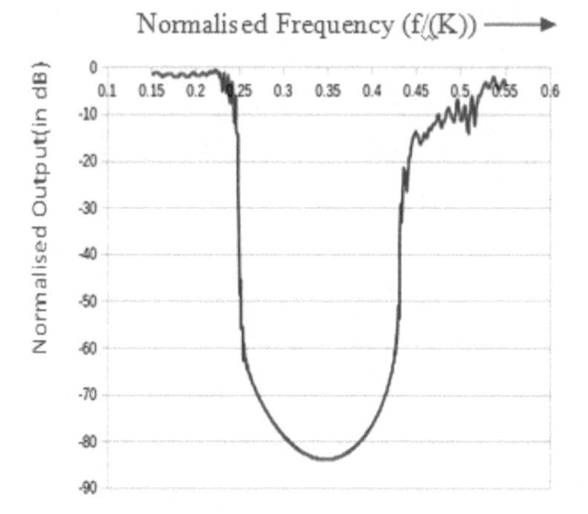

Fig. 1 . **a** Hexagonal lattice of holes in the slab. **b** Transmission spectra of (**a**)

PBG structure provides basic information about the crystal. It gives relationships of frequencies and wave numbers, i.e. modes. The band gap consists of a range of all the frequencies which are not allowed to propagate through the crystal. When defects, such as point, collection of points, and lines, etc. are introduced in photonic crystals, they subsequently cause defect modes in the band gap [3].

In this study, a 2D hexagonal lattice of air holes in a dielectric slab is chosen and two parallel line defects are created as shown in Fig. 1a. The two line defect rows are on the either side of a middle no defect row. The width of the line defect can be varied by separating the two sides of photonic crystal about the middle no defect row. The two variable line defects and the region in between constitute the waveguide.

2 Computation Methods

Finite difference time domain (FDTD) has been used to compute the real-time behavior of electromagnetic waves in photonic crystals using a time-domain approach. Using FDTD, by applying a pulsed field as the source and by taking the Fourier transform of the response obtained, the output is obtained over a wide frequency range, with the PML (perfectly matched layer) condition being taken into consideration.

Plane Wave Expansion Method (PWEM) is a method used to compute the band structures of the photonic crystals using a frequency-domain approach.

The band structure of the crystal depends upon various parameters, such as difference in refractive index and lattice structure etc. A basic hexagonal lattice of air holes in a dielectric slab is shown in Fig. 1.

Table 1 Width of waveguide

Figure	Width
1	0
2	0.2
3	0.4
4	0.6
5	0.8
6	1

3 Structure Parameters

The structures shown in the figures can be described by the following parameters:

Period of lattice = 1 unit (μm) (same for all figures)
Radius of holes = 0.42 units
Dielectric constant of slab = 12
Height of slab = infinite.

Width of each of the two line defects for the figures is given in Table 1 as a fraction of one lattice period.

Photonic crystal parameters are mutually scalable. So period and frequency can be expressed as normalized and independent of units.

4 Units

The units of amplitude and frequency are normalized. The frequency value, when multiplied with ($2\pi c/a = M$), gives a frequency in SI units, where 'c' is the velocity of light in vacuum, and 'a' is the lattice periodicity in SI units respectively. Power obtained in the output spectra is normalized (i.e. output/input flux) and is plotted on a dB scale (10 * log (output/input flux)). The width of the line defect can be varied by separating the two sides of the photonic crystal about the middle no defect row. The two variable line defects and the region in between constitute the waveguide.

5 Structure Design

1. A dielectric slab of dielectric constant 12 is used.
2. Through the slab, holes of radius 0.42 units are drilled in the triangular lattice form.
3. The width of the waveguide can be changed as a fraction of 1 period unit of the lattice as shown.

4. Center frequency of the source is 0.33 normalised units.
5. Mode of propagation: TE mode.

When the central frequency is multiplied by M ($2\pi c/a$), and a value of a = 1 μm, then we obtain a wavelength of 482 nm.

6 PWEM Analysis

A basic band gap structure was analyzed for the structure generated as shown in Fig. 1.

The band gap generated is shown in Fig. 2. As seen in Fig. 2, a band gap centered on 0.3–0.4 is obtained. Hence for the transmission analysis the source is taken to be centered on 0.33 normalised units.

7 Transmission Spectra

Shown in Fig. 1b is the transmission spectrum for the structure in Fig. 1a.

Given below are various examples of structures and their corresponding transmission spectra.

8 Results and Discussions

1. Fig. 1a is the regular triangular lattice and Fig. 1b shows the band gap of the lattice.
2. In Fig. 3a the basic waveguide structure is shown. Figure 3b therefore shows the respective change that occurs in the spectra due to the above change.
3. Similarly in Figs. 4a, 5a, 6a, and 7b the width of the waveguide has been set to 0.4, 0.6, 0.8, and 1 unit respectively. And Figs. 4b, 5b, 6b, and 7b show the respective spectra changes.

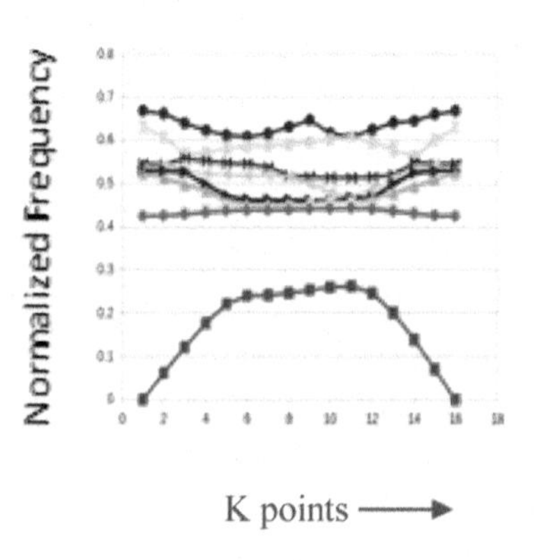

Fig. 2 Band structure for the structure in Fig. 1a

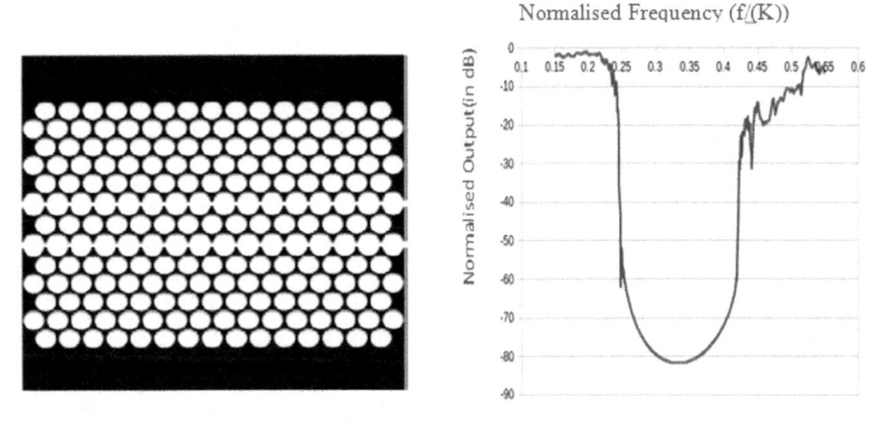

Fig. 3 **a** Structure with line defect = 0.2 units. **b** Transmission spectra of (**a**)

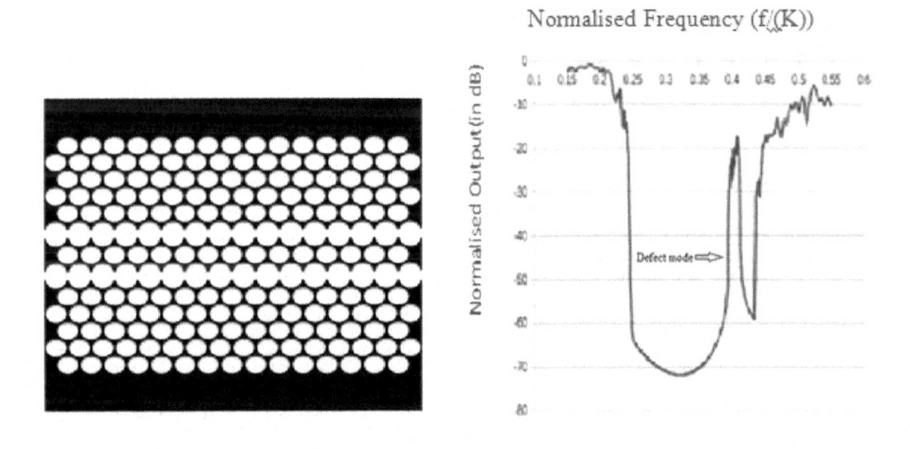

Fig. 4 **a** Structure with line defect = 0.4 units. **b** Transmission spectra of (**a**)

4. We can clearly see that as the width increases, the defect mode becomes wider in frequency and shifts toward the lower end of the band gap.
5. Fig. 6b reveals a closely spaced pair of defect modes.
6. Fig. 7b unlike previous cases, exhibits the possibility for a very narrow band frequency reject-type spectrum.

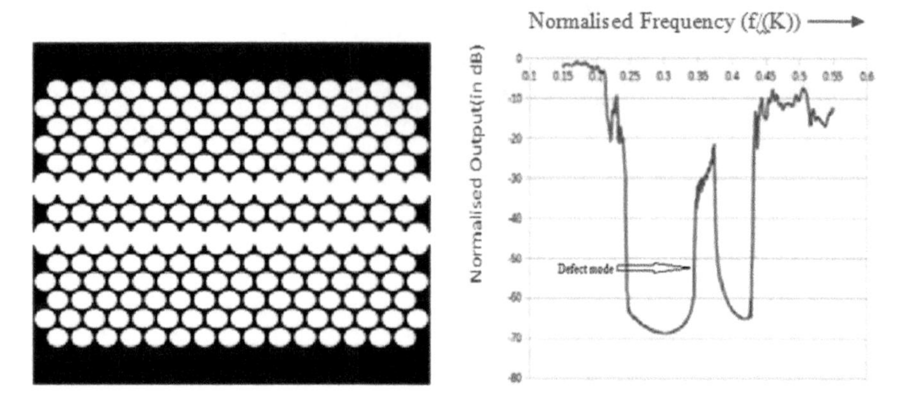

Fig. 5 **a** Structure with line defect = 0.6 units. **b** Transmission spectra of (**a**)

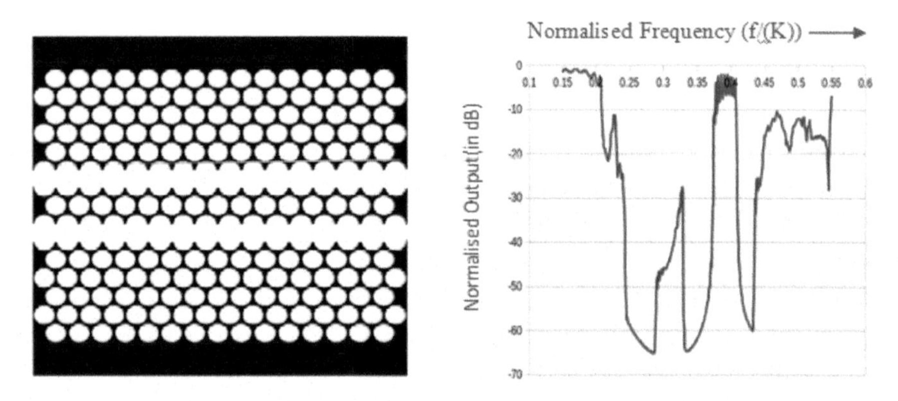

Fig. 6 **a** Structure with line defect = 0.8 units. **b** Transmission spectra of (**a**)

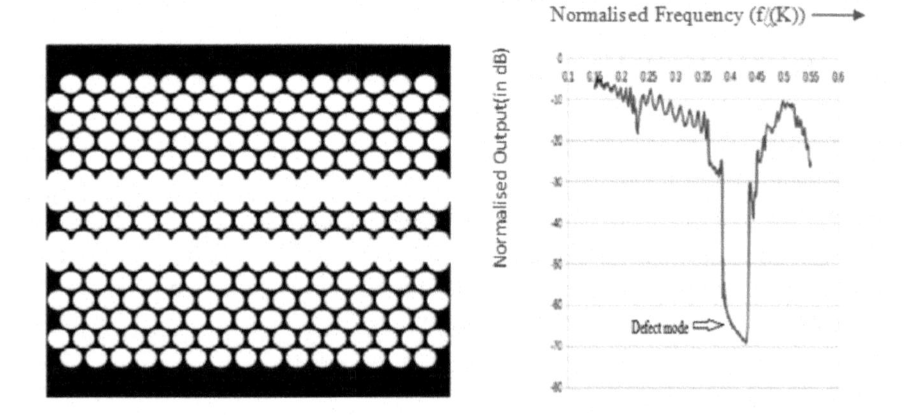

Fig. 7 **a** Structure with line defect = 1 unit. **b** Transmission spectrum of (**a**)

9 Applications

The above analysis can be used in optical communication devices and optical sensors. It can also be mapped and calibrated to measure physical properties like temperature and pressure [4].

10 Software Used

MEEP and MPB are the software used for analysis. Both of these have been developed by the Ab Initio Physics Research team at MIT (Massachusetts Institute of Technology). [5].

MPB is used for computing the band structures and electromagnetic modes of periodic dielectric structures using PWM.

MEEP is a FDTD simulation software used for real-time computation of transmission and reflection spectra, and frequency and field patterns in a dielectric substance.

References

1. Joannopoulos JD, Meade RD, Winn JN (1995) Photonic crystals: modelling the flow of light. Princeton University Press, Princeton, NJ
2. John S, Toader O, Busch K Photonic bandgap of material: a semiconductor of light. www.physics.utoronto.ca/~john/john/
3. Encyclopedia.pdf
4. Jha AK, Sandeep U, Samad SA, Srinivas T (2011) Analysis of line defects in 2D photonic band gap structures of dielectric rods in air for single waveguide modes, INDICON, Hyderabad
5. Ana M, Pinto R, Lopez-Amo M (2012) Photonic crystal fibers for sensing applications. J Sensors
6. Johnson SG, Joannopoulos JD (2001) Block-iterative frequency-domain methods for Maxwell's equations in a plane wave basis. Opt Exp 8(3):173–190
7. Tandaechanurat S, Ishida D, Nomura GM, Iwamoto S, Arakawa Y (2011) Lasing oscillation in a three-dimensional photonic crystal nanocavity with a complete bandgap. Nat Photon 5:91–94

An Anamnesis on the Internet of Nano Things (IoNT) for Biomedical Applications

Amruta Pattar, Arunkumar Lagashetty and Anuradha Savadi

Abstract This paper holds the data of broadly anamnesis and summery on internet of nano things (IoNT) for human services. This makes great possible to give the systematic and prognostic techniques and which in this way help in the medi cations of patients through correct bound pharmaceutical transport, tranquilize convey, tumor and various distinctive contamination's. The proposed study discusses the different network models of the IoNT and the architectural requirements for its implementation, which involves the different networking models, electromagnetic and molecular communication, channel modeling, information encoding, telemedicine aspects, and IoNT protocols.

Keywords IoNT · Molecular communication · Drugs delivery

1 Introduction

The recent developments in the area of the web known as the Internet of Things (IoT) have taken it to its next level of improvement by combining the nano sensors with IoT. Major developments and advancements in the field of nanotechnology, in combination with the internet, have arisen, Since the famous lecture on nanotechnology by Richard Feynman in 1959, the field has seen great progress and has also provided sophisticated devices with important applications, such as nanosensors, which helps in diagnosis at the molecular level and in turn can provide treatments, such as targeted drug delivery to tumor patients. The IoNT is one which relates in-depth to the internet of bio-nano things (IoBNT). Bio-nano things are those which can be defined as the special identification of basic structure and the functional unit that explains the work and the connectivity within the organic conditions.

A. Pattar (✉) · A. Lagashetty · A. Savadi
Appa Institute of Engineering and Technology, VTU, Kalaburgi, Karanataka, India
e-mail: amruta.pattar@yahoo.com

A. Kumar and S. Mozar (eds.), *ICCCE 2018*,
Lecture Notes in Electrical Engineering 500,
https://doi.org/10.1007/978-981-13-0212-1_22

Despite various papers published on nano gadgets every year, it is still not clear regarding how communication takes place between nano gadgets. There are two major and extensive aspects which is used by nanotechnology for the IoNT and they are electromagnetic communication, which uses radiowaves as the information carrier and molecular communication, where molecules are used as the information carriers. The molecular communication turns out to be better than electromagnetic communication as it shows a high energy efficiency, biocompatibility, and also has the competency to work in an aqueous medium which makes molecular communication much better for working on nano gadgets using the IoNT.

Molecular communication shows a good level of performance while using the IoNT as a communication media. Basically, molecular communication is all about invigorating the various molecules in the biological systems, since the communication is carried out by molecules. Molecular communication is one of the media which already exists in the natural world acting as an element between the nanoscales. Natural phenomena, such as intercellular and interbacterial communication are extremely helpful in providing essential information regarding the model of a nanonetwork. Figure 1 shows a simplified model of a molecular communication system. In this communication system, the encoded information of the transmitter is loaded on the molecules which are in turn called information molecules (proteins, ions, DNA, etc.). These information molecules are subsequently loaded onto the carrier molecules (molecular motors, etc.) and are then finally propagated to the receiver.

The main contribution of molecular communication is to provide a means to send, transport, and receive the molecules and it also adapts biological and artificially-created components such as sensors and reactors, which also facilitate communication between each other with the help of molecules. This is a major goal in the treatment of diseases at their molecular level, and helps in loading the drug at the particular area on the cancer cells.

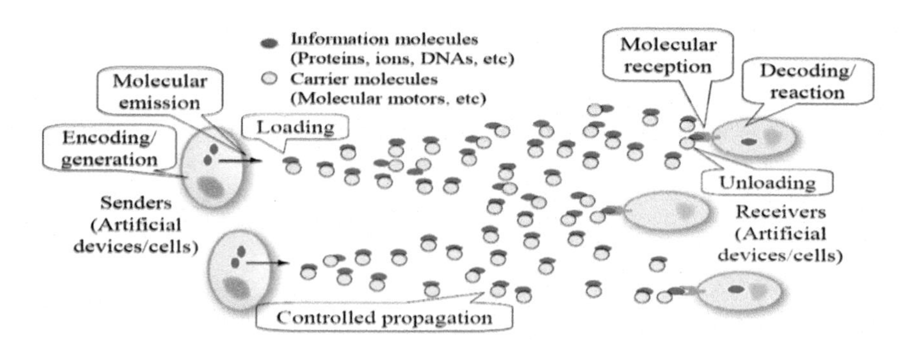

Fig. 1 A simple molecular communication system

2 Literature Review

The concept of the IoNT was initially plugged by the [1] were the author has described the summaried architecture for electromagnetic (EM) nanocommunication, which also embrace the basic concept of channel modeling, information coding, many other protocols. The author also describes the network architecture and discusses the interconnection of nanomachines with the accessible communication network system. Figure 2 characterize the basic introductory towards the IoNT which can retrieved in two distinctive manner, first the intrabody nanonetwork for healthcare, and second, the interconnected office.

The intrabody network is responsible in the facilitation of nanomachines, such as nanosensors and nanoactuators, which are deployed in the human body and can be operated with the help of remote, where the remote makes use of micro scale range by maintaining the major communication media as the internet. Whereas the interconnected office concerns each and every component regularly found in an office and an internal fragments are also equipped for a nanotransceiver which empowers them to be associated with the internet.

The paper entitled in [2] has come up with imaginative works of stack mode that allows the capture of unique characteristics of nanonetworks which is still in its early stages and is an active area of research. The author discusses the communication and networking aspects of the IoNT which involves the optimized version of layer-based models and non-layer-based models. The layer-based model is designed especially for nanonetworks and the protocol stack of this layer model is designed according to the following: application, transport, network, and finally the physical layer. It is maintained similarly from both the sender and the receiver sides. Whereas in the non-layer-based model, the protocol stack has the default assumption regarding nanonetworks, such as a multi-tiered, dynamic, and opportunistic hierarchical architecture that embraces nanomachines, nanorouters, and gateways. The author also also commented on the IoNT protocol stack,

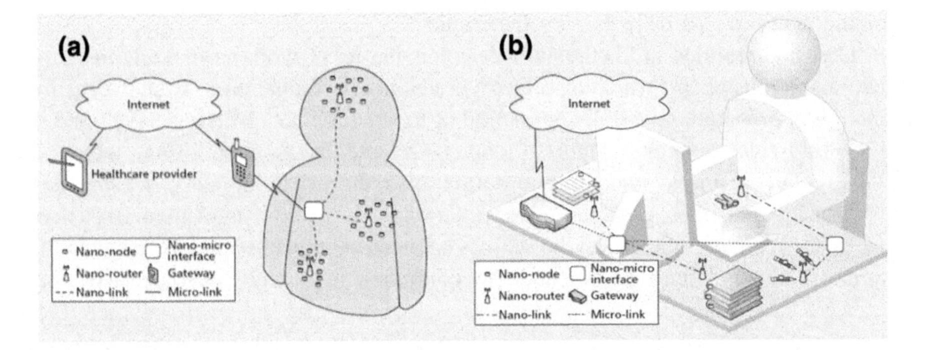

Fig. 2 Design of networks for the internet of nano things. **a** Intrabody nanonetwork for human services application. **b** The interconnected office

which includes an application layer, transport layer, network layer, medium access control layer, and a physical layer. The paper also tends to illustrate the importance towards significant applications, such as drug delivery and the effective detection of disease, for implementation in healthcare services.

The paper entitled in [3] describes the extensive study of networking and communication aspects which helps to understand the theoretical outcome of the physical implementation of molecular communication. The authors focused on the connectivity of molecular communication based on the IoNT in variable environmental conditions. The mathematical expression for connectivity is determined in terms of temperature (*temp*) and also the relative concentration of physical obstructions (x). The authors have also made use of MATLAB to reveal the state of physical obstruction when it gets implied with the change in accordance with the climate. Nanoscale network modeling is one of the major areas of study being carried out in the current research field, which also includes the major escalating of some of the parameters such as channel modeling, modulation and coding, receiver design, and reliability among others. For communication purposes, pheromones are used as signal carrier molecules, with the main drawback being that pheromones can be easily affected by changes in environmental conditions, such as temperature, or by physical obstacles resulting in a change in connectivity.

The paper entitled in [4] portrays avant-garde paradigm of the IoBNT, where the author discusses the origin of the IoBNT from a combination of synthetic biology and nanotechnology, which allows the engineering of computing devices with major help from biological components as shown in Fig. 3. The biological embedded computing device, which is based on biological cells and their functionalities in accordance with biochemical domains which incredibly promises the major purpose of sensing and actuation in the intrabody, and also helps in the environmental control of toxic agents and pollution. The author also describes the communication media used by the IoBNT, the major communication media used being molecular communication. In nature, the reciprocity of the information between cells is completely based on the synthesis, transformation, emission, propagation, and the receiving of molecules by making use of biochemical and physical processes. The challenges facing molecular communication in terms of engineering are also described by the author.

The paper entitled in [5] the author depicts the IoNT working in a telemedicine administrations chain, which also incorporates the investigation of frameworks for obtaining, preparation, and dissemination of therapeutic data which is enveloped in the worldwide creations. Figure 4 depicts remote health monitoring, which is nothing but smart healthcare. The author also discusses implants, sensors and nanosensors and their communication interface. The IBAN (implanted BAN) are one of the main specific grade, the sensors are used as implants and are placed in the human body and used as the node to communication with the outer world to improve monitoring.

As per the above figure, the architecture of remote health monitoring includes many classifications, such as

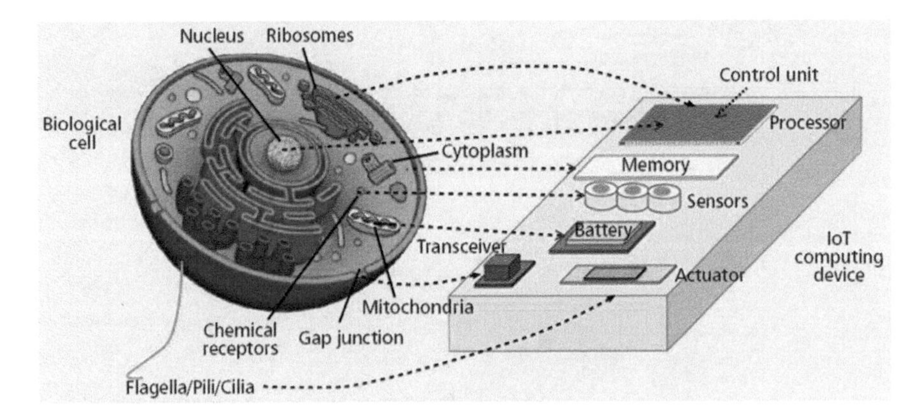

Fig. 3 Typical comonents of an IoT device and the elements of a biological cell

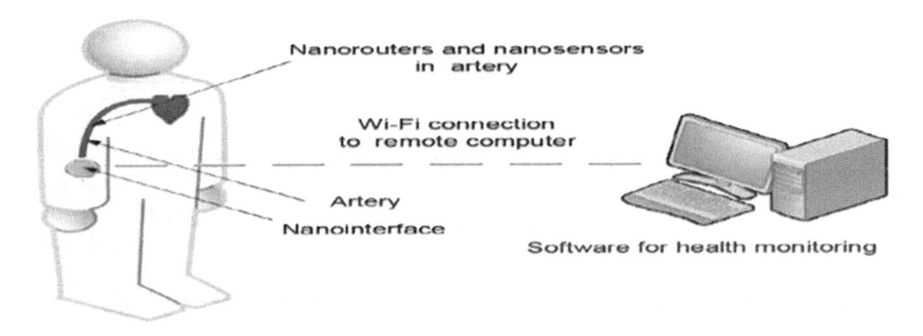

Fig. 4 Framework for observing wellbeing using WNSN and WBAN

- C-health—represents classical healthcare.
- E-health—electronic healthcare, which is also known as a subset of the c-health model using ICT.
- M-health—mobile healthcare, which makes use of mobile devices, and is also known as the subset of e-health.
- S-health—smart healthcare, helps in maintaining records, delivery data, and permissive prevention of health hazards.

As in [6] the author's main focus is to set a standard example of an IoBNT which has an extraordinary ability to set up the elimination of bio-nano devices from the internet source when it is required. This model imitates the apoptotic flagging pathways in living beings, where specific molecules are sent to cells to start their self-obliteration from the framework. Figures 5 and 6 represent the architectural model of the IoBNT. The main goal of the author is to focus on allowing the communication interface between the nanotransmitter and the other nanodevice by providing a deterministic model. The major work expressed here concerns the self-annihilation which explains the sending of the death command through the nanotransmitter, which has the capability to execute the received death command by using natural cells.

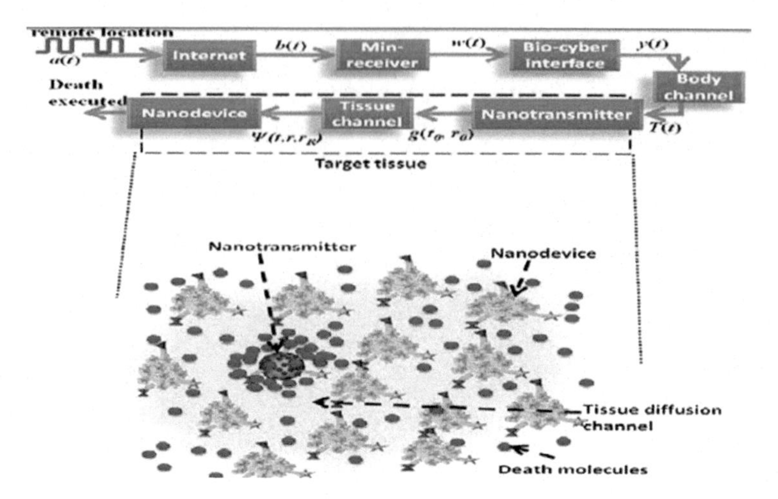

Fig. 5 Block diagram of diffusion-based IoBNT

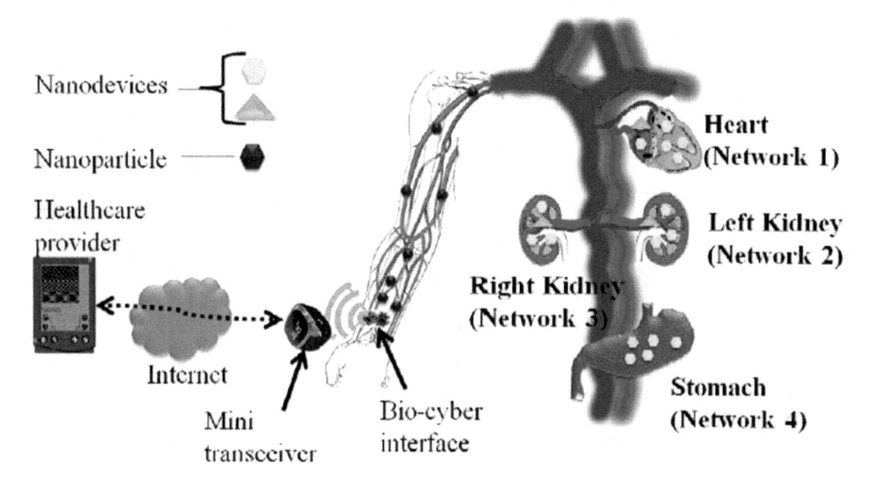

Fig. 6 Network architecture for the IoBNT

The paper entitled [7] the authors portray challenges, solutions, and applications in accordance with the IoNT which involves data collection, system architecture, routing technology, middleware, system management, data analysis, energy conversion, and other challenges, such as security and privacy. The main realization according to the author is the data collection regarding the IoNT, which in turn relates information from the nanosensors and their environmental condition according to which the system can handle the further process. The system architecture makes use of the data that has been collected by the device, which may not be so easy to process so the solution described by the author is to make use of *micro gateways.* Once the system architecture has completed the process, the next step is the routing technology. Routing technique is extremely important while the data is being transferred or received. This is made possible by the nanocommunication system which involves molecular and electromagnetic communication.

The paper entitled [8] depicts healthcare applications which incorporate different requirements, opportunities, and challenges. The author also introduces the basic classification of requirements relating to the generic application functionalities supported by the IoNT and also informs implementation and performance evaluation issues, particularly those which relate to deployment, communication, and coexistence with other networking paradigm. This paper describes the innovative smart technologies which were not previously feasible but are now easily accessible using wireless communication with high-end facilities. A body area network is used and this supports the near real-time sensing and also provides the reporting on the patient's various health conditions. This can be operated using mobile health (*M-health*) and also by the many health monitoring systems which provide for the monitoring of the health status of patients with the help of smartphones. The major aim of this paper is to convey the importance of the nanoscale in accordance with healthcare applications, such as diagnosing, treating, and monitoring, with the focus mainly on treating patients at the molecular level. The main advantage of the nanonetwork discussed by the author is that it has an ability to detect the availability of any imbalance regarding molecules, chemicals, or any kind of virus, and send an alert.

The paper entitled [9] depicts the current promises, future aspects and numerous applications in the area of the IoNT and also gives an extensive review of the IoT, IoE and IoNT, revealing the extreme extension of the IoT by implementing the IoNT. The main perspective is to focus on further studies which will mainly focus on industrial and biomedical areas. The author describes technologies which support the IoT mostly by the physical objects linked to the internet by the different methods of short-range wireless technologies such as ZigBee, RFID, sensor networks and also through location-based technologies. This paper focuses on describing the difference between these three techniques, which are an extension of one another which helps in the future technologies. The main concept which is been portrayed here is the improvement of the technology in terms of communication, such as person-to-person communication, person-to-machine, and now with the help of the IoNT, machine-to-machine communication within the body is possible.

The paper entitled [10] the paper emblematic the advancement of a novel radio channel demonstrate within the human skin at the terahertz, which will empower the association among potential nano-machines working in the bury cell regions of the human skin. The communication media used here is body-centric wireless communication (BCWC) which has been widely contemplated in the past at a range of frequencies. Since nanoscale technologies has attractive future potential to open up a large number of opportunities for making use of the latest nanomaterials, such as carbon tubes and graphene etc. As mentioned earlier, the main purpose of the author is to come up with unique idea of a channel model for use the skin, which considers all of the previously mentioned parameters and also undergoes blind testing for the analytical results which is carried out on the previous works done with the help of simulated data. The potential of the model suggested by author can also be evaluated by comparing it with measures of skin samples using THZ time-domain spectroscopy (THz-TDS).

3 Conclusion

This paper present a review on the IONT for healthcare applications. As per the requirement of new generation nanotechnology the communication between nanosensors and nanodevices within the human body is made possible by adopting the IOT technology such as molecular communication, electromagnetic communication, layer models, non-layer models, telemedicine, and embedded computing using biological cells. The paper also describes the major studies and novel thoughts of many author conveying the idea that the internet of nanothings looks as if it will be extremely helpful in the future, especially in the field of biomedicine, where the IoNT allows nanosensors to communicate with themselves and interface with the outer world according to human requirements.

References

1. Akyildiz IF, Jornet JM (2010) The internet of nano-things. IEEE Wireless Commun 17(6): 58–63
2. Najah AA, Wesam A, Mervat Abu E (2016) Internet of nano-things network models and medical applications. IEEE Wireless Commun 211–215. https://doi.org/10.1109/iwcmc.2016.7577059
3. Prachi R, Nisha S (2016) Study of environmental effects on the connectivity of molecular communication based internet of nano things. In: IEEE conference, pp 1123–1128. https://doi.org/10.1109/wispnet.2016.7566311
4. Akyildiz, IF, Pierobon M, Balasubramaniam S, koucheryavy Y (2015) The internet of bio nano things. In: IEEE conference Globecom workshop., vol 53, pp 32–40. https://doi.org/10.1109/mcom.2015.7060516
5. Jarmakiewicz J, Krzysztof P, Krysztof M (2016) On the internet of nano things in healthcare network. In: IEEE conference, pp 1–6. https://doi.org/10.1109/icmcs.2016.746572
6. Uche AK, Chude O, Reza M, Maharaj BT, Cholette CC (2016) Bio-inspired approach for eliminating redundant nanodevices in internet of bio-nano things. In: 2015 IEEE Globecom workshops (GC workshops), pp 1–6. https://doi.org/10.1109/glocomw.2015.7414163
7. Sasitharan B, Jussi K, Realizing the internet of nano things: challenges, solutions, and application. IEEE J Mag 46:62–68. https://doi.org/10.1109/mc.2012.389
8. Najah AA, Mervat A-E (2015) Internet of nano-things healthcare application: requirements, opportunities, challenges. In: IEEE conference, pp 9–14. https://doi.org/10.1109/wimob.2015.7547934
9. Mahdi HM, Maaruf A, Peter SE, Rich P (2015) A review on internet of things (IoT),internet of everything (IoE), and internet of nano things (IoNT). In: IEEE conference publications, pp 219–224. https://doi.org/10.1109/itecha.2015.7317393
10. Qammer HA, Hassan ES, Nistha C, Ke Y, Khalid AQ, Akram A (2016) Terahertz channel characterization inside the human skin for nano-scale body-centric networks. IEEE J Mag 6:427–434. https://doi.org/10.1109/2016.2542213

Minimization of the Size of an Antipodal Vivaldi Antenna for Wi-MAX and WLAN Applications

Sneha Tiwari, Trisha Ghosh and Janardhan Sahay

Abstract In this paper, miniaturization of the antipodal Vivaldi antenna is discussed. The antipodal Vivaldi antenna is a broadband antenna and thus it is suitable for use in many wireless applications. The proposed antenna has a center frequency at 3.6, 5.2, and 5.8 GHz and so it can be used in WiMAX as well as WLAN systems and to avoid potential interference from narrowband communication systems, it is advised to design a miniaturized broadband antenna with intrinsic band-notched characteristics which can be used in narrowband communication. The circular slots are applied on the edge at the width of the Vivaldi antenna and are etched properly and as a result this helps in the area minimization of the antenna. The circular slot with the greatest diameter is used to achieve the different center frequencies so that the proposed antenna can be used for wireless communication, i.e. for WiMAX and WLAN. The rectangles are used on the alternative circular slots so as it can be used as notched structure and the interference in between the two different wireless applications can be minimized. The bandwidth of the proposed antenna is 2.8–6.2 GHz. A comparison of a conventional Vivaldi antenna with an antipodal antenna is also undertaken. The simulation of the proposed antenna was performed using HFSS13.0 software. The return loss, gain, radiation pattern, z-parameters, and VSWR are shown.

Keywords High frequency simulation software (HFSS) · Ultra-wide band (UWB) · Voltage standing wave ratio (VSWR) · Wireless local area network (WLAN)

S. Tiwari (✉) · T. Ghosh · J. Sahay
BIT Mesra, Ranchi, India
e-mail: snehasandilya14@gmail.com

A. Kumar and S. Mozar (eds.), *ICCCE 2018*,
Lecture Notes in Electrical Engineering 500,
https://doi.org/10.1007/978-981-13-0212-1_23

1 Introduction

The features and design of the Vivaldi antenna was first discussed by Gibson in 1979 [1]. The Vivaldi antenna is a class of tapered UWB slot antenna. It is a special kind of aperiodic travelling wave antenna, with the slot used in the design being aperiodic in nature, and the em wave travelling into the slot before leaving the antenna structure. The design compromises an exponential slot whose guiding equation of the curve is given by $y = \pm c_1 e^{Rx} + c_2$, where R is the exponential factor [2]. This power factor dominates the beam width of the antenna and c_1 and c_2 are constants. The exponentially tapered design of the Vivaldi antenna produces a significant improvement in the antenna's performance parameters such as gain, efficiency, directivity, and bandwidth. The basic problem with the Vivaldi antenna is its large size [3–5]. Thus, different solutions have been proposed which can decreases the size of the Vivaldi antenna but the gain and other antenna parameters cannot be enhanced with these solutions. So, in this paper a proposal for the reduction of the area of the Vivaldi antenna with enhanced bandwidth is shown, which will be beneficial to the overall use of the antenna in wide-band situations. By making use of two types of slots in the design of the antenna a reduction in size as well as enhanced bandwidth can be easily achieved. In the proposed antenna the exponential slots are made in such a way that one part of the slot is designed on the ground of the antenna and the other part is made on the patch of the antenna thus giving it an antipodal structure. In the structure the circular slots are etched for two different purposes. The circular slot at the width edge of the structure is used for the minimization of the antenna. The slot at the length edge of the antenna is to obtain the center frequencies, i.e. for the application of the WiMAX and WLAN used in wireless communication. The diameter of this circular slot is higher than that of the other circular slot used at the width edge of the Vivaldi antenna [6]. The rectangles are used on the alternative circular slots so that it can be used as notched structure and the interference in between the two different wireless applications can be minimized.

2 Antenna Design

In Fig. 1 the geometry of the conventional Vivaldi antenna is shown. Here the length and width of the antenna are kept as 110 mm and 80 mm respectively. The substrate used is FR-4 with a relative permittivity of 4.4 and the height of the substrate is kept as 1.6 mm. The two patches of the Vivaldi antenna, i.e. above and below the substrate, are metalized using copper as a metal. The exponential length of the antenna is kept as 90 mm and the rectangular length is kept as 20 mm. The back-offset is kept as 25 mm. When this antenna is designed and simulated the operating frequency is between 5.8 and 6.8 GHz. Here, the size of the antenna is slightly large and thus in Fig. 2 the reduced size of the antenna is an area 50 *

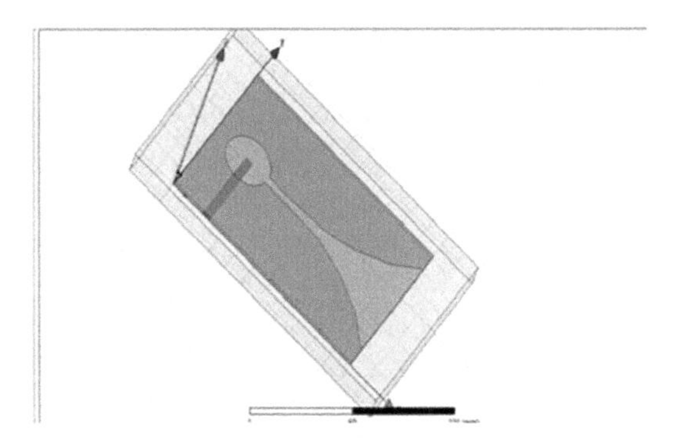

Fig. 1 Conventional Vivaldi antenna

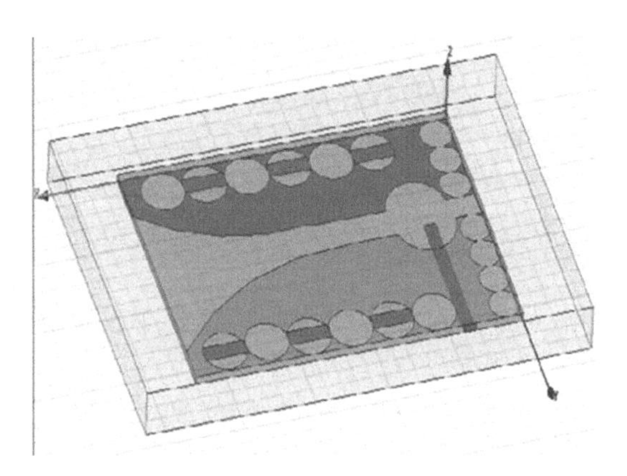

Fig. 2 Vivaldi antenna with reduced antipodal size

25 mm^2. In this modified antenna design the technique used for the reduction of the antenna is the use of circular slots on both the patch and the ground. The spacing of these slots are done by properly and the gap between the slots is the same in the ground as well as in the ground plane [7–9]. Etching of the slots on the patch and the ground makes the Vivaldi antenna a compact antenna. Even the mutual coupling between the different slots of the antenna could be reduced by using circles etched onto the exponential part of the antenna. The substrate used in the modified designing of the antenna is also FR-4 and the height of the substrate is also kept as 1.6 mm. The bandwidth of this antipodal structure with reduced area is 2.8–6.2 GHz.

2.1 Conventional Vivaldi Antenna Design and Antipodal Vivaldi Antenna Design

See Fig. 3 and Table 1.

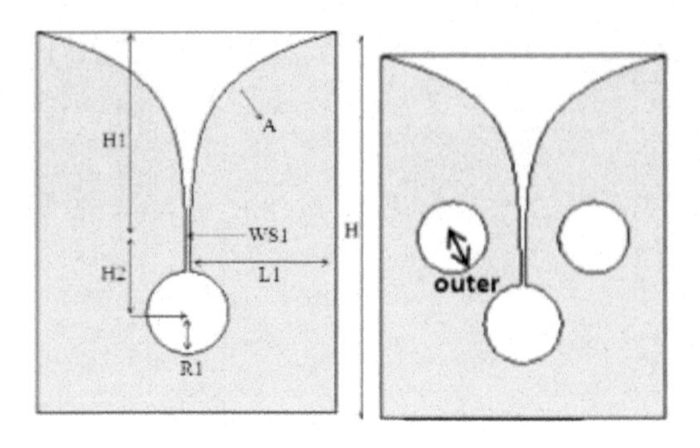

Fig. 3 Parametric structure of a Vivaldi antenna

Table 1 Dimensions used in conventional and antipodal antennae

Parameter	Conventional Vivaldi dimension	Modified Vivaldi dimension	Notation
Area	100 * 70 mm^2	50 * 25 mm^2	L * B
Dielectric	FR-4	FR-4	ε
Exponential height	70 mm	20 mm	H_1
Back off-set	20 mm	6 mm	$H-(H_1+H_2+R_1)$
Rectangular length	18 mm	4 mm	H_2
Feeding technique	Microsftrip feedline	Microstrip feedline	–
Radius of circle	10 mm	5 mm	R_1
Circular slot radius	–	5 mm	Outer 1
Ellipse slot radius	–	7 mm	Outer
Exponential rate	O.0445 * X (0.25 < X < 1)	0.0445 * X (0.5 < X < 1)	R
Exponential length	30 mm	10 mm	L_1
Gap between exponential slot	4 mm	1.5 mm	A

2.2 *Simulation and Results*

The simulation of the antipodal Vivaldi antenna with reduced size is designed. The software used for the simulation is HFSS13.0 (i). The cutoff or the center frequency of the antenna is taken to be 3.6–5.2 GHz.

Return loss

In Figs. 4 and 5, the return loss of the two different structures is shown. The bandwidth of the conventional Vivaldi antenna is 5.8–6.8 GHz and that of the antipodal Vivaldi antenna is 2.4–6.2 GHz.

VSWR

Figure 6 shows the VSWR (iii) of the antipodal Vivaldi antenna and it is seen that in its frequency range the value of VSWR is below 2 and at the cut-off frequency

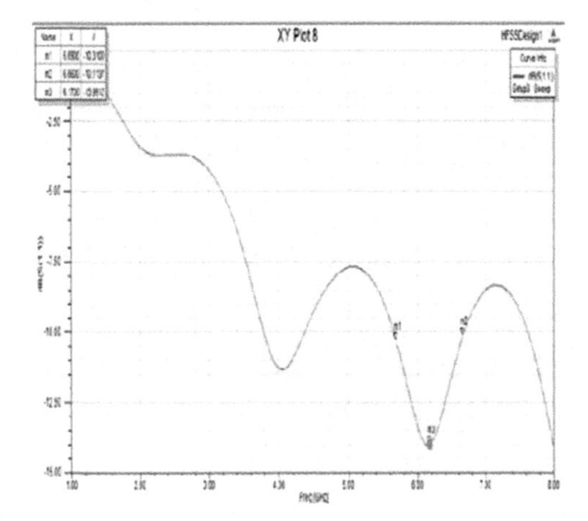

Fig. 4 Return loss of a conventional antenna

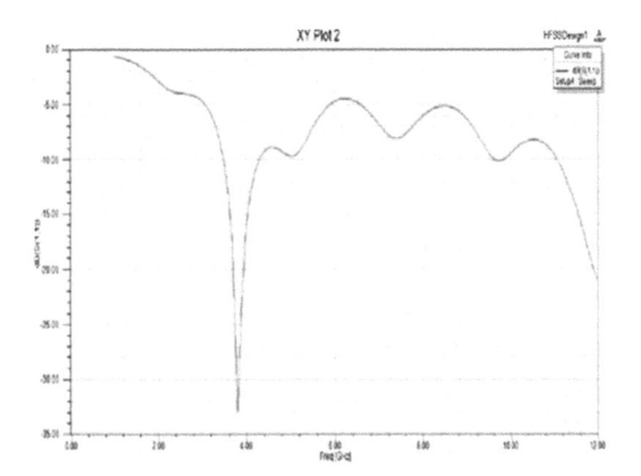

Fig. 5 Return loss of the antipodal Vivaldi antenna

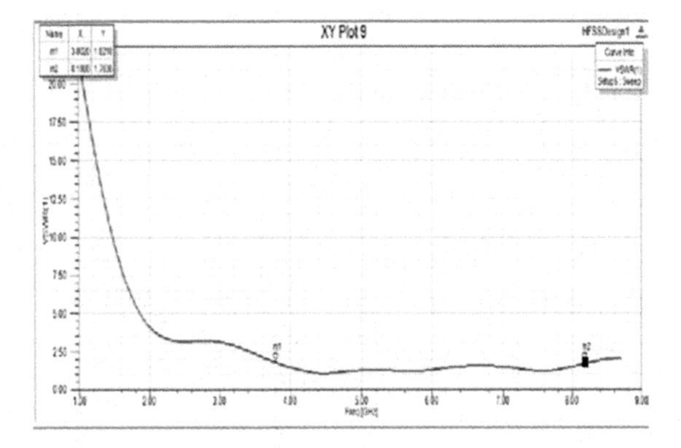

Fig. 6 VSWR of the antipodal Vivaldi antenna

the value of VSWR. At a frequency of 5.2 GHz the VSWR is nearly zero and as a result it can be used in WLAN communication.

Z-Parameter

See Fig. 7.

The impedance matching of the Vivaldi antenna is basically used to match the impedance between the feedline and the antenna [10, 11]. It can seen that perfect impedance matching occurs at 5.2 GHz with a Z-Parameter of 50 Ω.

Radiation Pattern

The radiation pattern of the antipodal Vivaldi antenna is directive in nature, and this can be seen in Fig. 8.

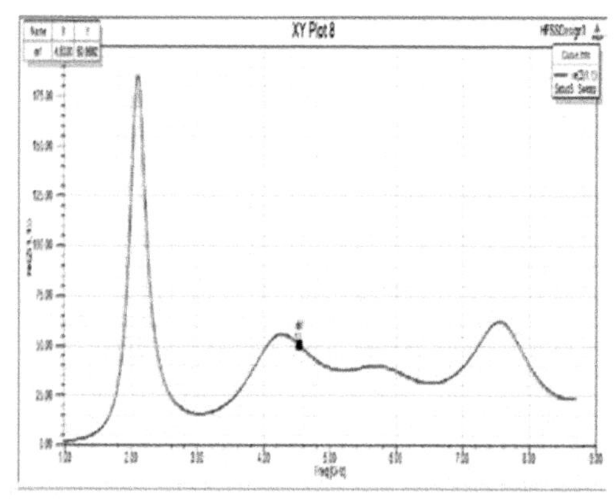

Fig. 7 Z-Parameter of the antipodal antenna

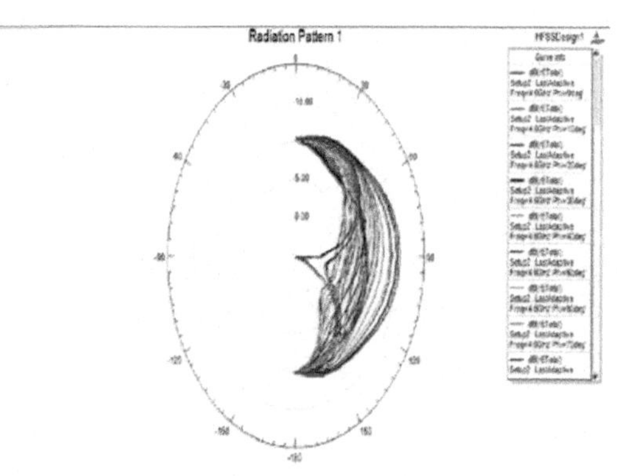

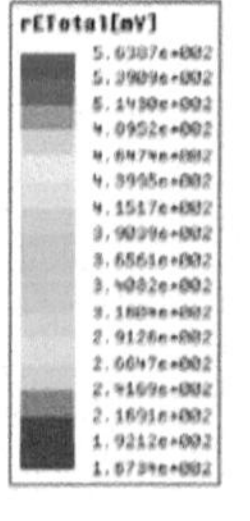

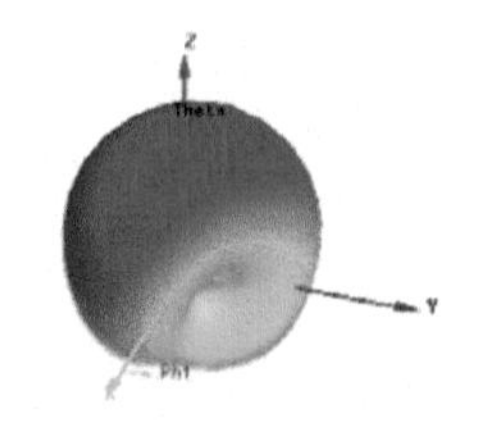

Fig. 8 Radiation pattern in 2-D and 3-D

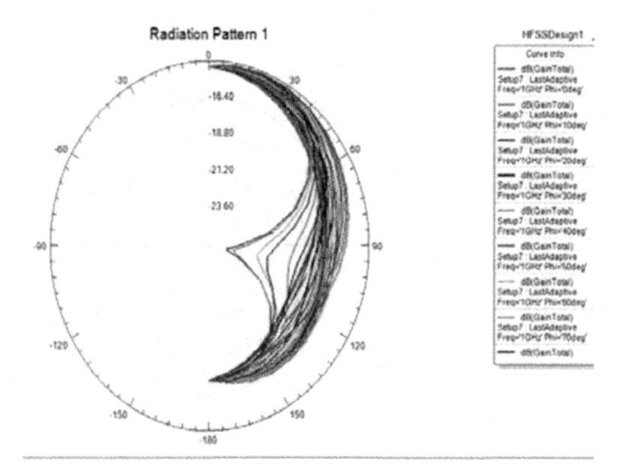

Fig. 9 Gain of the antipodal Vivaldi antenna with reduced size

From the above figure can clearly be seen that the gain of the Vivaldi antenna is −23 db, −21 db, −18.80 db and −16.40 db for the 3.6 GHz, 5.2 GHz, 5.8 GHz and 5.825 GHz cut-off frequencies respectively (Fig. 9).

2.3 Calculations

Different equations are used for the calculation of the different parameters of the antenna (Fig. 10).

Antenna Length:

$$F_{\text{max}} = 6.2\,\text{GHz},\ F_{\text{min}} = 3.4\,\text{GHz}$$

$$\lambda_{\text{min}} = 50\,\text{mm},\ \lambda_{\text{max}} = 150\,\text{mm}$$

There is the limiting condition on the length of the antenna, i.e. the antenna length should be greater than the average value of the maximum and minimum

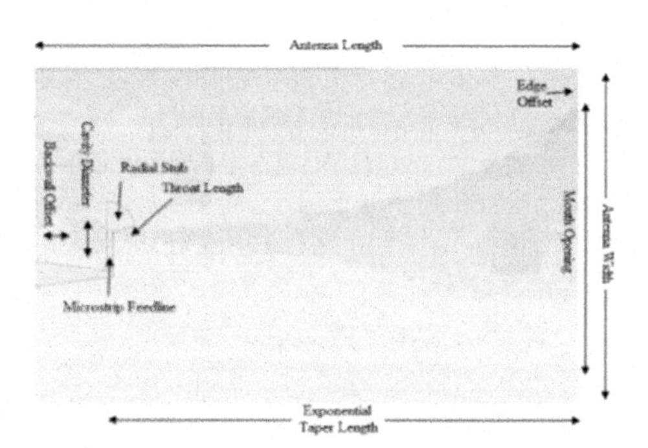

Fig. 10 Vivaldi antenna parameters

operating frequency [4, 12, 13]. The bandwidth of the antenna can be improved by using a longer length.

Antenna Width:

According to [14, 15] the antenna width should be greater than half of the average value of the maximum and minimum operating frequencies $W > \frac{\lambda_{min} + \lambda_{max}}{4}$.

Opening:

This is the outermost part of the antenna from which the em waves leaves the antenna structure [14]. The mouth opening should have a value inbetween $W_{min} = \frac{\lambda_g}{f}$ and $W_{max} = \frac{\lambda_g}{2}$.

Edge Offset:

This is the extra conducting area at the end of the exponential slot.

$$\text{Edge offset} = \frac{totallength - mouthopening}{2}$$

Exponential Slot:

The exponential slot is guided by the equation: $Y = C_1 e^{Rx} + C_2$ where $C_1 = \frac{y_2 - y_1}{e^{Rx_2} - e^{Rx_1}}$ and $C_2 = \frac{e^{Rx_2} y_1 - e^{Rx_1} y_2}{e^{Rx_2} - e^{Rx_1}}$ where R is the exponential rate and x_1,x_2,y_1 and y_2 indicate starting and ending points of the slot line [16].

Modified antipodal antenna design:
Radius of the etched circle = 5 mm
Radius of the etched ellipse = 7 mm
No. of circles used = 7
No. of ellipses used = 14.

2.4 Applications

1. This antipodal Vivaldi antenna is used in Bluetooth devices when a cut-off frequency of 2.4 GHz is used.
2. It can be used in WiMAX applications when a frequency band of 3.3–3.7 GHz is chosen or notched.
3. It can be used in WLAN devices when operating at 5.15–5.35 GHz or 5.725–5.825 GHz.

2.5 Conclusion

From the paper it can be concluded that the minimization of the Vivaldi antenna can be achieved using the etching of the circular slots at the edge of the antenna. Through the design of the antipodal Vivaldi antenna the bandwidth characteristics of the antenna can be enhanced. An etching ellipse with a diameter of 7 mm is used at the edge of the antenna on the long side and this helps in the achievement of the three different frequencies used in wireless communication, i.e. for WiMAX and for WLAN. The rectangular slots are kept on the alternative ellipse and thus act as a notched system preventing the interference between the different frequency bands used. Thus, this antipodal Vivaldi antenna can be used in a wide range of wireless communication applications.

References

1. DW (2014) Ultra-wideband slot-loaded antipodal Vivaldi antenna array. In: IEEE international symposium antenna and propagation. Li Y, Li W, Yu, W et al (2014) A small multi-function circular slot antenna for reconfigurable UWB communication applications. IEEE antennas and propagation society international symposium (APSURSI), pp 834–835
2. Ojaroudi N, Ojaroudi M (2013) Novel design of dual band-notched monopole antenna with bandwidth enhancement for UWB applications. IEEE Antennas Wirel Propag Lett 12:698–701
3. Bai J, Shi S (2011) Prathern (APSURSI). Washington, USA, July, Spokane, pp 79–81
4. Xianming Q, Zhi NC, Michael YWC (2008) Parametric study of ultra-wideband dual elliptically tapered antipodal slot antenna. Int J Antennas Propag 2008:1–9
5. Kollberg EL, Johansson J, Thungren T, Korzeniowski TL, Yngvesson KS (1983) New results on tapered slot endfire antennas on dielectric substrates. Presented at 8th international conference infrared millimeter waves
6. Yao Y, Cheng X, Yu J, Chen X (2016) Analysis and design of a novel circularly polarized antipodal linearly tapered slot antenna. IEEE Trans Antennas Propag 64(10):4178–4187
7. Gibson PJ The Vivaldi aerial. In: Proceedings of 9th European Microwave Conference, Sept 1979, vol 1, pp 101–105
8. John M et al (2008) UWB Vivaldi antenna based on a spline geometry with frequency band-notch. In: Proceedings of IEEE antennas propagation society international symposium, July 2008, pp 1–4
9. Mirshekar-Syahkal D, Wang HY (1998) Single and coupled modified V-shaped tapered slot antennas. In Proceedings of IEEE antennas propagations society international symposium, June 21–26, 1998, vol 4, pp 2324–2327
10. Ping L, Jia-ao Y (2011) Research on printed Vivaldi antennas. J Xidian Univ, 38(2):194–196. Feng Z, Yi-cail J, Guang-you F (2010) A compact Vivaldi antenna for 0.5-2 GHz. J Microw 26(6):54–57
11. Adamiuk G, Zwick T, Wiesbeck W (2008) Dual-orthogonal polarized Vivaldi antenna for ultra-wideband applications. In: Proceedings of 17th international conference microwaves, radar and wireless communications (MIKON'08), pp 1–4. Accessed 19–21 May 2008
12. Teni G, Zhang N, Qiu J et al (2013) Research on a novel miniaturized antipodal Vivaldi antenna with improved radiation. IEEE Antennas Wirel Propag Lett 12:417–420
13. Bai Jian, Shi Shouyuan, Prather Dennis W (2011) Modified compact antipodal Vivaldi antenna for 4–50 GHz UWB application. IEEE Trans Microw Theor Tech 59(4):1051–1057

14. Fei P, Jiao Y-C, Hu W, Zhang FS (2011) A miniaturized antipodal vivaldi antenna with improved radiation characteristics. IEEE Antennas Wirel Propag Lett 10:127–130
15. Wang Z, Yin Y, Wu J, Lian R (2016) A miniaturized CPW-fed antipodal Vivaldi antenna with enhanced radiation performance for wideband applications. IEEE Antennas Wirel Propag Lett 15:16–19
16. Zhou B, Cui TJ (2011) Directivity enhancement to Vivaldi antennas using compactly anisotropic zero-index metamaterials. IEEE Antennas Wirel Propag Lett 10: 3

Physical Layer Impairment (PLI) Aware Lightpath Selection in WDM/DWDM Networks

Vikram Kumar and Santos Kumar Das

Abstract The demand for high data rate with low bit error rate (BER) and large bandwidth is highly satisfied in wavelength division multiplexing/dense wavelength division multiplexing (WDM/DWDM) networks. However, the signal traveling inside the optical fiber can be affected by various physical layer impairments (PLIs). These impairments are caused due to fiber non-linearities and the non-ideal nature of optical components. Dispersion is one of the PLI constraint which affects signal quality. That needs to be compensated. This research work presents the approach of dispersion penalty (DP). It also suggests a PLI-aware lightpath selection algorithm based on DP.

1 Introduction

An optical network employing wavelength division multiplexing/dense wavelength division multiplexing (WDM/DWDM)forms the backbone of the next generation of communication systems. WDM/DWDM networks have an enormous bandwidth allowing them to satisfy emerging applications such as video on demand, medical imaging, and distributed central processing unit (CPU) interconnects. Optical network has evolved from traditional opaque networks toward all-optical network via translucent networks as a result of technology advancement [1]. In WDM/DWDM networks lightpath selection, after routing and assigning a wavelength to each connection, is termed as a routing and wavelength assignment (RWA) problem. RWA is a very complex problem which needs to be addressed before designing an optical networks. To maintain the transparency of optical networks, regenerators are removed. As a result distorted signals cannot be amplified during long-distance transmission.

V. Kumar · S. K. Das (✉)
Department of Electronics and Communication Engineering,
National Institute of Technology, Rourkela, Rourkela 769008, India
e-mail: dassk@nitrkl.ac.in

V. Kumar
e-mail: vikram.nit2015@gmail.com

A. Kumar and S. Mozar (eds.), *ICCCE 2018*,
Lecture Notes in Electrical Engineering 500,
https://doi.org/10.1007/978-981-13-0212-1_24

Since the transmitted signal remains in the optical domain, noise and distortion affects the quality of the received signal. This occurs due to the non-ideal nature of optical components and non-linearities of the optical fiber at the physical layer give rise to impairments like attenuation, dispersion etc. which affects the signal strength known as physical layer impairment [2]. The physical layer impairments (PLIs) is a major constraint for RWA decision to achieve longer distance with acceptable bit error rate (BER) [2]. PLI can be classified as linear and nonlinear impairments. Linear impairments (LIs) are static in nature, whereas non-linear impairment (NLIs) are dynamic in nature.

Many researchers are working on the RWA problem and suggested an algorithm to overcome the effect of various PLIs in order to set-up a lightpath. The authors in [3] have presented a comprehensive survey on various PLI-aware network design techniques, RWA algorithms, and PLI-aware failure recovery algorithms. This work also suggested that dispersion is the most serious issue for systems operating at a bit rate higher than 2.5 Gb/s. In [4], the authors have suggested an adaptive quality of transmission (QoT) aware routing technique incorporated with a new cost function based on the impairments. They have considered linear and non-linear impairments whose variance can be predicted. Zhao et al. [5] present a bidimensional quality of service (QoS) differentiation framework to improve network performance. In this framework, they have considered both PLIs and set-up delay as well as the impact of PLIs on QoT. In [6], a comprehensive survey on the impact of PLI on a transparent optical network is studied. It has presented a survey of various PLI-RWA algorithms discussed in the previous research in order to have a better understanding of optical networks. The authors in [7] have studied the static impairment-aware multicast RWA problem for transparent WDM networks. They have formulated this problem mathematically with the help of integer linear programming (ILP) considering various impairments present in physical layers, such as optical power, amplifier spontaneous emission (ASE) noise, crosstalk, and polarization effects. In [8], the authors have suggested a weighted mechanism for provisioning PLI-aware lightpath set-up in WDM networks. Dominant PLIs considered are self-phase modulation (SPM), cross-phase modulation (XPM), four-wave mixing (FWM), and total noise. They have proposed a novel weighted approach that (i) selects the optimum launch power, (ii) knows the current network state, and (iii) assigns weight to the wavelength based on PLIs.

In this chapter, the estimation and management of PLIs to provide efficient and qualitatively good lightpaths to end users is investigated. This chapter considers dispersion as one of the PLI constraints and suggests a dispersion penalty (DP) approach to compensate the signal distortion occurring inside the optical fiber. DP is defined as the increase in input signal power in order to achieve the same signal to noise ratio (SNR) as that of an ideal system. The routing algorithm will select those paths which have lower values for DP such that the impact of dispersion can be minimized. In other words, a lower value of DP guarantees less dispersion on that particular lightpath. The results indicate that only a few paths are suitable for wavelength assignment to ensure lower blocking probabilities.

2 System Model

The WDM/DWDM system model is shown in Fig. 1. It comprises of clients, connection requests, provider edge routers (PERs), core routers (CRs), a control manager (CM), and a central database (DB). This is a centralized system model, that has a data plane as well as the control plane. The data plane deals with data transmission, whereas the control plane deals with the management of network resources. The topology provides information such as (i) network connectivity, (ii) availability of wavelengths, and (iii) connection requests. The CM maintains a traffic matrix (TM) for all the clients. It records a database table for physical layer constraints such as routing information and DP matrices for all possible connections between any source-destination client pairs. CM performs a direct communication with optical CRs and updates its database.

The connection matrix $C(i,j)$ between any router pair i and j of the physical topology can be represented as:

$$T(i,j) = \begin{cases} 1 & \text{if there exist a link between}(i,j); \\ 0 & \text{otherwise} \end{cases}$$

Based on the system model, we have estimated and analyzed the DP.

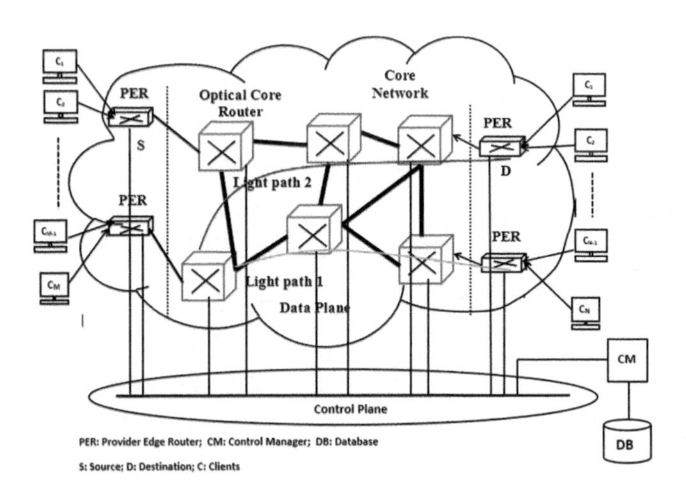

Fig. 1 System model

2.1 Estimation and Analysis of the Dispersion Penalty

Dispersion is defined as the broadening of light pulses as they travel along the fiber due to variation in the velocities of different spectral components. It can be categorized mainly as chromatic dispersion (CD) and polarization mode dispersion (PMD). CD is a phenomenon which degrades the signal quality caused by different spectral components traveling at their own velocities [9]. When a signal travels through different fiber links, the resultant dispersion is the summation of dispersion caused due to individual links. Mathematically CD can be expressed as [10]:

$$T_{cd}(i,j) = D_{cd}\sigma_{\lambda}L(i,j) \tag{1}$$

where, D_{cd} is the CD coefficient, σ_{λ} is the spectral width of the source (ranges from 40 to 190 nm for light emitting diode (LED)), and $L(i,j)$ is length of the link(i,j).

Fiber is made up of silica (SiO_2), which may contain some impurities due to the manufacturing process or environmental conditions like stress and temperature. These impurities act as an obstacle for the smooth movement of the signal inside the core of the fiber. As a result of this, different polarization of optical signals occurs leading to different group velocities. Hence, the pulse spreads in the frequency domain. This phenomenon is called PMD. The different group delay is proportional to the square root of link length $L(i,j)$, it can be represented as [11]:

$$T_{pmd}(i,j) = D_{pmd}\sqrt{L(i,j)} \tag{2}$$

where D_{pmd} is the PMD coefficient.

Total delay of pulses, caused due to dispersion, is calculated as a summation of the delay due to CD and PMD, which is written as follows [12]:

$$T_{total}(i,j) = \sqrt{(T_{cd}(i,j))^2 + (T_{pmd}(i,j))^2} \tag{3}$$

Now, the total delay for a source-destination (s,d) pair can be express as follows:

$$T_{total}(s,d) = \sum_{\forall(i,j)\in(s,d)} T_{total}(i,j) \tag{4}$$

The dispersion in the fiber is a major constraint for high-speed data transmission, as it increases the signal-to-noise ratio (SNR) or bit error rate (BER). As a result of this, intersymbol interference (ISI) occurs between the various channels inside the fiber that control the data rate. In order to maintain the same SNR and BER at high-speed transmission the system degradation should be compensated. For this, the signal power has to be increased to achieve the same SNR as that of the ideal system. This increase in power is known as DP. The DP for link (i,j) is expressed in terms of bit rate and total delay as [13]:

$$P_d(s,d) = -10\log_{10}[1 - 0.5(\pi \times B(s,d))^2\sigma_t^2(s,d)] \tag{5}$$

where $B(s,d)$ is the bit rate of a path and $\sigma_t(s,d)$ is total dispersion of a path. The maximum possible bit rate for a lightpath with source destination pair (s,d) can be computed as [13]:

$$B(s,d) = \frac{\varepsilon}{10^{-6} \times T_{total}(s,d)} \tag{6}$$

where the total dispersion of a fiber link $\sigma_t(i,j)$ can be expressed as [12]:

$$\sigma_t(i,j) = \sqrt{\sigma_c^2(i,j) + \sigma_n^2(i,j)} \tag{7}$$

where $\sigma_c(i,j)$ is the intramodal or chromatic broadening of pulses and $\sigma_n(i,j)$ is the intermodal broadening caused by delay differences between the various modes. The term $\sigma_c(i,j)$ consists of pulse broadening due to both material and waveguide dispersion and can be represented as follows:

$$\sigma_c(i,j) = \sigma_m(i,j) + \sigma_{wg}(i,j) \tag{8}$$

Since σ_{wg} is negligible compared to σ_m so:

$$\sigma_c(i,j) = \sigma_m(i,j) \tag{9}$$

and

$$\sigma_m(i,j) = \sigma_\lambda L(i,j)M \tag{10}$$

where σ_λ is the spectral width of an light emitting diode (LED) light source, $\approx$50 nm, and M is the material dispersion coefficient.

The intermodal dispersion σ_n is expressed as:

$$\sigma_n(i,j) = \sigma_s(i,j) \tag{11}$$

where $\sigma_s(i,j)$ is root mean square (rms) pulse broadening due to intermodal dispersion for a step index fiber. The expression for intermodal dispersion is expressed as:

$$\sigma_s(i,j) = \frac{L(i,j)(NA)^2}{4\sqrt{3}n_1c} \tag{12}$$

where $L(i,j)$ is the length of link, NA is the numerical aperture of the fiber, n_1 is the refractive index of the core, and c is the speed of light.

All the above-mentioned parameters are calculated for a single link, and for the calculation of an (s,d) pair we need to sum up all the links (i,j), represented as follows:

$$\sigma_s(s,d) = \sum_{\forall(i,j)\in(s,d)} \sigma_s(i,j) \tag{13}$$

$$\sigma_m(s,d) = \sum_{\forall(i,j)\in(s,d)} \sigma_m(i,j) \tag{14}$$

$$\sigma_t(s,d) = \sum_{\forall(i,j)\in(s,d)} \sigma_t(i,j) \tag{15}$$

Now, the dispersion penalty for a particular (s, d) pair is expressed as follows:

$$P_d(s,d) = -10\log_{10} \left[1 - 0.5\left(\frac{\pi\times\varepsilon}{10^{-6}}\right)^2 \times \left(\frac{1}{\sum_{\forall(i,j)\in(s,d)} \left(D_{cd}\sigma_\lambda L(i,j)\right)^2 + D^2_{pmd}L(i,j)}\right)^2 \times \left(\sum_{\forall(i,j)\in(s,d)} \sigma_\lambda^2 L^2(i,j)\, M^2 + \sum_{\forall(i,j)\in(s,d)} \frac{L^2(i,j)(NA)^4}{48n_1^2c^2}\right)\right] \tag{16}$$

3 Lightpath Set-up Algorithm

To compute an optimal path, an algorithm is given below which is used to get M possible lightpaths using the Floyd–Warshall approach [14]. This search algorithm is used to support multiple constraints. However, this computation uses two constraints, i.e., path length and threshold DP. In addition to the estimation of DP, comparison with the threshold value is also done in order to obtain lightpaths of better quality. Further, the availability of wavelength is checked and the lightpath is decided for setting up the connection. Lightpaths not satisfying the above criterion are blocked. Accordingly, the blocking probability is calculated for a set of source destination (s, d) pairs having load L and wavelengths λ, expressed as follows [15]:

$$P_{b(L,\lambda)} = \frac{\frac{L^\lambda}{\lambda!}}{\sum_{i=0}^{\lambda} \frac{L^i}{i!}} \tag{17}$$

where $P_{b(L,\lambda)}$ is the blocking probability and i is the ith link for an (s, d) pair.

Algorithm 1 DP Based path computation

1: connection request R arrives for a (s, d) pair
2: calculate K shortest path of R by the shortest path algorithm number the K shortest path as 1,2,....K in accordance with a ascending order of the length
3: calculate the threshold path length
4: **for** (i^{th} shortest path (i$\leq$ K))
5: **if** (path length $\leq$ threshold path)
6: **if** (DP $\leq$ threshold)
7: **if** (wavelength available)
8: optimal lightpath connection
9: established
10: **else**
11: Path is blocked
12: **end if**
13: **end if**
14: **end if**
15: Compute the blocking probability
16: **end for**

4 Numerical Results and Discussion

A National Science Foundation Network (NSFNET) is a North-American topology standard shown in Fig. 2 is used for numerical analysis, it consist of 10 nodes and 16 links. This is a new approach, not used previously in the literature. Therefore, we can not compare our result with any previous methods. All the numerical results were carried out using MATLAB, the system parameters considered are shown in Table 1 [10]. This chapter presents a lightpath selection mechanism based on quality parameters such as bit rate and DP for finding the best suitable connection. DP should

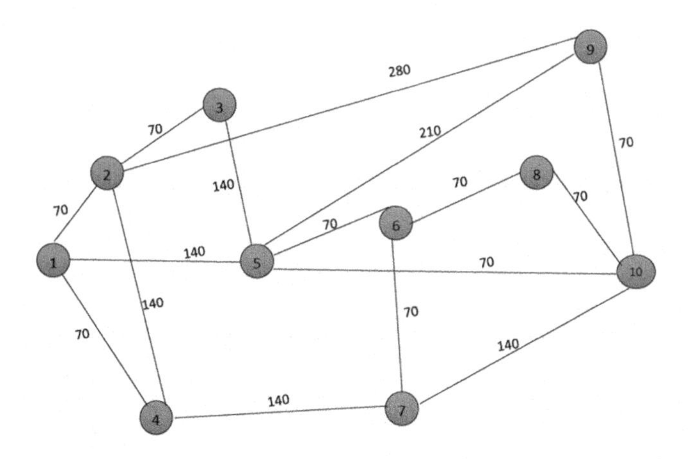

Fig. 2 An NSFNET topology

Table 1 Simulation parameters [11]

Parameters	Values
Pulse-broadening ratio, ε	0.182
Pulse-broadening factor, σ	0.1
PMD coefficient, D_{pmd}	$0.5\ \text{ps}/\sqrt{\text{km}}$
CD coefficient, D_{cd}	18 ps/nm km
Material dispersion coefficient, M	250 ps/nm km
Numerical aperture, NA	0.30
Core refractive index, n_1	1.50
Speed of light, c	2.9999×10^8 m/s
Source spectral width, σ_λ	50 nm

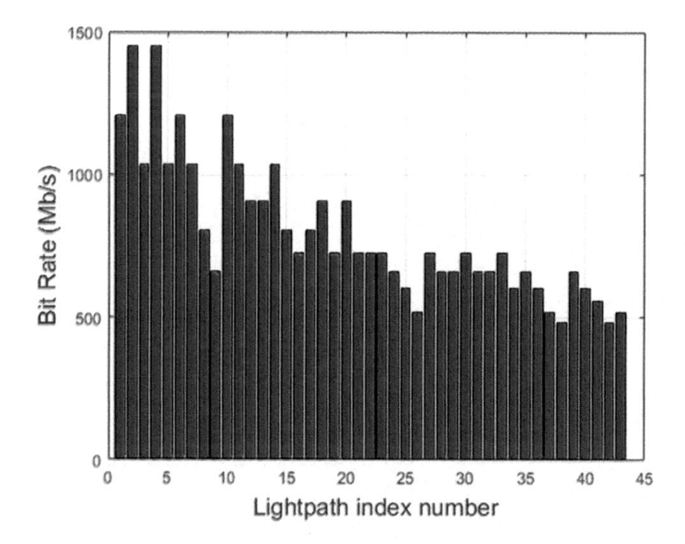

Fig. 3 Bit rate of all possible paths for source-destination pair (2, 8)

be as low as possible for better network performance. The Floyd–Warshall algorithm has been employed to calculate all possible paths between a particular set of (s, d) pairs having several links. For example, it considered node 2 as a source and node 8 as a destination, i.e., a (2, 8) pair.

There are total 43 lightpaths for this (s, d) pair. All these possible lightpaths are labeled as "lightpath index number" for the ease of representation for DP variation over individual lightpaths. Figure 3 shows the variation of bit rate for different set of paths. Bit rate is higher for the starting set of lightpaths compared to the later set of lightpaths. This depicts that the effect of impairments is dominant on longer paths. Now, using the literature mentioned in Sect. 3 DP has been calculated for all paths using (5). The plot for DP versus all possible paths for a (2, 8) pair is shown

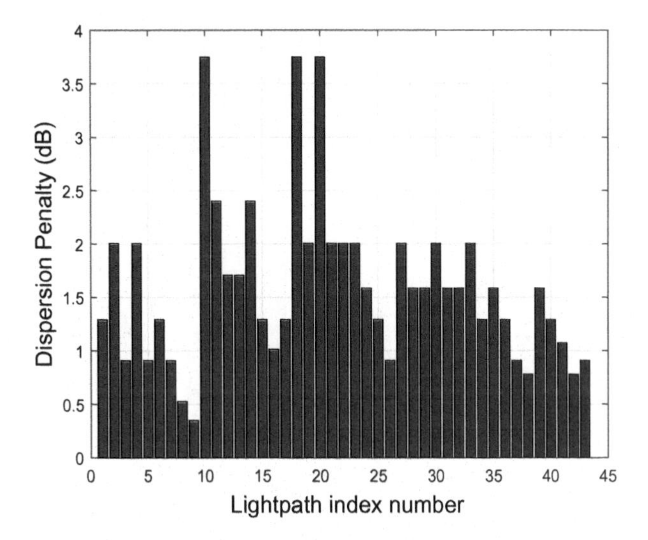

Fig. 4 DP of all possible paths for source-destination pair (2, 8)

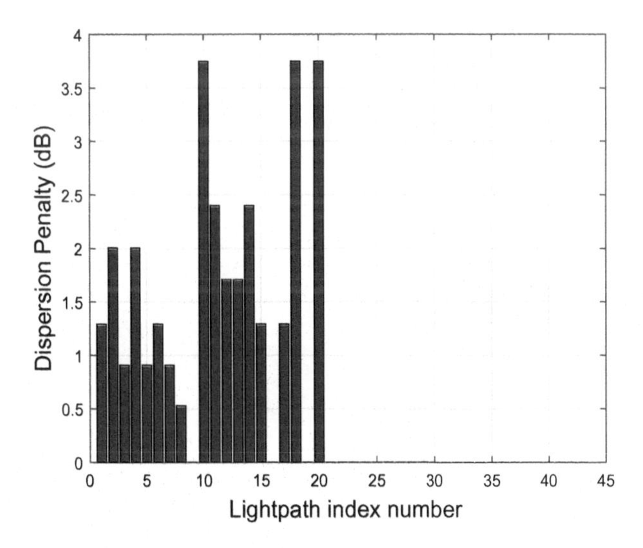

Fig. 5 DP $\leq DP(d_{th})$ for source-destination pair (2, 8)

in Fig. 4. In order to estimate the lightpath quality the threshold limit of DP is taken as 2 dB [13], so out of 43 paths only 29 paths fall under this condition, as shown in Fig. 6. Now the second condition is applied, i.e., the threshold path span (average path length). It is observed that only 17 paths satisfy the criterion, as shown in Fig. 5. Now this work considered both the constraints, i.e., DP and threshold path span, with only those paths selected which satisfy both. The optimal shortest path is obtained between the (2, 8) pair, which is suitable for wavelength assignment, as shown in

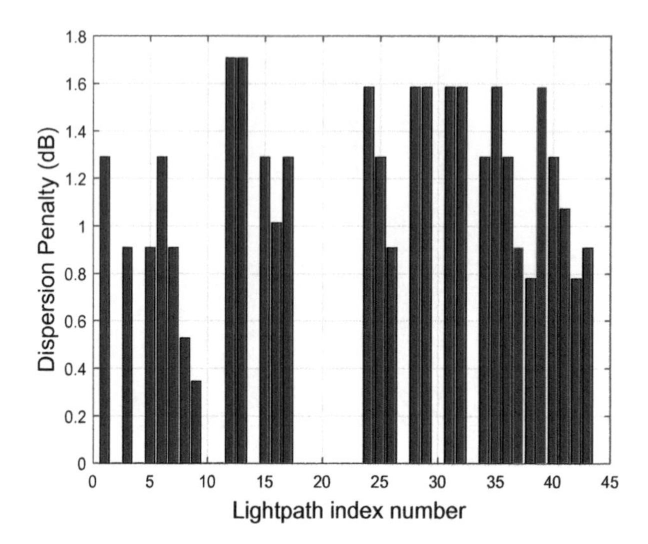

Fig. 6 DP ≤ 2 dB for source-destination pair (2, 8)

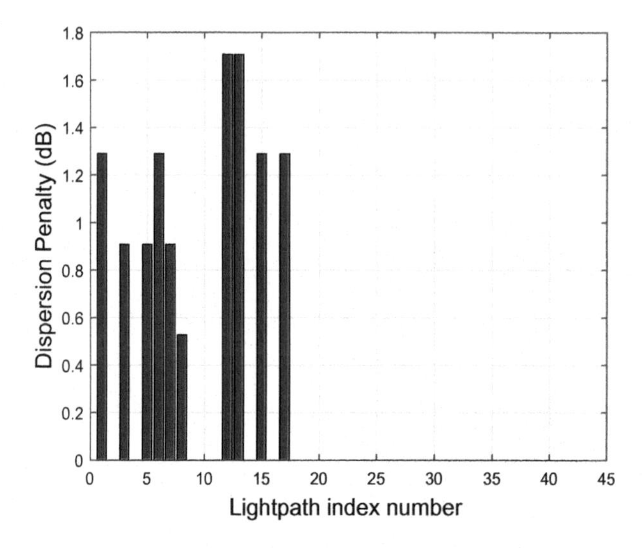

Fig. 7 Dispersion penalty plot of optimal paths for source-destination pair (2, 8)

Fig. 7. Therefore, out of 43 lightpaths only 10 lightpaths have small enough DP, desirable for larger bandwidth and high data rate transmission under the influence of PLIs. As per the proposed algorithm, the optimal lightpath connection will be 2-9-5-10-8 for the source-destination pair (2, 8).

The performance of the algorithm is analyzed by computing-blocking probability for the incoming connection requests. The blocking probability of the network, which is illustrated in (17), depends on the number of wavelengths, traffic load, and the

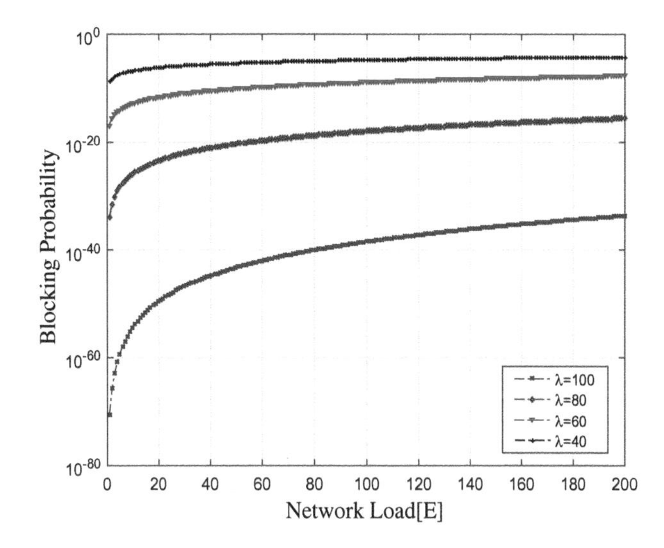

Fig. 8 Network-blocking probability for different number of wavelengths

number of nodes. In this chapter number of nodes = 10 and number of channels are varied. Figure 8 depicts the reduction in blocking probability with an increase in the number of wavelengths allocated per node. Traffic load is varied and a set of wavelengths, i.e., 40, 60, 80, 100 have been considered.

5 Conclusion

This chapter investigates the dispersion effect on the signal quality in transparent WDM/DWDM networks. It also presents an impairment aware lightpath quality estimation algorithm based on the dispersion. This algorithm focuses on optimal lightpath selection with a lower value of DP. It addresses the routing and wavelength assignment problem from the PLI point of view. Another advantage is that this approach can be effectively used to estimate the blocking probability, where the number of wavelengths is different on each link.

References

1. Azodolmolky S, Klinkowski M, Marin E, Careglio D, Pareta JS, Tomkos I (2009) A survey on physical layer impairments aware routing and wavelength assignment algorithms in optical networks. Comput Netw 53(7):926–944
2. Huang YG, Heritage JP, Mukherjee B (2005) Connection provisioning with transmission impairment consideration in optical WDM networks with high-speed channels. J light Technol 23(3):982–983

3. Saradhi CV, Subramaniam S (2009) Physical layer impairment aware routing (PLIAR) in WDM optical networks: issues and challenges. IEEE Commun Surv Tutor 11(4):109–130
4. He J, Brandt-Pearce M, Pointurier Y, Subramaniam S, "QoT-aware routing in impairment-constrained optical networks," in Global Telecommunications Conference, (2007) GLOBECOM'07. IEEE. IEEE 2007:2269–2274
5. Zhao J, Li W, Liu X, Zhao W, Maier M (2013) Physical layer impairment (PLI)-aware RWA algorithm based on a bidimensional QoS framework. IEEE Commun Lett 17(6):1280–1283
6. H. Dizdarevic, S. Dizdarevic, M. Skrbic, and N. Hadvziahmetovic, "A survey on physical layer impairments aware routing and wavelength assignment algorithms in transparent wavelength routed optical networks," *IEEE Information and Communication Technology, Electronics and Microelectronics (MIPRO)*, pp. 530–536, 2016
7. Panayiotou T, Manousakis K, Ellinas G (2016) Static impairment-aware multicast session provisioning in metro optical networks. IEEE Electrotechnical Conference (MELECON) 4:1–6
8. S. Iyer and S. P. Singh, "A novel launch power determination strategy for physical layer impairment-aware (PLI-A) lightpath provisioning in mixed-line-rate (MLR) optical networks," *IEEE Conference on wireless communication, signal processing and networking (WiSPNET)*, pp 617–622, 2016
9. R. Ramaswami, K. Sivarajan, and G. Sasaki, *Optical networks: A Practical Perspective*. Morgan Kaufmann, 2009
10. J. M. Senior and M. Y. Jamro, *Optical fiber communications: principles and practice*. Pearson Education, 2009
11. B. Mukherjee, *Optical WDM networks*. Springer Science Business Media, 2006
12. G. Keiser, *Optical fiber communications*. Wiley Online Library, 2003
13. G. Farrell, "Introduction to system planning and power budgeting," *Dublin Institute of Technology, December*, 2005
14. Khan P, Konar G, Chakraborty N, "Modification of Floyd-Warshall's algorithm for shortest path routing in wireless sensor networks", in India Conference (INDICON), (2014) Annual IEEE. IEEE 2014:1–6
15. Wason A, Kaler R (2007) Wavelength assignment problem in optical WDM networks. IJCSNS 7(4):27

Miniaturized MIMO Wideband Antenna with L-Shaped DGS for Wireless Communication

Trisha Ghosh, Sneha Tiwari and Janardhan Sahay

Abstract In this paper, a MIMO antenna is designed consisting of two planar symmetrical monopole antennas and the ground plane is slotted into L-shaped. The simulation results indicate that the antenna works well in the ultra-wide band and hence can be used in a wide range of applications. The return loss is below −62 dB at 7.1 GHz and below −32 dB at 3 GHz approximately, which is highly desirable. This antenna has a wide frequency range of 2.4–10 GHz. The overall antenna size is as low as 35 mm × 22 mm but the effect of the mutual coupling among the antenna elements is reduced and is below −10 dB over a wide range of frequencies, i.e. 2.4–8 GHz. Reducing the effect of mutual coupling is a challenge in MIMO antennas, and has been achieved in this case. The maximum gain achieved is approximately 2.2 dB. The design has been simulated using Ansoft HFSS software. The attributes of S-parameters, VSWR, gain, radiation pattern, and Smith chart are shown and its applications are discussed.

Keywords MIMO · Wideband · Multiband · Reflection coefficient WLAN · DGS · HFSS

1 Introduction

The MIMO technique enhances channel capacity and signal transmission. The incorporation of a number of antenna elements in both the transmitter as well as the receiver results in greater channel capacity. The presence of a number of paths in between the transmitter and the receiver ensures multipath propagation. The drawback of this multipath propagation is that it produces signal fading which is controlled by spatial diversity. Spatial multiplexing can be applied in the system. MIMO is the main technique used in advanced wireless communication systems, such as 4G and 5G, etc. The range and robustness of the whole system is also

T. Ghosh (✉) · S. Tiwari · J. Sahay
Birla Institute of Technology, Mesra, Ranchi, India
e-mail: putu.trisha@gmail.com

A. Kumar and S. Mozar (eds.), *ICCCE 2018*,
Lecture Notes in Electrical Engineering 500,
https://doi.org/10.1007/978-981-13-0212-1_25

augmented but at the same time this increases its complexity. In a MIMO antenna, it can be seen that in a single beam array, capacity increases even in the presence of high interference and high correlation between multipath signals, whereas in multi-beam arrays there is a decrease in capacity compared to normal antenna arrays. One of the biggest hurdles in MIMO antenna technology is mutual coupling among the antenna elements in the case of small sizes of antenna. In this case, the mutual coupling is reduced by taking an L-shaped lot in the ground plane. MIMO antennas with a miniature size have incredible future scope for use in variable portable devices as per users [1–4].

2 Experiments: Antenna Design

We have designed a MIMO antenna with two symmetrical planar monopole antennas that are separated by a distance of 9 mm, placed within the compact area of 35 × 22 mm^2. We have used Rogers as a substrate with a dielectric constant, $\in_r$ of 3.5, a loss tangent δ of 0.002 and a thickness of 1.6 mm.

In Fig. 1, two symmetrical monopole antennas designed with a square-shaped radiator of 8 mm are shown. The technique used for feeding both ports is microstrip feed. Impedance matching is effortless in the case of microstrip feed compared to other techniques. It is reliable and easy to construct. The ground plane has an L-shaped slot. Using the dimensions given in Table 1, the antenna is designed in Ansoft HFSS software [5–8] (Fig. 2).

In HFSS software, when we apply the radiation far field to the airbox in terms of software that enfold the antenna, the direction of propagation of radiation is observed as in Fig. 3.

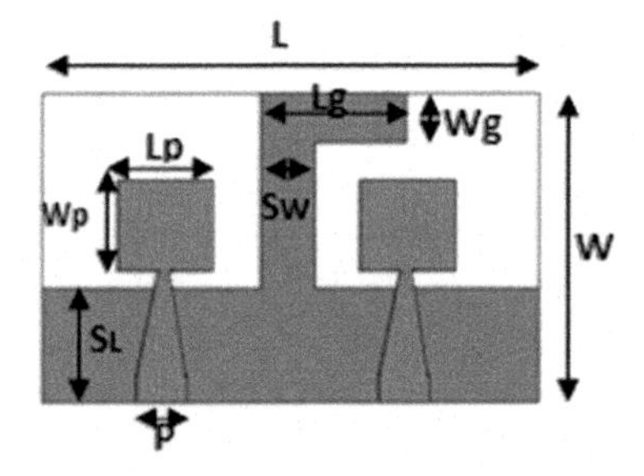

Fig. 1 Geometry of the MIMO antenna

Table 1 Dimensions of the MIMO antenna (in mm)

L	W	Lp	Wp	P	SL	SW	Lg	Wg
35	22	8	8	3.5	6	3.5	9	3

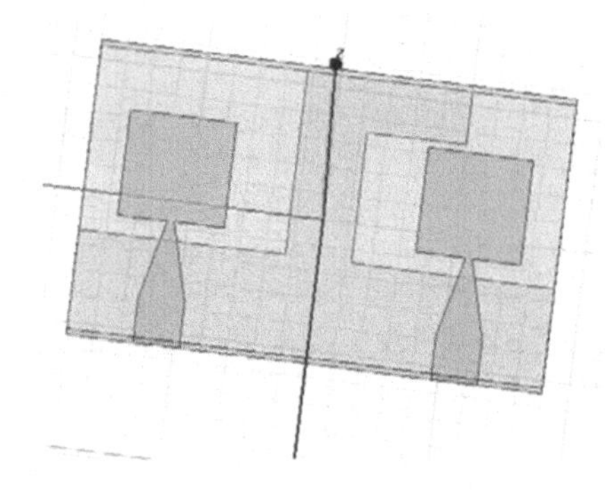

Fig. 2 Design of the MIMO antenna in a software interface using an L-shaped slot in the ground plane

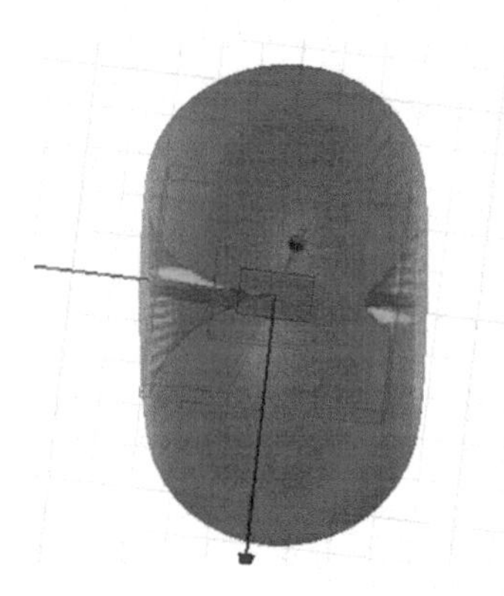

Fig. 3 Radiation far field applied in the software interface

2.1 Derivation and Explanation

All the parameters used in the geometry of the MIMO antenna are calculated using the following formulas:

Some waves travel in other substrates as well as in air, so an approach effective dielectric constant concept is introduced. The value of the effective dielectric constant, $\in_{reff}$ is given by:

$$\in_{reff} = \frac{\in_r + 1}{2} + \frac{\in_r - 1}{2}\left[1 + 12\frac{h}{W}\right]$$

where h: thickness of the antenna, W: width of the patch

Fringing effects increase the electrical length of the microstrip patch of the antenna. This makes the dimensions appear larger. Let the length of the patch be L and the length travelled by electric field be 2ΔL. Thus, the effective length is given by [9]

$$L_{eff} = \mathrm{L} + 2\Delta\mathrm{L}$$

The empirical values when pertained in the formula, imply that:

$$\frac{\Delta L}{h} = 0.412 \frac{\left(\in_{reff} + 0.3\right)\left(\frac{W}{h} + 0.264\right)}{\left(\in_{reff} - 0.258\right)\left(\frac{W}{h} + 0.8\right)}$$

For better radiation efficiency, the width of the radiator must be calculated by:

$$\mathrm{W} = \frac{c}{2f_r}\sqrt{\frac{2}{\in_r + 1}}$$

The actual length of the patch is given by:

$$\mathrm{L} = \frac{1}{2f_r\sqrt{\in_{reff}}\sqrt{\mu_0 \in_0}} - 2\Delta L$$

2.2 *Calculations*

The resonant frequency is taken as 4.4 GHz and accordingly the width is calculated first. Once the width (W) is calculated, we can easily get all other parameters by using the formulas mentioned above.

$$\mathrm{W} = \frac{3 \times 10^8}{2 \times 4.8}\sqrt{\frac{2}{3.5 + 1}} = 22.2\,\mathrm{mm} \approx 22\mathrm{mm}$$

In this design, the ground plane is cut into L-shaped so that the mutual coupling between the antenna elements can be controlled even when the size is miniaturized. The fabrication cost is estimated to be low because of the use of microstrip feed. All design parameters are calculated and implemented in the structure.

2.3 *Simulation and Results*

The reflection coefficient of the MIMO antenna is shown as:

In Fig. 4, the value of the reflection coefficient, S_{11} for this MIMO antenna can be seen as below −10 dB for a wide range of frequencies, i.e. 2.4–8 GHz for both ports but at 3 GHz and 7.7 GHz it made deep cuts close to −37.5 dB and −62.5 dB respectively. It can be observed that the port-1 output (colored red in the figure), gives better results compared to the port-2 output (in grey). Since there are two active ports, we get the return loss for both ports. In case of the MIMO antenna,

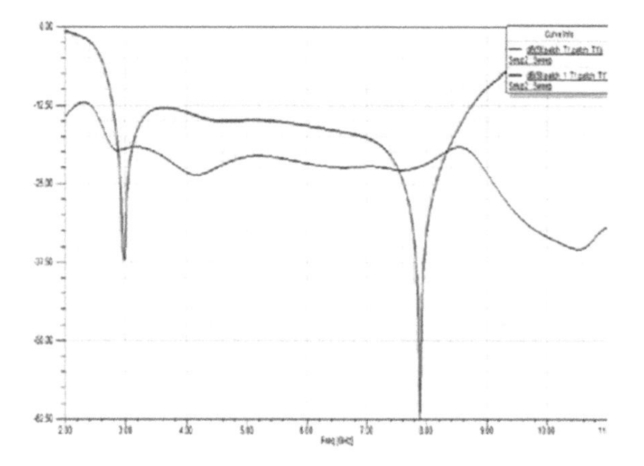

Fig. 4 Return loss S_{11} plot of the MIMO antenna

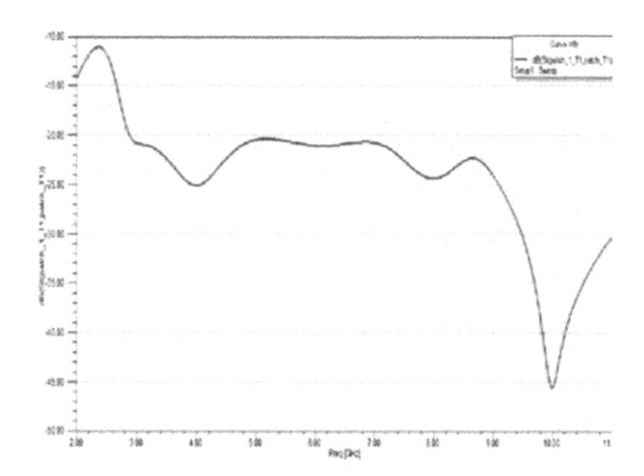

Fig. 5 Plot of S_{12} for the MIMO antenna

when a signal is transmitted port-1 gives a better response at 3 GHz and 7.7 GHz while port-2 remains below −10 dB throughout the bandwidth.

In Fig. 5, it can be seen that S_{12} is below −20 dB for a range of frequencies of 2.8–11 GHz and the maximum deep cut is of −45 dB at 10 GHz. This implies that the mutual coupling between the antenna elements is less and is controlled.

In Fig. 6, it can be seen that the VSWR for this particular MIMO antenna is less than 2 for the whole bandwidth. It is almost 1.6 which is desirable for an antenna and indicates that the impedance matching is good (Figs. 7 and 8).

The maximum gain is pragmatic at approximately 2 dB. This is seen in the two-dimensional plot and the three-dimensional figure.

In Fig. 9, the radiation pattern in the far field is shown. It is an omni-directional pattern, one which is desired for the MIMO antenna.

In the MIMO antenna, an omni-directional pattern is desirable since it is directional in one plane and isotropic in the other. The antenna is expected to be omni-directional and this will affect its application too (Fig. 10).

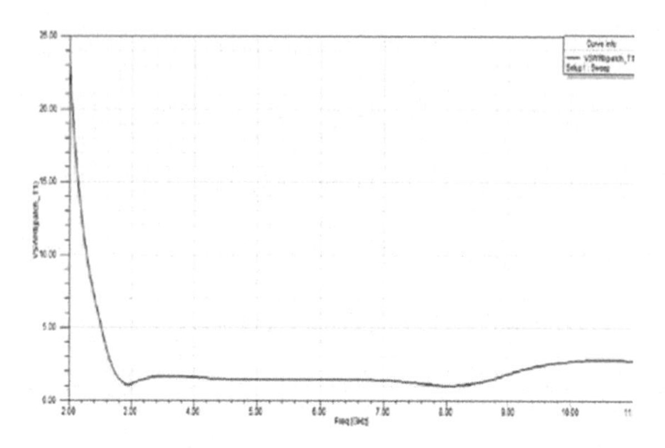

Fig. 6 VSWR for the MIMO antenna

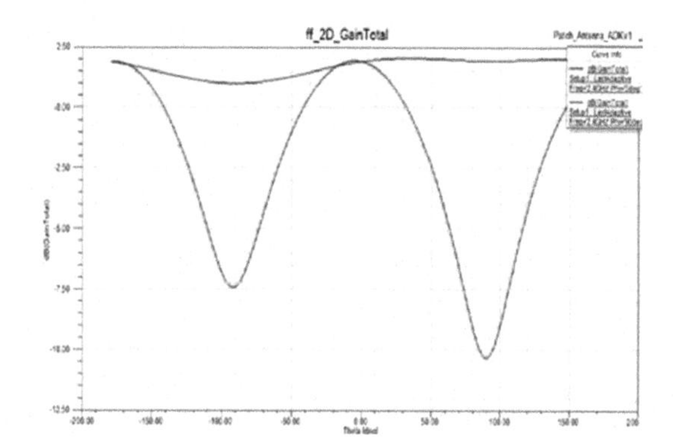

Fig. 7 2-Dimensional gain for the MIMO antenna

In the Smith chart, the centre represents 50 Ω. The aim of the designer is to reach to the centre of the Smith chart in the construction of the matching network, and this has been achieved in our design.

2.4 Discussion and Analysis

In the above experiment, a MIMO antenna is designed with the help of two symmetrical planar monopole antennas. The ground plane is cut into an L-shaped slot so that the mutual coupling which is undesirable in a MIMO antenna can be minimized. The dielectric material used as the substrate, i.e. Rogers, is easily available and hence easy to fabricate. The structure when simulated, gives S_{11} below −10 dB for a wide frequency range of 2.4–8 GHz and at 3 GHz and 7.7 GHz it made deep cuts close to −37.5 dB and −62.5 dB respectively. We take −10 dB as a reference point because it indicates that 90% of the radiation coming from antenna is radiated and only 10% is reflected back. We have observed that S_{11} made deep cuts up to −62.5 dB and this is highly desirable. If this structure is

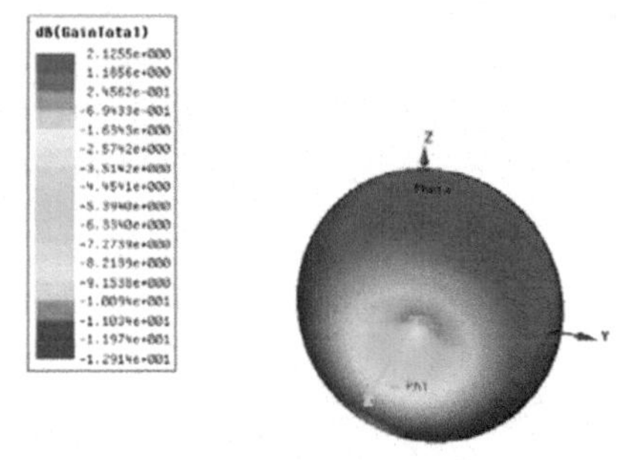

Fig. 8 3-Dimensional gain for the MIMO antenna

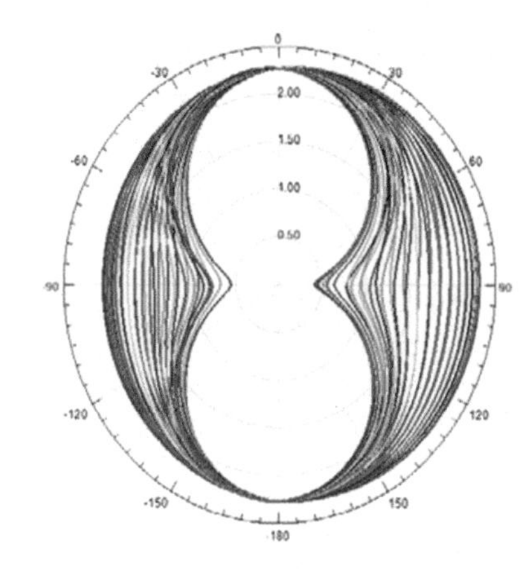

Fig. 9 Radiation pattern for the MIMO antenna design

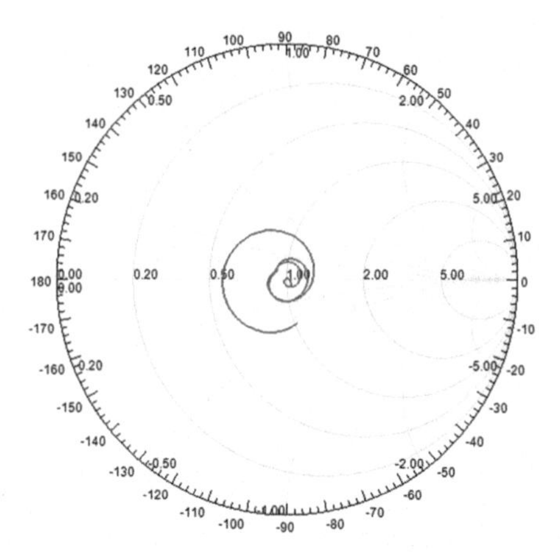

Fig. 10 Smith chart representing the path from port-1 to port-2

fabricated then even with a significant amount of fabrication error, it can be expected that S_{11} will have desirable values in the manufactured antenna. The plot of S_{12} represents power from port 2 delivered to port 1. There is superior amount of isolation between the ports for the MIMO antenna to work properly. The results show that it is below −20 dB over a wide range of frequencies (2.8–11 GHz) and even lower than that for the higher range, i.e. −45 dB at 10 GHz which shows that the mutual coupling is reduced between the antenna elements. The VSWR shows how good the impedance matching is. When the matching is perfect, more power is delivered to the antenna. The value of VSWR is generally greater than or equal to 1, but less than 2. In the graph shown above, the VSWR value is below 2, being at 1.6, which indicates that the antenna has good impedance matching. In respect to this structure, the maximum gain is approximately 2 dB. The radiation pattern at far field shows that it is omni- directional, which is highly desirable for a MIMO antenna. Impedance matching is good and therefore the Smith chart reading shows that the path travels to the center which means 50 Ω. All the parameters are verified and this MIMO antenna can be easily implemented for practical utilization [10–12].

2.5 Conclusion

Nowadays wireless communication plays a major role in society. The antenna is the main component of any wireless communication system. The performance of any wireless communication system depends completely on the high performance of the antenna design and implementation. In this MIMO antenna design, a rectangular radiator along with the L-shaped ground plane helps in achieving desirable parameters. The core issue in a MIMO antenna is reducing the mutual coupling when the antenna is of a compact size. This issue has been resolved by making a slotted ground plane. The reflection coefficient is −10 dB over a frequency range of 2.4–8 GHz. The reflection coefficient strikes deeper being −37.5 dB and −62.5 dB at 3 GHz and 7.7 GHz respectively. This makes it usable for various applications in wireless communication such as WLAN and WiMAX etc. All the parameters have been successfully verified by simulating the structure and the design can be fabricated easily.

References

1. Hossain E, Hasan M (2015) 5G cellular: key enabling technologies and research challenges. IEEE Instrum Meas Mag 18:11–21
2. Wang F, Duan ZY, Tang T, Huang MZ, Wang ZL, Gong YB (2015) A new metamaterial-based UWB MIMO antenna. In: IEEE IWS 2015. Shenzhen, China, pp 1–4
3. Rajagopalan A, Gupta G, Konanur AS, Hughes B, Lazzi G (2007) Increasing channel capacity of an ultrawideband MIMO system using vector antennas. IEEE Trans Antennas Propag 55(10):2880–2887
4. Manteuffel D (2009) MIMO antenna design challenges. In: Lough-borough antennas & propagation conference

5. Jusoh M, Jamlos MFB, Kamarudin, MR, Malek MFBA A MIMO antenna design challenges for UWB application. Prog Elect Res 36:357–371, ISSN 1937-6472
6. Hong W, Baek K-H, Lee Y, Kim Y, Ko S-T (2014) Study and prototyping of practically large-scale mm wave antenna systems for 5G cellular devices. IEEE Commun Mag 52:63–69
7. Amin Honarvar M, Naeemehsadat H, Bal SV (2015) Multiband antenna for portable device applications. Microw Optic Technol Lett 57(4)
8. Liu L, Cheung SW, Yuk TI (2013) Compact MIMO antenna for portable devices in UWB applications. IEEE Trans Antennas Propag 61(8)
9. Constantine AB Antenna theory—analysis and design, 3rd edn. Wiley publication
10. XueMing L, RongLin L (2011) A Novel dual-band MIMO antenna array with low mutual coupling for portable wireless devices. IEEE Antennas Wirel Propag Lett 10
11. Liu L, Cheung SW, Yuk T (2015) Compact MIMO antenna for portable uwb applications with band-notched characteristic. IEEE Trans Antennas Propag 63(5)
12. Dong L, Choo H, Heath RW Jr, Ling H (2005) Simulation of MIMO channel capacity with antenna polarization diversity. IEEE Trans Wirel Commun 4(4)

An Enhanced Reputation-Based Data Forwarding Mechanism for VANETs

Aman Kumar, Sonam Bhardwaj, Preeti Malik and Poonam Dabas

Abstract A vehicular ad hoc network (VANET) is a self-configuring and infrastructureless network connecting high mobility random vehicles by wireless links. Due to high mobility, data transmission between two vehicles may possible through other intermediate vehicles but it is difficult to transmit messages through these intermediate vehicles because intermediate vehicles may violate security by sending the wrong messages or by not forwarding messages. So transmission of messages using trust-based VANETs is a difficult task. Various techniques are proposed by researchers to forward packets in trust-based VANETs. Each technique has its own mechanism as well as its pros and cons. In this paper we propose an enhanced trust-based mechanism to select trusted nodes through which messages are transmitted. The proposed mechanism has been implemented using ONE (opportunistic network environment) simulator. Results shows that the proposed mechanism has a high delivery ratio and less message delay than existing reputation-based mechanisms.

Keywords VANET · Trust degree · Threshold · Prophet · ONE

A. Kumar (✉) · S. Bhardwaj (✉) · P. Malik · P. Dabas
Department of Computer Science and Engineering, University Institute of Engineering & Technology (UIET), Kurukshetra University, Kurukshetra, Haryana, India
e-mail: aman.sagwal92@gmail.com

S. Bhardwaj
e-mail: sbsonambhardwaj@gmail.com

P. Malik
e-mail: preetimalik71@gmail.com

P. Dabas
e-mail: poonamdabas.kuk@gmail.com

A. Kumar and S. Mozar (eds.), *ICCCE 2018*,
Lecture Notes in Electrical Engineering 500,
https://doi.org/10.1007/978-981-13-0212-1_26

1 Introduction

VANETs are developing requirement of users. It has many advantages over a traditional network as well as disadvantages too. There are lots of things to bear in mind while designing ad hoc networks. The foremost task of the researchers is to provide for safe transmission without incident. So to securely transmit messages in a VANET is challenging job because it is difficult to forward message through intermediate vehicles without trust [1]. Researchers have a done lot of work on this problem but each proposed mechanism has its own drawbacks. We have endeavored to propose a mechanism to transmit messages securely with trusted nodes. The proposed mechanism increases the delivery ratio and reduces the drop rate. When messages are transmitted without any trust mechanism, then intermediate vehicles may drop the packets or forward incorrect routing information and this will increase the packet drop rate and decrease the delivery ratio [2].

A. **Technical Challenges in VANET**

The technical challenges refers to the technical obstacles which should be overcome before the deployment of a VANET. Some of the challenges are given below:

i. Network management: Due to high versatility, the system topology and channel Condition change quickly.
ii. Congestion and impact control: The unbounded system measure likewise results in a Challenge. The movement load is low in country areas and at night even in urban areas. Because of this, system parcels regularly occur during surge hours, the movement load is high and consequently the system is congested and crashes occur.
iii. Environmental impact: VANETs utilize electromagnetic waves for correspondence. These waves are influenced by the Earth and in order to convey the VANET, environmental effects must be considered.
iv. Security: As VANETs gives the street wellbeing applications which is life basic in this way security of these messages must be fulfilled.

2 Related Works

Li et al. [3] proposed a reputation-based global trust establishment plot (RGTE). This plan takes into account three critical components which incorporate properties of a VANET, security, and proficiency in trust building. Hubs in the RGTE impart its trust to others by sending trust messages to a reputation management centre (RMC). An RMC is a bona fide framework that gathers trust from every single hub in VANET. Before figuring the reputation of a hub, the RMC ought to sift through suspicious trust messages with a measurable consistency. With the assistance of a

RMC any hub in the system, particularly new hubs, can gain exceptional trust data from the whole system in a secure manner.

Dhurandher et al. [4] proposed a reputation-based framework that looks to give a secure and robust vehicular system. The proposed algorithm set up wellbeing in a VANET through the accomplishment of trust levels for hubs in the system utilizing reputation and credibility checks. The algorithm has been planned principally for wellbeing related data that are communicated in a single bounce and transferred in multiple jumps through the middle of the road hubs. The bundles to be sent will be communicated safely and a unicasted parcel will be seen as malicious data. The algorithm follows an occasion-situated approach; that is a hub starts the correspondence when it watches an occasion through its sensors.

Ding et al. [5] planned an occasion-based reputation framework to sift through fake messages spread by malicious assailants in VANETs. As opposed to Nai-Wei's technique, they proposed a more mind-boggling model which orders all vehicles experience a similar movement occasion to various parts. Reputation capacities are intended for these diverse parts. Every part has a reputation assessment system to decide if an approaching activity message is trusted. A dynamic part subordinate status assessment system is introduced to channel fake cautioning messages. Reproduction tests demonstrate that huge execution additions can be achieved utilizing this structure.

Ma and Yang [6] proposed a trust-based directing convention. The trust a node has for a neighbor structures the essential building square of trust model. The proposed trust assessed system, which is executed by each node in the system freely, only utilizes neighborhood data in this way making it versatile. In addition, unlike GPSR and OLSR, this depends on area data that require a considerable measure of space and time for buffering parcels and calculating the separation between nodes. The drawbacks of this paper are that it does not provide any mechanism to compute trust value, with the reputation of node being totally dependent on the neighbor nodes. It is also possible that a neighbor node may provide an incorrect judgment of other nodes.

3 Existing Reputation-Based Mechanisms

In this section a standard reputation-based mechanism is presented. In a reputation mechanism messages are transmitted or forwarded to an intermediate node based on its reputation value. Nodes having a high reputation value have a greater chance of getting messages from the source node.

(a) *START*
(b) *Set Source S and destination*
(c) *Add address of destination node D into message sending by S.*
(d) *Check if D is neighbor of S?*

(e) *Now discovers neighbor of S.*
(f) *Computes trust value of all neighbor nodes.*
(g) *Search for a neighbor node with high trust value.*
(h) *Then forward message through the selected node.*
(i) *END.*

In existing reputation-based mechanisms each node gives its opinion of its neighbor node by checking its reputation value and classifying it as either a malicious node or a normal node. The drawbacks of this scheme are that it does not provide any mechanism to compute trust value, and the reputation of a node is totally dependent on its neighbor nodes. It is also possible that a neighbor node may provide an incorrect opinion of the reputation of other nodes. So to handle these kinds of drawbacks, we propose an enhanced reputation-based data forwarding mechanism that can overcome these obstacles [6–8].

4 Proposed Work

In this section, an enhanced reputation-based forwarding mechanism is presented in detail. Reputation-based Trust Management Scheme: In a reputation-based mechanism each node gives its opinion of its neighbor node by checking their reputation value and classifying it as either a malicious vehicle or a normal vehicle. The drawback of this scheme is that measuring the reputation of the vehicle is a difficult job due to the dynamic nature of vehicles and as a result it is based only on assumptions. To overcome this we propose an enhanced reputation-based mechanism.

Trusted Vehicles

i. If a vehicle delivers maximum packets then it is a trusted node.
ii. If a vehicle receives a number of packets but cannot forward some data then that node may behave as a malicious node.

Trust Degree

i. The trust degree indicates nodes message transmission prediction of whether a node is trustworthy to transmit messages or not. As higher the trust degree the chances of message transmission through that node is high.
ii. The trust degree can be computed by maintaining node routing information collected from the node buffer (Nbt).

Proposed Algorithm

1.//Parameter initialization//
Initialize number of vehicles with a unique ID in the network.
Set trusted authority.
Createdmsgs = number of messages created by

```
vehicle
Msgdelivereds = number of messages delivered
by vehicle
Msgreceiveds = number of messages received
by vehicle
2. for (i = 1; i <=n; i ++)
{
createdmsgs = null; msgdelivereds
= null; msgreceiveds = null;
msgdelivereds[i]=
createdmsgs[i] + msgreceiveds[i];
TrustDegree[i] = msgdelivereds[i]/
createdmsgs[i] + msgreceiveds[i];
Start
Initialize number of
nodes with unique Id
For (i = 1; i < N; i ++)
3. if (TrustDegree[i] > 0.5)
return true;
else
{
return false;
}
```

Description of Algorithm

In this algorithm, a mechanism to detect trusted vehicles is defined. If a vehicle successfully delivers messages created by it and received from other vehicles, then that vehicle is called a trusted vehicle otherwise vehicles are untrusted. To distinguish between them we provide a threshold value, i.e. 0.5. Vehicles whose trust degree greater than 0.5 are called trusted vehicles (Table 1).

Above table show routing information of vehicles during computed by data transmission. This routing information may helpful in measuring trust degree. The vehicles having trust value greater than 0.5 threshold value are trusted vehicles else they are untrusted vehicles. A threshold is a fix value used to find optimal solution (Fig. 1).

Table 1 Routing information of nodes created, delivered, and received by the start vehicle

Nodes	Created	Delivered	Recieved	Trust degree
N1	10	12	5	0.8
N2	12	13	6	0.72
N3	14	14	6	0.7
N4	16	16	7	0.6
N5	18	5	8	0.1
N5	9	3	22	0.09
N6	9	6	3	0.5

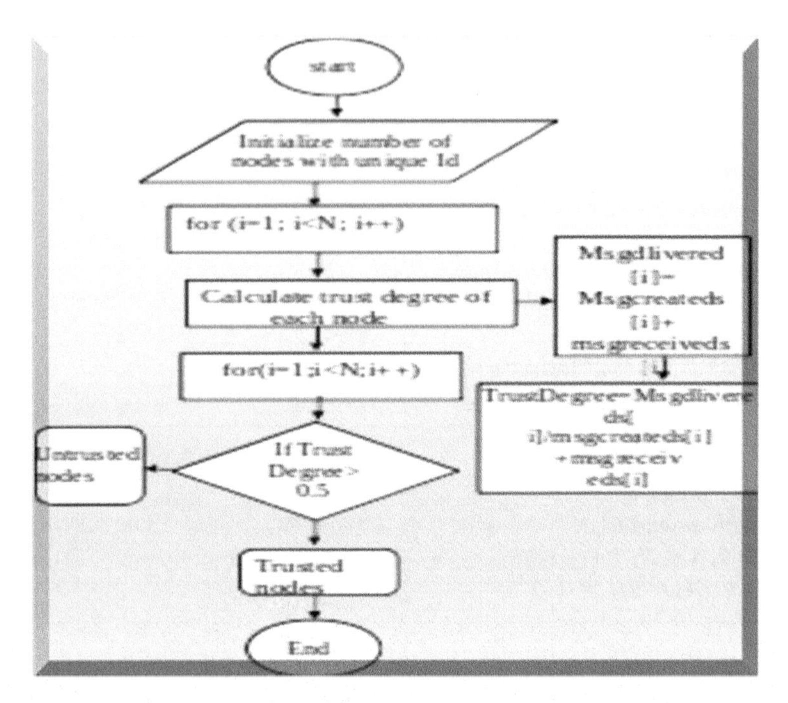

Fig. 1 Flow chart of the proposed mechanism

5 Results and Analysis

i. **Tool used:** A ONE simulator is used for simulating the proposed mechanism. It is an open source java-based simulator that runs on any platform [9].

ii. **Routing used:** To analyze the proposed mechanism a probability-based routing mechanism is used. In this routing the next hop is selected based on its highest probability of delivering messages to the required destination.

iii. **Performance metrics used:**

a) *Average Message Delay:* This is calculated by taking the average difference between the time of message delivery and the time of message creation.

b) *Number of Messages Delivered:* This is defined as the number of messages that are actually delivered to the destination.

c) *Delivery Ratio:* The delivery ratio is the ratio of the number of messages delivered to the number of messages created (Table 2).

Figure 2 shows the effect of the increasing number of vehicles on delivery ratio. As the number of vehicle increases, the delivery ratio also increases. In the proposed trust-based mechanism the delivery ratio is high compared to existing reputation-based mechanisms.

Table 2 Simulation parameters

Parameter description	Value
Simulation area	4500 m × 3400 m
Simulation time	20000 s
Mobility model	Map based movement
No. of groups	4
Transmission rage	10 m
Node speed	2 m/s
Warm-up period	1000 s
Time to live	300
Buffer size	5 M
Routing schme	Prophet

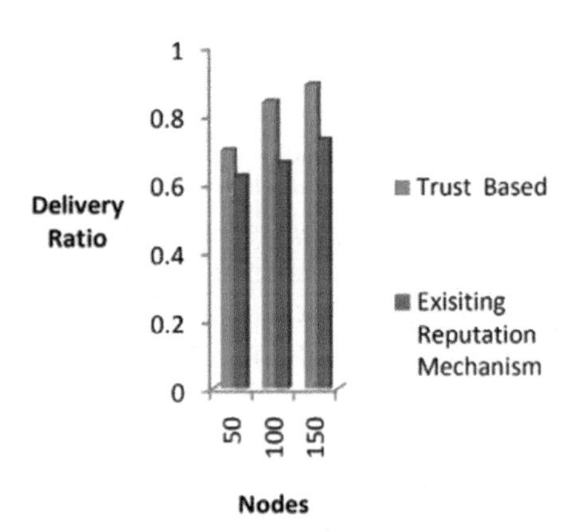

Fig. 2 Delivery ratio versus vehicles

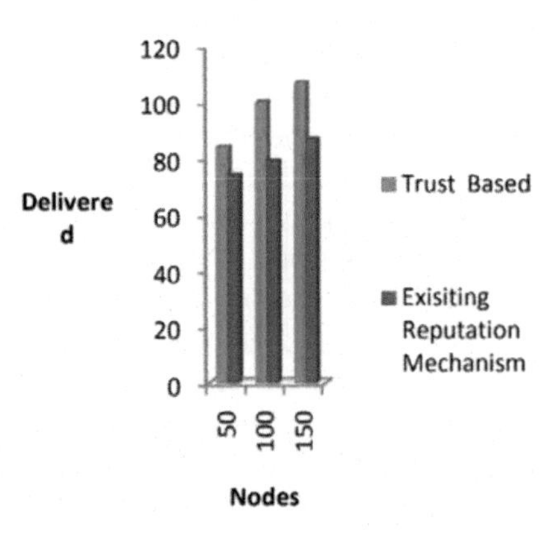

Fig. 3 Delivered versus vehicles

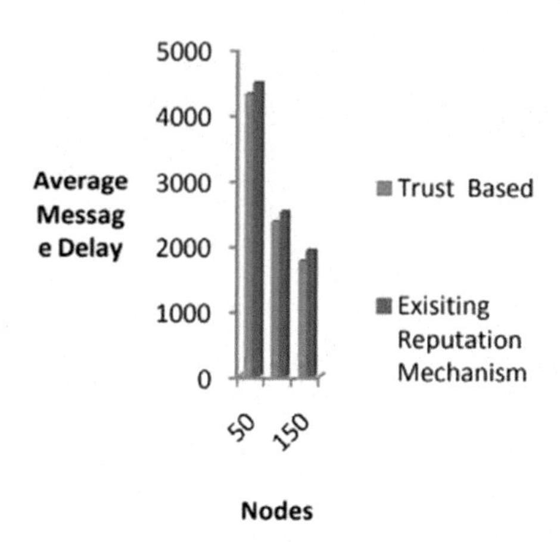

Fig. 4 Average message delay versus vehicles

In Fig. 3 the total numbers of messages delivered by vehicles are shown. In the proposed trust-based mechanism the delivery of messages is high compare to existing reputation-based mechanisms.

Figure 4 illustrates average message delay as it is affected by of variation in vehicles. As vehicles increase then the delay may also increases. In a trust-based mechanism the average message delay is slightly lower than in an existing reputation-based mechanism.

6 Conclusion

Security in VANETs is challenging. During data transmission, data or information may be accessed by attackers. As a result, trust-based transmission of messages in a VANET is a difficult task. In this paper, enhanced reputation-based data forwarding in a VANET has been proposed. In the proposed mechanism we attempt to detect trusted vehicles. If a vehicle successfully delivers messages created by it and received from other vehicles then that vehicle is called a trusted vehicle otherwise vehicles are untrusted. To distinguish between them we provide a threshold value, i.e. 0.5. Vehicles whose trust degree is greater than 0.5 are called trusted vehicles otherwise they are not trusted vehicles. A ONE simulator was used for simulation purposes. In the analysis of the simulation results, few performance metrics have been used, such as average message delay, delivery ratio, and numbers of packets delivered by vehicles. Simulation results show that the proposed mechanism has a higher delivery ratio and lower message delay than existing reputation-based mechanisms.

References

1. Sivasakthi M, Suresh S (2013) Research on Vehicular Ad Hoc Networks (VANETs): an overview. J Appl Sci Eng Res 2(1):23–27
2. Vumar V, Chand N (2010) Efficient data scheduling in VANETs. J Comput 2(8):32–37
3. Li X, Liu J, Li X, Sun W (2013) RGTE: a reputation based global trust establishment in VANETs.In: IEEE, 978-0-7695-4988-0/13 $26.00 ©2013, pp 210–214
4. Dhurandher SK, Obaidat MS, Jaiswal A, Tiwari A, Tyagi A (2010) Securing vehicular networks: a reputation and plausibility checks-based approach. In: 2010 IEEE, pp 1550–1554
5. Qing D, Jiang M, Li X, Zhou XH (2010) Reputation based trust model in vehicular ad hoc networks. In: IEEE, 978-1- 4244-7555-1/10/$26.00 ©2010, pp 1–6
6. Ma J, Yang C (2015) A Trust-based stable routing protocol in vehicular ad-hoc networks. Int J Secur Appl 9(4) pp 107–1165
7. J Secu Appl 9(4) 107–1165
8. Li Q, Malip A, Martin KM, Ng S-L, Zhang J (2012) A reputation based announcement scheme for VANETs. In IEEE,0018-9545/$31.00 ©2012, pp 4095–4108
9. Cao Z, Li Q, Wei Lim H, Zhang J (2014) A Multi-hop Reputation Announcement Scheme for VANETs. In: IEEE, 978-1-4799-6058-3/14/$3 1.00 ©2014, pp 238-243
10. Keränen A (2008) Opportunistic network environment simulator. Special Assignment Report, Helsinki University of Technology, Department of Communications and Networking, May 2008
11. Lindgren A, Doria A, Schelen O (2004) Probabilistic routing in intermittently connected networks. In: Proceedings of The first international workshop on service assurance with partial and intermittent resources (SAPIR), Aug 2004

Statistical Metric Measurement Approach for Hazy Images

T. Saikumar, K. Srujan Raju, K. Srinivas and M. Varaprasad Rao

Abstract A novel statistical metric measurement approach for the evaluation of enhancement of hazy images. Metric measurement plays a critical role in picture enhancement in hazy weather conditions and leads to a lessening in pixel resolution, a distortion in color, and gray images. In this paper hazy and foggy images are considered for evaluation using contrast-to-noise ratio (CNR) which dehazes the original hazy images. We propose a unique novel effective parameter based on an image filtering approach. The results demonstrated show a better CNR for dehazed images.

Keywords Metric measurement · Hazy images · CNR · Filtering approach

1 Introduction

The ability to view through air, irrespective of sunlight or moonlight is always greater with clean free air than with polluted air containing multiple tiny dust particles or droplets of water. There is an abundance of factors affecting the visibility of an image, including fog, mist, haze, and smoke. The ability physically distinguish between foggy and haze can be achieved by means of a visibility distance parameter. The degradation of images which are affected by haze, fog, and mist is caused by the scattering and absorption of light particles in the air or

T. Saikumar (✉) · K. Srujan Raju · M. Varaprasad Rao
CMR Technical Campus, Hyderabad, India
e-mail: tara.sai437@gmail.com

K. Srujan Raju
e-mail: ksrujanraju@gmail.com

M. Varaprasad Rao
e-mail: vpr_m@yahoo.com

K. Srinivas
Jyothishmathi Institute of Technology & Science, Karimnagar, India
e-mail: phdknr@gmail.com

A. Kumar and S. Mozar (eds.), *ICCCE 2018*,
Lecture Notes in Electrical Engineering 500,
https://doi.org/10.1007/978-981-13-0212-1_27

atmosphere [1–3]. A reduction in visibility is mainly caused by emission or scattering from light particles between an object and a predefined observer. Images or sequences of images acquired by camera system, such as those from video surveillance or digital remote sensing applications can be affected due to absorption of water droplets and light particles present in the atmosphere. In this paper, light particles and droplets of water from fog, mist, and haze are not ease [2]. Methods for enhancing the visibility of degraded open-air images or sequences of images are fall into two broad categories. There first category includes retinex theory and wavelet transformation which are well-known non-model-based methods. The second category of enhancement is defined as model-based methods, and these can achieve good results by means of modeling using scattering light particles, but frequently many additional assumptions of the image environment or image system have to be made, such as "for an estimate on depth of still scene" [4].

2 Degradation Model

Assume that the digital picture degradation model for a haze-affected weather environment is expressed mathematically as

$$\gamma = \chi + A\zeta \tag{1}$$

$$\chi = \varphi(x)\ t(x) \tag{2}$$

$$\zeta = (1 - \psi(x)) \tag{3}$$

where x is pixel location, φ is image radiance, A is atmospheric light, and ψ is transmission function.

In the present scenario, an effect on atmospheric layer depends up on a parameter of the depth of an images, which has to restorate its visibility with respect to evaluation of an image with color and their its physical properties of hazy and depth of an image mapping. Due to there being less knowledge regarding the structural element of images or the depth of images both gray and color images/image sequences, it is very difficult to distinguish between the two parameters ψ and A. The difficulty of the above Eq. (1) can be modified for the degradation model, which is not directly related to the enhanced contrast of images.

$$\xi = A\zeta \tag{4}$$

The above Eq. (1) is modified and written as

$$I(x) = \varphi(x) - \left[\frac{\rho(x)}{A}\right] + \xi(x) \tag{5}$$

where $\rho(x) = \varphi(x)\xi(x)$

From above Eq. (3) with a final modification after enhancement of image visibility is

$$\varphi(x) = [I(x) - \xi(x)] + \left[\frac{\rho}{A}\right] \tag{6}$$

assuming the values for A are isotropic throughout the demonstration. As expected from the results for which assessment of φ can be modified in term of $\xi(x)$ that results to haze images with a suppose of constant A (Fig. 1).

Enhanced version of the degradation image model in Eq. (6).

There are two steps for the enhancement of haze removal.

1. The primary step is to enhance the haze layer with respect to the pixel position X, where the density of the haze is directly proportional to the depth of a picture element.
2. By means of the physical properties attributes the layers of haze effected images have two significant contribution such as $0 \leq I(x)$ and a pixel of a gray or color picture with which minimal channel value is derived as:

$$[g_X(x)] = \min[I(x)] \tag{7}$$

First, compute G(x) by means of probability density function of an input picture $g_X(x)$ which as follows.

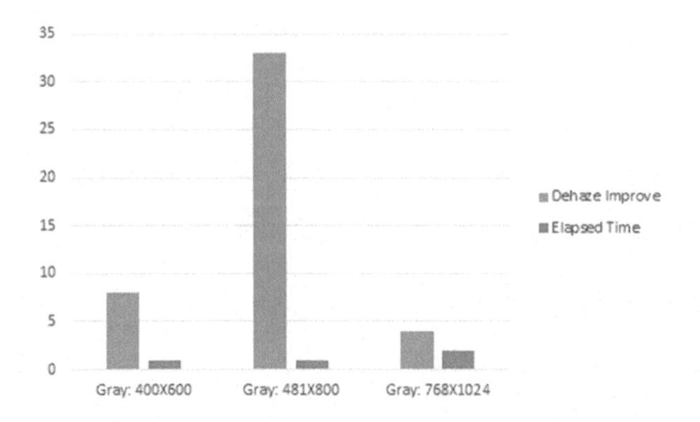

Fig. 1 Calculation of improvement by dehazing and elapsed time (unit: seconds)

Fig. 2 **a** Original images. **b** Enhanced images

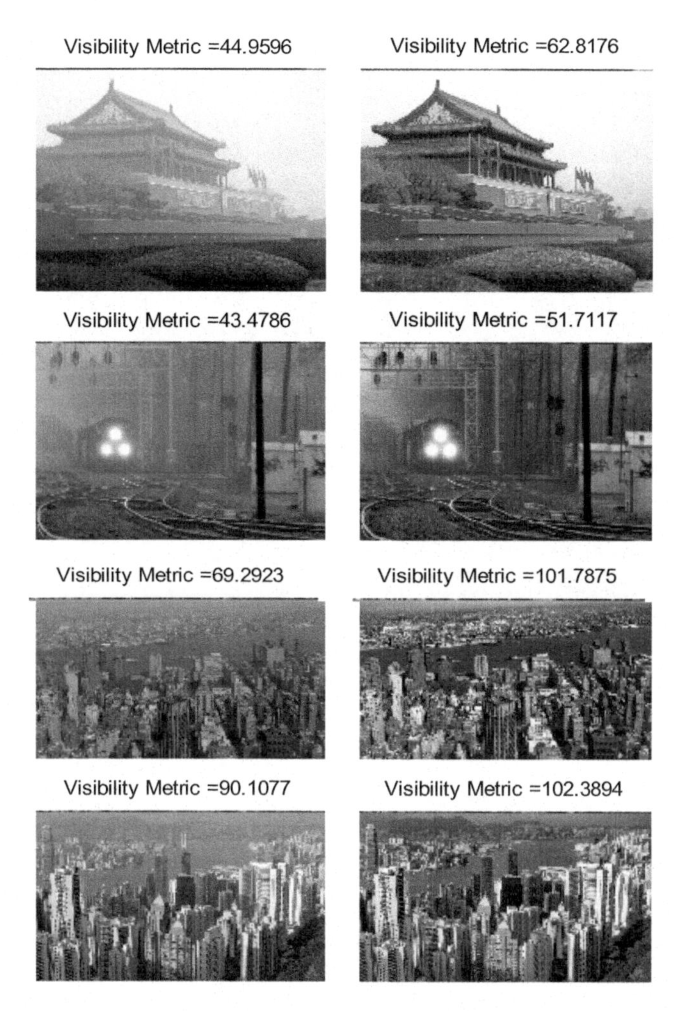

Fig. 2 (continued)

$$G_X(x) = median[g_X(x)] \tag{8}$$

where G_x denotes the local region at each digital pixel.

The mean value of the haze layer is smoothed by use of a mean filter (LPF) with changes in the depth of the images [2].

$$\overline{g_{x,y}}(m,n) = g_{x,y}(m,n) - \frac{1}{MN}\sum_{i=1}^{MN} g_{x,y}(m,n) \tag{9}$$

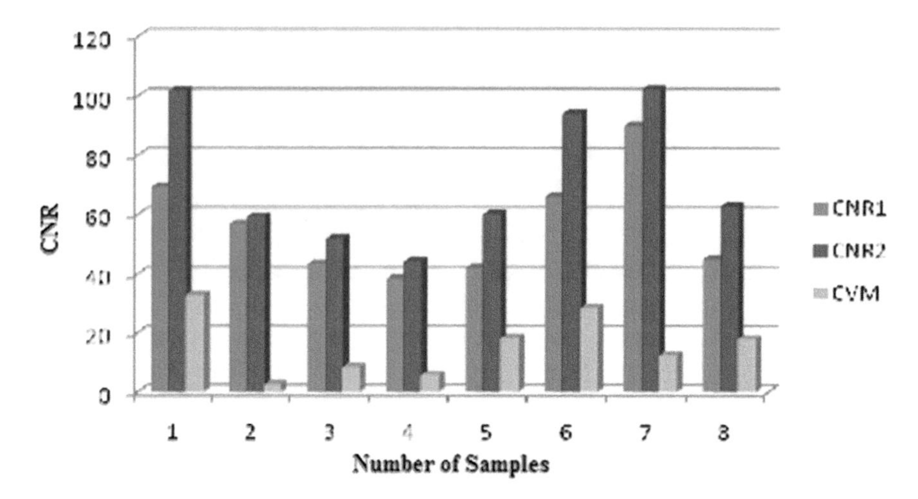

Fig. 3 Statistical analysis of different data set samples of hazy images versus contrast-to-noise ratio (CNR)

3 Experimental Results

The experimental setup consists of haze images which are evaluated and computed in terms of CNR Values (Figs. 2 and 3).

4 Conclusions and Future Work

The results of the experimental demonstrations relate to constant still images which are evaluated objectively and compared with the original CNR value in the images affected by haze. While comparing our proposed algorithm with a help of mean or average filtering which leads to a better enhancement in single haze images. However, the future scope of the proposed approach may see it able to works well for moving objects or videos affected by heavy fog, with few modifications being needed. It may even be suitable for use in video processing.

References

1. Narasimhan SG, Nayar SK (2003) Contrast restoration of weather degraded images. IEEE Trans Pattern Anal Mach Intell 25(6):713–724
2. Zhang Y-Q, Ding Y, Xiao J-S, Liu J, Guo Z (2012) Visibility enhancement using an image filtering approach. EURASIP J Adv Signal Process 1–6

3. Narasimhan SG, Nayar SK (2002) Vision and the atmospheric. Int J Comput Vision 48(3): 233–254
4. Saikumar T, Dwijendra B, Arika M, Nandita K, Mukesh A (2017) Visibility enhancement (VM) for haze images based on color channel. Int J Electron Electr Comput Syst (IJEECS) 6(5). ISSN: 2348-117X

Image Enhancement for Fingerprint Recognition Using Otsu's Method

Puja S. Prasad, B. Sunitha Devi and Rony Preetam

Abstract The internal surfaces of human hands and feet of have minute ridges with furrows between each ridge. Fingerprints have very distinctive features and have been used over a long period of time for the identification of individuals and are now considered to be a very good authentication system for biometric identification. For successful authentication of fingerprint, features must be extracted properly. The different types of fingerprint enhancement algorithms used in image processing all provide different performance results depending on external and internal conditions. External conditions include types of sensors and pressure applied by the subject etc. Internal conditions include the body temperature of a subject and skin quality etc. In this paper, we enhance an image using Otsu's method, which is one of the segmentation steps of image processing. This algorithm can improve the clarity of ridges and furrows of a fingerprint and enhances performance by reducing the total time for extraction of minutiae compare to other algorithms.

Keywords Minutiae · Gabor filtering · Ridge ending · Ridge bifurcation Wavelet domain · Otsu's method

1 Introduction

Experts use many details from a fingerprint for the authentication of a person. The process of fingerprint verification starts by investigating the quality of a finger image or input image taken by sensors and then proceeds by performing a number

P. S. Prasad (✉) · B. Sunitha Devi · R. Preetam
CMRIT, Medchal, Hyderabad, TS, India
e-mail: puja.s.prasad@gmail.com

B. Sunitha Devi
e-mail: sunithabigul@gmail.com

R. Preetam
e-mail: ronyjcpreetam@gmail.com

A. Kumar and S. Mozar (eds.), *ICCCE 2018*,
Lecture Notes in Electrical Engineering 500,
https://doi.org/10.1007/978-981-13-0212-1_28

of pattern search algorithms. Image enhancement as well as ridge segmentation is done by using local orientated ridge to filter parameters. After segmentation, a thinning process takes place to obtain a thinned image so that minute features can be extracted. Spurious minutiae are removed at the post-processing stage.

The way an image is captured or its machine representation, determines the success of any matching algorithm that works in decision module [1]. The performance of an algorithm actually depends on how accurate and reliable the results it gives are. The overall quality of an input fingerprint image plays a very significant role in the decision of the identification and verification algorithms used. This paper introduces a fingerprint enhancement method which is fast and actually improves the image quality of the valley and the ridge structures of input images, based on the orientation and frequency of the local ridges and thereby extracts the correct minutiae. The uniqueness of a person's fingerprint and its unchanging nature throughout an individual's lifespan make it a very important model for a biometric authentication system. A fingerprint consists of a unique pattern of what are called furrows and ridges. A ridge is a curved line or segment whereas a valley is the space between two adjacent ridges. A feature that is used for the identification of a person is called a minutia and is actually the discontinuity of a ridge segment. These discontinuities appear in the form of a bifurcation of a ridge, resembling a fork, whereas ridges end abruptly at a ridge ending. These minutiae are stored in the form of a template and are used for authenticating a person (Fig. 1).

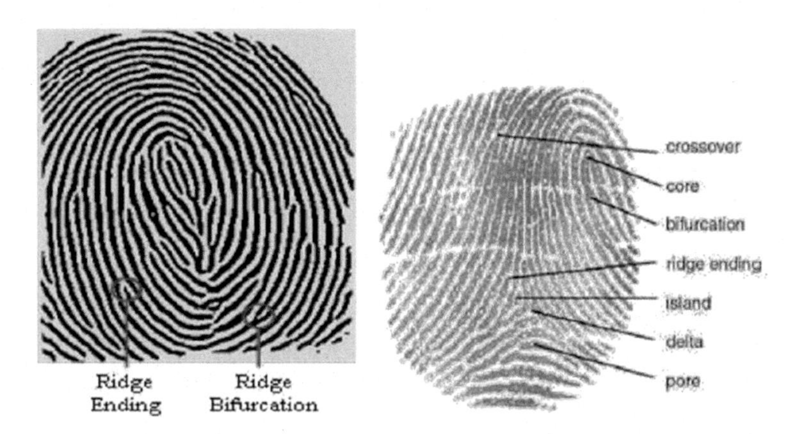

Fig. 1 Fingerprint showing minutia points: crossover, core, bifurcation, ridge ending, island, delta, pore etc.

1.1 Fingerprint Representation

The individuality of a fingerprint is determined by its ridge pattern and the occurrence of certain ridge anomalies called minutiae points. Normally, to make class of different kind of fingerprint the global design of ridges are used whereas the allocation of minutiae points is taken for matching between fingerprint and fingerprint template [2]. Automatic recognition and identification of a fingerprint matches query features against a large database of millions of different features stored with individual identifications. It depends on the pattern of ridges in the query image to refine their search in the database, a technique called indexing of fingerprints, and on the minutiae points to determine a precise matching fingerprint. The ridge pattern itself is hardly ever used for fingerprint matching.

- **Minutiae**

The local discontinuities in the ridge pattern called minutiae give the fingerprint features that can be used to authenticate a person's identity. Details such as the orientation, type and location of minutiae are taken as description when using minutiae as a fingerprint features. The two most important local ridge distinctiveness points are the ridge bifurcation and the ridge ending, and these are generally used for pattern recognition.

Different types of pattern are found in fingerprint minutiae and these are:

- Ridge endings—Ridge endings are the points where a ridge ends suddenly.
- Ridge Bifurcation—Ridge bifurcation is the characteristic where an individual ridge is cut into two ridges and looks like fork.
- Independent ridge or island—An independent ridge is an individual ridge that starts at some point and ends after travelling a short distance. It looks like an island.
- Ridge enclosures—In a ridge enclosure, a single ridge divides like a fork and then joins again abruptly afterwards and continue as an individual ridge.
- Spur—A spur is actually a bifurcation and a short ridge branching off a long ridge.
- Bridge or Crossover—This type of pattern involves a short ridge which crosses between two parallel ridges (Fig. 2).

Minutiae also refer to any small point that is distinct to an individual. Not all minutiae are used for verification, only ridge ending and ridge bifurcation (Fig. 3).

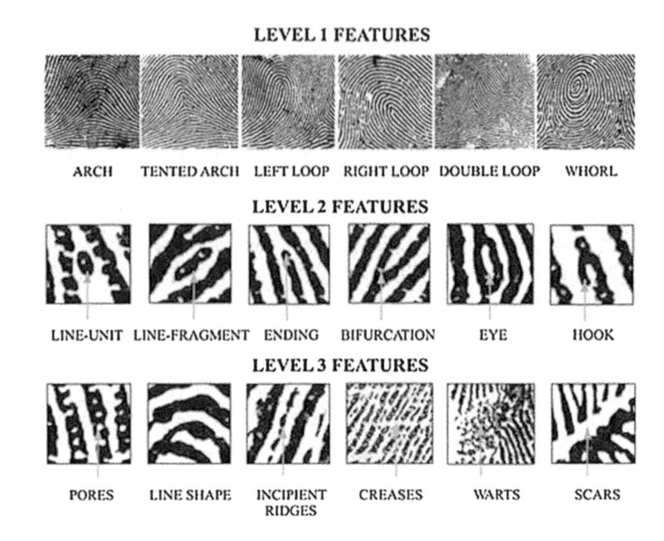

Fig. 2 Different minutia points showing crossover, spur, ridge bifurcation, and ridge ending that are used in fingerprint authentication systems

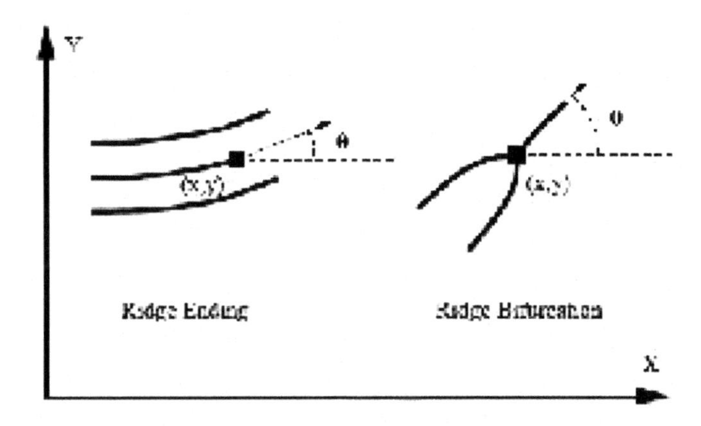

Fig. 3 Ridge ending and ridge bifurcation having its value using two points x and y with corresponding angle

2 Related Work

Sonavane and Sawant [3] offered a method by which an image of a finger is broken into a set of clean images and their orientation is estimated, giving a special domain fingerprint enhancement. Kukula et al. [4] proposed a method of applying different levels of force to investigate its effect on the performance of matching image scores in terms of quality, and the amount minutiae between capacitance and optical fingerprint sensors. Hsieh et al. [5] developed another method for fingerprint authentication where only ridge bifurcations are used and not ridge endings, using a different algorithm for ridge bifurcation that excludes unclear points. In addition, from a wide study into different research papers it is clear that there are a number of methods in use. A number of changes have occurred in different preprocessing techniques, such as segmentation by the use of external characteristics called morphological operations as well as an improved thinning process, different techniques for removing false minutiae, minutia marking using the triple branch counting method, breakdown of minutia unification into three terminations, and matching in the unified x-y coordinate system after a two-step conversion. An online fingerprint identification and recognition method using a hashing technique which is very fast and tolerant to distortion.

2.1 Finger Print Enhancement

A fingerprint enhancement algorithm takes its input from a fingerprint image. Fingerprint enhancement is either done on (i) a binary image or (ii) a gray level images. The database for this research consists of fingerprints which have been scanned to give an impression of the finger using an ink and paper technique because this method introduces a high level of noise to the image and helps to evaluate performance with different types of finger [6]. The main aim of using an image enhancement process is to enable the designing of an authentication system that works in the worst conditions as well as to enhance performance and so get the best results possible.

2.2 Direct Gray-Level Enhancement

Using a gray-level fingerprint image, ridges and valleys in a local neighborhood appear in the form of a sinusoidal-shaped wave, which has the properties of a well-defined orientation and frequency. There is a need to estimate these local orientations and frequencies to improve the quality of gray-level fingerprint images [2]. For the filtering, a Gabor filter is used which employs these orientation and frequency properties to enhance the image. Otsu's algorithm is a simple and popular thresholding method for image segmentation, which falls into the clustering category. The algorithm divides the image histogram into two classes by using a threshold such as the in-class variability being very small. This way, each class will be as compact as possible.

2.3 Algorithm

Algorithm: Thresholding segmentation using Otsu's method

Input: finger image (grayscale), overridden threshold value output: output image

```
1  Read (finger_ image)
2  N = finger_image.width × input    imnumberage.height    initialize
   variables
3  threshold, var max, sum, sumB, q1, q2, µ1, µ2 = 0
4  max intensity = 255
5  for i = 0; i <= max intensity; i++ do
6  histogram[value] = 0 accept only grayscale images
7  if num channels(input image) > 1 then
8  return error compute the image histogram
9  for i = 0; i < N; i++ do
10 value = input image[i]
11 histogram[value] + = 1
12 if manual threshold was entered then
13 threshold = overridden threshold
14 else auxiliary value for computing µ2
15 for i = 0; i < = m ax intensity; i ++ do
16 sum + = i × histogram[i]update qi(t)
17 for t = 0; t < = max intensity; t ++ do
18 q1 + = histogram[t]
19 if q1 == 0 then
20 continue
21 q2 = N − q1update µi(t)
22 sumB + = t × histogram[t]
23 µ1 = sumB/q1
24 µ2 = (sum–sumB)/q2 update the between-class variance
25 2 b (t) = q1(t)q2(t)[µ1(t) − µ2(t)]2update the threshold
26 if 2 b (t) > var max then
27 threshold = t
28 var max = 2b (t)build the segmented image
29 for i = 0; i < N; i++ do
30 if finger_image [i] > threshold then
31 output image[i] = 1
32 else
33 output image[i] = 0
34 return output image
```

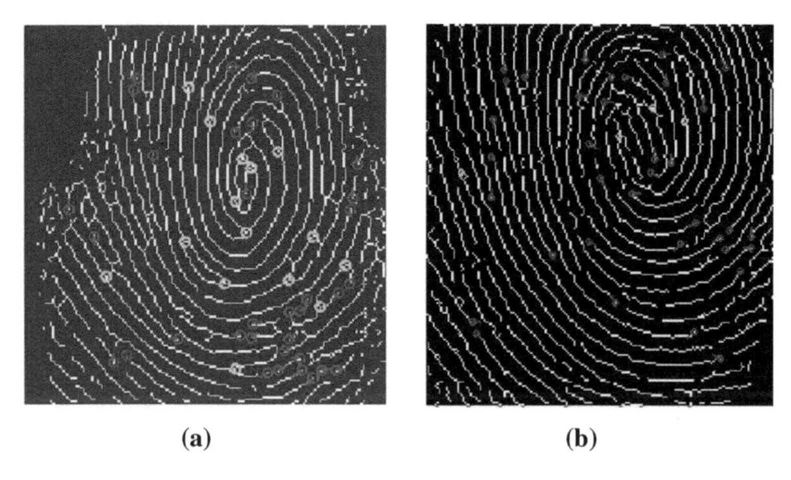

Fig. 4 **a** Minutiae extraction with enhancement (green dots show true minutiae, red false minutiae. **b** Minutiae extraction without enhancement (green dots show true minutiae, red false minutiae)

3 Experimental Results and Discussion

Fingerprint enhancement plays an important role in authentication systems because the performance of the system depends upon the false acceptance rate and false rejection rate. So minutiae must be extracted properly. By using Otsu's method for the segmentation phase, the performance of the system increases compared to its performance without the enhancement process (Fig. 4).

Different minutiae extraction algorithms [7] are present in the literature which we use a CN technique that is able to accurately detect all valid bifurcations and ridge endings from the skeleton image. The false acceptance rate and the false rejection rate determine the performance of the system. We evaluate the performance of both enhanced images and non-enhanced images and it can be seen that the matching rate is low, or that the false acceptance rate (FAR) is greater in non-enhanced images compared to enhanced images. We use a neural network for decision making by selecting a threshold value (Fig. 5).

4 Conclusion

Although there are different types of algorithm available for segmentation in the image enhancement process, the performance of the system is greatly affected by applying the enhancement algorithm. Otsu's method efficiently enhances the clarity of the minutiae (ridge structures). Using Otsu's method, ridge ending and ridge bifurcation are efficiently calculated during the image enhancement process and the noise level is reduced. A number of thresholding algorithms are available for the

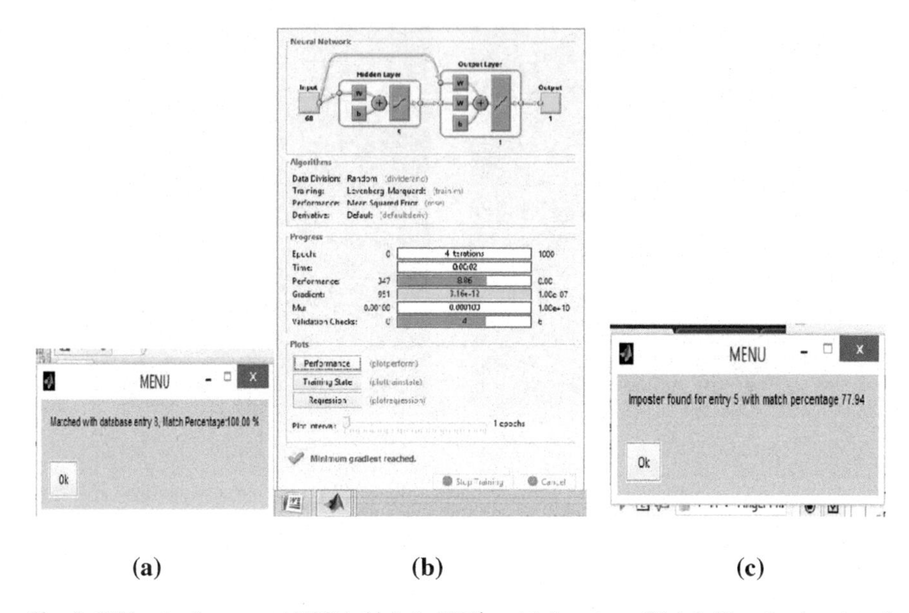

Fig. 5 Without enhancement FAR is high (**a** 100% match is not possible). **b** Neural network tool for matching minutiae. **c** With enhancement FAR is reduced

segmentation process, but Otsu's method is easily implemented and the time taken to extract minutiae also decreases due to inaccurate estimation of the orientation as well as to ridge frequency parameters. The performance of a biometric authentication system is evaluated by the false acceptance rate (FAR) and the false rejection rate (FRR). Without enhancement FAR increases compared to enhanced images at different threshold values. Gabor filter is also less effective due to presence of noise.

References

1. Handbook of Fingerprint Recognition by David Maltoni (Editor), Dario Maio, Anil K. Jain, Salil Prabhakar
2. Hong L, Wan Y, Jain AK (1998) Fingerprint image enhancement: algorithm and performance evaluation. IEEE Trans Pattern Anal Mach Intell 20(8):777–789
3. Raju Sonavane, Sawant BS (2007) Noisy fingerprint image enhancement technique for image analysis: a structure similarity measure approach. J Comput Sci Net Secur 7(9):225–230
4. Kukula EP, Blomeke CR, Modi SK, Elliott SJ (2008) Effect of human interaction on fingerprint matching performance, image quality, and minutiae count. International Conference on Information Technology and Applications, pp 771–776
5. Hsieh CT, Shyu SR, Hu CS (2005) An effective method of fingerprint classification combined with AFIS. EUC 2005: Conference Paper, Embedded and Ubiquitous Computing – EUC pp 1107–1122

6. Hong L, Wan Y, Jain A, Fingerprint image enhancement: algorithm and performance evaluation. East Lansing, Michigan
7. Fingerprint Minutiae Extraction, Department of Computer Science National Tsing Hua University Hsinchu, Taiwan 30043
8. Hong L, Jain A, Pankanti S, Bolle R (1996) Fingerprint enhancement. Pattern Recognit 202–207
9. Jain AK, Hong L, Pantanki S, Bolle R (1997) An identity authentication system using fingerprints. Proc IEEE 85(9):1365–1388
10. Garris MD, Watson CI, McCabe RM, Wilson L (2001) National institute of standards and technology fingerprint database, Nov 2001
11. Guo Z, Hall RW (1989) Parallel thinning with two-subiteration algorithms. Commun ACM 32(3):359–373
12. Jain AK, Farrokhnia F (1991) Unsupervised texture segmentation using Gabor filters. Pattern Recogn 24(12):167–186
13. Jain AK, Hong L, Bolle RM (1997) On-line fingerprint verification. IEEE Trans Pattern Anal Mach Intell 19(4):302–314
14. Gunjan VK, Shaik F, Kashyap A, Kumar A (2017) An interactive computer aided system for detection and analysis of pulmonary TB. Helix J 7(5):2129–2132. ISSN 2319–5592

Estimation of Success Probability in Cognitive Radio Networks

Chilakala Sudhamani, M. Satya Sai Ram and Ashutosh Saxena

Abstract In this paper, we considered a cooperative spectrum sensing over fading and non-fading channels. We proposed a model of a Rayleigh fading channel and a non-fading additive white Gaussian noise channel. Total error rates and the optimal number of cooperative secondary users over the non-fading channel and the success probability over the fading channel are calculated and the simulation results plotted. The simulation results convey that the optimal number of secondary users is five in both cases. We hope that our results will be useful in improving energy efficiency in identifying the unutilized spectrum.

Keywords Cognitive radio · Cooperative spectrum sensing · Success probability · Total error rate

1 Introduction

The widespread use of wireless technology has inevitably resulted in an increased need for spectrum resources. This leads to a spectrum scarcity issue caused by unutilization of the spectrum [1, 2]. Hence cognitive radio (CR) technology has been proposed to identify the unutilized spectrum and to avoid disagreement between spectrum underutilization and spectrum scarcity [3]. In order to identify the unutilized spectrum, spectrum sensing techniques have been used. In spectrum sensing techniques, the secondary user (SU) will sense the primary user's (PU) licensed spectrum by identifying the received signal strength, noise, and the number

C. Sudhamani (✉) · A. Saxena
CMR Technical Campus, 501401 Hyderabad, Telangana, India
e-mail: sudhamanich@gmail.com

A. Saxena
e-mail: saxenaaj@gmail.com

M. Satya Sai Ram
Chalapathi Institute of Engineering and Technology, 522034 Guntur, AP, India
e-mail: msatyasairam1981@gmail.com

A. Kumar and S. Mozar (eds.), *ICCCE 2018*,
Lecture Notes in Electrical Engineering 500,
https://doi.org/10.1007/978-981-13-0212-1_29

of users using that spectrum [4]. The availability of free spectrum depends on the availability of spectrum holes that change according to time and place. This shifts the challenge to identification and detection of the PU [5–7]. A cooperative spectrum sensing (CSS) technique has been proposed as an effective method to improve detection performance [8, 9].

In CSS all the SUs will sense the channel and forward their local decisions to the fusion center (FC). The FC will combine all local decisions using fusion rules and make a final decision. The performance of CSS in CR networks mainly depends on channel imperfection due to fading effects. The performance of CSS over a Rayleigh fading channel in terms of probability of misdetection has been studied in [10, 11]. In these papers the authors considered that the reporting channels between the SU and the FC are imperfect and they observed that the misdetection probability decreases only by increasing the reporting channel signal-to-noise ratio (SNR). As the SNR increases, channel imperfection decreases, and this automatically reduces the probability of misdetection. In [12], the probability of a false alarm and the probability of misdetections are calculated for fading channels in CSS and comparisons will be made in the future purpose. CSS for hard fusion rules is compared in [13] for a Suzuki fading channel. CSS using energy detection in log-normal shadowing was explained in [14]. This explains the performance of CSS over a large fading channel and requires more cooperation among SUs to improve detection performance. For indoor [15] and outdoor [16] environments, Weibull fading has been used. The performance of a single CR user spectrum sensing is best in a Weibull fading channel when compared to other channels, such as Rayleigh and Nakagami.

However, existing works have only examined the AWGN channel. In this paper, we consider the fading channel rather than the non-fading channel. We used the Rayleigh fading channel for calculating the success probability of CSS. The rest of the paper is organized as follows: The system model of a CSS is defined in Sect. 2. Success probability is calculated in Sects. 3 and 4 provides the results of simulation. Finally, conclusions are drawn and future directions discussed in Sect. 5.

2 System Model

Consider a CSS with a K number of SUs, one PU and one FC. In this system, each SU senses the local spectrum independently and then forwards its binary local decisions (1 or 0) to the FC. The FC combines all the local decisions and makes a final decision to identify the presence or absence of the PU. In a CR network, the absence and presence of the PU is given by hypothesis testing as H_0, H_1. Under these two hypotheses the received signal strength of the *i*th SU is given as [17]

$$Y_j = X_j + n_j \colon H_0 \tag{1}$$

$$Y_j = X_j + n_j + S_j \colon H_1 \tag{2}$$

where X_j is the complex-valued channel input, Y_j is the complex-valued channel output, and S_j is the transmitted signal of the PU.

3 Success Probability

For a non-fading channel, the probability of detection and the probability of false alarm at the *j*th SU are given as [18]

$$P_d^{(j)} = Q\left(\sqrt{2\gamma}, \sqrt{\lambda}\right) \tag{3}$$

$$P_f^{(j)} = \frac{\Gamma(U, \lambda/2)}{\Gamma(U)} \tag{4}$$

where U is the time bandwidth product, λ is the detection threshold, γ is the signal-to-noise ratio, $Q(.)$ is the Marcum Q-function, Γ (.) is the incomplete gamma function, and Γ (.) is the gamma function.

Success probability is defined as identifying the presence or absence of the primary user correctly by the secondary user. In CSS, in order to detect the presence or absence of the PU, K SUs will sense the PU channel and forward their local decisions to the FC. The FC will combine all the local decisions according to soft, quantized soft, and hard fusion rules. In soft combining, all the CRs will send their total sensing information to the FC. In the quantized soft method, CR users quantize the sensed data and then forward those quantized samples for soft combining. In the hard combining method, each CR user make a one-bit local decision and forwards that local decision to the FC for hard combining [19]. In all three methods, we are using a hard combining method to reduce the channel overhead.

We have three hard fusion rules AND, OR, and MAJORITY. From Figs. 1 and 2, it can be seen that the MAJORITY fusion rule is the optimal solution for calculating success probability. As a result we used MAJORITY rule in this paper. Hence by using the MAJORITY fusion rule, the probability of detection and the probability of false alarm at the FC are given as [20] respectively.

$$Q_{d,Maj} = \sum_{\left(l=\frac{K}{2}\right)}^{K} \begin{bmatrix} K \\ l \end{bmatrix} P_{d,j}^{l} \left(1 - P_{d,j}\right)^{K-l} \tag{5}$$

$$Q_{f,Maj} = \sum_{\left(l=\frac{K}{2}\right)}^{K} \begin{bmatrix} K \\ l \end{bmatrix} P_{f,j}^{l} \left(1 - P_{f,j}\right)^{K-l} \tag{6}$$

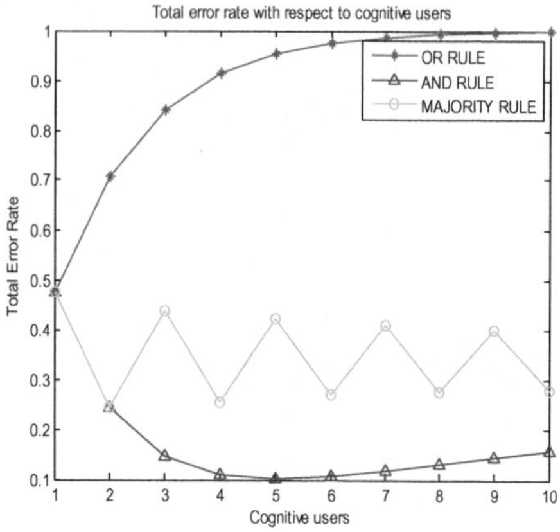

Fig. 1 Total error rate for AND, OR, and MAJORITY fusion rules with respect to cooperative secondary users

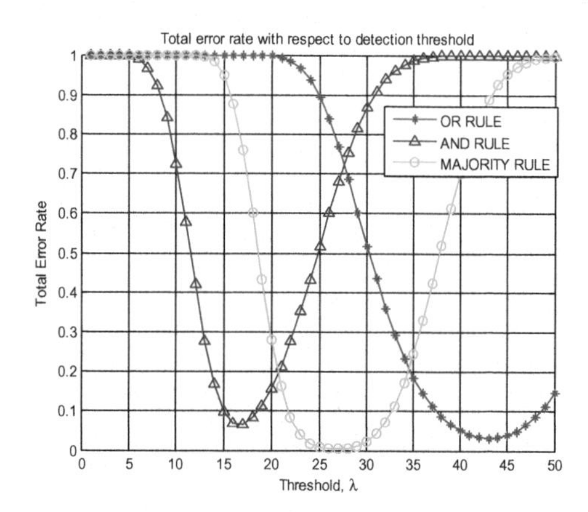

Fig. 2 Total error rate for AND, OR, and MAJORITY fusion rules with respect to detection threshold

$$Q_{m,Maj} = 1 - Q_{d,j} \quad (7)$$

In this paper, we estimated the optimal number of SUs by calculating the total error rate and success probability. Therefore, total error rate is defined as the sum of false alarm probability and misdetection probability, and the success probability is the detection probability.

4 Simulation Results

Consider a cognitive radio network with K = 10 SUs and assume that all the SUs are uniformly distributed around the PU. We assumed the different parameters of signal-to-noise ratio = 10 dB, path loss exponent = 2, and noise variance = 1. Total error rate is defined as the sum of the probability of misdetection and the

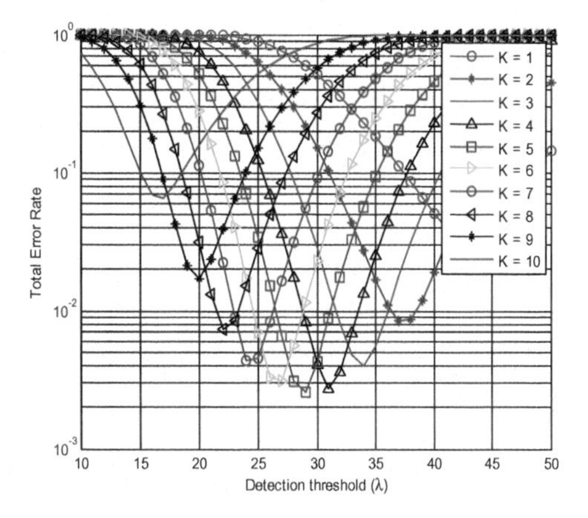

Fig. 3 Total error rate according to detection threshold for K = 1, 2, 3, 4, 5, 6, 7, 8, 9, 10

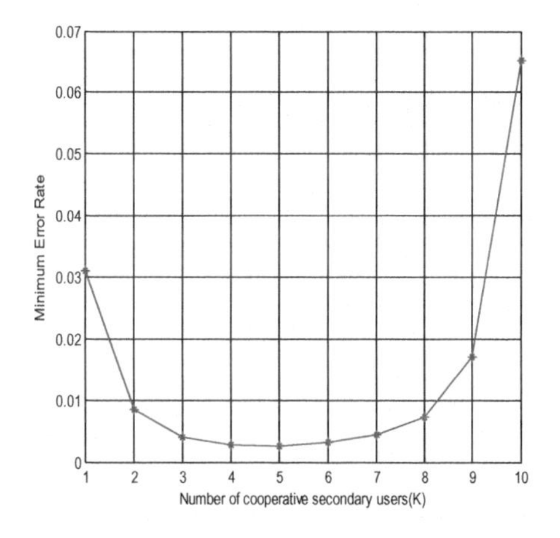

Fig. 4 Minimum error rate according to number of cooperative secondary users

probability of false alarm. The graph in Fig. 3 shows the variation of total error rate with respect to the detection threshold as the number of SUs increases from 1 to 10. We observe that as the detection threshold increases, the error rates first decrease to an optimal value and then increase based on the number of SUs. As the number of SUs varies from 1 to 10, the minimum error rate first decreases gradually up to K = 5 and then increases from K = 5 to K = 10. From Fig. 4 it is can be seen that the optimal number of cooperative SUs is 5 for a minimum error rate. The graph also suggests that beyond 7 SUs the error rates drastically increases. This is in a sense also true in real-time systems because as the number of SUs increases, the energy required for spectrum sensing and reporting sensing results to the FC increases. This reduces the energy efficiency because energy efficiency is defined as the ratio of average channel throughput to average energy consumption.

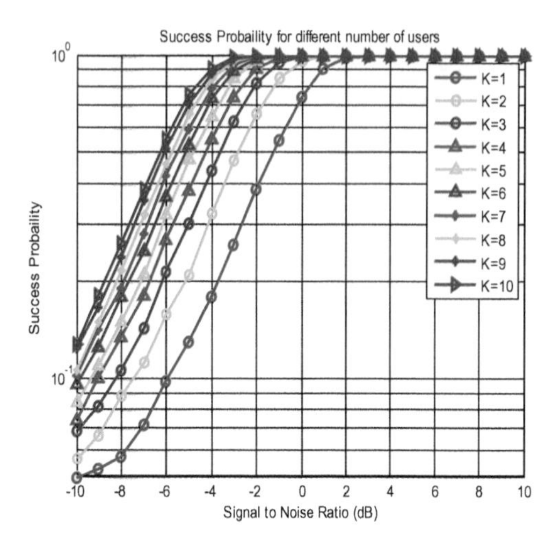

Fig. 5 Success probability with SNR (dB) for K = 1, 2, 3, 4, 5, 6, 7, 8, 9, 10

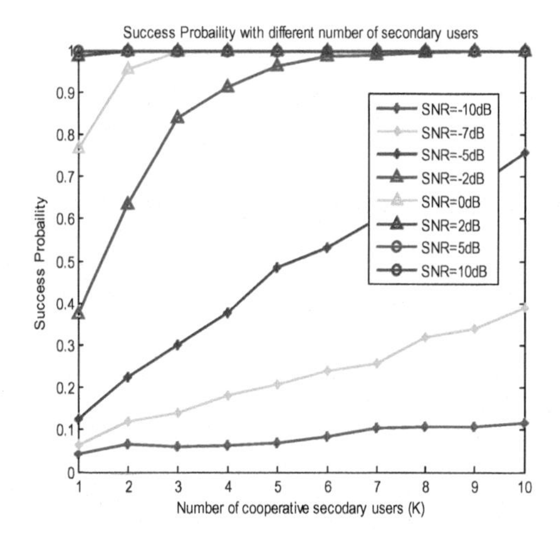

Fig. 6 Success probability with number of secondary users for different SNR values

In CSS, most of the work is done by assuming perfect channels between cooperative SUs. We considered imperfect channels between cooperative SUs for sensing and reporting their sensed information to the FC. Graphs are plotted for different SUs with varying signal-to-noise ratio (SNR). For a Rayleigh fading channel, the success probability for different number of SUs with SNR is shown in Fig. 5. With the increase in number of SUs, the success probability increases along with the SNR. For small values of SNR, the success probability increases and then maintains a constant value of one. The same can be observed in Fig. 6, where the SNR increases as the probability of detection increases.

5 Conclusion

In this study, we looked at CSS in a cognitive radio network over a non-fading AWGN channel and a Rayleigh fading channel. The total error rate for a non-fading AWGN channel was calculated for K = 1–10 SUs and it was found that the optimal number of cooperative SUs to be K = 5. For a Raleigh fading channel, the success probability was calculated to be K = 1, 5, and 10 SUs and it was found that the optimal number of cooperative SUs was K = 5. From this we observed that the optimal number of SUs is the same for both the fading and non-fading channels. This optimal number decreases the false alarm probability and improves the detection probability by using SU cooperation. As the number of cooperative SUs increases, the energy consumption required for spectrum sensing and for reporting sensing results to the FC increases, and this reduces energy efficiency. To improve energy efficiency, we need to reduce the cooperative SUs. Hence, energy efficiency is improved by knowing the optimal number of cooperative SUs in the case of both fading and non-fading channels. To continue our research we also plan to explore the success probability for Ricean and Nakagami fading channels and then present a comparative study on them. We hope that our results will be useful for improving energy efficiency in identifying the unutilized spectrum.

References

1. Spectrum Policy Task Force Federal communications commission. Rep. ET Docket, Washington, D.C., USA, Nov 2002
2. Soy H, Zdemir Z, Bayrak M, Hamila R, Al-Dhahir N (2013) Decentralized multiuser diversity with opportunistic packet transmission in MIMO wireless sensor networks. AEU Int J Electron Commun (Elsevier) 76(2), 910–925
3. Haykin S (2005) Cognitive radio: brain-empowered wireless communications. IEEE J Sel Areas Commun 23(2):201–220
4. Bhowmick A, Roy SD, Kundu S (2015) A hybrid cooperative spectrum sensing for cognitive radio networks in presence of fading. In: Twenty first national conference on communications (NCC), vol 16. https://doi.org/10.1109/NCC.2015.7084887
5. Larsson EG, Skoglund M (2008) Cognitive radio in a frequency-planned environment: some basic limits. IEEE Trans Wirel Commun 7(12):4800–4806. https://doi.org/10.1109/T-WC.2008.070928
6. Sahai A, Hoven N, Tandra R (2004) Some fundamental limits in cognitive radio. In: Proceedings of alert on conference on communications, control and computing, pp 131–136
7. Shafie AE, Al-Dhahir N, Hamila R (2015) Exploiting sparsity of relay-assisted cognitive radio networks. In: IEEE wireless communications and networking conference, WCNC 2015, New Orleans, LA, USA
8. Ganesan G, Li YG (2007) Cooperative spectrum sensing in cognitive radio part I: two user networks. IEEE Trans Wirel Commun 6(6):2204–2213
9. Ganesan G, Li YG (2007) Cooperative spectrum sensing in cognitive radio part II: multiuser networks. IEEE Trans Wirel Commun 6(6):2214–2222
10. Nallagonda S, Roy SD, Kundu S (2011) Performance of cooperative spectrum sensing with censoring of cognitive radios in Rayleigh fading channel. In: Proceedings of IEEE INDICON

11. Nallagonda S, Roy SD, Kundu S (2012) Cooperative spectrum sensing with censoring of cognitive radios in Rayleigh fading channel. In: Proceedings of IEEE eighteenth national conference on communications
12. Duan J, Li Y (2010) Performance analysis of cooperative spectrum sensing in different fading channels. In: Proceedings of IEEE international conference on computer engineering and technology (ICCET'10), pp 64–68
13. Kyperountas S, Correal N, Shi Q, Ye Z (2008) Performance analysis of cooperative spectrum sensing in Suzuki fading channels. In: Proceedings of IEEE international conference on cognitive radio oriented wireless networks and communications (CrownCom'07), pp 428–432
14. Hashemi H (1993) The indoor radio propagation channel. In: Proceedings of IEEE, pp 943–968
15. Adawi NS (1988) Coverage prediction for mobile radio systems operating in the 800/900 MHz frequency range. IEEE Trans Veh Technol, pp 3–72
16. Singh A, Bhatnagar MR, Mallik RK (2016) Performance of an improved energy detector in multi hop cognitive radio networks. IEEE Trans Veh Technol 732–743
17. Sethi R, Bala I (2013) Performance evaluation of energy detector for cognitive radio networks. IOSR J Electron Commun Eng 46–51
18. Akyildiz IF, Lo BF, Balakrishnan R (2011) Cooperative spectrum sensing in cognitive radio networks: a survey. Phys Commun J (Elsevier) 40–62
19. Kyperountas S, Correal N, Shi Q, Ye Z (2008) Performance analysis of cooperative spectrum sensing in Suzuki fading channels. In: Proceedings of IEEE international conference on cognitive radio oriented wireless networks and communications (CrownCom07), pp 428–432
20. Ghasemi A, Sousa ES (2005) Collaborative spectrum sensing for opportunistic access in fading environments. In: Proceedings of 1st IEEE symposium new frontiers in dynamic spectrum access networks, pp 131–136

Analysis of Road Accidents Through Data Mining

N. Divya, Rony Preetam, A. M. Deepthishree and V. B. Lingamaiah

Abstract There is currently a great deal of interest relating to road accidents that result in the loss of life or harm to an individual. GIS is capable of storing information regarding road accidents like vehicle accidents, hour wise accidents, day wise accidents. Apart from this, road accidents are also addressed by road traffic database. In this research on the city of Hyderabad, road traffic databases is taken into considerations where road accidents impact on the socioeconomic growth of society. A data mining technique is used to discover hidden information from the warehouse to handle road accident analysis. We implement algorithms, such as prediction and classification in Weka version 3.7. We use k-Madrid to form a cluster of related information. Different attributes are subjected to analysis with the conclusion that prediction is the most suitable and accurate algorithm.

Keywords GIS · Data mining · K-medoid · Prediction

1 Introduction

In the modern world lot of accidents happen on the road due to human negligence and traffic tampering. Information is collected through GIS and the best algorithm to minimize road accidents is built. We implement a data mining technique in order to categorize the information gathered through GIS and apply a data mining

N. Divya (✉) · R. Preetam · A. M. Deepthishree · V. B. Lingamaiah
CMRIT, Kandlakoya, Hyderabad, India
e-mail: n.divya38@gmail.com

R. Preetam
e-mail: ronyjcpreetam@gmail.com

A. M. Deepthishree
e-mail: Deepthiathni@gmail.com

V. B. Lingamaiah
e-mail: lingamaiah.vb@gmail.com

A. Kumar and S. Mozar (eds.), *ICCCE 2018*,
Lecture Notes in Electrical Engineering 500,
https://doi.org/10.1007/978-981-13-0212-1_30

prediction algorithm. We aggregate similar information to form a cluster. A report is compiled relating to the road accident information in a particular area of the city. Data mining concepts are implemented in order to uncover hidden information or patterns in the data.

The major motivation behind the use of data mining algorithms or techniques is that they provide an output for any given input, without the need for human effort, through use of the Weka tool. We choose fuzzy set from data mining. This is also known as possibility theory. It describes information in the form of categories or sets and then applies GIS to breakdown the threshold or boundaries for each category set. Truth values are used to represent degree of membership that a certain value abide in a given category.

Recent reports regarding road accidents are in the form of continuous values rather than categorical values. So we use a data mining technique that involves classification and prediction. Classification is a data mining technique that assigns an object to its predefined class based on attributes or training sets. Prediction is a data mining technique where we find the value of one variable based on predicted variables that are independent. We use numeric prediction, which is also known as regression. Regression analysis is used to model the relationship between one or more independent or predicted variables. There are two types of regression, linear and non-linear.

Krishnaveni and Hemalatha [1] studied the statistical properties of four regression models: two conventional linear regression models and two Poisson regression models in terms of their ability to model vehicle accidents and highway geometric design relationships. Roadway and truck accident data from the highway safety information system (HSIS) have been employed to illustrate the use and the limitations of these models. Abdel-Aty [2] used the fatality analysis reporting system.

Usually accidents occur due to the negligence on the part of the driver, lack of awareness of other drivers, and animals becoming obstacles in the road. Intoxicated drivers could also cause accidents and the consequences of these accidents may be minor injury, major injury, permanent damage to any body organ, or even death. Everyday database is updated as accidents occur, so the data stored increases steadily. Data mining would be used to discover new patterns from these databases [2]. Discovering a pattern from the database or repository is a difficult task so data mining is a useful tool by which to discover hidden information. Even if the data is stored across different organizations, data mining should be capable of handling and bringing up useful information.

Weka stands for wikato environment for knowledge analysis. Weka knowledge explorer makes use of a GUI with the help of Weka software. The major Weka software packages include classifiers, filters, clusters, associations and attribute selection with a visualization tool. We can work with any of the technique above via open source software under GNU (general public license). Datasets should be in ARFF format if the file is not in ARFF format. The pre-processor has the facility to receive data from a database as a CSV file using a filtering algorithm.

The other reason for using a database was to explain road accidents in different formats and to minimize the effort invested by the researches and different users in

Table 1 Data set of road accidents

Road	Speed limit (A, B, C)	Whether (X, Y, Z)	Pedestrian distance (L, M, N)	Accident type
Road1	A	X	L	X1
Road1	A	Y	M	X1
Road1	B	Y	L	Y2
Road1	C	X	N	X1
Road1	C	Z	N	Y2

order to collect accident reports. By doing this it can help decision makers to better formulate traffic safety control policies. With more vehicles and traffic the risk of accidents will be increased. There is a relationship between driver, road, car, and accident occurrence. One cannot get improved safety without successfully relating frequency.

Road traffic accident (RTA) analysis objectives include (1) to check, in particular, underlying road-related variables; (2) different data mining techniques are used to check the severity of accident prediction; and (3) the task that models standard classification comparison. Most data mining related studies analyze RTA data locally and globally, and obtain regular results (Table 1).

2 Literature Survey

In recent years, many researchers have undertaken RTA analysis using different methodologies and algorithms.

Tesema et al. [3] developed mining rules for RTAs. They used a clustering technique in order to aggregate the data and divide them into subsets and then used the subsets for classification. CART was most the efficient for analyzing predictive model. When the pre-processing technique was completed, the final dataset used for modelling contained 4,658 records using 16 different attributes of which 13 were base attributes and 3 were derived.

Hirasawa [4] developed a statistical approach to model a traffic accident analysis system. A digital map is used to indicate data on accident conditions. The goal was to use a GIS system to analyze factors contributing to road traffic accidents. GIS is an application developed in order to analyze road traffic analysis.

Nabi et al. [5] proposed some behaviours contributing to RTAs. The best predictors were exceeding limits on normal roads and highways, use of a cellphone while driving, and tired or drunk drivers were the main reasons. This research shows a negative attitude towards traffic safety and the rules of the road. Enforcing users to change their approach such as reducing speed or prohibiting alcohol could improve road traffic rules.

Krishnaaveni and Hemalatha [1] proposed a method for analyzing RTAs using data mining techniques. They used classification methods to predict the severity of injury during accidents. They compared J48 decision tree part naive Bayesian classification. The final result shows that J48 is better.

Sachin et al. (2015) proposed a framework for the city of Dehradun. Indian road accident (11.5 + 4) that took place during 2009 and 2014 using a K-modes clustering technique and association rule mining. The analysis made by the combination of the above techniques showed that the result will be effective if no segmentation has been performed with respect to generation association rules.

Ralambonetrainy (1995) implemented a k-means algorithm through data mining to club categorical data which transform different category attributes into binary numeric attributes. In data mining these attributes range from hundreds to thousands and k-means uses these attributes in order to compute space costs.

Sachin Kumar et al. (2016) recommend implementing a k-means algorithm and ARM technique to find solutions to the severity of traffic accidents problem. They divided the data set of traffic accidents into different levels, namely (a) high, (b) moderate, and (c) low frequency in order to discover the hidden information behind the data set and establish preventive actions prior to accident location.

Sowmya (2015) undertook a study related to traffic accident data provided by the government transport department in a certain country. The baseline techniques that were used are naïve Bayes, J48, adeboost, M.PART, and random forest classification in order to predict classification accuracy and analyze performance.

Krishnavani (2011) analyzed traffic accidents in Nigeria using different classification models and compared naïve Bayes Bayesian classifier [3]. The research also used an artificial neural networks approach and decision tree analysis to reduce deaths on the road. The data that was used for research was classified into two types (a) continuous and (b) categorical.

ANN is exclusively used to deal with continuous values of data, whereas decision has the ability to deal with categorical data. The results of the analysis revealed that a decision tree algorithm or approach was efficient when compared to ANN. A decision tree approach had a lower error rate and a higher accuracy rate

3 Existing Model

We focus on the predicting possibilities in a particular area using machine learning techniques such as SMO, J48, and IBK.

SVM: This uses the concept of decision trees and planes that define the boundaries of any decision. A decision plane separates objects or set of objects having different class memberships. The main task of SVM is to construct hyperplanes in multidimensional space. It can be categorized like regression and classification that can handle continuous or categorical values.

Decision Tree: This is one of the classification techniques and takes the form of a flowchart that selects labels for any given input. It consists of a root node, an internal node, and a leaf node. Decisions are made on the internal node. A leaf node cannot be further divided into any of the nodes. A root node contains a condition for any input values and selects a branch based on certain features.

Let S(i|t) denote function of records categorized under the class 'I' at any node 't'. We find three different equations to calculate impurity measures.

$$\text{Entropy} = -S(i|t)\log_2 S(i|t) \quad (1)$$

$$\text{Gini Index} = 1 - S(i|t)^2 \quad (2)$$

$$\text{Classification error} = 1 - \max[S(i|t)] \quad (3)$$

Here 'c' represents number of classes and $0\log_2(0) = 0$, and the calculation is computed for a given node 't'.

Association mining rule is a technique that explains how data are correlated in any transaction. An example is market basket analysis. It defines the underlying rules that produce pattern in any data set.

Let 'D' be given any data set which consists of n transactions where each transaction TϵD. Let I be the set of items in any transaction I = {I1, I2, …., In}.

An item set A will occur in T if and only if AcT. A-> B is an association rule such that AcI. IcI and AUB = o.

Association rule mining mainly focuses on supper and confidence.

The support of a rule A-> B states the percentage of A and B occurring together in a data set. It is also called the frequency constraint. Frequent item sets are generated if they support the minimum support threshold.

Confidence: A-> B defines the ratio of the occurrence of A and B together to the occurrence of A only. The higher the confidence value of a rule A-> B, the higher the probability of occurrence of B with the occurrence of A.

Lift: Lift for a rule A-> B calculates the expected occurrence of A and B together. It is also the ratio of the confidence of the expected confidence of c rule.

$$\text{Support}(S_p) = P(AuB) \quad (4)$$

$$\text{Confidence (Cf)} = P(AUB)/P(A) \quad (5)$$

$$\text{Lift (Lf)} = P(AUB)/P(A)*P(B) \quad (6)$$

4 Proposed System

Data mining consists of different techniques that are used to discover useful information from a large repository. Wherever the information is discovered should be used to make decision for any enterprises.

Discretize information into categories based on number of accidents accused and we can classify it into two categories, i.e. number of year and number of accidents.

Prediction is a data mining technique which allows us to predict continuous values or ordered values for any given input. If we wanted to find out the number of road accidents that has occurred for a particular year, prediction can be employed. Prediction is also known as regression in statistical methodology. Regression analysis can be used to model a relationship between one or more independent and dependent variables. In general, independent variables are also known as predictor variables, whose values are known to us. We use linear regression to solve the problem. Linear Regression involving response variables 'z' and a single predictor variable 'x' is a simple form of regression and models response variables as the linear function of the predictor variable 'x'.

$$z = c + wx \tag{7}$$

'z' is assumed to be constant, and 'c' and w are regression coefficients indicating intercept and slope of the line.

Keeping the weights in consideration, the regression coefficients can also be found using the formula

$$Z = a_0 + a_1 x \tag{8}$$

where a_0 and a_1 are the weights. Weight can be computed by

$$a_1 = (x_i - x)(y_i - y)/(x_i - x)^2 \tag{9}$$

where 't' is equal to training set are those whose class label is known.

$$a_0 = xy - w_1 x \tag{10}$$

Case 2: Non-linear: This involves transformation of the variables from a non-linear to a linear model.

Case 3: K-Medoid: We introduce an object-based representative technique known as k-medoid, where instead of taking the mean value of objects in a cluster as the reference point k-medoid will actually pick an object to represent the clusters. This object is used as a representative object in the cluster. The remaining objects within cluster if they are similar to the most representative object and then they are all clustered together.

Absolute error criterion is used to investigate the above road accident problems.

$$F = |k - R_i| \tag{11}$$

where 'F' is the sum of the absolute errors for all objects in the data set; 'k' is the point in space representing a given object in the cluster 'j'; and r_i is the representative object of r_i.

5 Results and Conclusion

See Fig. 1.

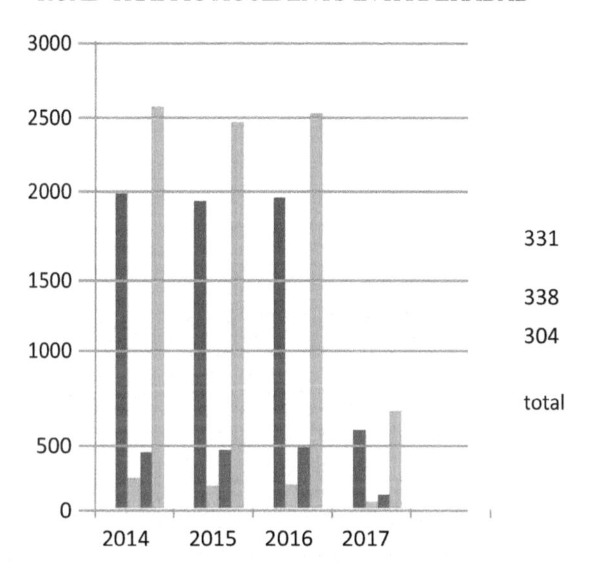

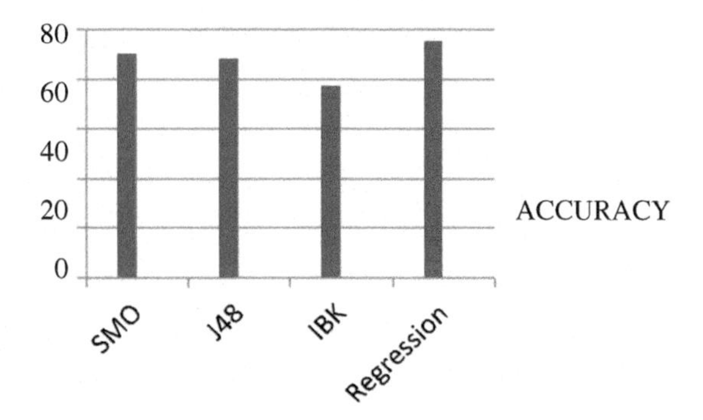

Fig. 1 Comparison of SMO, J48, IBK, and Regression algorithms

6 Conclusion

We explore different data mining techniques and their applications in Hyderabad to analyze road traffic accidents using a prediction model. Prediction may help traffic officers to make decisions in their control activities especially regarding behaviors like accident mode, time, and causes. Based on the results, systems can upgrade or enhance traffic safety policies using a k-medoid algorithm to combine most familiar information the output generated suggests that linear regression performs better than SMO, J48, and IBK.

References

1. Krishnaveni S, Hemalatha M (2011) A Perspective analysis of traffic accident using data mining techniques. Int J Comput Appl (0975–8887) 23(7)
2. Gartner Group Advanced Technologies & Applications Research Note 2/1/95
3. Tesema T, Abraham A, Grosan C (2005) Rule mining and classification of road traffic accidents using adaptive regression trees. I. J Simul 6(10):80–94
4. Hirasawa M (2005) Development of traffic accident analysis system using GIS. Proc East Asia Soc Transp Stud 10(4):1193–1198
5. Nabi H, Salmi LR, Lafont S, Chiron M, Zins M, Lagarde E (2007) Attitudes associated with behavioral predictors of serious road traffic crashes: results from the GAZEL cohort. Inj Prev 13(1), 26–31; Jacobs IS, Bean CP (1963) Fine particles, thin films and exchange anisotropy. In: Rado GT, Suhl H (eds) Magnetism, vol. III, Academic Press, New York, pp 271–350
6. Gupta M, Solanki VK, Singh VK (2017) A novel framework to use association rule mining for classification of traffic accident severity. J Ing Solidar 13(21). p-ISSN 1900-3102, e-ISSN 2357-6014
7. Gupta M, Solanki VK, Singh VK Analysis of data mining technique for traffic accident severity problem: a review. In: Second international conference in research in intelligent & computing in engineering

An Assessment of Niching Methods and Their Applications

Vivek Sharma, Rakesh Kumar and Sanjay Tyagi

Abstract Populace-based metaheuristics have been demonstrated to be especially powerful in taking care of MMO issues if furnished with particularly planned decent variety saving systems, commonly known as niching strategies. This paper provides a fresh review of niching techniques. In this paper, an assessment of niching methods is presented along with their real-time applications. A rundown of fruitful applications of niching techniques to genuine issues is used to show the capacities of niching strategies in giving arrangements that are hard to other enhancement techniques to offer. The critical viable benefit of niching techniques is clearly exemplified through these applications.

Keywords Niching methods · Multi-modal optimization · Metaheuristics
Multi-solution methods · Evolutionary computation · Swarm intelligence

1 Introduction

The two ideas of niche and species can be found in regular biological systems, where singular species must contend to get by going up against various parts [1]. Diverse species or organisms evolve to fill distinctive niches (or subspaces) in an environment that can support a diverse array of life. As commented in [2] "A niche can be characterized by and large as a subset of assets in the earth. A species, then again, can be characterized as a sort or class of people that exploit a specific niche. In this way, niches are divisions of a domain, while species are divisions of the

V. Sharma (✉) · R. Kumar · S. Tyagi
Department of Computer Science and Applications, Kurukshetra University, Kurukshetra, Haryana, India
e-mail: vivek.radaur@gmail.com

R. Kumar
e-mail: rakeshkumar@kuk.ac.in

S. Tyagi
e-mail: tyagikuk@gmail.com

A. Kumar and S. Mozar (eds.), *ICCCE 2018*,
Lecture Notes in Electrical Engineering 500,
https://doi.org/10.1007/978-981-13-0212-1_31

population" [3]. In science, a species is characterized as a gathering of organisms of comparative natural components fit for interbreeding among themselves, but not with organisms from an alternate gathering. Since every niche has a limited number of assets which must be shared among species, individuals involving that niche, after some time distinctive niches and species develop normally in the earth [4]. Rather than developing a solitary population of people apathetically, regular biological communities evolve into distinctive species (or subpopulations) to fill diverse niches [5].

2 Niche Genetic Algorithm

The basic idea of the niche genetic algorithm is that in biology different species have a tendency to live with species that have comparable characteristics and mate with them to various relatives. The condition that the species rely upon is called niche. The genetic algorithm copies species. Applying these ideas, we can influence the person of the genetic algorithm to advance in the particular living condition. So we can introduce the idea of the niche into the genetic algorithm, and as a result the niche genetic algorithm appears [6].

The existing niche genetic algorithm: The current niche genetic algorithm for the most part comprises of a swarming niche genetic algorithm, a sharing niche genetic algorithm and a disengagement niche genetic algorithm. These algorithms enhance the decent variety of the population and it is a successful strategy to settle the multi-modular capacity improvement issue [7, 8]. Nonetheless, on account of the multifaceted nature of the multi-modular capacity enhancement issue, the current niche genetic algorithm is hard to comprehend it [9, 10]. The problems are as follows:

(1) The number and position of the peaks of the capacity are indeterminate.
(2) The ranges' span of the peak is hard to decide in light of the fact that the width and the stature of the peaks are extraordinary.

3 Related Work

In [11] creators incorporate niche innovation and a PSO calculation into the traditional FastSLAM. The niche-PSO process was executed before ascertaining the significance weight of every molecule, with the goal that the molecule set can be nearer to the genuine condition of the versatile robot before resampling, and the resampling procedure was more effective. The distinction of niches ensures the distinction of particles utilized for the state estimation, so the decent variety of particles was kept up and the molecule consumption was maintained a strategic distance from, which protect improve the accuracy of the state estimation. Since the single molecule was viewed as a niche, the computational multifaceted nature can be decreased and the framework state can be evaluated with few particles.

In [12], by coordinating the supervision data and the neighborhood structure of heterogeneous information, a novel strategy named hetero-complex regularization (HMR) was proposed to learn hash capacities for proficient cross-modular pursuit.

In [13], to conquer the untimely deformity of customary subterranean insect province calculation, another enhanced niche subterranean insect settlement calculation (niche subterranean insect state calculation in light of the wellness sharing standard) was proposed by joining the wellness offering strategy to niche subterranean insect settlement calculation and connected to the multi-modular capacity improvement issue.

In [14], MEDAs were created to find different worldwide optima for multimodal streamlining issues. Dissemination estimation and niching are successfully used to understand the proposed calculations. Uniquely, the grouping-based niching strategies for swarming and speciation are joined, prompting swarming-based and speciation-based MEDAs, named MCEDA and MSEDA, individually. Further, they are improved with nearby pursuit, framing LMCEDA and LMSEDA, individually. The niching techniques for MEDAs are enhanced from those in the writing through building up a dynamic bunch measuring system to bear the cost of a potential harmony amongst investigation and misuse, whereby easing MEDAs from the affectability to the group estimate. Varying from traditional EDAs to appraise the likelihood circulation of the entire population, MEDAs concentrate on the estimation of appropriation at the niche level, and all people in every niche take an interest in the estimation of conveyance of that niche. Further, the option utilization of Gaussian and Cauchy dispersions to produce posterity takes the upsides of the two dissemination and possibly offers a harmony amongst investigation and abuse. At long last, the arrangement precision is upgraded through another neighborhood seek plot in view of Gaussian circulation with probabilities self-adaptively decided by wellness estimations of seeds.

In [15] a niche with population relocation procedure was proposed to understand the multi-modular capacity advancement issue. They approve the adequacy of their calculation by three standardized one-dimensional multi-modular capacities. Through contrasting and the trial information of applicable written works' calculations, the proposed system has favorable circumstances in the accuracy of the count and the support of the assorted variety of the population.

In [16] gives a refreshed study on niching strategies. The paper initially returns to the principal ideas about niching and its most illustrative plans, and at that point surveys the latest advancements in niching strategies, including novel and half-and-half techniques, execution measures, and the benchmarks for their appraisal. In addition, the paper provides a study on past work that looked at utilizing the capacities of niching to encourage different improvement errands (e.g., multi-goal and dynamic advancement) and machine learning undertakings (e.g., grouping, include determination and learning gatherings). A rundown of useful applications of niching techniques to certifiable issues was introduced to show the abilities of niching strategies in providing arrangements that are hard for other streamlining strategies to offer. The huge down to earth benefit of niching techniques was unmistakably exemplified through these applications. Finally, the paper offers difficulties and research conversation starters on niching that are yet to be fittingly tended to.

In [17] an area-based transformation was proposed and coordinated with different niching DE calculations in order to tackle multimodal improvement issues. Neighborhood change could limit the generation of posterity inside a neighborhood an indistinguishable niche from their folks. This strategy guarantees that the calculations are faster and have a high degree of accuracy. They showed that the area transformation can instigate stable niching conduct. The area-based DE calculation could find numerous worldwide optima and look after them. The after effects of exploratory examinations recommend that the proposed calculations can give a superior and more predictable execution than various best in class multimodal streamlining calculations for an expansive number of test issues.

In [18] expands upon that, additionally investigating distinctive methodologies towards complex system examination based versatile component. Initially, the arrangements were positioned by centrality just, and the substitution was constantly proficient by random re-initialization. In this form, distinctive methodologies can be picked. The elitism has been fused into the versatile instrument, i.e. the arrangement's quality is additionally considered when choosing whether or not to expel it from the population. Besides, the productivity of applying ABC's change administrator as opposed to supplanting the arrangement by a randomly created one was investigated, enhancing misuse rather than investigation. Three variations of Adaptive ABC calculation with expanded properties and included control parameters are exhibited.

In [19] proposed a methodology versatile memetic swarming DE (SAMCDE) which consolidates CDE with system versatile and fine pursuit strategy to tackle multimodal improvement issues. The proposed SAMCDE tackled the issue of choosing appropriate trial vector era system and control parameters. The fine hunt strategy upgrades the neighborhood seeks capacity of the proposed calculation which enhances the merging pace and exactness. One more stride to broaden the ebb and flow work includes the procedure adjustment and fine inquiry method to other DE-based niching calculations. Besides, it was intriguing to self-adjust population measure as this parameter likewise has a pivotal impact on niching calculations.

In [20] proposed DESBS a disseminated differential transformative calculation with species and best vector determination strategy. The choice weight for choosing the people as best people of a population was applied at two diverse levels first to give the opportunity to every people to perform and second to calculate the edge. The DESBS execution was greatly improved in multi-demonstrate capacities. In uni-show capacities, DESBS perform substantially more like SDE, yet the capacity assessments are on the higher side as a contrast with different calculations, in light of conveyed nature of DESBS. The idea of DESBS makes it more explorative in nature and SDE which was the posterity era component in DESBS make it exploitative.

In [21], a strategy in view of CLONALG was depicted for the computerized IIR channel plan and its execution was contrasted with that of GA and TS calculations. Keeping in mind the end goal to upgrade the worldwide optima seek ability of CLONALG in taking care of multi-modular capacity streamlining issues, various improvements are made to the calculation to enhance its execution and a novel multi-modular invulnerable advancement algorithm (MIOA) was introduced with the premise of incorporating the attributes of the Chaos and CLONALG in the paper. The correlative analyses after effects of multi-modular capacity advancement demonstrate that MIOA has quick joining speed and capable hunt ability.

4 Real-World Applications

i. In the Femtosecond laser pulse shaping problem, a (CMA-ES)-based niching strategy was utilized to understand the issue in the field of quantum control [22]. A separation metric was fittingly characterized between two possible arrangements, keeping in mind the end goal to find numerous one of a kind heartbeat profiles of high caliber. For this situation, distinctive niches speak to the same theoretical plans [23]. The (CMA-ES)-based niching strategy accomplished preferable arrangement comes about over the standard advancement technique strategy.
ii. Job shop scheduling problem (JSSP): This is a great improvement issue considered broadly in writing. The author speaks to one of the not very many examinations on JSSP with an emphasis on recognizing numerous arrangements. JSSP are normally multi-modular, displaying a perfect case for applying niching techniques. Their examinations recommend that not exclusively do niching techniques help to find numerous great arrangements, yet in addition to saving the decent variety more successfully than utilizing a standard single-ideal looking for hereditary calculation [24].
iii. Resource-constrained multi-project scheduling problems (RCMPSP): Here, different undertakings must be performed and finished utilizing a typical pool of rare assets. The trouble is that one needs to organize each venture's assignments to streamline target work without abusing both intra-project priority limitations and between venture asset requirements. A chief can profit by picking between various suitable planning arrangements, rather than being restricted to just one. Moreover, it is likewise substantially quicker than rescheduling [25].
iv. Seismological inverse problem: A niching GA was connected to a reversal issue of teleseismic body waves for the source parameters of a quake. Here a separation metric for waveform reversal was used to measure the likeness between arrangements. The niching GA appeared to be more proficient than a

matrix in recognizing a few worldwide and neighborhood optima over a range of scales, speaking to the blame and assistant planes.

v. Real-time tracking of body motion: A niching swarm filtering (NSF) calculation was produced to address the issue of the constant following of unconstrained full-body movement. In this situation, numerous critical worldwide and nearby arrangements of the design circulation are found.

vi. Competitive facilities location and design: In this office area issue, ordinarily numerous worldwide arrangements should be gotten. A niching strategy named the Universal Evolutionary Global Optimizer (UEGO) was appeared to essentially beat mimicked toughening and multi-begin techniques.

vii. Solving systems of equations: One of the primary niching PSO calculations were produced to settle frameworks of straight conditions. Niching calculations are appropriate to explain frameworks of conditions because of frameworks of conditions having different arrangements. As of late, it has appeared in that frameworks of nonlinear conditions can likewise be unraveled utilizing niching methods [26].

5 Assessment

Table 1 shows an assessment of different real applications with different performance metrics, such as interaction rate, average delay, and average execution time by applying niching methods. The assessment shows solving TSP problems and RCMPSP problems using niching methods provides a better interaction rate and less delay with optimal execution time.

Table 1 Assessment of different real-time applications with niching

Real-time applications	Interaction rate	Average delay	Average execution time
Job shop scheduling problem	Low	High	High
Resource-constrained multi-project scheduling problems (RCMPSP)	Medium	Medium	High
Travelling salesmen problem	High	Low	Medium
Seismological inverse problem	Medium	Medium	High
Real-time tracking of body motion	Low	High	High

6 Conclusion

Niching techniques are capable search strategies that can deliver various answers for a chief to look over. In this paper, we have returned to exemplary niching strategies in EAs and looked into reent improvements of niching techniques obtained from other metaheuristics. We have appeared through some true application cases that looking for numerous great arrangements is a typical errand over various disciplinary regions, and niching techniques can assume a critical part in accomplishing this undertaking. These cases of niching applications exhibit a more all-encompassing photo of the effect by niching techniques, and ideally, this will give an awesome driving force to a considerably more broad utilization of niching strategies.

References

1. Ward A, Liker JK, Cristiano JJ, Sobek DK (1995) The second toyota paradox: how delaying decisions can make better cars faster. Sloan Manag Rev 36(3):43
2. Boyd S, Vandenberghe L (2004). Convex optimization. Cambridge university press
3. Goldberg DE, Richardson J (July 1987) Genetic algorithms with sharing for multimodal function optimization. In: Genetic algorithms and their applications: proceedings of the second international conference on genetic algorithms. Lawrence Erlbaum, Hillsdale, NJ, pp 41–49
4. De Jong KA (1975) An analysis of the behavior of a class of genetic adaptive systems (Doctoral dissertation)
5. Mahfoud SW (1992) Crowding and preselection revisited. Urbana 51:61801
6. Beasley D, Bull DR, Martin RR (1993) A sequential niche technique for multimodal function optimization. Evol Comput 1(2):101–125
7. Harik GR (July 1995). Finding multimodal solutions using restricted tournament selection. In: ICGA, pp 24–31
8. Bessaou M, Pétrowski A, Siarry P (2000) Island model cooperating with speciation for multimodal optimization. Parallel problem solving from nature PPSN VI. Springer, Berlin/Heidelberg, pp 437–446
9. Yin X, Germay N (1993) A fast genetic algorithm with sharing scheme using cluster analysis methods in multimodal function optimization. In: Artificial neural nets and genetic algorithms, pp 450–457
10. Parsopoulos KE, Plagianakos VP, Magoulas GD, Vrahatis MN (2001) Objective function “stretching” to alleviate convergence to local minima. Nonlinear Anal Theory Methods Appl 47(5):3419–3424
11. Parsopoulos KE, Vrahatis MN (2004) On the computation of all global minimizers through particle swarm optimization. IEEE Trans Evol Comput 8(3):211–224
12. Pétrowski A (May 1996). A clearing procedure as a niching method for genetic algorithms. In: Proceedings of IEEE international conference on evolutionary computation, 1996. IEEE, pp 798–803
13. Li JP, Balazs ME, Parks GT, Clarkson PJ (2002) A species conserving genetic algorithm for multimodal function optimization. Evol Comput 10(3):207–234
14. Engelbrecht AP (2007) Computational intelligence: an introduction. Wiley
15. Horn J (1995) The nature of niching: genetic algorithms and the evolution of optimal. Cooperative Populations, University of Illinois, Urbana-Champaign, Illinois

16. Zhirong Z, Zixing C, Baifan C (May 2011) An improved FastSLAM method based on niche technique and particle swarm optimization. In: Control and decision conference (CCDC), 2011 Chinese. IEEE, pp 2414–2418
17. Zheng F, Tang Y, Shao L (2016) Hetero-manifold regularisation for cross-modal hashing. IEEE Trans Pattern Anal Mach Intell
18. Zhang X, Wang L, Huang B (August 2012) An improved niche ant colony algorithm for multi-modal function optimization. In: 2012 international symposium on instrumentation & measurement, sensor network and automation (IMSNA), vol 2. IEEE, pp 403–406
19. Yang Q, Chen WN, Li Y, Chen CP, Xu XM, Zhang J (2017) Multimodal estimation of distribution algorithms. IEEE Trans Cybern 47(3):636–650
20. Wang ZR, Ma F, Ju T, Liu CM (December 2010) A niche genetic algorithm with population migration strategy. In: 2010 2nd international conference on information science and engineering (ICISE). IEEE, pp 912–915
21. Li X, Epitropakis M, Deb K, Engelbrecht A (2016) Seeking multiple solutions: an updated survey on niching methods and their applications. IEEE Trans Evol Comput
22. Qu BY, Suganthan PN, Liang JJ (2012) Differential evolution with neighborhood mutation for multimodal optimization. IEEE Trans Evol Comput 16(5):601–614
23. Metlicka M, Davendra D (July 2016) Complex network based adaptive artificial bee colony algorithm. In: 2016 IEEE congress on evolutionary computation (CEC). IEEE, pp 3324–3331
24. Liang JJ, Ma ST, Qu BY, Niu B (June 2012) Strategy adaptative memetic crowding differential evolution for multimodal optimization. In: 2012 IEEE congress on evolutionary computation (CEC). IEEE, pp 1–7
25. Khaparde AR, Raghuwanshi MM, Malik LG (May 2015). A new distributed differential evolution algorithm. In: 2015 international conference on computing, communication & automation (ICCCA). IEEE, pp 558–562
26. Hong L (October 2008) A multi-modal immune optimization algorithm for IIR filter design. In: 2008 international conference on intelligent computation technology and automation (ICICTA), vol 2. IEEE, pp 73–77

A Novel Method for the Design of High-Order Discontinuous Systems

G. V. K. R. Sastry, G. Surya Kalyan and K. Tejeswar Rao

Abstract A new procedure is suggested for the design of high order discontinuous systems using an order reduction technique. The method is computationally very simple and straightforward. The proposed method is based on an improved bilinear Routh approximation method and illustrated using typical numerical examples.

Keywords Large scale systems · Modelling · Discontinuous systems

1 Introduction

Very few methods are available for the design of discontinuous systems [1–11]. Some of the existing methods of reduction have serious drawbacks, such as generating an unstable reduced order model for stable original higher order systems, and computational complexity etc. [1–5]. In this paper, an attempt is successfully made to suggest a procedure for the design of high order discontinuous systems using modeling which is computationally very simple and straightforward and overcomes the limitations and drawbacks of some of the familiar methods available in the literature.

A reduced order model is generated using a Routh approximation method and this model is used to design the controller for the high order original discontinuous

G. V. K. R.Sastry (✉)
EEE Department, GIT, GITAM University, Visakhapatnam, India
e-mail: profsastrygvkr@yahoo.com

G. Surya Kalyan · K. Tejeswar Rao
EEE Department, Chaitanya Engineering College, JNT University, Visakhapatnam, India
e-mail: sukaga80@gmail.com

K. Tejeswar Rao
e-mail: tejeshk222@gmail.com

A. Kumar and S. Mozar (eds.), *ICCCE 2018*,
Lecture Notes in Electrical Engineering 500,
https://doi.org/10.1007/978-981-13-0212-1_32

system. The new method is computationally superior over many of the existing methods for the design of discontinuous systems. The proposed method is illustrated by considering typical numerical examples available in the literature. The results are compared with different well-known methods available in the literature.

2 Proposed Procedure

Consider the original n-th order discontinuous system defined as:

$$G_n(z) = \frac{b_0 + b_1 z + b_2 z^2 + \cdots + b_{n-1} z^{n-1}}{a_0 + a_1 z + a_2 z^2 + \cdots + a_n z^n}; \quad a_0 \neq 0$$

Applying bilinear transformation, find G(s) as:

$$G(s) = (z+1) * G(z)|_{z=(1+s)/(1-s)} \tag{1}$$

The α- and β-parameters are obtained from the α-table and β-table given below:

α-table:

$$\begin{array}{llll}
 & a_0^0 = a_0 & a_2^0 = a_2 & a_4^0 = a_4 \\
\alpha_1 = a_0^0/a_0^1 & & & \\
 & a_0^1 = a_1 & a_2^1 = a_3 & a_4^1 = a_5 \\
\alpha_2 = a_0^1/a_0^2 & & & \\
\cdot & a_0^2 = a_2^0 - \alpha_1 a_2^1 & a_2^2 = a_4^0 - \alpha_1 a_4^1 & \\
\cdot & & & \\
\cdot & a_3^0 = a_2^1 - \alpha_2 a_2^2 & & \\
\cdot & & &
\end{array}$$

and so on.

β-table:

$$\begin{array}{lll}
 & b_0^1 = b_1 & b_1^1 = b_3 \\
\beta_1 = b_0^1/a_0^1 & & \\
 & b_0^2 = b_2 & b_2^2 = b_4 \\
\beta_2 = b_0^2/a_0^2 & & \\
\cdot & b_0^3 = b_2^1 - \beta_1 a_2^1 & b_2^3 = b_4^1 - \beta_1 a_4^1 \\
\cdot & & \\
\cdot & b_0^4 = b_2^2 - \beta_2 a_2^2 & \\
\cdot & & \\
\cdot & &
\end{array}$$

and so on.

Let the impulse energy value of the original system be

$$I = 1/4 \sum_{i=1}^{n} \left(\beta_i^2/\alpha_i\right) \tag{2}$$

For a reduced order system of order "k," the values of α'_k and β'_k are defined as

$$\frac{\beta_k^{'}}{\alpha_k^{'}} = \frac{\beta_k}{\alpha_k}$$

$$1/4\left[\sum_{i=0}^{k-1} \left(\beta_i^2/\alpha_i\right) + \beta_k^{'2}/\alpha_k^{'2}\right] = I \quad \text{where } k = 1, 2, 3, \ldots n \tag{3}$$

The numerator and the denominator of the reduced order model can be found by the algorithm by substituting the values of α_k with α'_k and β_k with β'_k respectively and applying an inverse bilinear transformation.

$$G_k(z) = \frac{B_m(z)}{A_m(z)}; \quad k = 1, 2, 3, \ldots n$$

where

$$B_m(z) = \alpha_m(z+1)B_{m-1}(z) + (z-1)^2 B_{m-2}(z) + \beta_m (z-1)^{m-1} \tag{4}$$

$$A_m(z) = \alpha_m(z+1)A_{m-1}(z) + (z-1)^2 A_{m-2}(z) \quad \text{where } m = 1, 2, 3, \ldots k \tag{5}$$

$$B_{-1}(z) = B_0(z) = 0; A_{-1}(z) = \frac{1}{(z-1)}; A_0(z) = 1$$

Design Procedure

The lower order system transfer function R(z) of order 'k' is obtained using the above proposed procedure for the given original uncompensated high order discontinuous unity feedback system with transfer function G(z). The low order model R(z) will then be used in place of the original system transfer function G(z) to find the controller D(z).

Let the transfer function of the compensated system be defined as:

$$H(z) = \frac{D(z) \cdot R(z)}{1 + D(z) \cdot R(z)}$$

where D(z) is the transfer function of the controller to be designed and incorporated in the forward path of the original system.

The controller is to be designed such that the overall compensated transfer function has the damping ratio (ζ) and natural frequency (ω_n) as per the requirement and also the steady state error must be zero.

$$\text{Let} \quad H(z) = \frac{b_0 + b_1 z^{-1} + b_2 z^{-2}}{A_1 + A_2 z^{-1} + A_3 z^{-2}}$$

where $A_1 = 1$, A_2 and A_3 are obtained from given values of ζ and ω_n.

Then the equations corresponding to the above conditions are:

$$H(z)\big|_{z=\infty} = 0; \quad H(1) = 1 \quad \text{and} \quad \frac{dH}{dz} = 1$$

By solving the above equations, $b_1 = A_1 - A_3$ and $b_2 = A_2 + 2A_3$

$H(z)$ is obtained by substituting the above values of b_1 and b_2.

Therefore, the controller transfer function is

$$D(z) = \frac{1}{R(z)} \cdot \frac{H(z)}{1 - H(z)}$$

3 Numerical Examples

Example 1 Consider the system given by

$$G(z) = \frac{0.1625\, z^7 + 0.125\, z^6 - 0.0025\, z^5 - 0.00525\, z^4 + 0.00262\, z^3 - 0.000875\, z^2 + 0.003\, z - 0.000412}{z^8 - 0.62075\, z^7 - 0.415987\, z^6 + 0.076134\, z^5 - 0.059152\, z^4 + 0.190593\, z^3 + 0.097365\, z^2 - 0.016349\, z + 0.002226}$$

Applying the bilinear transformation using the equation

$$G(s) = (z+1)G(z)\big|_{z=(1+s)/(1-s)}$$

$$G(s) = \frac{0.5711222 + 3.49336\, s + 9.168182\, s^2 + 12.753322\, s^3 + 10.13658\, s^4 + 4.619746\, s^5 + 0.97931\, s^6 + 0.09480472\, s^7}{0.255402 + 2.084249\, s + 17.66303\, s^2 + 50.41484\, s^3 + 73.36688\, s^4 + 66.04546\, s^5 + 36.66708\, s^6 + 9.835014\, s^7 + s^8}$$

By using the proposed procedure, the reduced order model obtained is:

$$R(z) = \frac{0.409429\, z - 0.294367}{1.235681\, z^2 - 1.948545\, z + 0.815774}$$

Controller Design

It is proposed to design a controller for the original system using the proposed reduced order model in order to obtain a required performance of the compensated system as per the parameters: $\omega_n = 2$, $\zeta = 0.7$, $\tau = 1$ s.

The transfer function of the compensated system is obtained as

$$H(z) = \frac{0.9392\,z - 0.0514}{z^2 - 0.0702\,z + 0.0608}$$

Therefore, the compensator of the system is

$$D(z) = \frac{1}{R(z)} \cdot \frac{H(z)}{1 - H(z)}$$

The time responses of the compensated system are compared with that of the uncompensated system in Fig. 1.

Comparison with Other Methods:

Example 2 Consider the fourth order discrete time system given by its transfer function [6].

$$G(z) = \frac{0.547377\,z^3 - 0.40473\,z^2 + 0.319216\,z - 0.216608}{z^4 - 1.36178\,z^3 + 0.875599\,z^2 - 0.551205\,z + 0.282145} \quad (\textbf{Original})$$

By using the proposed procedure, the reduced order model obtained is:

$$R(z) = \frac{0.727128\,z - 0.530431}{1.299757\,z^2 - 1.90185\,z + 0.798393}$$

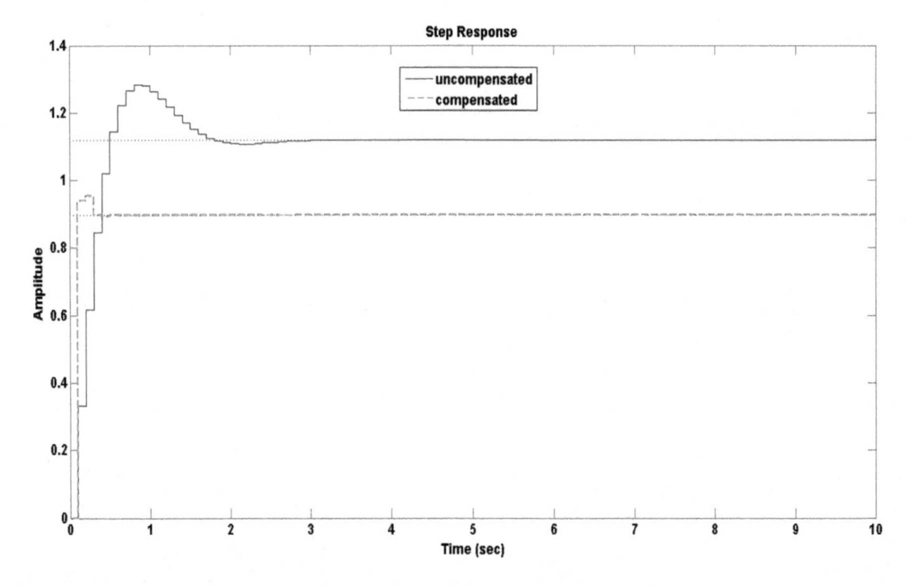

Fig. 1 Comparison of time responses

Therefore, the transfer function of the compensated system is obtained as

$$H(z) = \frac{0.63212\,z - 0.05014}{z^2 - 0.7859\,z + 0.367880} \quad \textbf{(Proposed)}$$

The reduced 2nd order model obtained by the Mukherjee method [6] is:

$$G_2^1(z) = \frac{0.4795\,z - 0.4035}{z^2 - 1.5025\,z + 0.6584}$$

Therefore, the transfer function of the compensated system obtained by the Mukherjee method is

$$H^1(z) = \frac{0.63212\,z - 0.05014}{z^2 - 0.7859\,z + 0.367880} \quad \textbf{(Mukherjee method)}$$

The reduced 2nd order model obtained by the Hsieh method [7] is:

$$G_2^2(z) = \frac{0.5531\,z - 0.4035}{z^2 - 1.4694\,z + 0.6187}$$

Therefore, the transfer function of the compensated system obtained by the Hsieh method is

$$H^2(z) = \frac{0.851\,z + 0.9833}{z^2 + 0.6855\,z + 0.1489} \quad \textbf{(Hsieh method)}$$

The time responses of the compensated system obtained by the proposed method, Mukherjee method, and Hsieh method are compared in Fig. 2.

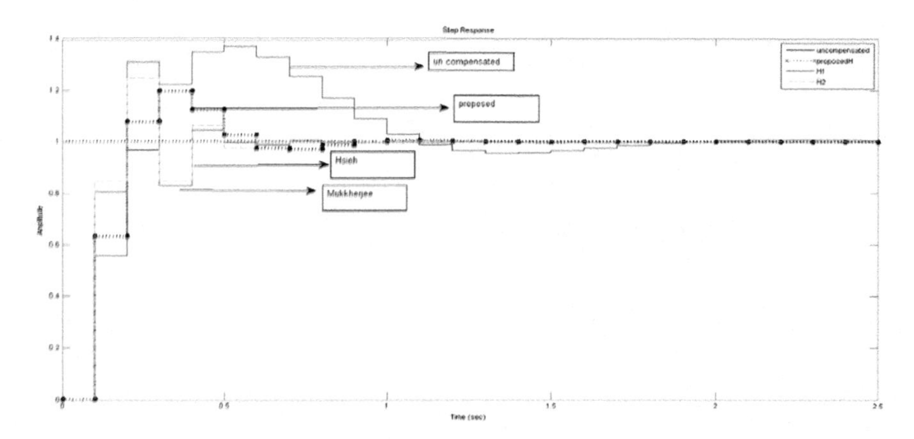

Fig. 2 Comparison of responses using proposed method and other methods

4 Conclusion

A procedure for the design of high order discontinuous time systems using modeling is suggested to overcome the limitations and drawbacks of some of the familiar methods available in the literature. The new procedure is successfully illustrated through typical numerical examples. The results are compared with other familiar methods available in the literature.

References

1. Farsi M, Warwick BSK (1986) Stable reduced order models for discrete time systems. IEEE Proc 133(3)
2. Chen TC, Chang CY (1979) Reduction of transfer function by stability equation method. J Frankl Inst
3. Young C et al (1985) Algorithm for biased continued fraction expansion of Z-transfer function. IEE
4. Unnikrishnan R, Gupta A (1990) Model reduction for discrete systems. Electron Lett
5. Choo Y (2001) Improved bilinear Routh approximation method for discrete-time systems. J Dyn Syst Meas Control 123
6. Mukherjeee S, Kumar V, Mitra R (2004) Order reduction of linear discrete system using an error minimization technique. Inst Eng 85:68–76
7. Hwang C, Hsieh C (1990) Order reduction of discrete time system via bilinear Routh approximation. ASME J Dyn Syst Meas Control 112:292–297
8. Sastry GVKR et al (2016) A new method for order reduction of high order multi variable systems using modified Routh approximation technique. IJCTA 9(5):129–136
9. Sastry GVKR, Tejeswara Rao K (2015) Application of Mihalov criterion for reduction of high order systems. IJAER 10(23):20740–20742
10. Jamshidi M (1983) Large scale system modeling and control series, no 9. Tata McGraw Hill Series
11. Sastry GVKR, Raja Rao G (2001) A simplified CFE method for large-scale systems modelling about s = 0 and s = a. IETE J Res

Efficient Integration of High-Order Models Using an FDTD–TDMA Method for Error Minimization

Gurjit Kaur, Mayank Dhamania, Pradeep Tomar and Prabhjot Singh

Abstract In this research paper, we have developed a hybrid FDTD–TDMA technique which is used to focus on solving the time domain equations of non-uniform structure applications by exploiting the finite difference time domain method. A mathematical model has been designed for the proposed FDTD–TDMA technique. The proposed technique shows better performance than existing FDTD–ADI method in terms of error minimization. The simulated comparison showed good results and an agreement between the two methods, which confirms the theory and validates the proposed FDTD–TDMA method.

Keywords Error correction · FDTD · FDTD–TDMA · Time step

1 Introduction

This finite difference time domain (FDTD) is an important computational technique which wide application for simulating electromagnetic behaviors [1, 2]. FDTD differential equations can be formed directly from Maxwell's equations and are important in mathematics for the simulation of complex dielectric media [3–5]. FDTD is of

M. Dhamania · P. Tomar
School of Information and Communication Technology,
Gautam Buddha University, Greater Noida 201312, UP, India
e-mail: mayank.dhamania@gmail.com

P. Tomar
e-mail: parry.tomar@gmail.com

P. Singh
Salesforce, San Francisco, CA, USA
e-mail: prabhjot27@gmail.com

G. Kaur (✉)
Department of Electronics and Communication Engineering,
Delhi technological University, Delhi, India
e-mail: gurjeet_kaur@rediffmail.com

A. Kumar and S. Mozar (eds.), *ICCCE 2018*,
Lecture Notes in Electrical Engineering 500,
https://doi.org/10.1007/978-981-13-0212-1_33

immense interest to researchers and engineers due to its wide range of applications [6–10]. The FDTD method has been widely used in the past for electromagnetic dosimeter and numerical calculation of electromagnetic grounds to couple error correction and numerical stability. Recently, several techniques were discussed in the literature. In [11], a Z transform was used to reduce complex integrals to simple equations in a LOD–BOR–FDTD method. The reported method had the disadvantage of a high error for dispersive media. Similarly, SLOD–FDTD [12] was used for electromagnetic channel modeling. This method had a drawback which resulted in errors for impulse responses and path loss and could only be applied indoors. Another method known as CLOD–FDTD [13] was introduced to model the curved boundaries, but the main drawback was its minor effect on numerical dispersion.

The present alternating-direction implicit (ADI) and finite-difference time-domain (FDTD) methods typically depend on explicit time addition systems. Hence, we propose a FDTD–TDMA implicit time as a simulation system, as a suggested approach to overcome this restriction.

2 Materials and Methods

The traditional FDTD method is used to statistically resolve the time-domain using Maxwell's calculations with the open space radiation condition. We then make use of design variables, such as the conductivity value at every Yee edge in the design area by using two FDTD solutions, namely the objective function and the gradient vector that can be computed for a number of design variables inside the design domain Ω. After introducing this concept, we make use of Maxwell's basic equations (1)–(2) which are listed below. These equations are compulsory time-dependent curl equations and we explain them in detail

$$\frac{\partial D}{\partial t} = \nabla \times H \tag{1}$$

$$\frac{\partial B}{\partial t} = -\nabla \times E \tag{2}$$

After defining the basic Maxwell equations, we take the dominant difference approximations, for the time and space co-ordinates respectively, hence developing the FDTD formulations given below in (3) and (4)

$$\frac{D_x^{n+\frac{1}{2}}(k) - D_x^{n-\frac{1}{2}}(k)}{\Delta t} = \frac{H_y^n\left(k - \frac{1}{2}\right) - H_y^n\left(k + \frac{1}{2}\right)}{\delta x} \tag{3}$$

$$\frac{B_y^{n+1}\left(k + \frac{1}{2}\right) - H_y^n\left(k + \frac{1}{2}\right)}{\Delta t} = \frac{E_x^{n+\frac{1}{2}}(k) - E_x^{n+\frac{1}{2}}(k)}{\delta x} \tag{4}$$

where δx is the space increase and Δt is the time step. These equations have both the electric and magnetic field in the double-positive (DPS) region; though the formulation varies in the double-negative (DNG) area.

$$D = \varepsilon(\omega) \times E \tag{5}$$

$$B = \mu(\omega) \times H \tag{6}$$

μ (magnetic permeability) and ε (electrical permittivity) are such chosen so that at a particular frequency, these parameters become negative, hence modeling the layer DNG. The base method leading to our proposed work is described by first defining the CN (Crank Nicolson) method.

We make use of the CN arrangement which averages the right-hand sides of the discretized Maxwell's equations at time step n and time step $n+1$. Using a 2-D Yee's mesh, the CN scheme can be expressed as shown in Eqs. (7)–(11), where Δt is the time step size; $a_1 = \Delta t/2\varepsilon$, $a_2 = \Delta t/2\mu$; and ε and μ are the permittivity and the permeability of the material respectively. Δx and Δy are the spatial meshing sizes beside the x and y axes; i and j are the integer-number guides of the computational cells; and n is the time step index. We have defined the medium to be linear, isotropic, non-dispersive, and lossless to maintain simplicity.

$$\begin{aligned} Ex^{n+1}\left(i+\frac{1}{2},j\right) = Ex^{n}\left(i+\frac{1}{2},j\right) + \frac{a_1}{\Delta y}\Bigg(&Hz^{n+1}\left(i+\frac{1}{2},j+\frac{1}{2}\right) - Hz^{n+1}\left(i+\frac{1}{2},j-\frac{1}{2}\right) \\ &+ Hz^{n}\left(i+\frac{1}{2},j+\frac{1}{2}\right) - Hz^{n}\left(i+\frac{1}{2},j-\frac{1}{2}\right)\Bigg) \end{aligned} \tag{7}$$

$$\begin{aligned} Ey^{n+1}\left(i,j+\frac{1}{2}\right) = Ey^{n}\left(i,j+\frac{1}{2}\right) - \frac{a_1}{\Delta x}\Bigg(&Hz^{n+1}\left(i+\frac{1}{2},j+\frac{1}{2}\right) - Hz^{n+1}\left(i-\frac{1}{2},j+\frac{1}{2}\right) \\ &+ Hz^{n}\left(i+\frac{1}{2},j+\frac{1}{2}\right) - Hz^{n}\left(i-\frac{1}{2},j+\frac{1}{2}\right)\Bigg) \end{aligned} \tag{8}$$

$$\begin{aligned} Hz^{n+1}\left(i+\tfrac{1}{2},j+\tfrac{1}{2}\right) = Hz^{n}\left(i+\tfrac{1}{2},j+\tfrac{1}{2}\right) &+ \tfrac{a_2}{\Delta y}\Big(Ex^{n+1}\left(i+\tfrac{1}{2},j+1\right) - Ex^{n+1}\left(i+\tfrac{1}{2},j\right) \\ &+ Ex^{n}\left(i+\tfrac{1}{2},j+1\right) - Ex^{n}\left(i+\tfrac{1}{2},j\right)\Big) \\ &- \tfrac{a_2}{\Delta X}\Big(Ey^{n+1}\left(i+1,j+\tfrac{1}{2}\right) - Ey^{n+1}\left(i,j+\tfrac{1}{2}\right) \\ &+ Ey^{n}\left(i+1,j+\tfrac{1}{2}\right) - Ey^{n}\left(i,j+\tfrac{1}{2}\right)\Big) \end{aligned} \tag{9}$$

The three electromagnetic field mechanisms in Eq. (7) are the final equations. We use Eq. (7) directly as a computational technique, where a large sparse matrix must be resolved at each time step, which is more appropriate than Yee's FDTD, which is not applied to many real problems. In order to solve Eq. (9) conveniently, the field components must be decoupled.

We have used the estimated decoupling technique for decoupling the coupled calculations Eqs. (7)–(9). Next we used the concept of TDMA in our design process where the frame time is divided into U slots and one slot is assigned per wavelength. Each derivative matrix includes the L matrix. The finite-difference vector arrival rate in TDMA is given as:

$$\mathrm{Puar} = \lambda + A * T_f * \bar{E} \tag{10}$$

where

Puar	Wavelength arrival rate [wave per length per second]
λ	Total length arrival rate for the system [waves per length per frame]
A	Message advent degree [message per length unit time]
T_f	Frame time [s]
$\bar{E}$	Average number of waves per message

It can be seen that in a TDMA system each length transmits only one wave in each time frame so we can say that the service rate is equal to 1 ($\mu_{TDMA} = 1$). Because of this process, we have defined the utilization factor or the traffic intensity (ρ) in our proposed work as:

$$\begin{gathered} \rho = \frac{\text{total user arrival rate [packets per user per frame]}}{\text{service rate}} \\ \rho_{TDMA} = \frac{\lambda_{TDMA}}{(\mu_{TDMA} = 1)} = A_{(TDMA)} {*} T_{f(TDMA)} {*} \bar{E}_{(TDMA)} \end{gathered} \tag{11}$$

This can lead to errors which occur in bits due to noise and outer boundary imperfections, but if an error is obtained in the transmission of a wave then it will be retransmitted until it is correctly received. Retransmission is made possible by increasing the average number of waves of each message from $\bar{E}_{outer}$ to $\bar{E}_{inner}$

$$\bar{E}_{inner} = \frac{\bar{E}_{outer}}{P_c^{(TDMA)}} \tag{12}$$

where

$\bar{E}_{new}$ is the new average number of waves per message. [Waves per message]; $\bar{E}_{old}$ is the old average number of waves per message. [Waves per message]; P_c is the probability of correct detection of a wave. In fact, Pc is the ratio of the correctly received waves to the total of transmitted waves.

$$Pc = \frac{\text{Correctly received waves}}{\text{Total transmitted waves}}$$

So we can say that the useful throughput or useful phase velocity for each length and errors polluting the numerical phase velocity is

$$\tilde{\rho} = \rho{*}P_c^{(TDMA)} \tag{13}$$

where $\tilde{\rho}$: is the useful phase velocity.

We now introduce our proposed concept where two approaches are developed for an efficient integration of high-order models (stencils) using FDTD and TDMA techniques together.

Next we consider a TDMA scheme in which time is partitioned into slots, and a central controller decides which node gets to transmit in which slot. Also, we considered another scheme, an RA scheme, in which nodes use a common carrier-sensing approach and random backoffs before deciding to transmit a wireless frame in the medium. After defining the concept of TDMA, we make use of phase velocity in our proposed work where the general expression for a wave is given as

$$Y = A\cos(\omega t - kx)$$

We define Y as movement at any instantaneous t; A is fullness of vibration, $\omega = 2\pi\nu$ is the angular incidence and $k = \frac{2\pi}{\lambda}$ is the wave vector or wave number.

Now the above concept has been utilized in defining phase velocity, which is the velocity of the wave when the phase is constant, i.e. $\omega t - kx$ = constant or, $kx = \omega t$ + constant or, $x = \frac{\omega t}{k}$ + constant. Hence, the formula for phase velocity used in designing our proposed work is

$$vp = \frac{dx}{dt} = \frac{\omega}{k}$$

Next, we define the group velocity, which is essential in the design and simulations of our proposed work. The very useful de-Broglie waves are characterized by a wave packet and hence we have related "group velocity" with them. Group velocity is termed as the velocity with which the wave packet traverses within a particular period of time.

Now we consider two waves having the same amplitude but having slightly different frequencies and wave numbers represented by the equations

$$Y1 = A\cos(\omega t - kx) \tag{14}$$

$$Y2 = A\cos[(\omega + \Delta\omega)t - (k + \Delta k)x] \quad (15)$$

The resultant displacement due to the superposition of the above two waves is,

$$Y = Y1 + Y2 \quad (16)$$

$$Y = A\cos(\omega t - kx) + A\cos\ [(\omega + \Delta\omega)t - (k + \Delta k)x]$$

$$Y = 2A\,\mathrm{Cos}\left\{\left(\frac{2\omega + \Delta\omega}{2}\right)t - \left(\frac{2K + \Delta k}{2}\right)x\right\}$$

As the difference in frequency of the two waves is very small, we can assume that:

$$2\omega + \Delta\omega \approx 2\omega \text{ and } 2k + \Delta k \approx 2k \quad (17)$$

$$Y = 2A\,Cos\left\{\left(\frac{\Delta\omega}{2}\right)t - \left(\frac{\Delta k}{2}\right)x\right\}cos(wt - kx) \quad (18)$$

The speed of the subsequent wave (group velocity) is specified by the quickness with which an orientation point, which is the determined amplitude point, changes its position. Considering the amplitude of the subsequent wave as constant, we have,

$$2A\,cos\left(\frac{\Delta\omega}{2}\right)t - \left(\frac{\Delta k}{2}\right)x = constant$$

$$\text{or}\left(\frac{\Delta\omega}{2}\right)t - \left(\frac{\Delta k}{2}\right)x = constant$$

$$\text{or, } x = \frac{\Delta\omega t}{\Delta k} + \text{constant} \quad (19)$$

Group velocity $vg = \frac{dx}{dt} = \frac{\Delta\omega}{\Delta k}$.
When $\Delta\omega$ and Δk are very small,

$$vg = \frac{d\omega}{dk}$$

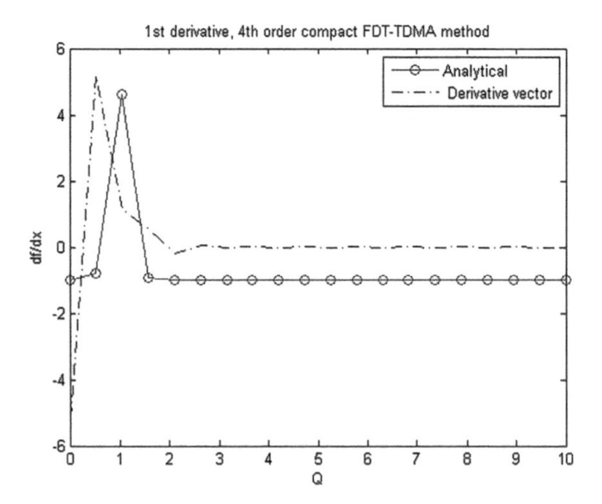

Fig. 1 Fourth orders dense/ compact FDT–TDMA for analytical and derivative order

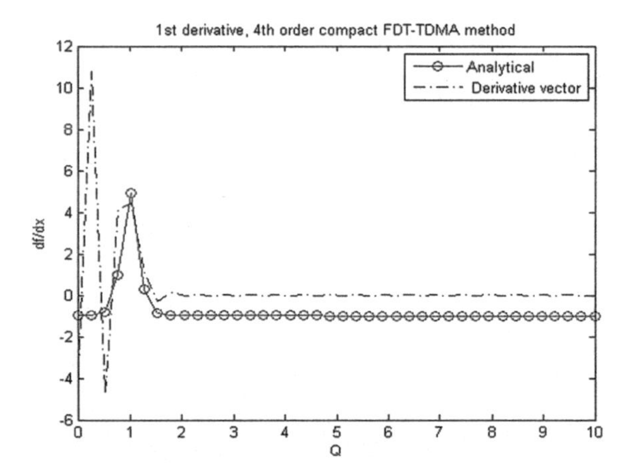

Fig. 2 Fourth orders dense/ compact FDT–TDMA for analytical and derivative order with a 40 × 40 grid

3 Simulation Results

The design of the proposed work is shown in Fig. 1. It consists of a time series difference using FDTD–TDMA. This section introduces a novel way to analyze and derive the vector estimate which involves finite-difference and time-domain schemes. FDTD–TDMA systems are estimates of results that increase in accuracy as the size of the time step goes to Q = 10, except that new errors can be seen in the approximated results.

Figure 1 shows that the maximum analytical value is 4 when Q = 10. A parallel-plate waveguide is demonstrated. The plate distance and its original phase velocity Ph is set at 4 and equivalent simulations are performed for iterations of

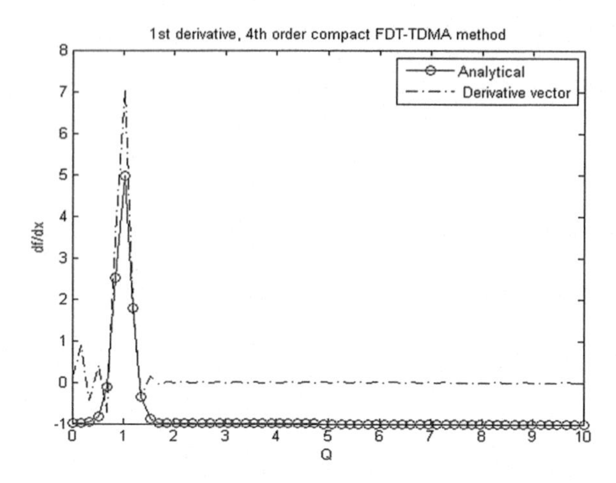

Fig. 3 Fourth orders dense/compact FDT–TDMA for analytical and derivative order with a 60 × 60 grid

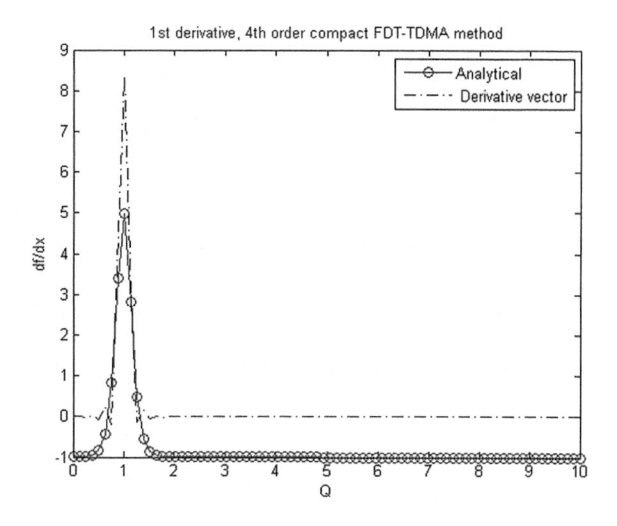

Fig. 4 Fourth orders dense/compact FDT–TDMA for analytical and derivative order with a 80 × 80 grid

multiples of 10, n = 20, 40, 60, 80, etc. The time step is determined by Q = 10 and Q= 5. The analysis is performed for a grid size of 20 × 20 .

The simulations are examined in a similar manner, where a parallel-plate waveguide is demonstrated. In Fig. 2, plate distance and its initial phase corresponding models are achieved for 10 iterations, taking a value of $n = 40$. The time step is determined by, Q = 10, and each grid position is calculated by analyzing the plane, respectively.

The next case when $n = 60$ is considered and shown in Fig. 3; the time step is determined by $Q = 10$ and it would be proceeds from 40 grid positions to the analytical plane of 60, meaning the time difference represents 20 iteration.

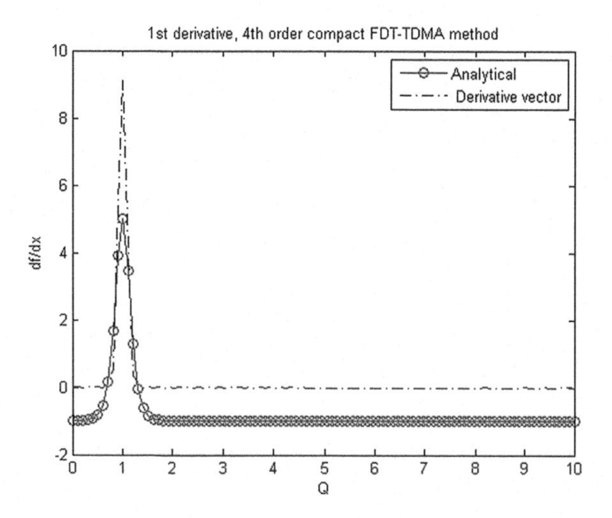

Fig. 5 Fourth orders dense/compact FDT–TDMA for analytical and derivative order with a 100 × 100 grid

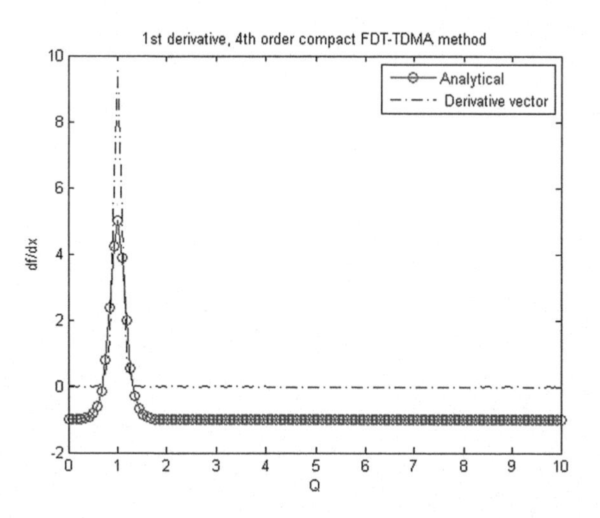

Fig. 6 Fourth orders dense/compact FDT–TDMA for analytical and derivative order with a 120 × 120 grid

Similarly in Fig. 4, we take the next iteration value as n = 80. The time step is determined by Q = 10 and proceeds from 80 grid positions to the analytical plane of 80, which means time difference represents 20 iterations. Considering n = 100 in Fig. 5 as next iteration value, the time step is determined by Q = 10 and proceeds from 100 grid positions to the analytical plane of 100.

Similarly in Fig. 6, the next iteration value is n = 120. The time step is determined by Q = 10 and proceeds from 120 grid positions to the analytical plane of 120.

In Fig. 7, n = 140. The time step is determined by Q = 5 proceeds from 140 grid positions to the analytical plane of 140. In Fig. 8, n = 160 which is taken as the next

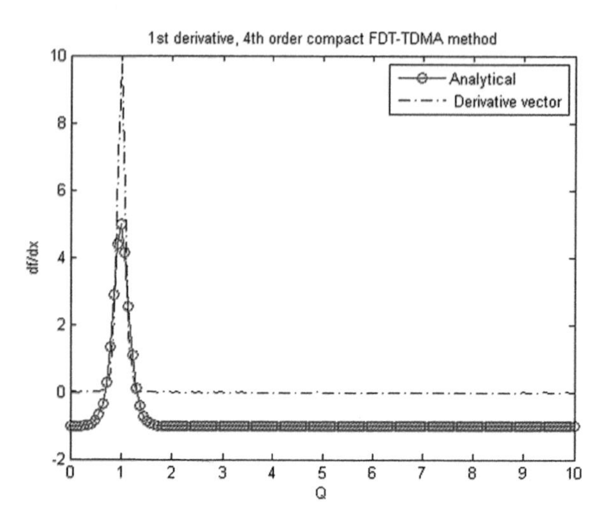

Fig. 7 Fourth orders dense/compact FDT–TDMA for analytical and derivative order with a 140 × 140 grid

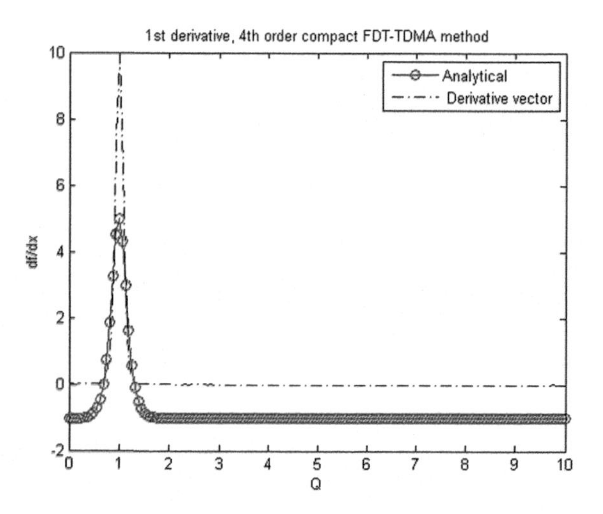

Fig. 8 Fourth orders dense/compact FDT–TDMA for analytical and derivative order with a 160 × 160 grid

multiple iteration. The time step is determined by Q = 5 and proceeds from 160 grid positions to the analytical plane of 160.

As in Fig. 9, considering n = 180, the time step is determined by Q = 5 and proceeds from 180 grid positions to the analytical plane of 180, which means the time difference starts with 160 positions covering 20 iterations.

In Fig. 10, n = 200. The time step is determined by Q = 5 and it has an analytical plane value of 200, which means the time difference starts with 180 positions covering 20 iterations, hence we find that the derivative vector specification becomes the same at different phase velocities.

After presenting our results we find that characteristics of wideband investigations are confirmed by the excited waveform. More precisely, the consistent

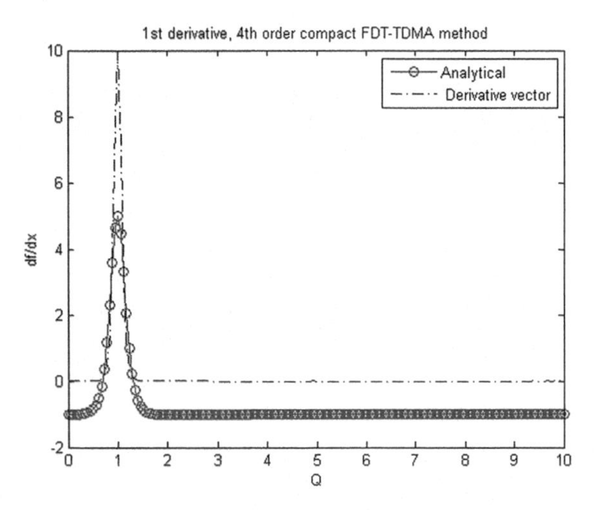

Fig. 9 Fourth orders dense/compact FDT–TDMA for analytical and derivative order with a 180 × 180 grid

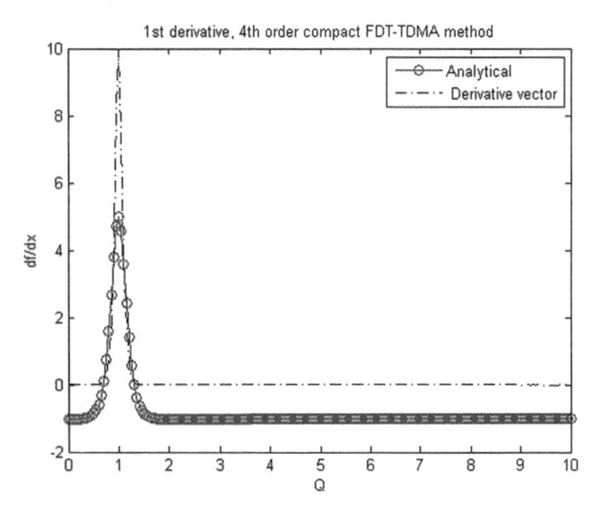

Fig. 10 Fourth orders dense/compact FDT–TDMA for analytical and derivative order with a 200 × 200 grid

excitation incidences are set to 4.1 and 8.2 GHz. The main principles of the L2 errors are shown in Figs. 1, 2, 3, 4, 5, 6, 7, 8, 9 and 10, for different grid sizes. As expected, the incorporation of standard 'fourth-order' spatial operators does not provide notable improvement over the proposed method using FDT counterparts, as they are not specific to different logarithmic distances representing a major difference compared to ADI–FDTD errors.

In Table 1 are the simulation results for a 4 × 4 cavity derived from Fig. 11, where overall error is designed taking account of both the mesh density (20–200 cells per wavelength) and parameter Q (1–10 range). For all cases, a project point that runs parallel to 50 cells per wavelength is measured for Log(Δx) with a −1 distance with respect to the time-varying ADI–FDT–TDM (error rate) of 0.58, and

Table 1 Error detecting in a 4 × 4 cavity in ADI–FDT–TDM

Log(Δx) (distances) for a 4 × 4 cavity	ADI–FDT–TDM (error rate)	FDTD–ADI (error rate)
−1.4	No error	No Error
−1.2	0.67	1.5
−1	0.58	1.52
−0.8	0.3	1.51
−0.6	0.8	1.4
−0.4	0.82	1.42
−0.2	No error	No error
0	No error	No error

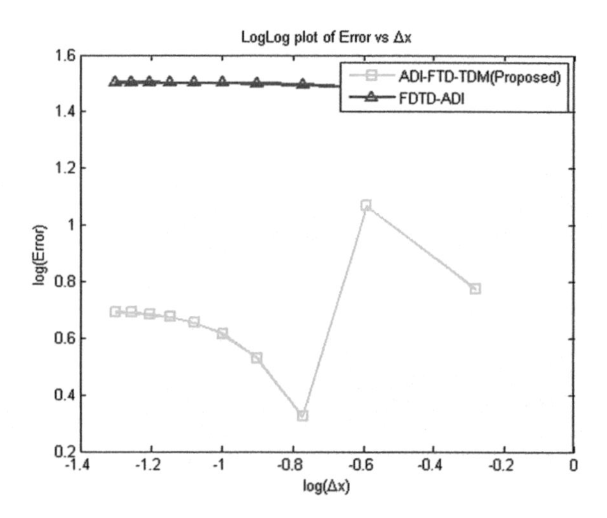

Fig. 11 Error detection at different distances

1.52 for ADI–FDTD (existing), which is obviously due to the surface's angle points.

We have arrived at a result showing that the cross-section of the presented surfaces with Q = constant gives a mean difference of (1.52 − 0.58)/2, i.e. 0.47, giving rise to an improvement in error for the proposed ADI–FDT–TDM over conventional ADI–FDTD, of 47%.

4 Conclusion

A hybrid FDTD–TDMA method for error improvement has been presented with measured and simulated results. Four-point spatial stencils using time-difference multiple access have been efficiently performed to incorporate artificial anisotropy up to tenth-order operatives as observed in the framework of the two-step leapfrog with the same grid position. We have compared this with existing methods by creating different machinists for different field mechanisms and correctly approving the dispersion effects. The proposed procedure is practical to each process with different time difference domains. The main finding of our proposed work is that for the same order, i.e., a 4 × 4 cavity size for constant phase velocity, the trigonometric developments of the subsequent error is improved compared to the ADI–FDTD method. The extracted approximations have been originated to overtake their 'fourth-order' complements and measured results demonstrate 47% improvements in error compared with the existing method. Both simulated and measured results have been found to be in good agreement. Furthermore, for future reference we need to introduce a dynamic mesh position, designed for greater robustness.

References

1. Aggarwal SK, Saini LM, Kumar A (2009) Electricity price forecasting in deregulated markets: a review and evaluation. Int J Electr Power 31(1):13–22
2. Lixia Y, Yijun W, Yingtao X, Gang W (2011) Modified uniaxial perfectly matched layer absorbing boundary condition for anisotropic dispersion media. High Power Laser Particle Beams 23(1):156–160
3. Qian W, Seng-Tiong H (2011) Implementation of perfectly matched layer boundary condition for finite-difference time domain simulation truncated with gain medium. J Light Wave Technol 29(10):1453–1459
4. Ramadan O. Systematic split-step perfectly matched layer formulations for modeling dispersive open region finite difference time domain applications. IET Microw Antenna 5 (9):1062–1066
5. Namiki T, Ito K (2001) Numerical simulation of microstrip resonators and filters using the ADI-FDTD method. IEEE Trans Microw Theory Tech 49:665–670
6. Lim JW, Lee YT, Pandey R, Yoo T-H, Sang B-I, Ju B-K, Hwang DK, Choi WK (2014) Effect of geometric lattice design on optical/electrical properties of transparent silver grid for organic solar cells. Opt Express 22(22):26891–26899
7. Ha S-G, Cho J, Choi J, Kim H, Jung K-Y (2013) FDTD dispersive modeling of human tissue based on quadratic complex rational function. IEEE Trans Antennas Propag 61(2):996–999
8. Chung H, Cho J, Ha S-G, Ju S, Jung K-Y (2013) Accurate FDTD dispersive modeling for concrete materials. ETRI J 35(5):915–918
9. Chung H, Jung K-Y, Tee XT, Bermel P (2014) Time domain simulation of tandem silicon solar cells with optimal textured light trapping enabled by the quadratic complex rational function. Opt Express 22(S3):A818–A832
10. Goyal P, Gupta S, Kaur G, Kaushik B (2017) Performance analysis of VCSEL using finite difference time domain method. Opt Int J of Light Electron Opt Elsevier, 2017. https://doi.org/10.1016/j.ijleo.2017.11.201 ISSN:0030-4026
11. Sullivan DM (1992) Frequency-dependent "FDTD" methods using "Z transforms,". IEEE Trans Antennas Propag 40(10):1223–1230
12. Masud Rana Md, Motin MA, Anower MS, Ali MM. Indoor propagation modeling for UWB communications using LOD 'FDTD' method. J Electr Eng
13. Hemmi T, Costen F, Senior Member, IEEE, Garcia S, Member, IEEE, Himeno R, Yokota H, Mustafa M. Efficient parallel LOD-'FDTD' method for Debye-dispersive media

Bearing Fault Detection and Classification Using ANC-Based Filtered Vibration Signal

Sudarsan Sahoo and Jitendra Kumar Das

Abstract The defective bearing in a rotating machine may affect its performance and hence reduce its efficiency. So the monitoring of bearing health and its fault diagnosis is essential. A vibration signature is one of the measuring parameters for fault detection. However, this vibration signature may get corrupted with noise. As a result this noise must be removed from the actual vibration signature before its analysis to detect and diagnose the fault. ANC (adaptive noise control)-based filtering techniques are used for this noise removal and hence to improve the SNR (signal-to-noise ratio). In our study an experimental setup is developed and then the proposed work is executed in three stages. In the first stage the vibration signatures are acquired and then ANC is implemented to remove the background noise. In the second stage the time (statistical) and the frequency analysis of the filtered vibration signals are done to detect the fault. In the third stage the statistical parameters of the vibration signatures are used for the classification of the fault present in the bearing using random forest and J48 classifiers.

Keywords ANC · Fault detection · Frequency analysis · J48 classifier ORF · Random forest classifier · SNR · Statistical analysis

1 Introduction

The bearing of a piece of rotating machinery in healthy condition carries its own vibration signature. When a defect occurs in the bearing then the vibration signature gets changed. The vibration signals acquired from both conditions can be compared in the time domain and frequency domain in order to detect the fault. However, the noise present in the measured vibration signature may affect the analysis. The noise is produced by the other elements of the rotating machinery. This background noise must be filtered before its analysis. As the noise is non-stationary in nature an

S. Sahoo (✉) · J. K. Das
KIIT University, Bhubaneswar, India
e-mail: sudarsan_iisc@yahoo.in

A. Kumar and S. Mozar (eds.), *ICCCE 2018*,
Lecture Notes in Electrical Engineering 500,
https://doi.org/10.1007/978-981-13-0212-1_34

adaptive noise cancellation technique is the best choice to filter the noise. Then this filtered signal is used for further processing to analyze the fault present in the bearing.

Many studies can be found in literature relating to fault diagnosis of bearings and gears in rotating machinery. P. K. Kankar et al. showed the fault diagnosis of ball bearings using continuous wavelet transform [1]. In some literature spectrum images of a vibration signal are used to diagnose the fault in the bearings [2]. The time, frequency and time-frequency analysis is used in some studies to detect the fault in the bearings [3]. In some studies, multi-scale morphological filters are used for fault detection in the gears [4]. In the other dimension of the work there are many examples in the literature where adaptive filters are used for SNR improvement of the measured signatures. Albarbar et al. [5] showed the acoustic monitoring of engine fuel injection based on adaptive filtering techniques. In some literature FIR (Finite Impulse Response)-based adaptive algorithms are used for adaptive noise cancellation [6]. The application of LMS (Least Mean Square) in adaptive filtering is found in some studies [7]. Young Wook Cho et al. showed the use of machine learning in a safety monitoring system [8]. In some studies the application of machine learning techniques can be found for fault diagnosis in rotating machinery [9, 10]. In the present study, vibration signals from healthy and defective bearings are acquired. Then ANC is implemented for noise filtering. After that the statistical parameters are calculated from the filtered vibration signals. Then FFT of the acquired vibration signals are determined and compared on the basis of ORF (outer race fault frequency). Statistical and frequency analysis is used to detect the fault in the bearings. Then the machine learning technique based on statistical data is used for the classification of the bearing faults.

2 ANC Implementation and Selection

In this section the ANC techniques are implemented to filter the noise. Three ANC algorithms are employed for this purpose. The algorithms used are LMS de-noising, EMD de-noising and wavelet de-noising. To compare the performances of the ANC techniques, the vibration signal from a defective bearing (Type-1) is taken and the ANC is implemented on it. In the de-noising process the vibration signal from the healthy bearing is used as the reference signal. The de-noising principle is shown in Fig. 1. The vibration signal acquired from the defective bearing before ANC and after ANC is shown in Fig. 2. The performances of the ANC techniques are compared on the basis of the SNR (signal-to-noise ratio) and the MSE (mean square error). The comparison is shown in Table 1. From the comparison EMD is found to be better. So an EMD algorithm is used to filter the noise from all the acquired vibration signals at the pre-processing stage. Then the filtered signals are used in further processing to analyze the defects.

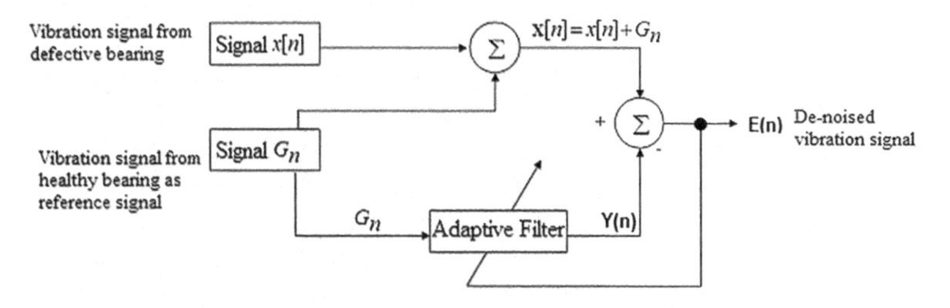

Fig. 1 Schematic of the de-noising principle for noise filtering

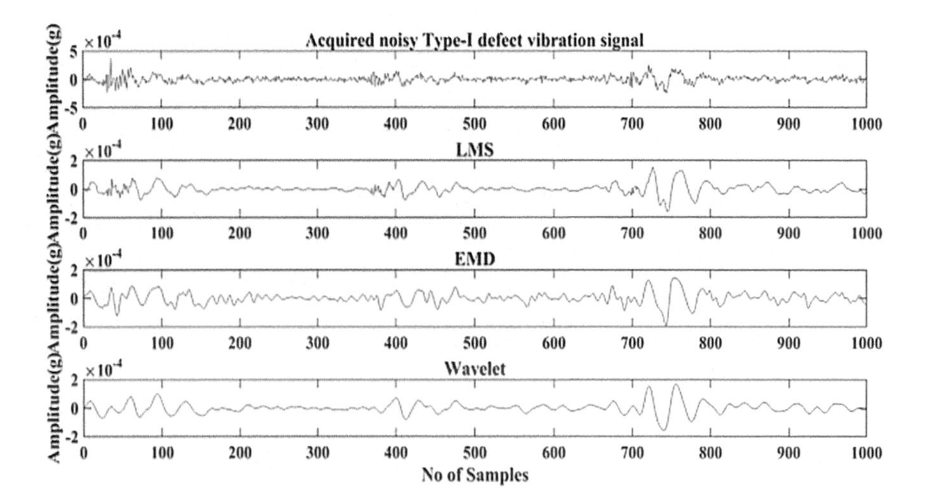

Fig. 2 Vibration signal before and after the ANC

Table 1 Performance comparison of ANC techniques

ANC techniques	SNR	MSE
LMS	11.068	0.0283
EMD	14.863	0.0210
Wavelet	13.061	0.0264

3 Experimental Work

To implement the proposed technique of fault diagnosis an experimental setup is developed. It consists of a single phase induction motor. The experimental setup and its model are shown in Figs. 3 and 4 respectively. To carry out the experiment three bearings are taken; one healthy bearing and two different defective bearings. In the type-1 defect bearing there is a cut at the outer race and in the type-2 defect bearing there is a hole at the outer race of the bearing. The healthy, type-1 and type-2 defect bearings are shown in Figs. 5, 6, and 7 respectively. The setup also

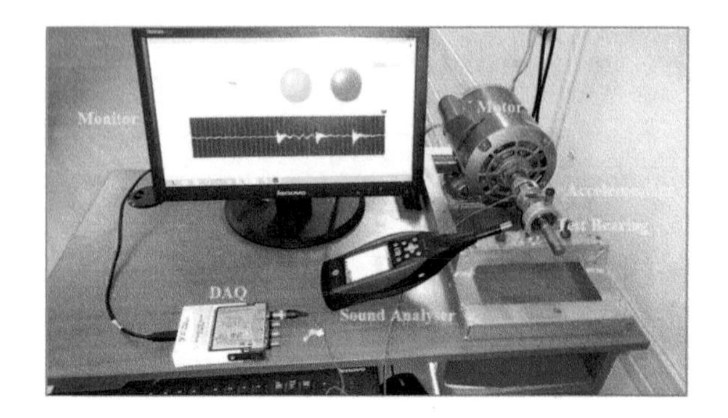

Fig. 3 The experimental setup

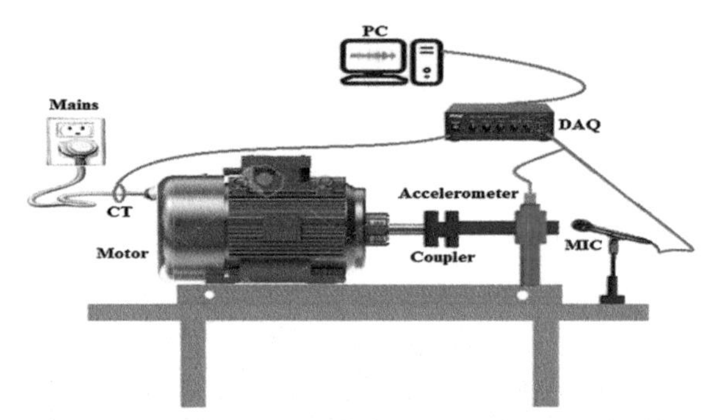

Fig. 4 The experimental setup model

Fig. 5 The healthy bearing

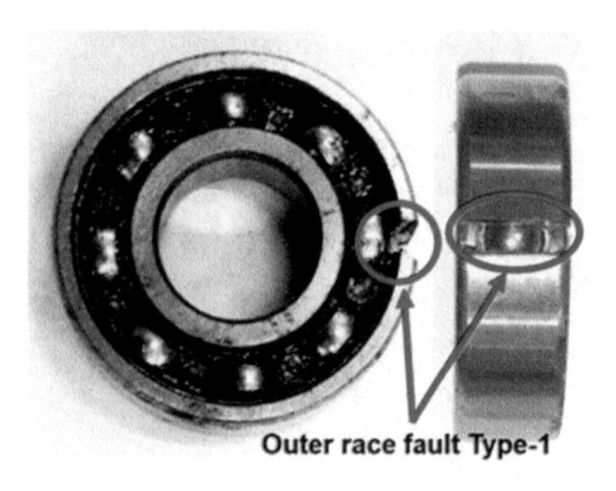

Fig. 6 The type-1 defect bearing

Fig. 7 The type-2 defect bearing

Table 2 Bearing specification

Model	6203z
No of balls (N)	8
Ball diameter (D)	6.74 mm
Pitch diameter (D)	30 mm
Contact angle (α)	0

Table 3 Motor specification

Type	Induction motor
RPM	8
Power	0.5 hp
AMP	2 AMP
Make	Crompton

consists of a data acquisition system to acquire the vibration signal from the bearings. The data acquisition system consists of an accelerometer (PCB 325c-03), which is a vibration sensor, a 4-channel DAQ card (NI-9234) and a PC with LabVIEW software. The specification for the motor and for the bearings are shown in Tables 2 and 3 respectively.

Initially the healthy bearing is mounted on the shaft of the motor and the corresponding vibration signature is acquired. Then the type-1 and type-2 defect bearings are mounted and the corresponding vibration signals are acquired. ANC using an EMD algorithm is implemented on all the acquired vibration signals and the filtered vibration signals from the healthy, type-1 and type-2 defect bearings are shown in Figs. 8, 9, and 10 respectively.

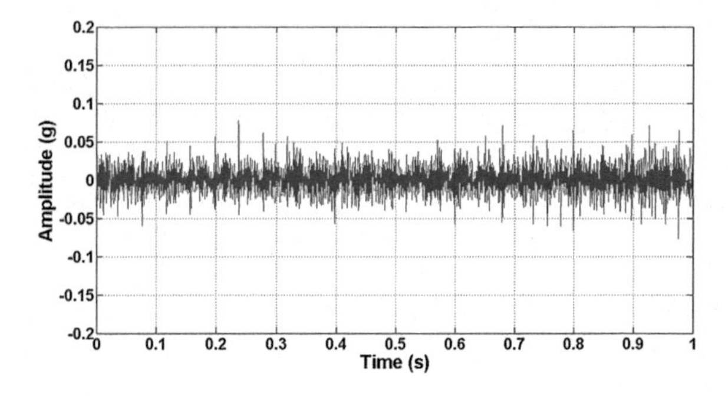

Fig. 8 Filtered vibration signal of a healthy bearing (B1)

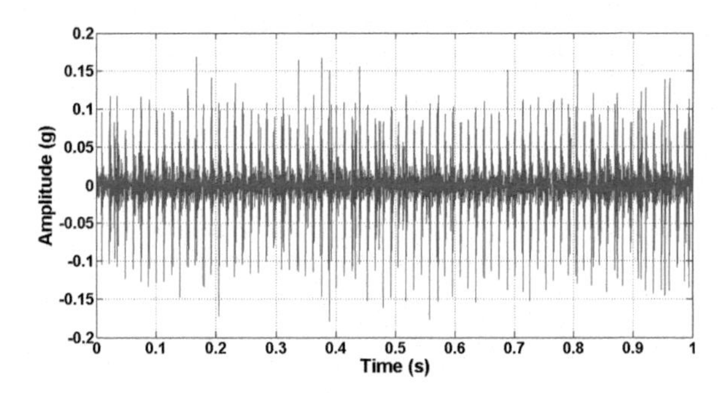

Fig. 9 Filtered vibration signal of a type-1 defect bearing (B2)

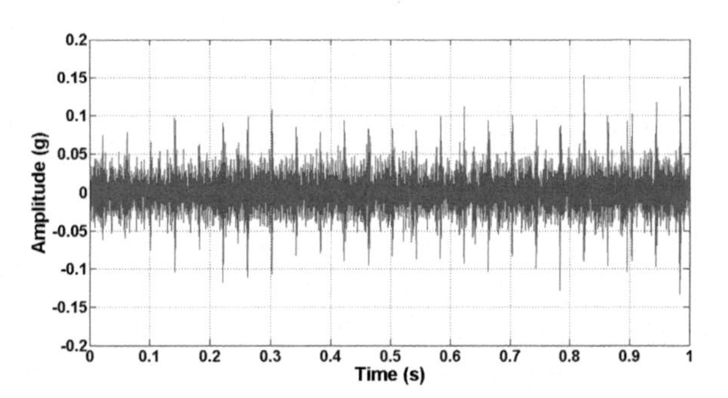

Fig. 10 Filtered vibration signal of a type-2 defect bearing (B3)

4 Results and Discussion

This section shows the statistical and frequency domain analysis of the vibration signatures used to detect the faults in bearings and also the performance of the machine learning algorithms used to classify the faults in the bearings.

Table 4 Statistical parameter comparison

Sl. no	Static parameters	Healthy bearing	Type-I defect bearing	Type-II defect bearing
1	Root mean square (RMS)	0.0134	0.0261	0.0203
2	Mean	0.6613	1.695	1.06
3	Peak value	0.0773	0.1734	0.1432
4	Crest factor	5.7737	6.6469	7.0564
5	Skewness	−0.0052	−0.2564	0.1228
6	Kurtosis	3.8532	9.0461	5.7542
7	Variance	0.0786	6.8005	4.1148
8	Standard deviation	0.0134	0.0261	0.0203
9	Clearance factor	716.0322	665.9750	624.7892
10	Impulse factor	7.4381	10.7450	9.4580
11	Shape factor	1.2883	1.6165	7.4381

4.1 Time (Statistical) Domain Analysis

The statistical parameters are computed from the vibration signals acquired from the healthy and defective bearings and the parameters are shown in Table 4. From the table it can be seen that the parameter values are changed when it is in a defective condition as compared to its healthy condition.

4.2 Frequency Domain Analysis

FFT is used for the frequency domain analysis. The FFT spectrum of the healthy and defective bearings are determined. The frequency spectra of the healthy bearing (B1), the type-1 defect bearing (B2), and the type-2 defect bearing (B3) are compared and shown in Figs. 11 and 12 respectively. After that the bearing characteristic frequency (BCF), which is the outer race defect frequency (ORDF) is calculated from the geometric configuration of the bearing by using the formula in Eq. 1.

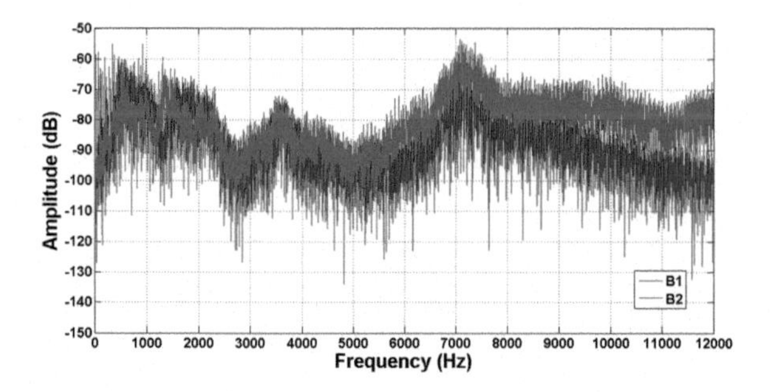

Fig. 11 FFT comparison of a healthy bearing (B1) and a type-1 defect bearing (B2)

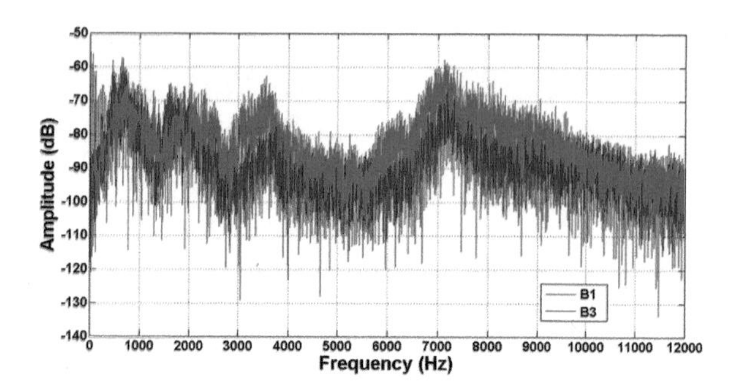

Fig. 12 FFT comparison of a healthy bearing (B1) and a type-2 defect bearing (B3)

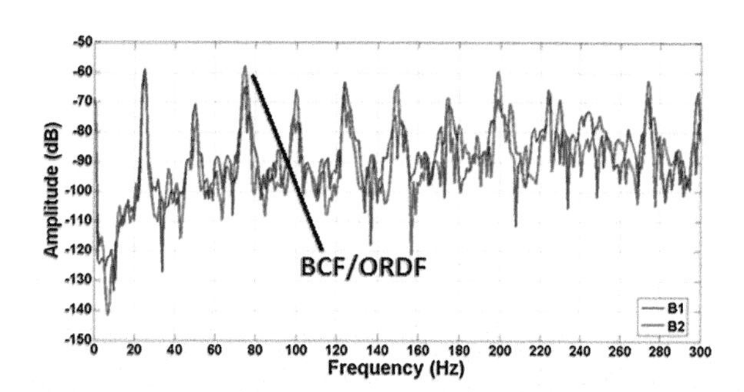

Fig. 13 FFT comparison of a healthy (B1) and a type-1 defect bearing (B2) at BCF

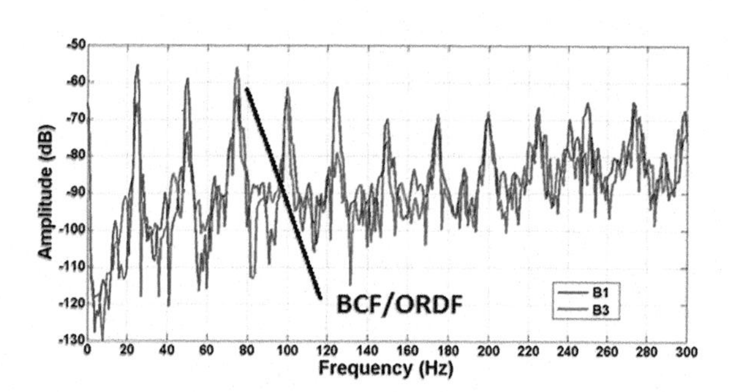

Fig. 14 FFT comparison of a healthy (B1) and a type-2 defect bearing (B3) at BCF

$$\mathrm{ORF(ourter\ race\ fault)} = \frac{\mathrm{N}}{2}\omega_{\mathrm{n}}\left(1 - \frac{\mathrm{d}}{\mathrm{D}}\cos\,\alpha\right) \tag{1}$$

Then the FFT spectra of the healthy and defective bearings are compared at the ORDF. An FFT comparison of a healthy bearing with a type-1 defect bearing is shown in Fig. 13 and with a type-2 defect bearing in Fig. 14. From the spectrum analysis the fault in the bearing is detected.

Table 5 Confusion matrix for the random forest classifier

Random forest classifier			
Defect class	Healthy	Type-1 defect	Type-2 defect
Healthy	*55*	0	*1*
Type-1 defect	0	*56*	0
Type-2 defect	*1*	0	*55*

Table 6 Confusion matrix for the J48 classifier

J48 classifier			
Defect class	Healthy	Type-1 defect	Type-2 defect
Healthy	*41*	0	*15*
Type-1 defect	0	*56*	0
Type-2 defect	*9*	1	*46*

4.3 Fault Classification Using Machine Learning Technique

For the classification of faults the random forest classifier and J48 machine learning algorithms are used. The statistical parameters are the feature vectors used in the algorithm to classify the faults. The vibration signals are segmented with a 1 s window consisting of 168 training samples out of which 56 are of a healthy bearing and 56 each of type-1 and type-2 defect bearings. Five features, namely kurtosis, mean, variance, RMS, and crest factor are extracted from all 168 samples. The algorithms are trained with the testing samples with 10-fold cross validation. The confusion matrix for the two classifiers are listed in Tables 5 and 6.

From the above confusion matrix it has been shown that in the case of the random forest algorithm, the classification of the type-1 defect is 100% while for the rest of the two classes, i.e. for the healthy and Type-2 defects it is misclassified for one test sample. Similarly, for J48 the classification accuracy is high for the Type-I defect but the misclassification rate of the J48 algorithm is found to be higher in the area of fault detection. It can also be seen that the random forest algorithm is found to be better than the J48 for classifying faults in bearings.

5 Conclusion

The present work showed the application of an adaptive noise control technique at the pre-processing stage of vibration signal processing. In the statistical analysis many parameters are computed to provide the information used to detect the fault. In addition, the frequency spectrum comparison of healthy and defective bearings at the bearing character frequency (BCF) is more informative for detecting the fault. The selected statistical parameters are used as the input to the machine learning algorithms to classify bearing faults. J48 and random forest classifiers are used to classify the faults and the random forest classifier is found to be better than the J48 classifier for classifying faults in bearings.

References

1. Kankar PK, Sharma SC, Harsha SP (2011) Fault diagnosis of ball bearings using continuous wavelet transform. Appl Soft Comput 11:2300–2312
2. Li W, Qiu M, Zhu Z, Wu B, Zhou G (2016) Bearing fault diagnosis based on spectrum images of vibration signals. Meas Sci Technol 27:035005 (10 pp)
3. Shakya P, Darpe AK, Kulkarni MS (2013) Vibration-based fault diagnosis in rolling element bearings: ranking of various time, frequency and time-frequency domain data-based damage identification parameters. Int J Cond Monit 3
4. Li B, Zhang P, Wang Z, Mi S, Zhang Y (2011) Gear fault detection using multi-scale morphological filters. Measurement 44:2078–20895
5. Albarbar A, Gu F, Ball AD, Starr A (2009) Acoustic monitoring of engine fuel injection based on adaptive filtering techniques. Appl Acoust 70:247–255
6. Liao CW, Lin JY (2007) New FIR filter-based adaptive algorithms incorporating with commutation error to improve active noise control performance. Automatica 43:325–331
7. Troparevsky MI, D'Attellis CE (2004) On the convergence of the LMS algorithm in adaptive filtering. Signal Process 84:1985–1988
8. Cho YW, Kim JM, Park YY (2016) Design and implementation of marine elevator safety monitoring system based on machine learning. Indian J Sci Technol 9:109889
9. Abraham Siju K, Sugumaran V, Amarnath M (2016) Acoustic signal based condition monitoring of gearbox using wavelets and decision tree classifier. Indian J Sci Technol 9:101335
10. Kankar PK, Sharma Satish C, Harsha SP (2011) Fault diagnosis of ball bearings using machine learning methods. Expert Syst Appl 38:1876–1886

A New Approach to Securing Online Transactions—The Smart Wallet

K. L. Anusha, G. Krishna Lava Kumar and Aruna Varanasi

Abstract For many years, two-factor authentication has been the only means of preventing cyber attacks and providing cyber security to online transactions. However, it seems to be vulnerable due to weak spots through which hackers are able to easily find ways of either intercepting message codes or exploiting account recovery mechanisms. Most of the available systems provide a onetime text password as an SMS to the registered mobile number of the user, while a few of them deliver it via telephone call leaving users to worry about its misuse through their phone being stolen or the SMS being seen by a hacker who can easily hack the SIM network provider and read the message, or by specific calls being diverted to his/another mobile number without the knowledge of the original recipient. The result was a huge detriment to the user, leaving him to worry about his hard-earned money. As a result, we are presenting this Smart Wallet approach.

Keywords Two-factor authentication · WinAuth by Google Authenticator Smart Wallet approach

1 Introduction

It is a fact that two-factor authentication doesn't seem to be as secure as it was intended to be. An attacker doesn't need your physical authentication information if they can easily trick your phone company and your network provider. However, it is true that nothing is perfect in this world and something is better than nothing.

K. L. Anusha · A. Varanasi
Sreenidhi Institute of Science and Technology, Ghatkesar, Hyderabad, India
e-mail: lakshmianusha@sreenidhi.edu.in

A. Varanasi
e-mail: arunavaranasi@sreenidhi.edu.in

G. Krishna Lava Kumar (✉)
CMR Institute of Technology, Medchal, Hyderabad, India
e-mail: krishna.lavakumar@cmritonline.ac.in; lavakumar.gopu@gmail.com

A. Kumar and S. Mozar (eds.), *ICCCE 2018*,
Lecture Notes in Electrical Engineering 500,
https://doi.org/10.1007/978-981-13-0212-1_35

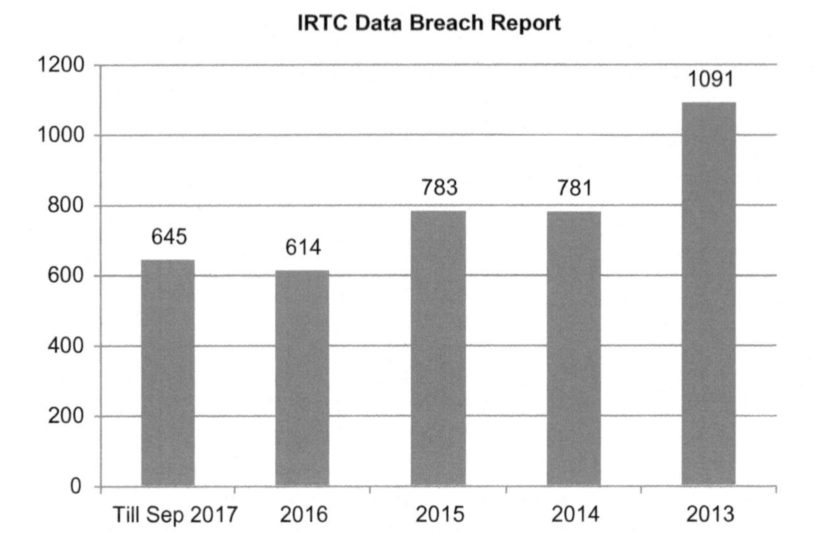

Fig. 1 ITRC (Identity Theft Resource Center) Data Breach Report from 2013 to Sept 2017 [1, 2]

Still, that is no reason to allow someone the chance to exploit us. For this reason we must always keep trying to improve on existing measures and develop new measures to provide security for users' hard-earned money.

In earlier days, many banks provided a security question and answer mechanism to process a transaction, with details given by the user at the time of account registration. A few banks provided a different password for completing a transaction which is different to the one that is used at the time of login. If incorrect details are supplied then the transaction was declined and some banks would even block the account if the transaction failed two to three subsequent times. This was inconvenient to the user simply trying to conduct a transaction and they would have to visit the bank to address the issue. This has led to the development of new, secure, and better ways of authenticating transactions made online. However, the shortfalls of these approaches and some of the loopholes identified have led us to propose this Smart Wallet approach (Fig. 1).

2 Two-Factor Authentication

Two-factor authentication was considered to be the solution to securing online transactions as well as to recognizing the authenticated person and log them into a system or application and many current and new companies are racing to deploy it. The trouble is that two-factor authentication is loaded with difficulties that can mislead companies trying to secure their environments, and even consumers who believe that their security is as good as it can be. There has been a lot of expectations surrounding two-factor authentication which have resulted in several misconceptions.

The first and foremost thing about two-factor authentication systems on many websites is that they work by sending a text message to a use's phone via SMS when someone tries to log in. Some of them may even allow you to remove the two-factor authentication protection from your account after confirming you have access to a phone number you configured as a recovery phone number. That sounds fine, but the underlying point that everyone must accept is that every cellphone has a phone number and a physical SIM card inside the phone that ties it to that phone number with your cellphone provider. Unfortunately your phone number isn't as secure as you think because of the fact that hackers can easily breach to your data by hacking your network provider [3].

In general, everyone transfers their existing phone number to a new SIM card after losing their phone or purchasing a new phone. A hacker will just call your cellphone company's customer care department and pretend to be you. They will simply ask about your phone number and ask for a few personal details which are very easy for a hacker to discover through getting into databases where you have provided such information to someone in the past. That is how a hacker can attempt to get your contact number moved to his phone and it is then easy to understand how he can easily read your passwords.

There are other easier ways as well, such as using call forwarding set up on the phone company's end so that all the incoming voice calls to your phone number are forwarded to their phone and never reach you in the case of one-time passwords on voice calls. This is the reason that in 2017, two factor authentication isn't no longer sufficient for performing costly transactions online [4].

3 WinAuth by Google Authenticator

Everyone knows about Google Authenticator, but the good thing is that the Google Authenticator app is built on a very well-documented algorithm which means that any developer can create their own apps on any platform that can generate time-based one-time passwords just like Google Authenticator. WinAuth by Google Authenticator is one such application for a Windows PC. It is a service that sends a second-factor login prompt directly from its servers to Android phones or to the Google search app for iOS. Even more secure still are systems that don't require any message to be sent at all. Apps like Google Authenticator and tokens like those sold by RSA generate one-time-password codes that change every few seconds. Those same exact codes are generated on the servers run by services like Slack, WordPress, or Gmail, so that the user can match up the code to prove their identity without it being sent over the internet at all. The math behind this WinAuth by

Google Authenticator system is very intelligent: Whenever the user logs into any of the Google services, the WinAuth app by the Google Authenticator and the server both start with a seed value that is transformed into a long, unique string of characters with a hash value which is a mathematical function that cannot be reversed by any attacker on the system as it is not transported over the internet. Again, that string of characters is hashed, and the results are once again hashed to make it more secure in a process repeating every few seconds. Finally, only a few digits of those characters are displayed as the login code in order to prevent anyone who glances at a user's phone from starting their own hash chain [5].

Unfortunately, some services like Twitter still only offer two-factor authentication that depends on the security of SMS. Twitter says that they are exploring additional ways to make sure that users' accounts remain secure. In other words, Twitter, like other services that store your sensitive data, may soon be offering a second-factor option other than the rickety telephone line that SMS represents [6].

4 The New Approach—A Smart Wallet

A Smart Wallet is a smart case for cards—for both debit and credit cards that can read a fingerprint from the user/owner of the card. It is a perfect combination of both hardware and software. It protects the card from being misused and mishandled. It can be termed as a cyber-security weapon to protect a user's money and bank information. A Smart Wallet is built to make users' banking transactions all the more secure while being difficult to hack. It serves the purposes of all bank users who are worried about their money being stolen. The stunning number of hacks that were made to bank accounts—even after two factor authentication and many other additional measures taken by the banks—also prompted us to propose this Smart Wallet approach.

The working of the Smart Wallet approach can be made very simple and user friendly. It uses biometrics (i.e. a fingerprint) to detect the identity of the cardholder and only then proceeds to make the transaction, thereby preventing other people who are pretending to be the cardholder from using the card. The idea is to take the fingerprint of each cardholder and store it in the cloud and at the time of every transaction, it should ask for the card details and then ask the cardholder to provide the fingerprint through the smart wallet which can then be matched against the cloud storage, and only if the prints match will the transaction be processed else the transaction will be declined. With this approach, we can protect users/customers hard earned money by making sure that no-one can use the card by stealing it or by stealing the account details and/or hacking the one-time passwords sent over the mobile phone network (Fig. 2).

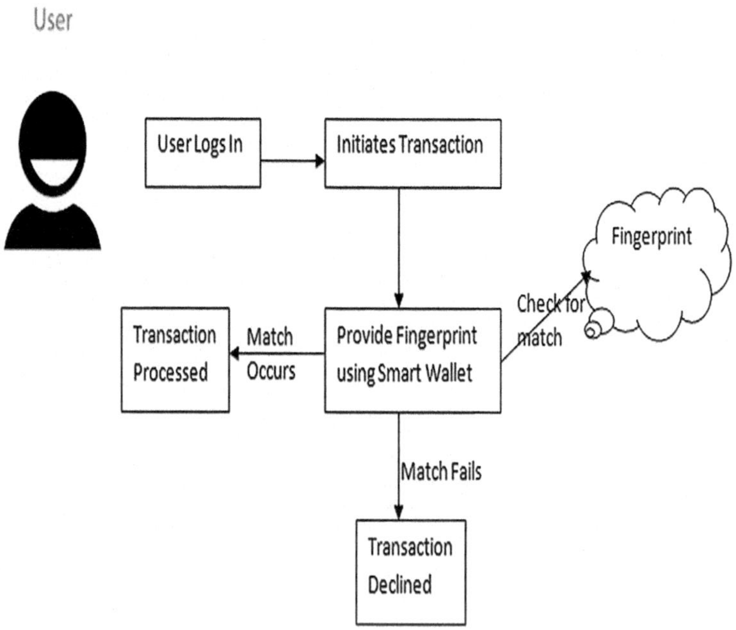

Fig. 2 Implementation of the Smart Wallet process

5 Conclusion

The challenge ahead is to build a Smart Wallet that scans biometrics rapidly and concludes transactions without delay. It must also look attractive and be user friendly. It integrates biometrics in a way that is reliable and requires minimal fault maintenance. The Smart Wallet is a product that can be used for any card, anywhere in the world, and can make that card secure.

Further biometric scanners, such as retina scanners, palm scanners, and face recognizers etc., can also be integrated to enhance a user's options. Our aim is to see that no person's hard earned money is lost to thieves who steal cards or to hackers who steal information and perform transaction pretending to be the legitimate user/owner of the card.

References

1. https://revisionlegal.com/data-breach/2017-security-breaches
2. http://www.idtheftcenter.org/Data-Breaches/data-breaches
3. https://twofactorauth.org/#banking
4. https://duo.com/assets/ebooks/Duo-Security-Modern-2FA.pdf
5. https://winauth.github.io/winauth/index.html
6. https://www.wired.com/insights/2013/04/five-myths-of-two-factor-authentication-and-the-reality

A Global Dispatcher Load Balancing (GLDB) Approach for a Web Server Cluster

Kadiyala Ramana, M. Ponnavaikko and A. Subramanyam

Abstract With the volatile expansion of the internet, numerous innovative online applications and services are in development. In conventional internet architecture, the innovative disputes are imposed by the fashionable applications. By using multiple servers, web server performance is improved and the effectiveness of a simulated web server system depends upon the process of distributing client requests. The distribution of client requests must occur in a way that is transparent to users among multiple server nodes, which affects availability and scalability in the distributed web server system. Thus, in this study, an efficient load balancing architecture called global dispatcher-based load balancing (GDLB) is proposed, which uses both domain name system and dispatcher. With this approach, performance is estimated to be better than with existing approaches. To analyze performance, a JMeter testing tool is used for dynamic load generation and performance measurement in a real-life internet scenario.

Keywords World Wide Web · Response time · Load balancing
Web server · Cluster

1 Introduction

Over the past two decades the World Wide Web (WWW) has seen implausible developments and provided enormous opportunities for the exploitation of a wide range of web services, making them accessible to a massive number of online users.

K. Ramana (✉)
School of Computer Science and Engineering, SRM University, Kattankulathur,
Chennai 603203, Tamil Nadu, India
e-mail: ramana.it01@gmail.com

M. Ponnavaikko
Vinayaka Missions University, Chennai, India

A. Subramanyam
AITS, Rajampet, India

A. Kumar and S. Mozar (eds.), *ICCCE 2018*,
Lecture Notes in Electrical Engineering 500,
https://doi.org/10.1007/978-981-13-0212-1_36

To sustain a large number of concurrent clients reliably, responsively, and economically, popular web services need to be ascendable due to the exciting character and extraordinary scale of the internet. Such availability and scalability requirements place a great challenge on both networking capacity and processing power. In particular, web cluster architecture is gaining in popularity due to its relevant features, such as efficient and inexpensive approaches. Along with one billion pages of index search, 6,000 Linux or Intel personal computers, and 12,000 disks were used by the Google search engine to serve an average of 1,000 queries per second in December, 2000 [1]. In 2011 [2], Google used more than one million servers as per an estimation, but this was based on an energy consumption and the company is probably running about 90,000 servers. In the WWW, the Google search engine found more than 30 trillion unique URLs per day, and had 20 billion site crawls and 100 billion searches to process every month. In 1999, to crawl and build an index of about 50 million pages too Google one month, whereas in 2012 to accomplish the same task it took less than 1 min [3].

Many well-known websites necessarily have to upgrade their server capabilities because of the rapid evolution of internet traffic. The popular way method is to provide a list of alternative, or equivalent mirrored servers in different locations. The mirrored servers are not transparent to the users and it is hard to provide fault tolerance and load balancing [4]. Even so, the effective load balancing architectures are required to improve response times. In many of the available load balancing architectures, the mirrored website has all the servers in only one cluster. However, to provide a service under heavy loads, multiple clusters which are geographically distributed are needed. In this scenario, the first step is to choose a cluster and then choose a server available in that cluster, which changes the load balancing problem.

The performance of proposed schemes which are available for load balancing under different load conditions are investigated and evaluated. It is very hard to build analytical models of real-life internet scenarios. It is also very hard to understand the behavior of servers, DNS, and networks, and also simulations of the same.

In this study, a geographically distributed web server architecture is proposed and experiments performed in order to estimate and compare the performance of various load balancing architectures.

2 Existing Architectures

The classification of existing architectures is split into five classes based on which component dispatches the incoming client request between servers. In these five classes, the first requires modification of software at the client side, and in remaining four one or more elements will be affected in the web server cluster. The five approaches are:

- Client-based approaches
- DNS-based approaches
- Dispatcher-based approaches
- Server-based approaches
- Dynamic dispatcher-based approaches (using both DNS and dispatcher).

2.1 Client-Based Approaches

In these approaches, a web client entity (either a web browser or proxy servers at the client) will be responsible for selection of a server. No processing will be done at the server side for selection of the same. The dispatching of client requests to various replicated servers will be done using client software.

- **Web clients**: In replicated web server architecture, all web clients are aware of the existence of the servers. Netscape's Approach [5] and Smart Clients [6] are the two schemes used for selection of server at the client side.
- **Client's DNS resolver**: For a I2-DSI System, Beck and Moore [7] proposed this scheme. Here, at the client side they used a DNS resolver which issues probes for the servers and chooses the server based on earlier access information or response time from the client.
- **Client-side proxy**: The proxy server is similar to a web client which redirects a client request to web server nodes. Baentsh et al. proposed an approach which incorporated server replication and caching [8]. In a client-side proxy, by implementing web location and information service they recorded replicated URL addresses and redirect requests to the selected server.

The above approaches reduce the load on servers by performing dispatching at the client side. However, the drawback was limited applicability as the user must know that the architecture is distributed.

2.2 DNS-Based Approaches

In these approaches, an authorized DNS is used at the server side which maps the domain name to an IP Address of any one server in the cluster by using numerous scheduling strategies. The selection of the server will be made by the server side DNS which does not suffer from the problems faced by client-based approaches. The authorized DNS has limited control over the requests which reach the server cluster. To control network traffic between the client and DNS, many caching techniques, such as web browsers, DNS resolvers, and intermediate name servers etc. will be used.

DNS not only provides the IP addresses of server nodes, it also includes a validity period called the time-to-live (TTL) value in the name-resolving process. When this value expires, the mapping request is sent to the authorized DNS otherwise it is resolved by any of the caching techniques mentioned above. This value cannot be set as low or as zero because it doesn't work for non-cooperative name servers and caching at the client side. This will increase network traffic and can become a bottleneck to itself. Some of the DNS-based approaches are elucidated in [9, 10]. Based on the scheduling algorithms which are used for a selection of servers and TTL values, DNS-based approaches are classified below.

- Constant TTL algorithms: Based on the server and client state information (location, load, etc.) these algorithms are classified as system stateless algorithms [11], server state-based algorithms, client state-based algorithms [12] and server and client state-based algorithms.
- Dynamic TTL algorithms: In these algorithms, the TTL value is dynamically changed when a URL maps to an address [9]. These are classified as variable TTL algorithms and adaptive TTL algorithms.

In all the above approaches, when a replicated object changes from one place to another, a change in mapping is required. Hence all the approaches mostly support static replication schemes rather than dynamic replication schemes. These approaches also have limited control of requests because of the mapping performed at different levels. Because of packet size limitations in UDP, these approaches cannot handle beyond 32 web servers for a public URL [10].

2.3 *Dispatcher-Based Approaches*

These approaches provide full control to the server-side entity over client requests. In these, the DNS will return the address of a dispatcher, which dispatches the client requests to one of the servers available in the cluster. At the server side, the dispatcher acts as a centralized scheduler which controls the distribution of all client requests. This approach is much more transparent because to the outside world it looks like a single IP address. These mechanisms were characterized as packet single rewriting [13], packet double rewriting [14], and packet forwarding [15]. Various dispatcher-based approaches are elucidated in [12, 16].

In this approach the dispatcher is the single decision entity. Whenever the request rate increases rapidly, it will lead to a bottleneck at the dispatcher. Furthermore, this system will fail because of its centralized nature. Performance is also degraded because of the modification and rerouting of each request through the dispatcher.

2.4 *Server-Based Approaches*

In these approaches, dispatching will be done on two levels. First at the cluster DNS then at each server (the request is received at any of the servers in the cluster if it is required). The problem of client request non-uniform load spreading and inadequate control of DNS was solved using this approach. Some of the server-based approaches are elucidated in [13, 17, 18].

These approaches increase the latency time observed by the clients due to redirection mechanisms.

2.5 *Dynamic Dispatcher-Based Approach*

This approach is based on DNS and dispatcher. A DNS server will initially communicate with the server and converts the URL to an IP address. One dispatcher is associated with all of the web servers available in the cluster. Every dispatcher is associated with the dispatcher selector. Each dispatcher is comprised of a load collector, which gathers the load of every web server, and an Alarm Monitor, which monitors the load and provisionally stops the services of web servers with very high loads. Every server comprises a load checker and a request counter which compute and direct the information about the load on the web server.

In this approach, the client request will first be sent to the DNS. The DNS will forward the request to the dispatcher selector who forwards the request to the dispatcher having a minimum-loaded web server in the cluster. The dispatcher analyses the load collector which receives the data from the load checker, Request counter and also checks the alarm monitor component for the least-loaded web server among all the servers in the cluster. The dispatcher forwards the load information about the minimum loaded server to the dispatcher selector. The dispatcher selector forwards the IP address of the minimum load dispatcher to the DNS, which returns this client. Then the client sends the request to the web server and gets a response directly from the web server [19].

The proposed method has been designed in a way such that it yields a better response time, throughput, and number of requests served in a better way when compared to the approaches mentioned in the literature survey.

3 Proposed Architecture

To analyze different schemes for request dispatching in a web cluster, an architecture which simulates real internet scenarios and follows all steps in the HTTP request service is considered and employed. In this architecture, all standard components used in the WWW like BIND for DNS and the Apache web server are used. A JMeter testing tool is used to generate HTTP client requests.

3.1 Design Goals

This architecture is intended to provide for an easy measurement of numerous parameters of web server cluster performance, such as average response time for requests, CPU utilization, error rate, and throughput of the web server system. By keeping the following goals in mind, this architecture has been designed.

1. The architecture should be scalable and it must permit a large number or clients and servers.
2. It should emulate real internet scenarios.
3. It should be general, i.e. diverse load balancing schemes can be instigated.
4. It should permit an easy configuration of load balancing schemes.
5. It should allow dynamic schemes that make their choice based on system state information.

3.2 Architecture of the Web Server Cluster System

To implement different load balancing schemes, a generalized system model is chosen, as shown in Fig. 1, for the web cluster system. In the system model the group of clusters are represented as replicas and individual clusters have a few servers and dispatchers. In the processing sequence, by the use of DNS the request is first forwarded to one of the clusters by the reciprocating dispatcher IP address of that particular cluster. Second, the request is directed by the user (client) to the dispatcher of that particular cluster. Based on the request, the dispatcher decides the server that should serve and directs the request.

This framework allows the enactment of both dispatcher- and DNS-based structures and as well as both combination structures. For excellent or superior services to clients in various locations, it allows a state where servers are situated in diverse geographical regions. Instead it is also probable the model where it has no dispatcher, one cluster, and a single server system.

For request spreading a variety of diverse schemes are allowed at two locations, i.e. dispatchers and DNS. Based on the desired structures, the cluster IP address can be selected at the DNS location. In practical scenarios the three methods such as DNS-based [20], dispatcher-based [20] and dispatcher-based dynamic load balancing [19] are implemented. In the proposed framework, new structures and methods can be instigated very easily. However, every new TCP connection from a client can be scheduled on the desired server at the dispatcher's location. At the

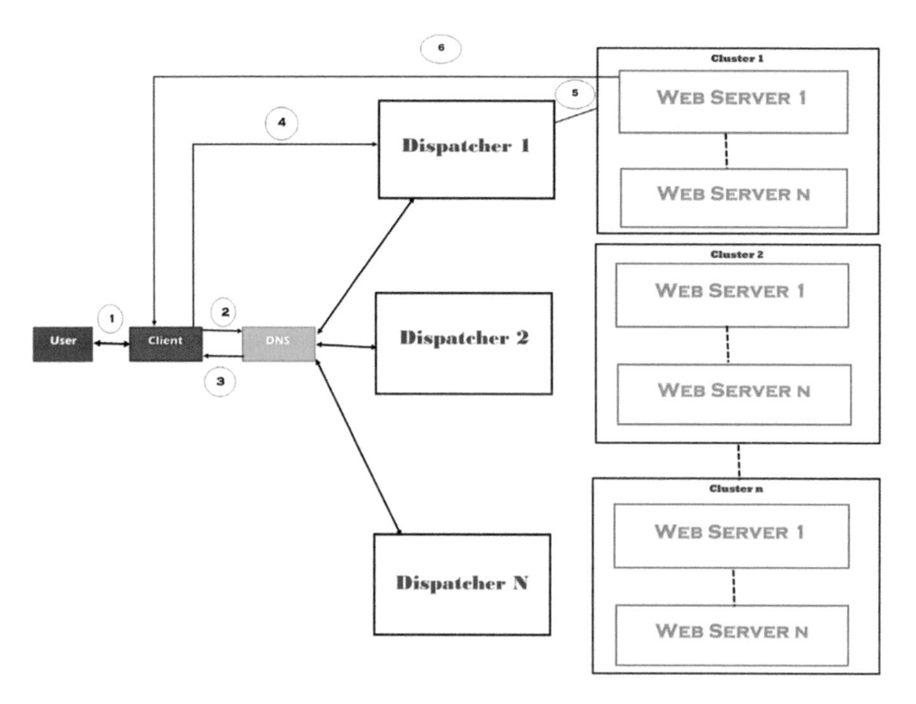

Fig. 1 Distributed web server cluster architecture

DNS, the dispatcher of the cluster will be selected based on the geographical location of the client. At the dispatcher, server selection will be random.

3.3 *Pseudo Code*

Step 1. User opens a web browser (invokes a client).

Step 2. User requests through the client by entering the website URL. Client forwards the request for mapping the URL into an IP address at the DNS.

Step 3. Based on the client's geographical location information, the DNS forwards the IP address of the dispatcher to the client. (DNS contains a list of IP addresses of the dispatchers.)

Step 4. The client sends the web page request to the dispatcher with the received IP address.

Step 5. The dispatcher records the relevant information for session entry into the database and forwards the requests of clients to the randomly selected web server. (The dispatcher contains a list of IP addresses of available web servers in the cluster).

Step 6. The selected web server responds and serves the web request to the client directly.

4 Implementation

PCs running on Linux for all components have been used in this architecture. The request generation and collecting of statistics is done by JMeter. The DNS server has been enhanced to return one of the possible IP addresses based on the geographical location of the client. Finally, dispatcher software was used to distribute requests transparently and randomly within a cluster. All components have been implemented in a way that means the new schemes can be implemented as modules that can be added to existing architecture easily.

4.1 Client-Side Software

To scrutinize the performance of the proposed system, a JMeter testing tool has been taken as a load testing tool for analyzing and measuring the performance of a variety of services, with a focus on web applications. JMeter is designed for testing web applications and can be further extended to test other functions.

An Apache JMeter may be used to test the performances of both static and dynamic resources. It can be used to simulate an overloaded web server, network or object to test its strength and analyze overall performance under different load types.

4.2 DNS Software

As discussed earlier, DNS-based schemes for load-balancing require that the DNS returns the IP addresses of the server or cluster, based on the state information. A current application of the domain name server (BIND-9.1) provides such support. It supports random and round-robin selections of IP address. In this approach, the BIND was extended to provide flexibility.

4.3 Server Software

All server machines can run Apache web server. However, one could use any other software without necessitating any change in the architecture.

4.4 *Dispatcher Software*

The dispatcher is responsible for distributing requests within a cluster. Depending on the scheme, it can take into account loads on various servers and previous request rates of the clients, in order to choose a particular server.

5 Performance Evaluation

The performance of the proposed GDLB web cluster system has been analyzed based on the factors below and the results has been compared with existing approaches.

5.1 *Number of Requests Served*

A general and broadly accepted definition of performance is to observe the system output that represents the number of successfully served requests from a total of input requests.

5.2 *CPU Utilization*

A web cluster system can be busy in the overload condition and may be required to balance the load dynamically. CPU utilization has been taken as the decisive factor in the selection of minimum loaded web server. In the proposed system, a dispatcher collects the utilization of each web server and CPU utilization is calculated by the following equation for "n" number of requests [21].

$$CPU\ Utilization = \sum_{i=0}^{n} \frac{(Current\ CPU\ Utilization - Initial\ CPU\ Utilization)}{(Current\ Time - Initial\ Time)}$$

where i = number of requests.

5.3 *Response Time*

Response time measures the performance of an individual request, transaction or query on the web cluster system. Response time is the amount of time from the moment that a user sends a request until the time that the application indicates that the request has completed.

The dispatcher acts as an intermediate request system which calculates the response time of each web server. The response time has been calculated using the following equation [21].

$$Average\,Response\,Time = \sum_{i=0}^{n} \frac{(End\,Time\,of\,Request - Start\,Time\,of\,Request)}{Total\,Number\,of\,Requests}$$

where i = number of requests.

5.4 Error Rate

Error Rate is a significant metric because it measures "performance failure" in the application. It tells us how many failed requests are occurring at a particular point in time during our load test. The value of this metric is most evident when we can easily see the percentage of problems increase significantly as the higher load produces more errors. In many load tests, this rise in error rate will be drastic. This rapid rise in errors tells us where the target system is being stressed beyond its ability to deliver adequate performance.

5.5 Throughput

Throughput indicates the number of transactions per second an application can handle, and the amount of transactions produced over time during a test. In a single packet rewriting approach, a web server is responding to a client over a period of time and then throughput is calculated based on the number of successful requests sent in a particular period.

6 Performance Analysis

To analyze the performance of the proposed GDLB web cluster system, a JMeter testing tool has been taken as a load testing tool to evaluate and compute the performance of a range of services, with particular attention placed on web applications.

To test the proposed GDLB web cluster system using a JMeter testing toolkit, a master system and numerous slave systems are considered. A master system generates multiple threads on each slave system and each thread is capable of sending

Table 1 Comparative analysis of serving requests with the proposed GDLB approach

Number of requests generated	Number of requests served			
	DNS-based web server system	Dispatcher-based web server system	Dynamic dispatcher-based web server system	Global dispatcher-based web server system
10,000	8551	8992	9523	9999
20,000	12,376	15,771	16,329	19,964
30,000	18,349	19,336	20,931	28,031

different types of requests on the system under test conditions. The proposed GDLB web cluster system has been tested with multiple requests generated by a master system and distributed over the slave systems to evaluate performance.

The test is performed on DNS-based [20], dispatcher-based [22], dynamic dispatcher-based [19], and the proposed web cluster system with two clusters of two servers and three servers respectively, with Apache Tomcat web server software installed on it and a simple web page hosted on it. Every web server can handle a limited number of client requests per IP address simultaneously and can provide a response to the maximum number of users. However, it becomes unresponsive when a web server approaches or exceeds its limit.

Table 1 shows that the number of serving requests with the proposed web cluster system is greater when compared to the DNS-, dispatcher- and dynamic dispatcher-based web server systems. The proposed GDLB web cluster system has served more requests in comparison to the DNS, dispatcher and dynamic web

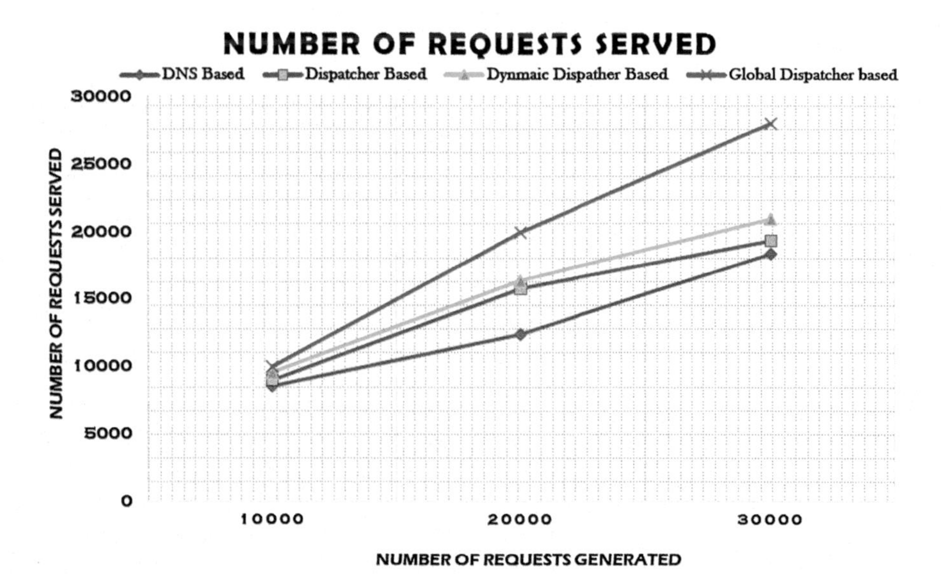

Fig. 2 Comparative analysis of serving requests with the proposed GDLB approach

Table 2 Comparative analysis of CPU utilization (%) with the proposed GDLB approach

Number of requests generated	CPU utilization (%)			
	DNS-based web server system	Dispatcher-based web server system	Dynamic dispatcher-based web server system	Global dispatcher-based web server system
10,000	85.51	89.92	95.23	99.99
20,000	61.88	78.86	81.65	99.82
30,000	61.16	64.45	69.77	93.4

server system as shown in Fig. 2. It also shows that growth of serving requests decreases as the number of generating requests increases in the proposed system, but by more than the DNS-, dispatcher- and dynamic dispatcher-based web server systems.

Based on the serving requests and the number of requests generated, the percentage of CPU utilization has been calculated as shown in Table 2. It can be observed that the proposed GDLB cluster system shows a high percentage of utilization in comparison to the DNS-, dispatcher- and dynamic dispatcher-based web server systems.

Figure 3 shows that the percentage of CPU utilization decreases as the number of generating requests increases, but in comparison to the DNS- dispatcher- and dynamic dispatcher-based web server systems, the proposed web server system shows a high percentage of CPU utilization.

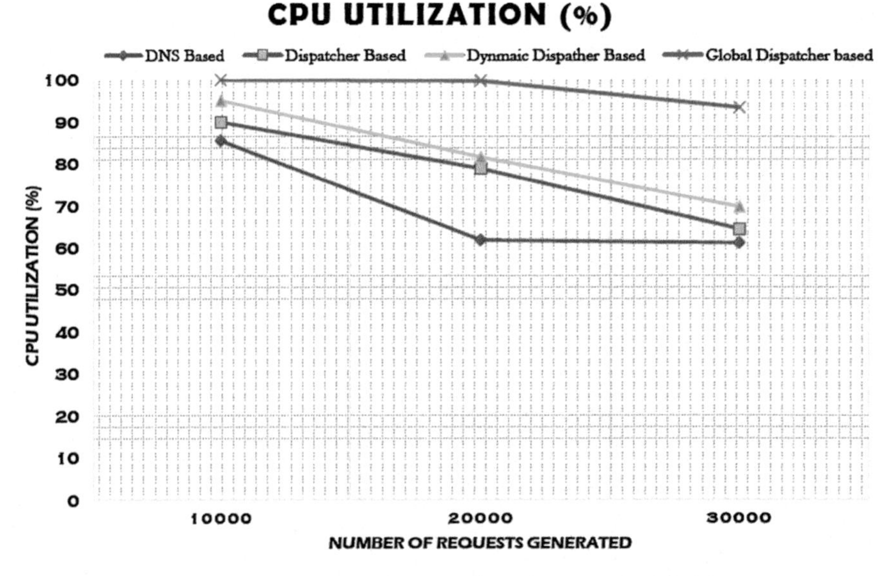

Fig. 3 Comparative analysis of CPU utilization (%) with the proposed GDLB approach

Table 3 Comparative analysis of average response time with the proposed GDLB approach

Number of requests generated	Average response time (ms)			
	DNS-based web server system	Dispatcher-based web server system	Dynamic dispatcher-based web server system	Global dispatcher-based web server system
10,000	555	459	371	386
20,000	1130	903	855	738
30,000	1487	1351	1254	1026

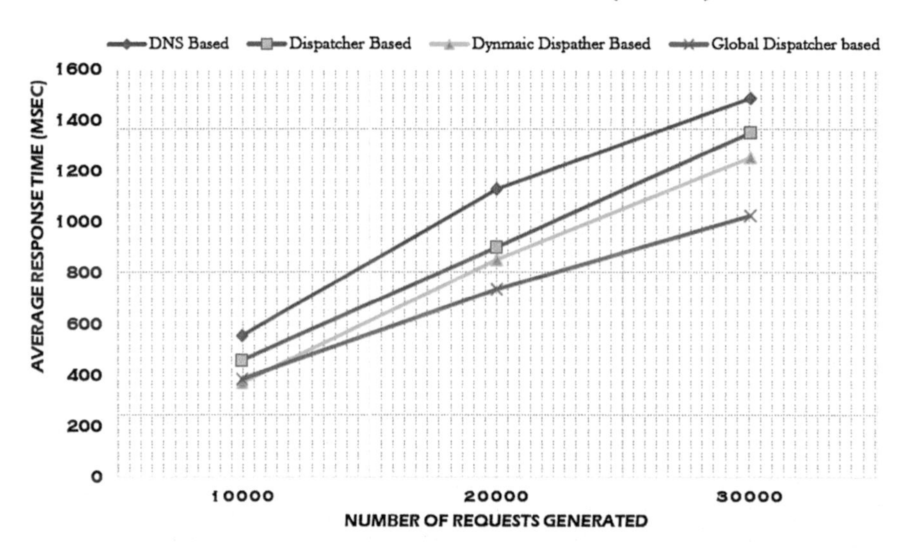

Fig. 4 Comparative analysis of average response time with the proposed GDLB approach

The average response time of generating requests has been measured on the DNS-, dispatcher-, and dynamic dispatcher-based webs systems, and the proposed web server system as shown in Table 3.

The average response time of the GDLB web server system increases as the number of generating requests increases. The rate of increase is low in the proposed web server system in comparison to the DNS-, dispatcher- and dynamic dispatcher-based web server systems as shown in Fig. 4.

As the number of requests generated on the web server systems increases, it serves the requests as per the availability and some requests may not be served. These errors of the web server system are also measured, as shown in Table 4.

Table 4 Comparative analysis of error rate (%) with the proposed GDLB approach

Number of requests generated	Error rate (%)			
	DNS-based web server system	Dispatcher-based web server system	Dynamic dispatcher-based web server system	Global dispatcher-based web server system
10,000	14.49	10.08	4.77	0.01
20,000	38.12	21.14	18.35	0.18
30,000	38.84	35.55	30.23	6.6

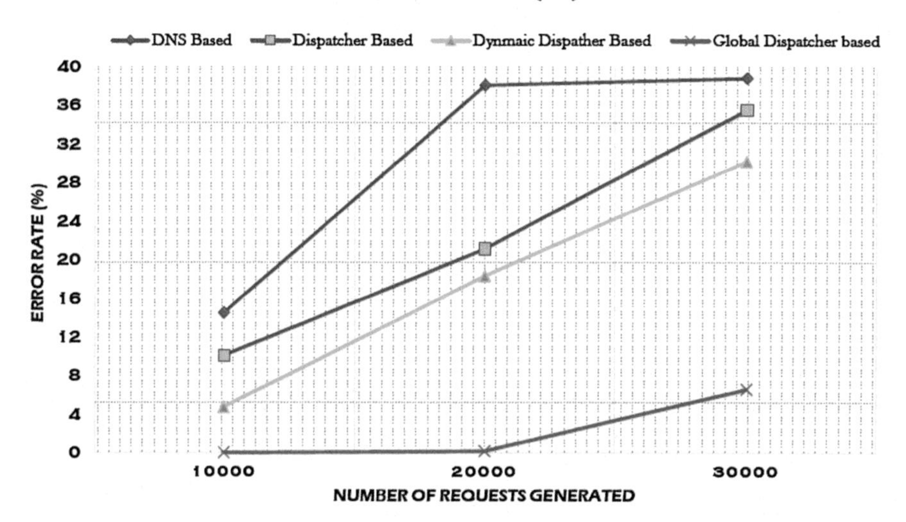

Fig. 5 Comparative analysis of error rate (%) with the proposed GDLB approach

In Fig. 5 the error percentage of unserved requests in the proposed web server system is lower in comparison to the DNS-, dispatcher- and dynamic dispatcher-based web server systems, but it increases as the number of generated requests increases.

Further throughput of the proposed GBLB web cluster has been also calculated for the same number of generating requests and it can be seen from Table 5 that the throughput is increasing more in comparison to the DNS-, dispatcher- and dynamic dispatcher-based web server systems. In Fig. 6, the proposed GDLB web cluster system shows a higher throughput in comparison to the DNS-, dispatcher- and dynamic dispatcher-based web server systems.

Based on the comparative analysis of DNS-, dispatcher- and dynamic dispatcher-based web server systems with the GDLB web cluster system, it can be

Table 5 Comparative analysis of throughput with the proposed GDLB approach

Number of requests generated	Throughput (Req/s)			
	DNS-based web server system	Dispatcher-based web server system	Dynamic dispatcher-based web server system	Global dispatcher-based web server system
10,000	130.5	137.3	147.8	155.5
20,000	138.9	174.6	180.2	189.0
30,000	177.6	185.0	203.5	209.6

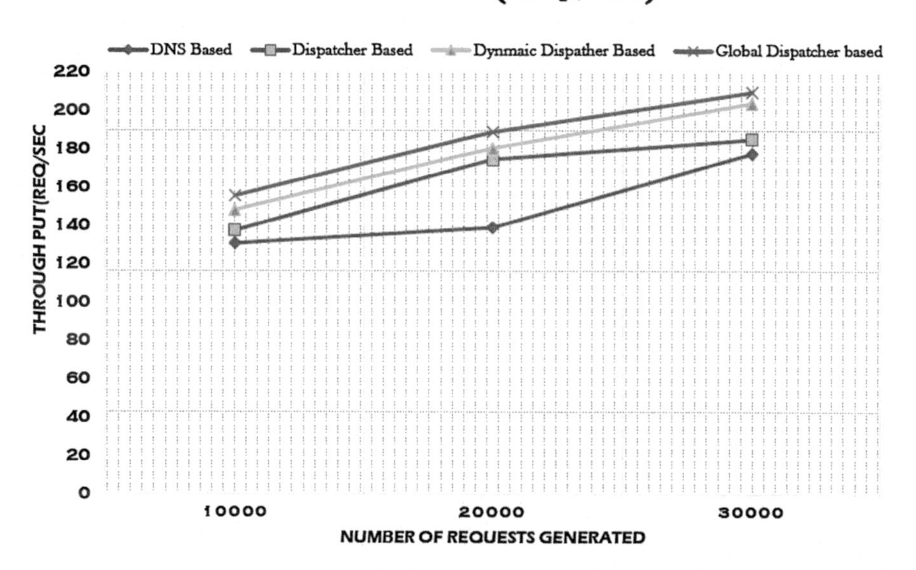

Fig. 6 Comparative analysis of throughput with the proposed GDLB approach

seen that the proposed GDLB approach designed for dynamic load balancing shows a better response time, CPU utilization, and throughput. It also provides a robust and efficient approach to dynamic load balancing for a web cluster system.

7 Conclusion

In this study, the performance of the proposed GDLB approach is measured where the request dispatching will be performed at both the DNS and the dispatcher. To generate client requests a JMeter tool was used. The parameters concerned in the experimental investigation are number of requests served, CPU utilization, average response time, error rate and throughput. Thus, the experimental outcomes show that the anticipated GDLB approach achieves an improved performance over the DNS-, dispatcher- and dynamic dispatcher-based web server systems.

References

1. Hennessy JL, Patterson DA (2003) Computer architecture: a quantitative approach. Morgan Kaufmann Publishers Inc., San Francisco, CA
2. Koomey J (2011) Growth in data center electricity use 2005 to 2010. Analytics Press, Oakland, CA. http://www.analyticspress.com/datacenters.html
3. Amith Singal, Senior Vice President, Google (2012). http://www.google.com/zeitgeist/2012/#the-world
4. Ramana K, Ponnavaikko M (2015) Web cluster load balancing techniques: a survey. Int J Appl Eng Res 10(19):39983–39998
5. Mosedale D, MCool R (1997) Lessons learned administering Nets ape's site. Internet Comput 1(2):28–35
6. Yoshilakawa C, Chun B, Eastham P (1997) Using smart clients to build scalable services. In: Proceedings of Usenix 1997, January 1997
7. Beck M, Moore T (1998) The Internet-2 distributed storage infrastructure project: an architecture for Internet content channels. In: 3rd International WWW caching workshop, Manchester, UK, June 1998. http://www.ahe.ja.net/events/workshop/18/mbeck2.html
8. Baentsh M, Baum L, Molter G (1998) Enhancing the Web's infrastructure: from caching to replication. Internet Comput 1(2):18–27
9. Colajanni M, Yu PS, Cardelini V (1998) Dynamic load balancing on geographically distributed heterogeneous web servers. In: IEEE 18th International conference on distributed computing systems, pp 295–302, May 1998
10. Cardelini V, Colajanni M, Yu PS (1999) Dynamic load balancing on web server systems. IEEE Internet Comput 3(3):28–39
11. Kwan TT, McGrath RE, Reed DA (1995) NCSA's World Wide Web server: design and performance. IEEE Comput 11:68–74
12. Cisco Systems Inc (1997) Distributed director white paper. http://www.cisco.com/warp/public/cc/cisco/mkt/scale/distr/tech/d_wp.htm
13. Sanghi D, Jalote P, Agarwal P, Jain N, Bose S (2004) A testbed for performance evaluation of load-balancing strategies for Web server systems. Softw Pract Exp 34(4):339–353
14. Anderson E, Patterson D, Brewer E. The Magi router: an application of fast packet interposing. http://s.berkeley.edu/~eanders/projects/magirouter/osdi96-mrsubmission.ps
15. Hunt GDH, Goldzsmit GS, Mukherjee R (1998) Network dispatcher: a connection router for scalable internet services. In: Proceedings of 7th international World Wide Web conference, April 1998
16. Damani O, Chung P, Kintala C (1996) ONE-IP: techniques for hosting a service on a cluster of machines. In: Proceedings of 41st IEEE computing society international conference, pp 85–92, February 1996
17. Andersen D, Yang T, Holmedahl V, Ibarra OH (1996) SWEB: towards a scalable World Wide Web-server on multi computers. In: Proceedings of 10th IEEE international symposium on parallel processing, Honolulu, pp 850–856, April 1996
18. Akamai Inc. How FreeFlow works. http://www.akamai.com/service/howitworks.html
19. Singh H, Kumar S (2011) Dispatcher based dynamic load balancing on web server system. Int J Grid Distrib Comput 4(3)
20. Hong YS, No JH, Kim SY (2006) DNS-based load balancing in distributed web-server systems. In: Proceedings of the the fourth IEEE workshop on software technologies for future

embedded and ubiquitous systems, and the second international workshop on collaborative computing, integration, and assurance (SEUS-WCCIA'06), pp 251–254, 27–28 April 2006
21. Jain R (2010) The art of computer systems performance analysis-techniques for experimental design, measurement, simulation, and modelling. Wiley, London
22. Pao TL, Chen JB (2006) The scalability of heterogeneous dispatcher based web server load balancing architecture. In: Proceedings of the 7th international conference on parallel and distributed computing, application and technology, pp 213–216

Automation of Railway Crossing Gates Using LabVIEW

N. Nagaraju, L. Shruthi and M. S. D. Hari

Abstract Train accidents at level crossings have always been a concern for the railways. The aim of this paper is to propose an automatic gate system at unmanned level crossings to replace gates operated by gatekeepers. By using this automatic railway gate management system, the arrival of the train is detected by the proximity detector placed with reference to the gate. Hence, the time that the gates are closed is less than with hand-operated gates and the amount of human labor needed is reduced too. This system can be employed at unmanned crossings wherever the probabilities of accidents is high and reliable operation is needed. It provides safety to road users by reducing accidents that occur as a result of carelessness on the part of either road users or gatekeepers. Mistreatment IR sensors and servo motors work automatically to open and shut the gates depending on the presence of train before the crossing. With relevance to the paper that used a single IR sensor to trace the train's position by employing a controller, we used two IR sensors on either side of the track. And later we programmed them by connecting them to myRIO through National Instruments LabVIEW platform.

Keywords myRIO · LabVIEW · IR sensors

N. Nagaraju (✉) · L. Shruthi · M. S. D. Hari
Department of Electronics and Communication Engineering, Institute of Aeronautical Engineering, Dundigal, Hyderabad 500043, Telangana, India
e-mail: smyle_nag@yahoo.co.in

L. Shruthi
e-mail: lakkireddyshruthi@gmail.com

M. S. D. Hari
e-mail: harimsd8@gmail.com

A. Kumar and S. Mozar (eds.), *ICCCE 2018*,
Lecture Notes in Electrical Engineering 500,
https://doi.org/10.1007/978-981-13-0212-1_37

1 Introduction

Railways are the most affordable mode of transportation and are preferred over other modes for that reason. There are two main types of level crossing, namely manned level crossings and unmanned level crossings. The history of level crossings reflects their placement, however, typically early level crossings had a flagman in a booth close to the tracks that would, on the approach of a train, wave a red flag or lamp to stop all traffic and clear the tracks. Nowadays, Asian countries have the world's largest railway network. It has become somewhat risky and dangerous for general public due to the accidents that are happening regularly. Many accidents at railway crossing gates occur thanks to the impatience of the crosser waiting for the train to cross the gate and those who try to cross the track in spite of the gate being closed. The lack of knowledge concerning the probability of the train crossing the railway gate from the time the gate is closed ends results in this parlous situation. The main aim of this paper is to develop an automatic control system that acquires information from infrared sensors and uses it to prevent accidents between trains and road users. The non-inheritable information from the IR sensors initiates the logic designed to regulate the gate and alerts the crosser regarding the train's location and keeps them secure.

An automatic control system is a meeting of physical parts connected in such a way that it doesn't need any manual interaction, and in recent years automatic management systems have rapidly become important in all fields of engineering. The benefits of an automatic system are: it provides protection from accidents, elimination of human error, saves energy, increases safety and leads to an economy of operation. It also frees people from having to do the thinking. Even the most advanced technology can't guarantee accident free and 100% safe operating conditions. However scientific investigations are often used simply to form progressive enhancements to a theory, method or current system. Putting a railway over a bridge is straightforward to do and the price is attractive compared to previous alternative methods [1]. Balog et al. [2] gave the chance of mistreatment of RFID technology by railway transport observance. The major part of the article describes the application of RFID technology in the railway industry of Slovakia. It also describes the principles of the RFID technology and its implementation in railways. Sankaraiah et al. [3] discussed the commonest types of RFID tags and readers, and the principles of RFID technology. The IR-based automatic railway gate system contains a microcontroller, motors, and sensors. The sensors are placed on the track with reference to the distance away from the track. Once the train arrives, appropriate signals will be sent from the sensors. Once the train arrives, at an equivalent time an associated IR transmitter senses generates a signal. An IR receiver receives the signal and generates an interrupt signal [4–7].

2 Methodology

This design consists of four major components:
(1) IR sensors (2) Servo motors (3) NI myRIO (4) LabVIEW software (Fig. 1).

2.1 IR Sensor

The basic principle of the IR sensor is based on an IR emitter and an IR receiver. The IR emitter will emit infrared continuously when power is supplied to it. On the other hand, the IR receiver will be connected and perform the task of a voltage divider.

2.2 Servomotor

A servomotor may be a mechanism or linear actuator that enables precise management of angular or linear position, speed, and acceleration. It consists of an acceptable motor coupled to a sensing element for position feedback. It is a special kind of motor that is mechanically operated up to a bound limit for a given command with facilitation of error-sensing feedback to correct performance.

2.3 NI myRIO

The myRIO is a time period embedded analysis tool created by National Instruments, RIO abbreviates to "Reconfigurable I/O" device. It is used to show and implement multiple style ideas with one reconfigurable I/O (RIO) device. Applications can be developed that utilize its onboard FPGA and silicon chip. myRIO

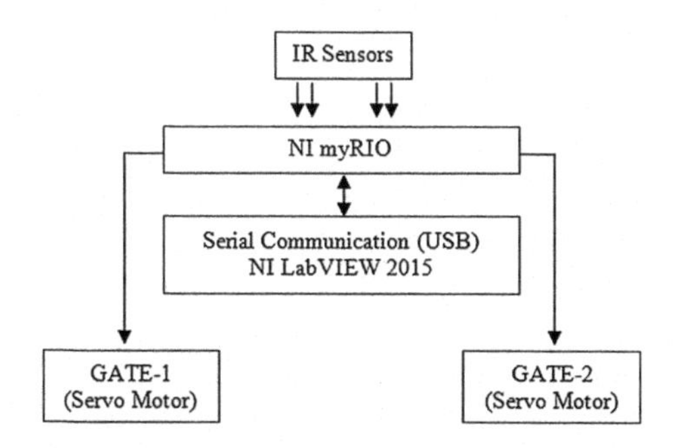

Fig. 1 Block diagram representation of the model

provides analog input (AI), analog output (AO), digital input and output (DIO), audio, and power output in a compact embedded device. It is supported by four components: a processor, a reconfigurable FPGA, inputs and outputs, and graphical design software.

2.4 NI LabVIEW-2015

LabVIEW, short for Laboratory Virtual Instrument Engineering Workbench, is a system-design platform and development setting for an evident language from National Instruments. LabVIEW is usually used for info acquisition, instrument management, and industrial automation on a growing number of operational systems, alongside Microsoft Windows, various versions of OS, Linux, and Mac OS. However, LabVIEW is more than a language. It is an academic degree and an advanced certification in the perspective of jobs. LabVIEW can create programs that run on those platforms, such as Microsoft Pocket PC, Microsoft Windows CE, Palm OS, and a growing number of embedded platforms, alongside field programmable gate arrays (FPGAs), and digital signal processors (DSPs).

3 Proposed Design

In this paper, a pair of IR sensors on either facet of crossing and is separated by a minimum distance of 25 m, i.e. the length of one compartment of a traditional Indian train. Detector activation time is therefore adjusted by calculating the time taken at a particular speed to cross a minimum of one compartment of an ordinary train of the Indian railway. In our study, 5 s was the length of time used. Sensors are located at 1 km on either side of the gate. The sensors were termed according to the direction of the train as "foreside sensors" or "after facet sensors."

3.1 Case-I

When both of the foreside IR receivers are activated, the gate motor is turned on in one direction at an angle of 0° and the gate is closed and stays closed until the train crosses the gate and reaches the after-facet sensors. The gates are closed as we introduced a delay element in the logic. The length of the delay depends on the length of train and the distance from the gate to the sensor. After this delay ends, the gates will be opened and returned to their initial 90° position. When train reaches the after side IR sensors, the logic won't take the inputs and the gates remain open.

3.2 Case-II

In the case of a train coming from the other end, the after side IR sensors detects a signal and logic takes inputs and the gates will be closed, i.e. they move to 0°. As soon as the gates close, the delay is ON, and the gates remain closed until the train crosses the crossing. At the end of the delay, the gates return back to their initial 90° position. After this, logic won't be taking inputs from the foreside sensors and the gates remain open for the rest of the time. A buzzer can be implemented so as to alert the road users in case they do not notice the arrival of train and to give time to drivers to clear the gate area in order to avoid being trapping between the gates. The buzzer stops sounding after the train has passed. Once all the hardware connections have been setup we have to form the logic in LabVIEW and connect RIO to that logic for execution. All the connections from the sensors to RIO, VCC, and GND must be proper and without any floats. Even if one GND connection is not proper, a precise output may not be produced. The complete logic was implemented in a LabVIEW platform (Fig. 2).

4 Results and Discussion

Virtual IR sensors can be added to the block diagram. IR sensors are considered as a digital input to logic. So, to add it from the function palette, right-click on the block diagram and scroll down to find myRIO > click digital input (DI). A pop-up window appears to add more pins and we can view a detailed description. It shows the number of digital pins actively connected to RIO. Add pins 11 and 13 for IR sensors 1 and 2 connected to channel A. Add one more digital input for IR sensors 3 and 4 connected to channel B. The digital inputs from RIO are active low signals and in order to make them high for further processing add NOT gates to each of the

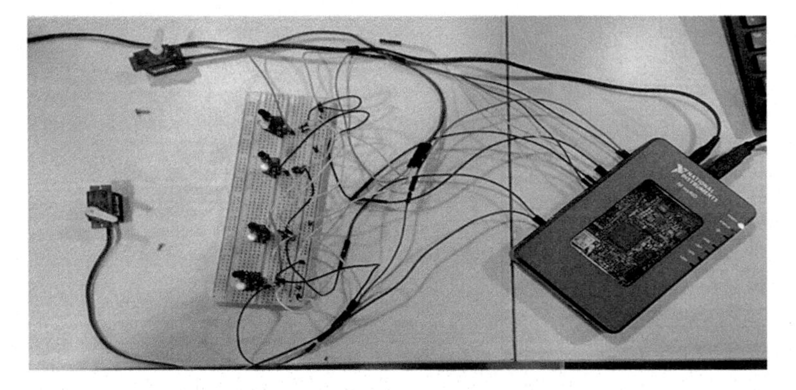

Fig. 2 Integration of all components with myRIO and a personal computer

terminals of the digital inputs. Both pins from channel A are connected to an AND gate. IR sensor 2 i.e. is connected to case structure which has two cases, either TRUE or FALSE, depending on the input. In the TRUE case a specific delay element is placed with a delay of 5 s (Fig. 3).

Pins from channel B are also connected in the same fashion as channel A and sensor 4, i.e. pin 13, is connected to the delay through a case structure. The outputs from both AND gates are connected to an OR gate. This OR gate output is given to another case structure with two cases. The TRUE case consists of 0, which means 0° and the FALSE case consists of 90, which means 90°. The output of the case structure is connected to PWM.

Servomotors are considered as PWM signals. Add two PWM functions from the function palette to the block diagram, one for pin 27 and the other for pin 29, both in channel A. Here a constant duty cycle of 0.1 is employed so that a servomotor rotates smoothly. The output of the case structure is given to the PWM output. As a result, the gates are rotated based on the value obtained from the case structure which has 0° and 90°. Before starting the process, the servomotors must be in a neutral position of 90° and that should be made a default value. Controls for the four IR sensors and a knob for the gate are connected so as to view their status on the front panel. Place the whole code in a while loop to automate the whole logic, and connect a STOP button to a Boolean control. Whenever a loop is used in the logic, we shouldn't click the second button, or in other words, run it continuously as it consumes more power. Also, never use the abort button in order to restore default values. Initially, the gates, i.e. the knob, are at 90°, as all the IR sensors are FALSE.

4.1 *Test Case-I*

When the first two IR sensors detect a signal, IR 1 and IR 2 glow and the servomotors reach 0° as shown on the front panel. The train is moving from left to right. Hence, IR sensors (1 and 2) detect the presence of train and the servomotor (the

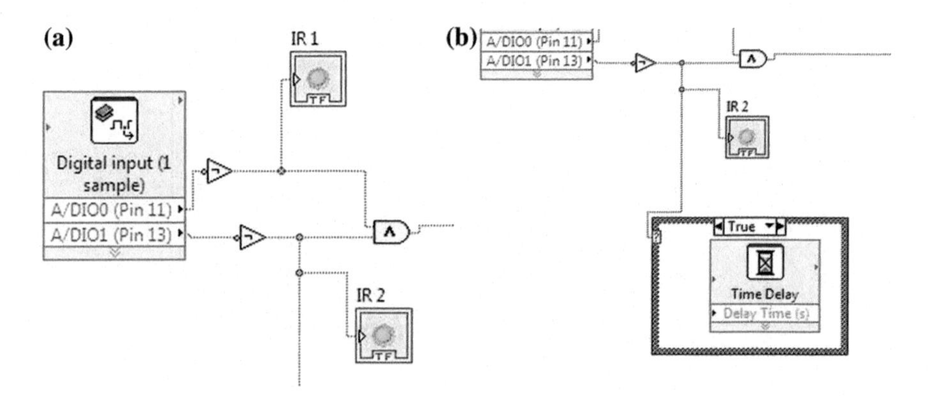

Fig. 3 **a** Inputs taken from an IR sensor through myRIO **b** delay element to IR sensor 2

knob) moves to 0°. The delay is now active and as soon as the train crosses the crossing, the delay will become zero and the gates move to neutral, i.e. 90°. When IR 4 is detected again, another delay associated with IR 4 is activated and hence the gate remain at 90° even when IR 4 and IR 3 are detected in the flow (Fig. 4).

4.2 *Test Case-II*

Here the train is moving from right to left. Hence, IR Sensors (3 and 4) detect the presence of the train and the servomotor (the knob) moves to 0° (Fig. 5).

In case of a train arriving from the opposite direction, IR 3 and IR 4 sensors are detected and glow and the servomotor reaches 0°. Here, the delay becomes active again and as soon as train crosses the crossing the gates reach 90°. When IR 2 detects a signal again, another delay associated with IR 2 is activated and the gate remains at 90° even when IR 1 and IR 2 are detected in the flow.

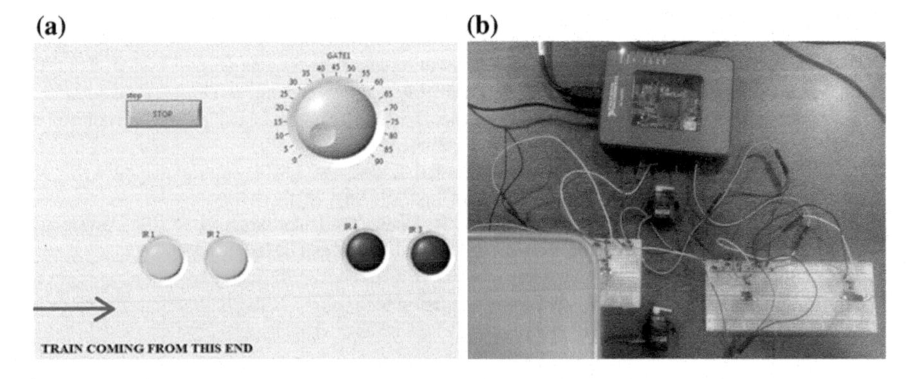

Fig. 4 **a** Display when the train is moving from the left **b** practical output from test case I

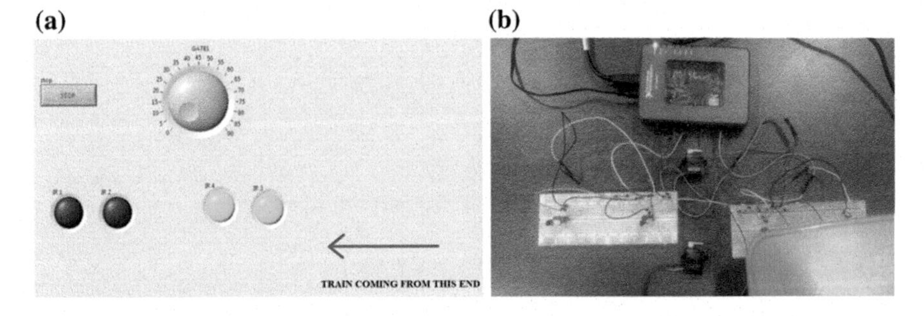

Fig. 5 **a** Display when the train is moving from the right **b** practical output from test case-II

5 Conclusion

An automatic railway gate system is proposed with the aim of reducing human involvement in the operation of the crossing gates that either allow or prevent cars and humans from crossing the railway tracks. One of the benefits of this project is that IR sensors are positioned at a minimum distance of 24 m. One limitation is that once the gates are closed they can't be opened again for the crossing of cars and vehicles. We can vary the location of IR sensors as per our need, however, it is better to use load sensors or perhaps proximity sensors. Our intention for future work is to implement our system in real time by fixing the minor issues regarding technologies.

References

1. Sharad S, Sivakumar PB, Ananthanarayanan V (2016) An automated system to mitigate loss of life at unmanned level crossings. Procedia Comput Sci 92:404–409
2. Balog M, Husár J, Knapcíkova L, Soltysova Z (2015) Automation monitoring of railway transit by using RFID technology. Acta Tecnología 9–12
3. Sankaraiah S, Mirza SAB, Narsappa R (2015) Automatic railway gate control system using IR and pressure sensors with voice announcement. Int J Innov Technol (IJIT) 3(3):0355–0357
4. Krishnamurthi K, Bobby M, Vidya V, Edwin B (2015) Sensor based automatic control of railway gates. Int J Adv Res Comput Eng Technol (IJARCET) 4(2)
5. Kottalil AM, Abhijith S, Ajmal MM, Abhilash LJ, Babu A (2014) Automatic railway gate control system. Int J Adv Res Electr Electron Instrum Eng 3(2)
6. Banuchandar J, Kaliraj V, Balasubramanian P, Deepa S, Thamilarasi N (2012) Automated unmanned railway level crossing system. Int J Mod Eng Res (IJMER) 2:458–463
7. Ilie-Zudor E, Kemeny Z, Egri P, Monostori L (2006) The RFID technology and its current applications, Proceedings of the modern information technology in the innovation processes of the industrial enterprises. MITIP 5(7)

RF Energy Harvesting Using a Single Band Cuff Button Rectenna

R. Sreelakshmy and G. Vairavel

Abstract In this paper a small-sized cuff button rectenna is proposed. The proposed antenna structure is a circular patch antenna of cuff button shape. It is fabricated and tested for suitability to power wearable electronics at 2.4 GHz. The wearable antenna, which resembles a cuff button, is made of a PTFE taconic ceramic substrate. The substrate has a permittivity ε_r equal to 10. RF-DC conversion is achieved by a diode rectifier and a DC-DC step-up converter. The measured efficiency is 51% at 2.4 GHz or 0 dBm. This rectenna can be used in wireless power transmission systems that transmit power by radio waves.

Keywords Wearable rectenna · Cuff button antenna · Taconic ceramic RF-DC conversion

1 Introduction

Electronic gadgets that can be worn are referred as wearable technology. These gadgets are becoming increasingly popular and important. The clothing material in which electronic devices and sensors are integrated into and worn on the body are referred to as e-textiles wearable technology (e-textiles) [1]. Major importance is placed on miniaturization and low power consumption in these wearable devices. The major challenge in this field is that the devices are battery driven and require frequent charging Wearable technology is a combination of both textile and electronic technology. In this paper a wearable cuff button antenna is designed to be operated at 2.4 GHz. RF energy harvesting involves RF-DC conversion and could be utilized for powering wearable sensors and other associated electronics.

R. Sreelakshmy (✉) · G. Vairavel
Department of Electronics and Communication Engineering, Vel Tech Rangarajan Dr. Sagunthala R&D Institute of Science and Technology, Chennai 600062, India
e-mail: lakshmysuneesh@gmail.com

G. Vairavel
e-mail: vairavel@veltechuniv.edu.in

A. Kumar and S. Mozar (eds.), *ICCCE 2018*,
Lecture Notes in Electrical Engineering 500,
https://doi.org/10.1007/978-981-13-0212-1_38

In energy harvesting techniques, energy is captured with minimum loss and maximum efficiency from natural sources like the sun, heat, wind, and radio frequencies.

The RF energy harvester comprises two sections—an antenna and a rectifier circuit. They are collectively called a rectenna [2, 3]. In this paper a single band cuff button antenna is designed and fabricated. The design of the wearable antenna is important as its operation on the human body may cause electromagnetic coupling and antenna characteristics may change, resulting in the reduction of radiation efficiency. In order to avoid this the designed antenna has a ground plane which extends long the entire bottom patch. The designed antenna structure is lightweight and of a small size. With a diameter of only 22.3 mm it is suitable for wearable applications. RF energy received by the single band cuff button antenna will be converted into DC by the half-wave rectifier circuit. This obtained DC voltage is fed to a LTC3105 DC-DC step-up converter.

2 RF Energy Harvester-Rectenna Design

The main principle of the electromagnetic energy harvester is to convert RF power density incident on the single band cuff button antenna into DC power which could drive wearable electronics. This harvester is a combination of an antenna and a rectifier, and so it is termed a rectenna [2, 3]. The basic block diagram is shown in Fig. 1.

2.1 Single Band Cuff Button Antenna Design

The single band cuff button antenna is designed with a circular radiating patch and the entire bottom patch has a ground plane. This is to reduce the effect of the human body on antenna performance and to reduce the effect of radiation on the human body [4, 5]. The antenna is fed by a microstrip transmission line of 50 Ω impedance. The antenna is designed with a PTFE taconic ceramic substrate of 3.18 mm thickness and $\varepsilon_r = 10$ is a square-shaped with filleted edges. This substrate has very good thermal and electrical conductivity.

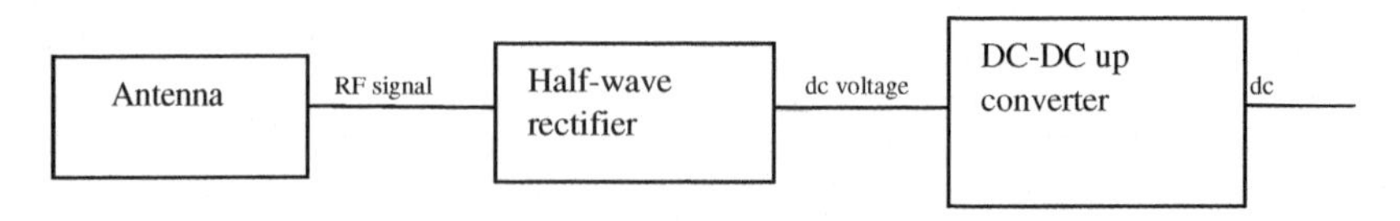

Fig. 1 Basic block diagram of the RF energy harvester

Circular Patch Radius and Effective Radius

Since the dimension of the patch is treated as a circular loop, the actual radius of the patch is given by

$$a = \frac{F}{\sqrt{\left\{1 + \frac{2h}{\pi \varepsilon_r F}\left[ln\left(\frac{\pi F}{2h}\right) + 1.7726\right]\right\}}} \quad (1)$$

$$F = \frac{8.791 \times 10^9}{f_r \sqrt{\varepsilon_r}} \quad (2)$$

The effective radius of patch used is given by

$$a_e = a\sqrt{\left\{1 + \frac{2h}{\pi \varepsilon_r a}\left[\ln\left(\frac{\pi a}{2h}\right) + 1.7726\right]\right\}} \quad (3)$$

Hence, the resonant frequency for the dominant TM^z_{110} is given by

$$f_{r110} = \frac{1.8412\, v_0}{2\pi a_e \sqrt{\varepsilon_r}} \quad (4)$$

where v_0 is the free space speed of light.

Figure 1 shows the geometry of the wearable antenna at 2.4 GHz, ISM band. The substrate used is PTFE taconic ceramic of permittivity 10 [6]. This proposed antenna structure has a diameter of 22.3 mm and a thickness of 3.18 mm. The ground and radiator are made of copper with a thickness of 60 μm. A 50 Ω microstrip feed, with a SMA connector of height 5 mm is used as an antenna feed. This antenna is designed using ANSYS HFSS software (Fig. 2).

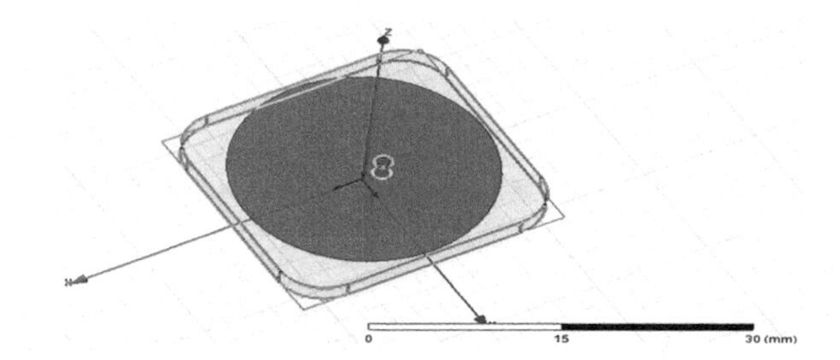

Fig. 2 Cuff button antenna structure

Properties: buttonantenna - HFSSDesign1

Local Variables

Value Optimization Tuning Sensitivity Statistics

Name	Value	Unit	Evaluated Value	Type	Description	Read-only	Hidden
subx	patchx+2mm		23.7321268427...	Design			
suby	patchy+2mm		23.0513mm	Design			
feed_length	20.83	mm	20.83mm	Design			
subh	3.18	mm	3.18mm	Design			
patchx	21.732126...	mm	21.7321268427...	Design			
patchy	21.0513	mm	21.0513mm	Design			
feedx	-1.8	mm	-1.8mm	Design			
feedy	0.00377078	mm	0.00377078mm	Design			

Show Hidden

Add... Add Array... Edit... Remove

Fig. 3 HFSS design parameters for the cuff button antenna

2.2 *Rectification Circuit*

A vital part of the RF energy harvester is the rectifier circuits. The most commonly used rectifier circuits for RF energy harvesting are half-wave rectifier and bridge rectifier. The energy of electromagnetic wave is inversely proportional to frequency. This proposed antenna works at 2.4 GHz and so a half-wave rectifier of low threshold voltage is selected. In this paper, a HSMS2850 half-wave rectifier of 150 mV threshold voltage is chosen (Fig. 3).

2.3 *DC-DC Step-up Converter*

The half-wave rectifier circuit shown in Fig. 4 produces a low voltage DC output. This low voltage DC is quite insufficient to drive wearable electronics. Hence it has to be boosted to obtain a high DC voltage to drive the components. In order to achieve this a DC-DC step-up converter is used. The major goal of energy harvester

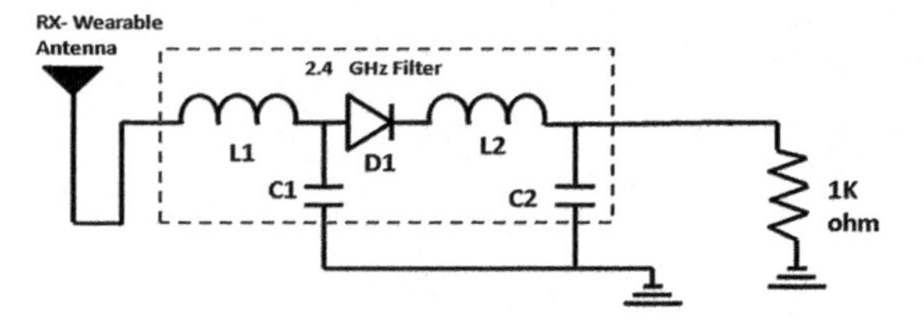

Fig. 4 Half-wave diode rectifier circuit

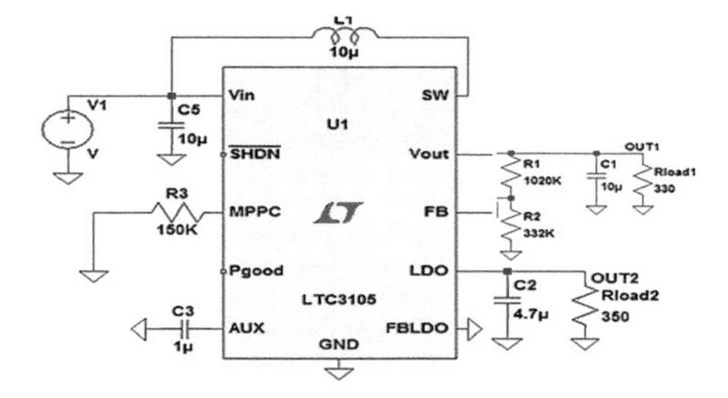

Fig. 5 LTC3105 step-up converter diagram

is to ensure higher efficiency [7]. In the RF energy harvester, a LTC3105 DC-DC step-up converter of 225 mV operating voltage and with maximum power point control is used. The pin configuration of the LTC3105 is shown in Fig. 5.

3 Results and Discussion

The antenna part of this rectenna is designed and simulated by a HFSS 13.0 tool. The antenna structure is designed and fabricated using a PTFE taconic ceramic of 3.18 mm thickness and a permittivity of 10. The structure has a return loss of −18 db at 2.4 GHz. The structure was fabricated and tested at CUSAT laboratory. This produces a high gain and a VSWR of 1.5 at 2.38 GHz. All the measured results and simulated results show close agreement and prove that the designed antenna is 94% efficient. The use of a half-wave rectifier using L and C components also helps in improving the efficiency of the harvester (Figs. 6 and 7).

The fabricated structure was tested and the test results show a close agreement with results from the simulation. The gain of the antenna is 5 db and the measured VSWR is 1.5 at 2.38 GHz. A slight deviation of resonant frequency is due to mechanical abrasions. All vital parameters that determine antenna performance were measured and all proved to be good. The simulation output and experimental result match well and can be considered as a strong indication to consider this as a

Fig. 6 Fabricated cuff button antenna structure

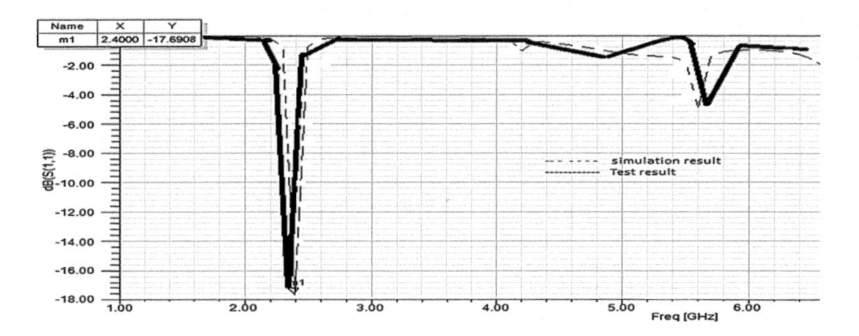

Fig. 7 Return loss, S11 parameter at 2.4 GHz

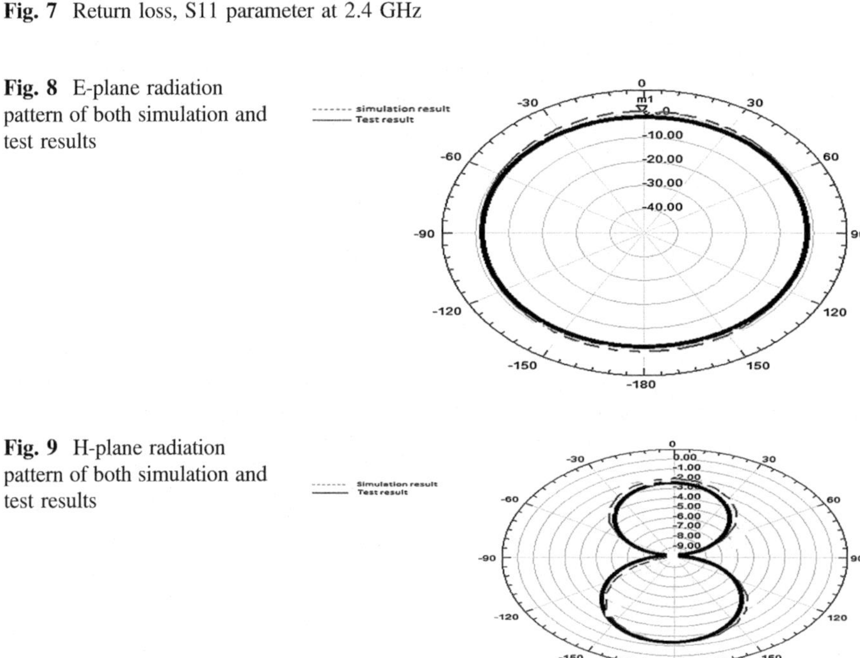

Fig. 8 E-plane radiation pattern of both simulation and test results

Fig. 9 H-plane radiation pattern of both simulation and test results

novel design [8]. The VSWR value of the antenna structure determines its radiating ability. Usually, an antenna with a voltage standing wave ratio of less than 3 is considered to be good. This wearable antenna structure provides a VSWR of 1.5 at 2.38 GHz and 2.6 at 2.4 GHz in the ISM band. So it proves that the proposed structure is radiates well (Figs. 8 and 9).

The results show that the radiation has a broad 360° coverage and so the range is very high. It is known that electromagnetic body modes are influential in guiding signals around the user and also to elevated external access points (Fig. 10).

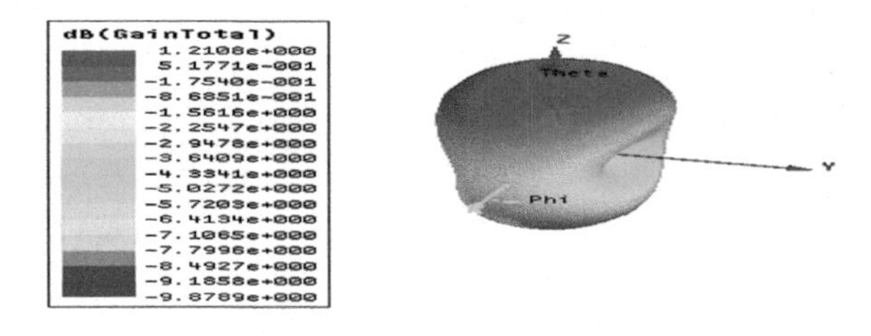

Fig. 10 Directivity of the fabricated structure

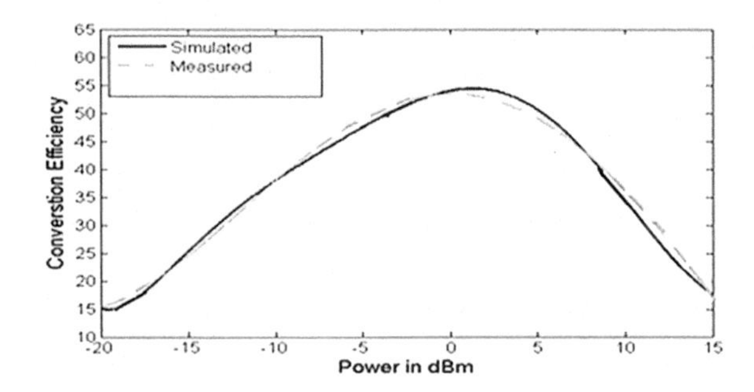

Fig. 11 Measured and simulated overall efficiency versus input power at 2.4 GHz

The micro power at the rectenna terminals can be determined using the Friis transmission equation

$$P_{rx} = P_{tx} G_{tx} G_{rx} \frac{C^2}{(4\pi D_r f_o)^2} \tag{5}$$

where P_{tx} is the transmitting power, G_{rx} is the receiving antenna gain, C is the velocity of light, and f_o is the frequency of the microwave. The overall efficiency of the rectenna can be calculated by Eq. 6

$$\eta_{EH} = \frac{P_{outDC}}{P_{rx}} = \frac{\frac{V^2_{outDC}}{R_L}}{P_{rx}} \tag{6}$$

From the above equation it is clear that the square of the DC output voltage is directly proportional to efficiency and DC output power, and is inversely proportional to load resistor R_L [2, 9]. The RF energy harvester has a measured efficiency of 60% at 2.4 GHz. The testing of the step-up converter only at maximum power operating points of the low power level half-diode rectifier maximized the harvesting efficiency. The DC-DC converter is operated with an input voltage. The maximum efficiency operating voltage at 2.4 GHz is 8 V (Fig. 11).

4 Conclusion

In this study, a single band cuff button rectenna at 2.4 GHz is proposed and presented at a compact size of 22.3 mm diameter. The design and implementation of the wearable antenna are explained as is its application in wearable electronics. Future research should focus on increasing the harvesting efficiency and miniaturization of the antenna. Research can be done on the bending effect and how it affects harvesting efficiency. In this paper, a novel cuff button antenna at 2.4 GHz is designed and the associated rectification circuitry provides a DC voltage which can be boosted up to a maximum of 4 V by the DC-DC step-up converter.

References

1. Stoppa M, Chiolerio A (2014) Wearable electronics and smart textiles: a critical review. Sensors 11957–11992
2. Song C et al (2015) A broadband efficient rectenna array for wireless energy harvesting. In Proceedings of EuCAP 2015, Lisbon (Portugal), pp 1–5
3. Stein R, Ferrero S, Helfield M, Quinn A, Krichever M (1998) Development of a commercially successful wearable data collection system. In: Second international symposium on wearable computers
4. Sanz-Izquierdo B, Batchelor JC, Sobhy MI (2010) Button antenna on textiles for WLAN on body application. IET Microwav Antennas Propag 4:1980–1987
5. Volakis JL, Chen JL, Fujimoto K. Small antennas: miniaturization, techniques & application, 409. Kellomäki T (2012) Analysis of circular polarization of cylindrically bent micro strip antennas. Int J Antennas Propag 2012:1–8
6. Pun K, lslam MN, Ng TW (2014) Foldable substrates for wearable electronics. In: International conference on electronic packaging technology (ICEPT)
7. G. TR25.914. Measurements of radio performances for UMTS terminals in speech mode
8. Sreelakshmy R, Ashok Kumar S, Shanmuganantham T (2017) A wearable type embroidered logo antenna at ISM band for military applications. Microw Opt Technol Lett 59(9):2159–2163
9. Babić M, Galoić A, Tkalac A, Žanic D (2016) Wearable energy harvester. In 2016 IEEE AP-S student design contest, presented at APS/URSI 2016, Fajardo, Puerto Rico

Hexagonal Intersection-Based Inner Search to Accelerate Motion Estimation

P. Palaniraj and G. Sakthivel

Abstract The computational complexity of motion estimation (ME), increases proportionally with the number of search points. As a result, rapid ME techniques reduce complexity by using different search patterns. Of these techniques, hexagonal search (HS) with a small diamond search pattern (SDSP) significantly reduces complexity compared to other fast ME algorithms. The proposed hexagonal intersection search (HIS) algorithm improves the inner search of HS. The new algorithm cares the inner points near to the intersection of a hexagon rather than a SDSP or a full search. The proposed algorithm reduce 7–15% search points compared with HS. Compared with other popular algorithms, the HIS algorithm made the lowest number of average search points with negotiable PSNR loss.

Keywords Motion estimation · Hexagonal search · Hexagonal intersection search

1 Introduction

Motion estimation (ME) is technique used in motion detection and video codecs. In a video codec ME analyzes the similarities between a successive frame with the candidate frame and identifies the best match position by search points in different patterns and the best position is called the motion vector (MV). A video codec's ME is the most complex process and requires greater computational cost; for MPEG4 computational cost for ME was 30–60% [1]. So search points play a very important role in the reduction of computational cost, especially in video encoding engines. Statistics [2] shows that 40% of overall internet traffic is occupied by video data so the use of videos has drastically increased over last decade. And the use of codecs has expanded into almost every gadget, especially battery-operated devices.

P. Palaniraj (✉) · G. Sakthivel
Annamalai University, Chidambaram 608002, Tamil Nadu, India
e-mail: palaniraj_p@yahoo.co.in

G. Sakthivel
e-mail: gsauei@gmail.com

A. Kumar and S. Mozar (eds.), *ICCCE 2018*,
Lecture Notes in Electrical Engineering 500,
https://doi.org/10.1007/978-981-13-0212-1_39

So a reduction in ME computational costs will accelerate the increase in power efficiency of video codecs.

There are many types of ME search patterns used for MV detection, e.g., logarithmic search, three step search, diamond search, four step search, and hexagonal search etc. Research to improve the MV has featured many aspects. ME [3] was improved in the h264 codec by reducing complexity through reuse of SAD (sum of absolute difference) values. Complexity was reduced to 12% in the h264 [4] by applying the hierarchical block matching method. On other hand, the acceleration of ME was also promoted in the design of the MV engine in the VLSI by improving the power and speed and by reducing and reusing the logic gates.

2 Hexagonal Search Pattern

The hexagonal pattern search made remarkable improvements in search points. It reduced search points so that the complexity was also reduced. The search pattern is shown in Fig. 1a. The search only involves the seven positions denoted by red dots. Then the mean of absolute difference (MAD) value was calculated, if the minimum MAD value was other than the center, the minimum point was taken as the center and a further hex search was continued with new search points. For example, if point “2” is the minimum, it is taken as the center and a new search starts as shown in Fig. 1b. Positions 4, 5, and 6 are already searched, so the search applies only to 2, 3 and 7. This expanded search continues until it reaches the center as the minimum. Once it reaches the center then the inner search is carried out for the eight positions around the center as per Fig. 1c or in the SDSP shown in Fig. 1d. Finally, the position of the minimum MAD from the nine points is considered as the motion vector.

The simulated results of various videos of different sizes are tabulated. The search boundary was −7 to 7, all video sequences are 16 frames and the block size was 16. Table 1 contains the average number of search points per block (ANSP) for various fast search patterns such as TSS (three step search), FSS (four step search), DS (diamond search), and HS (hexagonal search). HS had the fewest search points and an average reduction of 22% search points compared with the DS.

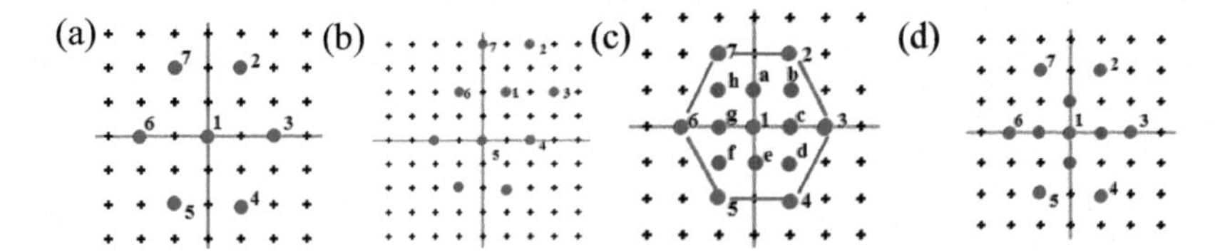

Fig. 1 **a** HS pattern, **b** extended HS, **c** SDSP inner search, **d** full inner search

Table 1 Average search points per block

	Caltrain		Missa		Salesman		Trevor		Surfside	
	AN SP	PS NR	AN SP	PS NR	AN SP	PS NR	AN SP	PS NR	AN SP	PS NR
FSS		31.3		37.8		37.8		32.9		36.3
TSS	23.8	30.1	22.9	37.5	22.7	37.5	21.3	32.9	24.7	36.1
FSS	19.2	30. 6	17.7	37.5	16.0	37.5	14.7	32.8	24.3	35.9
DS	16.1	31.2	15.9	37.5	13.0	37.5	11.5	32.8	25.9	35.9
HS	13.1	30.5	11.7	37.0	10.6	37.0	9.6	32.8	17.7	35.9

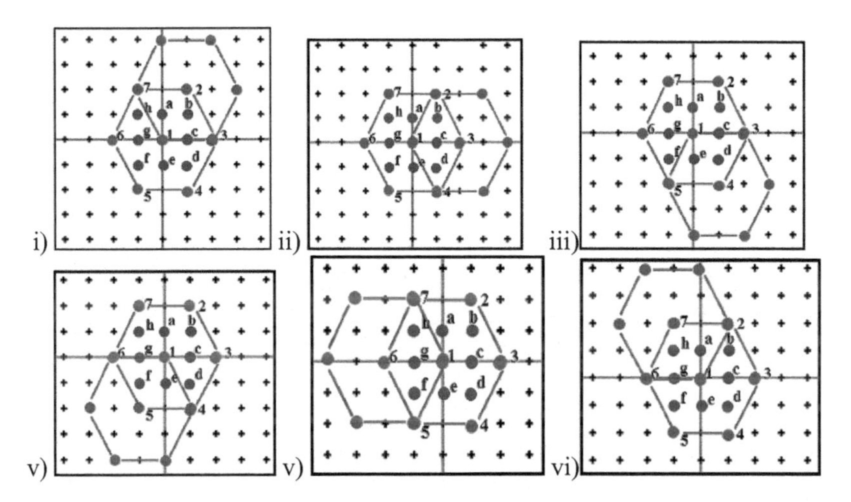

Fig. 2 Inner points on the intersection of the hexagonal area (i) points a, b, (ii) points b, c, d, (iii) points d, e, (iv) points e, f, (v) points f, g, h, (vi) points a, h

The PSNRs for various fast search patterns are shown in Table 1. The full search (FS) result is also included in the table for comparison. It shows that when comparing the HS with the FS, there was an average 0.48 db loss. This is an acceptable loss.

3 Hexagonal Intersection Search Pattern

The HS-based simulation shows an improvement in the ME. The simulation was carried out with a small diamond (SD) inner search. The inner search plays a significant role in HS-based ME and it shows a marked improvement. The role of the inner search was further analyzed, and it was found that if a full eight point inner search was executed, it only improved the PSNR by 0.3 db. This means that a ME in high frequency lies outside of the first hexagonal area. In the proposed new

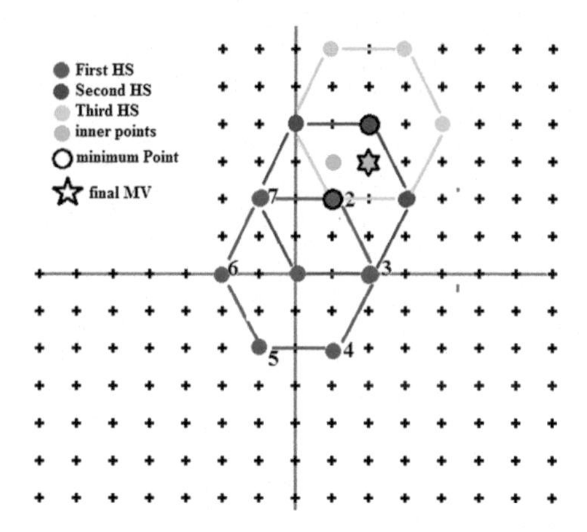

Fig. 3 Example of a HIS search

inner search position, selection is based on the intersection of the final and pre-final hexagonal area. The six intersection of the hexagonal area and the possible inner search is shown in Fig. 2.

The intersections created by the points and corresponding inner search points are 2-a, b, 3-b, c, d, 4-d, e, 5-e, f, 6-f, g, h, and 7-a, h as shown in Fig. 2. This process reduces the number of search points from eights to two or three, and as a result it significantly reduce the complexity of the ME process.

4 Algorithm for the Proposed Search

The algorithm for the hexagonal intersection search is as follows:

Step 1: Calculate the MAD for the first hexagonal area. If the minimum was point 1 then perform a SDSP and goto step 4.
Step 2: Take the minimum point as the center point for the next hexagonal search. Calculate the MAD and continue the process until reaching the center as the minimum.
Step 3: Do the search process on inner points corresponding to the intersection of the final and pre-final hexagonal area.
Step 4: Take the minimum MAD as the MV.

5 Example of a HIS Search

Figure 3 shows an example of the proposed search. In this example, in the first HS the minimum position was found at point ‘2’. Then the second HS was carried out by taking the first minimum as the center. Again the minimum was at point ‘2’.

The process was continued and in the third HS the center of point '1' was the minimum. At this stage the HS stopped and the third hexagon was consider as final and the second hexagon was considered as pre-final. So for the inner search we took the points of intersection and the final MAD from the intersecting points and center point. The position corresponding to the minimum MAD value is considered to be the MV.

6 Results and Discussions

The HIS was implemented in MATLAB 2010b. The simulation was performed with different sizes and types of image sequence. The simulated results of the HIS were compared with fast searching algorithms, such as FS, TSS, FSS, DS, and HS. The following image sequences were used for the simulation: Caltrain (16 frames 400 × 512), Missa (16 frames 288 × 360), Salesman (16 frames 288 × 360), Trevor (16 frames 256 × 256), and Surfside (16 frames 1024 × 2048).

The video sequences of Caltrain and Missa were simulated and the results are shown in the Figs. 4, 5, 6 and 7. In the Caltrain sequence, motion happens in nearly 60% of the frame, and in the Missa sequence in nearly 40%. From the results, it can

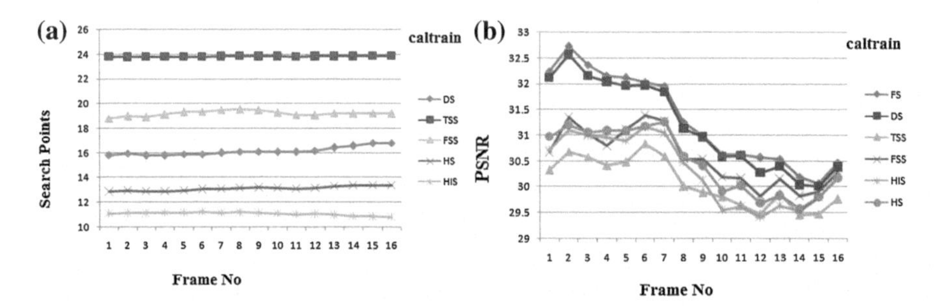

Fig. 4 **a** Average search points per block for Caltrain, and **b** PSNR for Caltrain

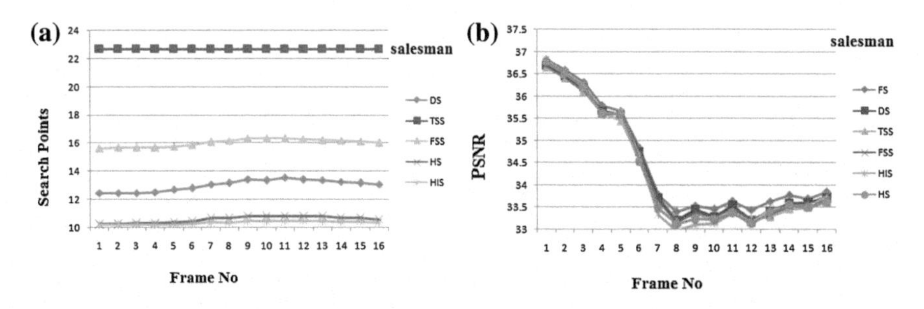

Fig. 5 **a** Average search points per block for Salesman, and **b** PSNR for Salesman

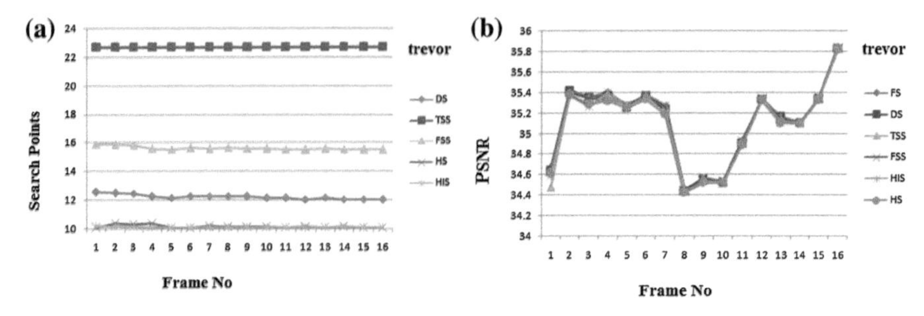

Fig. 6 **a** Average search points per block for Trevor, and **b** PSNR for Trevor

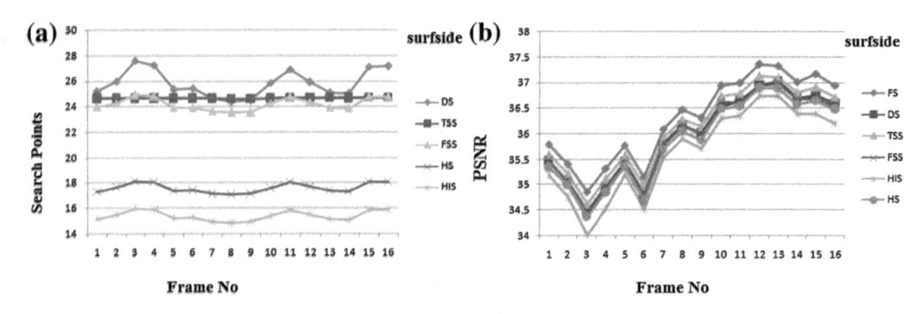

Fig. 7 **a** Average search points per block for Surfside, and **b** PSNR for Surfside

be seen that the search points are reduced to 15.75% in Caltrain and 7.12% in Missa. The PSNR loss value of Caltrain was 0.17 dB and in Missa 0.21 dB. This loss is negligible so the quality remains almost the same as with the HS.

The simulated results of both the Salesman and Trevor image sequences are shown in Figs. 5 and 6. In both sequences motion occurs in less than 25% of the frame. The possibilities that the MV lies outside the first hexagon was low, so the average search points of the HIS reduced by 2 to only 5%, almost equal to the HS. PSNR is also very close to the HS.

The Surfside image sequence had more than 90% of motion and the results are shown in Fig. 7. The HIS reduced the search points of 12.37% over HS and so compare the HIS with DS 40.28% less. The PSNR of the HIS was 0.21 dB less than the HS and its loss was negligible.

7 Conclusion

The HIS accelerates the ME process with negligible PSNR loss. The HIS algorithm improves the search process by up to 15% compared to a conventional HS. The results also show that the HIS performs well in high frequency motion of the image sequence. In low frequency motion the results are similar to the HS. Furthermore, the HIS technique can be combined with a group-based inner search [5], an efficient inner search [6], or a point-oriented inner search [7]. There are, therefore, possibilities to reduce the search points and this should lead to better acceleration in the ME process.

References

1. Chouliaras VA, Nunez JL, Mulvaney DJ, Rovati FS, Alfonso D (2005) A multi-standard video accelerator based on a vector architecture. IEEE Trans Consum Electron 51(1):160–167. https://doi.org/10.1109/TCE.2005.1405714
2. Cisco visual networking index: global mobile data traffic forecast update, 2016–2021
3. Jeon H-S, Yoo C-J (2007) Efficient fast motion estimation method for H.264. IJCSNS Int J Comput Sci Netw Secur 7(5)
4. Cai C, Zeng H, Mitra SK (2009) Fast motion estimation for H.264. Elsevier Signal Process Image Commun 24:630–636. https://doi.org/10.1016/j.image.2009.02.012
5. Zhu C, Lin X, Chau L-P (2004) An enhanced hexagonal search algorithm for block motion estimation. IEEE Trans Circuits Syst Video Technol 14(10). https://doi.org/10.1109/tcsvt.2004.833166
6. Su C-Y, Hsu Y-P, Chang C-T (2005) Efficient hexagonal inner search for fast motion estimation. In: IEEE international conference on image processing, 2005, ICIP 2005, 14 Sept 2005. https://doi.org/10.1109/icip.2005.1529945
7. Po L-M, Ting C-W, Wong K-M, Ng K-H (2007) Novel point-oriented inner searches for fast block motion estimation. IEEE Trans Multimed 9(1). https://doi.org/10.1109/tmm.2006.886330

A Comprehensive Study of 1D and 2D Image Interpolation Techniques

V. Diana Earshia and M. Sumathi

Abstract Image interpolation plays an important role in converting a low resolution image into a high resolution image. This paper provides a comprehensive study of perdurable image interpolation techniques, such as nearest neighbor, bilinear, bicubic, cubic spline, and iterative linear interpolation. The usage of a Lagrange polynomial and a piecewise polynomial gives a better fitting curve for interpolated pixel values. The parameters of interest are the signal-to-noise ratio, peak signal-to-noise ratio, mean square error and processing time. Experiment results are used to analyze the performance of interpolation algorithms. These results help us to choose an appropriate algorithm for better usage.

Keywords Image interpolation · Image scaling · Image magnification
Image interpolation algorithms

1 Introduction

Image interpolation [1, 2] is widely used in many areas of modern electronics, such as digital cameras and computer graphics etc. Image interpolation is a method of estimating new data points with the range of discrete set of known data points. It is a process of finding a value between two points on a line or curve.

The objective of interpolation is to retain the qualitative characteristics of a reproduced image from artifacts such as blurring, checkerboard effects, edge discontinuities, and jagging etc. Interpolation techniques were developed to convert low resolution images to high resolution images, with low computational complexity and high accuracy. The various interpolation techniques that were

V. Diana Earshia (✉) · M. Sumathi
Department of Electronics and Communication Engineering,
Sathyabama Institute of Science and Technology, Chennai, Tamil Nadu, India
e-mail: earshy@gmail.com

M. Sumathi
e-mail: sumagopi206@gmail.com

A. Kumar and S. Mozar (eds.), *ICCCE 2018*,
Lecture Notes in Electrical Engineering 500,
https://doi.org/10.1007/978-981-13-0212-1_40

developed are nearest neighbor interpolation [3, 4], bilinear interpolation [5, 6], bicubic interpolation [7, 8], cubic spine interpolation [9, 10] and iterative linear interpolation [11, 12].

Interpolation is the problem of approximating the value of a function for a non-given point in a space when given the value of that function at points around that point. In computer graphics and digital imaging, image scaling refers to the resizing of a digital image. In video technology, the magnification of digital material is known as upscaling or resolution enhancement [1, 2]. Photo interpolation is the process by which the number of pixels comprising an image is increased to allow printing enlargements that are of higher quality than photos that are not interpolated. Interpolation is commonly used to make quality large prints from digital photos and film-scanned images.

2 Interpolation Techniques

The various interpolation techniques considered for comprehensive study are as follows.

2.1 Nearest Neighbor Interpolation (NNA)

In NNA [3, 4] neighboring pixel points are chosen to calculate the value of the interpolated pixel. Now consider the interpolated pixel as $A = f(x, y)$. Let the neighboring pixels be at (i, j), $(i, j + 1)$, $(i + 1, j)$, $(i + 1, j + 1)$. Its pixel value is manipulated based on the algorithm described below.

2.1.1 Methodology

If $(y - j) < ((j + 1) - y)$ then the value of the interpolated pixel will be one of the top two pixel values, else any one of the bottom two pixel values will be chosen.

If $(\mathrm{x} - \mathrm{i}) < ((\mathrm{i} + 1) - \mathrm{x})$ then the value of the interpolated pixel will be one of the left two pixel values, else any one of the bottom two pixels will be chosen.

Figure 1b depicts how the interpolated pixel value $A(x \cdot y)$ is obtained from the actual image shown in Fig. 1a. The actual image is divided into, say, four pixels. The scaled image has sixteen pixels. The interpolated image should have similar feature characteristics to the actual image.

This method is simple and easy to implement and has flexible features choices. The major drawbacks are its speed and poor interpolated image generation with aliasing and blurring effects.

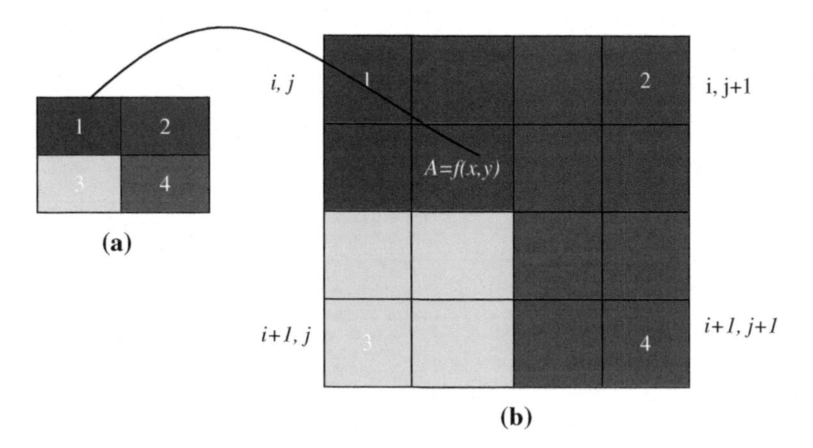

Fig. 1 **a** Actual image **b** image block describing the interpolated pixel $A(x, y)$ using NNA [4]

2.2 *Bilinear Interpolation (BLI)*

In BLI [5, 6] four adjacent pixel points on a rectilinear 2D grid are used for calculating the weighted value of the interpolated point in a scaled image. Linear calculation functions are used in two directions, namely horizontal and vertical, to find the interpolated pixel value.

2.2.1 Methodology

To find the value of P:

$$f(i, j+y) = [f(i,j+1) - f(i,j)]y + f(i,j) \tag{1}$$

To find the value of Q:

$$f(i+1,j+1) = [f(i+1,j+1) - f(i+1,j)]y + f(i+1,j) \tag{2}$$

To find the value of A:

$$\begin{aligned} f(i+x,j+y) &= (1-x)(1-y)f(i\cdot j) - (1-x)yf(I,j+1) + x(1-y) \\ &\quad f(i+1,j) + xyf(i+1,j+1) \end{aligned} \tag{3}$$

Figure 2 shows the pictorial representation of the process of obtaining the weighted value of the interpolated pixel using BLI. Equations (1), (2), and (3) help to obtain the weighted value of $f(x, y)$. It uses a weighted average of the four nearest cell centers. The closer an input cell center is to the output cell center, the higher the influence of its value is on the output cell value.

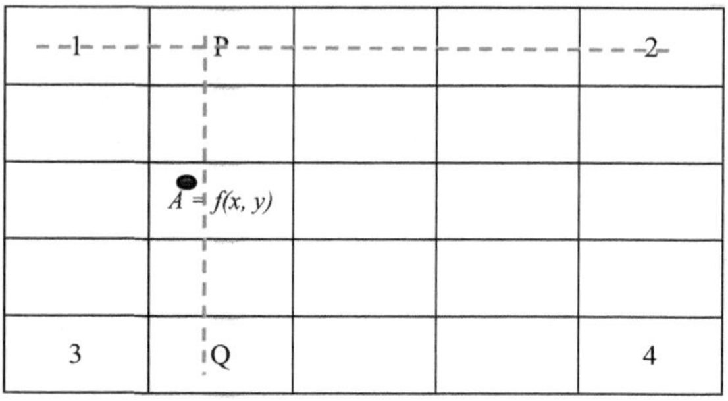

Fig. 2 Image depicting the prediction of weighted value for the interpolated pixel using BLI [5]

This means that the output value could be different to the nearest input, but is always within the same range of values as the input. Since the values can change, BLI is not recommended for use with categorical data. Instead, it should be used for continuous data, such as elevation and raw slope values [13].

The major advantage is that the visual distortion caused by fractional zoom is reduced. The block uses the weighted average of two translated pixel values for each output pixel value. The local gradients for the test images can be investigated [14] by applying an iterative procedure. A perceptually optimum value for each interpolated pixel can be obtained from the local mean value of the inverse gradients.

2.3 *Bicubic Interpolation (BCI)*

In BCI [7, 8] Lagrange polynomials are used for calculating the weight of the interpolated point A = f(x, y) in a scaled image. A 4 × 4, i.e. 16, adjacent pixels are used. For every single dimension calculation, the values of four pixels positioned at i, j, i − 1, and j + 1 are used. Figure 3 depicts the image for predicting the interpolated pixel value using BCI.

2.3.1 Methodology

The weighted sum of the interpolated pixel can be found by manipulating Eq. (4):

$$f(x,y) = \frac{1}{16}\sum_{l=-1}^{2}\sum_{m=-1}^{2} f(i+l,j+m)u(dx)u(dy) \tag{4}$$

where, $f(i+l,j+m)$ gives the gray value of that pixel, $u(dx)$ gives the variation on the x axis, and $u(dy)$ gives the variation on the y axis.

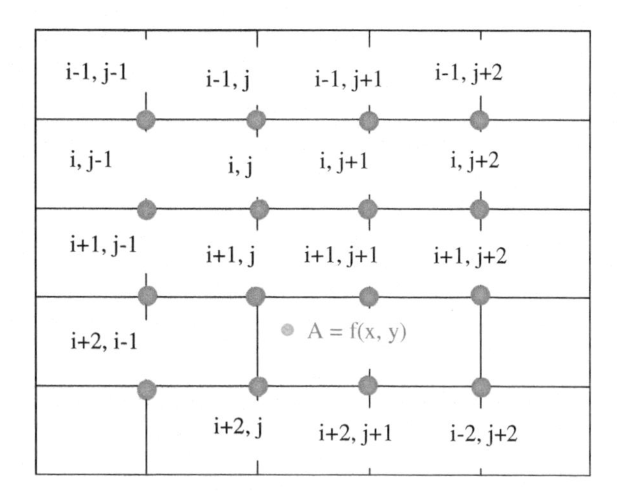

Fig. 3 Image depicting the prediction of the weighted value for the interpolated pixel using BCI [8]

The above figure shows how the 16 pixel values are considered for calculating the weighted average of the final interpolated image. It gives sharper images than the methods described above, but requires more computational time.

2.4 Cubic Spline Interpolation (CSI)

This [11] interpolation technique uses a special type of piecewise polynomial called a spline. A cubic spline is a spline constructed of piecewise third-order polynomials which pass through a set of control points. The second derivative of each polynomial is commonly set to zero at the endpoints, since this provides a boundary condition that completes the system of equations. The interpolation curves are produced as a smooth curve.

CSI is widely used in curve fitting because of its ability to work with both low and high degree polynomials [15]. The depth of the interpolation is variable, and can be set to depend on an absolute or relative error tolerance. It is meant to be easy to interpolate expensive functions that take a lot of time.

2.4.1 Methodology

Given a function *f*(*x*) defined on [a, b] and a set of nodes [12],

$$a = x_0 < x_1 < \cdots < x_n = b \tag{5}$$

Equation (5) describes the set of nodes. Considering *f*(*x*) is a piecewise cubic polynomial, a cubic spline interpolant, S, is given in Eq. (6),

$$S(x) = \begin{cases} a_0 + b_0(x - x_0) + c_0(x - x_0)^2 + d_0(x - x_0)^3 & \text{if } x_0 \le x \le x_1 \\ a_1 + b_1(x - x_1) + c_1(x - x_1)^2 + d_1(x - x_1)^3 & \text{if } x_1 \le x \le x_2 \\ \cdot & \\ \cdot & \\ \cdot & \\ a_{n-1} + b_{n-1}(x - x_{n-1}) + c_{n-1}(x - x_{n-1})^2 + d_{n-1}(x - x_{n-1})^3 & \text{if } x_{n-1} \le x \le x_n \end{cases} \tag{6}$$

2.5 *Iterative Linear Interpolation (ILI)*

ILI [4] adopts the fuzzy gradient model to estimate gradients of the target point according to its neighbor sample points in different directions by weighing the gradients using fuzzy membership grades. It estimates the difference between the target point and its neighboring sample points and finally obtains the target point. In 1D signal reconstructions, it uses only three multipliers. Bidirectional interpolation is composed of multiple 1D interpolations. To approximate 2D signal, five 1D ILIs are used, which cost only eight multipliers to obtain similar peak signal-to-noise ratios (PSNR). Further exploiting, multiple 1D interpolation has moderate PSNR performance but better robustness.

2.5.1 Methodology

Fuzzy sets and membership functions are chosen first. The pictorial description of membership functions is shown in Fig. 4.

Consider uniform sampling is done. The below equations show that only two adders and two multipliers are required for calculating the value of interpolated pixel. The left side and right side gradient values are defined in Eqs. (7) and (8).

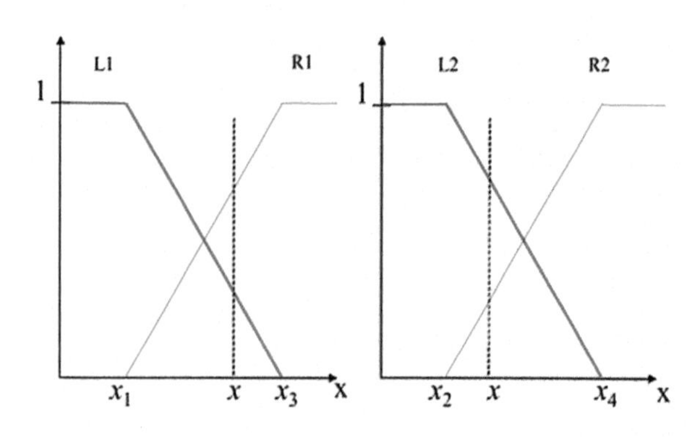

Fig. 4 Image depicting the fuzzy sets and membership functions [4]

$$\hat{f}_L(x) = f(x_2) + (x - x_2) \times q_L$$
$$\hat{f}_R(x) = f(x_3) + (x - x_3) \times q_R. \tag{7}$$

where,

q_L describes the left to right gradient functioinal value.
q_R describes the right to left gradient functional value.

$$q_L = \frac{q_{12} \times L_1 + q_{23} \times R_1}{L_1 + R_1}$$
$$q_R = \frac{q_{23} \times L_2 + q_{34} \times R_2}{L_2 + R_2}. \tag{8}$$

3 Results and Discussion

The following are the pictures considered for comprehensive analysis. These pictures are captured using a Nokia 310 (flower), Samsung GT-S5360 (teddy), and Samsung GT-18552 (vase and bird).

The performances of the algorithms discussed above are compared based on signal-to-noise ratio (SNR), peak signal-to-noise ratio (PSNR), mean squared error (MSE), and processing time. The results were obtained with the help of Matlab R2011a and are tabulated in Table 1.

From Fig. 5, we can see the SNR of various interpolation algorithms, when acted on interpolating images of sizes from 256 × 256 to 1024 × 1024 as tabulated in Table 2.

From Fig. 6, we can observe the PSNR of various interpolation algorithms, when acted on interpolating images of sizes from 256 × 256 to 1024 × 1024 as tabulated in Table 3.

From Fig. 7, we can see the processing time of interpolating techniques and their mean squared error for image sizes from 256 × 256 to 1024 × 1024.

Table 1 Comparison of interpolation techniques based on SNR (dB) for image sizes from 256 × 256 to 1024 × 1024

Interpolation type	Teddy	Flower	Vase	Bird
Nearest neighbor	18.05	21.61	19.12	19.15
Bilinear	22.32	26.37	20.23	22.13
Bicubic	22.56	25.66	20.98	22.24
Cubic spline	22.74	24.35	18.65	21.37
Iterative linear	22.68	25.59	20.11	22.58

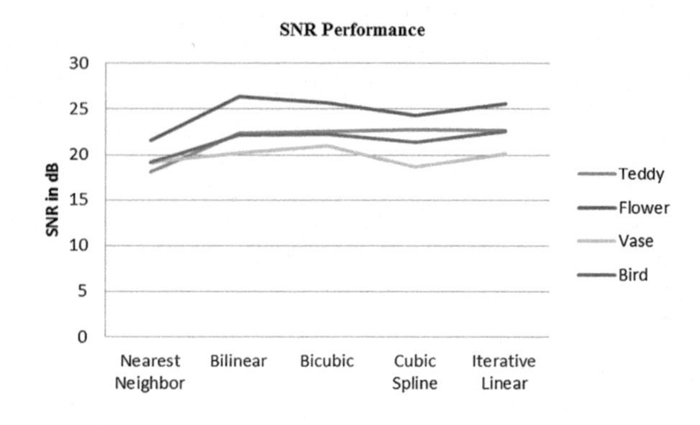

Fig. 5 SNR performance of interpolation techniques

Table 2 Comparison of interpolation techniques based on PSNR (peak signal-to-noise ratio) for image sizes from 256 × 256 to 1024 × 1024

Interpolation type	Teddy	Flower	Vase	Bird
Nearest neighbor	33.50	29.16	23.77	25.18
Bilinear	33.47	29.17	23.84	25.15
Bicubic	32.57	28.84	23.23	23.92
Cubic spline	30.75	29.78	23.89	25.26
Iterative linear	27.96	29.25	23.71	24.65

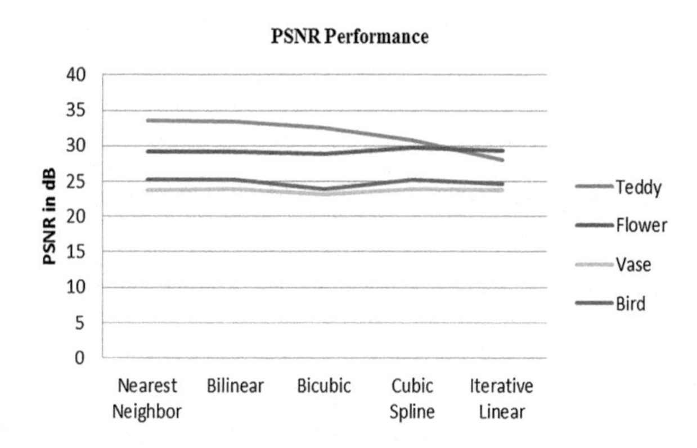

Fig. 6 PSNR performance of various interpolation techniques

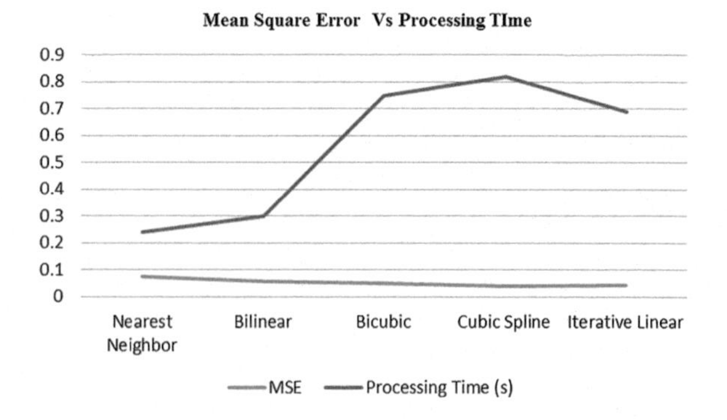

Fig. 7 Mean square error versus processing time for image sizes from 256 × 256 to 1024 × 1024

Table 3 Comparison of interpolation techniques based on MSE (mean square error) and processing time for images of sizes from 256 × 256 to 1024 × 1024

Interpolation type	MSE	Processing time (s)
Nearest neighbor	0.0761	0.24
Bilinear	0.0556	0.3
Bicubic	0.0484	0.75
Cubic spline	0.0395	0.82
Iterative linear	0.0412	0.69

4 Conclusion

In this paper, various interpolation techniques were discussed for their better usage in the field of image interpolation. This survey will assist us in developing a new technique for interpolating images with improved results. As a result future algorithms developed in the future will be optimized for any sort of imaging.

References

1. Gonzalez RC, Woods RE (2008) Digital image processing. Nueva Jersey, p 976
2. Pratt WK (2001) Processing digital image processing, vol 5, no 11
3. Parker JA, Kenyon RV, Troxel DE (1983) Comparison of interpolating methods for image resampling. IEEE Trans Med Imaging MI-2:31–39
4. Hanssen R, Bamler R (1999) Evaluation of interpolation kernels for SAR interferometry. IEEE Trans Geosci Remote Sens Part 1 37(1):318–321
5. Lehmann TM, Gonner C, Spitzer K (1999) Survey: interpolation methods in medical image processing. IEEE Trans Med Imaging 18(11):1049–1075
6. Han D (2013) Comparison of commonly used image interpolation methods. In: Proceedings of the 2nd international conference on computer science and electronic engineering (ICCSEE 2013), pp 1556–1559
7. Haifeng Z, Yongfei Z, Ziqiang H (2010) Comparison of image amplifying method. Mod Electron Tech (24):33–36
8. Amanatiadis A, Andreadis I (2008) Performance evaluation techniques for image scaling algorithms. In: Proceedings of the IEEE international workshop on imaging systems and techniques, pp 114–118 (2008)
9. McKinley S, Levine M (1998) Cubic spline interpolation. Coll Redw 45(1):1049–1060
10. Hou HS, Andrews HC (1978) Cubic splines for image interpolation and digital filtering. IEEE Trans Acoust Speech Signal Process ASSP-26(6):508–517
11. Chen C, Lai C (2014) Iterative linear interpolation based on fuzzy gradient model for low-cost VLSI implementation. 22(7):1526–1538
12. Muresan D, Parks T (2004) Adaptively quadratic image interpolation. IEEE Trans Image Process 690–698
13. Sen Wang, Kejian Yang (2008) An image scaling algorithm based on bilinear interpolation with VC++. J Tech Autom Appl 27(7):44–45
14. Hwang JW, Lee HS (2004) Adaptive image interpolation based on local gradient features. IEEE Signal Process Lett 11(3):359–362
15. Huang JJ, Siu WC, Liu TR (2015) Fast image interpolation via random forests. IEEE Trans Image Process 24(10):3232–3245

Reinforcement Learning-Based DoS Mitigation in Software Defined Networks

A. VishnuPriya

Abstract A software defined network (SDN) is an OpenFlow-based network that initiates innovative traffic engineering and also simplifies network maintenance. Network security is still as stringent as that of traditional networks. A denial of service (DoS) attack is a major security issue that makes an entire network's resources unavailable to its intended users. Blocking the flows based on the number of flows per port threshold was the most common method employed in the past. At some occasions legitimate traffic also takes the huge flow will punish by default rules. In order to address this issue, I proposed a reinforcement learning-based DoS detection model that detects and mitigates huge flows without a decline in normal traffic. An agent periodically monitors and measures network performance. It also rewrites the flow rules dynamically in the case of rule violation.

Keywords Denial of service • Software defined networks • Sflow
Reinforcement learning

1 Introduction

A denial of service (DoS) is a serious attack performed by cyber attackers creating bots in commercial networks. These bots are software that runs independently on host network/machines to launch frequent requests. The objective of these attacks is financial gain and to corrupt the availability of network resources to intended users. According to Verisign DDoS trends report Q2014 the average attack has increased to 7.39 Gbps, or 15% higher than previously. There are different categories of DoS attack vector, such as volumetric, fragmentation, TCP state-exhaustion and

A. VishnuPriya (✉)
Department of Electronics & Communication Engineering,
Vel Tech Rangarajan Dr. Sagunthala R&D Institute of Science and Technology,
Chennai, India
e-mail: vishnu.priya@veltechuniv.edu.in

A. Kumar and S. Mozar (eds.), *ICCCE 2018*,
Lecture Notes in Electrical Engineering 500,
https://doi.org/10.1007/978-981-13-0212-1_41

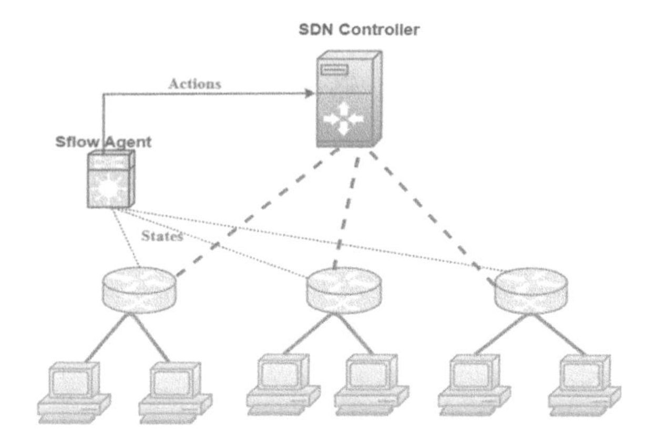

Fig. 1 A reinforcement-based SDN model

application layer attacks. Freely accessible toolkits like Pandora, Dereil, ad HOIC make attacks more serious.

A SDN is a three plane network comprising a data plane, control plane, and application plane. Each plane has been distressed by flooding fake flows. The attacker specifically blocks the controller interface and makes the controller unreachable, and as a result the entire network performance is severely damaged. In the data plane, it works by initiating fake flows that cause the flow table to overflow. In case of the application plane, a malicious application program can modify the authenticate flow rules by flooding commands.

Here, a novel mechanism is proposed that can detect a DoS by use of a reinforcement learning method. We allotted a reinforcement agent that has close interaction with network environment as shown in Fig. 1. These agents sense metrics from the network and take appropriate actions. Sensed metric values are compared with threshold values and necessary actions will be taken.

This paper is organized as follows: Sect. 2 reviews related work. Section 4 discusses the proposed model. Section 5 presents the implementation and results. Finally, Sect. 6 is the conclusion.

2 Related Works

Braga et al. [1] proposed self-organizing map (SOM) techniques to detect a Distributed Denial of Service (DDoS) attack. A SOM is trained by collecting flow metrics such as average packets per flow, average bytes per flow, and average duration per flow. The performance of SOM improves when more statistics are gathered. However, it fails to find spoofed packets. In 1985 Solnushkin [2] proposed a fat-tree network, which is tree-like in structure with a bottom layer connected with a processor. The link gets bigger towards the top of the tree, and more links are present at the root switch of the tree compared to other switches. The problem with this approach is that every packet of a flow follows a single predefined path via a network. In case of any switch link failure, packets tend to drop or other switches should be configured manually to choose a different path. As the network grows this becomes a difficult task.

Cai et al. [3] suggest using multi-core processors to improve the controller computing performance in the case of frequent requests, but don't recommend any prevention techniques. Nayana et al. [4] prefer threshold-based DDoS detection and load balancing techniques in terms of hard blocking. So there is the possibility of false positives and false negatives. Son [5] proposed queue-based early detection of DDoS techniques. These segregate the request based on priority and provide service accordingly. No mitigation possibilities were suggested.

Balyaev and Gaivoronski [6] introduced two-level load balancing techniques during DDoS attacks, which improved the survival time of the controller during attacks. It initiated a dynamic load balancing algorithm when imbalance occurred. Chin et al. [7] recommends selective packet inspection to detect a TCP SYN flooding attack. They implemented this algorithm in a GENI cloud infrastructure. DoS attack possibilities in different SDN planes and related work appear in Table 1.

Table 1 Related work on DoS attacks and mitigation methods in different SDN layers

Attack type	Related work	Methods
DoS in control plane (controller)	Line switch [8]	It employs a probabilistic proxy and blacklisting of network traffic to prevent attack
	Avantguard [9]	A connection migration tool reducing data-control plane interaction and actuating trigger to install flow rules
	Lightweight DDoS [1]	Statistical information with self-organizing maps to classify traffic as normal or malicious
	CONA [10]	Rate and pattern-of-content requests are analysed to detect DDoS attacks
	Enhance security in Openflow [11]	Secure transport layer and add IDS as middle boxes
	SDN-guard [12]	Add three different modules: 1. Flow management 2. Flow aggregator 3. Monitoring modules Instead of dropping malicious flow it redirects to a longer routing path
Data plane	ProtoGENI [13]	Priority-based resource allocation and user authentication
Application layer	FortNOX [14]	System has a conflict detection engine to find flow rule conflicts
	ROSEMARY [15]	Resource utilization monitoring and application authentication
	LegoSDN [16]	Follows network application abstraction to avoid application conflicts
	Fleet [17]	Fleet controller to detect the malicious administrator

3 Reinforcement Learning-Based DDoS Mitigation

3.1 SFlow-rt

Sampled flow is a standard for packet export in layer 2 of the OSI model. It is an industry standard technology for monitoring high speed networks [18]. It gives complete visibility into the use of a network enabling performance, optimization, usage, and defense against security threats. It is an asynchronous analytic system to support SDN apps. Its functions are:

- It receives real-time telemetry streams from agents which are assigned in switches and converts them as an action through RESTflow API.
- With the aid of network visibility it can support scale services in the cloud infrastructure.
- The agent receives metrics and fixes the thresholds. RESTcall occurs via an external application program written in JavaScript.

3.2 Reinforcement Learning

This is an area of machine learning technique that has close interaction with the real-time environment and also measures metrics online. An agent obtains knowledge from the observed metrics. To operate the agent, a parametric model of the particular environment is used. This environment may be stationary or non-stationary. If it is non-stationary the actual state space may not be fully observed then the state and action spaces are discrete.

Learning Approaches:

- *Indirect Learning*
  Learn the estimated model and find the optimal solution for the same.
- *Direct Learning*
  Learn the optimal policy of the model before learning the estimated model.

– Policy space search: Genetic algorithm, policy gradient, etc.
– Dynamic programming principles: Temporal difference learning (TDL), Q-learning, etc.

4 Proposed Solution

DoS mitigation is implemented based on the impact of the reinforcement direct learning strategy. The agents are set in Openflow switches that obtain the global knowledge of the entire network and network devices. Initially, it receives details regarding the IP address, MAC address, and port number of all network devices. After start the sflow agents, it begins to retrieve the number of permissible packets per destination and compares it with the threshold value. The threshold value can be fixed based on network capacity and administrative perspectives. Here we fixed the threshold at 100 packets per port. If the packet rate is higher than threshold, block flow action space will start to work. The attacker performs this type of flooding by spoofing the IP address. This means that we have created mapping between the MAC and corresponding IP address log. Our proposed technique verifies the IP and MAC of the network sources from the stored log and then it tests the threshold comparison. If it fails in both such condition, sflow starts the block flow action via RESTAPI to the controller. Sflow changes the flow rule and pushes the comments to the static flow pusher API of the controller. The detailed algorithm and flowchart are explained in Fig. 2.

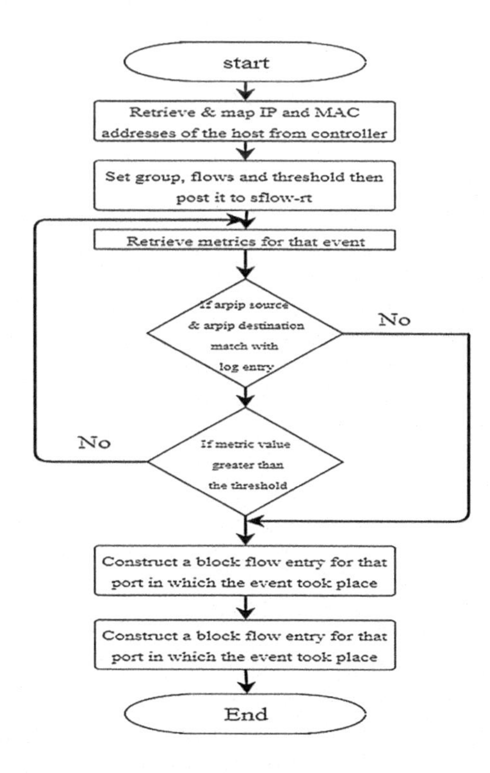

Fig. 2 Flow diagram of the proposed model

DoS mitigation procedure

Create a basic network topology with Si (i = 1, 2, 3 … n) switches and Ni (i = 1, 2, 3 … n) nodes
Connect Si switches to controller C and start
Assign Agent Ai in Openflow switches and start
For each Agent Ai
    Retrieve IP and MAC address of network devices and create IP log
    Fix the threshold value based on network condition
For each event from Node Ni
    Collect metrics for each event
    Check IP-MAC combination and also verify the threshold value
    If it fails in both conditions
        Block the flow through RESTAPI of Sflow

5 Implementation and Results

Experiments were conducted for various attack scenarios to show the feasibility of the proposed solution. Tests were performed using a HP 15-r204TX running on Windows 10 with an Intel® Core™ i5-4210U CPU @ 1.70 GHz 2.40 GHz with an installed RAM of 8.00 GB. An Ubuntu Virtual Machine (VM) with 2 cores and 4 GB of RAM with Intel VT enabled was also used. Mininet, a network emulator was used to create the testbed. It creates a virtual network of hosts, switches, controllers, and links according to a given topology. These switches support Openflow and are orchestrated by a SDN controller. The Mininet emulator is installed on a Ubuntu 14.10 Linux VM that runs on VMware Workstation Pro. Floodlight is used as the SDN controller.

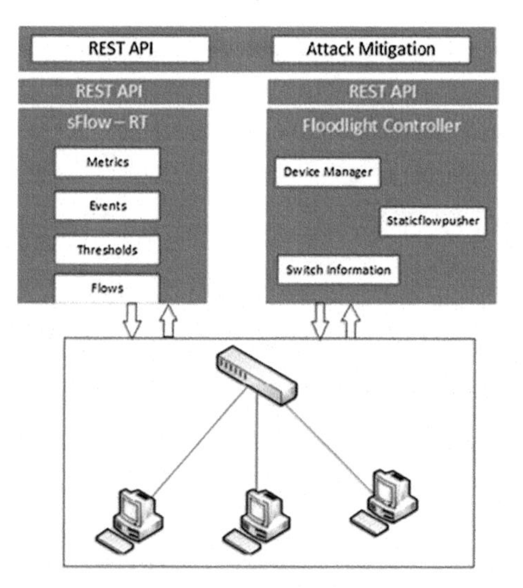

Fig. 3 Network topology

Fig. 4 Retrieved metric from Sflow-rt

ID	Time	Name	Metric	Threshold	Value	Agent	Data Source
1	Tue Nov 14 2017 03:26:54 GMT-0500 (EST)	dos	dos	100	104.54	172.17.107.104	6
0	Tue Nov 14 2017 03:26:54 GMT-0500 (EST)	dos	dos	100	104.54	172.17.107.104	7

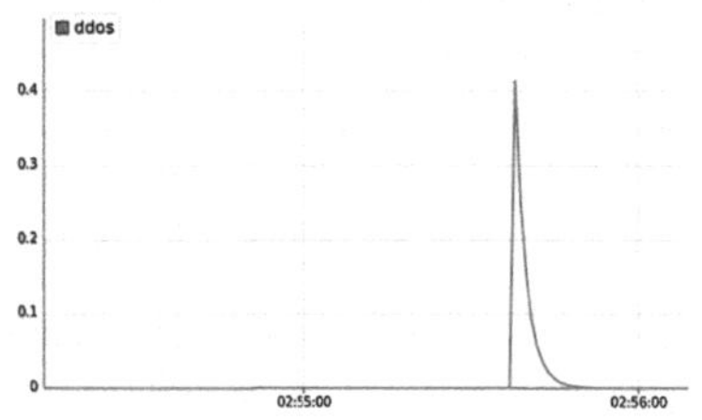

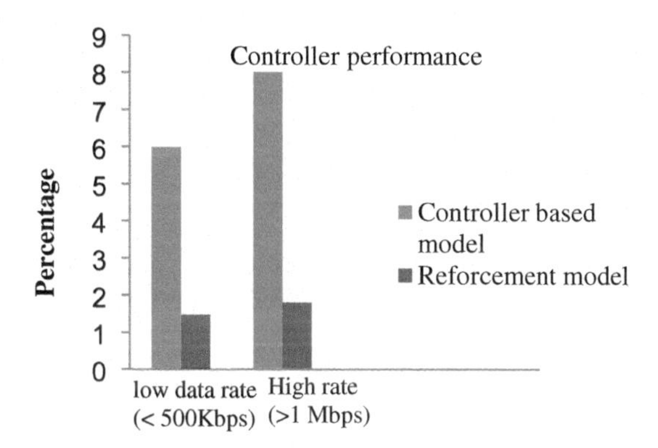

Fig. 5 Comparison of controller performance

A synthetic and real traffic based data sets was created and used in the experiments with the network topology as shown in Fig. 3. Mitigation structure has to be implemented on the top of SDN controller by using new DoS detection method. Combining all these areas should allow the network to react quickly to a DoS attack and also increases the ability to filter malicious traffic. The output screen of our proposed model is shown in Figs. 4 and 5. In this model, we have not added any application module to the controller. Mitigation runs on an external controller which improves controller performance compared to existing models as shown in Fig. 5.

6 Conclusion

We framed the reinforcement impact DoS mitigation model in a SDN and also tested it in a Floodlight controller. This is an agent-based solution which reduce the overheads of a central controller and provides a better performance. The performance of the hosts were measured using an iperf tool before and after mitigation.

CPU usage was found to be high during the attack scenario and the response time was quite low but was found to be comparatively impressive after allowing the mitigation script to run in the background. The identified attackers were blocked and their communication with the rest of the network was blocked successfully.

References

1. Braga R, Mota E, Passito A (2010) Lightweight DDoS flooding attack detection using NOX/OpenFlow. In: Proceedings of the conference on local computer networks, LCN, pp 408–415 (2010). https://doi.org/10.1109/lcn.2010.5735752
2. Solnushkin KS (2013) Automated design of two-layer fat-tree networks
3. Cai Z, Cox A, Ng TSE (2011) Maestro: a system for scalable OpenFlow control. Rice University
4. Nayana Y, Tech JM, Girish L (2015) DDoS mitigation using software defined network. 24:258–264
5. Son NH (2016) A mechanism for early detecting DDOS attacks based on M/G/R PS queue. Int J Netw Secur Appl 8:17–24
6. Belyaev M, Gaivoronski S (2014) Towards load balancing in SDN-networks during DDoS-Attacks. In: SDN NFV next generation of computational infrastructure—2014 international science technology conference—modern networking technologies, MoNeTec. https://doi.org/10.1109/monetec.2014.6995578
7. Chin T, Mountrouidou X, Li X, Xiong K (2015) Selective packet inspection to detect DoS flooding using software defined networking (SDN). In: Proceedings of the 2015 IEEE 35th international conference on distributed computing systems workshops, ICDCSW 2015, pp 95–99. https://doi.org/10.1109/icdcsw.2015.27
8. Ambrosin M, Conti M, De Gaspari F, Poovendran R (2017) LineSwitch: tackling control plane saturation attacks in software-defined networking. IEEE/ACM Trans Netw 25:1206–1219
9. Shin S, Yegneswaran V, Porras P, Gu G (2013) AVANT-GUARD: scalable and vigilant switch flow management in software-defined networks. In: ACM SIGSAC conference on computer and communications security (CCS 2013), pp 413–424. https://doi.org/10.1145/2508859.2516684
10. Suh J et al (2010) Implementation of content-oriented networking architecture (CONA): a focus on DDoS countermeasure. In: 1st European NetFPGA developers workshop, pp 1–5
11. Chellani N, Tejpal P, Hari P (2016) Enhancing security in OpenFlow, pp 1–10
12. Dridi L, Zhani MF (2016) SDN-guard: DoS attacks mitigation in SDN networks. In: Proceedings of the 2016 5th IEEE international conference on cloud networking, CloudNet 2016, pp 212–217. https://doi.org/10.1109/cloudnet.2016.9
13. Li D, Hong X, Witt D (2013) ProtoGENI, a prototype GENI under security vulnerabilities : an experiment-based security study, pp 1–11
14. Porras P et al (2012) A security enforcement kernel for OpenFlow networks. In: Proceedings of the first Workshop on hot topics in software defined networking—HotSDN '12, p 121. https://doi.org/10.1145/2342441.2342466
15. Shin S et al (2014) Rosemary: a robust, secure, and high-performance network operating system. In: Proceedings of the 2014 ACM SIGSAC conference on computer and communications security—CCS '14, pp 78–89. https://doi.org/10.1145/2660267.2660353

16. Chandrasekaran B, Benson T (2014) Tolerating SDN application failures with LegoSDN. In: Proceedings of the third workshop on hot topics in software defined networking—HotSDN '14, pp 235–236. https://doi.org/10.1145/2620728.2620781
17. Matsumoto S, Hitz S, Perrig A (2014) Fleet: defending SDNs from malicious administrators. In: Proceedings of the second ACM SIGCOMM workshop on hot topics in software defined networking—HotSDN '14, pp 103–108. https://doi.org/10.1145/2620728.2620750
18. Systems BC (2009) Brocade sFlow for network traffic monitoring, p 12

Design of a Low Power Full Adder with a Two Transistor EX-OR Gate Using Gate Diffusion Input of 90 nm

J. Nageswara Reddy, G. Karthik Reddy and V. Padmanabha Reddy

Abstract A full adder is the one of the main parts of an arithmetic logic unit (ALU). In this paper a full adder is developed using gate diffusion input (GDI) to perform fast arithmetic operations. The main aim of this paper is the design of a two transistor XOR gate-based full adder using a gate diffusion input (GDI) technique. A two transistor (2T) EX-OR gate is a suitable gate in the design of a full adder. The intention behind the novel method of a 2T EX-OR gate-based full adder design is to reduce power and improve speed in an optimized area with a lower transistor count compared with CMOS technology. A GDI approach is the one of better methods available for the design of digital logic circuits and tends to run the improved conditions. The proposed technique is then applied to a full adder design. The complete work is carried out using the 90 nm technology of a cadence tool to calculate power, delay, and area for the 2T EX-OR gate. The resulting analysis shows that the proposed method is better than conventional CMOS technology.

Keywords GDI · CMOS · EX-OR · Cadence tool

1 Introduction

Currently, due to the circuit complexity and increased number of transistors embedded on a chip, electronics are characterized by various factors, such power consumption, speed, and area, which are the major issues in VLSI (very large-scale integration) design. The main trend towards research into low-power VLSI has

J. Nageswara Reddy (✉) · G. Karthik Reddy
ECE Department, CMR College of Engineering & Technology, Hyderabad, India
e-mail: nag_ece424@yahoo.com

G. Karthik Reddy
e-mail: gandhari.karthik@cmrcet.org

V. Padmanabha Reddy
ECE Department, Institute of Aeronautical Engineering, Dundigal, Hyderabad, India
e-mail: vpr2008@gmail.com

A. Kumar and S. Mozar (eds.), *ICCCE 2018*,
Lecture Notes in Electrical Engineering 500,
https://doi.org/10.1007/978-981-13-0212-1_42

increased with the increase in transistor count on a chip, leading to increased temperature in the IC and increased power consumption, which in turn directly affect battery durability in portable devices. In CMOS VLSI circuits there are three types of power consumption with the primary one being static power consumption due to unwanted leakage current when the transistor is used in the off state. The second is dynamic power consumption due to the switching of parasitic components, and third is short-circuited power consumption due to variation of current between V_{dd} to V_{ss}.

Up to now, the design of a full adder has usually employed one logic method that is standard static CMOS circuit, pass transistor logic and transmission gate logic methods in the conventional full adder.

The few advantages having compare with CMOS technology by using pass transistor logic (PTL). Just because of a small area the low power interconnecting effect in VLSI circuit [1, 2], low power consumption and higher speed due to a lower number of transistors [3]. However, one of main disadvantages in PTL is the threshold drop across each transistor resulting in reduced sink and source current and leading to a reduction in operating speed at low supply voltages and a high input voltage at the reforming inverters is not V_{dd} [4–6]. The full adder with 10 transistors faces the problem of double threshold loss [7], and the design of a high-speed and low-power adder has become one of the most important aims of research. Here, the main issue is static power dissipation which occurs due to the PMOS transistor (which is not in the fully off condition) in the inverter. To overcome this problem, the proposed method is implemented for a high-speed and low-power digital circuit design known as GDI [8].

The aim of the proposed method is to design a low power full adder with a 2T EX-OR gate using a gate diffusion input of 90 nm. The following section explains basic GDI technology briefly. The third section describes the conventional method full adder. The fourth section discusses the proposed method of a 2T EX-OR-based full adder designed with a 6T transistor using the GDI technique. Section 5 shows the results and comparison. Section 6 is the conclusion.

2 Gate Diffusion Input (GDI)

GDI is a new approach for low power digital combinational circuits. It always increases the speed and reduces the power consumption and area of digital combinational circuits to maintain the low complexity of the logic circuit design.

A simple basic GDI cell is shown in Fig. 1. The GDI cell has three inputs. G is for the common gate input for nMOS and pMOS, P is the input for the source or drain for pMOS, N is the input for the nMOS source or drain, the pMOS substrate is made to connect with Vdd and the nMOS substrate with the ground. Different inputs can be connected to the P, N and G terminals. Compared with CMOS technique, GDI occupies less silicon area because of the lower number of the transistor count, and as area occupation is less the node capacitance value is lower.

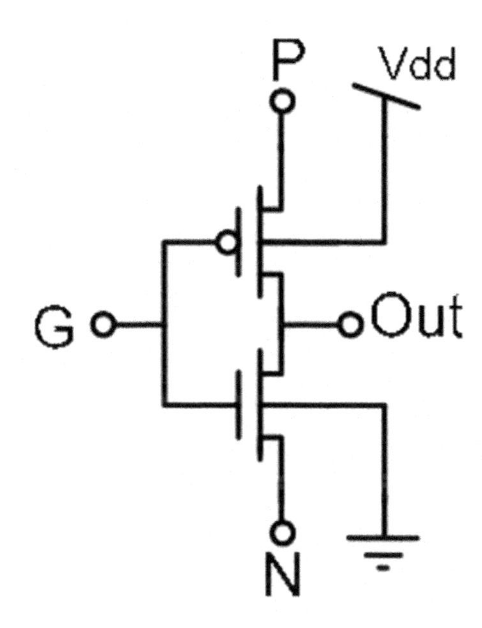

Fig. 1 A basic GDI cell

These all imply a high operating speed, which shows that the GDI logic method is an efficient method to use in the design of an adder.

The threshold voltage depends on the source-to-bulk voltage [9]. The bulk terminal of the nMOS and pMOS should be connected to their diffusion to minimize the bulk effect. Variation in threshold due to a change in VSB is called the body effect. This effect directly shows the impact of the threshold voltage when not connected to a source.

3 8T CMOS Full Adder

The circuit below in Fig. 2 is a full adder designed with 8T having the inputs A, B, and C in, and outputs SUM and Cout. In this paper, a full adder designed with 2 EX-OR gates and one 2:1 mux. SUM is generated at the output of the second EX-OR gate and carry is generated multiplexers (COUT) output as shown in Fig. 2. In the conventional method the full adder is implemented by using CMOS technology.

4 Proposed 6T Full Adder Design

The performance of the full adder circuit can be explained here as follows: The addition of two single bit inputs A and B with input carry C_{in} gives the two single bit outputs Sum and carry out C_{out},

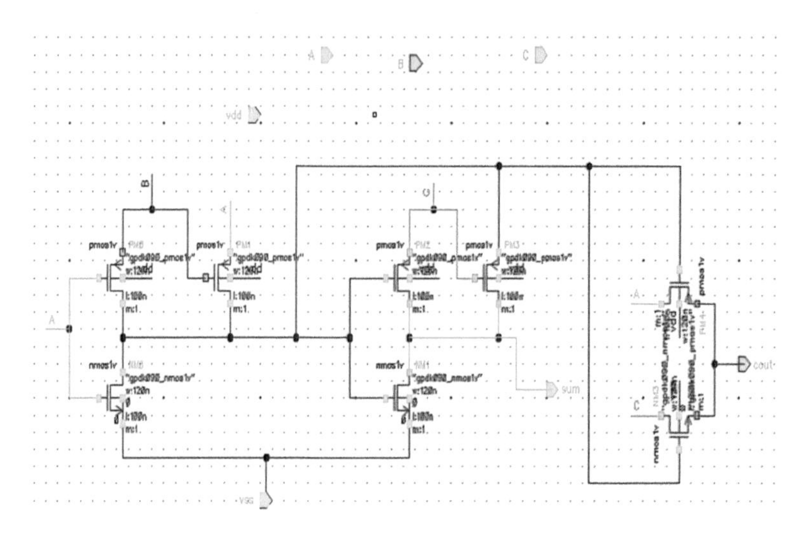

Fig. 2 CMOS full adder design

where

$$\text{Sum} = (\text{A XOR B}) \text{ XOR C} \tag{1}$$

$$\text{Cout} = (\text{A} \cdot \text{B}) + (\text{A XOR B}) \cdot \text{C} \tag{2}$$

Table 1 deals with the functional approach of the full adder. The standard way to build a full adder using XOR gates and a multiplexer is as shown in Fig. 3. The XOR gate and multiplexer are two basic components in the design of a full adder circuit. Arithmetic and logical operations of the full adder are completely based on the XOR gate and multiplexer blocks. The XOR gate design should have a low number of transistors for low power dissipation. The reason behind adopting the multiplexer circuit in our proposed design is to generate C_{out}. A transmission gate is used as a multiplexer because it speeds up the carry propagation and improves the output voltage swing as level restoring circuit. The proposed full adder circuit needs two XOR gates, one multiplexer, and only six transistors.

Table 1 Truth table for the full adder

A	*B*	C_{in}	*SUM*	*Carry*
0	0	0	0	0
0	0	1	1	0
0	1	0	1	0
0	1	1	0	1
1	0	0	1	0
1	0	1	0	1
1	1	0	0	1
1	1	1	1	1

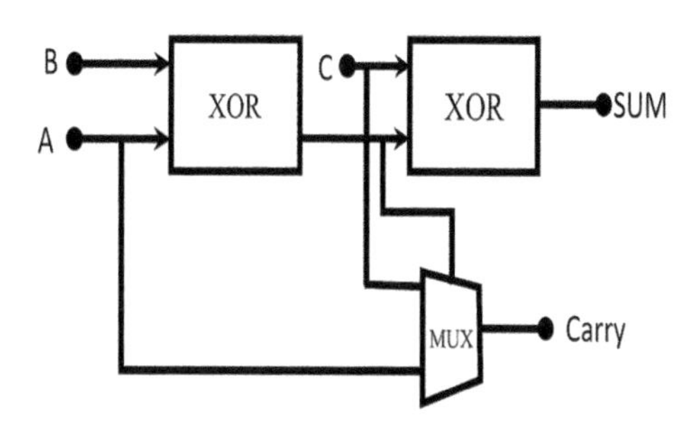

Fig. 3 Block diagram of the full adder

Figure 4 shows the design of the XOR gate using the GDI technique. It has two transistors, pMOS and nMOS, with *A* and *B* inputs. When *A*, and *B* are both at logic low pMOS is in the ON state and nMOS is in the OFF state so that the output is low. When input *A* is low and *B* is high, then pMOS is in the off state and nMOS is in the on state so that output is high. If *A* is high and *B* is low then pMOS is in the on state and nMOS is in the off state so that the output is at high. When both inputs are high then output is low, hence the above circuit acts as a XOR gate.

Figure 5 represents the proposed full adder circuit with six transistors, namely M1, M2, M3, M4, M5 and M6. M1 and M2 transistors act as the first XOR gate. Input *A* is connected to the drain of M2, A *bar* is connected to the source of M1and *B* is connected to the gate terminal of M1 and M2. M3 and M4 act as a second XOR gate. Input C_{in} is connected to the drain of M4 and C_{in}*_bar* is connected to the source of M3. The output of the first XOR gate is applied as the input to the gate terminals of M3 and M4, and the second XOR gate produces *SUM* as the output.

Two AND gates and one OR gate of an existing full adder are replaced with the multiplexer in the proposed circuit. Here the output of the first XOR gate is applied

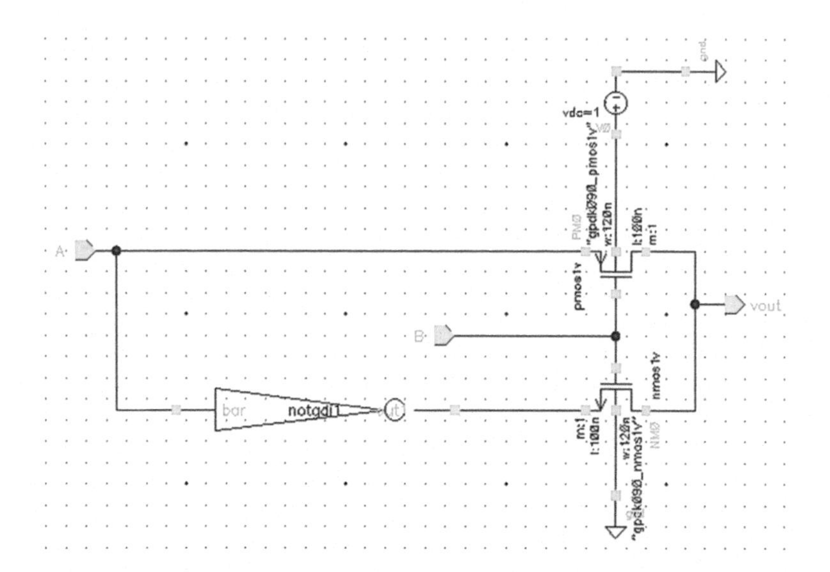

Fig. 4 A 2T XOR gate

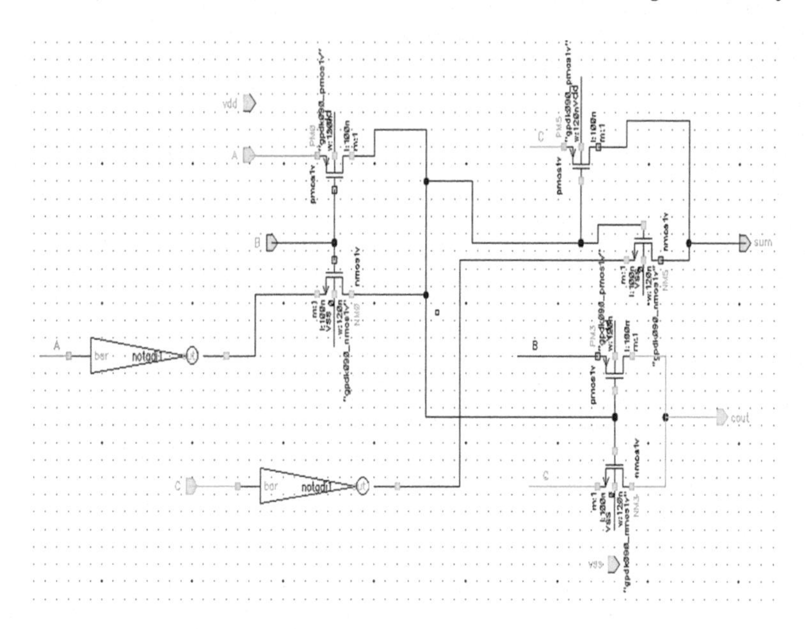

Fig. 5 The proposed full adder design

to the gate terminal of M5 and M6, input B is connected to drain of M6, input C_{in} is connected to source of M5, and the multiplexer produces $C_{out.}$

5 Result and Discussion

Table 2 shows the improvement in power and delay analysis of the 90 nm technology of the proposed system over the conventional method. The consideration of area shows the transistor count is comparatively low next to the conventional CMOS technique [10]. If the number of transistors is reduced, the complexity of the circuit will decrease, speed will increase, and the utilization of power will reduce. Figure 6 shows the timing delays and power analysis tabulated in Table 2. In the proposed method, power reduces up to 32% and delay reduces up to 10% compared

Table 2 Comparison table

Adder style	Area (transistor count)	Power (w)	Time delay (s)	Area (transistor count)	Power (w)	Time delay (s)
	CMOS 90 nm			GDI 90 nm		
Not	2	166.2×10^{-9}	235×10^{-12}	2	13.1×10^{-9}	201.8×10^{-12}
EXOR	8	388.6×10^{-9}	168×10^{-9}	2	40.47×10^{-9}	20.07×10^{-9}
Full adder	8	592.3×10^{-9}	50.14×10^{-9}	6	146.9×10^{-9}	30.03×10^{-9}

with the conventional CMOS full adder. Time delay and power are calculated using a cadence tool.

Figure 6 shows the time delay due to different carrier mobility associated with the pMOS and nMOS devices (Fig. 7).

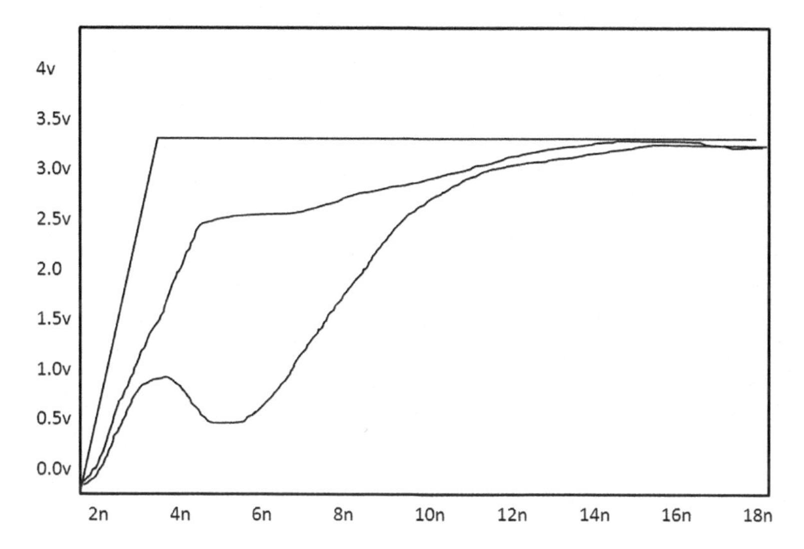

Fig. 6 Time delay for the full adder

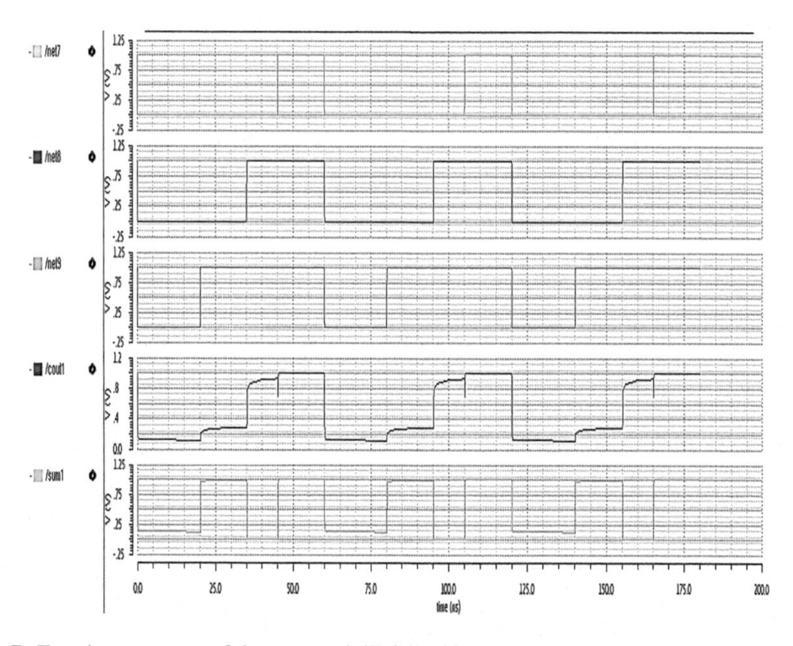

Fig. 7 Transient response of the proposed 6T full adder

6 Conclusion

The proposed full adder designed using 90 nm has an improved performance in terms of area, delay, and power consumption compared with a conventional full adder. Design complexity is also reduced when used the GDI technique. Considering all these factors, the proposed technique can also be used to design combinational and sequential circuits.

References

1. Sakurai T (1993) Closed-form expressions for interconnection delay, coupling, and crosstalk in VLSI's. IEEE Trans Electron Devices 40(1):8–124
2. Adler V, Friedman EG (1997) Delay and power expressions for a CMOS inverter driving a resistive-capacitive load. Analog Integr Circuits Signal Process 14:29–39
3. Segura J, Hawkins CF (2004) CMOS electronics: how it works, how it fails. Wiley-IEEE, 132 pp. Chandrakasan R, Brodersen W (1995) Minimizing power consumption in digital CMOS circuits. Proc IEEE 83(4):498–523
4. Al-Assadi W, Jayasumana AP, Malaiya YK (1991) Pass-transistor logic design. Int J Electron 70:739–749
5. Yano K, Sasaki Y, Rikino K, Seki K (1996) Top-down pass-transistor logic design. IEEE J Solid-State Circuits 31:792–803
6. Lin J-F, Hwang YT, Sheu M-H, Ho C-C (2007) A novel high-speed and energy efficient 10 transistor full adder design. IEEE Trans Circuits Syst I: Regular Papers 54(5):1050–1059
7. Tung CK, Hung YC, Shieh SH, Huang GS (2007) A low-power high-speed hybrid CMOS full adder for embedded system. IEEE No. 14244-1161
8. Morgenshtein A, Shwartz I, Fish A (2010) Gate diffusion input (GDI) logic in standard CMOS nanoscale process. In: IEEE 26th convention of electrical and electronics engineers, Israel
9. Kang S-M, Leblebici Y (2002) CMOS digital integrated circuits. McGraw-Hill Professional
10. Wairya S, Nagaria RK, Tiwari S (2012) Comparative performance analysis of XOR-XNOR function based high-speed CMOS full adder circuits for low voltage VLSI design. Int J VLSI Des Commun Syst (VLSICS) 3(2)

Analysis of e-Recruitment Systems and Detecting e-Recruitment Fraud

M. Niharika Reddy, T. Mamatha and A. Balaram

Abstract Recruitment is the one of the major tasks for the human resource department of any organization. At present, human resource management recruits employees using a manual procedure. This manual procedure means that employees must attend interviews and this is time consuming for the organization as well as for the candidates who attend interviews. To overcome this limitation, several e-recruitment tools are available. In this paper we analyze these e-recruitment tools along with potential fraud detection tools. We also look at the advantages and disadvantages of e-recruitment.

Keywords Human resource · Organization · Recruitment · Fraud detection

1 Introduction

e-Recruitment structures have visible expanded in recent years, allowing human resources (HR) groups access to a very large audience for a small cost. This state of affairs might be overwhelming to HR companies that need to allocate human assets to manually assess candidate resumes and evaluating their suitability for the positions in question. Automating the technique of studying applicant profiles to determine those that fit the position's specification could lead to an enhanced overall performance. For instance, SAT telecom noted a 44% financial saving and a drop in the average time taken to fill a vacancy from 70 to 37 days after deploying an e-recruitment device. Several e-recruitment structures have been proposed with

M. Niharika Reddy (✉) · T. Mamatha
Sreenidhi Institute of Science and Technology, JNTUH, Hyderabad, India
e-mail: niharika.hridaya@gmail.com

T. Mamatha
e-mail: mamathat@sreenidhi.edu.in

A. Balaram
CMR Institute of Technology, JNTUH, Hyderabad, India
e-mail: balaram.balaram@gmail.com

A. Kumar and S. Mozar (eds.), *ICCCE 2018*,
Lecture Notes in Electrical Engineering 500,
https://doi.org/10.1007/978-981-13-0212-1_43

the goal of speeding up the recruitment process, main to a higher normal patron revel in. E-Gen gadgets evaluate and categorize unstructured interest offers (i.e. in the form of unstructured text documents) to obtain a ranking of applicants. A common framework applies semantic web technologies inside the discipline of HR management. In this framework, the candidate's personal tendencies, decided via a web questionnaire which is crammed-in through manner of the candidate, are taken into consideration during recruitment. In order to match applicants to positions those systems generally integrate strategies from classical IR and recommender structures, along with relevant comments, semantic matching and an analytic hierarchy process. Another approach proposed makes use of NLP era to mechanically represent CVs in a well-known modelling language. These techniques, even though beneficial, are beset by the problems associated with inconsistent CV formats, forms, and contextual records. Furthermore, they may be now not be able to investigate a few secondary traits associated with CVs, which include style and coherence, which may well be essential in CV evaluation.

Hiring can be modelled as a multistep process that begins with composing and advertising and ends with successful hiring. To accomplish their assignment more effectively, hiring managers depend on a significant range of cloud-based solutions, specifically applicant tracking systems (ATS). Unfortunately, the growing adoption of ATS has attracted scammers. In most instances, this phenomenon (a) jeopardizes jobseekers' privacy; (b) results in financial loss, and (c) vitiates the credibility of organizations. Today, as distinct in section three, venture frauds have become rather ingenious. Phony content is difficult to identify from benign content, so countermeasures are commonly advert-hoc and their practical fee is regularly questionable. Furthermore, the peculiarities of this precise concern render the utility of cutting-edge solutions designed to address relevant problems hard to develop. Specifically, lots of work has gone into addressing problems, such as e-mail, unsolicited mail, phishing, Wikipedia vandalism, cyber bullying, trolling, and opinion fraud. Nevertheless, in terms of employment scams, the proposed answer has been insufficient in practice.

2 Related Work

Recruitment consists of those practices used by a company with the primary motive of identifying and attracting potential employees (Breaugh and Starke 2000). It has developed into a complicated interactive engine with the capacity to automate each side of the hiring system faithfully (Joe Dysart 2006). The internet can simplify the choice of employees, especially where lengthy distances are concerned (Galanaki 2005). e-Recruitment has grown swiftly during the last ten years and is now widely utilized by recruiters and jobseekers the world over (Cober and Brown 2006). These rapid advances have dramatically changed the manner the enterprise is conducted and the increasing use of technology is seen in the variety of groups and people that utilize the internet and e-mail (Erica 2007). In phrases of human useful resource management, the internet has modified the manner of recruitment from the perspectives of both organizations and jobseekers (Epstein 2003). Online recruitment is a money saving and time saving means of selecting candidates. It reduces utility charges additionally recommended elevated applications from below qualified process seekers, triggering the increased effort of corporations to improve screening mechanisms (Freeman and Autor 2002). It facilitates the employer to control their database electronically. There are many process portal vendors for each agency&employee, where the employees can put their CVs free of charge but the employers are charged for filtration of the CVs.

As Bill Gates stated, one of the key advantages of the internet is that it reduces inefficiencies inside the market by improving the data exchange between customers dealers. For company staffing managers in all sectors of the economy, this gain has been adopted through the growing use of online recruiting ("e-recruitment") as a primary approach for advertising jobs in an increasing number of global labor markets. Indeed, critiques of job postings featured on online placement services and company websites illustrate a developing reliance on those resources. For instance, research shows that online resources now maintain 110 million jobs and 20 million unique resumes (which include 10 million resumes on Monster.Com alone), and that US online recruitment revenues topped $2.6 billion in 2007 (Li, Charron, Roshan, and Flemming 2002). Similarly, records show that actually all Fortune 100 corporations now use some shape of e-recruiting strategy (Lee 2005) and that 94% of Global 500 agencies use their websites for recruitment, as compared to only 29% in 1998 (Greenspan 2003). These developments are still rising due to the fact that e-recruiting has changed recruiting from "batch mode" to the more efficient "continuous mode" (Lee 2005) and has reduced hiring costs by about 87% compared to traditional recruiting thru newspapers magazines ($183 as opposed to $1,383, respectively). Also contributing to this growth are positive reviews from prominent employers including Dow Chemical, which reduced its hiring cycle from 90 to 34 days whilst reducing its associated costs by 26% (Gill 2001).

3 e-Recruitment System Architecture and Analysis

3.1 Proposed e-Recruitment System

e-Recruitment is getting an increasing number of well-known businesses growing their personal websites and organizations with online paintings forums. Now an afternoon's recruitment via social media is also in. Top organizations refer social money owed to research the behaviour and abilities of a candidate. The increased competiveness in the recruitment market has brought about a greater amount of time, efforts, and resources spent on developing recruitment techniques and expanding the variety of marketing techniques. The utilization of the internet in this area has transformed the recruitment system. Connecting the jobseeker and the

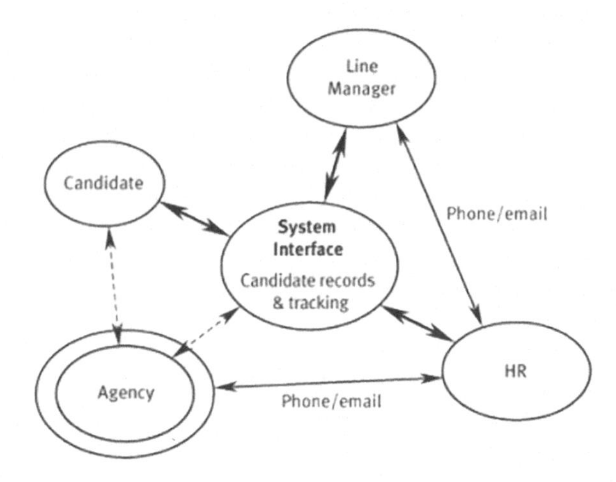

Fig. 1 Architecture of the e-recruitment system

corporation has become easy, fast, and cheap; internet tools enable employees/ recruiters to reach a much wider pool of candidates and to fill positions faster in competitive surroundings. It's completely at the enterprise a way to take the concept of e-recruitment in addition. Technology has permit company websites extra sophisticated, interactive and to attach usually (Fig. 1).

3.2 Use of Online Interviews in the Present Era

Regardless of the present day monetary climate, the common variety of interviews is increasing as staffing actions greater toward transient staffing and people on common live shorter in a single task. This trend creates several issues and ends in improved pressure on recruitment technique efficiency. In the hunt for a more flexible and efficient process, technological developments offer businesses new opportunities in terms of recruiting skills. The use of online video interviews already affords many organizations with a solution to the recruitment troubles they have to deal with. Below are four reason why you must not forget the usage of web-based video interviews for as part of you recruitment process.

Saving Time and Resources: Screen interviews are often a high-priced and time-consuming venture. The candidate has to travel and take day off. On the other hand, managers spend their precious time on lengthy interviews despite often making a decision inside the first five minutes regarding whether or not to employ the candidate. An internet-based method in the first round reduces fees and receives rid of tour problems. With video interviews managers and candidates can conduct them at any place and time and this is convenient and doesn't cost valuable time or assets.

Better Candidate Experience: The candidate experience is becoming increasingly critical. The preferences and needs of the candidate are changing and cannot be ignored in today's job market. Employers want to create an enjoyable experience for the candidate to make a very good first impression. Showing some recognition of the candidate's wishes is consequently an excellent way to construct an effective image of your company's ethos. With video interviews applicants don't have to travel and take a day off so people do not drop out of the first round of interviews due to scheduling troubles or cost issues. Most jobseekers also still greatly prize non-public communication. For many, a CV, letter, or phone conversation isn't always an appropriate alternative to appraise them. Online interviews offer an efficient means of doing so without losing privacy that candidates appreciate.

Better Screening, Better Matches: The use of online video interviews gives a better first spherical screening and thereby increases the chances of choosing a suitable candidate. With online interviews it's feasible to base the first choice now not simply on a CV (information and revel in), however also at the "tender skills." By changing the time-eating interviews at the workplace for a good deal faster first introductions thru the webcam, there's time to see extra applicants. Finally, options such as recording conversations and ranking the candidate's responses make it much less difficult to evaluate and review applicants.

Reduce Time-to-Hire: The time that it takes to first face-to-face conferences are one of the primary causes of delays in an organization's hiring procedure. Today's candidate needs a quick and efficient process. Candidates frequently reject a system because another one is quicker. Companies who can invite candidates quickly have an advantage. Reducing the time-to-hire by means of the use of video interviews gives organizations an alternate to live in advance of the opposition and recruit the excellent expertise

3.3 Advantages of an e-Recruitment System

Cost Effectiveness: Online recruitment strategies can be much more cost-effective than conventional strategies. Cost effectiveness is the most important reason for using e-recruitment.

Speed and Efficiency: The recruitment process can be made quicker via online postings of jobs, filling the online software shape and e-mailing CVs is straightforward. Applications can be processed inside minutes, saving both recruiters jobseekers valuable time.

Employer Branding: It is also a key purpose for groups to undertake e-recruitment to sell their image as a progressive employer. As more information can be provided by an organization's website than in a newspaper advert, it improves the image of business or enterprise and gives an indication of the organization's culture.

Create a Wider Candidate Pool: It creates a wider pool via offering 24/7 access to jobseekers from around the globe. It affords a better chance of locating the most suitable candidate. It also attracts a larger or more numerous applicant pool.

Quality of Candidates: Online recruitment improves the pleasant of the candidates due to blended finding. It additionally widens the pool of applicants and decreases the time to recruit. Online recruitment can help businesses to compete for the most suitable candidates.

Benefits of Video Interviews:

- Screen more than one candidate simultaneously.
- Search analyze responses—all interview statistics are recorded.
- Get insights quickly—via study of verbal cues and body language.
- Recruiters don't lose candidates because of scheduling issues—applicants can document their solutions at a time that works best for them.
- Bring group paintings and collaboration for your hiring selections.

3.4 Disadvantages of an e-Recruitment System

- Screening checking the content and authenticity of millions of resumes is a problem and a time-consuming exercise for businesses.
- There is low internet penetration, limited access and a lack of knowledge of the internet in many locations throughout India.
- Organizations cannot depend entirely on internet-based recruitment methods.
- In India, employers and personnel still prefer a face-to-face interaction rather than sending e-mails.
- Poor segmentation of the marketplace.
- Lack of human contact.
- Net beaten.
- Discrimination closer to non-customers.
- Privacy issues.

4 Detection of Fraudulent Candidates

Applicant Tracking Systems: The hassle of detecting employment scams can be defined because the system of distinguishing the subset some of the sum of content material of an ATS that targets at being used for fraudulent activities as opposed to valid recruiting. Such a system is usually finished with the aid of correlating information approximately the textual, structural and contextual attributes of that content. One can easily be aware that employment rip-off detection stocks not unusual characteristics with applicable troubles inclusive of email spam, phishing, Wikipedia vandalism, cyber bullying, trolling and opinion fraud.

Before we delve into the hassle of employment scam, we should remember that it is essential to discuss the context around ATS. An ATS, also called a candidate management system, is software program designed to assist organizations recruit employees more successfully by way of enabling the electronic management of recruitment phases. Such systems are similar to customer relationship management (CRM) systems, but are tailor-made to meet the unique requirements of the recruitment procedure.

5 Conclusion

In this paper, we discussed e-recruitment systems and fraud detection during online interviews in an e-recruitment system. By using e-recruitment system, we can reduce the manual procedure facing organizations and jobseekers who are called for interview. This proposed e-recruitment system is a time efficient system. By using fraud detection methods and tools we can improve the efficiency and security of e-recruitment systems.

References

1. Sharma N (2015) Recruitment strategies: a power of e-recruiting & social media. Int J Core Eng Manag (IJCEM) 1(5)
2. Vidros S, Kolias C, Kambourakis G, Akoglu L (2017) Automatic detection of online recruitment frauds: characteristics, methods, and a public dataset
3. Maurer SD, Liu Y (2007) Developing effective e-recruiting websites: insights for managers from marketers. Bus Horiz 50:305–315
4. Faliagka E, Ramantas K, Tsakalidis A, Tzimas G (2012) Application of machine learning algorithms to an online recruitment system
5. Allen DG, Van Scotter JR, Otondo RE (2004) Recruitment communication media: Impact on prehire outcomes. Pers Psychol 57(1):143–171
6. Barber AE (1998) Recruiting employees: individual & organizational perspectives. Sage Publications, Thousand Oaks, CA
7. Vidros S, Kolias C, Kambourakis G (2016) Online recruitment services: another playground for fraudsters. Comput. Fraud Secur. 201:8–13
8. Moon WJ (2007) The dynamics of internet recruiting: an economic analysis. Issues in Political Economy, vol. 16, pp 1–8
9. Othman RM, Musa R (2007) E-Recruitment practice: pros & cons. Public Sector ICT
10. Panayotopoulou L, Vakola M, Galanaki E (2007) E-HR adoption & the role of HRM: evidence from Greece. Personal Rev 36(2):277–294

Issues in Wireless Sensor Networks with an Emphasis on Security

Kartik Sharma and Sheeba Sharma

Abstract Wireless sensor networks (WSNs) are spatially dispersed systems with self-configuring sensor nodes connected via a wireless medium. The applications of WSNs are growing exponentially regardless of their known shortcomings. The very first applications of WSNs were military and surveillance based and even now military applications form a major portion of WSN applications. Due to the aforementioned reason and the sensitive nature of the data collected by WSNs, security has become a prime concern. This paper is focused on various issues present in WSNs with a minor emphasis on security.

Keywords WSN · WSN issues · Threats · Wireless sensor networks Security · Security threats in WSNs

1 Introduction

Wireless Sensor Networks (WSNs) were first limited to use military applications only but now their scope is being broadened with their employment in civilian applications including healthcare and environmental monitoring etc. [1]. WSNs like any other technology are not immune to flaws. They have numerous issues especially with respect to energy supply and miniaturization. WSNs need to be autonomous, which is why they need a higher degree of fault tolerance. In the not-so-distant future, we can expect ourselves to be surrounded by WSNs regardless of the type of applications they serve. The issues in WSNs have been researched extensively. The intent of this paper is to explore the issues faced by WSNs while focusing on the security of these networks.

K. Sharma (✉) · S. Sharma
University Institute of Engineering and Technology, Kurukshetra University, Kurukshetra, India
e-mail: kartik_s@outlook.in

S. Sharma
e-mail: sheebasharma10@gmail.com

A. Kumar and S. Mozar (eds.), *ICCCE 2018*,
Lecture Notes in Electrical Engineering 500,
https://doi.org/10.1007/978-981-13-0212-1_44

2 Issues in Wireless Sensor Networks

2.1 *Energy Consumption*

In WSNs, the nodes are supposed to have a degree of mobility. For that very reason, they depend on a power source packed within their microelectronic circuit [2]. That makes power the scarcest resource for any WSN. Nodes may fail due to battery depletion caused by normal use or by short circuits leading to incorrect sensor readings.

2.2 *Volatile Nature of the Wireless Medium*

The wireless medium's efficiency largely depends on the environment's static as well as dynamic nature. The noise in the wireless medium is a result of the geographical location, the timing, and various other factors.

2.3 *Scalability and Versatility of the Architecture*

The size of the sensor network is something which we can't truly foresee while designing the communication protocols. These protocols should be versatile enough to handle a tiny 3–4 node sensor network as well as another network with virtually infinite nodes.

2.4 *Quality of Service*

Quality of service (QoS) is crucial to any type of communication. The cardinal nature of QoS makes it hard to ignore while designing WSNs and their protocols. The basic purpose of QoS is to deliver data in a timely manner with no compromise in the integrity of the data. It is difficult to maintain in WSNs because of their obvious dynamic nature [3].

2.5 *Real-Time Mode*

In order to achieve a real-time mode in WSNs, the nodes' system must fall in line with the basic QoS requirements of the network.

2.6 Data Collection

Data collection is not as simple as it sounds. It comprises of gathering data from different sensors while also dealing with the issue of redundancy. The collected information must be delivered to the terminus with no loss or compromise of any kind.

2.7 Scheduling

Scheduling is the least-mentioned issue of WSN but that does not reduce its importance. It refers to the pattern which determines if/when the various sensor nodes are in active mode, standby mode or sleep mode. A decent scheduling algorithm can affect the energy consumption of the network in a positive way [4].

2.8 Node Deployment

In order to set up a WSN, we need to set up various nodes in an arbitrary manner. The deployment of nodes depends on the motive behind setting up the network. Some applications may require a sparse topology while some may work more effectively with a dense topology.

2.9 Dynamism of Sensor Nodes

The basic architecture of any WSN warrants the network to be topology free. The WSN may face the failure of nodes or the addition of new nodes, which is why the network needs to be compliant.

2.10 Network Congestion

Owing to the numerous simultaneous transmission attempts made by a sensor node, the network may become congested. The communication protocols which are supposed to govern the in-network communication are supposed to deal with this problem and manage the nodes' transmissions accordingly [5].

2.11 Unsupervised Operation

WSNs of a topology-less nature with no proper infrastructure. Once the network is put in working order, it must be able to work in an autonomous way. The nodes themselves are responsible for reconfiguration in case of any substantial change in the network. The network must be able to sustain itself regardless of the addition/ failure of its member nodes [6].

2.12 Production Cost

The production cost is a key aspect in the realistic applications of WSNs. For a feasible WSN, production costs of the sensor nodes needs to be kept low.

2.13 Synchronization

For all the nodes to work competently, the sensor nodes first need to synchronize their local clocks. In some WSN applications, a global clock solves the problem. Network time protocol and global positioning systems are energy hungry equipment which may cause excessive power consumption. To achieve a higher degree of accuracy, we need to employ more resources. The trade-off between synchronization accuracy and number of resources is something that needs to be dealt with in accordance with the actual applications of the network.

2.14 Limited Resources

In WSNs, each node requires data to be stored individually and hence the need arises for computational power and memory size which are other meager resource [6].

2.15 Security

The security of any network is paramount, but due to the battlefield applications of WSNs, it is even more vital in their case [7]. Since there is no identity assignment in case of any sensor nodes, the network becomes vulnerable to attacks mostly by the practice of spoofing [7–10].

3 Security in Wireless Sensor Networks

WSNs are more susceptible to attacks in comparison with normal wireless networks. Some of the basic flaws of the WSNs aid the attacker's capability to infiltrate the network. Timely detection of attacks in WSNs is a tough task as attacks often seem like network failures. To achieve resilience in the networks, we first need a way to distinguish between a network failure and an attack. Secondly, we need to employ reactive measures in the form of protocols to deal with that attack.

Reactive measures are available in a variety of arrangements. The simplest option is to shut down the whole network by sending a termination/hold signal to all the sensor nodes. Disabling all communication in the network for a specific period makes it very hard for an adversary to pursue the attack. As a negative effect, it introduces a down time in our network until adversary gets tired of waiting and leaves. On a positive note, this reactive measure preserves the energy resources of the nodes during the attack. Another reactive measure that can be taken is to keep the network in a working condition and to disregard any type of transmission until the attacker is taken care of. Enactment of these measures in the network is a complicated task because of the various known WSN constraints. A description of the most common types of attacks follows.

3.1 Denial of Service (DoS)

WSNs are more prone to denial of service attacks than other networks. The simplest means of subjecting a WSN to a DoS attack is by jamming the network. Jamming the network is a relatively easy task due to the communication medium being wireless. The attacker can easily introduce another wireless signal to interfere with the network's own wireless signal rendering the communication meaningless. This type of DoS attack is known as a physical layer attack.

On the data link layer, a DoS attack can be executed by breaching the communication protocols of the network, which may then require all the nodes to initiate the same transmission, again and again, leading to the exhaustion of the data source without being able to complete any transmission. In the same way, the network layer and the transport layer can also be subjected to a DoS attack by violating their respective protocols [11].

3.2 Compromising Information En Route

Since the very first applications of WSNs were military ones and they still form a considerably large portion of WSN applications, the information being transmitted by the sensor nodes is usually of a highly sensitive nature. The transmission between the sensor nodes needs to be immune to monitoring, eavesdropping, compromise, traffic analysis, camouflaged adversaries, and modification. In order to achieve this, we need to employ a set of privacy protocols which deal with the issues of confidentiality, authentication, and integration etc. [9, 10].

3.3 Node Replication

In a node replication attack, the attacker attempts to insert a new sensor node into the network. The node to be added originates from an actual node which has been cloned to form the new one. Cloning a node refers to copying the data/metadata from an authorized node to a new node in order for the new node to appear as an authorized node as well. The new node can be modified after cloning to allow transmission of data from it to the attacker. Node replication transforms into an even more serious issue when a base station is cloned [9].

3.4 Network Layer Attack

A network layer attack on a WSN is focused on the routing protocols of the WSN. A routing protocol attack may be in different forms such as a node using false routing information or transmitting only the data that suits its own purpose. Selective forwarding, hello flood attacks, sybil attacks, wormhole attacks, false routing information attacks, and sinkhole attacks are the most common forms of network layer attack. The worst of these is the false routing information attack, which possesses the capability to transform the whole network according to its own needs.

3.5 Desynchronization Attack

Desynchronization attacks deal with the attacker counterfeiting the messages with bogus sequence numbers and control flags requiring the nodes to retransmit the missed/compromised frames. This results in a substantial loss of power in the legitimate nodes. These attacks can be dealt with using the help of a proper authentication protocol requiring all packets to contain authentication.

4 Conclusion

WSNs have become an integral part of our technological environment. Like any other technology, WSNs have their shortcomings. Various threats to the secure operation of WSNs have been explored in this paper. WSNs are largely vulnerable to insertion of false information and nodes, which can be dealt with properly but not without compromising the energy efficiency of the system. Almost every issue or its solution in turn adversely affects the energy efficiency of the system. According to our observations, we don't just need a security framework for WSNs but we require a security framework which is capable of dealing with these issues without undermining the efficiency of the system.

References

1. Akyildiz I, Su W, Sankarasubramaniam Y, Cayirci E (2002) A survey on sensor networks
2. Tiande, Wenguo Y, Guo (2010) The non-uniform property of energy consumption and its solution to the wireless sensor network. In: 2nd international workshop on education technology and computer science
3. Wang Y, Attebury G, Rammurthy B (2006) A survey of security issues in wireless sensor networks. In: IEEE communications survey, 2nd quarter
4. Takai M, Bagrodia R, Tang K, Gerla M (2001) Efficient wireless networks simulations with detailed propagations models. In: Kluwer Wireless Networks, pp 297–305
5. Sharma P, Tyagi D, Bhadana P (2010) A study on prolong the lifetime of wireless sensor network by congestion avoidance techniques. In: Int J Eng Technol 2(9)
6. Jangra A, Swati, Richa, Priyanka (2010) Wireless sensor network (WSN): architectural. Int J Comput Sci Eng 2
7. Gupta K, Sikka V (2015) Design issues and challenges in wireless sensor networks. Int J Comput Appl (0975–8887) 112(4)
8. Danier EB (2011) Security of wireless sensor networks. In: Eighth international conference on information technology: new generations
9. Stallings W (2003) Cryptography and network security- principles and practice, 3rd edn. Prentice Hall
10. Tanenbaum A (2003) Computer networks, 4th edn. Prentice Hall
11. Shi E, Perrig A (2004) Designing secure sensor networks. In: J IEEE Wirel Commun 11(6)

Evaluation of Selected Tree- and Mesh-Based Routing Protocols

T. Harikrishna and A. Subramanyam

Abstract This paper researches various routing protocols, problems and necessities comparatively in MANET routing and layout concerns which include classifications primarily based on layers and other aspects. The layout and implementation of PUMA is a declarative constraint-fixing platform for coverage-based total routing and channel selection in multi-radio wi-fi mesh networks. PUMA integrates a high-performance constraint solver with a declarative networking engine. PUMA achieves a high data delivery ratio with very restricted manage overhead, which is almost constant in a large range of community situations. PUMA uses an unattached manipulate packet format for querying the receivers while ODMR has separate manage packets for querying exclusive manipulate information. The outcomes from a huge range of eventualities of varying mobility, organization members, a wide variety of senders, traffic load, and a wide variety of multicast organizations show that PUMA attains higher packet delivery ratios than ODMRP and MAODV, whilst incurring some distance less manipulate overhead.

Keywords PUMA protocol · Multicast routing · Control packet
MANET routing

1 Introduction

Construction of pleasant viable multicast timber and retaining the organization connections in sequence is challenging even networks under stress. Though in MANETs multicast is one of the hard surroundings, the implementations of the

T. Harikrishna (✉)
Rayalaseema University, Kurnool, India
e-mail: thk.itd@gmail.com

A. Subramanyam
Annamacharya Institute of Technology and Sciences, Rajampet, India
e-mail: smarige@gmail.com

A. Kumar and S. Mozar (eds.), *ICCCE 2018*,
Lecture Notes in Electrical Engineering 500,
https://doi.org/10.1007/978-981-13-0212-1_45

tree-based routing techniques are less complicated compared to mesh-based total routing strategies. In tree-based routing the simplest unattached path connects the source node and free-moving spot node, while in mesh-based routing a couple of routes connect the supply node and vacation spot node. Primarily tree-based routing protocols are afflicted by an inferior overall performance in terms of mobility. If a link breakdown occurs then the statistical messages are lost until a brand new dimension is constructed. Multicast tree structures are easily damaged and have to be readjusted continuously as connectivity changes. Furthermore, normal multicast trees regularly require a global routing substructure. Frequent modifications of routing vectors or link nation table's reasons continuous changes in topology. As a result, it generates excessive channel and processing overheads. Limited bandwidth, limited strength, and mobility of community hosts make the design of multicast tree-based total routing protocols particularly difficult. For this reason it is a major problem for researchers to increase primarily tree-based routing protocols. To remedy the problems inherent in tree-based techniques, a new topology referred to as mesh has been proposed. Mesh topology is characterized by the fact that it affords multiple paths among a source and a receiver which lets in multicast records grams to be brought even if a hyperlink fails. A schematic way of layout and experimentation of NS2 configuring for crucial analysis. Here we take into consideration the protocols MAODV and ADMR of the tree-based class and two other protocols, PUMA and ODMR, of the mesh-based elegance type to illustrate the protocol rating/ordering technique. However, this contribution has a technological value and no longer having a lot thought nature. The performances of the protocols referred to above are considered for the QoS parameters which might be crucial in evaluating the worthiness of novel routing protocols which might be developed through this work.

2 Selected Tree-Based Routing Protocols

2.1 Multicast Ad-hoc On-Demand Distance Vector Routing (MAODV) Protocol

MAODV is an improved model of AODV. It is a dynamic, self-beginning, multi-hop routing protocol. MAODV creates a shared tree that connects more than one source and receiver in the multicast institution. The root of each institution tree is both one of the multicast supply or receiver of the institution that has been designed as a group leader. The root is the primary member of a multicast group. When a utility on a node troubles to be part of request for a multicast organization, this node floods the RREQ packet within the complete network. If no reaction is obtained from the group then the RREQ packet is repeated and the requested node becomes the group leader for that institution. When a brand new supply wants to send packets to a collection, it follows identical steps. This group leader takes on the task of keeping the multicast group sequence variety. MAODV uses the unique collection range to recognize the multicast organization. The multicast group leader initializes the sequence variety and increases it at normal periods through a timer. By the use of a modern day collection number it generates routes for multicast businesses.

After this, the group leader floods the network through a group hello packet to broadly inform the community of the lifestyles of this group and its modern series quantity. By the usage of the group hello packet, the individuals within the institution update their request desk and distance to the group leader. The MAODV discovers multicast routes on-demand by using broadcast discovery mechanisms, i.e. route request and route reply. If nodes are asked to enroll in the institution or nodes want to ship packets to the multicast organization, then those nodes are required to obtain a group hello packet from the group leader and unicast a RREQ packet to the group leader.

Once the group leader receives the RREQ packet it uncasts a RREP packet again to the originator of the RREQ which responds with a multicast activation (MACT) packet. The MACT packet establishes multicast forwarding country between the newly joined receiver and the shared tree. If a source node does not receive a MACT within a certain timeframe then it broadcasts any other RREQ. After the quantity of RREQs, the supply assumes that there aren't any different members of the tree that can be reached and announces itself as the group leader. Damaged links are detected with the assistance of periodic hello packets transmitted through every node inside the community and nodes use the increasing ring search mechanism to reconnect the shared tree.

2.2 *Adaptive Demand Driven Multicast Routing (ADMR) Protocol*

ADMR is a receiver initiated multicast tree. If at the least one supply and one receiver are lively for the organization it creates a tree through the use of an on-demand mechanism. ADMR helps receivers to receive multicast packets dispatched through any sender. As well as receivers might also is part of a multicast institution dealt on behalf of unique senders. The multicast source does not now understand who the receivers are and in which community they may be located. The receivers want now not realize who are the assets and wherein network they are placed. ADMR works with the nodes which can flow at any time within the complete community and any packet can be lost inside the network. To be part of a multicast organization, an ADMR receiver transmits a MULTICAST SOLICITATION packet to the entire community. Once a source receives this packet, it replies through sending a unicast KEEP-ALIVE packet to that receiver and additionally confirms that the receiver has joined that source. The receiver replies to the KEEP-ALIVE through sending a RECEIVER JOIN packet in the reverse direction which units up forwarding state along the shortest paths. Additionally, in order to the receiver's join mechanism, a source floods the RECEIVER DISCOVERY packet periodically inside the entire community. The receivers which exist in the community and if there is no longer already related to the multicast tree then they get this packet and reply to it with a RECEIVER JOIN packet. In order to locate broken hyperlinks within the tree, the supply monitors the packet forwarding charge to determine if the tree has been damaged or the supply node has emerged as silent. When the hyperlink break occurs then the node initiates a repair. If the source node stops sending packets then any forwarding country is silently eliminated. Receivers screen the packet reception rate and can join the multicast tree if intermediate nodes have been not able to reconnect the tree. The receivers ensure a restore by means of broadcasting a new MULTICAST SOLICITATION packet. On the other hand a node at the multicast tree transmits a REPAIR NOTIFICATION packet down its sub-tree to cancel the restore of downstream nodes. The most upstream node sends a hop-constrained flood of a RECONNECT packet. If any forwarder receives this packet then it forwards the RECONNECT up the multicast tree to the source. The supply responds to the RECONNECT packet through sending a RECONNECT REPLY as a unicast message that follows the route of the RECONNECT returned to the repairing node. Thus, it performs both direction discovery and route renovation features on-demand.

3 Selected Mesh-Based Routing Protocols

3.1 On-Demand Multicast Routing (ODMR) Protocol

This is a mesh-based total multicast routing protocol. To create a mesh for each multicast organization, the ODMR protocol [1] makes use of the method of a forwarding group. This protocol establishes multicast routes and organizations on-demand and it is brought to the source. The ODMR protocol makes use of a course request and reply phase. In the network, if a source node has packets to forward then it periodically declares a member advertising and marketing packet called a join query. If a node receives a join query then it collects the upstream node ID and additionally rebroadcasts the packet. If the join query packet reaches a multicast receiver then the receiver creates or updates the supply entry in its member table. When legitimate entries are present inside the member table, join replies are broadcast periodically to the nearest node. If a node receives a join query it tests if the next node ID is one of the entries matches its own ID. If the node realizes that it is in the direction of the supply and a phase of the forwarding institution then it broadcasts its own join reply. Every forwarding group member sends the join reply until it reaches the multicast source via the shortest path. In the forwarding institution this manner may be used to create or replace the routes taking off from sources to receivers and paperwork a mesh of nodes. After the formation of the institution alongside the direction production procedure a multicast supply sends packets to receivers via desired routes and forwarding businesses. Periodic control packets are brought handiest whilst outgoing records packets are nevertheless there. While receiving a multicast information packet a node sends packets simplest while it isn't always a replica. To join or depart the institution no specific manipulate packets are required to be sent. If a multicast supply desires to leave the institution it stops sending join query packets immediately because it does not have any multicast statistics to forward to the institution. From a particular multicast group a receiver which no longer wants to receive, the receiver gets rid of the subsequent entries from its Member Table and want not transmit the Join Reply for that organization. Sample Heading (Third Level). Only two levels of headings should be numbered. Lower level headings remain unnumbered; they are formatted as run-in headings.

3.2 Protocol for Unified Multicasting Thru Announcements (PUMA)

PUMA is a mesh-based routing protocol which supports a source node to transmit multicast packets addressed to a known multicast group without having the information of the way the group is. Within the community it selects one of the receivers of a collection as the center of the organization. In addition it informs every router as at the least one subsequent-hop closer to the selected center of each organization. Every node on the shortest route connecting any receiver and the core, forms a mesh. A sender transmits a facts packet to the group using any of the shortest paths connecting the sender and the core. Once the statistics packet reaches a mesh member it floods the mesh. In addition, nodes maintain a packet ID cache to drop duplicate packets.

PUMA makes use of a single control packet for each function, i.e. a multicast announcement packet (MAP). Every MAP has a series variety, group ID (address of the organization), core ID (deal with of the center), distance to the middle, mesh member flag, and determine to choose a neighbor to reach the middle. Succeeding MAPs contain a better collection number than advance multicast announcements dispatched by way of the equal center. Using this information, nodes pick cores and locate routes for sources past a multicast and additionally provide the information concerning joining or leaving the mesh of a set and preservation of the mesh. A node in the organization at the core of a collection periodically transmits the multicast announcement. When the multicast statement travels for the duration of the community, it creates a connectivity listing at each node inside the community. By method of connectivity lists nodes want to create a mesh and also direction the statistics packets from senders to receivers. A node collects the facts from each multicast announcement and it accepts from its nodes in the connectivity list. A new multicast statement from a neighbor that's having higher collection wide variety overwrites with that of a lower sequence wide variety for the identical institution.

For a given institution, a node has only one access in its connectivity list from a specific neighbor and it keeps most effective those facts with the modern day sequence range for a given core.

Every access in the connectivity listing collects the records concerning the multicast declaration, the time when it is received, and the neighbor from which it changed into received. The node creates its personal multicast declaration primarily based on the excellent entry within the connectivity listing. For the similar center ID and maximum sequence variety the multicast announcements via smaller distances to the center are considered pleasant. Connectivity list and document work a recent list that's constrained to the new core. If each and each subject is equal then the multicast declaration that arrived formerly is taken into consideration. After figuring out the satisfactory multicast declaration packet, the node creates its own multicast declaration packet. The connectivity list collects records concerning all the routes that exist to the center. If a core change takes place for a particular institution then the node removes old entries.

4 Performance Evaluation Through Simulation

In order to evaluate the performance of a routing protocol we can apply a community simulator like NS-2, Qual Net, OPTNET and GloMoSim etc. In this study we used NS-2 to create the simulation environment to implement the protocols and to compare their overall performance with each other.

4.1 Performance Metrics

To evaluate the overall performance of the selected protocols we recollect that throughput, packet shipping ratio and postpone throughput are defined as the whole common variety of facts packets acquired by way of the destination in bytes per second. The packet delivery ratio is the ratio of the data packets obtained at the destination to the variety of records packets transmitted by the supply. The average end-to- end delay represents the common time, i.e. the transmission put off of facts packets which might be introduced correctly. This postponing consists of propagation delays, queuing delays at the interfaces, and buffering delays at some point of path discovery.

4.2 Simulation Scenario and Scenario Setup

The simulator for evaluation of the proposed routing protocol is carried out using Network Simulation (NS2) in Linux. The network size is of 50–200 nodes incrementing via 50 nodes placed randomly over a 1100 m × 1100 m area.

The transmission variety for each node is 250 m and the channel potential is 2 Mbits/s. The mobility version of the nodes within the simulations is the random waypoint version. Every node begins at a random position in the simulation region and stays deskbound for an interval of pause time. The node then generates a uniformly dispensed latest location, which is a random destination within the simulation area. The mobility speed is uniformly dispensed between precise mobility speeds of 0–10 m/s, with a pause time of 0 s equivalent to a regular motion of 10 s.

4.3 *Channel and Radio Model*

The propagation models in NS2 are a loose area version, an X-ray floor mirrored image model and a shadowing version. The free area model assumes the precise scenario in which handiest one clean line-of-sight route some of the transmitter and receiver. The two ground reflection models consider the direct path and floor mirrored image course together. The shadowing version consists of parts i.e. the first component is direction loss example, mobile nodes may not move in straight lines at regular speeds for the whole simulation due to the fact that actual cell nodes can no longer pass in such a confined way. There are unique sorts of mobility fashions which might be to be had i.e. random mobility model, group mobility version, temporal mobility model, and spatial mobility model. Of the random type, random stroll mobility, random waypoint mobility, and random direction mobility versions etc. are available. In the proposed protocols we consider the widely used random point route mobility model for the overall performance of the protocols.

4.4 *Random Waypoint Mobility Model*

The random route point mobility model comprises the pause times which entail modifications in direction and/or velocity. A node starts off evolved in staying at one function intended for a particular timeframe, i.e. the pause time. When this time expires the cellular node selects a random destination within the simulation vicinity and the velocity that's uniformly distributed the various most pace and minimal speed. The cellular node moves in the direction of the newly selected destination with the aid of the chosen pace. On arrival, the cellular node stops for a selected time period prior to beginning the system again.

In most of the performance evaluations that used the random waypoint mobility model, the mobile nodes are at first disbursed randomly at some stage in the simulation vicinity. The random distribution of model and the second one element reflects the version of the received strength at unique distance. The two-ray ground reflection model is used for simulation inside the proposed protocol.

4.5 MAC Protocol

The IEEE 802.11 MAC protocol thru allotted coordination feature (DCF) is used as the MAC layer. DCF uses a RTS/CTS/DATA/ACK for all unicast packets. Multicast data packets are sent without using ACK in the series.

4.6 Simulation Parameters and Traffic Scenario

The selected protocols are evaluated with the use of a network simulator (NS2) of 50–200 nodes incrementing through 50 nodes. The mobility version chosen is a random way point model. In this mobility model, a node randomly selects a destination and it moves in the direction of the vacation spot with a velocity uniformly chosen among the minimum pace and maximal velocity. After it reaches the destination, the node remains there for a pause time after which it actions again. Each node moves randomly with a pace of 0–10 m/s and stays at the identical region with a pause time of 0–10 s. The distributed coordinated function (DCF) of IEEE 802.11 for wi-fi LANs is believed as the MAC layer protocol. The two-ray ground version is selected for propagation. A bandwidth of 2 Mbps with a radio variety of 250 m is taken into consideration. We have elected to use CBR as the form of verbal exchange and the maximum interface queue period is 250. The overall performance metrics considered are throughput, average end-to-end delay and the packet delivery ratio.

5 Simulation Results

We evaluated and compared the overall performance of the tree-based total routing protocols, i.e. ADMR and MAODV routing protocol, and the mesh-based total routing protocols, i.e. ODMR and PUMA. We took into consideration the various node mobilities, various node densities and their group sizes. We decided on a network with a node mobility starting from 2 to 10 m/s incrementing in steps of two, node density degrees from 50 to 200 which increments within the steps of fifty and their institution sizes in with increments of thus the feasible mixtures for comparing above 4 protocols is i.e. No. of distinct densities taken into consideration extended without distinct businesses considered. The consequences for different overall performance metrics plotted for different parameters and node mobilities are illustrated in Fig. 3.1. In the overall performance analysis, throughput, packet shipping ratio, and end-to-end delay comparisons are detailed below.

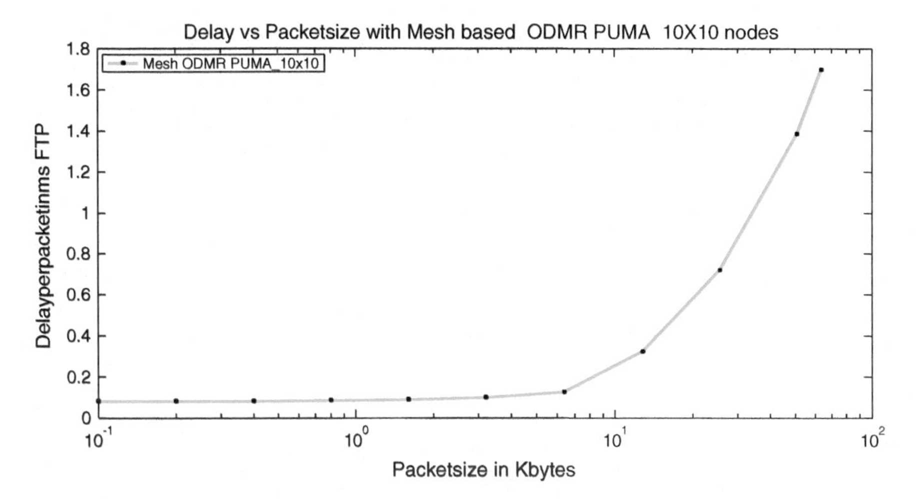

Fig. 1 Delay versus throughput dimension

5.1 Throughput

From the graphs plotted in Fig. 1 it can be seen that once the wide variety of agencies is equal to at least one, PUMA is slightly higher than ODMR and in turn ODMR is better than MAODV and ADMR. As the organization dependence is increased from one to two, PUMA is a good deal higher than ODMR. This behavior is because of the distinction within the format of the manipulate packets used. PUMA uses a single manipulate packet format for querying the receivers while ODMR has separate manipulate packets for querying specific control information. Thus the manage packets exchanged between senders and receivers are greater compared to the records packets in ODMR than PUMA. Hence the result of throughput is lower in ODMR. PUMA offers higher throughputs because it focuses on mesh redundancy in the region.

5.2 Packet Delivery Ratio

From the graphs plotted in Fig. 2 you could word the risky transport ratios exhibited by using ADMR for the diverse densities and institution counts considered except the density equals to 200. However, its performance is very poor compared to the other three protocols because of the greater number of link breakages that came about in ADMR. The ODMR presentations good following ratios with organization be counted equals to at least one in comparison to

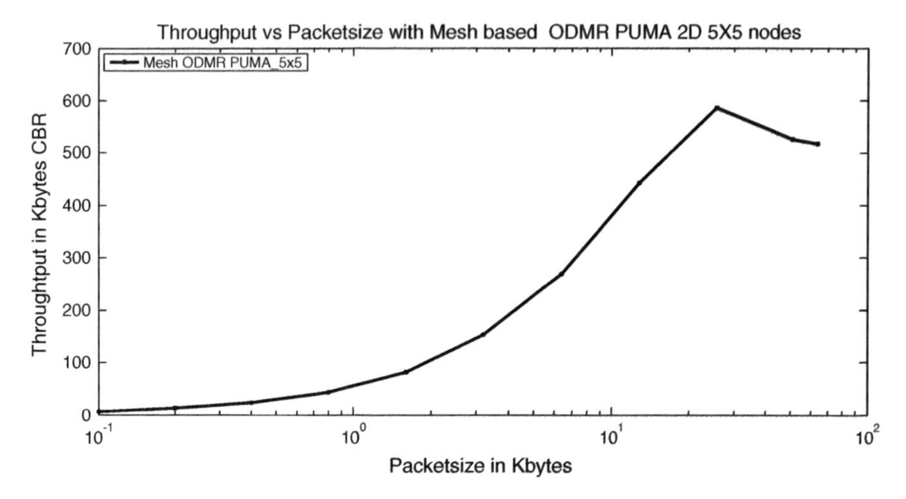

Fig. 2 Throughput versus packet size

MAODV. However, with the number of companies equal to 2, the transport ratios of ODMR and MAODV are almost identical with ODMR being only slightly better.

5.3 *End-to-End Delay*

Figure 3 conveys the reality that the give up-to-stop delay of ADMR is higher in comparison to MAODV except in a few instances where both of them overlapping in all respects. This is because of the fact that during ADMR the receiver needs to

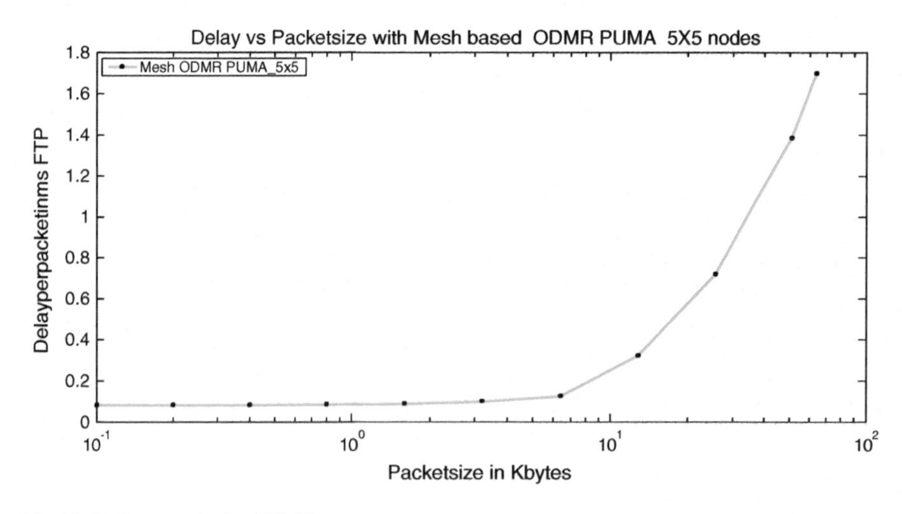

Fig. 3 Delivery ratio for PUMA

send an affirmation to every supply. The give up-to-cease delays in PUMA are lower in comparison to the other three protocols and ODMR is clearly better than MAODV and ADMR. However, ODMR shows higher delays compared to the other three protocols. PUMA nevertheless presents regular quit-to-give up delays.

References

1. Nagaratna M, Raghavendra Rao C, Kamakshi Prasad V (2011) Node disjoint split multipath protocol for unified multicasting through announcements (NDSM-PUMA). In: Proceedings of international conference on computer science and information technology CSIT-2011, Bangalore, India, pp 373–380
2. Agarwal DP, Zeng QA (2005) Introduction to wireless and mobile systems. Brooks/Cole
3. Anderson RGT, Bershad B, Wetherall D (2000) A system architecture for pervasive computing. In: Proceedings of 9th ACM SIGOPS European workshop, Kolding, Denmark, pp 177–182
4. Chlamtac I, Redi J (1998) Mobile computing: challenges and opportunities. In: Ralston A, Reilly E (eds). International Thomson Publishing
5. Verma AK, Dave M, Joshi RC (2007) Secure routing in mobile networks: a review. Int J Syst, Cybern Inf (IJCSI). ISSN 0973-4864
6. Perrig A, Canetti R, Tygar JD, Song D (2000) Efficient authentication and signing of multicast streams over lossy channels. In: Proceedings of IEEE symposium on security and privacy
7. Corson S, Macker J (1999) Mobile ad hoc networking (MANET): routing protocol performance issues and evaluation considerations, RFC 2501
8. Wang Z, Liu L, Zhou MC (2004) Protocols and applications of ad-hoc robot wireless communication networks: an overview. Int J Intell Control Syst 10(4):296–303, December 2005; Architectures and protocols. Prentice-Hall
9. Perkins CE (2008) Ad hoc networking. Pearson Education Inc
10. Royer EM, Perkins CE (1999) Multicast operation of the ad-hoc on-demand distance vector routing protocol. In: Proceedings of the 5th ACM/IEEE international conference on mobile computing and networking, Mobicom'99, pp 207–218
11. Jetcheva JG, Johnson DB (2001) Adaptive demand-driven multicast routing in multi-hop wireless ad hoc networks. In: Proceedings of the 2nd ACM international symposium on mobile ad hoc networking and computing (MobiHoc'01), pp 33–44
12. Vaishampayan R, Garcia-Luna-Aceves JJ (2004) Protocol for unified multicasting through announcements (PUMA). In: Proceedings of the IEEE international conference on mobile ad-hoc and sensor systems (MASS'04)
13. Dahill B, Levine B, Belding Royer E, Shields C (2001) A secure routing protocol for ad hoc networks. University of Massachusetts, Technical Report, pp 01–37
14. Toh CK (2002) Ad Hoc mobile wireless networks: protocols and systems. Prentice-Hall, Englewood Cliffs, NJ, USA
15. Paul S (1989) Multicasting on the internet and its applications. Kluwer Academic Publishers, Norwell, Mass, USA, 1998; Stojmenovi´c I (ed) Handbook of wireless networks and mobile computing. Wiley, New York, NY, USA, 2002
16. Badarneh OS, Kadoch M (2009) Multicast routing protocols in mobile ad hoc networks: a comparative survey and taxonomy. Hindawi Publishing Corporation EURASIP J Wirel Communications Netw, Article ID 764047. https://doi.org/10.1155/2009/764047
17. Junhai L, Liu X, Danxia Y (2008) Research on multicast routing protocols for mobile ad-hoc networks. Comput Netw 52(5):988–997

18. Chiang CC, Gerla M, Zhang L (1998) Adaptive shared tree multicast in mobile wireless networks. In: Proceedings of the IEEE global telecommunications conference (GLOBE-COM'98), vol 3, pp 1817–1822
19. Shen CC, Jaikaeo C (2005) Ad hoc multicast routing algorithm with swarm intelligence. Mob Netw Appl 10(1):47–59
20. Park S, Park D (2004) Adaptive core multicast routing protocol. Wirel Netw 10(1):53–60
21. Toh CK, Guichal G, Bunchua S (2000) ABAM: on-demand associativity-based multicast routing for ad hoc mobile networks. In: Proceedings of the IEEE vehicular technology conference (VTC'00), vol 3, pp 987–993
22. Ozaki T, Kim JB, Suda T (2001) Bandwidth-efficient multicast routing for multihop, ad-hoc wireless networks. In: Proceedings of the 20th annual joint conference of the IEEE computer and communications societies (INFOCOM'01), vol 2, pp 1182–1191
23. Pompili D, Vittucci M (2006) PPMA, a probabilistic predictive multicast algorithm for ad hoc networks. Ad Hoc Netw 4(6):724–748
24. Xie J, Talpade RR, McAuley A, Liu M (2002) AMRoute: ad hoc multicast routing protocol. Mob Netw Appl 7, 132(6), 429–439
25. Ge M, Krishnamurthy SV, Faloutsos M (2006) Application versus network layer multicasting in ad hoc networks: the ALMA routing protocol. Ad Hoc Netw 4(2):283–300
26. Ye Z, Krishnamurthy SV and Tripathi SK (2003) A frame work for reliable routing in mobile ad hoc networks. In: Proceedings of the 22nd annual joint conference of the IEEE computer and communications societies (INFOCOM), vol 1, pp 270–80
27. Mueller S, Ghosal D (2005) Analysis of a distributed algorithm to determine multiple routes with path diversity in ad hoc networks. In: Proceedings of the third international symposium on modeling and optimization in mobile, ad hoc and wireless networks(WIOPT), pp 277–285
28. An H-Y, Zhong L, Lu X-C, Peng W (2005) A cluster-based multipath dynamic source routing in MANET. In: Proceedings of the IEEE international conference on wireless and mobile computing, networking and communications (WIMOB), vol 3, pp 369–376
29. Wisitpongphan N, Tonguz OK (2003) Disjoint multipath source routing in ad hoc networks: transport capacity. In: Proceedings of the IEEE 58th vehicular technology conference (VTC), vol 4, p 2207–2211
30. Bagrodia R, Gerla M, Hsu J, Su W, Lee SJ (2000) A performance comparison study of ad hoc wireless multicast protocols. In: Proceedings of the 19th ComSoc IEEE Conference, pp 565–574

Reduction of Kickback Noise in a High-Speed, Low-Power Domino Logic-Based Clocked Regenerative Comparator

N. Bala Dastagiri, K. Hari Kishore, G. Vinit Kumar and M. Janga Reddy

Abstract The comparator is the most significant element in the design of ADCs. Also there is a lot of demand for low-power, high-speed VLSI circuits. Therefore to maximize power efficiency and speed in ADCs, there is a desire to design high-performance clocked regenerative comparators. The regenerative latch of the comparator is responsible for taking decisions quickly and accurately. Normally, the accuracy of an ADC is degraded due to disturbance in the input voltage called kickback noise, which usually occurs with large variations of voltage at coupled regenerative nodes. This paper describes an analysis of the minimization of kickback noise in a clocked regenerative double tail comparator. To improve further on the double tail comparator, a new domino logic-based regenerative comparator is realized with high speed, low power and reduced kickback noise at low supply voltages. The simulated results using 130 nm CMOS technology confirm the theoretical results. The analysis of the proposed design demonstrates that kickback noise, power, and delay are considerably reduced. The simulation work was carried out using Mentor Graphics tools.

Keywords Clocked regenerative comparators · Kickback noise
Domino logic · CMOS technology

1 Introduction

A latched comparator compares the voltage levels at the inputs and a corresponding voltage is generated at the output. An open-loop op-amp is the simplest approach for designing a comparator, but the slow response time and input offset voltage of the op-amp induces errors in the comparator and limits accuracy and resolution.

N. Bala Dastagiri (✉) · K. Hari Kishore
Koneru Lakshamaiah Education Foundation, Guntur, A.P, India
e-mail: baluece414@gmail.com

G. Vinit Kumar · M. Janga Reddy
CMR Institute of Technology, Hyderabad, Telangana, India

A. Kumar and S. Mozar (eds.), *ICCCE 2018*,
Lecture Notes in Electrical Engineering 500,
https://doi.org/10.1007/978-981-13-0212-1_46

Due to the presence of device mismatches an input offset voltage is generated in the comparator. Comparators have found widespread use in many applications, such as ADCs, data transmission, power regulators and so on [1]. Comparators with low power dissipation, high speed and a small die area are mostly used in ADC's. These high-speed comparators suffer from supply voltages in sub-micron technologies. Hence, designing power-efficient and speed-efficient comparators at lower supply voltages is very challenging [1, 2]. Techniques like supply boosting methods [3, 4] and methods employing body-driven transistors [5, 6], and current mode designs [7] use dual-oxide processes for handling higher supply voltages at lower supply voltages. Boosting and bootstrapping are the two techniques used to address the switching and offset problems at lower supply voltages [5]. Also in [8–11] dynamic regenerative comparators suffer from a noise called kickback noise which is generated due to large voltage variations across regenerative nodes. This noise degrades the accuracy of the comparator [1]. In [8–11] an additional circuit is added to improve performance at lower supply voltages. The proposed topology of the clocked regenerative comparator with domino logic has a supply voltage of 0.8 V.

In this paper, the transient analysis of conventional dynamic comparators is presented and kickback noise calculated for various architectures. A new dynamic comparator is based on double-tail architecture, which does not require higher voltages. After modification the resultant comparator topology arrives at high speed, low power and low kickback noise voltage when compared with traditional comparators.

The preceding sections in this paper are organized as follows: Sect. 2 investigates the working of various traditional comparators and their advantages and disadvantages. In Sect. 3 the new comparator topology is proposed. Section 4 addresses the simulation results and comparisons followed by conclusion and the scope of future work in Sect. 5.

2 Traditional Clocked Comparators

2.1 Dynamic Latched Single Tail Comparator

The conventional latched single tail comparator shown in Fig. 1 operates in two stages [1]. The immediate stage is followed by a regenerative stage with two cross-coupled inverters, where the outputs of each inverter are connected as the input to the next stage. The two modes of operation are the interface mode and the regenerative mode. The comparator shown is a single tail comparator with a NMOS transistor as a tail, and is connected to the ground. During the interface mode CLK becomes '0' which in turn turns off the tail transistor M9. The output reaches to ground or VDD depending on the voltages applied to the input nodes. If inn is less than inp the output of Vn discharges more quickly than the output Vp and vice versa. In the regenerative mode, as CLK equals VDD, transistor M9 turns on and

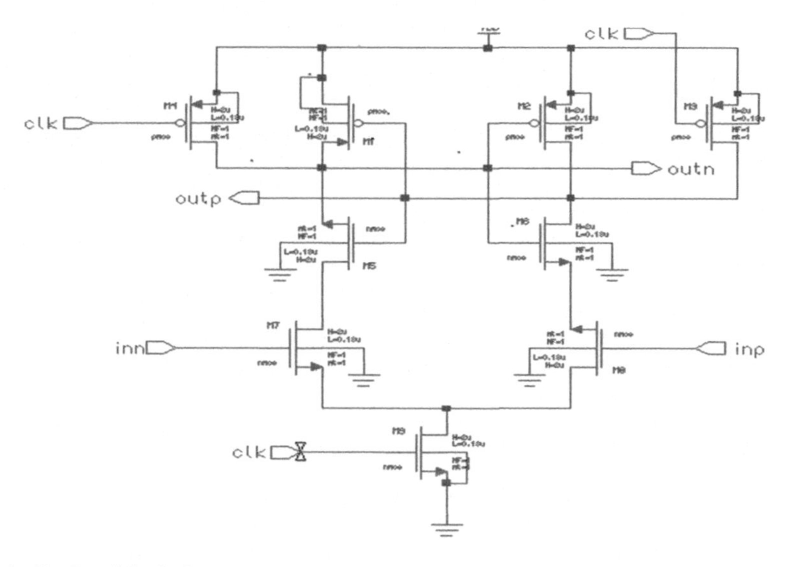

Fig. 1 Single tail latched comparator

the comparator outputs discharge to ground. At the nodes where the drains of transistors M7 and M8 connect together a noise called kickback noise originates and degrades the accuracy.

The important limitation of the single tail comparator is that the tail transistor M9 has only one path to distribute currents for differential amplifier and regeneration latch. Tail transistor M9 mostly operates in the triode region, which is not favourable for the working of the latch.

2.2 *Conventional Latched Double Tail Comparator*

A conventional latched double tail comparator with less stacking of transistors is shown in Fig. 2 [1]. Due to its architectural features, a double tail comparator can be operated with low supply voltages when compared with the single tail comparator.

The double tail comparator is operated in two modes, i.e. the evaluation mode and the decision-making mode. In the reset mode, when clk = '0' both the tails Mtail1 and Mtail2 are in the off state. Both the output nodes Outp and Outn are pulled to VDD as the transistors M3 and M4 are pre-charged which in turn discharges the output nodes are discharged to ground. During the evaluation mode, CLK equals VDD or "1", and turns on the transistors Mtail1 and Mtail2. As both tail transistors are turned on, the transistors M3 and M4 are turned off resulting in the output nodes reducing by a ratio of $I_{mtail1}/C_{OUTN(p)}$. The input and output protection obtained from the cross-coupled latch reduces power and delay.

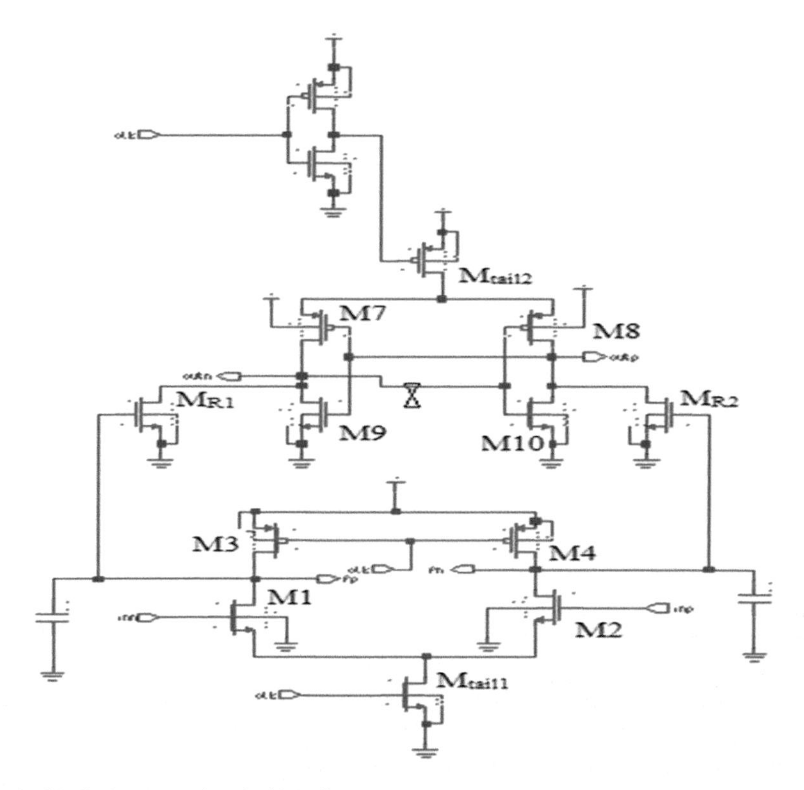

Fig. 2 Clocked regenerative double tail comparator

In this double tail comparator, the intermediate stage transistors have an inactive role in improving the latch effectiveness as they will be mostly in the cut-off region. However, during the reset phase, regenerative nodes should be at VDD which results in greater power consumption.

3 Proposed Dynamic Regenerative Comparator

The requirements of the upcoming challenges facing VLSI designs are new digital logic design techniques with high speed, a robustness with respect to noise, and energy efficiency. Such circuits are designed using domino logic gates because of their high speed. On the other hand, domino logic has the disadvantage of a high noise sensitivity. As double tail dynamic comparators exhibit better performance at low supply voltages [1], the comparator proposed is designed using domino logic and double tail architecture to improve the performance in terms of speed. However, kickback noise still exists and needs to be suppressed. The idea behind this structure is to improve the regeneration speed of the latch and thereby cancel the kickback noise originating at the regenerative nodes.

3.1 Operation of the Proposed Comparator

The architecture of the proposed domino logic double tail comparator is depicted in Fig. 3. The schematic shown is an extension of a double tail comparator with a domino logic circuit connected to the output [10, 11]. The operation of this architecture is same as that of a double tail structure. The proposed double tail domino logic comparator operates in two modes, i.e. reset mode and comparison mode.

In reset mode, CLK equals '0' and both the tails Mtail1 and Mtail2 are off. Here both the output nodes Outp and Outn are pulled to VDD since the transistors M3 and M4 are pre-charged and the output nodes are discharged to ground and in turn transistors Mr1 and Mr2 are affected. In the evaluation phase of operation, CLK

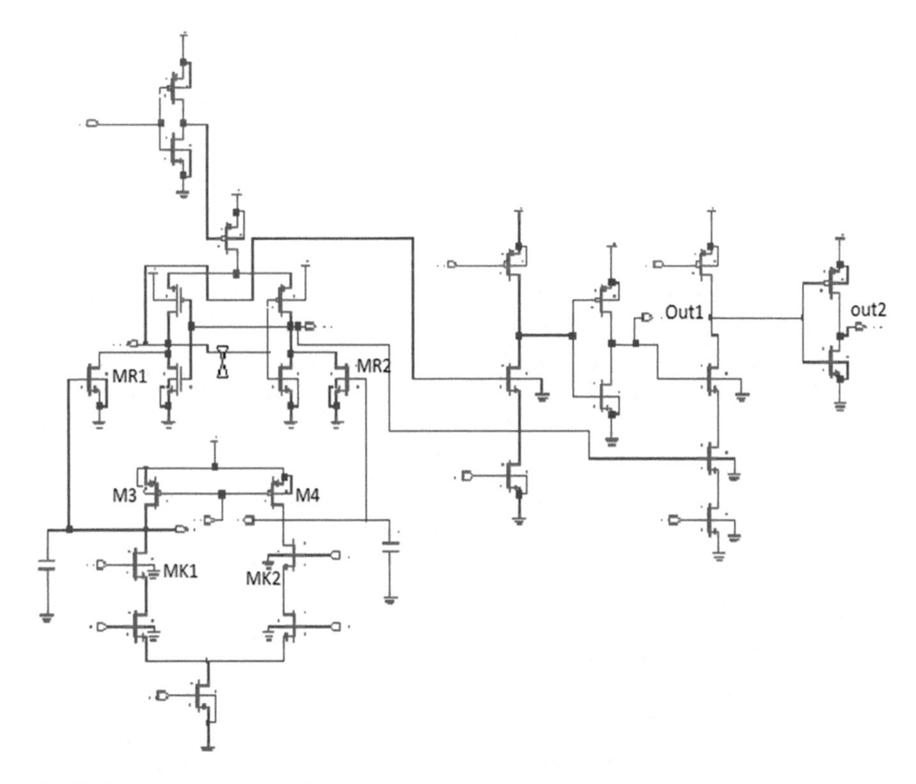

Fig. 3 Proposed dynamic double tail comparator

equals VDD or "1" and turns on the transistors Mtail1 and Mtail2. As the both tail transistors are ON, the transistors M3 and M4 are OFF resulting in the output nodes reducing by a ratio of $I_{mtail1}/C_{OUTN(p)}$. In this double tail comparator, the intermediate stage transistors do not play any role in improving the effectiveness of the latch as they will be mostly in the cut-off region. Moreover, the domino logic increases the speed of latch regeneration. The input and output protection obtained from the cross-coupled latch reduces power and delay.

3.2 *Reduction of Kickback Noise*

In dynamic latched comparators, wide voltage variations are seen at regeneration nodes. Due to this, the voltage at the input node is disturbed, which in turn degrades the accuracy of the comparator, usually called kickback noise. In general, it is the more power efficient and higher speed comparators that suffer from high kickback noise.

In order to reduce the effect of kickback noise at the input nodes, two transistors Mk1 and Mk2 are introduced. These transistors separate the input from the voltage variations at regenerative nodes. The parasitic capacitances of the transistors Mk1 and Mk2 attenuate the voltage variations at the drains of the input transistors. Thus kickback noise can be present and this may increase the transistor count with a slight reduction in power dissipation. Most importantly, noise can be minimized.

4 Simulation Results and Comparison

The simulated transient responses of various dynamic comparators along with the proposed topology is presented in Figs. 4, 5 and 6. The proposed architecture, along with single tail and double tail comparators is simulated using 130 nm technology with VDD = 0.8 V. The simulated results also show the characteristic parametric

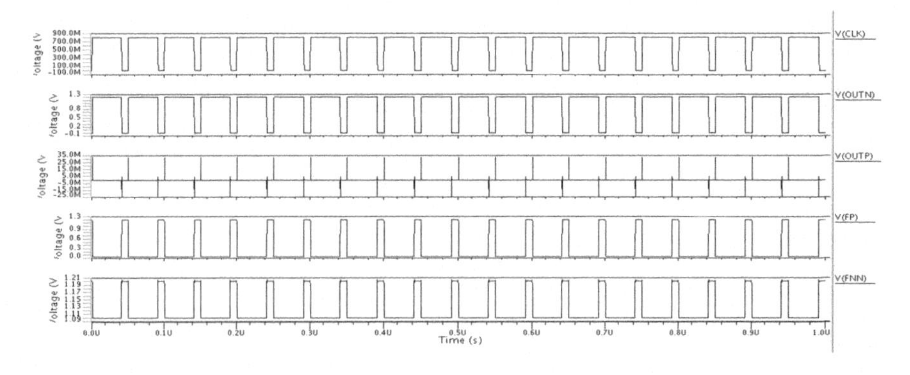

Fig. 4 Transient response of a single tail comparator

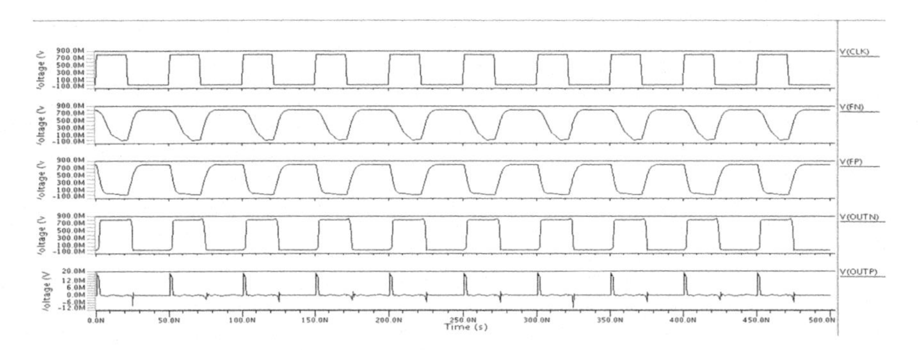

Fig. 5 Transient response of a dynamic latched double tail regenerative comparator

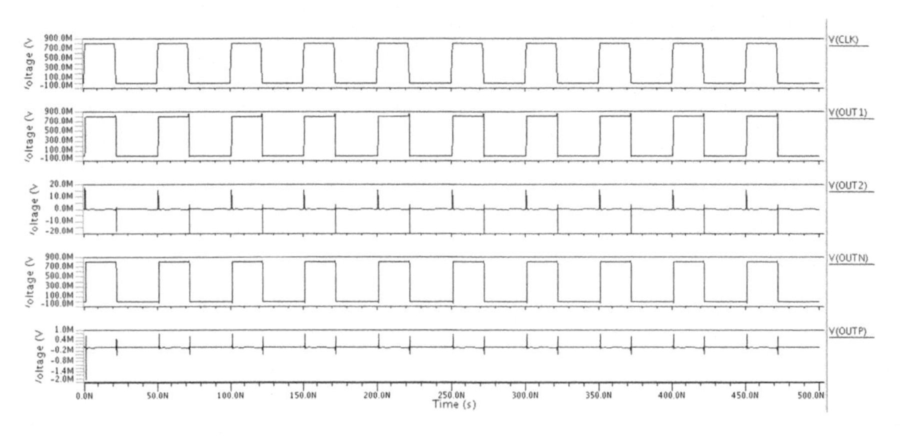

Fig. 6 Transient response of the proposed domino logic transistor induced double tail comparator

changes related to speed and delay. The comparator circuits discussed above are simulated using Mentor Graphics software tools. A comparison table shows the values of propagation delay, power dissipation, slew rate, and input offset voltage (Table 1).

The table shows the comparison results of various traditional and proposed architectures of dynamic regenerative comparators in terms of parameters such as power dissipation, propagation delay, and kickback noise voltage. It can be seen

Table 1 Performance comparison

Dynamic comparator type	Power dissipation (W)	Delay (Ns)	Kickback noise (uv)
Dynamic latched	0.4798 N	50.0912	972.09
Conventional double tail	1.1822 N	48.5015	472.90
Proposed double tail	1.1556 N	36.7820	154.75

that the kickback noise has been considerably reduced using the proposed architecture. The power dissipation and delay have also been reduced.

5 Conclusion

In this paper we present an analysis for various topologies of dynamic regenerative comparators in terms of power, speed, and kickback noise. The results show that the proposed dynamic comparator has an improved performance compared to existing comparators in terms of speed, power consumption, and kickback noise. Due to the modified technique, the noise is reduced by more than 60% when compared with the other comparator topologies.

Acknowledgements The authors gratefully acknowledge Annamacharya Institute of Technology & Sciences, Rajampet, India and KL University, Vijayawada, India for the support given to carry out this work. The authors would like to extend their gratitude to Prof. B. Abdul Rahim, Mr. S. Fahimuddin, and Dr. G. Vinith Kumar for the stimulating discussions.

References

1. Babayan-mashhadi, Lotfi (2014) Analysis and design of a low-voltage low-power double-tail comparator. IEEE Trans Very Large Scale Integr (VLSI) Syst 22(2)
2. Goll B, Zimmermann H (2009) A comparator with reduced delay time in 65-nm CMOS for supply voltage down to 0.65 v. IEEE Trans Circuits Syst II, Exp Briefs 56(11):810–814
3. Mesgarani, Alam MN, Nelson FZ, Ay SU Supply boosting technique for designing very low-voltage mixed-signal circuits in standard CMOS. In: Proceedings of IEEE international midwest symposium circuits systems
4. Ay SU A sub-1 volt 10-bit supply boosted SAR ADC design in standard CMOS. Int J Analog Integr Circuits Sig Process 66
5. Maymandi-Nejad M, Sachdev M 1-bit quantiser with rail to rail input range for sub-1 V modulators. IEEE Electron Lett 39
6. Blalock BJ (Feb 2000) Body-driving as a low-voltage analog design technique for CMOS technology. In: Proceedings IEEE southwest symposium mixed-signal design, pp 113–118
7. Okaniwa Y, Tamura H, Kibune M, Yamazaki D, Cheung T-S, Ogawa J, Tzartzanis N, Walker WW, Kuroda T (2005) A 40 Gb/s CMOS clocked comparator with bandwidth modulation technique. IEEE J Solid-State Circuits 40(8):1680–1687
8. Goll B, Zimmermann H (Feb 2007) A 0.12 μm CMOS comparator equiring 0.5 V at 600 MHz and 1.5 V at 6 GHz. In: Proceedings IEEE international solid-state circuits conference, digital technical papers, pp 316–317
9. Goll B, Zimmermann H (Feb 2009) A 65 nm CMOS comparator with modified latch to achieve 7 GHz/1.3 mW at 1.2 V and 700 MHz/47 μW at 0.6 V. In: Proceedings of IEEE international solid-state circuits conference digital technical papers, pp 328–329

10. Goll B, Zimmermann H (2007) Low-power 600 MHz comparator for 0.5 V supply voltage in 0.12 μm CMOS. IEEE Electron Lett 43(7):388–390
11. Shinkel D, Mensink E, Klumperink E, van Tuijl E, Nauta B (Feb 2007) A double-tail latch-type voltage sense amplifier with 18 ps Setup + Hold time. In: Proceedings of IEEE international solid-state circuits conference digital technical papers, pp 314–315

Two-Level Intrusion Detection System in SDN Using Machine Learning

V. Vetriselvi, P. S. Shruti and Susan Abraham

Abstract Software Defined Networking (SDN), the new paradigm in network architecture is changing how we design, manage, and operate an entire network, making networks more agile, flexible, and scalable. Such admirable features arise from the design factor that, in SDN, the control plane is decoupled from the data plane and instead resides on a centralized controller that has complete knowledge of the network. As SDN continues to flourish, security in this realm remains a critical issue. An effective intrusion detection system (IDS), which can monitor real-time traffic, detect and also identify the class of attack would greatly help in combating this problem. This work aims to heighten the security of SDN environments by building an IDS using the principles of machine learning and genetic algorithms. The proposed IDS is divided into two stages, the former to detect the attacks and the latter to categorize them. These stages reside in the switches and the controller of the network respectively. This approach reduces the dependency and the load on the controller, as well as providing a high attack detection rate.

Keywords Intrusion detection system · Software defined networks
Machine learning · Iterative dichotomiser 3 · Genetic algorithm
Distributed denial of service · Programming protocol-independent packet processors

V. Vetriselvi (✉) · P. S. Shruti · S. Abraham
Department of Computer Science and Engineering,
College of Engineering Guindy, Chennai, India
e-mail: vetri@annauniv.edu; kalvivetri@gmail.com

P. S. Shruti
e-mail: shrutisekaranp@gmail.com

S. Abraham
e-mail: susan.abraham199423@gmail.com

A. Kumar and S. Mozar (eds.), *ICCCE 2018*,
Lecture Notes in Electrical Engineering 500,
https://doi.org/10.1007/978-981-13-0212-1_47

1 Introduction

SDN, the new approach to networking, is completely redefining the arena, giving it new expectations in terms of agility and flexibility. This dynamic architecture is ideal for a wide range of today's applications, due to its very design. The architecture decouples the control plane from the data plane, giving the centralized controller absolute knowledge of the network, including where the hosts connect to the network and what the network topology connecting all of the hosts together looks like. Such an omniscient central controller acting as the brains of the network, allows network engineers to implement unique, adaptable forwarding policies as well as allowing them to devise a number of interesting network applications, such as load balancing, WAN management, and network monitoring. An overview of SDN's architecture and related security issues has been presented by Scott-Hayward et al. [1].

In traditional networks, devices residing at all planes have some control over forwarding decisions. Though this seems more flexible, separating the control plane from the data plane, as is done in SDN has numerous advantages. It enables the controller to allow or block specific packets or flows, assign priorities, and decide the path of packet flow in the network, etc. The controller, which is the control plane device and the switches which are the data plane devices share a master-slave relationship. The controller sends instructions to the various switches, telling them how to handle a particular packet or flow, and the switches execute these instructions, acting as simple forwarding devices.

SDN's myriad benefits make it ideal for emerging technologies such as the IoT, and cloud computing as well as for improving existing applications. There are few in the networking community who have escaped the impact of SDN. However, SDN's main benefits, i.e. control logic centralization and network programmability, introduce new fault and attack planes. As more and more SDN devices and systems hit the market, security in SDN remains a cause for concern. SDN as well as traditional networks, suffer from a diverse range of attacks such as DoS, viruses, IP spoofing, etc. Due to the major differences in architecture, SDN has additional security issues, and it is also difficult to apply the solutions created for traditional networks in SDN. As the controller is the "brain" of the network, attacks on and vulnerabilities in controllers are probably the most severe threats to SDN architecture.

This project addresses such issues by building an efficient, effective IDS for an SDN environment. IDS come in a variety of "flavors" and approach the goal of detecting suspicious traffic in different ways, one of which has been described by Sayeed et al. [2]. In this project, we propose to build a two-stage IDS based on machine learning. The first stage, deployed in the switches extracts information and detects attacks, whereas the second stage using ID3 and a genetic algorithm, is deployed in the controller to further identify the category of the attacks. This is based on the work of Sarvari et al. [3]. Such a system can work robustly in the detection of threats and provides additional information to launch countermeasures, all the while balancing the load on the controller and maintaining the efficiency of the network.

The remainder of the paper is organized as follows. Section 2 discusses the related work of this project. Section 3 deals with the workflow of the proposed system, a detailed architecture diagram, building blocks, the algorithms used in each module, and detailed design diagrams of the individual modules. Section 4 describes the implementation details, test cases and their results, and evaluation metrics for the proposed system, along with a comparison of the proposed resource provisioning system performance with existing systems. The conclusion and future work are discussed in Chap. 5.

2 Related Work

SDN is becoming one of the most important technologies in the market. To give a brief picture about what SDN does and who benefits from it, a comprehensive survey of the research related to the security of SDN was presented by Scott-Hayward et al. [1] which discusses both the security enhancements to be derived from using the SDN framework and the security challenges introduced by the framework.

The main focus of SDN is to separate the control plane and data plane in order to make networks programmable and scalable. To make the networking environment flexible and simple, OpenFlow is used in SDN. However, in the current scenario, security is greatly underexploited. An intrusion detection mechanism for OpenFlow-based SDN has been described by Sayeed et al. [2]. The study focuses on developing a packet filtering firewall over a SDN controller, namely floodlight, and the application of association rules to find patterns among the data passing through the firewall. The patterns recorded serve as the motivation behind the development of an anomaly-based intrusion detection mechanism.

In SDN the major advantage is the provision of central control over the network. But if it is made unreachable by a distributed denial of service (DDoS) Attack, the entire network crashes. Using the central control of SDN an efficient method is proposed to detect attacks, by Mousavi and St-Hilaire [4]. It also shows how DDoS attacks can exhaust controller resources and provides a solution to detect such attacks based on the entropy variation of the destination IP address. This method is able to detect DDoS within the first five hundred packets of the attack traffic.

The complexity of network management and network security is continually increasing. To overcome this, SDN emerged, which has the potential to replace the existing network infrastructure. To address the lack of security in SDN, a signature-based firewall and a statistical-based network intrusion detection system has been proposed by Mantur et al. [5] which has an easy to use interface. Mininet has been used for the creation of network topology and an opendaylight controller for controlling the topology. A centralized firewall and network intrusion detection system in SDN (CFNIDSS) has been created, which is used to provide security to SDN from various types of attacks. Various experiments have been performed by sending different types of malformed packets and CFNIDSS which consist signature-based firewall and statistical-based NIDS has been successfully tested.

The different features in the attack on a network eat up time in terms of training and prediction setups leading to higher dimensionality of data that has to be analyzed for detecting attacks. This remains one of the key issues concerning IDS. A hybrid method of a support vector machine (SVM) and a genetic algorithm (GA) has been suggested by Sarvari et al. [3]. These methods are used for reducing the number of features from 41 to 11 using the KDD Cup'99 dataset. Using the GA the features are classified as three priorities. The feature distribution is done in such a way that four features are placed in the first priority, five in the second, and two in the third.

There are several machine-learning paradigms including neural networks, linear genetic programming (LGP), support vector machines (SVM), and fuzzy inference systems (FISs) etc., which have been explored for the design of IDS. A hybrid method of C5.0 and SVM used to investigate and evaluate the performance of their proposed method with a DARPA dataset has been developed by Golmah [6].

The performance of the hidden markov model (HMM) and SVM has been analysed by Jain and Abouzakhar [7] for anomaly intrusion detection. The proposed techniques discriminate between normal and abnormal behavior of network traffic. The specific focus of this study is to investigate and identify distinguishable TCP services that comprise of both normal and abnormal types of TCP packets, using a J48 decision tree algorithm.

Some of the existing works in the various fields of SDN, which have been instrumental in the development of this system have been described. These works not only gave us a deeper insight, but also provided us with a broad, intensive perspective of their respective areas. The following section deals with the detailed design and the architecture of the proposed system.

3 Proposed System

The overall architecture of the proposed system is illustrated in Fig. 1. The system comprises the SDN controller in the control plane, the switches in the data plane, and various hosts. The anomaly detection and anomaly classification processes reside in the controller as modules, and in the switches as rules.

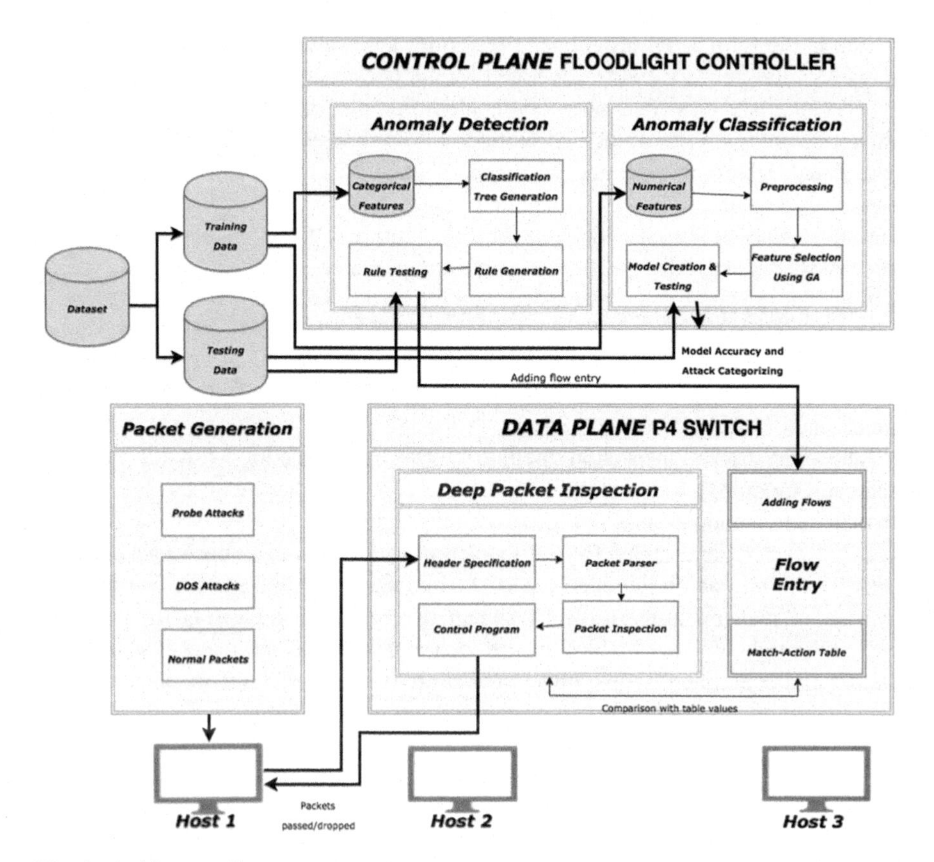

Fig. 1 Architecture diagram

In the depiction of the proposed system, only one controller and one switch are elaborated upon for simplicity and ease of experimentation and evaluation. All the hosts of the network are connected to this P4-based switch. This network can be further scaled up as per requirement. The network traffic first passes through the P4-based switch, which ex-tracts necessary features from the packet and compares these features with rules formulated by the anomaly detection module. These rules determine whether the packet is malicious or normal. If it is found to be normal, it is forwarded to the intended destination. Otherwise, it is sent to the SDN controller for further analysis in the anomaly classification module.

The anomaly detection module is developed based on the principles of ID3 trees. The algorithm takes the categorical features of the dataset, and builds a classification tree using these to determine which values of the features signify an attack. This tree, when parsed gives the detection rules which are then embedded into the switch in the form of P4 programs. These rules are evaluated by the testing dataset as well as the network traffic generated by the packet generation module.

The anomaly classification module identifies the class of attack a malicious packet belongs to. The numerical features of the dataset are used. Each feature is considered a chromosome and each record in the dataset is considered an individual. A genetic algorithm is applied to these individuals to select the most important features for classification. In this algorithm, each individual in the population is evaluated for its fitness value, after which the fittest, selected using tournament selection, are crossed over and mutated to create the next generation of the population. Fifteen such generations are created, after which the chromosomes in the final population indicate the selected features. These selected features are used to build a multi-class classifier, which can categorize each packet based on its attack class.

The deep packet inspection module works by interacting with the anomaly detection module. It extracts the features and compares them with the rules in the switch. This module uses P4 programming language to define the necessary headers, parsers, and tables for the extraction of features from each packet in the network traffic. Each rule generated by the anomaly detection module is translated as a set of match-actions in the P4 switch. Every packet traversing the switch is passed through a series of tables, where each table extracts and matches values of a feature and consequently determines the nature of the packet and decides a course of action.

The packet generation module creates various attacks on the system, in order to test the rules created by the anomaly detection module. Scapy is used as a tool to set the necessary fields to either create an attack or a normal packet. Few of the attacks generated include SYN flood, Land and WinNuke. The created packets are transmitted from a host connected to the P4 switch, which examines and decides to forward or drop them.

4 Anomaly Detection Module

The purpose of this module is to analyse the categorical data and generate rules to determine if a packet is malicious or not. The ID3 classification tree algorithm is used to generate a classification tree. Rules are created by parsing the tree, which shall be tested against the test dataset. These rules will then be deployed in the switches of the network.

Algorithm 1. GenerateTree

```
Input : Categorical data
Output : ID3 Tree

Create root node N
If all instances in D are positive then
return N with label = 0
If all instances in D are negative then
return N with label = 1
If size of A = 1 then
return N with label = most common value of the Y in D
Otherwise
BestAttribute = Attribute in A that best classifies in-
stances in D
  Decision Tree attribute for N = BestAttribute
  for each possible value vi of BestAttribute do
    Add a new tree branch below N, correspomding to the
test BestAttribute= vi
    Let D(vi) be the subset of instances in D that have
the value vi for BestAttribute
    If D(vi) is empty then
      below this new branch add a leaf node with label =
most common target value of Y in D
    else below this new branch add the subtree generate-
Tree(D(vi),A-{BestAttribute},Y)
  end for
return N
```

5 Anomaly Classification Module

The purpose of this module is to analyse the numerical data and classify the attacks detected. Feature selection is performed to determine the most important features using a genetic algorithm. The initial population generated from numerical features undergoes fitness evaluation, mutation, and crossover after which the best individuals are selected as the next generation of the population. This is repeated to produce 15 generations. Using 10-fold cross validation, the accuracy of the model is examined for 15 generations.

Algorithm 2. FeatureSelection

```
Input : Numerical features extracted from dataset
Output : Selected subset of features

Initialize population X
Initialize database D
for each chromosome Xi do
  k = evaluateFitness(Xi)
  Store Xi and k in D
end for
Sort D on Fitness score
Generate possible combinations C of pairs of top 4 chro-
mosomes
for each combination Ci in C do
  create chromosome temp1 and temp2
  temp1 = performCrossover(Ci)
  temp2 = performMutation(temp1)
  k = evaluateFitness(temp2)
  store temp2 and k in D
end for
Sort D on Fitness score
Generate new population X
if termination condition is not met repeat step 3
return first record in D
```

6 Packet Generation

We generate the following attacks to test the proposed system.

- A SYN Flood is a form of DoS attack in which an attacker sends a succession of SYN requests to a target's system in an attempt to consume enough server resources to make the system unresponsive to legitimate traffic.
- A WinNuke is a DoS attack where the URG flag in the packet is set and sent to specific ports. This can cause fragment overlapping and lead to OS crashes.
- A port scan attack occurs when an attacker sends packets to a particular IP address, varying the destination port. Such probes can give the attacker an idea about which services are being used, and which ports are available to stage further attacks.

Other attacks which are demonstrated are violations of the TCP protocol by misuse of the flags set.

- SYN FIN flags set: The SYN flag indicates the opening of a TCP connection while the FIN indicates its close. So in no scenario can both flags be set simultaneously.
- Only FIN flag set: As the FIN flag indicates closing a TCP connection, it's usually accompanied by the ACK flag as well. With just the FIN flag it becomes a violation.
- No flag bits set: TCP Protocol specification does not allow that no flag-bits are set. This abnormal situation may lead to replies which reveal information about the operation system.

7 Deep Packet Inspection

The general mechanism of extraction of packets and the flow of a P4 program is explained below. First the control program decides the overall flow, i.e. they determine the series of match action tables through which details of the packet must be compared against. Each table requires extraction of certain features. This extraction is performed by the parsers. The parsers extract all fields for a particular value as defined in the header program. Finally, after matching the extracted values in the table, an action is decided and performed as defined in the action program. Possible actions are modifying the fields, adding a value to a field or dropping the entire packet. A fraction of the program used in this system is shown. As the selected features in the first module are protocol and service flag, these are extracted from each packet. Headers are defined for all three separately and the fields are extracted in different parsers in order of their levels, i.e. service, given by port is level 5, and protocol and flags are level 4. The rules generated by the tree are translated as match action tables as required by P4. Hence, based on an exact match of these values by the match action tables, the P4 switch either drops or forwards the packet to the appropriate destination.

The overall architecture of the system, its various modules and their interaction with each other has been depicted. Each section in the above section describes the various algorithms designed and the intermediate output of each module. Following this, the final implementation of the system, the various tools used, and the intermediate results are detailed in the next section.

Sample Code. Deep Packet Inspection

```
Input : Network packet
Output : Packet dropped / forwarded

Header
header_type tcp_t
{
     fields
     {
          srcPort : 16;
          dstPort : 16;
          seqNo : 32;
          ackNo : 32;
          dataOffset : 4;
          res : 3;
          ecn : 3;
          ctrl : 6;
          window : 16;
          checksum : 16;
          urgentPtr : 16;
     }
}     header tcp_t tcp;

Parser
parser parse_tcp
{
     extract(tcp);
     return ingress;
}

Table
table http_pac
{
     reads
     {
          tcp.dstPort : exact;
     }
     actions
     {
          route;
          _drop;
     }
}

Control
control ingress
{
     apply(count_table);
     apply(ipv4_lpm);
     apply(forward);
     apply(http_pac);
     apply(app_packet);
}

Action
action route()
{
     add_to_field(ipv4.ttl, -1);
}
```

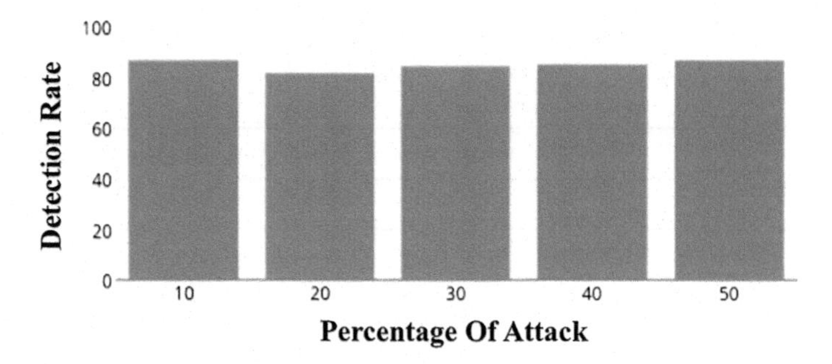

Fig. 2 Detection rate for various percentages of attack

8 Experimental Results

The proposed system is implemented in a small-scale simulation environment using several tools such as mininet, floodlight, eclipse, scapy, weka, and P4. In this project several metrics were selected for analyzing the performance of the IDS.

The rate of correct detection for varying percentages of attack packets can be seen in Fig. 2. Values found for three iterations have been averaged for each percentage of attack packets. It can be concluded from the graph that the detection rate is better for this model.

The misclassification rate for varying percentages of attack packets can be seen in Fig. 3. Values found for the three iterations have been averaged for each percentage of attack packets. It can be concluded from the graph that though the misclassification rate also increases slightly due to an increase in the number of attacks in the network traffic, it is better for this model.

The performance of IDS in terms of accuracy of the model can be visualized in Fig. 4. This graph shows that the model detects attacks at a better rate with a good level of accuracy. The precision of the model for varying percentages of attack can be seen in Fig. 5. From the graph, it can be concluded that the IDS has good precision with a small number of wrong predictions.

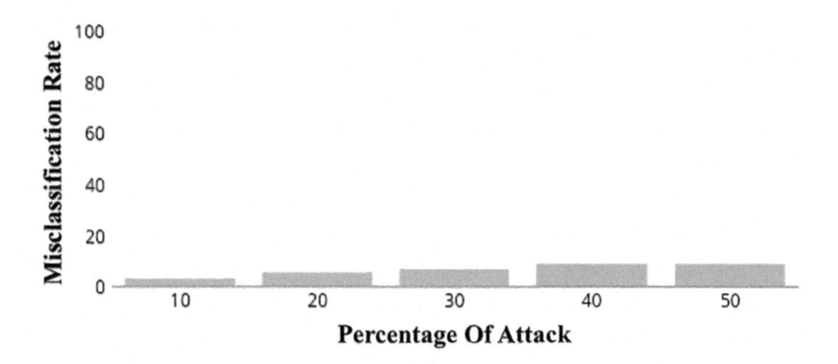

Fig. 3 Misclassification rate for various percentages of attack

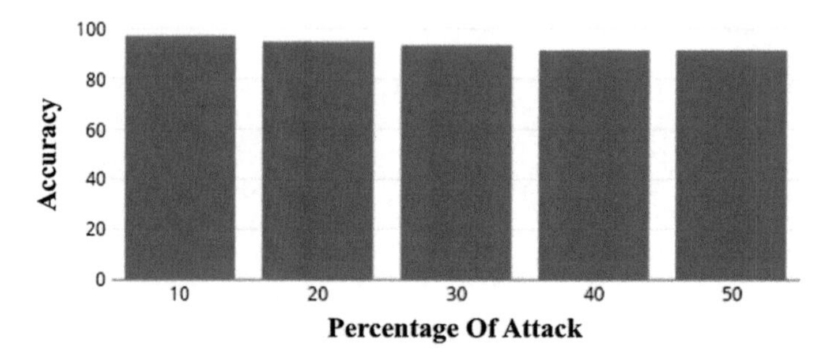

Fig. 4 Accuracy for various percentages of attack

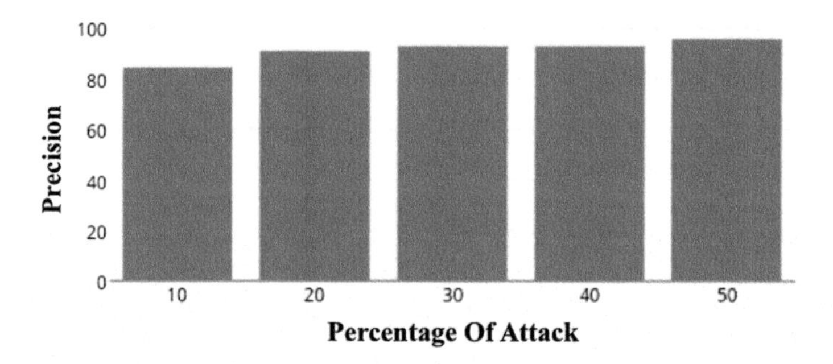

Fig. 5 Precision for various percentages of attack

9 Conclusion

In this machine-learning approach, several ideas were introduced for better development of an IDS to exploit SDN capabilities in order to secure the SDN environment. While a deep evaluation is currently in progress, preliminary results are promising, proving that there is a chance to build an effective IDS with extraction of limited features from the packets. Before extending this work to include more classes of attack, as well as increasing the real-time accuracy, more sophisticated techniques must be considered. For example, in the case of numerical features, their extraction instead of selection may prove to better distinguish the patterns of various attack classes. Therefore, there remains scope for many such extensions and enchantments in this area.

References

1. O'Callaghan G, Scott-Hayward S, Sezer S (2013) SDN security: a survey. In: IEEE SDN for future networks and services (SDN4FNS), Nov 11–13, pp 1–7. IEEE

2. Sayeed A, Sayeed MA, Saxena S (2015) Intrusion detection system based on software defined network firewall. In: 1st international conference on next generation computing technologies (NGCT), Sept 4–5. Dehradun
3. Ahmad I, Barati M, Muda Z, Sarvari S (2015) GA and SVM algorithms for selection of hybrid feature in intrusion detection system. Int Rev Comput Softw (I.RE.CO.S.) 10(3):265–270
4. Mousavi SM, St-Hilaire M (2015) Early detection of DDoS attacks against SDN controllers. In: International conference on computing, networking and communications, Feb 16–19. California
5. Mantur B, Desai A, Nagegowda KS (2015) Centralized control signature-based firewall and statistical-based network intrusion detection system (NIDS) in software defined networks (SDN). Emerg Res Comput Inf Commun Appl 497–506
6. Golmah V (2014) An efficient hybrid intrusion detection system based on C5.0 and SVM. Int J Database Theory Appl 7(2):59–70
7. Abouzakhar NS, Jain R (2013) A comparative study of hidden Markov model and support vector machine in anomaly intrusion detection. J Internet Technol Secured Trans (JITST) 2 (3):607–615

Geometric Programming-Based Automation of Floorplanning in ASIC Physical Design

N. Bala Dastagiri, K. Hari Kishore, Vinit Kumar Gunjan, M. Janga Reddy and S. Fahimuddin

Abstract The ASIC physical design process is a complex optimization problem with various objectives such as minimum chip minimum wire length, area, minimum of vias. The main aims of optimization are to improve the performance and reliability etc., of the ASIC design process. The objectives mentioned can be achieved through the effective implementation of floorplanning before other steps are implemented. In this study, a pseudo code is developed for floorplanning using geometric programming to achieve global optima. This study uses simulations performed using MATLAB GGP toolbox.

Keywords ASIC physical design · Floorplanning · Geometric programming

1 Introduction

ASIC is an acronym which emerged in the late 1970s and stands for Application Specific Integrated Circuit. There has been tremendous development in the field of ASIC design accompanied by a wide variety of design styles. The continuing increase in the transistor count of VLSI chips has encouraged ASIC developments which have brought about new architectures. The design flow of ASIC is demonstrated in Fig. 1. Steps 1–4 constitute logical design and steps 4–9 show the physical design flow. The process of physical design starts from deciding the locations of logical blocks, the placing of logical cells inside logical blocks, and their interconnections [1].

N. Bala Dastagiri (✉) · K. Hari Kishore
Koneru Lakshamaiah Education Foundation, Guntur, Andra Pradesh, India
e-mail: baluece414@gmail.com

V. K. Gunjan · M. Janga Reddy
CMR Institute of Technology, Hyderabad, Telangana, India

S. Fahimuddin
AITS, Rajampet, Andra Pradesh, India

A. Kumar and S. Mozar (eds.), *ICCCE 2018*,
Lecture Notes in Electrical Engineering 500,
https://doi.org/10.1007/978-981-13-0212-1_48

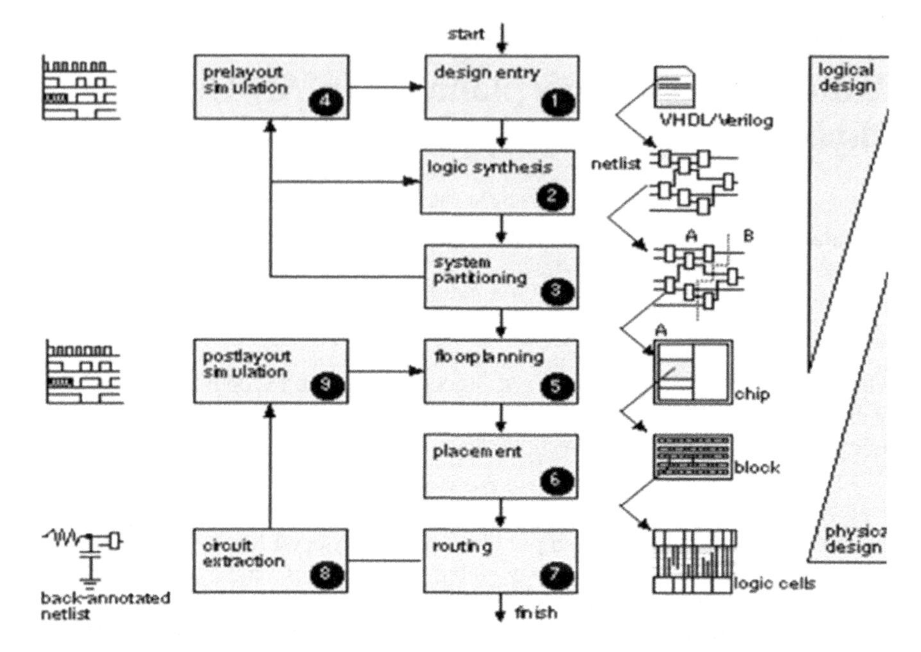

Fig. 1 ASIC design flow

The goals of various steps involved in physical design are as follows:

Floorplanning: calculate the sizes of all the blocks and assign them locations.
Placement: assign the interconnecting areas and the location of all the logic cells within the flexible blocks.
Routing: determine the location of all interconnections and completely route all of them.

Among the steps involved in physical design, floorplanning plays a crucial role as it can able to evaluate the performance characteristics of ASIC and thus provide a shorter delay and a higher fanout. With the effective implementation of floorplanning the designer can also reduce the cost.

In order to optimize the physical design process there are different automation algorithms available for implementing floorpanning in ASIC and they are broadly classified as follows:

- Evolutionary algorithms and metaheuristic optimization algorithms, such as genetic algorithms, particle swarm optimization, and simulated annealing.
- Linearly constrained optimization like integer programming for basic analog cell design.

All the above methods consume a lot of implementation time and they can stuck in local optima because of it uses a simulation tool as a part of the optimum loop [2].

In this study we will be using geometric programming for the design automation of floorplanning.

This work is organized as follows:

- Geometric programming optimization is explained in Sect. 2.
- The design procedure and the pseudo code of the floorplanning problem are explained in Sect. 3.
- In Sect. 4 results are discussed and followed by a conclusion in Sect. 5.

2 Geometric Programming (GP)

Geometric programming-based work started in the 1980s. In recent times GP has been used in the optimization process of physical design. GP is also used to solve non-linear problems [3–8].

Consider x to be a vector $(x_1, x_2 \ldots x_m)$ of m real variables and positive variables.

The GP optimization problem is of the form minimize $f_0(x)$ such that

$$\begin{aligned} f_i(x) &\leq 1,\ i=0,1,2,3 \ldots m \\ g_i(x) &= 1,\ i=0,1,2,3 \ldots p \\ x(i) &> 0,\ i=0,1,2,3 \ldots n. \end{aligned} \tag{2.1}$$

where $f_0(x)$ denotes the objective function and $f_i(x)$ and $g_i(x)$ are inequality constraints respectively. $f_0, \ldots, f_m$ are posynomial functions and $g_1, \ldots g_p$ are monomial functions. The posynomial function is in the form

$$f(x_1, \ldots, x_m) = \sum_{k=1}^{t} c_k x_1^{a_{1k}} x_2^{a_{2k}} \ldots x_m^{a_{mk}} \tag{2.2}$$

where $c_j \geq 0$ and a_{ij} are real variables. In the given example, f is a monomial function. GP is converted into a convex optimization problem by changing the variables and taking the logs of the function. The convex property is liable to provide the guarantee of a globally optimal solution.

The problem in (2.1) is not a global convex optimization problem. Therefore it is converted to a convex problem by applying "log" as $y_i = log\, x_i$ for all $i = 1, 2, \ldots, n$.

For a monomial g defined in (2.1)

$$g(x_1, x_2, \ldots, x_n) = g(y_1, y_2, \ldots, y_n) = ce^{y_1\alpha_1} e^{y_2\alpha_2} \ldots e^{y_n\alpha_n} = e^{a^{\sim T} y + b} \tag{2.3}$$

where $a^{\sim T} = [\alpha_1, \alpha_2, \ldots, \alpha_n]$ and b,

For a posynomial defined in (2.2),

$$f(x_1, x_2, \ldots, x_n) = f(y_1, y_2, \ldots, y_n) = \sum_{k=1}^{m} c_k e^{y_1 \alpha_{1_k k}} e^{y_2 \alpha_{2k}} \ldots e^{y_n \alpha_{nk}} = \sum_{k=1}^{m} e^{a^{\sim T} y + b_k} \quad (2.4)$$

where $\alpha_k^{\sim T} = [\alpha_{1k}, \alpha_{2k}, \ldots, \alpha_{nk}]$, and $b_k = log\, c_k$ for all k = 0, 1, 2, ..., m.

Therefore GP in (2.1) is rewritten as an optimization of the variable $y \ni R^n$ as minimize

$$\sum_{k=1}^{m} e^{a_{ik}^{\sim T} y + b_{0k}} \text{ such that } \sum_{k=1}^{m} e^{a_{ik}^{\sim T} y + b_{ik}} \leq 1;\ i = 1, 2, \ldots, m \quad (2.5)$$

$e^{a_{ik}^{T} y + b_{ik}} = 1;\ i = 1, \ldots, p$

Applying the logarithms to the objective and constraint functions in (2.5),

$$\begin{aligned} \text{minimize} \quad & f_0(y) = \log\left(\sum_{k=1}^{m} e^{a_{0k}^{T} y + b_{0k}}\right) \\ \text{such that} \quad & f_i(y) = \log\left(\sum_{k=1}^{m} e^{a_{ik}^{T} y + b_{ik}}\right) \leq 1);\ i = 1, \ldots, m \end{aligned} \quad (2.6)$$

where f_i is convex function in their argument y. h_i is an affine function.

Thus, Eq. (2.6) is a convex optimization problem.

The formulation in (2.6) is equivalent to the standard GP in (2.1)

3 Floorplanning Problem

Consider a floorplanning problem as follows: Here the problem is to maximize the volume of a logic block with a height h, depth d, and width w subject to constraints as we need to limit the total area of the logic block to 2(hw + hd), and logic block area as wd, with lower and upper bounds on the aspect ratios h = w and w = d. This leads to the GP as maximize volume hwd subjected to constraints, i.e. 2 (hw + hd) ≤ Albs; wd ≤ Albf; α ≤ h/w ≤ β, γ ≤ d/w ≤ δ.

Where h, w, and d are the given optimization variables, and Albs and Albf are the problem parameters and α, β, γ, δ are the lower and upper limits of the aspect ratios.

The pseudo MATLAB code for the above problem is as follows:

```
% Given problem data
Albs = 1000; Albf = 1000;
alpha = 0.4; beta = 3; gamma = 0.6; delta = 4;
% GP variables in the given problem
gpvar h w d;
% objective function is the logic block
volume = h * w * d;
% set of constraints
constr = [2 * (h * w + h * d) <= Albs; % wall area limit
w * d <= Albf; % floor area limit
% aspect ratios
alpha <= h/w; gamma <= d/w;
h/w <= beta; d/w <= delta;];
% solving the GP
[max_volume, solution, status] = gpsolve (volume, constr, 'max')
% no semicolon after the gpsolve command, so
% maximum volume, solution, and status
assign (solution);
```

This pseudo code is simulated in the MATLAB GGP toolbox. The given code solves the problem of placing logic cells such that they don't overlap and thus the area is minimized. The code provide us with the maximum volume of the logic block and its dimensions so as to accommodate a greater number of logic cells in it.

4 Simulation Results

As the code is simulated using the MATLAB GGP toolbox we get the following dimensions for the logic block as

```
max_volume = 6.8041e+003

solution =

'd' [40.8248]
'h' [8.1650]
'w' [20.4124]

status = Solved
```

In the graph it can be seen that the area is minimized with an increase in aspect ratio and vice versa (Fig. 2).

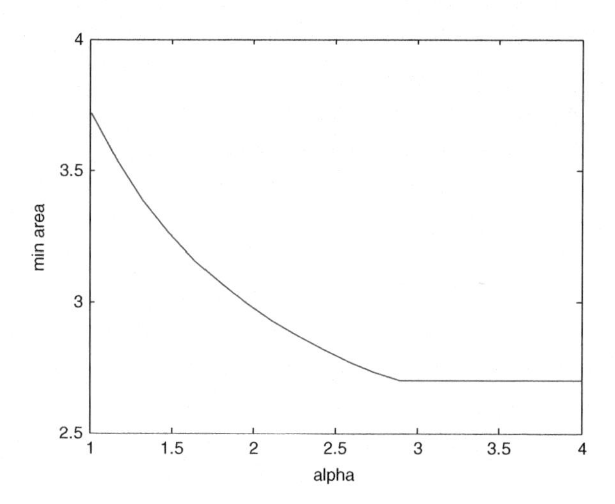

Fig. 2 Min. area versus aspect ratio

5 Conclusion

In this paper we presented GP as a method of solving non-linear problems and developed a pseudo code for floorplanning using GP. The simulation was performed using the MATLAB GGPLAB toolbox and the results plotted. In future we intend to use GP to solve other problems in ASIC design flow and extend it to solve issues in mixed signal CMOS circuits.

References

1. Guo PN, Takahashi T, Cheng C-K, Yoshimura T (2001) Floorplanning using a tree representation. IEEE Trans Comput Aided Des Integr Circuits Syst 20:281–289
2. Shafaghi S, Farokhi F, Sabbaghi-Nadooshan R (2013) New Ant colony algorithm method based on mutation for FPGA placement problem. Smart Electr Eng 2.1. Winter, 2251-9246. F 53–60
3. Boyd S, Kim S-J, Vandenberghe L, Hassibi A (2007) A tutorial on geometric programming. Springer Science + Business Media, pp 67–127
4. Posser G, Flach G, Wilke G, Reis R (2011) Gate sizing minimizing delay and area. In: IEEE computer society annual symposium on VLSI, ISVLSI 2011. Chennai, India
5. Chen W, Hseih C-T, Pedram M (2000) Simultaneous gate sizing and placement. IEEE Trans Comput Aided Des Integr Circuits Syst 19(2):206–214
6. Chen Y-C, Li Y (2010) Temperature-aware floorplanning via geometric programming. Math Comput Model 51(7-8):927–934
7. Boyd S, Kim S-J, Patil D, Horowitz M (2005) Digital circuit optimization via geometric programming. Oper Res 53(6):899–932
8. Chu C, Wong D (2001) VLSI circuit performance optimization by geometric programming. Ann Oper Res 105:37–60

Design of a Power Efficient ALU Using Reversible Logic Gates

B. Abdul Rahim, B. Dhananjaya, S. Fahimuddin and N. Bala Dastagiri

Abstract Today's VLSI design technology is moving very quickly into low power, high speed and micro areas of development. Reversible logic has played an important role in this, notably in quantum computing and DNA computing, and presently moving into optical computing also. It is also found that under some ideal conditions it can produce zero power dissipation. A known fact that an arithmetic logic unit (ALU) is one of the core components of a the CPU in a computer. The design of an ALU using different reversible logic gates is proposed. The proposed reversible logic-based ALU is implemented using a Mentor Graphics tool in 130 nm technology for power efficiency. The power dissipation of two proposed ALU designs and a conventional area-based ALU have been compared. The conventional ALU dissipates the power 10% reversible logic-based ALU.

Keywords Quantum computing · Arithmetic logic unit · CMOS 130 nm technology

1 Introduction

Nowadays reversible logic has become attractive and popular because of its inherent ability to reduce the consumption of power, particularly in quantum computing machines in the field of low power VLSI circuits. A reversible circuit input is a unique regain from its output. The indispensible value of reversible computing is its electric charge on a storage cell. In which case it can be recyclable through reversible computing, which can reduce energy consumption. This logic is

B. Abdul Rahim (✉) · S. Fahimuddin · N. Bala Dastagiri
Department of Electronics and Communication Engineering,
Annamacharya Institute of Technology and Sciences, Rajampet, A.P., India
e-mail: abdulrahimbepar@gmail.com

B. Dhananjaya
Department of Electronics and Communication Engineering,
Bheema Institute of Technology and Sciences, Adoni, A.P., India

A. Kumar and S. Mozar (eds.), *ICCCE 2018*,
Lecture Notes in Electrical Engineering 500,
https://doi.org/10.1007/978-981-13-0212-1_49

also useful in mechanical applications of nanotechnology. Reversible circuits also improve speed performance and lead to improvements in energy efficiency [1].

Basically, a reversible logic gate is an n-input n-output logic gate. It consists of equal number of inputs and outputs. These gates are very helpful in differentiating the outputs from the inputs and also the inputs can be uniquely from the outputs. This concept is also known as one-to-one mapping. These circuits can be synthesized. Fan-out in reversible circuits can be achieved using additional gates. Any reversible circuits should be implemented with fewer reversible gates. The controlled V− and V+ gates are basic 2 × 2 reversible logic gates.

An arithmetic logic unit (ALU) is a digital electronic circuit that performs both arithmetic as well as logical operations. An ALU is the heart of a computer processor, working as a data processing unit. This allows the computer to execute arithmetic operations and logical operations. It is a combinational circuit that can have one or more inputs, but only one output. The design and function of an ALU may vary between different processors [2–4].

2 Implementation of Different Reversible Logic Gates in CMOS

2.1 Feynman Gate

A Feynman gate is a fundamental reversible gate also known as controlled NOT gate [9]. It can also be used for fan-out applications. A, B and P, Q are the inputs and outputs of a Feynman gate. The quantum cost for this gate is one. Figure 1 shows the logical representation of a Feyman gate and Fig. 2 is the CMOS representation of the gate. Table 1 shows the corresponding truth table of the gate.

2.2 Fredkin Gate

The Fredkin gate is a reversible 3 × 3 logic gate where I(A, B, C) and O(P, Q, R) are the respective inputs and outputs of the gate [10]. The relationship between the outputs and inputs is represented as $P = A$, $Q = A'B \oplus AC$ and $R = A'C \oplus AB$. It is also defined as a controlled swap gate. The quantum cost for a Fredkin gate is 5.

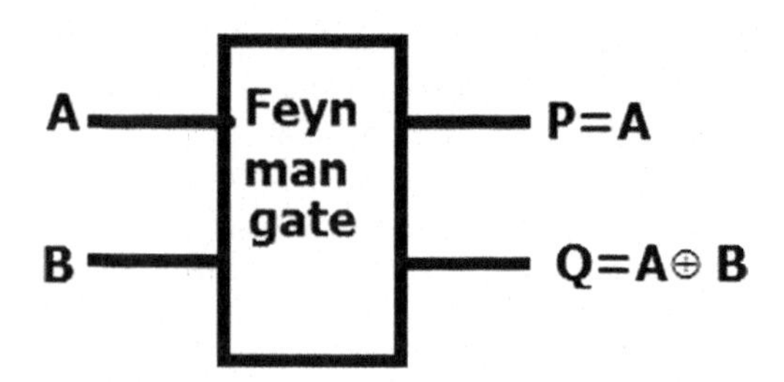

Fig. 1 Feynman gate block diagram

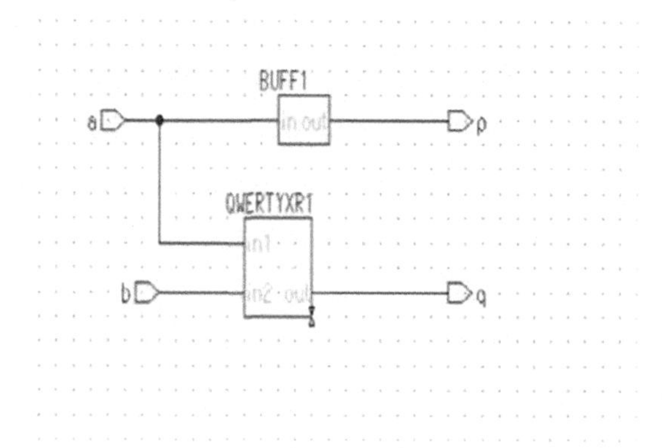

Fig. 2 Implementation of a CMOS Feynman gate

Table 1 Feynman gate truth table

'A'	'B'	'P'	'Q'
'0'	'0'	**'0'**	**'0'**
'0'	'1'	**'0'**	**'1'**
'1'	'0'	**'1'**	**'1'**
'1'	'1'	**'1'**	**'0'**

Figure 3 shows the block diagram and Fig. 4 shows its corresponding CMOS implementation of a Fredkin gate. The truth table of this gate is shown in Table 2.

2.3 MRG Gate

An MRG gate is a 4 × 4 programmable reversible gate where I(A, B, C, D), and O (P, Q, R, S) are the respective inputs and outputs of the concerned gate. Morrison and Ranganathan designed this gate [11]. The output and input relationship is as shown as $P = A$, $R = (A \oplus B) \oplus C$, $Q = (A \oplus B)$, and $S = (AB \oplus D) \oplus ((A \oplus B) \oplus C)$. The quantum cost for an MRG gate is 6.

Figure 4 shows the block diagram and Fig. 5 shows its corresponding CMOS implementation of a MRG gate. The truth table of this gate is shown in Table 3 (Fig. 6).

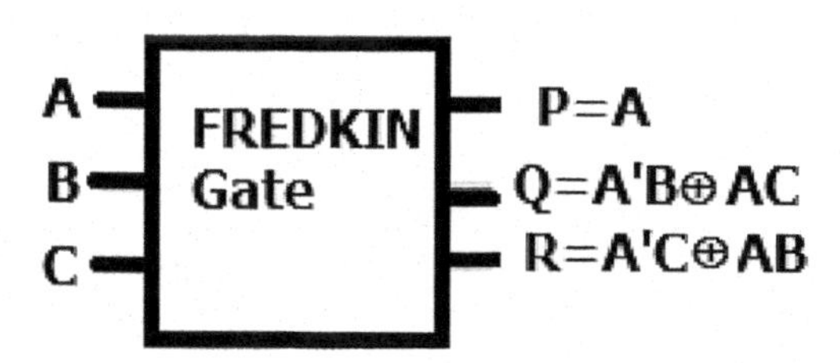

Fig. 3 Fredkin gate block diagram

Fig. 4 CMOS implementation of a Fredkin gate

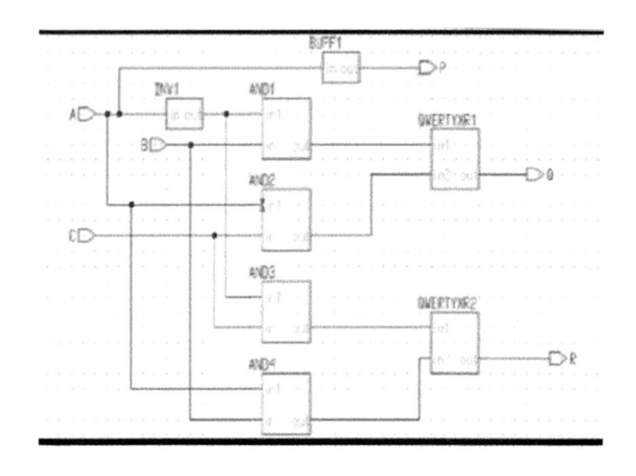

Table 2 Fredkin gate truth table

'A'	'B'	'C'	'P'	'Q'	'R'
'0'	'0'	'0'	**'0'**	**'0'**	**'0'**
'0'	'0'	'1'	**'0'**	**'0'**	**'1'**
'0'	'1'	'0'	**'0'**	**'1'**	**'0'**
'0'	'1'	'1'	**'0'**	**'1'**	**'1'**
'1'	'0'	'0'	**'1'**	**'0'**	**'0'**
'1'	'0'	'1'	**'1'**	**'1'**	**'0'**
'1'	'1'	'0'	**'1'**	**'0'**	**'1'**
'1'	'1'	'1'	**'1'**	**'1'**	**'1'**

2.4 HNG Gate

HNG gates were designed by Hagparast et al. [2]. The relation between the inputs and outputs in an HNG gate is defined as $P = A$, $R = (A \oplus B) \oplus C$, $Q = B$, and $S = (A \oplus B)\ C \oplus (AB \oplus D)$. The quantum cost for this gate is also 6.

Figure 7 shows the block diagram and Fig. 8 shows its corresponding CMOS implementation of an HNG gate. The truth table of this gate is shown in Table 4.

2.5 PAOG Gate

The Peres And Or gate is an extension of a Peres gate [6]. This is a 4 input 4 output gate. The input and output vectors are I(A, B, C, D), and O(P, Q, R, S) respectively.

Fig. 5 MRG gate

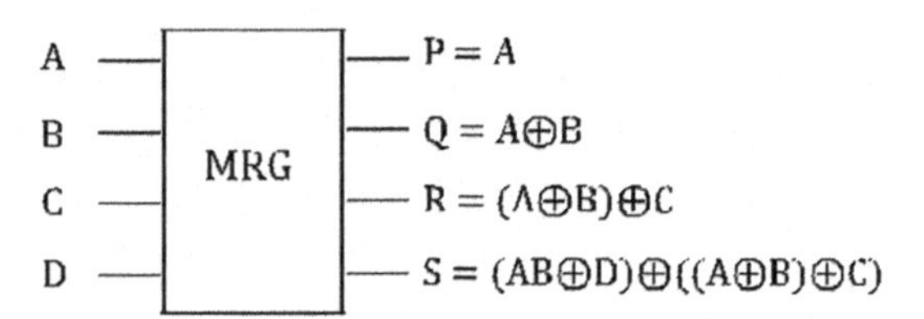

Table 3 MRG gate truth table

A	B	C	D	P	Q	R	S
0	0	0	0	**0**	**0**	**0**	**0**
0	0	0	1	**0**	**0**	**0**	**1**
0	0	1	0	**0**	**0**	**1**	**1**
0	0	1	1	**0**	**0**	**1**	**0**
0	1	0	0	**0**	**1**	**1**	**1**
0	1	0	1	**0**	**1**	**1**	**0**
0	1	1	0	**0**	**1**	**0**	**0**
0	1	1	1	**0**	**1**	**0**	**1**
1	0	0	0	**1**	**1**	**1**	**1**
1	0	0	1	**1**	**1**	**1**	**0**
1	0	1	0	**1**	**1**	**0**	**0**
1	0	1	1	**1**	**1**	**0**	**1**
1	1	0	0	**1**	**0**	**0**	**1**
1	1	0	1	**1**	**0**	**0**	**0**
1	1	1	0	**1**	**0**	**1**	**0**
1	1	1	1	**1**	**0**	**1**	**1**

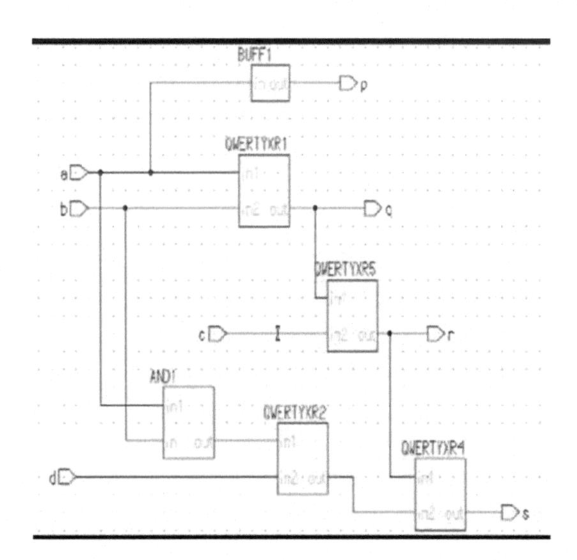

Fig. 6 CMOS implementation of a MRG gate

The relation between inputs and outputs of the concerned gates are P = A, R = AB ⊕ C, Q = A ⊕ B, and S = ((A ⊕ B) ⊕ D) ⊕ (AB ⊕ C).

Figure 9 shows the block diagram and Fig. 10 shows its corresponding CMOS implementation of a PAOG gate. The truth table of this gate is shown in Table 5.

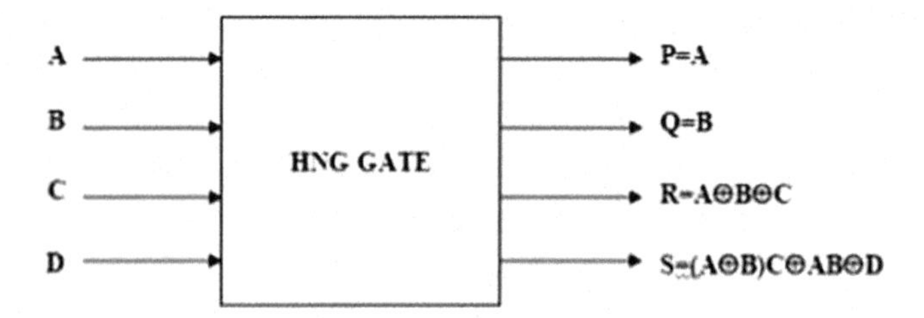

Fig. 7 HNG gate block diagram

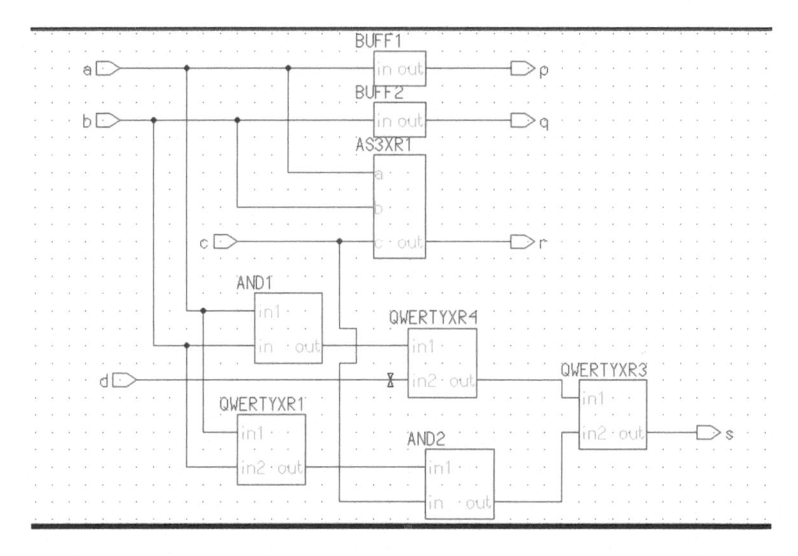

Fig. 8 CMOS implementation of an HNG gate

Table 4 HNG gate truth table

'A'	'B'	'C'	'D'	'P'	'Q'	'R'	'S'
'0'	'0'	'0'	'0'	**'0'**	**'0'**	**'0'**	**'0'**
'0'	'0'	'0'	'1'	**'0'**	**'0'**	**'0'**	**'1'**
'0'	'0'	'1'	'0'	**'0'**	**'0'**	**'1'**	**'0'**
'0'	'0'	'1'	'1'	**'0'**	**'0'**	**'1'**	**'1'**
'0'	'1'	'0'	'0'	**'0'**	**'1'**	**'1'**	**'0'**
'0'	'1'	'0'	'1'	**'0'**	**'1'**	**'1'**	**'1'**
'0'	'1'	'1'	'0'	**'0'**	**'1'**	**'0'**	**'1'**
'0'	'1'	'1'	'1'	**'0'**	**'1'**	**'0'**	**'0'**
'1'	'0'	'0'	'0'	**'1'**	**'0'**	**'1'**	**'0'**
'1'	'0'	'0'	'1'	**'1'**	**'0'**	**'1'**	**'1'**
'1'	'0'	'1'	'0'	**'1'**	**'0'**	**'0'**	**'1'**
'1'	'0'	'1'	'1'	**'1'**	**'0'**	**'0'**	**'0'**
'1'	'1'	'0'	'0'	**'1'**	**'1'**	**'1'**	**'1'**

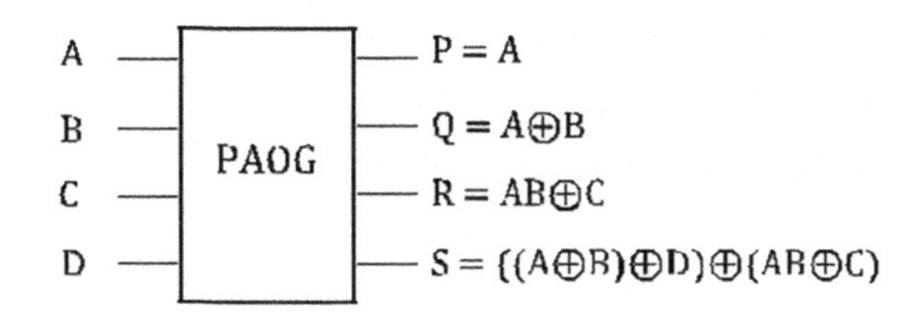

Fig. 9 PAOG gate

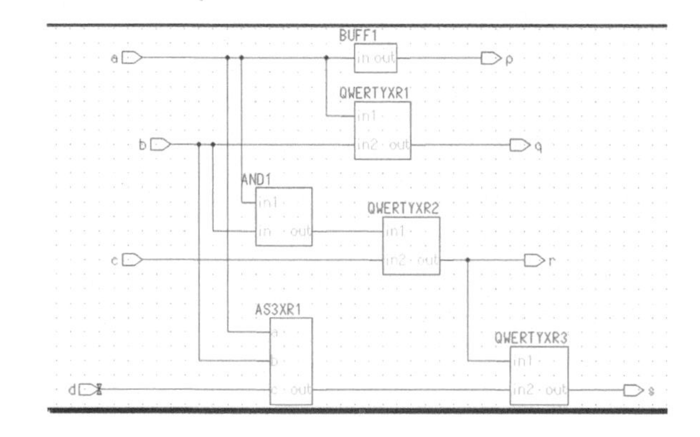

Fig. 10 CMOS implementation of a PAOG gate

Table 5 PAOG gate truth table

'A'	'B'	'C'	'D'	'P'	'Q'	'R'	'S'
'0'	'0'	'0'	'0'	**'0'**	**'0'**	**'0'**	**'0'**
'0'	'0'	'0'	'1'	**'0'**	**'0'**	**'0'**	**'1'**
'0'	'0'	'1'	'0'	**'0'**	**'0'**	**'1'**	**'1'**
'0'	'0'	'1'	'1'	**'0'**	**'0'**	**'1'**	**'0'**
'0'	'1'	'0'	'0'	**'0'**	**'1'**	**'0'**	**'1'**
'0'	'1'	'0'	'1'	**'0'**	**'1'**	**'0'**	**'0'**
'0'	'1'	'1'	'0'	**'0'**	**'1'**	**'1'**	**'0'**
'0'	'1'	'1'	'1'	**'0'**	**'1'**	**'1'**	**'1'**
'1'	'0'	'0'	'0'	**'1'**	**'1'**	**'0'**	**'1'**
'1'	'0'	'0'	'1'	**'1'**	**'1'**	**'0'**	**'0'**
'1'	'0'	'1'	'0'	**'1'**	**'1'**	**'1'**	**'0'**
'1'	'0'	'1'	'1'	**'1'**	**'1'**	**'1'**	**'1'**
'1'	'1'	'0'	'0'	**'1'**	**'0'**	**'1'**	**'1'**
'1'	'1'	'0'	'1'	**'1'**	**'0'**	**'1'**	**'0'**
'1'	'1'	'1'	'0'	**'1'**	**'0'**	**'0'**	**'0'**
'1'	'1'	'1'	'1'	**'1'**	**'0'**	**'0'**	**'1'**

3 ALU Design

An ALU is an important component of the Central Processing Unit (CPU). It performs both logical and arithmetic operations. Implementation of ALU using conventional logic gates is shown in Fig. 11.

3.1 Proposed ALU Design

Two 1-bit ALUs are implemented here. One ALU can be implemented using MRG, HNG, Feynman, and Fredkin gates as shown in Fig. 12.

The input for this reversible ALU has 3 data inputs and 5 fixed selection lines. CMOS implementation for this ALU is shown in Fig. 13. This design produces six logical operations: NOR, SUB, XOR, ADD, OR, and XNOR [11].

Another ALU can be implemented using PAOG and HNG gates to produce the same six logical operations [11]. The block diagram for this circuit and its CMOS implementation are shown in Figs. 14 and 15 respectively.

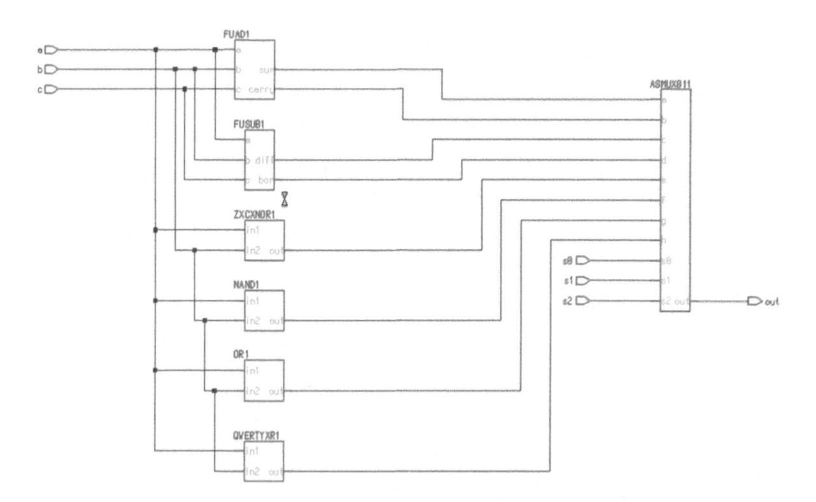

Fig. 11 CMOS implementation of a conventional ALU

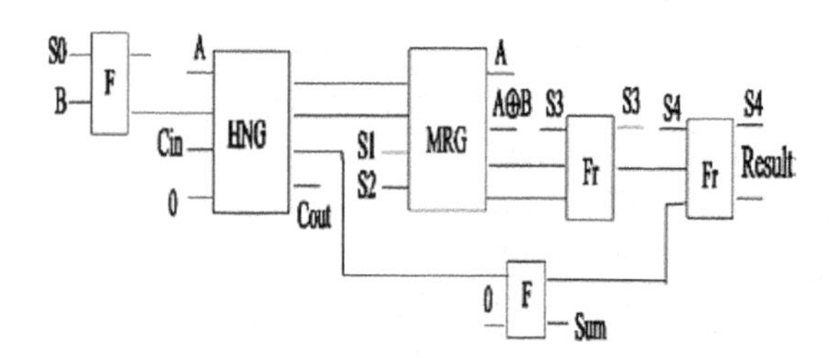

Fig. 12 Design of an ALU using MRG and HNG gates

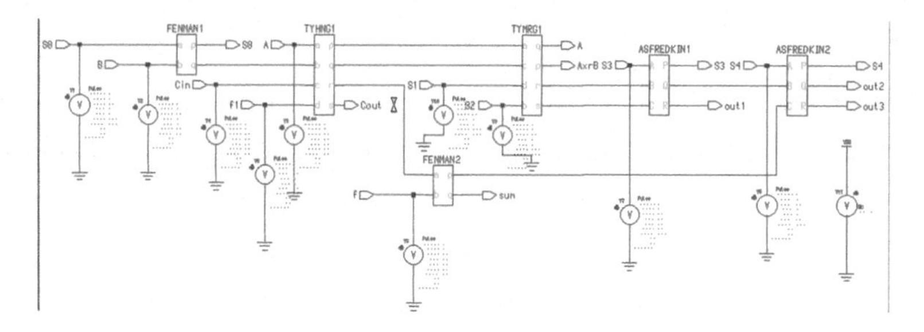

Fig. 13 CMOS implementation of the ALU using HNG and MRG gates

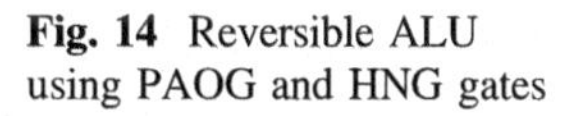
Fig. 14 Reversible ALU using PAOG and HNG gates

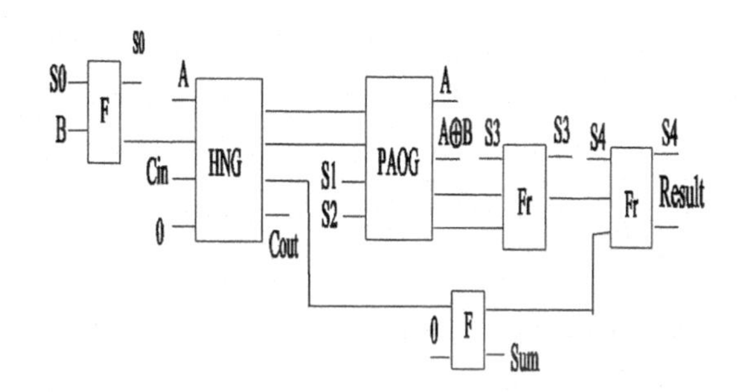

4 Comparison Results

Here both conventional and proposed ALU designs were implemented using a Mentor Graphics tool. The proposed ALU designs dissipate less power than the conventional ALU. The total power dissipation for the conventional ALU is 50.80 nW whereas the proposed ALU designs have a total power dissipation of 45.39 nW and 47.42 nW respectively (Figs. 16 and 17, Table 6).

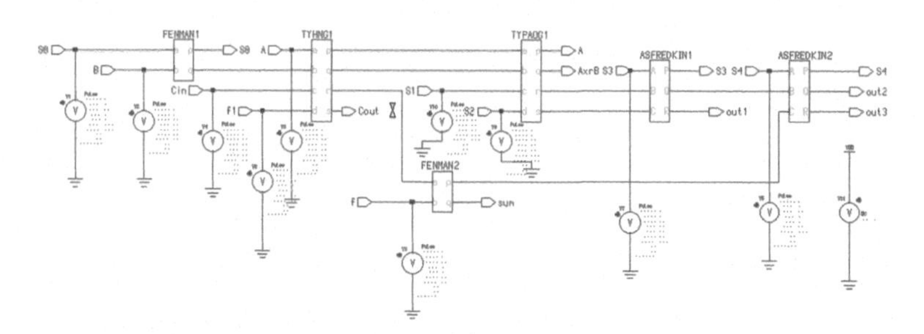

Fig. 15 CMOS implementation of the ALU using HNG and PAOG gates

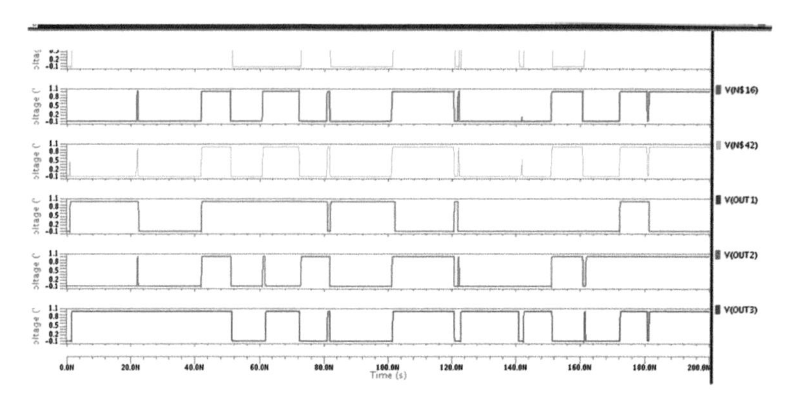

Fig. 16 Transient response of the proposed ALU using PAOG and HNG gates

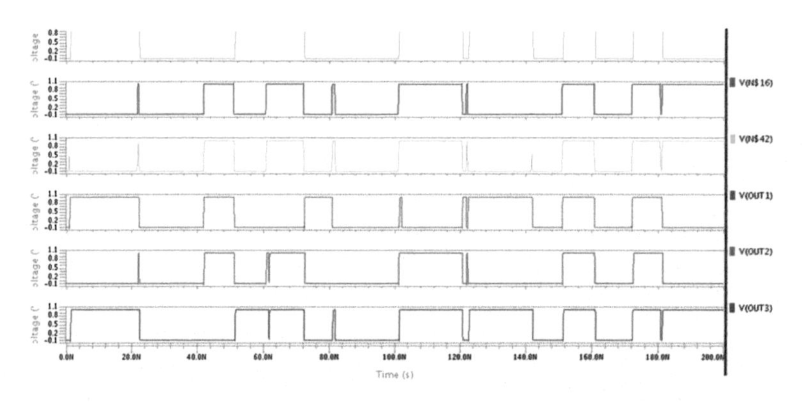

Fig. 17 Transient response of the proposed ALU using HNG and MRG gates

Table 6 Comparison results truth table

Parameter	Conventional ALU	Proposed ALU using HNG and PAOG gates	Proposed ALU using HNG and MRG gates
Total power dissipation (nW)	50.80	45.39	47.42

5 Conclusion

Two types of ALU approaches are represented using different reversible logic gates. The simulation was carried out using a Mentor Graphics Tool with 130 nm technology. The proposed design is compared with the conventional ALU. When compared to both designs, the conventional ALU is less power efficient than the reversible ALU. The conventional design dissipates power as 50.80 nW whereas the proposed reversible ALUs dissipate power as 45.39 and 47.42 nW.

Acknowledgements This work is sponsored and assisted by Annamacharya Institute of Technology and Sciences, Rajampet, India and we are thankful to that organization.

References

1. Mohammadi M, Eshghi M (2009) On figures of merit in reversible and quantum logic designs. Quant Inf Process 8(4):297–318
2. Haghparast M, Jassbi SJ, Navi K, Hashemipour O (2008) Design of a novel reversible multiplier circuit using HNG gate in nanotechnology. World Appl Sci J 3:974–978
3. Nachtigal M, Thapliyal H, Ranganathan N (2010) Design of a reversible single precision floating point multiplier based on operand decomposition. In: Proceedings of 10th IEEE international conference on nanotechnology (To appear)
4. Thapliyal H, Ranganathan N (2010) Design of reversible sequential circuits optimizing quantum cost, delay, and garbage outputs. ACM J Emer Technol Comput Syst
5. Toffoli T (1980) Reversible computing. Technical Report MIT/LCS/TM-151
6. Peres A (1985) Reversible logic and quantum computers. Phys Rev 32(6):3266–3276
7. Feynman R (1982) Simulating physics with computers. Int J Theoret Phys
8. Smolin J, Divincenzo D (1996) Five two-bit quantum gates are sufficient to implement the quantum Fredkin gate. Phys Rev A 53:2855–2856
9. Feynman R (1986) Quantum mechanical computers. Found Phys 16(6)
10. Fredkin E, Toffoli T (1980) Conservative logic. Int J Theor Phys 21:219–253
11. Morrison M, Ranganathan N (2011) Design of a reversible ALU based on novel programmable reversible logic gate structures. In: 2011 IEEE computer society annual symposium on VLSI, 978-0-7695-4447-2/11 $26.00 © IEEE

Modelling and Mitigation of Open Challenges in Cognitive Radio Networks Using Game Theory

Poonam Garg and Chander Kumar Nagpal

Abstract Cognitive radio networks (CRNs) are being envisioned as drivers of the next generation of ad hoc wireless networks due to their ability to provide communications resilience in continuously changing environments through the use of dynamic spectrum access. However, the deployment of such networks is hindered by the vulnerabilities that these networks are exposed to. Securing communications while exploiting the flexibilities offered by CRNs still remains a daunting challenge. In this survey, we put forward concerns relating to security, spectrum sensing and management, and resource allocation and performance of CRNs and model mitigation techniques using game theory. Game theory can be a useful tool with its ability to optimize in an environment of conflicting interests. Finally, we discuss the research challenges that must be addressed if CRNs are to become a commercially viable technology.

Keywords Ad hoc CRN · Game theory · Security · Spectrum sensing Trust · Cooperation

1 Introduction

The rapidly growing volume of data transfer in wireless networks has rendered spectrum availability as a critical issue, making flexible spectrum utilization a mandatory requirement. To tackle the issue, the concept of CRNs was proposed [1] wherein unlicensed users (referred to as secondary users or SUs) can use the available vacant spectrum not currently being used by licensed users (referred to as primary users or PUs) provided they don't cause any interference in the environment [2, 3]. Sensing of the available vacant spectrum is a major activity done by

P. Garg (✉) · C. K. Nagpal
YMCA University of Science and Technology, Faridabad, India
e-mail: poonamgarg1984@gmail.com

C. K. Nagpal
e-mail: nagpalckumar@rediffmail.com

A. Kumar and S. Mozar (eds.), *ICCCE 2018*,
Lecture Notes in Electrical Engineering 500,
https://doi.org/10.1007/978-981-13-0212-1_50

SUs involved in the working of CRNs. Conventionally CRNs are dependent upon the information gathered by other SUs to ensure the accuracy of spectrum sensing making them vulnerable to security attacks and leading to the need for security mechanisms like cryptography and trust. Various proposals based on cryptography and trust exist in the literature. However, there are still various open issues related to security, spectrum sensing and management, resource allocation, and performance of CRNs which need to be addressed.

Organization of paper
The rest of the paper is organized as follows. Section 2 contains an overview of various open challenges in CRN. Section 3 contains an overview about game theory. Section 4 models and mitigates open issues using game theory. In Sect. 5, we put forward remaining challenges which seems difficult to handle by game theory that must be addressed to make cognitive radio networks commercially viable.

2 Open Challenges in Cognitive Radio Networks

In the recent past various research investigations have been made into spectrum management, resource allocation, and security for CRNs, mainly focused on spectrum selection and availability in networks. Gaps still exist in the area of security for CRNs and must be thoroughly investigated.

1. No defined security model: The current literature [4–10] assumes either a hierarchical or distributed security model. There is no quantized security model.
2. No distinguish between selfish and malicious users: Existing security mechanisms uses selfish, malicious and intruder term as same but these adversaries are very are different. Behaviour of nodes changes momentary and to deal with it is a great challenge.
3. No secure mechanisms: CRNs communicate through radio and the flowing information between these radios must be authenticate, authorize and protected. But because of their dynamic behaviour existing mechanisms are not suitable.
4. Improper learning mechanism: Various learning techniques to defend against long-term and short-term attacks are available in literature but no learning technique is appropriate.
5. Dynamic nature of CRN: The dynamic nature of CRN is due to different modulation schemes and working frequencies [11]. The attacker can use these powerful characteristics to hinder the working of CRN [11].It is still an open issue.
6. Asynchronous sensing: Cooperation of SU is by its wish that is why each SU has independent and asynchronous sensing. So there is more probability of false alarms due to missed cooperation in spectrum sensing [12]
7. High latency: Cooperation in CRN, raises network traffic it results in higher latency in collecting this information due to channel contention [13]. This is still an open issue in CRAHNs.

8. Resource limitations in CRNs: Resource limitations hinder securities in CRNs. CRNs are expensive in terms of power so an efficient secured technique design is a difficult task [11].
9. Spectrum sensing in multi-user network: In multi-user environment, it is difficult to sense the intruders, empty channels and interference. These factors slow down the whole network [14].
10. Trust management: Trust management in CRN is beyond the security design and traditional trust management of MANET cannot work for CRN [13]. Trust management in CRN is an advanced communication overhead.
11. Analog primitives: If primary base stations transmit analog signals in some frequency bands like TV band and all security mechanism (game theory also) works on digital domain then it is not possible to integrate [15].
12. Performance degradation during hand-off in multiple spectrum-bands: Severe performance degradation occurs in CRNs during hand-off [16]. Mobility and connection management are very difficult to handle in such a scenario. Hence, security mechanisms must be scalable to CRN with multiple spectrum bands to provide high communication efficiency [17, 18].
13. Interference due to spectrum sensing: Spectrum sensing is severely hindered by the variation in the count of SUs [16, 19].
14. Channel fading in spectrum sensing: Primary signal may be faded by heavy obstacles or a secondary user is cooperative and sharing the sensed report but not reached due to some reason. Both these factors affect the performance of CRN [16, 20].

3 Game Theory

Game theory, first introduced by J. V. Neumann and O. Morgensterrn in 1944, is a collection of modelling tools that aid in the understanding of interactive decision-making problems. It analyzes the strategic interactions amongst multiple independent decision makers on the basis of rationality. Its predictive ability in an environment of conflicting interests makes it a useful suitable tool to manage an ad hoc network in the presence of autonomous selfish, malevolent, malicious, and attacker nodes.

Game theory can be divided in two main categories:

- Cooperative game theory, in which players cooperate with each other to maximize overall utility.
- Non-cooperative game theory, in which players are selfish and take their actions independently with the aim of maximizing their own utility function.

Game theory provides a mathematical framework for modelling and analyzing decision-making problems that handles situations in which players with contradictory interests or goals compete with each other. A particular instance of such a game is described by a set of rational players, strategies available to the players, and payoffs to the players. A rational player has his own interests in mind and chooses strategies or actions that help it in achieving those interests. A player is assumed to be able to estimate the outcome or payoff of the game [21] depending upon the actions of all the participants included in it.

The basic components of a game in an ad hoc network environment are:

$N = \{1, 2, \ldots n\}$ set of all mobile nodes (i.e. game players) in a routing path
$A_k = \{C, D\}$ possible action set for the node k, C denotes cooperative and D denotes defective
a_k = action chosen by the kth node
$\boldsymbol{a}_{-k}$ = action chosen by rest of the nodes except kth= $(a_1, a_2, a_3, \ldots a_{k-1}, null, a_{k+1}, a_{k+2}, \ldots a_n)$
$\boldsymbol{a}$ = action chosen by all the nodes $= (a_k, \boldsymbol{a}_{-k}) = (a_1, a_2, a_3 \ldots, a_k, a_{k+1}, a_{k+2}, \ldots a_n)$
$U(\boldsymbol{a})$ = Utility function of all the nodes, utility means payoff/benefit
$U_k(\boldsymbol{a})$ = Utility function of kth the nodes
D = Cartesian product of all action sets of a node.

The game for a node k is defined by a tuple:

$$G = \langle N, D, U_k(a_k, a_{-k}) \rangle \tag{1}$$

The purpose of the game is to bring the network to a position which is best suited to the network as a whole. In *Nash Equilibrium* all the nodes in the network are rationally benefited with no preference to the individual. In ad hoc networks the problem of selfish nodes is quite prominent wherein to save its resources a node may adopt a selfish attitude. In such an environment, game theory can convert an ad hoc network into the form of a mutually beneficial society.

4 Mitigation of Open Issues Using Game Theory

Game theory provides an elegant means to model strategic interaction between agents which may or may not be cooperative in nature. By leveraging the mechanisms of game theory, we can model heterogeneous spectrum sharing in CRNs as a repeated game in which collocated CRNs in a given region are its players. The payoff for every player in the game is determined by the quality of the channel to which it is able to gain access.

4.1 Computation of Payoff

Interaction between SUs can be modelled as a cooperative game. Each SU can adopt one of the two strategies: to cooperate or not to cooperate. A cooperating SU will get the revenue for participating in the spectrum sensing process by incurring the cost of participation, e.g., energy for sensing, etc. On the other hand, a non-cooperative/defective SU gets revenue without incurring any cost even without participating in the spectrum sensing process. The payoff for a cooperative SU is

$$\text{Payoff\_C} = \text{Benefit\_C} - \text{Cost\_C} \tag{2}$$

The payoff for a non-cooperative SU is

$$\text{Payoff\_NC} = \text{Benefit\_C} \tag{3}$$

A selfish/non-cooperative/defective SU does not incur any cost (in terms of energy) due to non-participation in spectrum sensing but obtains same profits. Local spectrum sensing information may be erroneous due to noise and channel impairments, such as shadowing, fading, etc. Thus, the payoff functions of a cooperating and a non-cooperating user are affected by the probability of false alarm and the probability of misdetection. The final decision is shared with all SUs. The payoff of a SU is affected by following scenario: if it gives a false alarm or a missed detection then Payoff_C decreases. Hence the payoff function value detects the behavior of the SU node:

- Payoff continuously increases out of boundary: Surely a selfish/non-cooperative SU node
- Payoff continuously increase at a steady rate: Cooperative SU node
- Payoff toggles between high and low: Malicious SU node.

There are various rules for decision making with regard to sensed information like the OR rule, AND rule, K2 rule, and majority rule. However, the majority rule is preferred because it says at least half of the SUs in the network give the local sensing result "1" when the channel is actually occupied.

4.2 Repeated Game

SUs tend to be non-cooperative/defective continuously; the best way to discourage such non-cooperating behavior is in the "Prisoner's Dilemma" which repeats the game. Hence, spectrum sensing occurs during the entire lifetime of a SU.

In a time slot, if sensing results from one SU are not obtained, it is directly assumed to be non-cooperating or deviated. To check the behavior of a SU nodes, sliding window concept is used where size of the window is Ws slots. If sensing reports (T_R) are not received in the last window slot W_S then it is declared as non-cooperative. Moreover, if the sensed results deviate from the results obtained through the "majority rule" and this occurs in two continuous W_S, then that SU is declared as an intruder/defective. So, a punishment is triggered in both cases. Here T_R and W_S are thresholds which are utilized for decision making. Since collisions are a part of transmission, few collisions are considered in every time slot. Let P_C is the probability of a collision in a single slot. So, the expected number of collisions in W_S slots is ($W_S \cdot P_c$). Accordingly the threshold assumed is $T_R > W_S \cdot P_c$. Hence the incentives of cooperating SUs are sustained and defaulters are identified and punished. Figure 1 describes the complete process in a flow chart.

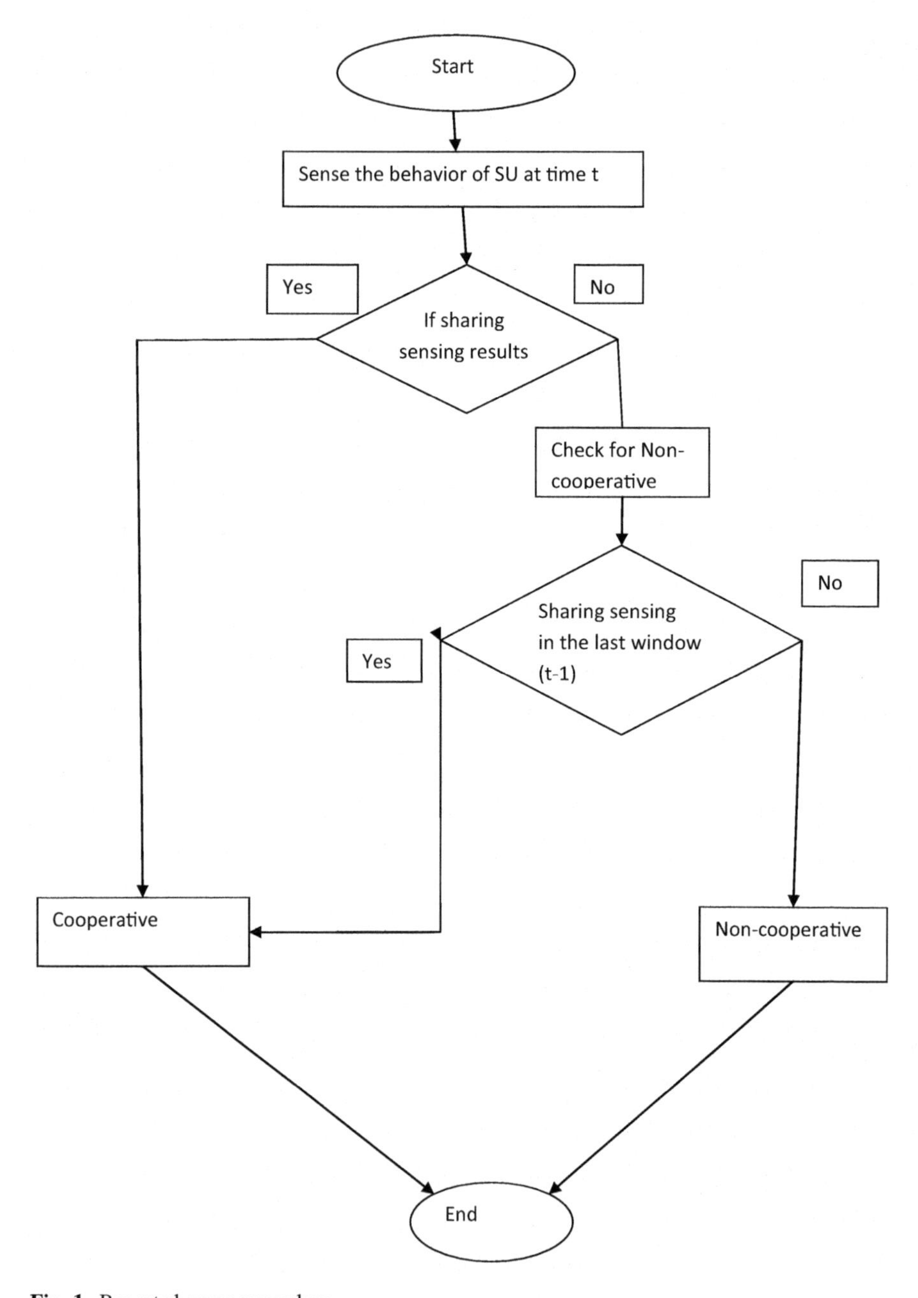

Fig. 1 Repeated game procedure

Almost all the mentioned issues relating to CRNs are resolved by using the above-mentioned techniques, such as differentiation between selfish and malicious nodes, enhancing cooperation, security mechanisms, fading, interference, latency, resource limitation, and trust management.

5 Conclusion and Scope for Future Work

This paper modelled and surveyed the mitigation of several issues related to security, spectrum sensing and management, and resource allocation and performance of CRNs using game theory. The ability to model individual, independent decisions makes it attractive for application in various fields of information technology, and in particular in the analysis of the performance of wireless networks. In this paper, we discuss how various interactions in cognitive radio-based wireless networks can be modelled as a game. Game theory provides incentives for individual users to behave in socially constructive ways. Gaps still exist in the working of CRNs due to the heterogeneous environment and must be thoroughly investigated with regard to the following issues:

1. Developing analog primitives
2. The dynamic natures of CRNs
3. Performance degradation during hand-off in multiple spectrum bands.

References

1. Mitola J III (2000) Cognitive radio: an integrated agent architecture for software defined radio. PhD thesis. Royal Institute of Technology (KTH), Sweden
2. Haykin S (2005) Cognitive radio: brain-empowered wireless communications. IEEE J Sel Areas Commun 23(2):201–220
3. Akyildiz IF et al (2006) Next generation/dynamic spectrum access/cognitive radio wireless networks: a survey. Comput Netw 50:2127–2159
4. Mathur CN, Subbalakshmi KP (2007) Security issues in cognitive radio networks. In: Cognitive networks: towards self-aware networks. Wiley Ltd., Chap 11
5. Zhang X, Li C (2009) The security in cognitive radio networks: a survey. In: Proceedings of international conference on communications and mobile computing, pp 309–313. ACM, Leipzig, German
6. Qin TYuH, Leung C, Shen Z, Miao C (2009) Towards a trust aware cognitive radio architecture. Newsletter ACM SIGMOBILE Mob Comput Commun Rev 13(2):86–95
7. Chen R, Park JM (2006) Ensuring trustworthy spectrum sensing in cognitive radio networks. In: 1st IEEE workshop on networking technologies for software defined radio networks. SDR '06, pp 110–119. Reston, VA, USA
8. Chen R, Park J-M, Reed JH (2008) Defense against primary user emulation attacks in cognitive radio networks. IEEE J Sel Areas Commun 26(1):25–37
9. Nhan N-T, Koo I (2009) A secure distributed spectrum sensing scheme in cognitive radio. In: Proceedings of the intelligent computing 5th international conference on emerging intelligent computing technology and applications, pp 698–707. Springer, Berlin

10. Jakimoski G, Subbalakshmi KP (2009) Towards secure spectrum decision. In: Proceedings of the IEEE international conference on communications, pp 2759–2763. Dresden, Germany
11. Araujo A, Blesa J, Romero E, Villanueva D (2012) Security in cognitive wireless sensor networks, challenges and open problems. EURASIP J Wirel Commun Netw 48(1)
12. Akyildiz IF et al (2009) CRAHNs: cognitive radio ad hoc networks. Ad Hoc Netw 7:810–836
13. Chen R, Park J-M, Hou YT, Reed JH (2008) Toward secure distributed spectrum sensing in cognitive radio networks. IEEE Commun Mag Special Issue Cognit Radio Commun 50–54
14. Akyildiz IF, Lee WY, Vuran MC, Mohanty S (2008) A survey on spectrum management in cognitive radio networks. IEEE Commun Mag 40–48
15. Mathur CN, Subbalakshmi KP (2007) Digital signatures for centralized DSA networks. In: Proceedings of 4th IEEE conference on consumer communications and networking, pp 1037–1041
16. Khalid L, Anpalagan A (2010) Emerging cognitive radio technology: principles, challenges and opportunities. J Comput Electr Eng 36(2):358–366
17. Fu X, Zhou W, Xu J, Song J (2007) Extended mobility management challenges over cellular networks combined with cognitive radio by using multi-hop network. In: Proceedings of ACIS international conference on software engineering, artificial intelligence, networking, and parallel/distributed computing, pp 683–688. Qing-dao, China
18. Atakan B, Akan O (2007) Biologically-inspired spectrum sharing in cognitive radio networks. In: Proceedings of IEEE wireless communications and networking conference, pp 43–48. Hong Kong
19. Sahai A, Hoven N, Tandra R (2004) Some fundamental limits on cognitive radio. In: Proceedings of alert on conference on communication, control and computing. Monticello, Chicago
20. Yau K-LA, et al (2012). Reinforcement learning for context awareness and intelligence in wireless networks: review, new features and open issues. J Netw Comput Appl 35(1):253–67
21. Avinash KD, Skeath S, David HR (2010) Games of strategy, 3rd edn. W.W. Norton & Company, New York.

On Control Aspects of Quality of Service in Mobile Ad Hoc Networks

C. Siva Krishnaiah and A. Subramanyam

Abstract Provisioning Quality of Service (QoS) in a MANET is a prominent research area due to the ongoing increasing range of MANET applications. The need to improve QoS in these networks has been vital due to the traits which include dynamically changeable network topology, be short of facts about state, unavailability of a primary controller, and insufficient availability of resources. To quantitatively evaluate QoS in a MANET several associated metrics are preferred. This paper explores QOS aspects and metrics, after which mentioned the scope and relevance of manipulated aspects in view of the divisible and non-divisible traffics in the network for QoS.

Keywords QoS metric · Manet · Ad hoc wireless networks
Quality of service

1 Introduction

A mobile ad hoc network (MANET) is an endlessly self-configuring network possessing a set of portable devices which can converse between them lacking infrastructure associated wirelessly. Each piece of equipment in a MANET is liberated to be in motion alone in any route, and will frequently change its associations to other devices. The most important challenge in building a MANET is to enable each piece of equipment to keep up the information requisite to properly en route traffic. Such networks may function by themselves or may be linked to the wider internet. QoS in a network is considered in terms of the provision of a definite amount of data which a network conveys from one point to another during a certain amount of time.

C. Siva Krishnaiah (✉)
Rayalaseema University, Kurnool, India
e-mail: sivacmca@gmail.com

A. Subramanyam
Annamacharya Institute of Technology and Sciences, Rajampet, India
e-mail: smarige@gmail.com

A. Kumar and S. Mozar (eds.), *ICCCE 2018*,
Lecture Notes in Electrical Engineering 500,
https://doi.org/10.1007/978-981-13-0212-1_51

To quantify QoS, numerous correlated components of the network service are considered along with well-known parameters of QoS, such as error costs, bit rate, throughput, communication delay, availability, and jitter. QoS is the potential to make available diverse priority to diverse programs, users, or data flows, or to promise a certain level of overall performance with respect to data flow.

There are a few former QoS routing algorithms like Ant-E, PBANT, and many more. Ant-E is built on Blocking-ERS to manage the operating cost and region retransmission to obtain increased reliability. It resumes its route discovery procedure to find a route to the end node from the point where it finished in the last round consequent a failure. The PBANT algorithm optimizes the route detection procedure by taking into consideration the point of the nodes which can be identified by the GPS receiver.

FQMM [1] model classifies nodes into ingress nodes (source), interior nodes (intermediate relay nodes) and ingress nodes (destination), and considers a per flow basis. It assumes that the proportion of flows required for each flow QoS is to a great extent lower than that of the low-priority flows that can be pooled into QoS classes. FQMM combines the reservation system intended for high-priority traffic through service differentiation for low-priority traffic. Therefore, FQMM provide the best QoS for per flow as well, overcoming the scalability crisis by categorizing low-priority traffic into service classes. This deals with the scalability crisis. However, it cannot resolve other problems, such as choice of traffic categorization, allocation of per flow, the aggregated service intended for the specified flow, quantity of traffic fit into per flow service, as well as scheduling or forwarding of traffic through the intermediate nodes.

HQF [2] enable to deal with QoS at three diverse levels: the physical interface level, the logical interface level, and the class level used for the QoS queue and decisive methods by means of the MQC interface to make available a fine QoS architecture.

AQuaFWiN [3] uses a hierarchical approach in which a cluster of portable end hosts are coupled to a base position. These are linked to a supervisory node associated to the wired communications. The changing conditions in wireless due to intervention and perhaps mobility enable that the real-time application need rigorous QoS ought to be adaptable. The framework uses a standard response apparatus to maintain adaptability at all levels of the wireless network.

IntServ framework implemented the RSVP to reserve bandwidth on routers level all along the pathway of a flow [4]. When a flow arrives by way of a QoS obligation, the ingress edge routers initiate the pathway establishment procedure by distributing a PATH message to the destined egress edge router. The egress edge router responds by shipping a RESV message turn around to the ingress router and attempts to reserve bandwidth requisite for the requested QoS along the pathway towards the source ingress edge router. Core routers along the pathway shape their traffic control mechanism such that every admitted flow is assured to accept the bandwidth reserved, and accordingly the requested QoS. In the course of this per-flow-based hop-by-hop signaling, IntServ delivers end-to-end QoS assurance.

In the DiffServ [5], flows are composed into a small extent of classes at the edge of a network and the routers within the network simply put into practice a set of scheduling/buffering mechanisms depending on the classes. This per-class makes a router's functionalities simpler as well as reducing the state that a router has to maintain. The per-class approach eliminates the operating cost of per-flow QoS signaling. DiffServ is more scalable than IntServ. On the other hand, if bandwidth is provisioned on a pathway used only by a single flow then it loses its advantages and degrades to IntServ.

2 QoS Metrics

QoS metrics and their possible applications (see Table 1) referred in [4] are:

2.1 Additive Metric

A metric whose value over a pathway is the summation of the values at every hop.

Example: Cost, delay, and jitter

An additive metric (A_m) is termed as

$$A_m = \sum_{k=1}^{N} L_k(m)$$

where, L_k (m) is the value of metric "m" over link L_k and $L_k \in$ P, "N" is hope length of path P, "m" is metric, "L_k" is link and "P" is path.

2.2 Concave Metric

This represents the minimum value over a path. The accessible resource resting on every single link should be at least identical to the requisite value of the metric. Example: Bandwidth.

Table 1 Applications and associated metrics [6]

Applications	KEY QOS metrics/parameters
Multimedia applications	Bandwidth, delay etc.
Military applications	Security, reliability, delay, bandwidth etc.
Emergency/realtime applications	Availability, reliability etc.
Hybrid wireless network applications	Maximum available link life, delay, bandwidth, channel utilization etc.
Sensor network applications	Battery life, energy conservation etc.

A concave metric (C_m) is termed as

$$C_m = \min(L_i(m)) \text{ where } L_i(m) \in P.$$

2.3 *Multiplicative Metric*

This is the product of QoS metric values.

Example: Reliability/Availability, Link-break probability.

$$M_m = \prod_{k=1}^{N} (L_k(m)). \text{ where } L_k(m) \in P.$$

3 Control Constrictions for Quality of Service

The network model [7] is a graph defined in terms of Q = (G, E) where "G" is the set of "N" nodes and "E" is a set of edges. Each node "k" has a bandwidth probable and delay is intended for node-pair (m, n). The bandwidth is indistinguishable for both transmitting and receiving communications. So as to this, total bandwidth for transmit data in addition to the whole bandwidth intended for receiving data at node "k" will at this time not go beyond. The delay is associated with the maximally permissible hop-matter for the node pair (m, n). From the network model, we can see that these requirements have an effect on the QoS provisions of the network. We want to discover an unbiased control mechanism that convenes the QoS provisions and has higher directional validity, minimized delay, and bandwidth conservation.

Let $\Omega_{m,\ n}$ denote the traffic call meant for the node pair (m, n), $\rho_{m,\ n}$ denote the maximally permissible hop-count figure meant for the node-pair (m, n) and furthermore, d (i, j) is the gap between node "I" and node "j". The topology control problem can be formally defined as observed. Given a node set "G" surrounded by their position and each one node "k" with B_k as well as D_k, further given $\Omega_{m,\ n}$ and $\rho_{m,\ n}$ for each node-pair (m, n), discover minimum bandwidth, put off used for $1 \leq k \leq n$, hence all the traffic necessities can be routed in the hop-rely basis.

There are some possible considerations such as end-to-end traffic consignment are not divisible, so as $\Omega_{m,\ n}$ for node-pair (m, n) be obliged to be routed at the equal route starting "m" to "n".

End-to-end traffic consignments are divisible, in order that $\Omega_{m,\ n}$ can be routed on dissimilar paths starting "m" to "n".

4 Consideration I: Non-divisible Traffic

Here, consider that the traffic consignment is not divisible, so within the online situation it isn't feasible to route the traffic among a node-pair through numerous paths on occasion or maybe from the synchronized requests.

Variables

1. $x_{k,l}$ Boolean variables $x_{k,l}=1$ if there is a connection from node 'k' toward node "l"; in any other case $x_{k,l}=0$.
2. $x_{k,l}^{m,n}$ Boolean variables, $x_{k,l}^{m,n}=1$ if there is a connection from node "m" toward node "n" lying on link (k, l); in any other case $x_{k,l}^{m,n}=0$.
3. B_k, the bandwidth and D_k is delay for node-pair (m, n) at node "k".
4. $B_{\max}$, the maximum bandwidth usage of node pair.
5. $D_{\max}$, the maximum delay of node pair.

Optimize

To optimize the bandwidth usage of node pair (m, n):

$$Min\, B_{\max} \tag{1}$$

To optimize the delay:

$$Min\, D_{\max} \tag{2}$$

Constrictions

1. Topology Constriction:

$$x_{k,l} \le x_{k,l^1} \text{ If } d(k,l^1) \le d(k,l)\, \forall k,\ l,\ l^1 \in V \tag{3}$$

2. Bandwidth Constriction:

$$\sum_{(m,n)} \sum_{l} x_{k,l}^{m,n} \lambda_{m,n} + \sum_{(m,n)} \sum_{j} x_{l,k}^{m,n} \lambda_{m,n} \le B_k\, \forall k \in V \tag{4}$$

3. Delay Constriction:

$$\sum_{(k,l)} x_{k,l}^{m,n} \le \gamma_{m,n}\, \forall\, (m,n) \tag{5}$$

4. Flow Maintenance Constriction:

$$\sum_{l} x_{k,l}^{m,n} - \sum_{l} x_{l,k}^{m,n} = \begin{cases} 1 & if \quad m=k \\ -1 & if \quad d=k\, \forall\, k \in V \\ 0 & otherwise \end{cases} \tag{6}$$

5. Route Legality Constriction:

$$x_{k,l}^{m,n} \le x_{k,l} \quad \forall k, l \in V \tag{7}$$

6. Binary Constriction
   a. For Directed Node Connectivity or Unidirectional:

$$x_{k,l} = 0\, or\, 1\, x_{k,l}^{m,n} = 0\, or\, B_k = 0\, or\, 1 \tag{8}$$

   b. For Undirected Node Connectivity or Bidirectional:

$$x_{k,l} = x_{l,k}\, for\, all\, k, l \in V \tag{9}$$

Observation 1: Constriction (3) convinced that nodes encompass broadcast capability. The transmission all the way through the node may be acquired via all of the nodes inside its transmission range. This correspond to the hyperlinks within the networks as, for node “k”, if there’s a hyperlink to “l” (i.e., $x_{k,l} = 1$), then there should be a hyperlink to any node l^1(i.e., $x_{k,l^1} = 1$) when $d_{k,l^1} \leq d_{k,l}$ while that’s constriction (3).

Observation 2: Constriction (4) convinced that the overall transmission and reception of data on a node do no longer cross past the bandwidth usage of that node. The first expression on the left side of inequality (4) corresponds to the entire departure communication load at node “k” (origin) and the second expression corresponds to all of the inward traffic at the destination. While this constriction does not guard against the scenario of concurrent communication at a node, it is relevant to the typical case where a node is geared up with the simplest set of transceivers and cannot put out and gets hold off on the same time.

Observation 3: Constriction (5) gives provable assurance that the hop-count for each node pair (m, n) does not go beyond the prespecified bound.

Observation 4: Constriction (6) is for go with the flow conservation, here the traffic is considered as non-divisible, so $x_{k,l}^{m,n}$ is both 0 and 1, and in lieu of that either the intact traffic of (m, n) goes through linkage (k, l) or nothing does. This constriction specifies that the aim of the whole traffic for (m, n) originates at node “m” and goes down at node “d”, along with at any intermediate node where the (m, n) communication data coming into this node should be the same as the data exiting this node.

Observation 5: Constriction (7) gives provable specification that the route legality intended for each pair of node, pointing out that there may be traffic glide from node “k” to node “l” that is possible to handle whilst there be a link (k, l).

4.1 Consideration II: Divisible Traffic

In this situation we should not forget that the traffic demands may be divisible, so within the online situation it is more viable to direct the traffic between a node-pair thru diverse paths once in a while, or maybe from the synchronized requests.

Variables

1. $x_{k,l}$ and B_{max}, D_{max} continue to be the same as in the non-divisible case.
2. $T_{k,l}^{m,n}$, variables specified intended for the quantity of traffics of the node pair (m, n) that under go all the way through link (k, l).
3. B_{max}, the maximum bandwidth usage of the node pair.
4. D_{max}, the maximum delay of the node pair.

Optimize

To optimize the bandwidth usage of the node pair (m, n):

$$Min\, B_{\max} \tag{10}$$

To optimize the delay:

$$Min\, D_{\max} \tag{11}$$

Constrictions

1. Topology Constriction:

$$x_{k,l} \leq x_{k,l^1} \text{ If } d(k, l^1) \leq d(k, l)\, \forall k, l, l^1 \in V \tag{12}$$

2. Bandwidth Constriction:

$$\sum_{(m,n)} \sum_{l} T_{k,l}^{m,n} + \sum_{(m,n)} \sum_{j} T_{l,k}^{m,n} \leq B_k\, \forall k \in V \tag{13}$$

3. Delay Constriction:

$$\frac{1}{\lambda_{m,n}} \sum_{(k,l)} T_{k,l}^{m,n} \leq \gamma_{m,n}\, \forall (m,n) \tag{14}$$

4. Flow Maintenance Constriction:

$$\sum_{l} T_{k,l}^{m,n} - \sum_{l} T_{l,k}^{m,n} = \begin{cases} \lambda_{m,n} & if \quad m = k \\ -\lambda_{m,n} & if \quad d = k \forall k \in V \\ 0 & otherwise \end{cases} \tag{15}$$

5. Route Legality Constriction:

$$T_{k,l}^{m,n} \leq T_{k,l}^{m,n}\, x_{k,l} \quad \forall k, l \in V, (s, d) \tag{16}$$

6. Binary Constriction
   a. For Directed Node Connectivity or Unidirectional:

$$\begin{aligned} &x_{k,l} = 0 \, or \, 1 \quad B_k \geq 0 \\ &T_{k,l}^{m,n} \geq 0 \forall k, l \in V, (m, n) \end{aligned} \tag{17}$$

b. For Undirected Node Connectivity or Bidirectional:

$$T_{k,l} = T_{l,k} for \, all \, k, l \in V \tag{18}$$

Observation 6: The purpose and specification of the constrictions are comparable to the non-divisible case.

Observation 7: The delay is calculated by constriction (14) as the average hop-count of multi-flows among nodes. The specification of delay constriction is justifiable, because in the divisible scenario, traffic sandwiched between a node pair can be routed through diverse paths and a bound on the average delay provides a high-quality delay guarantee for network applications.

Observation 8: Constriction (15) indicates the whole traffic for all the routes intended for node pair (m, n). The whole traffic for (m, n) (i.e. $\lambda_{m,n}$) is now divisible into numerous flows (i.e. $T_{k,l}^{m,n}$).

5 Performance Evaluation

To assess the performance, an NS-2 simulator is appropriate to create the simulation environment on the way to implementing the proposal, and additionally for comparing its overall performance with each other. To find the overall performance, we need to recollect throughput, packet shipping ratio, queue size, and postpone. The transmission put off of facts packets which might be introduced correctly. This postpone consists of dissemination delays, queue delays on the perimeter and buffering delays at some point of path discovery.

References

1. Xiao S, Chua KC (2000) A flexible quality of service model for mobile ad hoc networks. In: Proceedings of IEEE, pp 445–449
2. Cisco (2013) Hierarchical queueing framework configuration guide, Release 15M&T
3. Bobby R, Sonia J, Sudhir Adaptive QoS framework for multimedia in wireless networks and its comparison with other QoS frameworks
4. Murthy S, Manoj BS Ad Hoc wireless networks architectures and protocols, Pearson Education
5. Carlson M, Davies E, Wang Z, Weiss W, Blake (1998) An architecture for differentiated services, IETF RFC 2475
6. Siva Krishnaiah C, Subramanyam (2017) Framework on quality of service in mobile ad hoc networks. IJASTEMS 3(5). ISSN 2454-356X
7. Santi P (2005) Topology control in wireless ad hoc and sensor networks. Wiley, Chic ester

Securing CoAP Through Payload Encryption: Using Elliptic Curve Cryptography

M. Harish, R. Karthick, R. Mohan Rajan and V. Vetriselvi

Abstract The vision of the IoT is to not only make our everyday lives easier but also at the same time to ensure a secure environment. As the networking world moves closer towards an environment comprising of minimalistic ubiquitous nodes, IoT-based protocols cannot afford to accommodate security vulnerabilities. In this paper, we identify and mitigate the existing security flaws in the CoAP protocol of the IoT. A real-time system is developed to put the mitigated system into use and analyze the enhanced security. Additionally, we quantitatively look to evaluate the vulnerability of current implementation of CoAP and the magnitude of mitigation the method suggested in this paper provides. A secondary quantitative measure is used to prove that the overhead of the applied encryption is acceptable in terms of efficiency for the mitigation achieved.

Keywords IoT security · CoAP · Payload encryption

1 Introduction

CoAP is one of the major application layer protocols for constrained devices, and hence is employed in many Internet of Things (IoT) devices. Naturally, since IoT communication is a new area of networking, it is prone to security threats and subsequent attacks. Apart from networking, the IoT has spread across multiple dimensions, from networking computational applications to smartphones, making the implication of a possible security breach in its protocol much more dangerous.

As far as the future is concerned, the computing world is moving towards low cost, low computational power devices, able to communicate using internet protocols. In an environment that spans sensing devices and home smart devices that exclusively function on IoT protocols, and also traditional networking devices that

M. Harish · R. Karthick · R. Mohan Rajan · V. Vetriselvi (✉)
Department of Computer Science, Anna University, Chennai, India
e-mail: vetri@cs.annauniv.edu

A. Kumar and S. Mozar (eds.), *ICCCE 2018*,
Lecture Notes in Electrical Engineering 500,
https://doi.org/10.1007/978-981-13-0212-1_52

run sophisticated networking protocols on the web, it has become mandatory to handle HTTP or an equally secure protocol's request to an IoT device. The implication of such a request has an immense importance, especially when it involves sensitive data that the sensors or home devices handle. In some cases, such an implication involving the handling of http requests to an IoT node requires an additional proxy that manages the connection by translating packets arriving from an http protocol onto a packet form which an IoT node can sense and operate onward. Although the CoAP protocol was designed to accommodate simple integration with web protocols like HTTP, in some cases where the translation of protocol is not straightforward or not defined, the use of an additional proxy becomes mandatory.

The CoAP layer in the IoT has numerous security vulnerabilities. These get exposed when an HTTP request to an IoT device through the CoAP layer is involved. An alarmingly large number of applications involve handling HTTP requests to IoT devices, which intuitively has to be handled by the CoAP application layer. This implication has opened up a window for numerous attacks that inhibit the use of http requests to an IoT server (running CoAP).

The problem that is addressed in this paper coincides with the problems prevalent in the CoAP protocol, which is to completely mitigate the security flaw that arises whilst a HTTP connection is made to a CoAP node through a proxy. When an HTTP request needs to access a CoAP node, a proxy like 6LBR must be used as discussed by Raza et al. [7]. It is possible to instantiate an attack by modification of message packets at the proxy or by adding additional packets that either snoop data or modify data. Clearly, there is no concrete standard for proxies mapping HTTP requests to a CoAP server.

The crux of the suggested solution is in encrypting the payload inside the packet structure of the request that awaits handling by the proxy. Upon encrypting the payload before it reaches the proxy, the flaws concerning the pathway and insecure transition nodes are mitigated to a considerable extent.

The most obvious challenge of the system is the analysis and identification of the flaws existing in the current CoAP system. This is due to the fact that the solution can only be designed based upon the flaws identified in the existing CoAP in the sense that all the modules must work coherently to mitigate the identified flaw, the identification of which becomes the focus of the proposed solution.

2 Related Work

A translation of the message headers created by the (D)TLS record layer and the protocols lying on top of it must be implemented and this is discussed by Brachmann et al. [2]. Enabling security during transmission is an important criterion as the information sent should be secure enough to protect it from any kind of attack as explained by Granjal et al. [4] and Rahman et al. [6].

2.1 DTLS and CoAP Integration

The integration of DTLS and CoAP for the IoT devices can bring many more advantages than expected, as explained by Raza et al. [8]. DTLS header compression can significantly reduce the energy and power consumption, and this is turn can reduce the number of transmitted bytes maintaining DTLS standard compliance. A number of factors have to be considered, such as who initiates the connection, whether it is HTTP or CoAP, and whether the proxy is within the CoAP network or if it is trusted.

Solving the problems of end-to-end security and secure group communication is the key to ensure a secure IP-based IoT as explained by Heer et al. [5]. CoAP architecture is split into two layers, a message layer and a request/response layer. The first layer is responsible for controlling the message exchange over UDP between two end points. CoAP uses UDP which is why it uses DTLS as explained by Vŭcinić et al. [11]. The mapping from HTTP to CoAP requires a proxy which can be useful to translate packets. The proxy can be a 6LoWPAN border router. A trusted example of a 6LBR is required to ensure that no malicious code is added. The 6LBR is used to interconnect the WSNs with the IP world. The solution to this problem is to achieve a fully secure communication between an HTTP and a CoAP entity.

2.2 Proxy Vulnerabilities

The 6LoWPAN [7] concept originated from the idea that "the Internet Protocol could and should be applied even to the smallest devices," and that low-power devices with limited processing capabilities should be able to participate in the IoT. The resource constraints of the devices and the lossy nature of wireless links are among the major reasons that hinder applying general E2E security mechanisms to 6LoWPANs. An example is building control automation in order to directly control lighting, heating or security settings in a building through a mobile phone. The user's identity should be protected and the information which is exchanged must be secured to prevent any kind of attacks as explained by Roman et al. [10]. The network must be flexible enough to withstand DoS attacks and must be capable of repairing the damaged system in case of any attacks. End-to-end security is a must for devices involved in communication as it can prevent data leaking to man-in-the-middle attacks.

2.3 Multicast Problem

Multicast messages are used in CoAP to manipulate resources in a group of devices at the same time, as explained by Granjal et al. [4]. Unicast messages can be secured via DTLS with PSK. A solution is needed to ensure security when multicast messages are used and it can be achieved using a proxy/6LoBR which has to make translations from HTTP to CoAP, TLS to DTLS, and in this case a mapping from the unicast address in the destination field of the UDP header to a multicast address. A secure DTLS connection is required within a group of devices with a single session key. We need to consider different topologies and various cases have to be handled in order for secure connections when multiple devices are connected in a network.

2.4 Security Issues

The current state of the art in the IP-based IoT is discussed by Zhao et al. [14]. General security problems of communication network are a threat to data confidentiality and integrity. Key algorithms mainly include symmetric key and asymmetric keys. An asymmetric keys algorithm mainly uses Rivest-Shamir-Adleman and ECC. However, there are still some common threats, including illegal access networks, eavesdropping on information, confidentiality damage, integrity damage, DoS attacks, man-in-the-middle attacks, virus invasion, and exploit attacks etc., as explained by Xu et al. [12]. Cryptography technology can not only realize user privacy protection, but can also protect the confidentiality, authenticity, and integrity of the system in order to prevent countermeasures, as explained by Zhang et al. [13]. As an important part of the IoT perception layer, data is transmitted in free space. The attacker can easily intercept and analyze data. In the systematic approach of Riahi et al. [9], data encryption mechanisms prevent attackers from eavesdropping and tampering with data during its transmission, and encode data to ensure data confidentiality. An attacker sends a package, which has been received by the destination host in order to obtain the trust of the system as described by Gou et al. [3]. It is mainly used in authentication processing and destroys the validity of certification. This means that effective authentication technology must be used in order to prevent illegal user intervention. The introduction of IDS can monitor the behavior of network nodes in a timely manner, and discover the suspicious behavior of nodes.

Our Contribution The proxy vulnerabilities outlined in previous work have been thoroughly studied so as to get an understanding of the security flaws it imposes. From those works, for a real-time implementation of CoAP, the areas which needs to be dealt with are understood and packet translation is considered in detail. Deviating from the general problems facing multicast and translation security—

which are discussed in the earlier works—payload security of the packets is put under scrutiny, thereby enabling this project to contribute further regarding the unexplored parameters of security of the IoT.

3 Proposed System

In our proposed work, we secure HTTP connections to an IoT node, by handling the HTTP requests through a proxy and securing the CoAP layer by encryption of the payload through elliptic curve cryptography.

The general schema of the system involves an IoT device, which is controlled by an IoT controller. The IoT controller also handles requests to the IoT device from other nodes in the network. Intuitively, the IoT controller needs an inbuilt router to control the network traffic and oversee the packet movement across the bandwidth. The overall system architecture of the proposed system is explained further.

3.1 System Architecture

In order to implement the proposed solution, a wi-fi-based wireless network with IoT-based nodes sharing a common access point is considered. One of the IoT device nodes in the network is a passive infrared (PIR) motion sensor that detects motion around it. As previously mentioned, this IoT device is connected to an IoT controller which handles requests to the sensor data from other nodes in the network and also enables the sensor to communicate with other nodes. The architecture is an abstract representation of how IoT devices handle payload in a network and hence any specific device can replace the PIR sensor and the abstraction of the architecture reduces it to the payload that is generated from the IoT device. The network is chosen to be a wi-fi-based wireless network since most of the practical applications and subsequently the scope of attack of IoT devices are through wireless media.

Figure 1 shows the entire flow of the proposed implementation.

The overall design of the system is split into three modules as follows:

- Connection Establishment Module
- Packet Translation Module
- ECC Encryption and Decryption Module.

Connection Establishment Module The main function of this module is to initiate the connection from the client to the IoT device that the client wants access

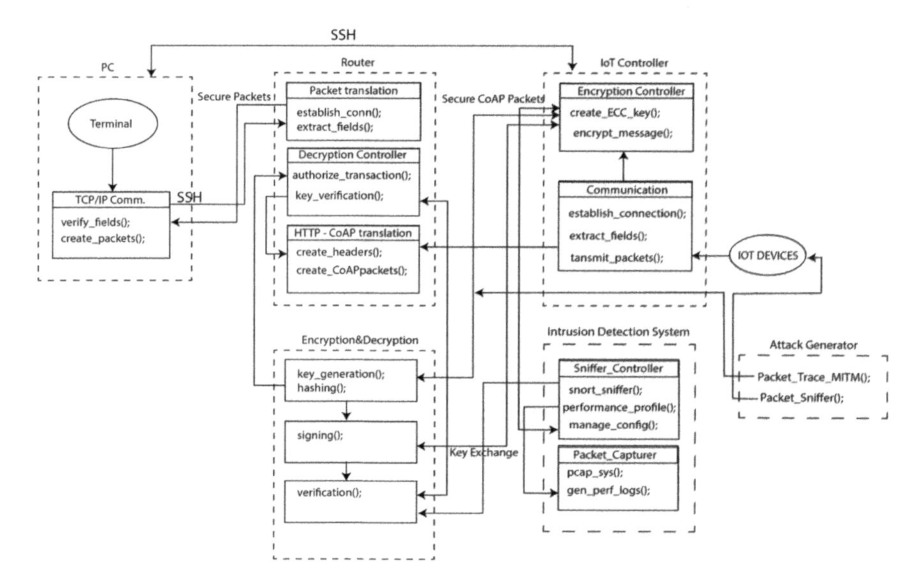

Fig. 1 Securing CoAP through payload encryption

to. Subsequently, the IoT device may also communicate or relay its value to the client. Any IoT node in the network that wants to communicate or access data with another node or the PIR sensor in the network is regarded as a client. Since the client cannot explicitly communicate with the PIR sensor, the request is routed through the IoT controller.

In order to implicate the security vulnerability of the CoAP protocol upon translation of HTTP requests, the initiation of the connection from the client to the sensor is through "http" and a proxy process is initiated in the IoT controller. This process is invoked in the controller whenever a client makes an http request to the PIR sensor. Alternatively, this process can be done by a separate node in the network instead of it residing in the IoT controller, if the controller has computing constraints. The working of the proxy process is explained in the next section.

The connection is built upon TLS and DTLS protocols for security in HTTP and CoAP respectively. This module encompasses the entire architecture and acts as the crux, since any communication-based system is based on connection establishment.

Packet Translation Module The actual transition from HTTP to CoAP (and subsequently from CoAP to HTTP if required), which essentially is the main function of the proxy, is achieved with this module. From the incoming packets the entity body must first be extracted. Upon verification of the http entity-body, the individual bytes are checked. CoAP supports only UTF-8 decoding. Therefore the bytes are converted if required. After successfully obtaining the required information for the CoAP packets to be formed in the correct character set decoded format, the CoAP packet is constructed and then returned.

The controller takes over the returned CoAP packet and routes it to the device.

Algorithm 1 HTTP content analysis and Packet translation

```
1: if httpMessage instanceof HttpResponse then
2:httpEntity=((HttpResponse)     httpMessage).getEntity()
3: end if
4: if httpMessage instanceof HttpEntityEnclosingRequest then
5:httpEntity=((HttpEntityEnclosingRequest)     httpMessage).getEntity()
6: end if
7: if httpEntity == null then
8: throw new IllegalArgumentException(" httpEntity == null" )
9: end if
10: payload = EntityUtils.toByteArray(httpEntity);
11:coapCharset=UTF-8
12:httpContentType= ContentType.getOrDefault(httpEntity)
13: if httpCharset != null && !httpCharset.equals(coapCharset) then
14:charset = changeCharset(payload, httpCharset, coapCharset)
15: end if
16: return payload
```

ECC Encryption and Decryption Module The encryption and decryption model is the most crucial aspect of the system that aims at safeguarding the payload throughout its transition from the client to the end CoAP node through a proxy.

After the payload is defined in the packet structure, the DTLS built upon the UDP layer is used for communication. After the DTLS is established and subsequently after the UDP is invoked, the packet is sent to the destination details as set using the UDP protocol using the '*coap://*' prefix denoting a conventional CoAP protocol.

Pertaining to this system, after the DTLS establishment, the payload is encrypted using the encryption module of elliptical curve cryptography (ECC) and at the receiving end, when the same stack is used to receive the CoAP packet, the decryption module of ECC is used to get the original data back. Therefore, an additional layer is effectively added before the core CoAP application layer in the communication protocol of IoT. The added layer encrypts and decrypts the payload in the sending and receiving nodes respectively.

The selection of ECC for encryption and decryption is justified by the fact that it requires much less memory compared to other similar public key encryption methods and hence is more suited to memory and computationally constrained IoT networks.

4 Evaluation and Results

The evaluation of the system is performed whilst considering the PIR sensor data being relayed across to the nodes in the network. Essentially, the payload comprises of the data that is derived from the PIR sensor, if and when it detects motion. The value of the PIR sensor is taken from a file written to the memory of the IoT controller. The choice of value that is to be written to the file for different states of the sensor can be programmatically altered.

For evaluation purposes, an environment with multiple sensor devices is considered. The controller primarily aggregates values from all these sensors into a single value and relays it to the nodes the sensor wants to communicate with.

Assuming an eavesdropping attacker passively has access to one or more sensors in the network, the probabilistic measure of how correctly he can figure out the actual value relayed to the node forms the basis for the quantifying parameter considered in this section.

Definition 1 *Eavesdropping Vulnerability* [1] Eavesdropping vulnerability is a probabilistic methodology that aims to quantify eavesdropping vulnerability in IOT networks. It is a function of distribution of sensor values (sensor network model), size of compromised nodes, aggregation function, and error tolerance.

The parameter is described for a sensor network which can be further extended for individual nodes in a network. Basically, this parameter quantitatively signifies how vulnerable a system is against eavesdropping attacks.

We choose to use Eq. (1) which brings in the attacker aggregation parameter S. In some cases of attacks through secured channels and masking, the size of nodes being attacked (SA) cannot be evaluated. Hence, we compute the expected eavesdropping vulnerability as,

$$\gamma_E = \sum_S (S_A = s) \cdot I(|\phi(S_C) - \phi(s_a)| \leq \delta) \tag{1}$$

where I is the indicator function which computes to the Boolean values of 1 or 0 if the relational logic it is applied to is true or false respectively, and δ is the error tolerance of the adversary. The aggregation function σ, can be derived from any of the following standard aggregation functions: min/max, sum, average or median. Choosing average as the aggregation function, we get the derived formula,

$$\gamma_E = \sum_S (S_A = s) \cdot I(|\sigma(S_C) - \sigma(s_a)| \leq \delta) \tag{2}$$

where σ denotes the average of the sensor nodes as the aggregation function.

Applying this logic and expanding Eq. (2), we get,

$$\gamma_E = \sum_{i=1}^{|S|} p(|S_A| = i) \cdot (\Phi|S_A|(u) - \Phi|S_A|(l)) \tag{3}$$

$$\gamma_E = \sum_{S} (S_A = s) \cdot I(|\phi(S_C) - \phi(s)| \leq \delta) \tag{4}$$

where p denotes the probability that a single node in the network is compromised and φ is the cumulative density function (cdf) of the distribution of S, and u, l are the upper and lower limits of error tolerance.

With the help of Eqs. (3) and (4), we can find the eavesdropping vulnerability of a protocol. This is applied to quantify how vulnerable the current implementation of CoAP is. Upon assuming that the probability that a node is compromised as 0.2 ($p = 0.2$), the CoAP with the encrypted payload and the conventional CoAP is compared with regard to the magnitude of eavesdropping vulnerability as shown in Table 1. The table shows that with the use of different sensors, and by comparing it with the values obtained from the device (from the CoAP and the encrypted CoAP), the aggregation function is calculated.

The vulnerability of the conventional CoAP protocol with varying ease of eavesdropping, can be analyzed by varying the probability of attack p, as shown in Table 2.

Intuitively, the magnitude of vulnerability of CoAP in its current form is higher when the node can be easily compromised and lower when it is difficult to attack a node. Assuming the probability that a single node is attacked as 0.2 for comparison purposes, a CoAP network is built for consideration. The system constitutes of 4 sensors whose values are aggregated and sampled by an IoT server. The sampled final value is requested by an http protocol running node in the same network. The

Table 1 Eavesdropping vulnerability probability

S. no	Probability of eavesdropping	Eavesdropping vulnerability
1	0.1	0.11
2	0.2	0.2496
3	0.3	0.4251
4	0.4	0.6496
5	0.5	0.9375
6	0.6	1.3056
7	0.7	1.7731
8	0.8	2.3616
9	0.9	3.0951

Table 2 Eavesdropping comparison

Data from nodes	Sensor 1	Sensor 2	Sensor 3	Sensor 4	Aggregation function (σ)	CDF (φ)
Actual values from device	0	1	3	5	$2.25(\sigma C)(SC)$	0.48777553
Data vulnerable to eavesdropping in CoAP	0	1	3	5	$2.25(\sigma A{-}coap)(SA{-}coap)$	0.48777553
Data vulnerable to eavesdropping in encrypted CoAP	0	39	117	195	$87.75(\sigma A{-}ECoap)(SA{-}ECoap)$	1

aggregated data values are then sent to the server with and without encryption, and the magnitude of eavesdropping vulnerability in both cases is illustrated in Table 2. Subsequent computations show that the main input to the indicator function (that needs to be less than the error tolerance) for the encrypted-CoAP is computed as 0.5122. This effectively means that the suggested CoAP can withstand an error tolerance of around 51% even when an attacker has complete access to the IoT network. To put things in perspective, this is with respect to the 0% error tolerance of current CoAP once an adversary attacks a single node in a network. This is further illustrated in Fig. 2.

$$|\varphi \cdot S_C\Sigma - \varphi \cdot S_{A-CoAP}\Sigma| = 0 \quad (5)$$

$$|\varphi \cdot S_C\Sigma - \varphi \cdot S_{A-ECoAP}\Sigma| = 0.51222447 \quad (6)$$

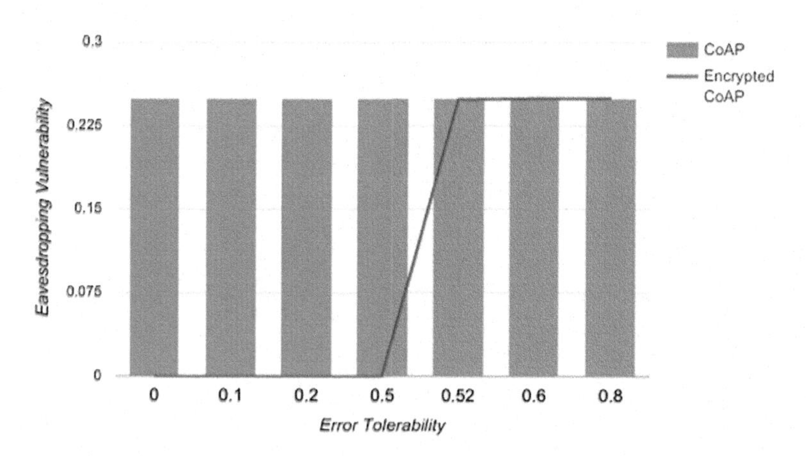

Fig. 2 Comparison of eavesdropping vulnerability

Assuming the probability that an attacker has access to a sensor, p = 0.2, and assuming the allowable error factor is 0, it is possible to derive Eqs. (7) and (8)

$$CoAP\colon \gamma_E = \sum_{i=1}^{4} p^i \cdot I(0 \leq 0) = 0.2496 \tag{7}$$

$$Encrypted\ CoAP\colon \gamma_E = \sum_{i=1}^{4} p^i \cdot I(0.5122 \leq 0) = 0 \tag{8}$$

Assuming an allowable error factor of 0.3, we can derive Eqs. (9) and (10)

$$CoAP\colon \gamma_E = \sum_{i=1}^{4} p^i \cdot I(0 \leq 0.3) = 0.2496 \tag{9}$$

$$Encrypted\ CoAP\colon \gamma_E = \sum_{i=1}^{4} p^i \cdot I(0.5122 \leq 0.3) = 0 \tag{10}$$

Assuming an allowable error factor of 0.6, we can derive Eqs. (11) and (12)

$$CoAP\colon \gamma_E = \sum_{i=1}^{4} p^i \cdot I(0 \leq 0.6) = 0.2496 \tag{11}$$

$$Encrypted\ CoAP\colon \gamma_E = \sum_{i=1}^{4} p^i \cdot I(0.5122 \leq 0.6) = 0.2496 \tag{12}$$

While this parameter basically signifies the efficiency of the distribution of encrypted values and hence the overall efficiency of the encryption method used in encryption of the payload, the system efficiency is a function of the cost incurred for encryption and the resulting output achieved. Feeding the output result to the forthcoming parameter, which effectively quantifies the system efficiency, was achieved by using the eavesdropping vulnerability.

Definition 2 *Performance Ratio* The eavesdropping vulnerability, γ_E, gives us the probability that an adversary can obtain a good estimate of the actual aggregate.

Obviously, to reduce the eavesdropping vulnerability γ, we will have to incur additional overheads. So, it becomes important to quantitatively measure the vulnerability reduction and cost incurred to achieve it. Ideally, a good system must have a high performance ratio.

We define the performance ratio of the adversary relative to a set of compromised nodes, ρ_A as,

$$\rho_A(\sigma, S, S_A, S_C, \delta, C) = \frac{\gamma(\sigma, S, S_A, S_C, \delta)}{C_r \cdot (S_A)} \tag{13}$$

The performance ratio is now calculated, as shown in Eq. (14).

$$\rho\left(\sigma, S, S_A, S_C, \delta, C, C'\right) = \frac{1}{\rho_A(\sigma, S, S_A, S_C, \delta, C)} \cdot \frac{C(S)}{C(S')} \tag{14}$$

where, C is the cost of initially obtaining sensor values (system cost) and C′ is system cost after mitigation of eavesdropping attacks. By varying the probability of eavesdropping, the performance ratio can be tabulated, as shown in Table 3.

Let us assume the energy values as follows, Cost of sensing = 0.015 J, Cost of transmitting and receiving through the medium = 0.025 J, Cost for encryption—0.010 J.

Since we have four sensors scrutinized for attack, the cost of eavesdropping, w.r.t. to the adversary, can be calculated as Cr(SA) = 4 * 0.025 = 0.1 J.

Overall CoAP system cost, C(S) = 4 * (0.015 + 0.025) = 0.16 J.

The overall encrypted CoAP system cost, C′(S) = 4 * (0.15 + 0.015 + 0.025) = 0.2 J.

Therefore, the formula to compute performance ratio of the adversary can be derived as

$$\rho_A \sum_{i=1}^{4} \frac{|\phi(S_C) - \phi(S_{A-ECoAP})| * p^i}{0.1\, i} = \sum_{i=1}^{4} \frac{0.5122 * p^i}{0.1\, i} \tag{15}$$

Assuming the probability of eavesdropping as p = 0.2 and with the help of the Eq. (16), we can calculate ρ_A

$$\begin{aligned} \rho_A &= ((0.51220.2^1) \cdot 1) + ((0.51220.2^2) \cdot 2) + ((0.51220.2^3) \cdot 3) \\ &\quad + ((0.51220.2^4) \cdot 4) = 1.1425 \end{aligned} \tag{16}$$

Table 3 Performance ratio

S. no	Probability of eavesdropping	Performance ratio
1	0.05	3.04
2	0.1	1.48
3	0.2	0.7
4	0.3	0.43
5	0.4	0.3
6	0.5	0.22
7	0.6	0.17
8	0.7	0.13
9	0.8	0.11
10	0.9	0.09

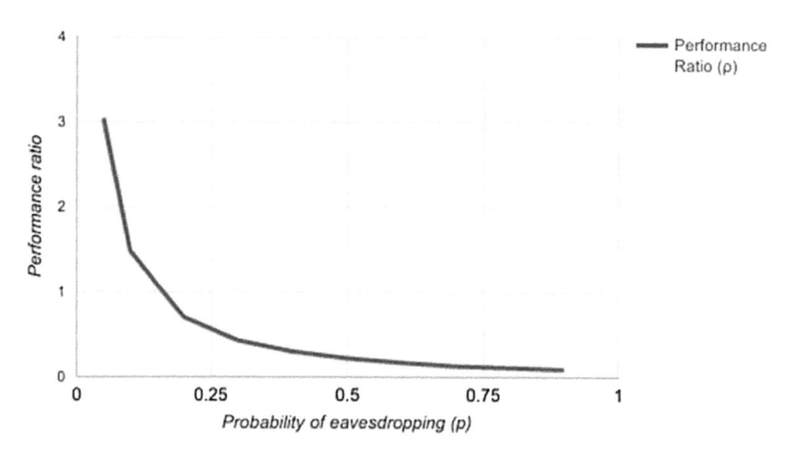

Fig. 3 Performance ratio

This illustration showcases the fact that the variance of vulnerability of a node, which in effect shows the variance in difficulty of attacking a single node in a network, does affect the effectiveness of the system but the system still functions exceedingly well for p¡0.3, which complements the fact that more system resource is required to deploy the suggested method in an insecure network where it is very easy to attack a node (p is close to 1) (Fig. 3).

5 Conclusion

The first step towards achieving the preset objectives was to fully understand the implementation of CoAP and understand the points at which it is vulnerable. After this was discovered, by virtue of understanding the implementation of the proxy, the transportation of the payload from the HTTP request to a CoAP endpoint is analyzed. From the assessed vulnerability, a system was developed with clients trying to access a PIR sensor wirelessly in an IoT network. In the developed system, the payload was encrypted and the system was put into use. Attacks such as man-in-the-middle and eavesdropping were eliminated by virtue of implementing this solution. DTLS and CoAP integration plays a major role in this implementation as it can secure the system against most of the attacks in IoT networks which have passive eavesdropping as their essence. Additionally, ECC is chosen so that less memory and less bandwidth are required. The amount of memory utilized by ECC is minimal when compared with other encryption algorithms.

Upon evaluation by quantifying the eavesdropping vulnerability of the system, it is found that an error tolerance of approximately 51% can be achieved. This means that even when half of the nodes of the system are attacked, it is probabilistically possible that the sampled values at the server are still safe and the system is fully secure. Furthermore, another quantifying measure is used to quantify the overhead involved in the system and the cost incurred for the additional security is studied and quantified and found to be efficient and more than acceptable for the security achieved.

6 Future Work

The IoT is a field which is very new and fast growing. As a result there is a never ending scope for future work. Pertaining to the system under consideration, the extension of the module where the PIR sensor sends out a signal to the ECC for encryption can be further discussed and enhanced through iteration of accepting and manipulating multiple devices.

Since every node connected contains a CoAP entity, the translation of the message from the IoT device along with the layers of the DTLS and the additional layers on top of it can be enhanced to decrease the size of each packet and a more optimal way of delivering packets with minimal load can be achieved by eliminating redundant layers in the stack for incoming devices. Another important topic that lends scope to future work is the dynamic expansion of the network. In a wireless environment, when devices enter and leave the network space at will, the configuration and encryption of newly entered devices through key exchange must be automated and made efficient in such a way that there is no latency for enabling key-exchange and encryption for newly registered devices in the network.

In addition, the IDS enacted in this scenario will initiate filtering only at periodic intervals specified by the client system or the host which setsup the system for sending out the packets. This can be further improved by instantiating the IDS as a primary gateway for the packet to pass. This in turn will make sure that the system will verify each and every packet which comes in or goes out of the system to stop altered messages from any attacker. Moreover, the IDS is not fully aware of the CoAP packets which are somehow new to some of the threats from attackers if the IDS is based on signatures.

References

1. Anand M, Ives Z, Lee I (2005) Quantifying eavesdropping vulnerability in sensor networks. In: Proceedings of the 2nd international workshop on data management for sensor networks, pp 3–9, ACM
2. Brachmann M, Garcia-Morchon O, Kirsche M (2011) Security for practical coap applications: issues and solution approaches. In: Proceedings of the 10th GI/ITG KuVS Fachgespraech Sensornetze (FGSN11), pp 15–16. Paderborn, Germany
3. Gou Q, Yan L, Liu Y, Li Y (2013) Construction and strategies in iot security system. In: 2013 IEEE international conference on green computing and communications and IEEE internet of

things and IEEE cyber, physical and social computing, pp 1129–1132. https://doi.org/10.1109/GreenCom-iThings-CPSCom.2013.195
4. Granjal J, Monteiro E, Silva JS (2015) Security for the internet of things: a survey of existing protocols and open research issues. IEEE Commun Surv Tutor 17(3):1294–1312
5. Heer T, Garcia-Morchon O, Hummen R, Keoh SL, Kumar SS, Wehrle K (2011) Security challenges in the ip-based internet of things. Wireless Pers Commun 61(3):527–542
6. Rahman RA, Shah B (2016) Security analysis of iot protocols: a focus in coap. In: 2016 3rd MEC international conference on big data and smart city (ICBDSC), pp 1–7. IEEE
7. Raza S, Trabalza D, Voigt T (2012) 6 lowpan compressed dtls for coap. In: 2012 IEEE 8th international conference on distributed computing in sensor systems, pp 287–289. IEEE
8. Raza S, Shafagh H, Hewage K, Hummen R, Voigt T (2013) Lithe: lightweight secure coap for the internet of things. IEEE Sens J 13(10):3711–3720
9. Riahi A, Challal Y, Natalizio E, Chtourou Z, Bouabdallah A (2013) A systemic approach for iot security. In: 2013 IEEE international conference on distributed computing in sensor systems, pp 351–355. https://doi.org/10.1109/DCOSS.2013.78
10. Roman R, Najera P, Lopez J (2011) Securing the internet of things. Comput Netw 44(9): 51–58
11. Vŭcinić M, Tourancheau B, Rousseau F, Duda A, Damon L, Guizzetti R (2015) Oscar: object security architecture for the internet of things. Ad Hoc Netw 32:3–16
12. Xu T, Wendt JB, Potkonjak M (2014) Security of iot systems: design challenges and opportunities. In: Proceedings of the 2014 IEEE/ACM international conference on computer-aided design, ICCAD'14, pp 417–423. http://dl.acm.org/citation.cfm?id=2691365.2691450
13. Zhang ZK, Cho MCY, Shieh S (2015) Emerging security threats and countermeasures in iot. In: Proceedings of the 10th ACM symposium on information, computer and communications security, pp 1–6. ACM
14. Zhao K, Ge L (2013) A survey on the internet of things security. In: Computational intelligence and security (CIS), 2013 9th international conference on, IEEE, pp 663–667

A Survey of Fingerprint Recognition Systems and Their Applications

Puja S. Prasad, B. Sunitha Devi, M. Janga Reddy and Vinit Kumar Gunjan

Abstract Recognition for authentication using biometrics is an intricate pattern recognizing technique. The process is really hard to architect and design, and choosing precise algorithms competent of fetching and extracting significant features and then matching them correctly, particularly in the cases where the quality of the fingerprint images are poor quality image capturing devices are used. Problems also occur where minutia are clearly visible on very small fingerprint area that are not exactly capture by camera. It is a false assumption that fingerprint recognition is a completely settled area regarding the authentication of a person just because it always give the correct identity of an individual. Fingerprint identification remains a very complex and intricate pattern-recognition system for authentication of a person.

Keywords Binarization · Minutia · Ridge · Bifurcation · Galton points

1 Introduction

This paper focuses on the various research and techniques applied by researchers that have been used in iris and fingerprint recognition, as well as the different techniques of fusion of two biometrics. A number of authors and researchers give a number of different techniques for fingerprint recognition, iris recognition, and the

P. S. Prasad (✉) · B. Sunitha Devi · M. Janga Reddy · V. K. Gunjan
CMRIT, Medchal, Hyderabad, TS, India
e-mail: puja.s.prasad@gmail.com

B. Sunitha Devi
e-mail: sunithabigul@gmail.com

M. Janga Reddy
e-mail: principalcmrit@gmail.com

V. K. Gunjan
e-mail: vinitkumargunjan@gmail.com

A. Kumar and S. Mozar (eds.), *ICCCE 2018*,
Lecture Notes in Electrical Engineering 500,
https://doi.org/10.1007/978-981-13-0212-1_53

fusion of different biometric along with their advantages, disadvantages, and gaps in the research. Human fingerprints are rich in complex feature details that are unique to each person and it is these minutiae which can be used for authentication purposes.

The objective of all researchers is to build a model for fingerprint authentication through extracting and then matching minutiae points, and to enhance the accuracy of the model. In order get good quality minutiae points, i.e. ridge endings and ridge bifurcations in fingerprints, a number of steps are undertaken. Preprocessing is first applied in the enhancement of an image and then binarization of the biometric takes place (fingerprints in this case) before evaluation. Hardware and software has been combined to build an overall fingerprint recognition technique extractor and a minutia matcher. False minutia points also add noises, so its elimination is also required to remove noises. A number of algorithms have been developed for minutia matching. These algorithms are capable of searching the relationship between the pattern of the extracted minutia and the database templates following a thorough search. Then evaluation of the performance of the developed system occurs using the template and fingerprints captured from different people.

1.1 Research in Fingerprint Recognition

Sir Francis Galton defined Galton points, which are the attributes of a particular finger and by which authentication has been in used for identifying fingerprints since the late nineteenth century. "Galton points" were the first steps taken in the study of fingerprint recognition, a field which has expanded and changed over the past century. Fingerprint recognition began to be mechanized in the late 1960s alongside the advances in computing technologies. Galton points are referred to as minutiae and have been used to build up automatic fingerprint recognition technology.

Ballan [1] gives the idea of fingerprint processing, which is directional by using smoothing fingerprint identification and classification based on single points, i.e. core points and delta points. These points are taken from histograms of an input fingerprint. According to this method, Lasso and Wirbel are the two main categories of this process which includes the formation of an image, image block representation, point detection, and the result.

Ratha et al. [2] projected a unimodal fingerprint authentication method using minutia extraction technique, which is based on graph theory. Hastings [3] proposed a method for enhancing the minutia point called a ridge by oriented dispersal using an adaptation of anisotropic diffusion for smoothing the image in the parallel ridge flow direction.

Prabhakar et al. [4] a developed fingerprint identification method using filter-based representation. This technique uses both global and local individuality in a fingerprint to enable recognition. In this process, each finger image is filtered in several directions resulting in the extraction of around 640 feature vectors in the middle region of the extracted fingerprint. Euclidian distance computed in matching stage between the template finger and the input finger. The process results in very fine matching with an excellent degree of accuracy.

Bazen et al. [5] proposed a correlation-based fingerprint verification and identification system. The advantage of the correlation-based system is that it uses the richer information of a grayscale image compared to normal minutia-based systems. This correlation-based technique first selects appropriate templates in the prime fingerprint, then that initially chosen fingerprint is used for template matching to set them in the secondary print, and then both fingerprints are compared using the minutia positions. In contrast, a system which uses other techniques like minutia based or a correlation-based method for fingerprint verification system is trained to deal with images that suffer from non-uniform shape distortion or images from which no minutiae can be extracted clearly. Compared with other techniques, the correlation-based technique gives improved results in terms of error rate as well as increased performance where images are not captured properly.

Pores and ridges used to match fingerprints for identification and verification are proposed by Jain et al. [6]. The detail of fingerprint ridge friction is described in a hierarchical order of three levels. Level 1 is a pattern, Level 2 is a minutia point, and level 3 includes pores and ridge contours. To increase the sensing power of the fingerprint scanner, it rarely enhances the performance of fingerprint matching unless all the extended features are used. They propose a hierarchical system of matching in which all the three levels are used. Level 3, which uses pores and ridges extracted using Gabor filters and Wavelets transform, matches fingerprints by using the iterative closest point algorithm. There is a reduction in error rates, i.e. error rates are equal when level 3 features are combined with level 2 and level 1. Some conventional studies have often reported a different category of error, i.e. false rejection rate and false acceptance rates that match score distributions.

Govindaraju et al. [7] present papers in feature extraction using the chain code method. They give two chain code algorithms, one for enhancing fingerprint images and the other for extracting minutia. For enhancing images, quick averaging occurs first and then binarization occurs before the chain code is generated. Chain code representation helps in calculating the direction field of given minutia. A filtering algorithm is also used for enhancing images. The quality of the image is very important as chain code representation is for binary images. The detected minutiae use a sophisticated ridge outline following procedure. Conducting numbers of experiment proves that the method is very efficient.

Hsieh et al. [8] developed a method for classification of fingerprint images. They use the thinned image of a fingerprint for extracting directional as well as a singular point from the region of interest (ROI) of the extracted images. The main benefit of this algorithm is that it reduces the computation time required to extract the features from the images.

Sonavane and Sawant [9] offered a method by which the fingerprint image is broken into a set of clean images and their orientation is estimated, giving a special domain fingerprint enhancement.

Girgis et al. [10] worked on the problem of incomplete fingerprint images. During capture of the image, most of the ridge endings and bifurcations get deformed due to the incomplete nature of the image. Their proposed genetic algorithm improves the quality of deformed ridges. Ambiguities in pairing a minutia are solved by using a dynamic programming which of Genetics algorithm and it shows the transformation that is globally optimized.

Lu et al. [11] proposed a well-organized algorithm that is very effective in extracting minutia and results in a reduction of the computation time needed to extract minutia thereby enhancing the performance of automatic fingerprint authentication systems [12]. This algorithm is very significant in that it extracts true minutiae while removing fake minutiae in the final post-processing stage. This planned new fingerprint image algorithm for post-processing makes a hard work to reliably discriminate bogus minutiae from accurate ones by making use of ridge number information, crafty and arranging various giving out techniques properly, referring to the original gray-level image, and also choosing various processing parameters carefully. The post-processing algorithm for bogus minutia removal works well.

Kumari and Suriyanarayanan [13] proposed a different method for enhancing performance for minutia extraction like searching the edges in a fingerprint using local operators, such as Roberts method, Prewitt method, Canny method, and LOG method. The segmentation of captured image is in advance done to extract individual segments from the image.

Kukula et al. [14] proposed a method that applies different levels of force to investigate its effect performance of matching image scores in terms of quality, and minutiae amount between capacitance and optical fingerprint sensors [12]. "Kaur et al. [15] has given the joint techniques to make an extractor that extracts a pattern present in the fingerprint and a minutia matcher. It uses segmentation process using morphological operation used to get the better removal of false minutiae, fingerprint minutia thinning.

Zhao et al. [16] worked on a method called a pore model for extracting fingerprint pores. Pores developed due to sweating are now the subject of ongoing research as they are considered to be one of the best variables for authenticating a person and give an extremely good result. On the other hand, the authors explain that the nature of pores is not always isotropic. They developed a direct approach to fingerprint matching by using fingerprint pores. A RANSAC algorithm is used to refine the correspondence between pores based on their local features. By using pore matching free from minutia matching, it facilitates the fusion of the pore and minutia match scores, which improves the efficiency of the fingerprint recognition system.

Tarjoman and Zarei [17] looked at competent methods of classification and at authenticating a person using fingerprints. In this important paper, the researcher introduces a new approach that is structural in nature and uses directional images for the classification so that the number of subclasses increases. In this process, they use pixels present in the same direction for image segmentation. The segmented image is used to construct a relational graph using relevant information.

Finally, they compare the obtained graph and the model graph for matching purposes. The increasing number of subclasses increases the accuracy and processing speed of the fingerprint recognition system, as explained by the authors.

Cui et al. [18] proposed an algorithm that detects edge, as the edge is an important parameter in the thinning process. This paper also gives some operators for edge detection and compares their characteristics and performance. This study also performs an experiment to show that each algorithm has its advantages and disadvantages. Finally, a suitable algorithm should be selected according to the characteristic of the images detected, so that it can perform as well as possible. Canny operator is not suitable for edge detection as it is very much affected by noise.

Edge detection is very important as it preserves a great deal of important data that is then further used for image processing. The advantage of canny operator is that it uses a smoothing effect to remove noise as well as having two different threshold values for detection of the weak edge and the strong edge. However, the complex algorithm used by canny operator make it time consuming. The sobel operator is very simple but its accuracy suffers from noise as well as images with a gray gradient. It does not give the location of the edge very accurately. The edge detection algorithm for binary images is very simple, detects the boundary of the image accurately, and the output-processed images do not have to go through the thinning process. Different algorithms have their advantages and disadvantages, and the selection of algorithm only depends on the quality and nature of the images taken.

Karna et al. [19] proposed cross-correlation, which is normalized-based fingerprint matching. To perform fingerprint matching using minutia pairings has been done but it is used fairly less time. However, there is a limitation with this technique as it is not very competent in recognizing the squat characteristic of fingerprints. A correlation technique is introduced to overcome this problem, providing an enhanced result. Due to their providing better results, correlation-based methods are in great demand in the biometrics field. This proposed technique for matching cross-correlation normalized method reduces error rates as well as reduces computational speed compare to another matching method and is effectively used for fingerprint identification and recognition system.

Vatsa et al. [20] proposed a combination technique where they combine ridges and pores with the aim of improving fingerprint recognition performance. The objective of this paper is to combine level 1 and level 2 features to enhance the performance of the matching algorithm. A ridge-tracing algorithm is used to extract minutia. By using level-2 minutiae and level-3 pore and ridge features, a speedy algorithm was developed. To register a fingerprint image, a two-stage process is used. In the first stage, coarse registration is performed on images based on a Taylor series transformation. In the second stage of fine registration, a thin plate spline transformation is used. The functional curve evolution called a Mumford–Shah curve is used to efficiently segment boundaries and extract ridge features, which is a feature extraction algorithm.

Nowadays a number of digital devices have an inbuilt feature, i.e. Bluetooth, which also makes them very suitable for indoor positioning. Subhan et al. [21] uses Bluetooth devices for indoor positioning by using the strength level of the signal to estimate the position of an object. However, the measurements are corrupted by a variety of environmental conditions, such as reflection, temperature, the human body, presence of obstacles, and other different communication signals. Filtering is therefore needed. The authors present an experimental relationship between the standard radio propagation model using the received power level.

Onyesolu and Ezeani [22] present work on fingerprint identification for enhancing the security of ATMs using fingerprint authentication. They use an instrument that consists of three section. In the first section, they deal with the user profile. The second section deals with its frequent uses and user reliability. Their study was carried out over four months and shows an efficient result.

Detecting the real life nature of a fingerprint and protecting the system from the use of an artificial fingerprint or the use of gelatin finger is a very difficult task. Fingerprint liveness detection methods have been developed as an attempt to overcome the vulnerability of fingerprint biometric systems to spoofing attacks. Fingerprint liveness detection is challenging and no experiment can produce detection that is 100% correct.

Ghiani et al. [23] provide a novel fingerprint liveness technique termed BSIF. This method of liveness detection, which is very similar to local binary pattern and local phase quantization-based representations, encodes the local fingerprint texture on a feature vector.

Arjona and Baturone [24] propose a global fingerprint feature called q finger map. This provides fuzzy information related to fingerprint images. A fuzzy rule is built that unites information from numerous q finger maps and it is used to list or enroll an individual in a database. There is not much more differences in the error and access rates in this fuzzy retrieval system compared to other systems that operate on similar rules but can be implemented in hardware platforms of vastly lower computational resources, also results in lower processing time but can be implemented in hardware platforms having lower computational resources, results in lower processing time.

2 Conclusion

A large number of studies have been carried out in the field of fingerprint recognition and authentication systems. Different researchers have worked on different aspects, with some working on image processing algorithms, such as enhancement of image and false minutia extraction. Others have worked on pattern recognition problems that a number of algorithms to match a minutia. Fingerprint recognition is one of the growing areas for authenticating a person and has its advantages and disadvantages. In this paper, we present a survey of the different research that benefits the new user in undertaking a project on the fingerprint authentication and recognition process. Extracting minutiae is one of the major steps in a fingerprint recognition system. Matching a pattern with new and existing minutiae is also of great importance for the overall authentication system.

References

1. Ballan M (1998) Directional fingerprint processing. In: International conference on signal processing, vol 2, pp 1064–1067
2. Ratha NK, Karuk K, Chen S, Jain AK (1996) A real-time matching system for large fingerprint databases. Trans Pattern Anal Mach Intell 18(8):799–813
3. Hastings R (2007) Ridge enhancement in fingerprint images using oriented diffusion. In: IEEE computer society on digital image computing techniques and applications, pp 245–252
4. Prabhakar S, Jain AK, Wang J, Pankanti S, Bolle R (2002) Minutia verification and classification for fingerprint matching. In: International conference on pattern recognition, vol 1, pp 25–29
5. Bazen AM, Verwaaijen GTB, Gerez SH, Veelenturf LPJ, van der Zwaag BJ (2000) A correlation based fingerprint verification system. In: Proceedings of workshop on circuits systems and signal processing, pp 205–213
6. Jain AK, Ross A, Prabhakar S (2004) An introduction to biometric recognition. In: IEEE transactions on circuits and systems for video technology, vol 14, no 1, January 2004
7. Govindaraju V, Shi Z, Schneider J (2003) Feature extraction using a chaincoded contour representation. In: International conference on audio and video based biometric person authentication, Surrey, UK
8. Hsieh C-T, Shyu S-R, Hu C-S (2005) An effective method of fingerprint classification combined with AFIS. In: Part of the lecture notes in computer science book series (LNCS, vol 3824)
9. Sonavane R, Sawant BS (2007) Noisy fingerprint image enhancement techniques for image analysis: a structure similarity measure approach. J Comput Sci Netw Security 7(9):225–230
10. Girgis MR, Mahmoud TM, Abd-Hafeez T (2007) An approach to image extraction and accurate skin detection from web pages. In: World academy of science, engineering, and technology, pp 27
11. Lu H, Jiang X, Yun Yau W (2002) Effective and efficient fingerprint image post processing. In: International conference on control, automation, robotics and vision, vol 2, pp 985–989
12. Wei L (2008) Fingerprint classification using singularities detection. Int J Math Comput Simul 2(2):158–162
13. Vijaya Kumari V, Suriyanarayanan N (2008) Performance measure of local operators in fingerprint detection. Acad Open Int J 23:1–7
14. Kukula EP, Blomeke CR, Modi SK, Elliott SJ (2008) Effect of human interaction on fingerprint matching performance, image quality, and minutiae count. In: International conference on information technology and applications, pp 771–776
15. Kaur M, Singh M, Girdhar M, Sandhu S (2008) Fingerprint verification system using minutiae extraction technique. In: Proceedings of world academy of science, engineering and technology, vol 36, pp 497–502
16. Zhao Q, Zhang D, Zhang L, Luo N (2010) Adaptive fingerprint pore modeling and extraction. Pattern Recogn 43(8):2833–2844
17. Tarjoman M, Zarei S (2008) Automatic fingerprint classification using graph theory. In: Proceedings of world academy of science, engineering and technology, vol 30:831–835
18. Cui W, Wu G, Hua R, Yang H (2008) The research of edge detection algorithm for fingerprint images. In: Published 2008 in World Automation Congress

19. Karna DK, Agarwal S, Nikam S (2008) Normalized cross-correlation based fingerprint matching. In: Conference on computer graphics, imaging and visualisation, CGIV'08
20. Vatsa M, Singh R, Noore A, Singh SK (2009) Combining pores and ridges with minutiae for improved fingerprint verification. In: Elsevier, signal processing, vol 89, pp 2676–2685
21. Subhan F, Hasbullah H, Rozyyey A, Bakhsh ST Indoor positioning in bluetooth networks using fingerprinting and alteration approach. In: Information science and applications (ICISA), 2011 international conference on information science and applications, 26–29 April 2011
22. Onyesolu MO, Ezeani IM (2012) ATM security using fingerprint biometric identifier: an investigative study, (IJACSA). Int J Adv Comput Sci Appl 3(4)
23. Ghiani L, Hadid A, Marcialis GL, Roli F (2013) Fingerprint liveness detection using binarized statistical image features. In: 2013 IEEE sixth international conference on biometrics: theory, applications, and systems (BTAS)
24. Arjona R, Baturone I (2015) A fingerprint retrieval technique using fuzzy logic-based rules. In: International conference on artificial intelligence and soft computing ICAISC 2015: artificial intelligence and soft computing, pp 149–159
25. Ratha NK, Chen S, Jain AK (1995) Adaptive flow orientation-based feature extraction in fingerprint images. Pattern Recogn 28(11):1657–1672

Iris Recognition Systems: A Review

Puja S. Prasad and D. Baswaraj

Abstract Recognition for authentication using biometric features is an intricate pattern-recognizing technique. The process is extremely hard to design and build, and choosing the exact algorithms competent to fetch and extract significant features and then match them correctly, particularly in cases where the quality of the captured images is poor or low-quality image capturing devices with very small capturing areas are used. It is a false assumption that biometric recognition is a completely settled area regarding the authentication of a person just because it always gives the correct identity of an individual. Iris identification remains a very complex and intricate pattern recognition system for authenticating a person. This paper focuses on the different techniques used for authentication.

Keywords Iris · Biometric · Hough transform · Laplacian
Canny edge detection

1 Introduction

This paper focuses on the various studies and associated techniques applied by researchers that have been used in iris authentication and recognition systems. A range of authors and researchers discuss a variety of different techniques for iris recognition, their fusion with different biometric, their advantages and disadvantages, and the gaps in research. The human iris has very rich and complex feature details which are unique to an individual. It is actually the floral part of iris that makes it very valuable in identifying a person and this can be used for authentication purposes.

The aim of all researchers in the field is to build a model for iris authentication through extracting and then matching the floral part or points and to enhance the

P. S. Prasad (✉) · D. Baswaraj
Department of CSE, CMRIT, Hyderabad, TS, India
e-mail: puja.s.prasad@gmail.com

D. Baswaraj
e-mail: d.baswaraj@cmritonline.ac.in

A. Kumar and S. Mozar (eds.), *ICCCE 2018*,
Lecture Notes in Electrical Engineering 500,
https://doi.org/10.1007/978-981-13-0212-1_54

accuracy of that model. To get good iris images, a numbers of steps are taken. Pre-processing is first stage and is applied in the enhancement of an image followed by binarization of the iris before evaluation. Hardware and software have been combined to build an overall iris recognition technique extractor and a pattern matcher. A number of algorithms have been developed for iris matching. These algorithms are capable of identifying the relationship between the extracted floral pattern and the database templates following a thorough search. Then evaluation of the developed system takes place to better understand its performance using the template and the iris images captured from different people.

2 Research in Iris Recognition and Authentication

John Daugman was the first person to use iris characteristics in the development of a model of an iris identification system. Daugman's system is a very popular iris recognition systems and is used in various organizations for authentication purposes. Daugman's iris system works in different stages, such as segmentation, normalization, then feature encoding, and iris code matching [1].

Wildes et al. [2] designed and implemented an automated recognition system for the iris, in which localization of iris region occurs by using histogram processing and then image filtering with band pass filters using band pass filter coated with isotropic substance is done. Following that, Wildes et al. [3] again proposed a model for personal authentication using iris characteristics called a machine-vision system for iris recognition. As the iris is an external body, its appearance is suitable to remote examination with the help of a machine-vision system. This system exhibited perfect performance in the evaluation of 520 iris images.

Wildes [4] proposed another modal that uses diffuse light sources in a personal identification system using the iris. In this model, he first extracted the boundaries of the iris using an edge detection method called canny edge detection. This was followed by a Hough transform circular function. By the use of a Laplacian Gaussian filter at multiple scales, he produced the templates used and in then in matching stage he computed the correlation.

Boles and Boashash [5] developed a new technique using different resolution levels by using zero-crossings of wavelet transform in different concentric circles of the iris, and thus this one dimensional image is used for identification using their wavelet transform. Matching was completed by taking into account two variation functions between the iris image and the iris template.

Zhu et al. [1] also works on the extraction of global features for personal identification using the iris by applying a Gabor filter and wavelet transform technique, by applying the multi-channel Gabor filter and wavelet transform technique. As global feature are less sensitive to noise. El-Barky [6] proposed a recognition system involving the iris, made quicker by the use of neural networks. Lim et al. [7] proposed an algorithm named "Efficient Iris Recognition through

Improvement of Feature Vector and Classifier". In this algorithm segmentation, normalization, and encoding feature method was taken same as the Daugman techniques. The matching stage has two methods of learning called winner selection and weight vector initialization for vector quantization (LVQ), and these were used to categorize the feature vectors. This system was used for the verification and identification of individual's personality. Ma et al. [8] employed another filter called circular symmetric filters (CFS) for iris recognition system. CFS or circular symmetric filters has certain variation from the Gabor wavelet. It uses a function called circular symmetric sinusoidal function for modulation. To detect boundaries they used a different edge detection function and a Hough transform function. This system focused on the fact that the center point coordinate of the pupil and iris were generally not the same.

Lye et al. [9], proposed a system that works using a different method. First, localization of the iris takes place and this is then used to determine different iris patterns. Using general camera for capturing images and then from that captured images relevant iris patterns are detected. Once the iris pattern is detected, that selected iris pattern is converted into a rectangular format which is then used for extracting features from the iris.

Sanchez-Avila et al. [10] used a wavelet transform called dyadic wavelet transform. They compared the results for different distances techniques, i.e. Euclidean, Hamming, and direct distance using zero-crossing, for classification and verification and they came to the conclusion that Hamming distance had better results. The iris recognition systems proposed by Daugman [11] were intended for both verification and identification, while the other systems were proposed for identification only. Boles and Boashash [5] worked on a one-dimensional signal whereas Daugman and Wildes worked on two-dimensional images. A zero crossing transformation is used for encoding a one-dimensional image at different resolution is done for recognizing iris pattern. Many projects are undertaken using the algorithm that Daugman proposed for use in iris recognition and authentication systems.

Lee et al. [12] introduced the distinct feature, which is actually a binary feature and is used as key iris recognition system. The variation in the quality of iris images is not important in their work. The keys for iris matching are actually obtained by the given pattern, which is then constructed as a lattice-structured image to symbolize a bit pattern for an individual iris image. Then the given reference patterns are put into a filter. Iris texture is shown according to the iris power spectrum in the corresponding frequency sphere. Chen et al. [13] proposed, method that is completely based on wavelet of iris images. Delivering good spatial adaptation and determining local quality measures for different regions of an iris image is the main aim of their work. The proposed quality index can reliably estimate the matching performance of an iris recognition and authentication system. By amalgamating local quality measures in the matching algorithm, they also saw a relative matching performance enhancement of about 21% and 11% at the equal error rate (EER) on the WVU and CASIA databases for irises respectively.

Cui et al. [14] proposed a matching algorithm called elastic blob to reduce the problem created by local feature-based classifiers in the iris recognition algorithm. The measurement of numbers of minutiae is greatly affected by the noise present during the capturing of an image, such as nonlinear deformation of the iris, and occlusions by eyelashes and eyelids. Local feature-based classifiers are combined with elastic blob for recognizing iris image efficiently. Intra-class comparison are used when local feature classifiers are uncertain about the decision.

Sun et al. [15] attempted to reduce the restrictions of local feature-based classifiers. In order to identify a number of iris images competently, a fresh chain scheme is proposed to combine the iris blob matcher and LFC. When there is uncertainty in decision due to poor quality iris images then comparison is done using intra-class value. The iris blob matcher is used to result the input iris identity as it has the capability to identify noisy images. Miyazawa et al. [16] give a matching algorithm which is actually phase based. To build an iris authentication system in a combined fashion with a simple matching algorithm they use a phase components using Fourier Transforms of two-dimensional iris images is used. Lili and Mei [17] proposed an algorithm which is different from other algorithms, this algorithm uses edge points for finding curve fitting. Fundamental iris image quality evaluation is important in an automatic system for iris recognition and authentication.

To reduce noise in an iris recognition system, an algorithm is given by Wang and Runtao [18], in which to stay away from noise, iris features preserving principle is apprehended. Result of by doing number of experiment resultant phase information for iris image is given and then phase preserving using complex Gabor wavelets filters is also discussed. To analyze the algorithm, white noise is added to iris images and after applying a de-noising algorithm, hamming distances between the iris images are calculated.

Yuan and Shi [19] have explain the iris images that works on structural attribute of our eyes, they come forward an mathematical values for iris location which is fast in nature. Gray projection is used to find the center of the iris and two points that are present at the right and left boundaries of the image respectively, and have finally find a point that are present at the lower boundary by using direction of edge detection operators, then they give the edge of pupil and possible points at the center. Using a Hough transform they were able to find the accurate pupil boundary and center in an iris image that go into image processing phases. The next step in the process is to search for two points that are located at the right and left boundaries between the iris and the sclera in a horizontal direction by using the precise center and direction edge detection operators. Finally they got at center of iris the horizontal coordinate and the two points are searched that are situated at upper and lower boundaries between iris and sclera start and at the horizontal coordinate of the center of iris along the directions and making plus and minus thirty respectively by using directionally edge detection operators, so they ensured the coordinate of the center and the boundary between the sclera and the iris. This method is much faster than existing methods in terms of processing speed, as they had already proved. A value is chosen and a window is centered on it.

Bolle et al. put forward approximation methods for authentication systems that use biometric for authenticating people. Authentication by using biometrics is becoming popular nowadays due to its growing reliability and ease of use. Performance analysis of systems using biometrics is an important factor in that it attempts to deal with two ideas of performance estimation that have been traditionally neglected.

Ziauddin and Dailey [20] proposed a hybrid technique for iris scanning that provides a more accurate result compared to other authentication techniques available and is one the most precise authentication technologies. However, even with their tremendously high accuracy in an ideal imaging environment, their performance degrades when the iris images captured are noisy or the enrollment environment and verification imaging situation are significantly different. To address this issue and enhance the performance of iris recognition in less than ideal conditions of image capturing. Introduction of weighted majority voting technique applicable to any of the biometric authentication and recognition system by comparing time taken during enrollment and verification bitwise of biometric templates. Experiments using CASIA database for iris, they find that their method performs existing reliable bit selection method. In their experiment, local frequency distribution is obtained for iris localization, which then uses multiple different techniques for the selection of two 64 $\times$ 64 corresponding iris regions. The iris gradient (iris texture change) is then obtained for the sclera and iris boundary. For pupil segmentation, a 20 $\times$ 20 iris region was recreated and then threshold intensity applied to localize the pupil region. Iris segmentation is mainly based on a circular Hough transform but requires several improvements.

Lagree and Bowyer [21] used another method called a log Gabor filter for iris recognition and authentication. The segmentation process takes place in which images are segmented to obtain an output for the preferred step. A linear Hough transform method is applied to the occluded iris portions of the upper and lower eyelids. ID Gabor bitmask is used to separate eyelashes by using eyelash and eyelid occlusion location, image filters and isolate collarette region, iris Complex segmentation and masking are exactly same as pattern called Iris Bee Pattern. In normalization, rubber sheet model but log Gabor filter is not used normalized patterns are actually vectors having texture feature. After converting polar coordinates into Cartesian filters, Spot detector and line detector created. A two-dimensional array is created which contains calculations for all the iris images. For each vertical and horizontal image dimensions of radial and angular dividing this previously normalized image array into a number of smaller resolutions correspondingly.

Ross et al. [22] gives five basics sections of directionality in an iris image. Using different image processing and pattern recognition algorithms, image acquisition is achieved and iris texture is extracted using segmentation, normalization, encoding, and matching as represented in a binary iris code. Multiple iris codes of the same eye are determined and then aligned. During image acquisition, a number of images are captured and different iris texture pattern classifications can be used to extract features from images. In the segmentation stage, a non-cooperative database is used and three regions are formed using iris images. Next, boundary detection for undesired areas like eyelids regions in right, left and bottom are removed from images. SIFT method is applied for most critical section of iris image to achieve leading orientation as well as feature point to increase accuracy. To improve the efficiency and accuracy of the proposed system, they present a new approach to making a feature vector compact and efficient by using wavelet transform, and two straightforward but efficient mechanisms for a competitive learning method such as a weight vector initialization and the winner selection [23, 24].

3 Conclusion

In this paper, we show how a person can be identified using their iris. There are a number of ways for identifying a person. Instead of having many keys or remembering passwords, we can use features that are present in our body like our iris, fingerprint, and ear lobe etc. These features are called biometrics. We constructed a biometric recognition system using physical characteristics or habits of any person for identification. Different researcher works on iris and this paper presents their work in brief. A number of techniques are used in iris recognition, such as image enhancement, image capturing, and image matching, and these use different algorithms. To complete the dream of Mr. Daugman, where iris recognition systems are deployed in many fields, a lot of improvements are still required in such systems.

References

1. Zhu Y et al (2000) Biometric personal identification based on iris patterns. In: IEEE Pattern Recognition. Proceedings, vol 2, Sept 2000, pp 801–804
2. Wildes RP, Asmuth JC, Green GL, Hsu SC, Kolczynski RJ, Matey JR, McBride SE (1994) A system for automated iris recognition. In: Proc IEEE Workshop Mach Vis Appl pp 121–128
3. Wildes R, Asmuth J, Green G, Hsu S, Kolczynski R, Matey J, McBride S (1996) A machine-vision system for iris recognition. Mach Vis Appl 9:1–8
4. Wildes R (1997) Iris recognition: an emerging biometric technology. Proceedings of the IEEE, vol 85, no 9. September 1997
5. Boles W, Boashash B (1988) A human identification technique using images of the iris and wavelet transform. IEEE Trans Signal Process 46(4):1185–1188. https://doi.org/10.1109/78.668573
6. El-Barky HM (2001) Human iris detection using fast cooperative modular neural nets, neural networks. Proceedings of international joint conference on IJCNN '01, vol 1, 2001. pp 577–582
7. Lim S, Lee K, Byeon O, Kim T (2001) Efficient iris recognition through improvement of feature vector and classifier. ETRI J, vol 23, no 2, Korea
8. Ma L, Tan T, Wang Y (2002) Iris recognition based on multichannel Gabor filtering. In: Proceedings of the international conference on asian conference on computer vision, pp 1–5

9. Lye WL, Ali C, Liau CF, Jamal AD (2002) Iris recognition using self organizing neural network. In: IEEE 2002 student conference on research and development proceedings, Shah Alam, Malaysia, pp 169–172
10. Sanchez-Avila C, Sanchez-Reillo R, de Martin-Roche D (2001) Iris recognition for biometric identification using dyadic wavelet transform zero-crossing. In: Proceedings of the IEEE 35th international carnahan conference on security technology, pp 272–277
11. Daugman J (2004) How iris recognition works. In: IEEE transactions on circuits and systems for video technology, pp 21–30
12. Lee J-C, Huang PS, Chiang C-S, Tu T-M, Chang C-P (2006) An empirical mode decomposition approach for iris recognition. In: Proceedings of the IEEE international conference on image processing, pp 289–292, 8–11 October, Atlanta, GA, 2006
13. Chen E-Y, Huang Y-P, Luo S-W (2002) An efficient iris recognition system. In: International conference on machine learning and cybernetics, pp 450–454
14. Cui J, Wang Y, Huang JZ, Tan T, Sun Z (2004) An iris image synthesis method based on PCA and super resolution. In: Proceedings of the 17th international conference on pattern recognition, Aug 23–26, pp 471–474. IEEE Explore Press, USA. https://doi.org/10.1109/icpr.2004.1333804
15. Sun Z, Tan T, Yang Y, et al (2005) Ordinal palmprint representation for personal identification. In: Proceedings of CVPR 2005, San Diego, pp 279–284
16. Miyazawa K, Ito K, Aoki T, Kobayashi K, Nakajima H (2005) A phase-based iris recognition algorithm. In: Zhang D, Jain AK (eds) Advances in biometrics. ICB 2006. Lecture notes in computer science, vol 3832. Springer, Berlin, Heidelberg
17. Lili P, Mei X (2005) The algorithm of iris image preprocessing. In: Fourth IEEE workshop on automatic identification advanced technologies (AutoID'05), 17–18 October 2005, Buffalo, New York, USA, pp 134–138
18. Wang J-M, Ding R-T (2005) Iris image denoising algorithm based on phase preserving. In: Sixth IEEE international conference on parallel and distributed computing, applications and technologies, PDCAT 2005, 05–08 December, 2005, Dalian, China, pp 832–835
19. Yuan X, Shi P (2005) Advances in biometric person authentication, Springer
20. Ziauddin S, Dailey MN (2009) A robust hybrid iris localization technique. In: Electrical engineering/electronics, computer, telecommunications and information technology, May
21. Lagree S, Bowye KW (2011) Predicting ethnicity and gender from iris texture. In: IEEE international conference on technologies for: Homeland Security
22. Ross A, Pasula R, Hornak L (2009) Exploring multispectral iris recognition beyond 900nm. In: IEEE 3rd international conference on biometrics: theory, applications, and systems (BTAS). Washington, DC
23. Lim S, Lee K, Byeon O, Kim T (2001) Efficient iris recognition through improvement of feature vector and classifier. ETRI J 23(2):61–70
24. Daugman J (2004) Iris recognition and anti-spoofing countermeasures. In: Seventh international biometrics conference. London
25. Wildes R, Asmuth J, Green G, Hsu S, Kolczynski R, Matey J, McBride S (1994) A system for automated iris recognition. In: Proceedings IEEE workshop on applications of computer vision, Sarasota, FL, pp 121–128

Efficient Image Segmentation Using an Automatic Parameter Setting Model

D. Baswaraj and Puja S. Prasad

Abstract Most image analysis methods perform segmentation as a first step towards producing the object description. In these methods both input and output are images only, but the output is an abstract representation of the input. Image segmentation is responsible for partitioning an image into multiple sub-regions based on a desired feature, such as edge, point, line, boundary, texture, and region. The most common segmentation methods use intensity-based images. Snakes or active contours (AC) are used extensively in computer vision and image processing applications, particularly to locate object boundaries. During a literature survey, it has been identified that many segmentation issues like gray values of pixels remain equal for a area, rapid change of image gradient, large gradient due to noise and identification of boundary between areas. The traditional adaptive distance preserving level set evolution method works fine for natural and synthetic images. However, it is necessary to set the performance parameters manually every time with respect to the type of input image. Hence, the proposed method of an automatic parameter setting model (APSM) improves the adaptive distance preserving level set evolution based on region by setting the performance parameters automatically with the help of an analysis of the image quality. The results show the effective performance of segmentation by reducing the number of iterations with an improved output quality.

Keywords Image analysis · Image segmentation · Active contours
Level sets · Image quality · Self performance parameter setting

---

D. Baswaraj (✉) · P. S. Prasad
Department of CSE, CMR Institute of Technology,
Medchal, Hyderabad, TS, India
e-mail: d.baswaraj@cmritonline.ac.in

P. S. Prasad
e-mail: puja.s.prasad@gmail.com

A. Kumar and S. Mozar (eds.), *ICCCE 2018*,
Lecture Notes in Electrical Engineering 500,
https://doi.org/10.1007/978-981-13-0212-1_55

1 Introduction

The digital computer used for scientific applications as well as in social media become more efficient and faster now a days. The processes of collecting and analysing information by humans are generally termed as sight and perception respectively. The processes of collecting and analysing information by a computer are termed image processing and image analysis respectively.

Most of the image analysis methods perform segmentation as a first step towards producing the object description. In this methods, both input and output are images only, but the output is an abstract representation of the input. The segmentation technique basically divides the spatial domain on which the image is defined into meaningful parts or regions. This method plays a vital role in many applications, such as acoustics, remote sensing, medical imaging, material analysis, and voice communication processes. Image segmentation classifies the segmentation as either supervised or unsupervised based on available theoretical information. The region homogeneity, region contrast, line contrast, and line connectivity are the important features [1] that play a vital role in unsupervised segmentation. The unsupervised segmentation is performed independently by the computer, whereas supervised segmentation requires human intervention.

Errors are measured using a reference in the case of supervised segmentation. The distinctions between the output of the original and reference image segmentation determine the performance of the segmentation algorithm. The probability of error is one of simplest measures of supervised segmentation [2] and can be used to determine the optimal threshold values. The errors can then be divided into under-merging and over-merging categories. In some cases, the quality of segmentation is measured using the approaches used in references [3, 4], which is based on differences of features determined from a reference and an original segmented image.

In the literature many techniques have been proposed [5, 6], of which the snakes or active contour models (ACMs) [1, 2, 5, 7, 8] are one of the most successful methods used to locate the boundaries of an object. A snake is an energy minimizing curve directed by external forces and influenced by image forces that pull it towards target features like lines and edges. The active contour minimizes its energy function and expose the dynamic behaviour always during run-time. Based on the constraints applied during segmentation, the existing ACMs can be categorized into edge-based [2, 5, 7, 8] and region-based models [1, 3, 4]. ACMs are largely partitioned into parametric and geometric models. In parametric models (see Fig. 1), the curve will be stored as vertices and each vertex is moved iteratively until the contour of a subject image has been detected.

In the geometric model (see Fig. 2), the curve will be stored as coefficients and sampled before iteration. Each sample is then moved iteratively to new coefficients until the contour of an interested object has been detected. Further, the geometric ACM classified to boundary and region model. In the boundary model, gradient

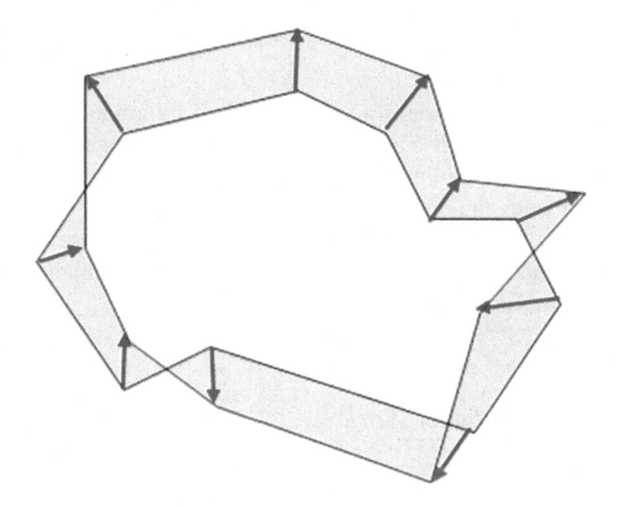

Fig. 1 Parametric active contours

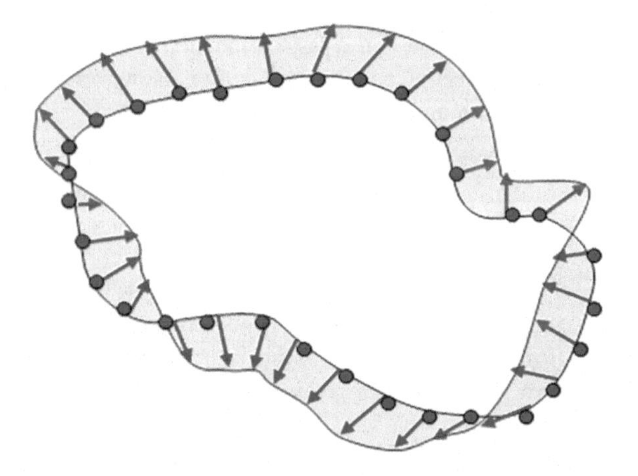

Fig. 2 Geometric active contours

information is considered to be the driving force for curve evolution. However, in the region model, gray level information is considered for curve evolution. Beside advantages of level set method which is based on curve evolution, it needs repeated initialization and a lot of calculation.

The objectives of the proposed segmentation methods are:

- To accomplish static segmentation by integrating both edge and region information of a given image.
- To acquire optimal segmentation that is not sensitive to the performance parameters like variation in operator size and thresholds.
- To get close contours of the object(s) existing within the foreground.
- To precisely segment the object(s) lying within the foreground from the background area.

2 Related and Proposed Work

The region model uses the gray level information inside and outside the region of the contour. It has a good effect on the image of the uneven gray-scale and the evolution continues beyond the noise. Although the level set method based on curve evolution has the advantages of free topological transformation and provides a high-precision closed partition curve, it needs repeated initialization and a lot of calculations. To effectively overcome the reinitialization problem, fast projection methods [9] are used.

2.1 Related Work

To alleviate the problem of repeated initialization and numerous calculations in the region model, the proposed method [10] mainly corrects the level set function and the deviation of signed distance function. In addition, it also uses a simple finite difference method and a large time step to solve the boundary model problem of partial differential equation evolution. However, the significant disadvantages of this approach are as follows:

- The curve evolution always follows one direction only (either inward or outward), not the root. According to the image information inward or outward movement, the user must manually select the initial contour movement direction.
- The external energy functional weight is constant and cannot be adjusted according to the image information and size.
- If the curve evolution speed is slow, then the edge stop function converges to zero, resulting in a weak boundary which may still have greater evolutionary speed.
- The real value of the currently adopted regularized Dirac function with compact support determines that the evolution equation is in the control of local action, thereby restricting the active contour target acquisition ability of the object boundary.

To overcome the shortcomings of evolution, the adaptive distance preserving level set evolution introduces an image-dependent weighting coefficient. This can speed up the evolution but has the following limitations:

- The level set cannot evolve to the edge of the target object in the area, where gray values are equal that make image gradient and weight coefficient zero.
- If the intensity changes because of fast changes in image gradient, then it influences the weighting coefficient to become either large or small, which leads the boundary to drip out or does not allow the image to separate correctly.
- Splitting is not permitted as the edge stop function is close to zero due to the large gradient of the gray image.

- The image gradient is large due to noise, it makes edge ending function to reach to local minimum.

To alleviate the above limitations, the proposed adaptive distance method modifies the image-dependent weighting coefficient and also combines the regional information to the design of the adaptive edge stop function.

The energy functional $E(\phi)$ can be expressed as:

$$\begin{aligned} E(\phi) &= \mu P(\phi) + E_m(\phi) \\ &= \mathbf{1/2}\mu \int_\Omega (|\nabla\phi| - 1)^2 dxdy + \lambda \int_\Omega g(\nabla\phi)\delta(\phi)|\nabla\phi|\, dxdy \\ &+ \int_\Omega v(I) g(\nabla\phi) h(-\phi)\, dxdy \end{aligned} \tag{1}$$

where $P(\phi)$—Internal energy functional and $E_m(\phi)$—External energy functional.

Equation (1) uses a Euler-Lagrange equation to minimize the energy functional as:

$$\frac{\partial\phi}{\partial \mathbf{t}} = \mu\left[\Delta\phi - div\left(\frac{\nabla\phi}{|\nabla\phi|}\right)\right] + \lambda\delta(\phi) div\left(g(\nabla\phi)\frac{\nabla\phi}{|\nabla\phi|}\right) + v(I)g(\nabla\phi)\delta(\phi) \tag{2}$$

where, $\delta(x)$—Dirac function, $h(x)$—Heaviside function, $I(x, y)$—Image function, $v(I)$—Variable weights, $g(x)$—Edge stopping function.

v(I) and g(x) are defined as:

$$v_n(I) = c \cdot \mathbf{sgn}(\Delta G_\sigma \times I)\, |\nabla(G_\sigma \times I)| \tag{3}$$

$$g(\nabla I) = \mathbf{exp}(-|\nabla(G_\sigma \times I)|/m) = \mathbf{exp}(-s/m) \tag{4}$$

where sgn $(\cdot)$—the sign function, c, m—constants.

$\Delta G_\sigma \times I$—the results of the effect which the Laplace operator has on the Gaussian filtered image

$\nabla G_\sigma \times I$—Gradient after the image goes through the Gaussian filter.

2.2 Analysis of Variable Weight Coefficient

The level set evolution uses image edge features to make curve evolution minimize the energy functional. In this method the weight coefficient $v_n(I)$ controls the evolutional direction of the zero level set and directly affects the magnitude of the stopping function $g(s)$. It determines the ability of the zero level set to detect the multi-contours of the object [11]. When the level set evolves equal to the gray value, its gray value in (3) is still equal after the Gaussian filtering and the magnitude of image gradient as well as weight coefficient becomes zero; the level set will stop the evolution, hence, is unable to segment the image.

When the level set moves into the uneven gray area, the image gradient changes leading to great changes in the weight coefficient [12]. Thus, boundary leak may happen. As the curve evolves into the deeply recessed area, the weight coefficient has to be smaller, and hence the curve cannot break through the deeply recessed area to reach the boundary of the target object.

2.3 Analysis of the Edge Stopping Function

Figure 3 shows how the edge stopping function $g(s) = e^{-s/m}$ changes with the parameters m and s. The $g(s)$ is a monotone decreasing positive function concerning the image gradient. With m under certain conditions, the stopping conditions of the level set evolution depend only on the gradient and the speed with which it converges. For a large image gradient the edge stopping function $g(s)$ is larger in the non-edge portion with homogeneous intensity, leading to curve evolution rather than stopping the evolution.

The $g(s)$ may enter into the local minimum for a larger gradient with noise. For example, if the gradient of a noise in Fig. 3 is 20 and $m = 2$, then $g(s) = 4.5$ $105 \approx 0$, leads to prevent the curve evolution, and thus it cannot segment the target image correctly.

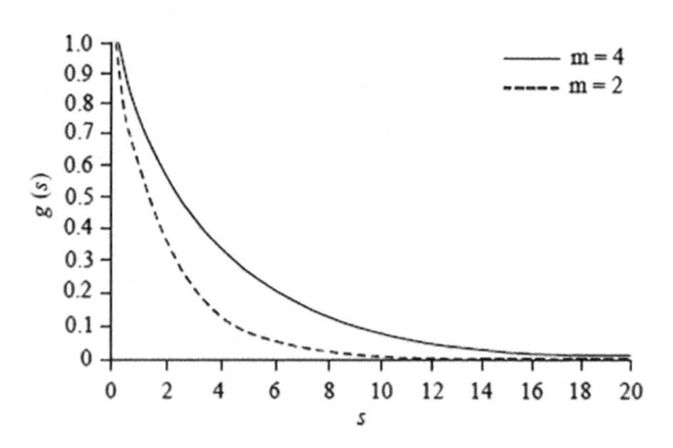

Fig. 3 Edge stopping versus the gradient of an image

2.4 Design of the Variable Weight Coefficient

To alleviate the problem of curve evolution in (3), the variable weight coefficient is modified [11] to satisfy the following conditions:

- Level set continues to evolve into the boundary in the area where the gray levels are equal.
- Detect the multi-layer profile where the gray level changes have been comparatively large.

Based on the above considerations, the variable weight coefficient is corrected and designed as:

$$v_n(I) = c \cdot \mathbf{sgn}(\boldsymbol{\Delta} G_\sigma \times I)(\boldsymbol{\alpha}|\nabla(G_\sigma \times I)| + \boldsymbol{\beta}) \tag{5}$$

where α and β are constants with values greater than zero. The analysis of (5) is as follows:

- The gray level equal to the area and it makes the value of $|\nabla(G_\sigma \times I)|$ and $(G_\sigma \times I)$ to zero. This ensures that the zero level set can continue evolving until it reaches the contours of the target object. The value in the range 0.5–1.3 can be assigned to the parameter α.
- Replacing $|\nabla(G_\sigma \times I)|$ with $\alpha|\nabla(G_\sigma \times I)| + \beta$ ensures that the curve penetrates the depth of the depression area to reach the target object boundary as the values of α and β can be properly adjusted to increase $v_n(I)$.
- Although the image gradient mode is relatively large when the gray scale changes greatly, α and β can still be properly adjusted to decrease $v_n(I)$ to ensure that the zero level set stays at the target boundary.

So the proper values for α and β control the level set evolution in different environments to reach the target boundary.

2.5 Design of an Edge Stopping Function

Consider that the value of the edge stopping function is high in (4), when the speed of the image gradient(s) converges to zero (see Fig. 2). As a result the proposed edge stopping function g introduces a coefficient M_ρ with regional information, to ensure that the function $g(I, M_\rho)$ adaptively changes. In this respect the stopping function has been defined as:

$$g(I, M_\rho) = \frac{1}{1 + |\nabla(G_\sigma \times I)/M_\rho|^2} \tag{6}$$

where, G_σ = a Gaussian kernel with a standard deviation, and a new adaptive varying coefficient M_ρ is defined as:

$$M_\rho = \rho \left| I - \frac{c_1 + c_2}{2} \right| + \theta \quad (7)$$

where, c_1—the average intensity values of pixels inside the evolution curve, c_2—the average intensity values of pixels outside the evolution curve of the image and ρ, θ—constants whose values can be in the range of 0–1.2.

The analysis of the edge stopping function $g(I, M_\rho)$ in (6) is given as follows:

- When the level set covers the object and is far away from the boundary, then the term $(\rho|I - \frac{c_1+c_2}{2}| + \theta) \gg \theta$, thus $M_\rho \gg \theta$ and $g(I, M_\rho) \approx 1$, despite the image $|\nabla(G_\sigma \times I)|$ being bigger in place with noises. The great change in gray scale, i.e. $M_\rho \gg \theta$, the $g(I, M_\rho)$ becomes larger, ensuring $g(I, M_\rho)$ does not fall into local minimum in the non-boundary, which increases the robustness of the algorithm.
- When the level set function evolves towards the object boundary, then $(|I - \frac{c_1+c_2}{2}|) \approx 0$ and $M_\rho = \theta$, as the gradient mode in the boundary is larger. If θ is close to zero, then $g(I, M_\rho)$ will converge to zero rapidly in the boundary of the object to make sure that the model falls into a global minimum value and improves the accuracy of the image segmentation.

The level set method proposed has the following significant advantages:

- The level set function can continue to evolve until the detected target object contour for the zero level has been set in the equal gray level area.
- Avoid edge leakage that is caused by severe intensity change.
- Restrict the noise in the level set by making it fall into local minimum.

2.6 Proposed Work: Automatic Parameter Setting Model (APSM)

The contour detection has to be efficient by setting the appropriate values of performance parameters like α, β, ρ, and θ based on the properties of the image tested. Hence, the proposed APSM considers the required behavior β of tuning parameters in the vicinity of two extreme local properties encountered in an image, namely edge regions and contours. The level set regularization reduces the domination of the noise magnification error. In this method, automatic selection of an image intensity-based weight coefficient reduces the above limitations. Finally, APSM effectively solves the problem that:

Table 1 Performance parameter space

Sl. no.	Performance parameters	Normal range	Special range
1	Alpha (α)	0.5–1.8	1.3–4.2
2	Beta (β)	0.3–1.8	0.7–2.8
3	Rho (ρ)	0.7–2.8	0.3–1.8
4	Theta (θ)	0.0–1.2	Nil

- The level set cannot continue to evolve in equal area for gray value.
- Inaccurate segmentation of the target image due to dramatic changes of image gradient by reducing noise sensitivity for model and it improves the robustness of the algorithm.

The numerical ranges for various performance parameters are shown in Table 1 and are fixed based on the results obtained from different types of image data sets. Based on an empirical analysis, an approximate range is assigned to each parameter. The parameters are depth of boundary, gray-scale changes, noise sensitivity, and evolution speed.

The following illustrations give more detail:

- To detect the depth of the depression boundary, the value α has to be in the range 1.3–4.2. Otherwise it has to be in the range 0.5–1.8.
- To detect the contours of target objects, whose gray scale change is relatively high, the value of β has to be in the range of 0.7–2.8 otherwise 0.3–1.8.
- The parameters ρ and θ are respectively used to control the noise sensitivity and the convergence speed to zero of $g(I, M_\rho)$, generally θ is in the range of 0–1.2.
- The larger ρ-value speeds up the evolution subject to no noise or little gray-scale change and
- For high-noise levels a relatively smaller ρ-value (between 0.3 and 1.8) ensures that the evolution curve skips the noise point and continues to evolve until it reaches the boundary of the object.

Usually the values are preferred to be from the normal range. However, with cases meant to detect the depth of depression boundary, a relatively high gray-scale level change to special range has to be preferred. Instead of selecting and setting the parameters manually, the APSM sets parameters automatically based on image properties, such as noise, gray level, gray density, and correlation between pixels.

2.7 Proposed Algorithm Design

The following guidelines help to design the proposed algorithm to automatically estimate the parameter values with the usage of MATLAB tools and library functions to detect the close contour of a given image.

- Create a gray level co-occurrence matrix (GLCM) by finding how frequently a pixel with gray level "I" occurs horizontally adjacent to a pixel with gray level "j" as: GLCM = *graymatrix*(I);
- Calculate the statistics of GLCM as: Stats = *graycoprops* (GLCM, properties); where properties indicate contrast, correlation, energy, or homogeneity of a given image.
- To determine the noise level in the given image, add noise using *imnoise* function first and then remove noise using linear or median filters using *imfilter or medfilt2* functions respectively.
- The larger ρ-value speeds up the evolution subject to no noise or little gray-scale change.
- Assign suitable values based on properties of an image to other performance parameters, such as α, β and θ.

Function: Automatic-parameter Setting

Input: Gray Level or Color Image

Output: Performance Parameter Values

- Determine GLCM for a given image using the *graymatrix* function.
- Determine the statistics of various properties such as contrast correlation, energy, and homogeneity of image using the *graycoprops* function.
- Determine noise level by adding and removing noise from an image using function *imnoise* and linear/median filters *imfilter/medfilt2* respectively.
- Based on image properties, statistics, and noise level, set the values as per Table 1 to the corresponding parameters.
- Return these performance parameter values to the curve evolution process.

3 Results and Discussion

Different case studies were used to evaluate the performance of the conventional and proposed methods discussed. From the different case studies, it can be observed that the results of both methods are effective for some cases based on the type of image. The experimental result on test images (see Figs. 4 and 7) shows that the proposed method completed the segmentation process in 60 iterations (see Figs. 6 and 9) as compared to 300 iterations (see Figs. 5 and 8) using traditional methods.

Fig. 4 Input test image

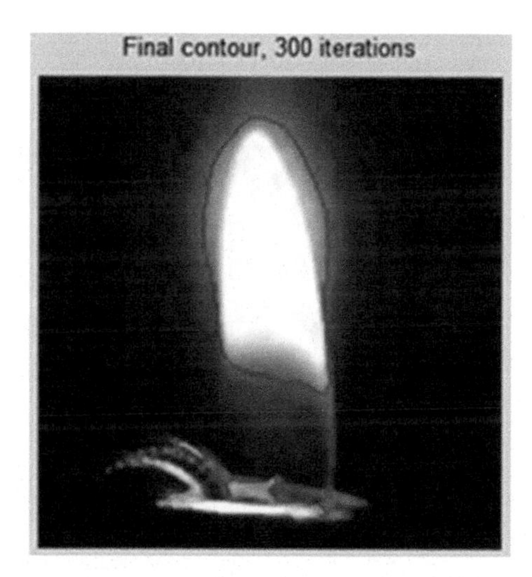

Fig. 5 Conventional region-based method

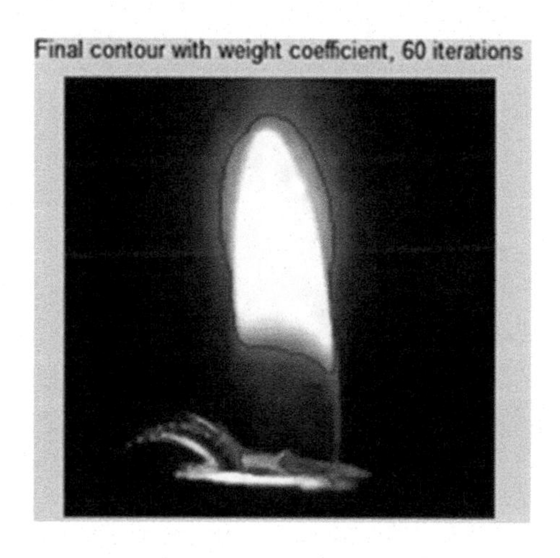

Fig. 6 Proposed method

Fig. 7 Input test image

Fig. 8 Conventional region-based method

Fig. 9 Proposed method

Case Study-1: Flame Image

Case Study-2: Flower Image

4 Conclusion

The proposed APSM uses the weight coefficient with self parameter setting to accelerate the evolution speed as well as optimizes the algorithm. This algorithm has been tested on a different variety of images captured from several clinical websites and scanners to identify whether the gray value is equal to the object boundary or not. The results show the effective performance of the segmentation by reducing the number of iterations with improved output quality.

The work presented in the APSM has been applied to detect the contour of an object whose gray level is uniform in some areas and also there are sudden changes in the gradient of image. However, it is unable to locate the contours of a low intensity or low luminous objects in a given image. This model can be extended further by analysing the quality of the image effectively, so that suitable values for performance parameters can be configured automatically.

References

1. Lavine MD, Nazif AM (1982) An experimental rule based system for testing low level segmentation strategies. In: Multicomputer and image processing algorithms and programs, pp 149–160. Academic Press
2. Yang L, Albregtsen F, Lonnestead T, Grottum P (1995) A supervised approach to the evaluation of image segmentation methods. In: Computer analysis of images and patterns: 6th international conference; proceedings/CAIP '95, pp 759–765. Springer
3. Gunjan VK, Shaik F, Kashyap A, Kumar A An interactive computer aided system for detection and analysis of pulmonary TB. Helix J 7(5):2129–2132, Sept 2017 (ISSN 2319–5592)
4. Zhang YJ, Gerbrands JJ (1992) Segmentation evaluation using ultimate measurement accuracy. In: Proceedings of society of photographic instrumentation engineers (SPIE), image processing algorithms and techniques III, vol 1657
5. Canny J A Computational approach to edge detection. IEEE Trans Pattern Anal Mach Intell June 1986, PAMI-8:679–698
6. Gonzalez RC, Woods RE (1993) Digital image processing. Addition Wesley
7. Xu Y, Olman V, Uberbacher EC (1998) A segmentation algorithm for noisy images: design and evaluation. Pattern Recogn Lett 19:1213–1224
8. Haralick RM, Shapiro LG (1985) Survey-image segmentation techniques. Comput Vision Graph Image Process 29:100–132
9. Zhou B, He C, Yuen Y (2012) Edge-based active contour model with adaptive varying stopping function. Appl Res Comput 29(1):366–368
10. Fu KS, Mui JK (1981) A survey on image segmentation. Pattern Recogn Lett 13:3–16
11. Li C, Xu C, Konwar KM, Fox MD (2006) Fast distance preserving level set evolution for medical image segmentation. In: Proceedings of the 9th international conference on CARV
12. Yong-qi Q, Tau Z (2013) Region-based method of adaptive distance preserving level set evolution. Res J Appl Sci Eng Tech 5(5)

Quantitative Evaluation of Panorama Softwares

Surendra Kumar Sharma, Kamal Jain and Merugu Suresh

Abstract Image stitching has been practiced in various computer vision and scientific study areas. Many different image stitching algorithms have been proposed by different research groups in the past, and there are many different image stitching software products available on the market. However, a comparison between different stitching software products and an evaluation of them has not been performed so far. Furthermore, most previous quality assessment approaches have not had an adequate number of performance matrices, while others have suffered from the adverse effects of computational complications. Our objective is to identify the best software for panoramic image stitching. In this paper we measure the robustness of different software products by assessing image quality of a set of stitched images. For the evaluation itself, a varied set of assessment criteria is used, and evaluation is performed over a large range of images captured in different scenarios using differing cameras. Results show that Autostitch performs relatively well for all types of scenes and for all types of dataset.

Keywords Image stitching · Panoramic image quality · Image quality assessment · Panorama software

1 Introduction

A panoramic image is a single image covering a wide field of view of the environment around the camera. Normally panoramas entirely surround the camera on the horizontal plane and can be approximately 120–180° in the vertical field of view [7]. Panoramic imagery is becoming more and more popular for 360° representations of landscapes, city squares, and indoor scenes. Panoramic images have wide

S. K. Sharma (✉) · K. Jain
Geomatics Engineering, Indian Institute of Technology Roorkee, Roorkee, India
e-mail: surendra123sharma@gmail.com

M. Suresh
R&D Centre, CMR College of Engineering & Technology, Hyderabad, India

A. Kumar and S. Mozar (eds.), *ICCCE 2018*,
Lecture Notes in Electrical Engineering 500,
https://doi.org/10.1007/978-981-13-0212-1_56

range of applications, such as computer vision, surveillance, 3D reconstruction, texture mapping in 3D GIS [17], and virtual reality being a few of them. They are also used in commercial applications such as entertainment, virtual tourism, and real estate. In the last 15 years, panoramic imaging systems have drastically improved. Due to the availability of a large number of software products not only experts can create panoramas, but also anyone having a computer and a digital camera is capable of creating panoramic images [7]. To generate a large panorama a number of overlapping images are stitched together using image stitching algorithms. Panoramic software uses an image stitching algorithm to generate a large panorama. Several image stitching applications have been developed over the last decade, but not a lot of work has been carried out to objectively evaluate and improve it. Image stitching applications in computer vision mostly rely upon assessing the quality of stitched results. In many of the cases of evaluating the quality of a stitching algorithm, human-based perception is selected as the evaluation framework [6].

However, as more accurate and faster software products have emerged in the past few years, it is not enough to depend entirely on visible inspection for the reason that it might go wrong when significantly differentiating the stitched images acquired through different kinds of software. Scientific analysis demands stitching quality to be measured quantitatively, instead of qualitatively [6]. This paper proposes a quantitative evaluation framework of many image stitching software products based on several matrices. We consider five commonly used panorama stitching software products tested on 30 real mosaic image pairs selected from different applications and from different cameras. There is a vast amount of literature published on the quality of images [3, 5, 10, 15, 16, 18, 20, 21] but for evaluating the quality of stitched images only a few references are available. Ghosh et al. [6] used different datasets but only one mosaicking algorithm (system) was used to evaluate the quality of stitched images. Also they only used two performance matrices PSNR (peak signal-to-noise ratio) and MI (mutual information). Dissanayake et al. [4] compared four stitching algorithms and for quality assessment they used subjective evaluation and objective evaluation. In their objective evaluation they used SSIM (structural similarity index) and a UIQ (universal image quality) index for geometric quality and for photometric quality they used SAM (spectral angle mapper) and IMR (image magnitude ratio) indices. Xu and Mulligan [19] worked on the color quality assessment of stitched images, in which they compared the quality of different color correction methods. Paalanen et al. [12] evaluated mosaicking performance using artificial video. First, they created an artificial video using an available image then a mosaicking algorithm was applied to the artificial video to create a mosaic. Finally the mosaicked image was compared with the original image. Authors tested this approach only for synthetic datasets and not for real datasets. This approach cannot be applied to real-world scenarios. A similar type of approach was also used by Boutellier et al. [1]. They also categorized the type of error in the stitched image and its cause.

Section 2 discusses panorama stitching software. We discuss commonly used panoramic stitching software products. Section 3 presents the image quality assessment metrics used to evaluate the quality of stitched images. Section 4 presents the dataset and methodology used in this work. Section 5 contains the results and discussion. Section 6 concludes the paper.

2 Panorama Stitching Software

We have considered five commonly used software products for panorama generation. In the following subsection five different examples of panorama software used in this work are described in alphabetical order (Table 1).

3 Evaluation Framework

There are two methods followed in evaluating image quality, the subjective and the objective method. The subjective evaluation method is considered expensive and time consuming due to the fact that we have to decide on a number of observers, show them a number of images, and then ask them to rate image quality based upon their own judgment. The objective image quality assessment (IQA) method employs automatic algorithms to evaluate image quality without human intervention. Based on the availability of the original image, objective image quality matrices are categorized into three different classes:

- **Full-reference**: Reference image is available.
- **Reduced-reference**: Complete reference image is not available but information about reference image is available in the form of image features which helps in the evaluation.
- **No-reference**: Reference image is not available. This is also known as "blind quality assessment."

In this paper we concentrate on the full reference objective image quality metric, as reference images are available to us.

Table 1 Software packages

Software	Brief description
Autostitch	Autostitch can create panoramas from unordered collections of images [2]
Hugin	Hugin is a cross-platform open source panorama stitching and HDR merging program [8]
Image composite editor	Image Composite Editor (ICE) is an advanced panoramic image stitching application. It can also create a panorama from a video [9]
Panorama maker	Panorama Maker (PM) turns photos and videos into panoramas [13]
Panorama plus	Panorama Plus (PP) seamlessly stitches together images and movie frames to create the final panorama. The stitching procedure works in two dimensions, creating vertical as well as horizontal panoramas [14]

The objective image quality measures used in this work to evaluate the quality of stitched panoramas are as follows:

3.1 Peak Signal-to-Noise Ratio (PSNR)

PSNR is used as a measurement of difference between two images. The PSNR of corresponding pixel values is defined as:

$$PSNR = \frac{10\ \log_{10}(max(G(i,j),\ O(i,j)))^2}{MSE}$$

where MSE is the mean square, G(i, j) and O(i, j) represent the (i, j)th pixel values in the input and stitched result respectively. MSE is defined as

$$MSE = \frac{\sum_i \sum_j (G(i,j) - O(i,j))^2}{N}$$

where N is the total number of pixels in each image.

If the difference between two images is lower, meaning that MSE is lower, this gives a higher PSNR between them.

3.2 Structural Similarity Index

A structural similarity (SSIM) index is a method for measuring the similarity between two images. The image which is being evaluated is termed the target image and the image to which the quality of the target image is compared is termed the reference image. The SSIM index can be seen as a quality measure of the target image provided the reference is of perfect quality [18]. It is an improved version of the universal image quality index.

The SSIM index is developed in such a way that it satisfies the conditions of symmetry, boundedness, and having a unique maximum. The luminance measurement is taken to be qualitatively consistent with Weber's law, which states that in the human visual system, a noticeable change in luminance is approximately proportional to the background luminance. Similarly, the contrast measurement is also consistent with the human visual system by noticing only the relative change in contrast, as opposed to the absolute contrast difference. The final SSIM index is a product of the above two values, together with a structural similarity component, which is also calculated based on luminance and contrast measurements [18]. SSIM between two images x and y is defined as

$$SSIM(x,\, y) = \frac{(2\mu_x\mu_y + C_1)(2\sigma_{xy} + C_2)}{\left(\mu_x^2 + \mu_y^2 + C_1\right)\left(\sigma_x^2 + \sigma_y^2 + C_2\right)}$$

where μ_x, μ_y, σ_x, σ_y, and σ_{xy} are the local means, standard deviations, and cross-covariance for images x, y.

The higher the value of the *SSIM* index, the less is the difference between the structure of x and y. $SSIM\ (x,\ y) = 1$ if there is no structural difference.

3.3 Feature Similarity Index

A feature similarity (FSIM) index is a full reference IQA method. It is based on the fact that the human visual system interprets an image according to its low-level features [21]. To find the FSIM index two features, phase congruency (PC) and gradient magnitude (GM) are evaluated. PC is used as a primary feature in FSIM and it is dimensionless measure of significance of a local structure. GM is considered to be a second feature. PC and GM are complementary in characterizing image's local quality. The FSIM measurement is separated between $f_1(x)$ and $f_2(x)$ into two components, each for *PC* or *GM*. The similarity measure in terms of $PC_1(x)$ and is defined as:

$$S_{PC}(x) = \frac{2PC_1(x) \cdot PC_2(x) + T_1}{PC_1^2(x) + PC_2^2(x) + T_1}$$

where T_1 is a positive constant.

The similarity measure in terms of *GM* values $G_1(x)$ and $G_1(x)$ is given as

$$S_G(x) = \frac{2G_1(x) \cdot G_2(x) + T_2}{G_1^2(x) + G_2^2(x) + T_2}$$

where T_2 is a positive constant depending on the dynamic range of GM values.

$S_{PC}(x)$ and $S_G(x)$ are combined to get the similarity $S_L(x)$ of $f_1(x)$ and $f_2(x)$ which is given as

$$S_L(x) = S_{PC}(x) \cdot S_G(x)$$

The FSIM index between f_1 and f_2 is defined as

$$FSIM = \frac{\sum_{X \in \Omega} S_L(X) \cdot PC_m(X)}{\sum_{X \in \Omega} PC_m(X)}$$

Where $PC_m(X)$ is max of $PC_1(x)$, $PC_2(x)$.

3.4 Visual Saliency Index

Visual saliency (VS) has been widely studied in psychology, neurobiology, and computer science during the last decade to explore, which areas of an image will draw most attention of the human visual system. Distortions can mostly impact VS maps of images. With this consideration, Zhang et al. [20] proposed a simple but very effective full reference image IQA method using VS called a visual saliency-induced (VSI) index. The VSI metric between image1 and image2 is defined as:

$$VSI = \frac{\sum_{X \in \Omega} S(X) \cdot VS_m(X)}{\sum_{X \in \Omega} VS_m(X)}$$

where Ω means the whole spatial domain.

$$S(X) = S_{VS}(X) \cdot [S_G(X)]^{\alpha} \cdot [S_C(X)]^{\beta}$$

$S_{VS}(x)$	is the visual saliency similarity component between two images
$S_G(x)$	is the gradient modulus similarity component between two images
$S_C(x)$	is the chrominance similarity component between two images
$VSm(x)$	max $(VS_1(x), VS_2(x))$
$VS_1(x)$ and $VS_2(x)$	are the visual saliency maps of image1 and image2 respectively

4 Dataset and Methodology

A detailed dataset which is captured using a smartphone and digital camera covering 30 image pairs belonging to three types of scene has been selected to evaluate the efficiency of stitching software. The dataset is versatile in nature and covers all possible real-world situations for panorama generation.

4.1 Dataset

Table 2 discusses the specifications of the smartphone and digital camera used in the study. Images are captured for three different scene categories to replicate possible real world situations.

Images captured for the outdoor scene category mostly containing building features are ranked in one category, while the image dataset containing natural scenery is ranked as a second category. A third category images contains indoor environments (classrooms interiors). These datasets are chosen to achieve different lighting scenarios, features, color types, and other constraints imposed due to site-dependent image capturing conditions (natural and manmade). Figures 1, 2, and 3 depict the image datasets for the three discussed situations.

Table 2 Camera devices

Model	Camera resolution (MP)	Lens
Canon PowerShot A2200	14	28–112 mm, f2.8–5.9
iPhone 6 Smartphone camera	8	29 mm, f/2.2

Fig. 1 Scene containing buildings (dataset 1)

Fig. 2 Scene containing buildings (dataset 2)

Fig. 3 Indoor scene (dataset 3)

4.2 Methodology

As shown in the Fig. 4, five image pairs are selected in each scene category. All the captured images are stitched using software selected for the current study and discussed in Sect. 2. The efficiency of the stitching software is evaluated by comparing the quality of both overlapping and non-overlapping regions of input and the resultant stitched image. As presented in Fig. 5, for each left image (L) and right image (R) both the overlapping and non-overlapping areas are compared.

For each image pair IQA indices are computed in both the overlapped and non-overlapped region. IQA indices results define the quality of the image stitching method used by the individual software. For the overlapping region, So is compared with Lo and Ro separately. Similarly, for the non-overlapping region SLnon is compared with Lnon, and SRnon is compared with Rnon (see Fig. 5). Finally, based on all the metrics (PSNR, SSIM, FSIM and VSI) a rank is assigned to each software product, first for individual datasets and then combined for all datasets. Rank computation and assigned ranks are described in Sect. 4.3. The complete workflow of this study is shown in Fig. 6.

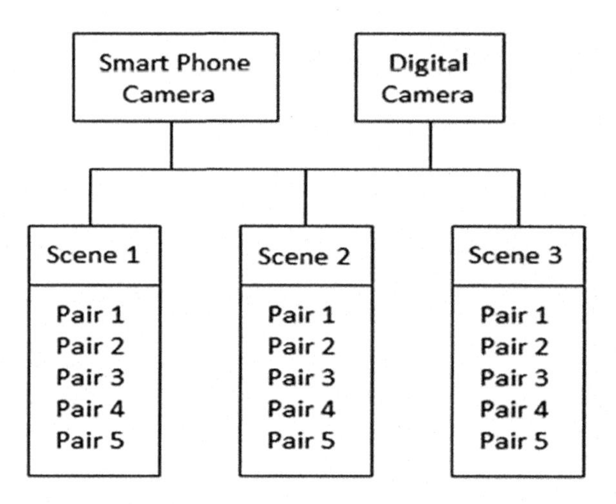

Fig. 4 Image capturing scenario

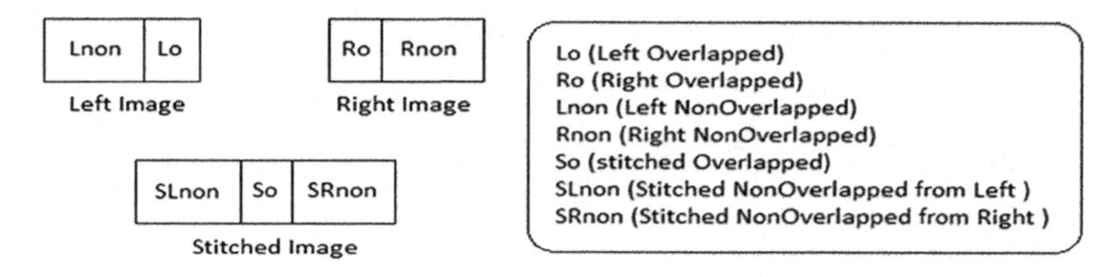

Fig. 5 Comparison of input and stitched images

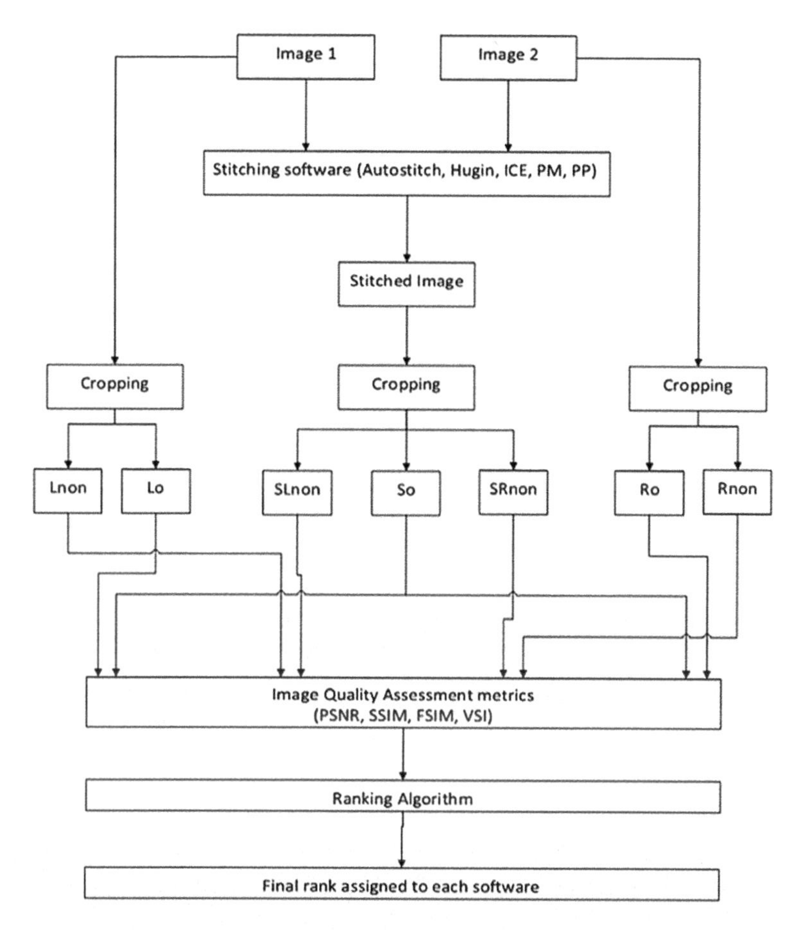

Fig. 6 Workflow adopted for panoramic image quality assessment

4.3 Rank Computation

For comparison, four types of metrics are available: PSNR, SSIM, FSIM and VSI for overlapping and non-overlapping areas. Individual metric ranks for each type of dataset are easy to compute. However, a single rank based on all metrics over all types of datasets would be beneficial.

The ranking methodology has been adopted from Mukherjee et al. [11]. The algorithm for ranking is described below:

1. For method m (representing a software program) with image area a (overlapping or non-overlapping area), camera type c (digital camera or smartphone camera), dataset d (dataset 1, 2, 3) and metric t, find the positional rank $P_{m,a,c,d,t}$ by sorting the metric values (a high value indicates better positional rank).

2. Compute the average rank of m for image area a, camera type c and dataset d over all metrics: $R_{m,a,c,d} = \frac{1}{N_t}\sum_t P_{m,a,c,d,t}$. Here, N_t represents the number of metrics. Find the positional rank $P_{m,a,c,d}$ of each method m for each dataset, image area and camera type based on the value of $R_{m,a,c,d,}$ by sorting the $R_{m,a,c,d}$ in ascending order (low $R_{m,a,c,d}$ represents better positional rank).
3. Compute the average rank of method m for camera type c and dataset d over all image areas: $R_{m,c,d} = \frac{1}{N_a}\sum_t P_{m,a,c,d}$. Here, N_a represents the number of image areas (overlapping, left non-overlapping and right non-overlapping). Find the positional rank $P_{m,c,d}$ of each method m for each camera type and each dataset based on $R_{m,c,d,}$ by sorting the $R_{m,c,d}$ in ascending order (low $R_{m,c,d}$ represents better positional rank).
4. Compute the average rank of method m for dataset d over all camera types: $R_{m,d} = \frac{1}{N_c}\sum_t P_{m,c,d}$. Here, Nc represents the number of camera types used (digital camera and smartphone camera). Find the positional rank $P_{m,d}$ of each method m for dataset based on $R_{m,d,}$ by sorting the $R_{m,d}$ in ascending order (low $R_{m,d}$ represents better positional rank).
5. Compute the average rank of method m across all datasets: $R_m = \frac{1}{N_d}\sum_t P_{m,d}$. Here, N_d represents the number of datasets. Finally, determine P_m, the positional rank across datasets by sorting R_m in ascending order.

5 Results and Discussions

This research compares five of the most prominent panorama image stitching application (Autostitch, Hugin, ICE, PM, PP) software programs using four IQA metrics. The final results are presented in two sections. In the first section a camera dataset is used and three varieties of results are compared for five software programs using four IQA matrices. When results are computed for both the combination of overlapping areas (So, Lo and So, Ro), it can be seen that the left image dominates in the stitching results for all the software products. Thus the "overlapping area" section in the table shows results for comparison of So and Lo only. The "Left non-overlapping area" section shows the results for comparison of Lnon and SLnon. Similarly the "Right non-overlapping area" section shows results for Rnon and SRnon.

Table 3 summarizes the IQA metrics values for camera dataset 1 which depicts an outdoor scene with building features. It can be seen that the IOA matrices values for the PP software are significantly higher in both the overlapping and non-overlapping area. For camera dataset 2, IQA values are presented in Table 4. Autostitch software has higher IOA metrics values and is suited to outdoor scenes which contain natural features such as trees. In the last camera dataset, which depicts indoor scenes, Autostitch software performs best in comparison to all the other stitching software (Table 5). In the second section the camera dataset is

Table 3 Camera dataset 1

Software name	Overlapping area				Left non-overlapping area				Right non-overlapping area			
	PSNR	SSIM	VSI	FSIM	PSNR	SSIM	VSI	FSIM	PSNR	SSIM	VSI	FSIM
Autostitch	14.1697	0.6403	0.8643	0.6128	14.7900	0.6710	0.8956	0.6587	14.3345	0.5787	0.8760	0.6475
Hugin	13.5505	0.6314	0.8539	0.5973	12.8251	0.6201	0.8862	0.6389	14.0541	0.5326	0.8648	0.6304
ICE	14.1074	0.6298	0.8557	0.5976	14.0342	0.5903	0.8929	0.6454	13.2475	0.5578	0.8682	0.6401
PM	13.0921	0.6007	0.8537	0.5822	12.2770	0.5841	0.8682	0.6275	11.1233	0.5346	0.8493	0.6012
PP	14.3038	0.6497	0.8663	0.6155	14.3206	0.6732	0.9035	0.6731	14.9200	0.5934	0.8920	0.6725

Table 4 Camera dataset 2

Software name	Overlapping area				Left non-overlapping area				Right non-overlapping area			
	PSNR	SSIM	VSI	FSIM	PSNR	SSIM	VSI	FSIM	PSNR	SSIM	VSI	FSIM
Autostitch	14.4912	0.5113	0.8839	0.6909	13.7820	0.4564	0.8612	0.6619	15.8101	0.5701	0.8975	0.6784
Hugin	13.9753	0.4716	0.8782	0.6689	13.4262	0.4048	0.8351	0.6283	14.7433	0.5397	0.8538	0.6292
ICE	14.0447	0.4994	0.8817	0.6827	13.4631	0.4467	0.8520	0.6482	15.2483	0.5579	0.8613	0.6364
PM	14.3161	0.5024	0.8824	0.6852	13.4970	0.4480	0.8583	0.6523	14.9552	0.5655	0.8782	0.6565
PP	14.0201	0.4904	0.8797	0.6765	12.5937	0.4441	0.8537	0.6460	14.4631	0.5469	0.8605	0.6354

Table 5 Camera dataset 3

Software name	Overlapping area				Left non-overlapping area				Right non-overlapping area			
	PSNR	SSIM	VSI	FSIM	PSNR	SSIM	VSI	FSIM	PSNR	SSIM	VSI	FSIM
Autostitch	18.0325	0.6651	0.9050	0.7350	22.0634	0.6477	0.9618	0.7912	18.0485	0.6230	0.9331	0.7485
Hugin	16.1442	0.5908	0.8881	0.7053	20.8888	0.6266	0.9358	0.7427	15.4757	0.5694	0.8892	0.6707
ICE	17.9168	0.6636	0.9044	0.7323	23.3518	0.6442	0.9460	0.7735	16.4559	0.6207	0.9122	0.7296
PM	17.4008	0.6260	0.8925	0.7147	20.5156	0.6381	0.9418	0.7511	15.0092	0.5859	0.8924	0.6905
PP	17.6210	0.6384	0.9001	0.7218	21.0936	0.6373	0.9434	0.7531	16.1984	0.6129	0.9015	0.6932

Table 6 Smartphone dataset 1

Software name	Overlapping area				Left non-overlapping area				Right non-overlapping area			
	PSNR	SSIM	VSI	FSIM	PSNR	SSIM	VSI	FSIM	PSNR	SSIM	VSI	FSIM
Autostitch	17.1579	0.5763	0.9276	0.6807	15.5227	0.5794	0.9140	0.6894	14.4222	0.5903	0.9062	0.7016
Hugin	16.9573	0.5678	0.9202	0.6712	13.2279	0.5227	0.9037	0.6693	14.5455	0.5620	0.9157	0.6907
ICE	16.9873	0.5772	0.9240	0.6733	14.0320	0.5602	0.9167	0.6843	15.7221	0.5894	0.9172	0.6914
PM	17.9083	0.5990	0.9308	0.6897	15.9321	0.5826	0.9201	0.6959	18.3993	0.6386	0.9319	0.7169
PP	17.9335	0.6229	0.9331	0.7076	16.8736	0.5834	0.9204	0.6960	19.3785	0.6809	0.9386	0.7431

Table 7 Smartphone dataset 2

Software name	Overlapping area				Left non-overlapping area				Right non-overlapping area			
	PSNR	SSIM	VSI	FSIM	PSNR	SSIM	VSI	FSIM	PSNR	SSIM	VSI	FSIM
Autostitch	14.3241	0.5246	0.8786	0.6726	13.1400	0.5059	0.8576	0.6441	14.5796	0.5488	0.8639	0.6447
Hugin	12.6626	0.4584	0.8480	0.6190	10.0106	0.4023	0.7825	0.5859	10.6321	0.4409	0.7942	0.5739
ICE	12.5755	0.4518	0.8536	0.6403	10.7969	0.4495	0.8345	0.6216	11.9116	0.4509	0.8176	0.6043
PM	12.5761	0.4819	0.8664	0.6548	12.4732	0.4586	0.8516	0.6345	12.7779	0.4875	0.8481	0.6236
PP	14.0309	0.5070	0.8719	0.6694	12.6876	0.4787	0.8540	0.6379	13.9763	0.5324	0.8491	0.6289

Table 8 Smartphone dataset 3

Software name	Overlapping area				Left non-overlapping area				Right non-overlapping area			
	PSNR	SSIM	VSI	FSIM	PSNR	SSIM	VSI	FSIM	PSNR	SSIM	VSI	FSIM
Autostitch	16.2034	0.7589	0.9069	0.7159	19.9430	0.7482	0.9064	0.7029	17.8754	0.6645	0.9065	0.7354
Hugin	14.5194	0.6609	0.8873	0.6899	17.2414	0.6552	0.8885	0.6737	14.1712	0.6018	0.8207	0.6238
ICE	15.5316	0.7122	0.8926	0.6995	18.1387	0.6793	0.8888	0.6777	16.3688	0.6373	0.8528	0.6576
PM	15.1732	0.6987	0.8876	0.6871	17.9720	0.6729	0.8913	0.6768	16.6107	0.6274	0.8382	0.6395
PP	15.8472	0.7371	0.8994	0.7060	19.3251	0.7119	0.9061	0.7009	17.1607	0.6584	0.9006	0.7328

replaced by a smartphone dataset and similarly for three varieties of datasets, results are compared for five software programs (Autostitch, Hugin, ICE, PM, PP) using four IQA metrics. Table 6 summarizes results for smartphone dataset 1 which depicts an outdoor scene with building features. It can be seen that the values of the IOA matrices are significantly higher for PP software in both the overlapping and non-overlapping areas. Results for smartphone dataset 2 are represented in Table 7. Autostitch software shows higher IOA matrices values and is suited to outdoor scenes with trees. In the last smartphone dataset which depicts indoor scenes, Autostitch software performs best in comparison to all other stitching software (Table 8).

We computed the rank of each method (software) *m* using the ranking algorithm described in Sect. 4.3. First, we calculated the rank of each method *m* for datasets 1, 2, and 3 individually then we calculated the rank of each method *m* for all datasets combined.

Table 9 (dataset 1 column) shows the ranking of software products for dataset 1. Panorama Plus ranks the highest, followed by Autostitch in second position. ICE and PM show the same type of results hence they have equal rank (rank 3). Hugin gives very poor results thus attains the lowest rank. Table 9 (dataset 2 column) shows the rankings for dataset 2. For dataset 2, Autostitch performs best therefore comes in first, Panorama Maker is in second position. Panorama Plus, the top ranker for dataset 1 shows average results for dataset 2 hence only attains third position. Hugin shows very poor results for dataset 2 also. Table 9 (dataset 3 column) shows the rankings for dataset 3. Again Autostitch is in first position for dataset 3, ICE and Panorama Plus are in second position. Again, Hugin performs very poorly hence finishes in last position.

We also calculated the overall rank of all the methods for all three datasets. Table 10 shows the rankings of all methods for all types of datasets. Autostitch obtains first rank which shows that Autostitch is best panorama software for all three datasets. Panorama Plus is in second position. Panorama Maker and ICE are third and fourth respectively. Hugin is in last position having performed very poorly for all datasets.

Table 9 Ranking of software products for each dataset individually

	Dataset 1		Dataset 2		Dataset 3	
Sl. no.	Method name	Rank	Method name	Rank	Method name	Rank
1.	PP	1	Autostitch	1	Autostitch	1
2.	Autostitch	2	PM	2	ICE	2
3.	ICE	3	PP	3	PP	2
4.	PM	3	ICE	4	PM	3
5.	Hugin	4	Hugin	5	Hugin	4

Table 10 Ranking of software products for all three datasets combined

Sl. no.	Method name	Rank
1.	Autostitch	1
2.	PP	2
3.	PM	3
4.	ICE	4
5.	Hugin	5

6 Conclusions

This paper has conducted a comparative study of five commonly used panorama software products. This study concludes that an objective image quality evaluation (IQA matrices) technique can be used as a very efficient way of compiling three different scenarios captured using a digital camera or smartphone for use in the five most popular panorama software products. When the results are compared for individual scenarios Autostitch performs relatively well for all types of indoor and outdoor scenes with both camera and smartphone datasets. This study recommends Autostitch as the best panorama image stitching software currently available.

References

1. Boutellier J, Silvén O, Tico M, Korhonen L (2008) Objective evaluation of image mosaics. Computer vision and computer graphics. Theory and applications. VISIGRAPP 2007. In: Communications in computer and information science, vol 21. Springer, Berlin, Heidelberg. https://doi.org/10.1007/978-3-540-89682-1_8
2. Brown M, Lowe D (2007) Automatic panoramic image stitching using invariant features. Int J Comput Vis 74(1):59–73. https://doi.org/10.1007/s11263-006-0002-3
3. Chandler DM (2013) Seven challenges in image quality assessment: past, present, and future research. ISRN Signal Proc 2013, Article ID 905685, 53 pp. https://doi.org/10.1155/2013/905685
4. Dissanayake V, Herath S, Rasnayaka S, Seneviratne S (2015) Quantitative and qualitative evaluation of performance and robustness of image stitching algorithms. In: International conference on digital image computing: techniques and applications (DICTA). https://doi.org/10.1109/dicta.2015.7371297
5. Eskicioglu M, Fisher PS (1995) Image quality measures and their performance. IEEE Trans Commun 2959–2965. https://doi.org/10.1109/26.477498
6. Ghosh D, Park S, Kaabouch N (2012) Quantitative evaluation of image mosaicing in multiple scene categories. In: International conference on electro/information technology (EIT). https://doi.org/10.1109/eit.2012.6220726
7. Gledhilla D, Tiana GY, Taylora D, Clarke D (2003) Panoramic imaging—a review. Comput Graph 435–445. https://doi.org/10.1016/s0097-8493(03)00038-4
8. Hugin (software). https://en.wikipedia.org/wiki/Hugin_
9. Image Composite Editor—Microsoft Research. https://www.microsoft.com/en-us/research/project/image/image-composite-editor
10. Liu A, Lin W, Narwaria M (2012) Image quality assessment based on gradient similarity. IEEE Trans Image Process 21(4):1500–1512. https://doi.org/10.1109/tip.2011.2175935

11. Mukherjee D, Jonathan Wu QM, Wang G (2015) A comparative experimental study of image feature detectors and descriptors. Mach Vis Appl 26:443–466. https://doi.org/10.1007/s00138-015-0679-9
12. Paalanen P, Ka¨ma¨ra¨inen JK, Ka¨lvia¨inen H (2009) Image based quantitative mosaic evaluation with artificial video. In: Sixteenth scandinavian conference on image analysis, pp 470–479. https://doi.org/10.1007/978-3-642-02230-2_48
13. Panorama-Maker. http://www.arcsoft.com/panorama-maker
14. Panorama-Plus. http://www.serif.com/panoramaplus/
15. Sheikh HR, Bovik AC (2006) Image information and visual quality. IEEE Trans Image Process 15(2):430–444. https://doi.org/10.1109/tip.2005.859378
16. Suresh M, Jain K (2015) Semantic driven automated image processing using the concept of colorimetry. Procedia Comput Sci 58(2015):453–460. https://doi.org/10.1016/j.procs.2015.08.062
17. Tiwari A, Jain K (2015) A Detailed 3D GIS architecture for disaster management. Int J Adv Remote Sens GIS 4(2015):980–989
18. Wang Z, Bovik AC, Sheikh HR, Simoncelli EP (2004) Image quality assessment: from error visibility to structural similarity. IEEE Trans Image Process 13(4):600–612. https://doi.org/10.1109/tip.2003.819861
19. Xu W, Mulligan J (2010) Performance evaluation for color correction approach for automatic multi-view image and video stitching. In: 23rd IEEE conference on computer vision and pattern recognition, pp 263–270. https://doi.org/10.1109/cvpr.2010.5540202
20. Zhang L, Shen Y, Li H (2014) VSI: a visual saliency induced index for perceptual image quality assessment. IEEE Trans Image Process 23(10):4270–4281. https://doi.org/10.1109/tip.2014.2346028
21. Zhang L, Zhang D, Mou X (2011) FSIM: a feature similarity index for image quality assessment. IEEE Trans Image Process 20(8):2378–2386. https://doi.org/10.1109/tip.2011.2109730

Emerging Trends in Big Data Analytics—A Study

G. Naga Rama Devi

Abstract Big data refers to exceptionally large datasets that are growing exponentially with time. The three key enablers for the growth of big data are (1) data storage, (2) computation capacity, and (3) data availability (Grobelnik M, Big-Data tutorial, 2012 [1]). This massive, heterogeneous, and unstructured digital content cannot be processed by traditional data management techniques and tools effectively, but this problem is overcome by using big data analytics. In this paper, we have discussed various big data services, languages, and data visualization tools. Big data helps organizations to increase sales and improves marketing results. It also improves customer service, reduces risk, and improves security. Both high storage and computation are important requirements for big data analytics. Information technology researchers and practitioners have faced the major challenge of designing systems for the efficient handling of data and its analysis for the decision-making process as the amount of data continues to grow. Big data is available in three forms, namely structured, unstructured, and semi-structured. The top ten big data technologies are (1) predictive analytics, (2) NoSQL databases, knowledge discovery and searching, (4) stream analytics, (5) data fabric for in memory computing, (6) distributed file stores, (7) virtualization of data, (8) integration of data, (9) preparation of data, and (10) quality of data. Amazon Elastic MapReduce, Apache Hive, Apache Pig, Apache Spark, MapReduce, Couchbase, Hadoop, and MongoDB are data integration tools used to manipulate big data accurately.

Keywords Big data trends and tools · Big data lambda architecture
Data storage and management · SQL-in hadoop · Data languages
Data visualization tools

G. Naga Rama Devi (✉)
Department of CSE, CMRIT, Kandlakoya, Medchal Road, Hyderabad, India
e-mail: ramadeviabap@yahoo.co.in; ramadevi.abap@cmritonline.ac.in

A. Kumar and S. Mozar (eds.), *ICCCE 2018*,
Lecture Notes in Electrical Engineering 500,
https://doi.org/10.1007/978-981-13-0212-1_57

1 Introduction

Big data is data of an extremely large size. It is generally said to occur in three forms. Structured data can be stored, accessed, and processed in the fixed format called "structured" data, e.g., a table definition in relational DBMS. Unstructured data are any data with an unknown form or where the structure is classified as unstructured data. A typical example of unstructured data is a "Google search," which contains heterogeneous data sources like simple text files, images, and videos etc. Semi-structured data contains both structured and unstructured forms of data. The characteristics of big data are described the 8 Vs and these are (1) volume, (2) velocity, (3) variety, (4) value, (5) veracity, (6) visualization, (7) viscosity, and (8) virality. The definition of big data involves the volume, velocity, variety, and veracity of information.

- Volume: The collection of massive data in the order of zettabytes.
- Velocity: Data in motion, i.e. streaming data. The data collection and analysis must be conducted rapidly and timely.
- Variety: Data available in many forms, such as structured (RDMBS), unstructured (videos, audio, webpages, etc.), and semi-structured (text, xml, etc.).
- Veracity: Data in doubt, i.e. uncertainty due to data inconsistency and incompleteness, ambiguities, latency, deception, and model approximations.

Big data technologies are used in different applications, such as (1) marketing, (2) finance, (3) government, (4) healthcare, (5) insurance, (6) retail, (7) telecommunications, and (8) gaming.

Broad and adaptive big data integration contains (1) the ability to access data formerly, process, combine, and consume it, (2) greater flexibility, (3) reduced risk, (4) support for the latest Hadoop distributions from Cloudera, Hortonworks, MapR, and Amazon Web Services, (5) ability to access data for preparation via SQL on Spark and to orchestrate existing Spark applications in Scala, Java, and Python, (6) integration with NoSQL stores including MongoDB and Cassandra, and (7) connectivity to analytic databases including HPE Vertical, Amazon Redshift, and SAP HANA.

The most popular big data tools for developers are (1) Splice Machine, (2) MarkLogic, (3) Google Charts, (4) SAP in Memory, (5) Cambridge semantics, (6) MongoDB, (7) Pentaho, (8) Talend, (9) Tableau, and (10) Splunk.

2 Lambda Architecture

Lambda Architecture (LA) [2] provides a hybrid platform by combining real-time data and data precomputed by the Hadoop. In Lamda Architecture we have discussed about importance of **Twitter's Summingbird and lambdoop**.

An LA framework contains any of the components which are shown in the successive Fig. 1 [3]:

1. **Twitter's Summingbird** [4]: Summingbird is a library that is used to write streaming MapReduce programs and execute them on distributed MapReduce platforms like Storm and Scalding. We can execute the Summingbird program in either of the following ways.
   a. Batch mode (using Scalding on Hadoop).
   b. Real-time mode (using Storm).
   c. Hybrid batch/real-time mode (offers attractive fault-tolerance properties).
2. **Lambdoop**: Lambdoop is a new big data middleware designed for data scientists and developers to build big data solutions combining streaming and batch data analytics.
   - All data entering into the system is dispatched to both the batch layer and the speed layer for processing.
   - The **batch layer** (**Apache Hadoop**) has two functions: (i) to manage the master dataset, and (ii) to precompute the batch views.
   - The **serving layer** (**Cloudera Impala**) indexes the batch views so that they can be queried in a low-latency, ad hoc way.
   - The **speed layer** (**Storm**, **Apache HBase**) compensates for the high latency of updates to the serving layer and deals with recent data only.

2.1 Properties of the Batch Layer and Saving Layer

Both layers satisfy all the properties, there is no concurrency issue and it scales insignificantly. The final layer, the speed layer will fixes the low latency updates.

- **Robust and fault tolerant**: In this, failover is handled by the Batch Layer using replication when machines go down and restarting computation tasks on other machines.

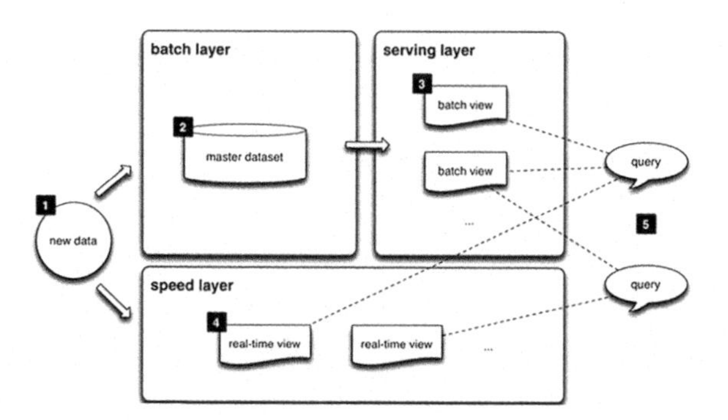

Fig. 1 Big data lambda architecture

- **Serving Layer**: To ensure availability, it uses replication when servers go down. Both are human fault tolerant, since, when a mistake is made, we can fix our algorithm or remove the bad data and recompute the views from scratch.
- **Scalable**: Both are easily scalable.
- **General**: Both have a general architecture. We can compute and update inconsistent views of both layers.
- **Allows ad hoc queries**: All of the data is available in one location and we are able to run any function we want on that data.
- **Minimal Maintenance**: Both maintain essential information to the large no of applications.
- **Debuggable**: Can be debugged when errors are found and it provides full information.

3 Data Storage and Management

The traditional systems unable to handle a large amount of data. A good storage provider should be available on which all analytical tools can be placed to store and query data [3].

3.1 Hadoop

Hadoop is open-source software. It is a distributed file system and MapReduce package. It is a software technology designed for storing and processing large volumes of data distributed across a cluster of commodity servers and commodity storage. It is used in analytics where there are terabytes of data to mine.

The base Apache Hadoop framework consists of the following core modules:

- Hadoop Common: This supports Hadoop modules.
- Hadoop Distributed File System (HDFS): This stores data on commodity machines, providing very high aggregate bandwidth across the cluster.
- Hadoop YARN: This provides a resource management platform for applications scheduling of users.
- Hadoop MapReduce: A programming model for large-scale data processing.

3.2 Big Data Analytics Tools [5–7]

There are many big data analytics tools that are available to handle the data effectively.

- **Ambari**: In Apache Ambari, Apache Hadoop clusters effectively.
- **Hive**: Apache *Hive* contains two modules, namely Apache Hadoop Distributed File System and MapReduce. It allows SQL developers to write *Hive* query language statements similar to standard SQL ones.
- **Pig**: Apache Pig is a scripting language. It provides a platform for analyzing large data sets. We can access and transfer data effectively.
- **Sqoop**: In Apache Spoop, data moves between Hadoop and RDBMS efficiently.
- **Flume**: In Apache Flume, data moves from log files into HDFS.
- **Mahout**: Apache Mahout is a library of scalable machine-learning algorithms. It is implemented on top of Apache Hadoop.
- **Tez**: Apache Tez builds high performance batch and interactive data processing applications. It is a data flow programming framework.
- **Spark**: Apche Spark runs on Hadoop, Mesos, standalone, or in the cloud. It can access diverse data sources including HDFS, Cassandra, HBase, and S3.
- **Zookeeper**: This is an Hadoop admin tool used for managing the jobs in the cluster and provides service for distributed applications.

3.3 *Cloudera*

Cloudera is a Hadoop with some extra services. It helps business people to build an enterprise data hub and provides better access to data storing. Cloudera is an open source element and helps businesses to manage their Hadoop ecosystem.

3.4 *MongoDB*

This is open source software. It is a NoSQL data store it manages unstructured or semi-structured data effectively as well as performing real-time processing effectively. It is a document-oriented and schema-less data store. MongoDB is used in a web application as a backend instead of MYSQL.

3.5 *Talend*

Talend is open source software. It is the open platform for data loading, data integration, data extraction, and data transformation. It is a rich ETL tool for extract, load, and transform processes.

4 Data Languages [8]

Data languages are used to understanding basic languages like R, Python, RegEx, XPath, and their functionalities.

- **R Programming**: R is an open source programming language. It performs statistical computation efficiently and displays the data graphically. It provides more debugging facilities for data validation. It can run easily run on Linux, Windows and Mac operating systems.
- **Python**: Python is an open source programming language for web development, game programming, desktop GUI's, scientific programming, network programming.
- **RegEX**: Regular expressions contain a set of characters that can manipulate data and perform pattern matching with strings or text effectively.
- **XPath**: This is a query language. It is used for data extraction and selecting certain nodes from XML documents effectively.

5 SQL-in-Hadoop

In this paper, we have discussed various SQL-in-Hadoop tools to manipulate big data and their functionalities as detailed in Table 1:

6 Data Visualization

Data visualizations tools are used to present complex data graphically and analyze it for perfect decision making [9]. Data visualization tools are Tableau, Silk, CartoDB, Chartio, and Plot.ly.

Tableau: Tableau is a data visualization tool used for business intelligence. Without programming we can create bar charts, maps, and scatter plots, and we can connect to databases and API using a data connector.

Silk: Silk is a simpler analytical and data visualization tool than Tableau. It builds interactive maps and charts quickly.

CartoDB: This is an open source tool and is used to facilitate making maps. It makes it possible to visualize and store geospatial data on the web. It can manage data files and types.

Chartio: Chartio is a visual query language that is easier to use than SQL. It does not require a separate data warehouse. It combines data sources and executes queries in-browser. Powerful dashboards can be created effectively.

Plot.ly: It allows the creation of stunning 2D and 3D charts without any programming knowledge.

Table 1 SQL-in-hadoop tools

Languages	Functionality
Apache Hive	1. It is used to querying and managing large datasets residing in distributed storage 2. It allows the map reduce programmers to plug in custom mappers and reducers
Impala	1. Cloudera's Impala is an open source massively parallel processing (MPP) SQL query engine 2. It runs natively in Apache Hadoop 3. Without requiring data movement or transformation, it enables users to directly query data stored in HDFS and Apache HBase
Shark	1. It is a data warehouse system and compatible with Apache Hive 2. It supports Hive's query language, meta store, serialization formats, and user-defined functions
Spark SQL	1. It allows relational queries expressed in SQL HiveQL, or Scala to be executed using Spark 2. Spark SQL is an alpha component
Apache Drill	1. Apache Drill, is an Apache incubation project 2. It provides ad hoc queries to different data sources 3. Inspired by Google's Dremel, Drill is designed for scalability and ability to query large sets of data 4. This project is backed by MapR
Apache Tajo	1. It is a big data relational and distributed data warehouse system for Apache Hadoop 2. Tajo is designed for low-latency and scalable ad hoc queries, online aggregation, and ETL (extract-transform-load process) on large-data sets stored on HDFS (Hadoop Distributed File System) and other data sources
Presto	1. Presto framework from Facebook, is an open source distributed SQL query engine for running interactive analytic queries against data sources of all sizes
Phoenix	1. It is an open source SQL query engine for Apache HBase 2. It is accessed as a JDBC driver and enables querying and managing HBase tables using SQL

Data wrapper: This is an open source tool. It creates embeddable charts and maps. It uses hands-on tables in key areas, enabling users to create high-quality charts much faster than with traditional workflows and tools [10].

7 Conclusion

In this paper we have discussed big data, Lambda architecture, data storage management, data languages, and data visualization tools. The big data emerging trends and tools survey and analysis may be useful to many practitioners and researchers. An effective and efficient system can be created by big data analytics. Handling big data is difficult using traditional methods, so it is overcome by using big data analytics and cloud security mechanisms.

References

1. Marko Grobelnik Big-Data tutorial, Jozef Stefan Institute Ljubljana, lovenia Stavanger May 8, 2012
2. http://www.databasetube.com/database/big-data-lambda-architecture/ https://www.mongodb.com/hadoop-and-mongodb
3. Massoud Sagiroglu S, Sinanc D (2013) Big data: a review. In: Proceeding of the 2013 international conference on collaboration technologies and systems (CTS). IEEE Computer Society, pp 42–47
4. Katal A, Wazid M, Goudar RH (2013) Big data: issues, challenges, tools and good practices, In: Sixth international conference on contemporary computing (IC3), pp 404–409
5. Louridas P, Ebert C (2013) Embedded analytics and statistics for big data. IEEE Softw 30 (6):33–39
6. Kambatla K, Kollias G, Kumar V, Grama A (2014) Trends in big data analytics. J Parallel Distrib Comput 74(7):2561–2573
7. Chen M, Mao S, Liu Y (2014) Big data: a survey. Mob Netw Appl 19:171–209
8. https://www.infoq.com/research/big-data-emerging-trends
9. Chen PCL, Zhang C-Y (2014) Data-intensive applications, challenges, techniques and technologies: a survey on big data. Inf Sci 275:314–347
10. Talia D (2013) Clouds for scalable big data analytics. Computer 46(5):98–101

A Novel Telugu Script Recognition and Retrieval Approach Based on Hash Coded Hamming

K. Mohana Lakshmi and T. Ranga Babu

Abstract Due to their many applications, optical character recognition (OCR) systems have been developed even for scripts like Telugu. Due to the huge number of symbols utilized, identifying the Telugu words is a very complicated task. Pre-computed symbol features are stored by these types of systems to be recognized or retrieved from a database. Hence, searching of Telugu script from the database is a challenging task due to the complication involved in finding the features of the Telugu word images or scripts. Here, we implement a novel Telugu script recognition and retrieval approach based on a method called hash coded hamming (HCH). Hash coding will be used as a feature extractor and the hamming distance will be utilized as a replacement for conventional Euclidean distance in order to measure the similarity between query and database images. Simulation analysis shows that the proposed scheme has a superior performance to the conventional approaches presented in the literature.

Keywords Optical character recognition (OCR) · Telugu script
Hash coding · MD5 algorithm · Hamming distance

1 Introduction

Document image retrieval is an emerging field of research with the continuous growth of interest in having data/information accessible in a digital format for effective access, safe storage, and long-term preservation. Scanned or digital

K. Mohana Lakshmi (✉)
Department of Electronics & Communication Engineering, CMR Technical Campus, Hyderabad, India
e-mail: mohana.kesana@gmail.com

T. Ranga Babu
Department of Electronics & Communication Engineering,
RVR & JC College of Engineering, Guntur, India
e-mail: trbaburvr@gmail.com

A. Kumar and S. Mozar (eds.), *ICCCE 2018*,
Lecture Notes in Electrical Engineering 500,
https://doi.org/10.1007/978-981-13-0212-1_58

documents don't contain searchable text regions inside the image and cannot be retrieved or searched by conventional search engines. Searching and retrieval for pertinent document images should be possible with the assistance of recognition-based and recognition-free approaches [1, 2]. A direct method of accessing these documents involves converting document images to their textual form by recognizing text from images. OCR alludes to a procedure of generating a character input by optical means, such as filtering, for recognition in ensuing stages by which a printed or handwritten text can be converted to a form which a PC can comprehend and control. The procedure of automatic reading of documents consists of different stages like image acquisition, pre-processing, object extraction, standardization or windowing, feature extraction, arrangement and post-processing. These are important stages of any word recognition system which is generic in nature. The vast majority of the Indian scripts are based on the Brahmi script through different transformations. The emphasis is on recognition-free approaches for retrieval of applicable documents from huge accumulations of document images. Like a recognition-based approach, recognition-free schemes likewise goes through many stages. When we get a document for analysis, it is preprocessed and then divided into words. At that point, features are extricated for portrayal of the fragmented word images. For a given inquiry, a similar methodology of highlight extraction is applied here. The issues that arises is of no too much significance here because the query images are exactly same in nature as present in the database images. In this area word searching, another method of optical character recognition (OCR), is lacking in ability to recognize and retrieve words from poor quality text documents. Google has developed digital libraries where users can search several thousands of printed documents to find the required text information. Conventional OCR approaches are poor in recognizing and retrieving text from poor quality documents. The main aim is to recognize the query images from images in the large database. The user feeds the query image and the system will find similar images and return the most relevant text images in the stored documents. A novel algorithm for OCR system to retrieve words is based on the extraction of features, such as shape, color, and geometric features etc. The overall structural procedure of Telugu word recognition and retrieval system has pre-computed symbol features which are stored by these types of systems so that words can be recognized or retrieved in a database.

2 Literature Review

There has been a lot of research and advancement in the range of machine learning and image handling in recent years. Powerful component descriptors like HOG [3], SIFT [4] and GLOH have been produced and utilized as a part of various applications like image recognition and image enrollment. The essential thought while choosing features is its invariance to perspective and lighting changes. These techniques can be connected to distinguish Telugu documents for preservation.

The character set of Telugu has various characters in it and has an exceptionally confounded structure when contrasted with Latin scripts. So in this paper, we have built up a calculation for distinguishing Telugu words. Telugu characters have certain primitive shapes and their recognition is a testing assignment. On the premise of SIFT features, recognition is done in view of BOW and SVM classifiers. Telugu dialect, not at all like English, is a dialect in which the written word system does not utilize vowels independently, rather, vowels are utilized as diacritics with a consonant in order to make a composite character. Therefore, there is a wide blend of composite characters which can be made. Rather than selecting singular character recognition it is henceforth better to accomplish word recognition with respect to simple character recognition. Telugu word image retrieval is a difficult and challenging task due to the fact that each word image has its own structure with single and multi-conjunct vowel consonants.On the other hand, characters in south Indian scripts like Telugu are composed of more than one object making it more complex to apply high level feature extraction techniques. Once in a while even consonants are joined to form composite characters termed ligatures. There are 16 vowels and 36 consonants in the Telugu dialect. A couple of learning models have been connected in the context of Telugu character recognition, for example, stochastic models [5] and machine-learning-based models [6]. SVMs have accomplished reasonable exactness in the recognition of individual characters, in digit recognition, and in character identification in Roman [7], Thai [8], English [9], and Arabic scripts [10]. For word identification, gradient-based features have appeared with a far superior performance when contrasted with texture-based features. A template coordinating technique will have the capacity to coordinate these features to related images. The techniques utilized for word order can be chiefly arranged into two sorts (i) structure-based and (ii) visual appearance-based methods. Recognition of words from scanned Telugu documents has not been investigated much in contrast to the customary scanned documents in English writing. As of late, some research has been done on script identification issues. Creator [11] proposed an approach in view of measurable script identification from images in different illumination conditions. The examination detailed by [12] uncovered that a better performance for word recognition has been accomplished by the utilization of Gabor features using closest neighbor and SVM classifiers. A proposition by [13] discussed the smoothness and cursiveness of lines to recognize the script. They considered only Tamil, English, and Chinese dialect scripts for their investigations. To perceive six different scripts, Creator [14] proposed spatial-gradient-features at the piece level by considering the separated text lines from the images for the trials and a normal arrangement rate of 82.1% was accounted for. Creator [15] examined and investigated three different features, which were Gabor, Zernike moment and a gradient with dimensionality of 400. The recognition and retrieval for the words was accomplished using support vector machine (SVM). These authors have clarified that the vital pre-handling methods are required to overcome the issues with the input or source. Sharma et al. [16] showed that to distinguish a word from an image; Gradient Local Auto-Correlation

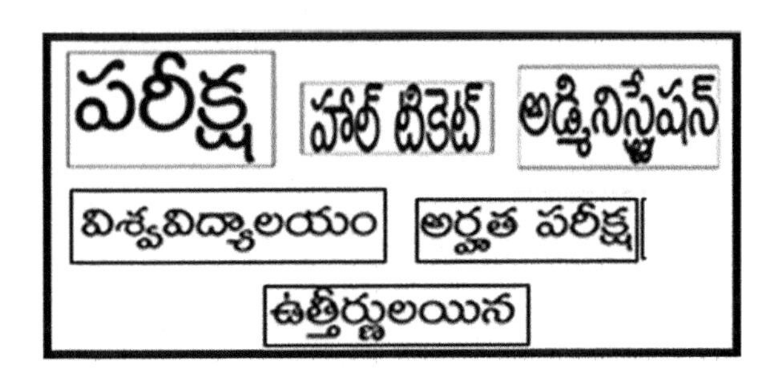

Fig. 1 Sample images from the dataset

(GLAC) feature is very efficient and effective and they found that for retrieving or recognizing the Telugu script, the gradient feature is computationally simpler and effective than the conventional texture features (Fig. 1).

A scheme termed template coordinating has been used for word recognition in [17] with the component identification as gradient angular features (GAF) applied to 760 words from six distinct scripts. In light of all the research papers discussed above and and from the literature review we can conclude that the method used for retrieving or searching words is the one which can distinguish words from each other based on features. Based on this, we have proposed a novel Telugu word recognition algorithm based on Hash coding capacity with similarity measured by using Hamming distance. We likewise introduced a comparative report and investigation of traditional techniques in order to comprehend the effectiveness and robustness of the proposed approach. Recently, author in [18] presented a HMM based correlation method for retrieving the Telugu word images from database. In [19], template matching scheme has been introduced for the retrieval of word images.

3 Proposed Approach

3.1 Hash Function

A hash function is a function which is used to map data of arbitrary size to data of settled size. The hash function returns certain values called hash values, hash codes, digests, and essentially, hashes. Data structure creates hash table using hash values as a part of PC programming for rapid data lookup. In a vast document hash functions accelerate table or database lookup by identifying copied records. Hash functions are identified with (and frequently mistaken for) checksums, check digits, fingerprints, lossy compression, randomization functions, error-correcting codes, and figures. Despite the fact that these concepts overlap to some degree, each has its own uses and requirements and despite the fact that these concepts overlap to some degree, each has its own uses and requirements and is defined and improved in an unconventional manner. Hash functions are occasionally called message process functions. Their motivation is to extricate a settled length bit string from a message (image, documents, and so forth.). Hash functions have discovered fluctuated applications in different cryptographic, compiler and database search applications. As of late, there has been a considerable measure of interest in utilizing hash functions in sight and sound applications both for security and ordering. A key element of ordinary hashing algorithms, for example, message digest 5 (MD5) and secure hash algorithm 1 (SHA-1) is that they are amazingly sensitive to the message, i.e. changing even one bit of the input will change the yield significantly. In any case, sight and sound data, for example, digital images, experience different controls, such as compression, improvement, trimming, and scaling. An image hash function ought to rather consider the adjustments in the visual area and deliver hash values in light of the image's visual appearance. Such a hash function would be valuable in recognizing images in databases, in which the image perhaps experiences coincidental changes, (for example, compression and format changes, normal flag handling operations, filtering or watermarking). A noteworthy use of a perceptual image hash could be for reliable image verification. In such cases, the hash must be invariant under perceptually irrelevant alterations to the image yet recognize malignant altering of image data. A few different applications can be distinguished in the territories of watermarking and information implanting in images.

3.2 MD5 Algorithm

The MD5 algorithm is a generally utilized hash function delivering 128-bit hash esteem. It can be utilized as a checksum to confirm data trustworthiness, the query word image gets affected with noise, occlusions and random distortions. MD5 processes a variable-length message into a settled length yield of 128 bits. The input message is separated into lumps of 512-bit blocks (sixteen 32-bit words); the message is cushioned with the goal that its length is distinct by 512. The padding fills in as takes after: initially a solitary bit, 1, is attached to the end of the message. This is followed by the same number of zeros as are required to bring the length of the message up to 64 bits less than a different of 512. The rest of the bits are topped off with 64 bits speaking to the length of the first message, modulo 264. The fundamental MD5 algorithm works on a 128-bit state, separated into four 32-bit words, signified A, B, C, and D. These are instated to certain settled constants. The principle algorithm at that point uses each 512-bit message hinder to thus alter the state. The handling of a message square comprises four comparative stages termed rounds. Each round is made out of 16 comparative operations in light of a non-direct function F, secluded expansion, and left rotation.

3.3 Algorithm

Step 1: Load the database script images
Step 2: Select and read a query image 'Q' from current directory
Step 3: Resize 'Q' to 128 × 128
Step 4: Now, apply hash coding to 'Q'
Step 5: Now, read all the script images from the database and resize the images with the size of 'Q'
Step 6: Apply hash coding to all the database script images
Step 7: Now, calculate the similarity distance between hash codes of 'Q' and database script images
Step 8: Display the most similar script image as a recognized script from the database

Hamming distances are positive integers that represent the number of pieces of data you would have to change to convert one data point into another. The Euclidean distance is the length of the line segment that connects two coordinates. When all is said and done, for expansive scale retrieval the most vital property is that the search time unpredictability will be directly proportional to the number of

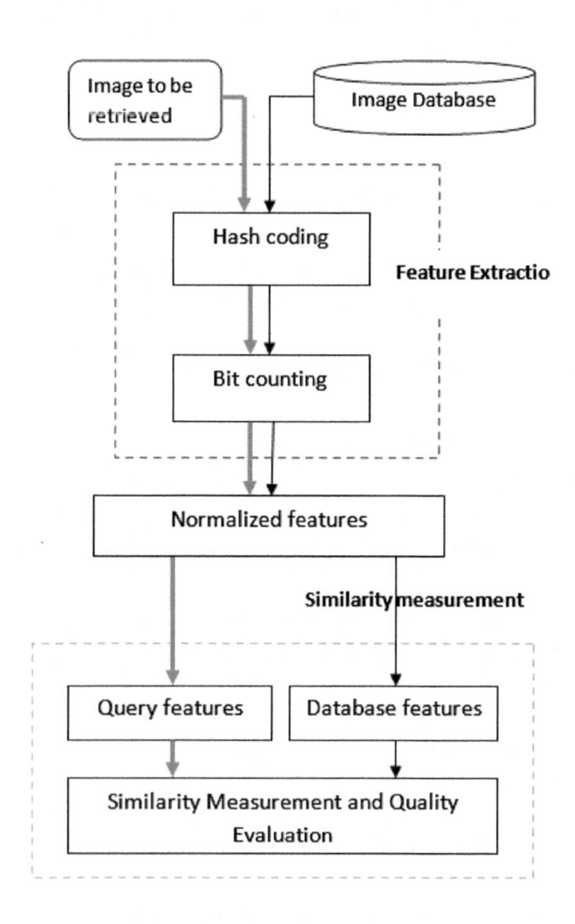

Fig. 2 Block diagram of the proposed system

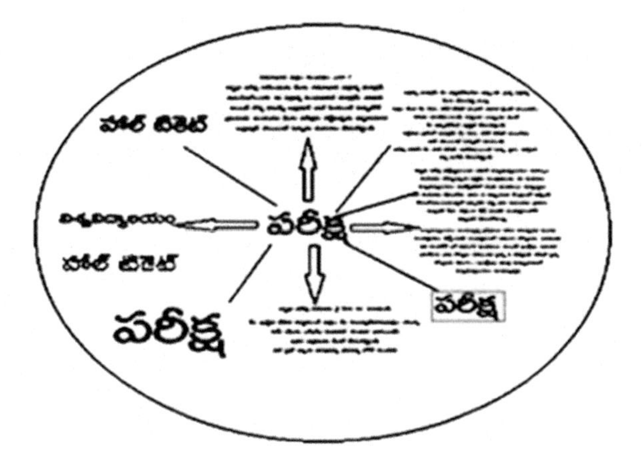

Fig. 3 Recognition of relevant Telugu script using HCH

database images. Also, given the dispersed idea of vast-scale figuring, the capacity to parallelize the search is vital for useful applications (Fig. 2).

In the specific context of parallel codes, as we consider here, retrieval includes finding all cases that have a zero or short hamming distance from the query, where the hamming distance between two double vectors is the quantity of bits that vary between them. To fulfill these requirements, we consider variations of hashing. Hashing is speedy and has insignificant storage requirements past the paired data vectors themselves. It uses all measurements of the twofold codes (bits) in parallel to perform retrieval. This is as opposed to tree-based algorithms, for example, kd-trees, where every case is found by settling on a progression of two ways to navigate the tree, every choice (bit) being restricted by the decisions above (Fig. 3).

4 Simulation Results

All the experiments were performed in a MATLAB 2016a environment with 4 GB RAM. We utilized various Telugu script images for training and testing purposes. First, a database of 60 images was trained and then a query Telugu script was given as an input to recognize and retrieve the relevant Telugu script from the database images. Figure 4 shows the images stored in the database for retrieval and recognition of query Telugu script.

Figures 5 and 6 show that the query script “PARIKSHA” and the recognized output image obtained by using the proposed algorithm and the obtained hash codes are shown in Table 1, where each retrieved image consists of values of size 1×16 and a data class with uint8 which indicates that the size of image is 128×128. The same is shown in Figs. 7 and 8 with a script called “HALLTICKET” and the respected hash codes are given in Table 2.

Precision is the fraction of retrieved words that are relevant to the query:

Fig. 4 Trained database

Query Script image

పరీక్ష

Fig. 5 Query sample 1

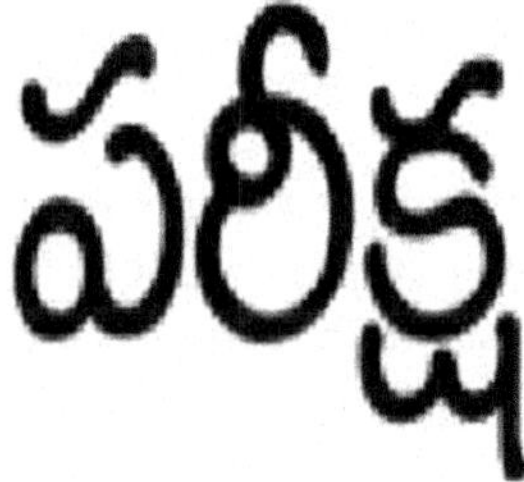

Fig. 6 Retrieved word image for query sample 1

Table 1 Hash table for retrieved images using the proposed approach with Telugu Script name "PARIKSHA"

R{1, 2, 3, 4}	89 31 245 13 71 217 119 222 25 14 26 252 224 123 87 63
R{5}	237 135 114 75 44 213 198 252 80 143 254 204 182 138 93 203
R{6, 7, 8, 9, 10, 11}	108 109 126 174 83 173 20 244 50 56 39 228 221 38 136 143
R{12}	135 19 124 26 40 227 124 152 80 146 26 74 90 216 105 147
R{13}	159 88 102 1 62 141 108 216 33 210 15 216 119 251 234 173

Fig. 7 Query sample 2

Fig. 8 Retrieved word image for query sample 2

Table 2 Hash table for retrieved images using the proposed approach with Telugu Script name "HALLTICKET"

R{1, 2, 3, 4, 5, 6}	108 109 126 174 83 173 20 244 50 56 39 228 221 38 136 143
R{7}	237 135 114 75 44 213 198 252 80 143 254 204 182 138 93 203
R{8}	191 252 174 214 220 129 148 203 6 180 113 174 214 151 89 34
R{9}	179 216 120 52 236 166 132 245 4 158 244 15 71 143 54 56
R{10}	228 133 14 187 193 198 177 153 119 107 25 56 77 179 4 13
R{11}	107 156 101 64 15 120 112 188 82 224 121 238 206 48 48 172
R{12, 13}	113 64 3 191 163 24 83 82 156 76 174 245 54 125 9 109

Table 3 Precision and recall values with respect to number of scripts

No. of scripts	Precision (in %)	Recall (in %)
50	97.77	92.12
90	96.52	90.45
150	94.98	88.78

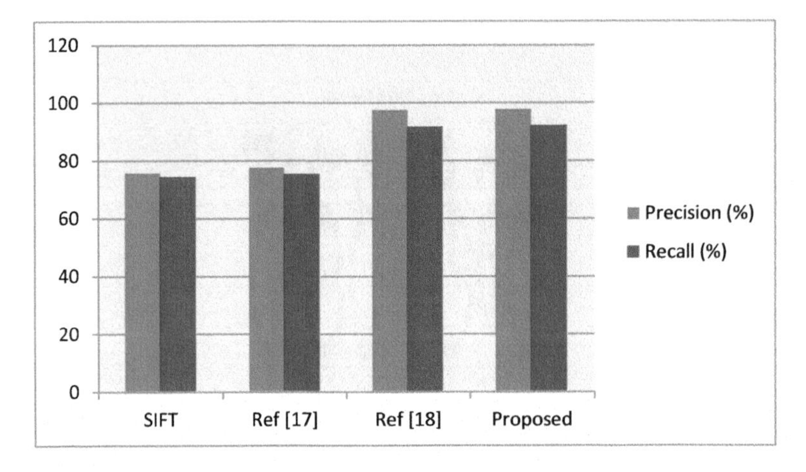

Fig. 9 Performance evaluation

$$P = \frac{|\{\textbf{relevant instances}\} \cap \{\textbf{retrieved words}\}|}{|\{\textbf{retrieved words}\}|} \tag{1}$$

Recall is the ratio of the relevant words that are successfully matched (Table 3 and Fig. 9).

$$R = \frac{|\{\textbf{relevant instances}\} \cap \{\textbf{retrieved words}\}|}{|\{\textbf{retrieved words}\}|} \tag{2}$$

5 Conclusions

A novel Telugu script recognition and retrieval scheme has been presented and evaluated with respect to conventional methods in terms of efficiency and accuracy. Our proposed HCH scheme is very simple and robust. It is very easy to implement even in real-time environments due to the usage of the application being in binary format. Hamming distance was utilized in of place conventional Euclidean distance for counting the number of bits instead of finding the length of the line segment;

this will produce more accurate measurements of similarity. Hence, it is very easy to find the most relevant database image for the query input. Finally, we achieved a maximum precision of 97.77% and higher recall of 92.12% by utilizing the HCH scheme.

6 Future Enhancement

Furthermore, the proposed scheme will be enhanced with respect to the meta database and the different types of noisy documents presented in the database as well as query images. These noises might be salt and pepper, Gaussian, and random etc. We will also consider the statistical parameters required to characterize an image by finding the ocular and texture features needed to get a better performance.

References

1. Rath TM, Manmatha R (2007) Word spotting for historical documents. IJDAR 9(2):139–152
2. Abidi A, Siddiqi I, Khurshid K (2011) Towards searchable digital Urdu libraries—a word spotting based retrieval approach. In: Proceedings of the 11th international conference on document analysis and recognition (ICDAR), pp 1344–1348
3. Dalal N, Triggs B (2005) Histograms of oriented gradients for human detection. In: IEEE conference on computer vision and pattern recognition, San Diego, CA, USA
4. Lowe DG Distinctive image features from scale invariant key points. Int J Comput Vis 60 (2):91–110
5. Li X, Plamondon R, Parizeau M (1998) Model-based online handwritten digit recognition. In: Proceedings of 14th international conference on pattern recognition, Brisbane, Australia
6. Sigappi AN, Palanivel S, Ramalingam V (2011) Handwritten document retrieval system for tamil language. Int J Comput Appl 31
7. Bunke H Recognition of cursive Roman handwriting—past, present, and future. In: 7th international conference on document analysis and recognition, vol 1, pp 448–459
8. Chanda S, Terrades OR, Pal U SVM based scheme for Thai and English script identification. In: 9th international conference on document analysis and recognition, vol 1, pp 551–555
9. Kortungsap P, Lekhachaivorakul P, Madarasmi S On-line handwriting recognition system for Thai, English, numeral and symbol characters. In: 3rd national computer science and engineering conference, Bangkok, Thailand
10. Lorigo LM, Govindaraju V Offline Arabic handwriting recognition: a survey. IEEE Trans Pattern Anal Mach Intell
11. Li L, Tan CL (2008) Script identification of camera-based images. In: Proceedings of the ICPR
12. Pati PB, Ramakrishnan AG (2008) Word level multi-script identification. Pattern Recognit Lett
13. Phan TQ, Shivakumara P, Ding Z, Lu S, Tan CL (2011) Video script identification based on text lines. In: Proceedings of the ICDAR
14. Zhao D, Shivakumara P, Lu S, Tan CL (2012) New spatial-gradient features for video script identification. In: Proceedings of the DAS
15. Sharma N, Pal U, Blumenstein M (2014) A study on word level multi-script identification from video frames. In: Proceedings of the IJCNN

16. Sharma N, Chanda S, Pal U, Blumenstein M (2013) Word-wise script identification from video frames. In: Proceedings of the ICDAR
17. Shivakumara P, Sharma N, Pal U, Blumenstein M, Tan CL (2014) Gradient-angular-features for word-wise video script identification. In: Proceedings of the ICPR
18. Nagasudha D, Madhavi Latha Y (2016) Keyword spotting using HMM in printed Telugu documents. In: International conference on signal processing, communication, power and embedded systems (SCOPES), pp 1997–2000
19. Mohana Lakshmi K, Ranga Babu T (2016) Searching for Telugu script in noisy images using SURF descriptors. In: IEEE 6th international conference on advance computing, pp 480–483

Comparison-Based Analysis of Different Authenticators

K. Kishore Kumar and A. M. Deepthishree

Abstract In the modern era, advances in technology are endless and information security has a vital role to play, in order to overcome or provide security related things. Authentication is an important factor when considering security. This paper focuses on evaluating different modes of security, such as passwords, biometrics, and security tokens etc. which we can state as authenticators or unique output for the combination. Here we focus on biometric techniques for the purpose of authentication. Every individual is recognized by parameters or characteristic features which are physiological in nature. In order to provide services to an individual, a verification system should be used or upgraded in order to avoid anonymous user and grant authenticated user based on authentication for any service. This paper details a review of authentication as it relates to different users and an evaluation based on source criteria that are unique in nature.

Keywords Biometric · Verification · Password · Individual authentication Control access

1 Introduction

The purpose of information security is to provide confidentiality, access to authorized users, integrity and uniformity of information, and availability etc., in all formats. In order to achieve these things different tools and techniques are used. In the computer era, we have no idea who is at the remote end of networks. It could be an eavesdropper, attacker, a friend or a machine. Over a network we can exchange information on

K. Kishore Kumar (✉) · A. M. Deepthishree
CMR Institute of Technology, Medchal, Hyderabad, India
e-mail: kishukanna@gmail.com

A. M. Deepthishree
e-mail: deepthianthni@gmail.com

A. Kumar and S. Mozar (eds.), *ICCCE 2018*,
Lecture Notes in Electrical Engineering 500,
https://doi.org/10.1007/978-981-13-0212-1_59

subjects such as finances, business plans, property, health, and many more that we may prefer to remain confidential. The world wide web adds complexities pertaining to our information, since fraudsters or attackers can access this information from online records without being physically present. In order to overcome such issues, we are being forced to switch to formal authentication techniques in our day-to-day lives and also if we want to communicate with a computer system over a network. As a result it is important to understand the underlying concepts of authentication techniques, how effective are they, and how they compare to each other.

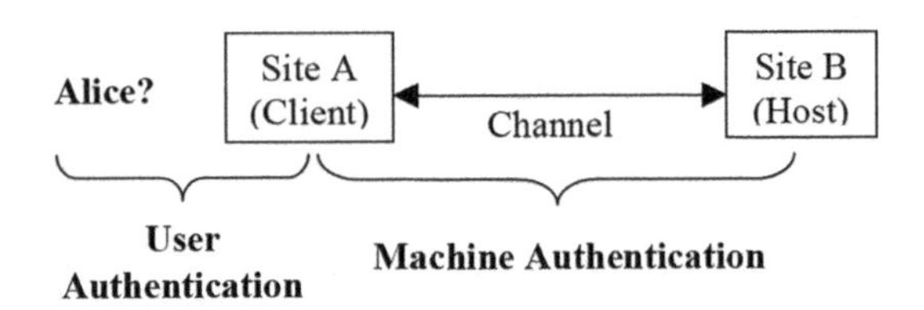

Authentications are of two types:

1. Machine by machine.
2. Machine by human (user authentication).

For example, if we wanted to secure a transaction over an internet we may prefer to use a "SSL" (secure socket layer) protocol. The main flaw with machine by machine authentication is that it verifies machine-based identities but there is no guarantee about the identity of any individual at the machine. Thus, we prefer user authentication for verification and validation of authorized users at every stage of information access. Let us consider the advanced encryption standard (AES) algorithm used for security purposes. This algorithm uses encryption in order to achieve its aim. The user in a AES system uses a private key to encrypt and decrypt information with 256 bits being the peak AES key length. If any attacker started guessing, there would be an average of over 1,076 guesses, which is highly time consuming. The major objective of this paper is to focus on the verification of different authenticators based on factors such as cost and security. Security to be addressed needs a better more of research, so the outline of the security is focused here.

Authenticators:

Pass-code: This is a private number, similar to a password except that it is generated by a machine and stored in a machine. It is random in nature, longer and changing.

Password: This is a string of characters or a word in which we may also include special characters or symbols in order to provide approval to achieve access to a particular resource. The problem associated with these types is that they can be easily guessed or cracked by a persistent attacker. So a random, longer or a changing password is difficult to guess and crack. A physical device such as access token, identity token. Securities help for authentication purpose.

Biometric: This is one of the most widely used features across the globe. A biometric is a feature extracted from an individual that distinguishes them from another individual for user authentication. It includes features like the eye, voice, hand, signature, and face. Futuristic biometrics are those such as smell and gait [1, 2]. For the eye, biometrics from either the iris or retina can be used for authentication. Biometric features can be forged/copied or counterfeited to source extent of difficulty in order to gain an access to a secured system.

Authentication types:

(a) ID-Based ("who you are?"): these are characterized based on the uniqueness of an individual and include passports, credit cards, and driving licenses. In the same manner biometrics include features such as iris scans or retina scans, face recognition, fingerprints, and voice prints. In the above cases, both for ID or biometrics, it is very difficult to forge documents but if lost or compromised, they are not easily replaced compared to passwords.
(b) Knowledge-Based (what do you know?): these are characterized based on secret or concealed passwords. They also include information that is less secure or obscure that can be loosely defined. The major drawback is that each time it is shared or used for authorization, it becomes less secret.
(c) Object-based: This is a physical possession. It differs from a cryptographic key. The drawback with it is that if it is lost, any user can directly enter the house. In order to avoid this problem, an additional password is attached with the token. The major advantage of physical possession is that the owner has the evidence of it.
(d) Multi-factor authentication is a scheme in which various authenticators are combined to upgrade security. A Boolean "AND" operation will be used for affirmation and as a result authentication should be satisfied at any cost. The best example of two-factor authentication is apple-id, where a password alone is not required to gain access to a system. We can blend a password with biometric data in order to make it more effective, so that if password is difficult to remember an alternate biometric factor will be used to gain access. This type of technique usually costs more than the usual type of password.

Types of Biometrics:

These can be of two types

(a) Physical type
(b) Behavioral type

Fingerprint, iris scans, and retina scans can be categorized as the physical type whereas the behavioral type includes gait (walking style of individual) and handwritten signatures.

We classify biometric signals rather than classifying the biometric itself.

- Alternate biometric signal
- Stable biometric signal.

Stable biometric signals are usually constant or remain invariant, including exceptions like plastic surgery, accidents caused.

For any feature of an individual, a template of smaller size is extracted [3]. Let S_B and S_T be the biometric template and biometric signal. The biometric template is desired from the biometric signal and the biometric signal in turn from. (B) Biometric of an individual. The template of the biometric signal is an unchanging constant BTS = f(B). For a fingerprint this is a biometric signal S_B and the extracted/derived feature make a biometric template (BTS).

With an alternate biometric signal there are two components:

(a) Variable u
(b) Stable biometric

Here again SB(x) is used to derive $S_T(x)$ and in turn they are brought together to yield signal. Therefore the template for an alterable biometric signal is given by BTn(x) = f(B, x). SB(x) be a speech signal of vocalization of variable, x be word/phrase, which evolved from stable vocal track filter B.

Comparison factors:

There are many factors through which we differentiate authenticators.

Entropy and keyspace: Any range of various possible values of a key is known as a keyspace. Any password with n characters and with each of these characters having c different values, will have a keyspace of

$$K_p = C^n$$

An entropy or statistical entropy would be defined as the measure of deviation, in terms of bits.

The keyspace password size Pk is proportional to the maximum entropy for any authentication/authenticator's number.

$$E_{max} = \log_2 P_k(\text{bits})$$

From this we come to the conclusion that both the entropy and keyspace should be at a maximum to minimize the probability of accurate guessing and avoid attacks which are brutal in nature. For any network there would be 3–5 attempts before any further attempt to enter a password becomes futile and the system locks users out. Similarly, any token of a two-factor authentication of lower entropy would just need a 4-digit PIN So that any attacker fails to—through the system. In the case of reading a smartcard or biometric scanning for authentication, it would take 2 or more seconds. There is also the probability of a brute force attack on it compared to password attacks. The password would possess higher entropy than these 2 factor case.

Effective keyspace of a biometric:

We cannot have fixed possible values for a biometric. If we take a fingerprint biometric into consideration. The fingerprints would be unlimited and if we try and measure continuous signal with infinite precision, then both of them can't be the same. Practically biometric is measured for discrete rather than continuous space. We can define the effective keyspace of a biometric for comparison purpose. If there is uniform distribution of password keyspace then the probability of guessing/predicting a password would be single. The probability of predicting a wrong password with the right one is analogous in nature. If we consider that for a given biometric the probability of it being the same as any other single biometric sample in a DB, is the false match rate for a single verification attempt, FMR(1), P(false, match) = FMR(1), as we know P(false, match) is analogous to P(correct guess) of any password, then $1/P_K$ would be analogous to FMR(1).

Host side security:

Static passcodes are usually located/stored at the host end in order to be matched against password/passcodes given/fed at the client end. It can be in any of the following formats:

1. Plaintext
2. Masked by reversible operation
3. Masked by irreversible operation.

Plaintext is highly unsecured when stored at the host machine because its essence is lost and is easily readable by the host administrators, meaning that secrecy is completely dependent on the host and how it manages. Usually hosts are sometimes not worthy, administrators would be unethical, plaintexts can be easily stolen.

Another approach to secure data is to mask the authenticators through end-to-end security or a reversible operation such as encryption, so that even if the authenticator file is lost from the host, the passcode can't be read directly. In order to decrypt the data or reveal the data, a decryption key should be available.

The problem associated with such a method is that the data can't be defended from untrustworthy or unethical host administrators. As the administrators already are aware of it. Another way of defending authentication from a host side attack is to use masking of an irreversible operation known as a hash function or a 1-way hash function. A hash function takes a variable-length message and converts it to a fixed-length string or hash code.

Token Protocol: in this type of authentication the token stores either a generated one-time password or a passcode that is static in nature. A sufficiently long random passcode is hashed H(R′), which in turn is combined with challenge random number and then forwarded to the host end. The user then accesses the passcode with password P′ from the token storage. One-time passcodes would be generated or a hashed user passcode H(R(U)) will be stored at the host.

Password Protocol: in two-way password protocol authentication the following sequence of steps are to be followed.

step 1: A user has to send a user identification for initiation or initialization.
step 2: In reply the host responds with a random number "r," which has the ability to identify a hash function H(), a challenge function F(), and a session.
step 3: The user sends the response consisting of the function involving the hash of the submitted password H(R′) and the random number r.
step 4: Authentication is only granted if the result matches with the result of the function with a random number and the hash of the genuine user H(P(U)). Any user information is not stored in the plaintext but in the hashed H(P(U)) to avoid theft.

Stable Biometric Protocol: at the client end the biometric B1 is obtained and is processed in order to obtain a biometric template TB1. This in turn is blended with the random number challenge r′ and then it encrypted E(). This entire thing is returned back to the host in response.

Alterable Biometric Protocol: here a challenge "h" is sent to the client by the host. This "h" would be a word, character, or a sequence of random numbers. These would be smaller than "r," because to yield the signal for the biometric $SB^1(w)$, the user only has to localize, type or write. We extract l1 and B1 for verification and they are then sent back to the host. The recognized h′ is compared with the originally stored h and similarly B′ is compared with B(u). The authentication is provided only if the r matches with r′ and B′ with B(U).

Cost and convenience: the cost can be categorized into three different criteria.

(a) **Administrative cost**: if the user forgets a password or a token is lost, then things have to be reset. This might take/require ongoing expenditure for a tri-labor force.
(b) **Infrastructure costs**: usually it may be a little expensive but can be minimized if the users are more in number per-client basis.
(c) **Per-user cost**: A password or passcode costs nothing to the user, but in case of a biometric, it is required to read a token as well as to read a biometric itself.

Security issues:

(a) **Watermarking techniques**: the inverse of a copying/theft attack would be a replay attack. In a copying/theft attack the attacker grabs the authenticator before the entry at client end, where in replay attack the attacker grabs the authenticator in the middle of the channel between the host and the client. In order to overcome this, a two-way protocol for authentication is needed. If the biometric is sent as a plaintext template format rather than hashed or combined in a response, then there is the probability of a replay attack. We can defend the replay attack by usage of a capture device that verifies the biometric.

(b) **Trojan horse attack**: in this the application impersonates a trusted application in order to obtain underlying information and gain entry to a system. In order to gain authentication, this attack is used to grab/steal a password biometric signal or a token passcode. We prefer the device to be machine authenticated when a decision is made at the client side.
(c) **Host attack**: this type of attack can be made on the plaintext or if the password is reachable at the host end. Sometimes credit card numbers are stolen in this way. best password files are stored in order to avoid it. One can still attack a hash file easily by undergoing a dictionary search attack, wherein hashed passwords are matched against hashed words or hashed combinations of words. To defend hashing we add a few random bits to each hashed password called salt, which in turn increases the duration of a dictionary search attack.
(d) **Denial of service attack (DOS)**: an example of a DOS attack is lockout by multiple failed authentications. A defense to this would be multi-factor authentication wherein a biometric or a password is combined with the token.
(e) **Theft, copying attacks, and eavesdropping**: sometimes a physical presence is needed for such kind of attacks and this restricts the opportunity for an attacker. A two-factor token is a good solution wherein the attacker has to steal both the physical token and a password. A biometric can be forged like a token can be stolen. A defense to such an attack is a liveness or anti-forgery check at the biometric capture point.
(f) **Client attack**: one of core features of a good authenticator is that it should surrender or give up easily for the guessing attacks or exhaustive search attack. A keyspace of larger size can be easily defended from such an attack. A token with high entropy passwords from the lower entropy biometric, password or passcode would be a good tool.

2 Conclusion

Token: when a password is combined with the token there are the following advantages. The first advantage is that it can generate or store multiple passwords, helps to remember only one single-password rather than multiple, changing passwords. The second advantage is that its absence is observed which in turn helps in compromised detection. The third advantage is that it provides a defense to DOS attacks with added protection. The major flaw with this method is the inconvenience and cost of the equipment to be costlier, than password.

Password: a single password would be effective and excellent authentication. The secret of the password would be good defense against theft. If it contains a higher keyspace then it is easy to defend from the host attack. High keyspacing and hashing prove reliable to safeguard against host attack. A password with low entropy is sensitive towards the dictionary search attack.

The major flaw with this method is that it doesn't provide good compromise detection and doesn't offer much defense against repudiation.

Biometric: The major strength of a biometric is that it cannot be given to any one or be stolen. As a result it has a strong defense against repudiation. A stable biometric can still be stolen by some means. Hence, the biometric shouldn't be deployed in a single mode factor. Furthermore, since a biometric operates best in verification mode, two-factor authentication would be a good choice to go with. The downside of a biometric is that the recognition rate of for speaker verification may not be high enough to provide security.

References

1. Jain A, Bolle R, Pankanti S (eds) (1998) Biometrics: personal identification in networked society. Kluwer Press, The Netherlands
2. Pankanti S, Bolle RM, Jain A (2000) Biometrics: the future of identification. Spec Issue Comput 33(2)
3. Ratha N, Karu K, Chen S, Jain A (1996) A real-time system for large fingerprint databases. IEEE Trans Pattern Anal Mach Intell 18(8):799–813
4. Rescorla E (2000) SSL and TLS: designing and building secure systems. Addison Wesley, Massachusetts
5. Federal Information Processing Standards Publication, FIPS-197, Specification for the Advanced Encryption Standard, NIST, Nov 2001. http://csrc.nist.gov/encryption/aes/
6. O'Gorman L (2002) Seven issues with human authentication technologies. In: IEEE workshop on automatic identification advanced technologies, Tarrytown, New York, Mar 2002, pp 185–186
7. Ratha NK, Connell JH, Bolle RM (2001) Enhancing security and privacy in biometrics-based authentication systems. IBM Syst J 40(3)
8. National Institutes of Science and Technology (1995) Secure hash standard. NIST FIPS Publication 180-1, US Department of Commerce, Apr 1995
9. Matyas SM Jr, Stapleton J (2000) A biometric standard for information management and security. Computers and Security 19(5):428–441
10. Blomme J (2003) evaluation of biometric security systems against artificial fingers. Master's thesis

Clustering Method Based on Centrality Metrics for Social Network Analysis

Siddapuram Arvind, G. Swetha and P. Rupa

Abstract The significance of a node in a social network is quantified through its centrality metrics, such as degree, closeness, and betweenness. However, many methods demonstrating the relevance of a node in the network have been proposed in the literature. In this digital smart world, the evolution of social networks occurs in various different directions at an unprecedented speed. A network evolution mechanism that provides the state of each node and its changes from its inception to its extinction over time will help in understanding its behavior. Often the strategy behind evolution is unknown and would not be reproduced in its totality. However, it is essential to understand behavior of the network as this can greatly facilitate its management before it becomes uncontrollable. A heuristics-based cluster method is proposed in this paper which combines centrality metrics and categorizes the entire network.

Keywords Social network analysis · Centrality metrics · Clustering

1 Introduction

In social network analysis (SNA), the study of communication patterns and structure of social networks is measured by centrality. In a social network the relative position of a node shows its centrality [1]. An object could be an actor or a thing that has a state and property. Centrality measures have been applied in various fields to influence investigation of patterns in inter-organizational networks, to

S. Arvind (✉) · G. Swetha · P. Rupa
CMRIT, Kandlakoya, Hyderabad, India
e-mail: scarvi@rediffmail.com

G. Swetha
e-mail: swetha9g@gmail.com

P. Rupa
e-mail: rupa537@gmail.com

A. Kumar and S. Mozar (eds.), *ICCCE 2018*,
Lecture Notes in Electrical Engineering 500,
https://doi.org/10.1007/978-981-13-0212-1_60

study the growth of an organization, to extract information from unethical and criminal networks, and to analyze employment opportunities etc. [2].

Degree centrality is the measure of the node knowing its previous status while calculating the number of links of that node with the remaining nodes in that network. High degree centrality nodes possess a higher probability of trans-receiving information that helps [3] in establishing rapid communication with neighboring nodes. Neighboring connections help to evaluate a node's importance (local measure). Messages sent through this network act as brokers. This measure is referred to as *betweenness* [4]. Closeness is another measure which is used to evaluate a node based on its nearest neighbor. Distant neighbors are reached through geodesic paths.

Metrics like betweenness, closeness, and eigenvector centrality are considered as *global* measures [5] since they evaluate the impact that a node has on the global structure or transmission of information within the network.

Determining groups in complex social networks is one of the issues in SNA. Disjointed collections of nodes linked with some sort of relation/interaction occupy different positions in a group. Some occupy central positions, others remain on the periphery or lie somewhere in between.

Though these measures possess their own significance in evaluating node level capabilities, individually they cannot determine the behavior of network evolution. Hence, it is necessary to combine both local and global measures. In this paper, the formulation of a cluster method based on heuristics is proposed that combines centrality metrics and demonstrates its applicability in categorizing the entire network.

2 Related Work

In the literature, various clustering algorithms have been proposed that hardly left any domain. Broadly speaking, clustering algorithms are divided into two groups [2]: (a) hierarchical method and (b) partitional method.

In recent years, many clustering algorithms that focus on categorical data based on various centrality metrics have been proposed. Freeman [1] proposed node betweenness centrality to help to detect communities in a network [4, 3] by transforming into relation among nodes to discover community structures; [4] associated the properties of resistor networks in discovering communities. Newman [6] described the community structure in networks based on eigenvectors of matrices [6, 7] proposed affinity propagation, which considers weighted similarity between pairs of data points as an input [7]. Rosvall and Bergstrom [8] proposed the probability flow of random walks in a network to break the network down into its components [8]. Ghaemmaghami [5] proposed a clustering method based on self-organizing feature map.

3 Proposed Method

This section describes a new cluster method based on heuristics that incorporates centrality metrics.

3.1 Clustering Method Based on Heuristics

Among the clustering types, most clustering algorithms fall into the category of intrinsic clustering. This takes place on the objects without supervision and which possess dissimilarities between them. If we know in advance about which objects are to be clustered, then we adopt extrinsic clustering.

Heuristics demonstrates its crucial role in solving problems typically through evolution strategy. It creates clusters based on a user-defined property. Heuristics "grows" a cluster from a specified root node in a network with the node's critical cluster and goes on increasing the network through the root node by adding its nearest neighbors. There is one significant decision to make which helps to identify the neighbor node to include in the cluster in the next iteration. A node must be assessed by the gain in the objective function's value for it to become part of the cluster.

Cluster size could be limited by one of the five approaches: Euclidean distance, network distance, population, area, or node count. To find the critical cluster for each node the algorithm runs sequentially in a network of a given size. The weight/cost value is estimated based on the total cluster weight over the number of exit lanes. Based on the classification of the network, the vulnerability level of an arc is defined as the worst-case vulnerability level of its two end nodes.

The following objective function is formulated based on heuristics for the present study.

1. MinMax rule

Clusters indicate vulnerability based on degree, betweenness, and closeness.

In the Min-Max rule the minimum of the maximum weighted deviations is sought.

$$Vulnerability = \left[\sum_{i=1}^{n} \cdot \beta^4 (1 - X)^4 \right]^{1/4}$$

where n = no. of parameters such as degree, betweenness, and closeness centrality measures;

β parameter weight
X parameter value

A heuristic is based on the deviation of values of each centrality metric from its ideal value, which is considered as 1. The importance of centrality metrics has been weighted in non-decreasing order of betweenness, closeness and degree.

4 Experiments and Results

A network of size 34 is considered for clustering and categorization based on the objective function explained in Sect. 3. Normalized values of degree, closeness, and betweenness measures of the 34 node network are shown in Figs. 1, 2, and 3 respectively. However, these measures are not useful in categorizing the network.

Categorization of the clusters is based on the objective function shown in Fig. 4 and extracts the strength of the links and orders them accordingly.

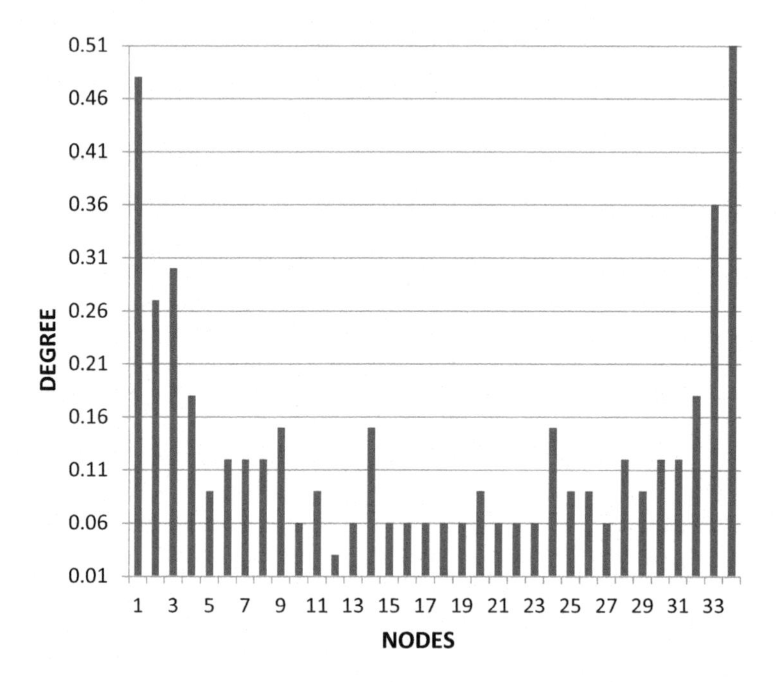

Fig. 1 Degree centrality

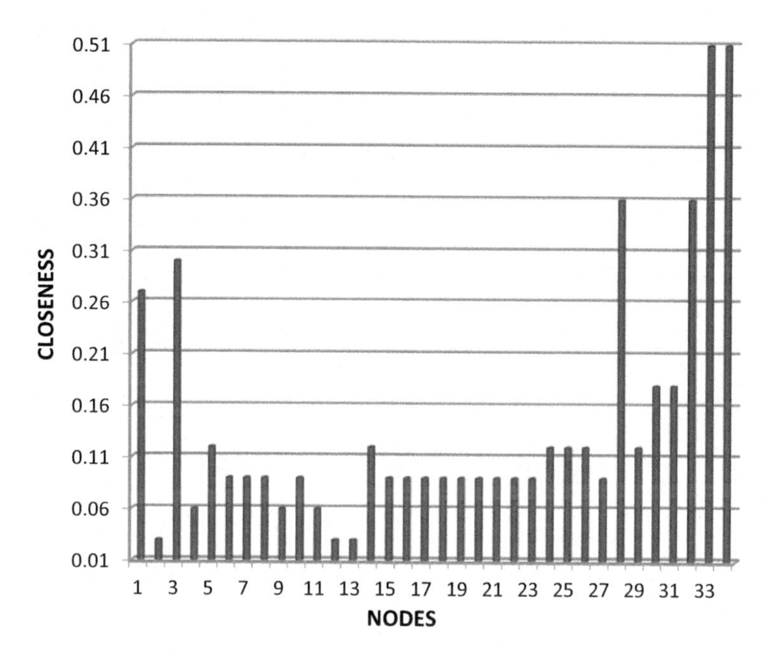

Fig. 2 Closeness centrality

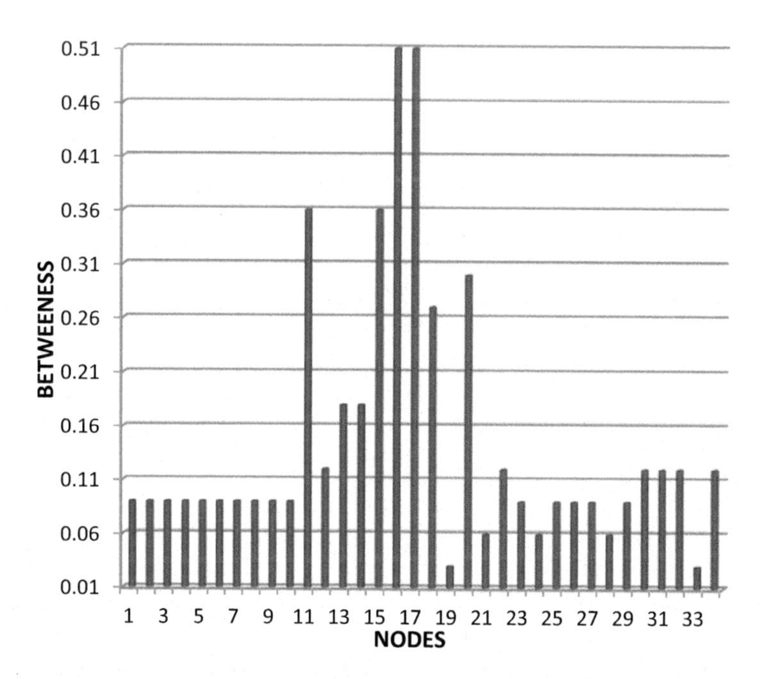

Fig. 3 Betweeness centrality

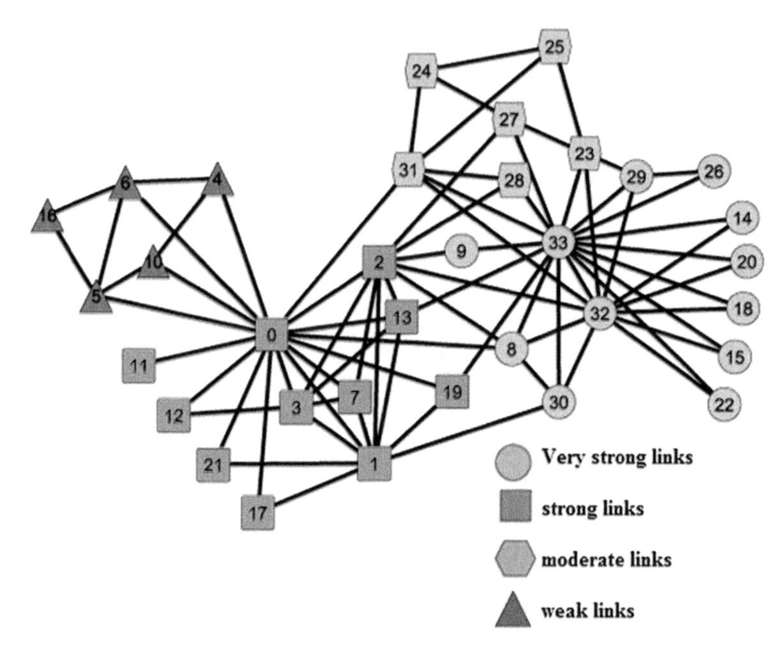

Fig. 4 Categorization of a 34 node network cluster based on a heuristic method

5 Conclusion

Individual centrality measures do not reveal hidden patterns and thus inhibit categorization of link strengths. A heuristic approach is devised for clustering and incorporated further for the categorization of the network. This methodology can be used to arrange the links in order of their influence in the network. It also helps to understand the highly influential links and remedies that slow down their impact can be incorporated accordingly. The weights used in the objective function can also be derived statistically or analytically although that was not a focus of this paper.

References

1. Freeman L (1977) A set of measures of centrality based upon betweenness. Sociometry 40:35–41
2. Jain AK (2009) Data clustering: 50 years beyond K-Means. http://dataclustering.cse.msu.edu/papers/JainDataClusteringPRL09.pdf
3. Newman MEJ, Girvan M (2004) Finding and evaluating community structure in networks. Phys Rev E 69:026113
4. Wu F, Huberman BA (2004) Finding communities in linear time: a physics approach. Eur Phys J B 38:331–338
5. Ghaemmaghami F (2013) Int J Comput Appl 84(5):0975–8887

6. Newman MEJ (2006) Finding community structure in networks using the eigenvectors of matrices. Phys Rev E 74:036104
7. Frey BJ, Dueck D (2007) Clustering by passing messages between data points. Science 315:972–976
8. Rosvall M, Bergstrom CT (2008) Maps of random walks on complex networks reveal community structure. Proc Natl Acad Sci USA 105(4):1118–1123

Future Aspects and Challenges of the Internet of Things for the Smart Generation

Chander Diwaker, Pradeep Tomar and Atul Sharma

Abstract The internet is now a basic necessity for human beings, especially in modern cities and metropolitan areas. Without the internet, an educated person feels helpless and unable to understand and follow events. At present, most people depend on machines. The field of computer engineering has helped the process of automation and the control of software as well as hardware devices. The internet of things (IoT) is a field of computer engineering that presents a synchronous behavior of components in a real-time system. Every piece of hardware and software that assists in accessing the internet or is used by the internet constitutes a main part of the IoT. The IoT includes the applications used in every field, e.g., healthcare, engineering, designing, inventory control, machine control, selling-purchasing, and the export-import of goods etc. In modern cities almost everyone uses the internet with individuals being linked to it through variable bandwidths and network ranges. People can access internet easily but they are not aware of the various issues, problems, and challenges of providing data to everyone at the same time on an unlimited number of topics. In this paper, the architecture of the IoT, the functioning of the IoT, the applications of the IoT in different fields, along with the research challenges and problems relating to the IoT are discussed.

Keywords IoT · Cloud computing · Applications · Security
Future aspects

C. Diwaker (✉) · A. Sharma
Department of CSE, UIET, Kurukshetra University, Kurukshetra, Haryana, India
e-mail: chander_cd@rediffmail.com

A. Sharma
e-mail: atulsharma2204@gmail.com

P. Tomar
Department of CSE, School of ICT, Gautam Buddha University, Greater Noida, Uttar Pradesh, India
e-mail: parry.tomar@gmail.com

A. Kumar and S. Mozar (eds.), *ICCCE 2018*,
Lecture Notes in Electrical Engineering 500,
https://doi.org/10.1007/978-981-13-0212-1_61

1 Introduction

The basis of the IoT is networking. The main components of networking are sensors; a finite set of heterogeneous devices that supports networking, storage devices, variable size of files, service providing, distributed system, and maintenance of networks etc. It is a collection of a heterogeneous system that is connected in a distributed manner to provide service to a customer for a particular query. Cloud computing also plays an important role in the IoT [1].

The revolution of the IoT started in 1970. At that time, there were low-speed processors, less RAM and cache memories, and low stage space. Currently, the speed of the processor, the capacity of RAM, the large data storage space, and small size of components help to perform high-speed data communication. The IoT helps in sending, receiving, and sharing information using the internet.

The IoT has applications in every field of engineering: from agriculture to flight systems, car manufacturing, education systems, software development, and real-time systems etc. The IoT includes the entire field in which software is used to manage and control things.

2 Architecture of the IoT

The IoT's architecture consists of three layers, i.e. perception layer, network layer, and application layer. The working of each layer is equivalent to an OSI model. Figure 1 shows the architecture for the IoT [2]:

(i) *Perception layer*: This is equivalent to the physical layer. It includes different types of sensors. Different sensors are used for different types of sensing information. The hardware components used are sensors, IP cameras, actuators, embedded communication and closed-circuit Television (CCTV), bluetooth, radio-frequency identification (RFID), and near-field Communication (NFC) etc.

(ii) *Network layer*: This uses a network device to find the optimal route for sending data packets. The network devices used are routers, bridges,

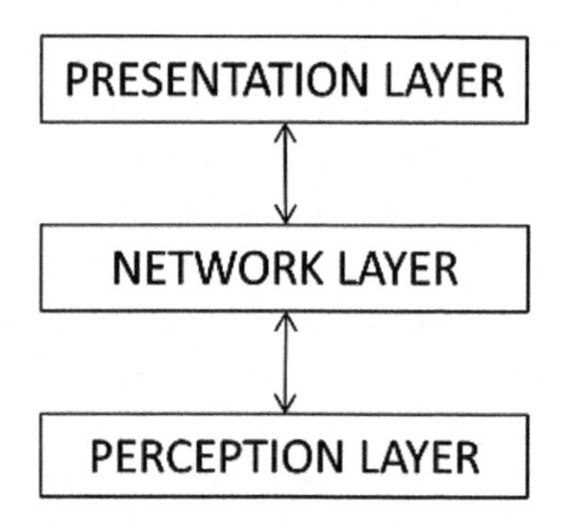

Fig. 1 Architecture of the IoT

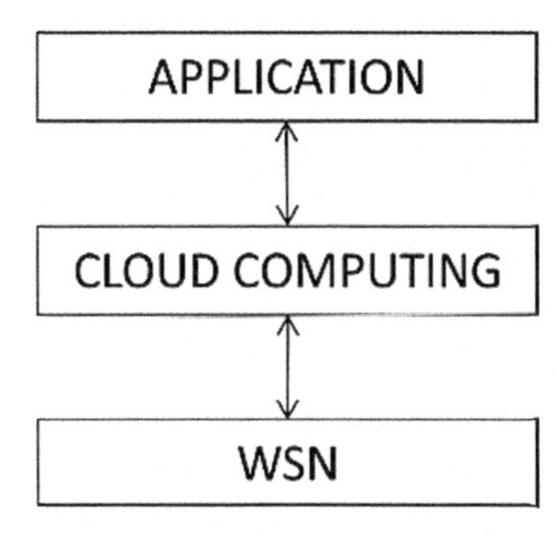

Fig. 2 Equivalent layers of the architecture of the IoT

gateways, hubs, and switches. The techniques used are 2G, 3G, 4G, and local area network (LAN) etc. Cloud computing and big data strategies are used to store and manage data.

(iii) *Presentation layer*: This uses different protocols for data transmission and to present that data in an understating and meaningful manner. This layer assists in monitoring and providing services to users.

The use of the IoT and its applications can be used based on the concept of 5As, i.e. anything, anytime, anywhere, anyhow, and anyway [3].

An equivalent layer architecture of the IoT is shown in Fig. 2. It includes the same layers as shown in Fig. 1 [4].

(i) *WSN*: WSN use set of protocols that provide the location of data, quality of service, and security to the network.
(ii) *Cloud Computing*: The architecture of cloud computing has three types: Software as a service (SaaS), infrastructure as a service (IaaS) and platform as a service (PaaS). These are responsible for data analytics, data storage, data visualization, and data computation.
(iii) *Application Layer*: The information provided by cloud computing is helpful in monitoring different activities at end users, such as building design monitoring, health monitoring, surveillance for security purpose, environmental monitoring, and transportation system monitoring etc.

3 Applications of the IoT

The IoT helps in developing smart devices, smart cars, smart cities, smart homes, and smartphones etc. The main research areas of the IoT are ad hoc networks, wireless sensor networks (WSN), cyber-physical systems (CPS), mobile computing (MC), and pervasive computing (PC) [5]. The interaction in these area helps in growing the IoT and helps its better utilization. Table 1 shows the general applications of the IoT.

Table 1 General applications of the IoT [4]

Healthcare services	Monitoring health activities
Emergency services, defence	Remote control, resource management, resource distribution, monitoring future and current disaster information
Crowd monitoring services	Monitoring crowds in public and private places
Traffic management services	Managing and monitoring real-time traffic conditions on the road air, in the air, and on the water, smart parking
Infrastructure services	Monitoring structural faults, accident monitoring
Water services	Quality, usage, waste management distribution, leakage, usage
Building management services	Temperature, activities for monitoring energy usage management, humidity control, heating, ventilation, and air conditioning (HVAC)
Environmental services	Waterways, air pollution, industry monitoring, noise monitoring

4 Challenges Facing the IoT

For successful implementation of the IoT, the prerequisites are [6]:

(i) *Dynamic demand*: The demand of accessing resources is increasing dynamically on the internet. Systems should be set up in a manner that enables them to handle and manage heavy load if resource demand increases.

(ii) *Real-time Systems*: A system should have the capability to tolerate faults and automatically repair faults.

(iii) Access to an open and interoperable cloud system.

(iv) Power usage of applications.

(v) Carrying out of the applications close to end users.

(vi) *Scalability*: Scalability is a concern with providing data as soon as possible and with minimum error. The addition of storage devices, processors, and low power consumption is a challenge.

(vii) *Multi-tenancy*: Multi-tenancy is a concern with shared IoT devices. Multi-tenancy minimizes faults as it uses shared machines. If one machine is not working properly or shows an error, the same work can be completed by another machine.
Multi-tenancy is necessary for monitoring health activities via pacemakers, MRI machines, ECG, CT Scans and other real-time systems.

(viii) *Network security*: This includes cryptography mechanisms, along with detection, prevention, and avoidance of attacks and intruders, and cyber crime etc.

(ix) *Low-power communication*: The addition of more memory, storage devices, and processors speeds up the whole system, but it consumes a lot of power and energy. The use of a greater number of electronic devices leads to an increase in electronic pollution.

(x) *Security Challenges*: Security in the non-real environment has a varied nature, i.e. the problem can be repaired or corrected after a period of time. However, in real time, high security must be implemented that can detect and avoid any mishaps while achieving a particular target, as observed in military operations, flight systems and healthcare systems [2]. The key factors of security are trust in another system, data privacy, and data confidentiality. Data Privacy and data protection are important factors that should be addressed with a high priority.

(xi) *Radio spectrum*: The use of different radio frequencies and spectra influence data transmission and reception. As the number of user increases, the demand also increases. The available frequency spectrum becomes less as demand rises. There is a need to increase the number of spectra for data transmission and receiving so that the demand of users can be fulfilled within an appropriate time period.

5 Related Work

Researchers and practitioners are focusing on sensors and energy generation mechanisms with reduced costs and less time-consuming mechanisms. The most recent work on the IoT found as follows.

Vyas et al. [7] discussed the applications, various application areas, challenges facing the IoT, and future aspects of the IoT. This paper focuses on open issues like naming, traffic characterization, QoS support, data integrity and data forgetting, security, and data management etc. It also looked at how big data can be used to manage a large database.

Perera et al. [8] discussed the necessity of IoT for a human body like of use of IoT-based wristwatches, socks, footwear, bands, gloves, helmets, and rings. This study included some of the trends in the IoT solutions based on domains, functionalities, and value. A survey compared approximately 70 different products with respect to variability, unit subscription, and service provided, such as Xively, PROJECT GRIZZLY, Smart Pile, NFC ring, SIGMO, Smart Things, and Where Dial etc. This survey took place during 2011–2103.

Ziegeldorf et al. [9] analyzed privacy issues in the IoT. Various privacy threats were classified and examined for identification, profiling, and tracking of known threats. The major threats to privacy were identified, i.e. violating interactions and presentations, inventory attacks, lifecycle transitions, and information linkage arise. As data accessing and storage demand increases, various threats make the management of big data a challenge.

Rad and Ahmada [10] focused on various applications of the IOT in different areas and explored the challenges and opportunities facing global industries. This study discussed the implementation and usability of the IoT on a global scale.

The application of the IoT in ways that are useful for both humans and traffic was also discussed.

Kaur and Kaur [11] focused on driver technologies and system design of the IoT. The relation of big data with IoT was also discussed. The use of the IoT was surveyed based on person to person (P2P), person to machine (P2M), and machine to machine (M2M) criteria. It was observed that 1% of the IoT is used for retail purposes and 41% of IoT-based applications are used for medical systems. A comparison of different protocols used in the application layer was also discussed. Parametric analysis was used in resolving challenges encountered in the IoT.

Gubbi et al. [4] presented a cloud-centric vision for the IoT. An implementation of cloud used Aneka based on the interaction of public and private clouds. It concluded that there is a need for convergence of the internet, WSNs, and distributed computing. This study presented the evolution of the IoT using a hype cycle of emerging technologies. The study focused on different groups in the city of Melbourne. Future technology will depend on machines rather than human beings.

Zeinab and Elmustafa [12] reviewed IoT applications and future possibilities relating to new technologies. The challenges and problems faced at the time of implementation were also discussed. The applications included smart cities, smart environments, smart energy and smart grids, smart manufacturing, and smart healthcare. The main issues in managing the IoT data are cloud computing, big data, security, privacy, distributed computing, and fog computing etc.

6 Future Aspects of the IoT

The future work is going on, How to use heat/energy released by different components, the vibration of components, movement of components, radio/other frequencies, wind energy, the temperature for providing high-speed networking, energy to hardware components, reusable electronic waste (e-waste), and other scratched material to generate energy and new products. Figure 3 shows, How the energy can be generated by using different resources that are easily available on earth. The main focus is on generating energy that can be sensed by a particular sensor through utilizing the following aspects:

- **Heat Energy**: Every machine that runs, with either a high speed or a low speed, consumes and exhausts heat energy. This heat can be used to increase the lifetime and to charge, or fulfill the requirement of energy to a particular machine. Those machines in which the engine/machine uses more displacement or horsepower release a huge amount of heat that can be used by applying a feedback mechanism to provide extra power to the machine. Alternatively this heat energy can be stored for the further use of the machine. The heart energy realized by a human being can also be used to power small machines, such as mobiles or watches.

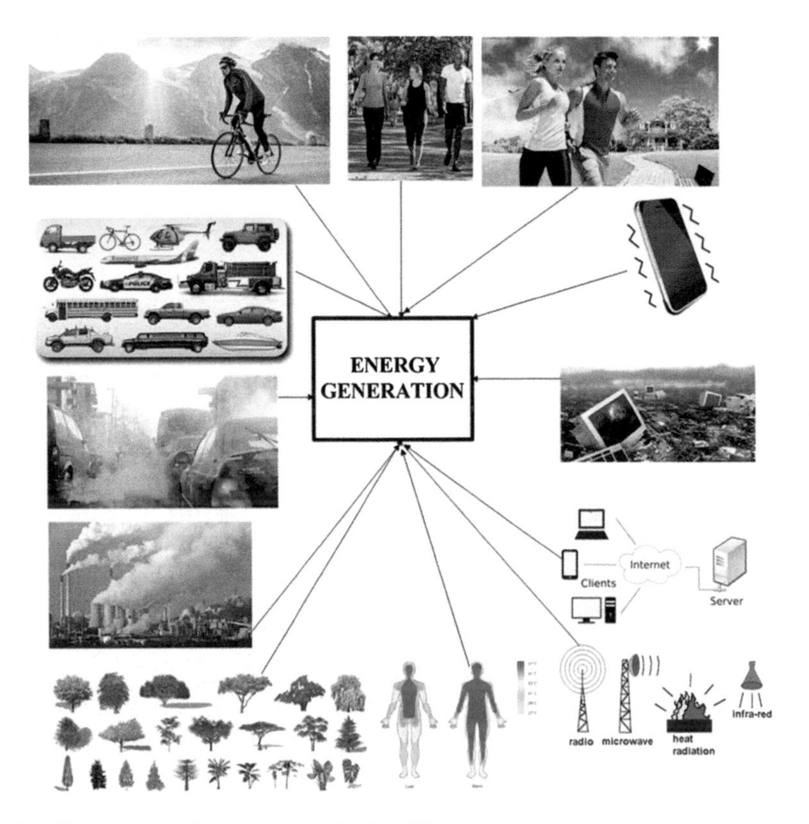

Fig. 3 Different ways of energy generation by different aspects

- **Vibration or Movement**: Energy can be generated by moving objects and vibration in objects. More vibrations and movement provide a greater generation of heat and energy.
- **Frequency**: High speed and high capacity frequency can be used for the generation of energy. However, these frequencies can only be used in particular ranges that don't affect human life.
- **Natural Resources**: Wind energy and solar energy can be major sources of energy. People need to be more aware of these natural resources in order to reduce pollution and increase economic wealth.
  Natural energy is a big source of energy production. It includes heat released by volcanos and the energy in deep layers of the earth. Other research is also stated: "How the plants survive in a different environment". If a plant can produce food using photosynthesis for survival, then a plant can be used to generate energy that can be used to run different kinds of applications.
- **Pollution/Smoke**: Researchers have experimented with using gases generated by the pollution of vehicles and industries to generate energy that can be used in different fields.

- **Other Components**: In computer networking there are a number of servers, client systems, and other hardware devices running for 24 h a day that release a lot of harmful gases and heat. These gases and heat can be used to generate and store energy that can then be used later.
- **Recycling of E-waste and others wastes**: E-waste is increasing daily. E-waste can be utilized to make/produce new products that will save time and money.

7 Conclusion

The IoT is becoming a part of human life. Without IT, it is difficult to live in the modern era or in a developing country. It has become a basic necessity of life. In this paper, an attempt has been made to present the highlights of the IoT, and the various research issues and challenges relating to it. The IoT in growing at an increasing rate every day and it is becoming a challenge to manage and provides services to end users. These challenges and problems can be minimized through the use of natural resources and the application of some logical mechanisms.

References

1. Alsaadi E, Tubaishat A (2015) Internet of things: features, challenges, and vulnerabilities. Int J Adv Comput Sci Inf Tech 4(1):1–13
2. Kumar JS, Patel DR (2014) A survey on internet of things: security and privacy issues. Int J Comput Appl 90(11):20–26
3. Madakam S, Ramaswamy R, Tripathi S (2015) Internet of things (IoT): a literature review. J Comput Commun 3(5):164–173
4. Gubbi J, Buyya R, Marusic S, Palaniswami M (2013) Internet of things (IoT): a vision, architectural elements, and future directions. Future Gener Comput Syst 29(7):1645–1660
5. Stankovic JA (2014) Research directions for the internet of things. IEEE Internet Things J 1 (1):3–9
6. Rose K, Eldridge S, Chapin L (2015) The internet of things: an overview. Internet Society, 1–50
7. Vyas DA, Bhatt D, Jha D (2016) IoT: trends, challenges and future scope. Int J Comput Commun 7(1):186–197
8. Perera C, Liu CH, Jayawardena S (2015) The emerging internet of things marketplace from an industrial perspective: a survey. IEEE Trans Emerg Top Comput 3(4):585–598
9. Ziegeldorf JH, Morchon OG, Wehrle K (2014) Privacy in the internet of things: threats and challenges. Secur Commun Netw 7(12):2728–2742
10. Rad BB, Ahmada HA (2017) Internet of things: trends, opportunities, and challenges. Int J Comput Sci Netw Secur 17(7):89–95
11. Kaur J, Kaur K (2017) Internet of things: a review on technologies, architecture, challenges, applications, future trends. Int J Comput Netw Info Secur 4:57–70
12. Zeinab KAM, Elmustafa SAA (2017) Internet of things applications, challenges and related future technologies. World Sci News 2(67):126–148

Impact of Node Mobility and Buffer Space on Replication-Based Routing Protocols in DTNs

Atul Sharma and Chander Diwaker

Abstract A delay-tolerant network (DTN) is a kind of network in which nodes are not directly connected with each other so they communicate through intermediate nodes. As the mobility of nodes is so high in DTNs it is difficult to deliver a message without the creation of duplicate copies for distribution in the network. In this paper the impact of node mobility and the impact of buffer spaces on replication-based routing techniques called epidemic routing and sprays and waits routing has been assessed. To evaluate performance metrics, measures such as delivery ratio, drop rate, overhead ratio, and the number of replications have been used. To simulate the above routing protocols ONE (opportunistic network simulator) simulation was used.

Keywords Delay-tolerant network (DTN) · Epidemic routing
Spray and wait routing · Movement models · Replication and ONE
(opportunistic network simulator)

1 Introduction

A DTN is a network intended to work successfully in profoundly tested conditions, where conventions received in associated networks (i.e. TCP/IP) come up short. Traditional networks accept low mistake rates, low engendering delays and, in particular, an associated end-to-end route between the source and destination. Be that as it may, a class of tested networks which damages at least one of these presumptions is increasingly being sought. Such networks for the most part

A. Sharma (✉) · C. Diwaker
Department of Computer Science and Engineering, University Institute of Engineering & Technology (UIET), Kurukshetra University, Kurukshetra, Haryana, India
e-mail: atulsharma2204@gmail.com

C. Diwaker
e-mail: chander_cd@rediffmail.com

A. Kumar and S. Mozar (eds.), *ICCCE 2018*,
Lecture Notes in Electrical Engineering 500,
https://doi.org/10.1007/978-981-13-0212-1_62

experience the ill effects of successive transitory segments and are alluded to as intermittently connected networks (ICNs) or delay-tolerant networks. This is especially obvious in provincial areas, for example, wild environments and towns that need essential frameworks. A DTN speaks to a class of networks that experience long postponements and irregular availability [1]. The DTN is a class of blame-tolerant network, where end-to-end associations are not required for the routing of messages from source to destination. In a DTN messages are transmitted jump by-bounce starting with one network node then onto the next until the point when the destination is reached.

2 Routings in DTNs

In challenging environments messages transfer through intermediate nodes. Choosing the next intermediate node for the transmission of a message in a network is a typical task. To transmit messages, various routing protocols like direct delivery routing [2], epidemic routing [3], and spray and wait routing [4] are available. In epidemic routing [3], messages are replicated over the whole network and delivered to an intermediate node. When messages are replicated over the network it increases the delivery ratio but also creates an overhead in the network. Figure 2 illustrates drop rate of both routing. In spray and wait routing [4], the messages are flooded over the nodes in the network and wait for the reply. After getting a reply from one node, the message would transfer through that node in the network but at same time too much delay was created in network. Epidemic routing (or straightforward flooding) occurs when zero-information of the network's topology is accessible. On the off chance that the node conveying the information package to be sent has no history information or learning of the portability examples of its neighbors or of the destination node, at that point the least difficult choice is to send the message (or a duplicate of it) to every node it experiences inside its transmission premises.

This, obviously, will bring about significant excess in the network. The network's consolidated buffer space will radically lessen in size since numerous duplicates of similar messages are being transmitted over the network. Moreover, interface data transmission is devoured by these re-transmissions.

Some investigations have demonstrated that epidemic routing is capable of conveying all transmitted messages. Such investigations expect limitless or adequately extensive buffer sizes at every node, which is a non-sensible supposition. Arbitrary routing may bring about the stopping of the entire network if the measure of traded/repetitive messages surpass the real buffer space constrains. At whatever point two nodes experience each other they will trade every one of the messages they as of now convey with each other. Toward the end of the process, both will have a similar arrangement of messages. As this procedure takes place, in the long run each node will have the capacity to send data to every other node. So the bundles essentially overflow through the network much like the spread of a viral epidemic. This speaks to the speediest conceivable route in which data can be

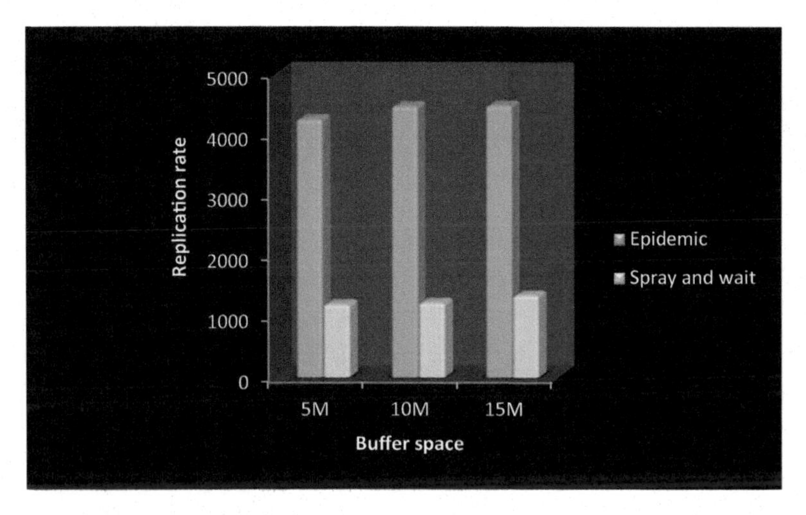

Fig. 1 Replication rate versus buffer space

spread in a network with boundless capacity and boundless transmission capacity requirements. This plan requires no learning about the network or the nodes. In any case, in most down to earth situations, such a plan will bring about wasteful utilization of the network's resources, for example, power, data transmission, and the buffer at every node. In addition, messages may keep on existing in the network even after they have been conveyed to the destination [3]. Epidemic routing fills in as the gauge for correlation for the greater part of the DTN routing plans. Epidemic routing is one of the easiest and most punctual routing plans for DTNs. In this routing strategy, whenever two nodes come in contact with each other, they exchange all the messages they currently carry at that point of time. In other words, the packets are spread like a viral epidemic. So this routing strategy is the fastest possible scheme. Figure 1 shows the functionality of epidemic routing. It is clear from the figure that whenever two nodes encounter each other the messages are flooded to the node.

3 Movement Models in DTNs

Random way point movement (RWPM) model: This is an arbitrary model for the movement of portable users, and how their area, speed and increasing speed change after some time. In the RWPM model portable users move randomly and openly without confinement. To be more specific, the destination, velocity, and course are all picked arbitrarily and freely for different users. In this development model users move arbitrarily in a subjective heading and no instrument is behind the hub portability.

Shortest path map-based movement model: Here users utilizes the idea of the shortest way as most limited accessible way is picked from among various different accessible ways in a guide-based environment. Here a Dijkstra calculation is utilized to determine the shortest of all the accessible ways and diverse hubs in hubs precede onward the premise of briefest way.

4 Simulation and Analysis

In this paper the impact of node mobility and the impact of buffer space on different routing protocols are analyzed. For simulation we use a ONE (opportunistic network environment) simulator in its latest version, 1.6.1. This section provides details of the GUI of the ONE simulator and detail of the performance metrics used for simulation [5].

A. ONE Simulator: ONE is open source software. It can be run on any operating system that supports java. ONE is specially designed for DTNs.
B. Performance metrics:
    i. Message drop rate: This is defined as the total number of messages dropped during data transmission.
    ii. Replication of messages: This is defined as the rate at which messages are duplicated in the network.
    iii. Overhead ratio: This is calculated as:
    (Total no of relayed message—total no of delivered messages)/total no of delivered messages.

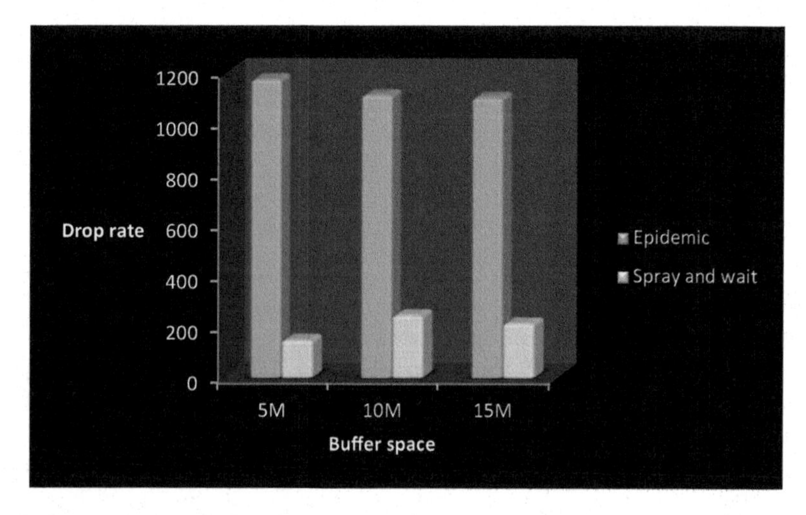

Fig. 2 Drop rate versus buffer space

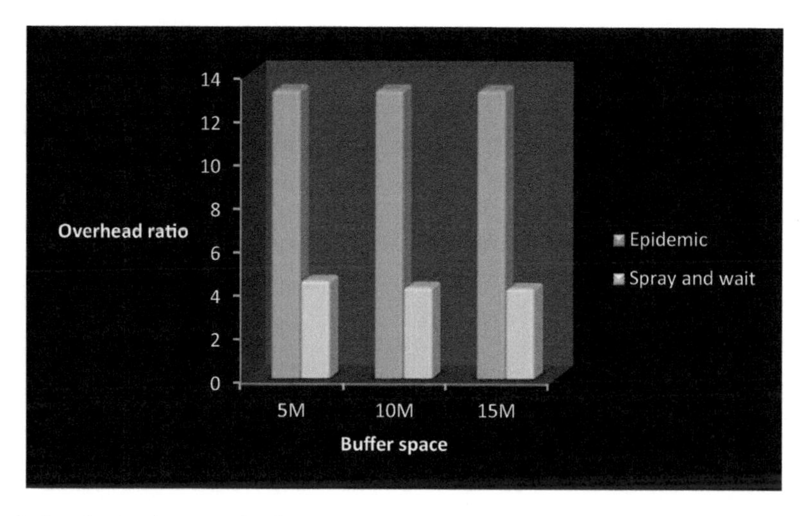

Fig. 3 Overhead ratio versus buffer space

iv. Delivery ratio: This is the successful message delivery rate over a communication channel. A physical or logical link can be used for data delivery or passage through a certain network node. The throughput is usually measured in bits per second.

C. Performance parameters used for performance analysis: Table 1 describes the parameter used in the simulation.
D. Results and Discussion:

a. Performance analysis with RWPM: In this section we present a performance analysis of epidemic routing and spray and wait routing from the perspective of the RWPM model. Here we varied buffer space from 5 to 15 MB.

Table 1 Simulation Parameters [6]

Parameter Description	Value
Simulation Area	4500 m * 3400 m
Simulation Time	30000 s
Mobility Model	Random way point
Routing protocols	[Epidemic; Spray and Wait]
Transmission Range	10 m
TTL(Time To Live)	300 s
Buffer Size	5MB; 10MB; 15MB
Warm Up Period	1000 s
Operating system used	Windows 10
No of groups	6
No of Nodes	126
Message size	500 KB to 1 MB

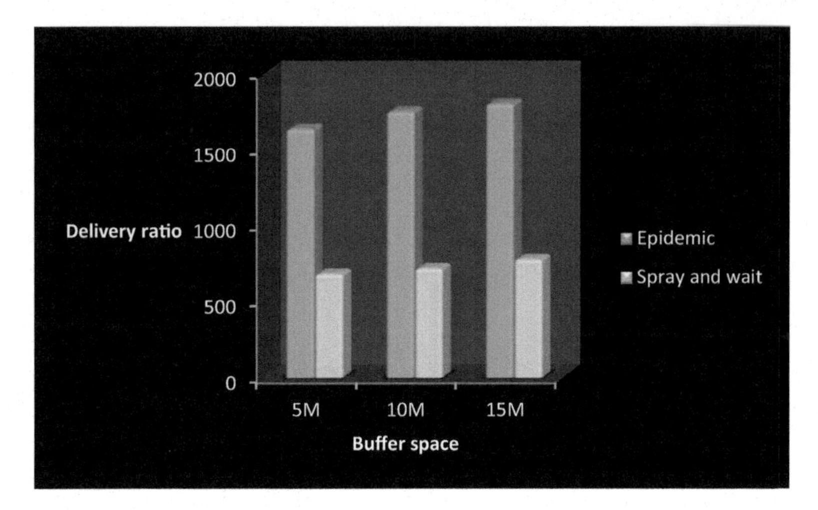

Fig. 4 Delivery ratio versus buffer space

Figure 1 illustrates the replication rates of both routing types. In spray and wait routing the replication rate is less when compared to epidemic routing. Figure 2 illustrates the drop rate of both routings. In spray and wait routing the drop rate is less when compared to epidemic routing. Figure 3 illustrates the overhead ratios of both routing types. In spray and wait routing the overhead ratio is less when compared to epidemic routing. Figure 4 illustrates the delivery ratio of both routing types. In spray and wait routing the delivery ratio is also less when compared to epidemic routing.

5 Conclusion

DTN provides the two most popular routing techniques, namely epidemic routing and spray and wait routing. These routing techniques create replication of messages in order to deliver them successfully. To deliver messages the mobility of nodes, as well as their buffer spaces, plays a crucial role in a DTN. In this paper an assessment of the two routing protocols with a movement model called random way point movement (RWPM) has been done by analyzing the impact of node mobility and buffer space. The results shows that in spray and wait routing the delivery ratio is less but at same time the replication of messages is low. In future it is intended to propose a novel routing mechanism in which a high delivery ratio is obtained, even with a low replication rate.

References

1. Fall K (2003) A delay-tolerant network architecture for challenged internets. In: proceedings of the ACM SIGCOMM, Karlsruhe, Germany, 27–34
2. Balasubramanian A, Levine BN, Venkataramani A (2007) DTN routing as a resource allocation problem. In: Proceedings of ACM SIGCOMM, Kyoto, Japan, pp 373–384
3. Spyropoulos T, Psounis K, Raghavendr CS (2005) Spray and wait: an efficient routing scheme for intermittently connected mobile networks. In: Proceedings of ACM WDTN, Philadelphia, PA, USA, pp 252–259
4. Spyropoulos T, Psounis K, Raghavendra CS (2007) Spray and focus: efficient mobility-assisted routing for heterogeneous and correlated mobility. In: Proceedings of IEEE PerCom, White Plains, NY, USA, pp 79–85
5. Sharma A (2016) Effect of node mobility on routing protocols in delay-tolerant networks. Int J Control Theory Appl 9:345–355
6. Chaintreau A, Hui P, Crowcroft J, Diot C, Gass R, Scott J (2006) Impact of human mobility on the design of opportunistic forwarding algorithms. In: Proceedings of IEEE INFOCOM, Barcelona, Spain, pp 1–5
7. Ott J, Kutscher D (2005) A disconnection-tolerant transport for drive-thru internet environments. In: Proceedings of the IEEE INFOCOM, vol 3, Miami, FL, USA, pp 1849–1862
8. Vahdat A, Becker D (2000) Epidemic routing for partially connected ad hoc networks. Department of Computer Science, Duke University. Technical Report CS: 1–6
9. Lindgren A Doria, Scheln O (2003) Probabilistic routing in intermittently connected networks. ACM Mob Comput Commun Rev 7(3):19–20
10. Burgess J, Gallagher B, Jensen D, Levine BN (2006) MaxProp: routing for vehicle-based disruption-tolerant networks. In: Proceedings of IEEE INFOCOM, Barcelona, Spain, pp 1–6
11. Jain S, Fall K, Patra R (2004) Routing in a delay-tolerant network. In: proceedings of ACM SIGCOMM, Portland, OR, USA, pp 145–157
12. Keränen A, Ott J, Kärkkäinen T (2009) The ONE simulator for DTN protocol evaluation. In: Proceedings of the 2nd international conference on simulation tools and techniques, Rome, Italy, pp 1–11
13. Project page of the ONE simulator. https://www.netlab.tkk.fi/tutkimus/dtn/theone

A New Surgical Robotic System Model for Neuroendoscopic Surgery

Velappa Ganapathy, Priyanka Sudhakara, Amir Huesin and M. Moghavvemi

Abstract During endoscopic surgery, the surgeon holds and manipulates the endoscope inside the operating area. Using a robotic handle for these tasks has beneficial points which have been covered by a rich literature. Most of the previous works have involved laparoscopy rather than neuroendoscopy which is fairly new in comparison. In this paper the difference between the two is discussed and the design of a suitable robotic handle for neuroendoscopy is proposed.

Keywords Surgical robots • Neuroendoscope • Degree of freedom (DOF)

1 Introduction

Neuroendoscopic surgery is a well-known minimally invasive procedure which is increasingly being used in various medical situations, such as skull base surgery, brain tumor management, and pediatric neurosurgery [1, 2]. An endoscope is the main tool in this kind of procedure. Basically there are two types of endoscope, namely rigid and flexible. In neuroendoscopy, there is a tendency towards using rigid endoscopes. This is due to rigid endoscopes posing less danger for the soft tissue of brain and also having a better image quality. Rigid endoscopes are made in

V. Ganapathy (✉) · P. Sudhakara
School of Computing, SRM University, Kancheepuram 603203, Tamil Nadu, India
e-mail: ganapathy.v@ktr.srmuniv.ac.in

P. Sudhakara
e-mail: priyanka.k@ktr.srmuniv.ac.in

A. Huesin · M. Moghavvemi
Department of Electrical & Electronics, University of Malaya, Kuala Lumpur, Malaysia
e-mail: amirhuesin@um.edu.my

M. Moghavvemi
e-mail: mahmoud@um.edu.my

A. Kumar and S. Mozar (eds.), *ICCCE 2018*,
Lecture Notes in Electrical Engineering 500,
https://doi.org/10.1007/978-981-13-0212-1_63

different sizes in terms of length and diameter, but the most suitable type for neuroendoscopy is 2.7–4.0 mm in diameter and 18 cm in length [1].

Previous work in this area has mainly been aimed at laparoscopic surgery [3–9]. There are some differences between laparoscopy and neuroendoscopy, which are not covered by these systems. The first difference is that in laparoscopy the space around the incision is bigger than in neuroendoscopy, meaning that the instruments and the endoscope can move with more freedom. This is due to having multiple incisions on the body to enter the instruments, whereas in some neurosurgery cases all of the instruments should enter from one incision. To save space the endoscope manipulator should be compact with the operating area. Figure 1 show a typical neuroendoscopy procedure.

The second difference is that unlike laparoscopy, in some neuroendoscopy cases (trans-nasal) the operation occurs through a canal like the nose. Here the pivot point of the endoscope can be some distance from the surface. Unlike in laparoscopy, this makes it difficult for the manipulator to have a resting point for the scope at the entering incision. Laparoscopic manipulators which have passive joints, and manipulators with not enough active axes of rotation are hard to use in neuroendoscopic cases.

AESOP by Computer Motion, Inc and EndoAssist by Armstrong Healthcare UK are two successful endoscope handles on the market. AESOP is a serial manipulator with six degrees of freedom (DOF) of which 2 DOF are passive joints. It mounts on the side rail of the surgery table and positions the scope on top of the patient's abdominal area (entry point) [8]. The entry point makes a pivot point which forms a spherical joint with 3 DOF.

In the EndoAssist system there is an infrared camera installed to track the surgeon's head movement to control the endoscope [9]. The production of EndoAssist subsequently transferred to Prosurgics Ltd. They replaced EndoAssist with the robot FreeHand system [10]. FreeHand uses a table mount instead of a floor stand and uses the same control system that EndoAssist uses. There are other works in this area that should be mentioned [11–13]. However, referring to clinical cases, all of the mentioned systems and robots use laparoscopy as their main endoscopic procedure. In some projects [4, 5] the authors tried to use a smaller and lighter manipulator for laparoscopic surgery by using a parallel manipulator. These

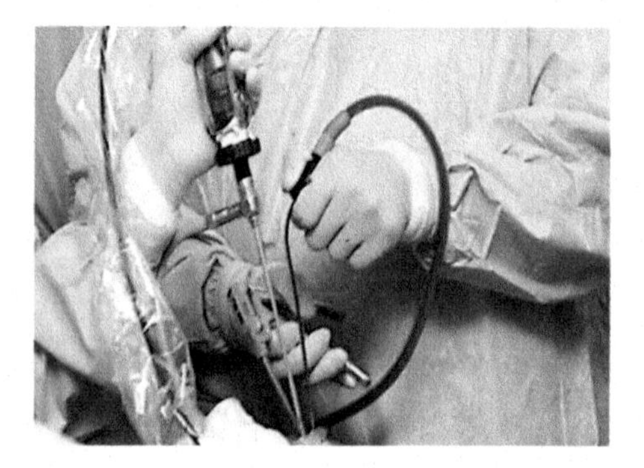

Fig. 1 A trans-nasal neuroendoscopy surgery. The space around the entry incision is severely limited

manipulators are small and light but still occupy lots of space around the entering incision. The system introduced in [14] is a capable neurosurgery robot developed by Renishaw Inc. This system can handle neuroendoscopy cases by assisting the surgeon for precision surgery.

This paper covers the design and simulation of a robotic system for manipulation of the endoscope in neuroendoscopic procedures. The paper continues with an analysis of the workspace, design, and simulation of this system.

2 Design

2.1 Workspace Analysis

Investigating the point of entry and trajectory of the endoscope, it can be said that there are two main categories. The first case involves entering from the skull and the second refers to trans-nasal cases in which the instruments enter from canals, such as the patient's nose. As illustrated in Fig. 2, during the operation the endoscope enters the surgery field from the entry point A and passes along the length, *L*. This length is affected by many factors such as the patient's skull specification, the placing of the entry point, and the nature of the surgery.

The movement of the endoscope should be a spherical movement with the point A as a pivot point. Placement of this point is dependent on many factors, such as anatomy of the patient. In laparoscopy the position is constant and acts as a spherical joint, but in neuroendoscopy the exact position of this point is not available as it differs throughout the procedure. The unavailability of this data prevents the idea of having passive joints to form a spherical joint at the surgery entry point. Figure 3 illustrates the movement of the endoscope while in operation.

The robot used for the endoscope must be robust enough when the endoscope is in the operating area. The workspace of the end effector (the lens of the endoscope) is inside the patient's brain so the robot should not have any additional movements.

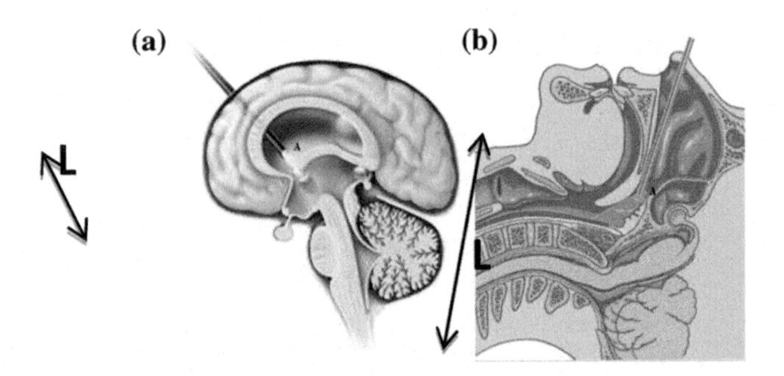

Fig. 2 The pivot point (A) in **a** a normal neuroendoscopy case, **b** in a typical trans-nasal case, where it is deeper and movement is restricted

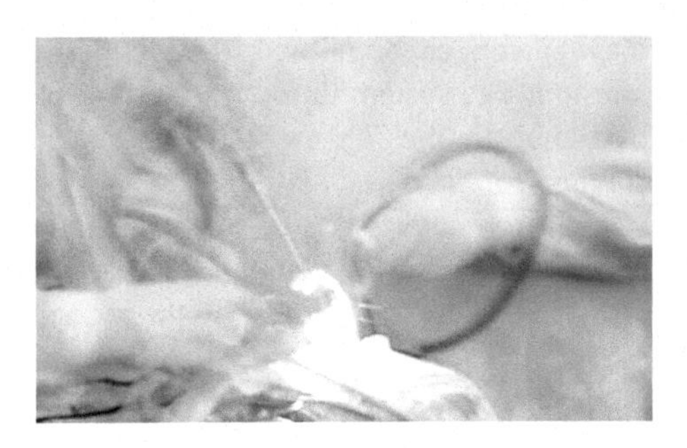

Fig. 3 Movement of the endoscope captured using a slow shutter camera

Furthermore, the robot should be able to compensate the gravitational force from the endoscope's weight.

The movement of the scope must be precise enough. The operation area is so tiny, and the surgeon needs to move the endoscope to have the best vision. Referring to the variety of sizes of instruments and following discussions with surgeons it has been decided to have millimeter precision for the end point movement. The entering hole to the operating area is very small. Outside of the operation area the surgeon's hands move closely together in a small space. As a result the griper and those parts of the robot which are attached to the endoscope must be small.

2.2 Topology Design

In parallel topology, the last joint (which is a plate) is connected to the previous one with multiple chains. This multi-link structure causes the moving plate to be big and bulky. As mentioned before the space near the endoscope is a critical variable in the design. In addition to the above problem, some other disadvantages should be mentioned [15], such as limited workspace, calibration difficulties, and passive joint non-linearity. It is fair to note that parallel robots have advantages, such as better accuracy [16, 17] and non-cumulative joint error [18]. The accuracy and error ratio of serial robots are good enough to ignore these advantages of parallel robots. In order to have a simpler kinematic in a serial manipulator, as lots of papers in this region [19] propose, the twisted angle between the manipulator joints should be either 0 or $|\pi/2|$. At least one of the link length and joint offset should be zero. Normally, to avoid mechanical complexity, having two actuators at one place should be avoided.

Here the design proceeds with a floor-mounted portable stand mounted on the table and the height robot can be adjustable. The portable base would give a freedom to put the robot anywhere around the operation table. Assigning the exact coordinate of the base is not possible due to its portable nature. The base should have a distance of between 200 cm and 50 cm from the operating table on the ground.

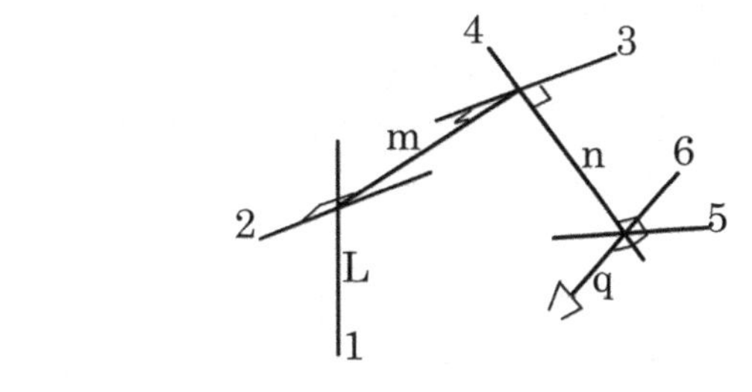

Fig. 4 Optimum structure of a 6 DOF serial manipulator

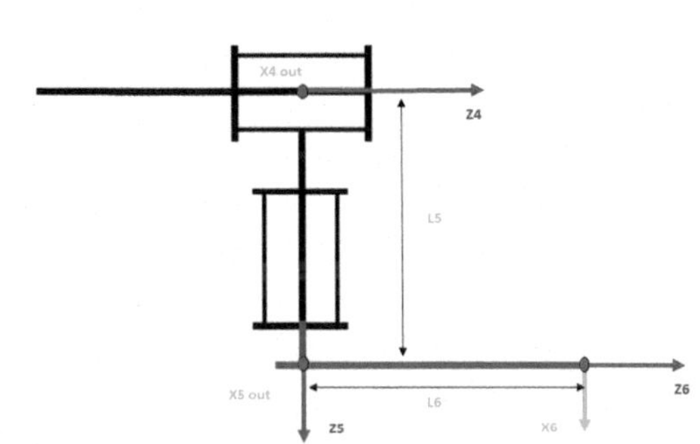

Fig. 5 The proposed 2 DOF wrist

The workspace of the manipulator can be decoupled into reachable and dexterous workspaces [20]. The aim is to have the maximum of both reachable and dexterous workspaces. However, in reality, manipulators with the existence of joint motion limitations and singularities means it is hard to maintain. As discussed in [21], Fig. 4 offers the best structure for a general purpose serial manipulator. This structure is based on 6 DOF. In a general case to overlay all the transport and orientation workspace the robot needs to have 6 DOF. However, in this paper the required orientation is a general case.

In Fig. 4, link L is the link between the base and second joint (elbow). Link m forms the arm, and n forms the forearm of the manipulator. Link q forms the distance from the wrist point to the tool gripper. To obtain the most dexterous workspace the spherical wrist is the best option [22]. One of the goals of this project is to have free space near the endoscope during the operation. Looking at the desired task, it can be seen that handling the endoscope has an axis of symmetry in the orientation space. It means if 6 DOF are used to maintain a specific orientation; in any given task there would be a joint axis that the desired orientation can be achieved with all of its angles. Eliminating this axis from the topology cannot affect the operation of the robot. The transporting structure needs to have at least 3 DOF, so a 2 DOF wrist instead of a 3 DOF spherical wrist is used (the axis 5 in Fig. 4 would be eliminated). This decision is supported by the axis of symmetry law [21, 23]. Figure 5 shows the diagram for the 2-DOF wrist.

The frames are attached according to D-H convention [23]. For ease of calculation, a passive and fixed joint have been added to the structure. The sixth frame belongs to the endoscope (Z6). If in any given angles the differential matrix doesn't

Table 1 The D-H parameters of the wrist

i	α_{i-1}	a_{i-1}	d_i	θ_i
4	0	0	0	θ_4
5	−90	0	0	θ_5
6	90	L6	L5	0

exist then there is a singularity at that point. This means that to avoid singularities, the points in which det J = 0 should be avoided. Table 1 shows the D-H parameters of the wrist

It is clear that the Jacobian matrix is

$$J = \begin{bmatrix} 0 & 0 & -Cos\theta_4 Sin\theta_5 \\ 0 & 1 & Sin\theta_4 Sin\theta_5 \\ 1 & 0 & Sin\theta_4 Cos\theta_5 \end{bmatrix}$$

det J = 0 means $Cos\theta_4 Sin\theta_5 = 0$ so in order to avoid singularities θ_4 should not be equal to ±90 and θ_5 should not be equal to 0. Singularities are called to the positions in which the manipulator is not able to move. The endoscope workspace is a cone rather than a hemisphere so θ_5 would not be equal to zero. The constraint $\theta_4 \neq \pm 90$ imposes limitations to the wrist movement but it can be avoided with proper frame assignment and a control algorithm.

The transfer structure consists of lengths *L*, *m* and *n*, and it has the responsibility of delivering and coincide the wrist point with the actual point in the workspace. Keeping in mind the endoscope workspace, the wrist point should have the workspace of a hemisphere around the patient's head. It can be concluded that the best geometrical configuration is $|m - n| = q$. In the best case q would be equal to zero; in that situation, m and n are equal. However, this is usually not possible due to the mechanical complexity involved.

To gather all the factors for the last design, the dexterous workspace for the proposed robot should be founded. The hachured area in Fig. 6 is the concluded workspace for this robot. The suitable workspace is between radii r_2 and r_3.

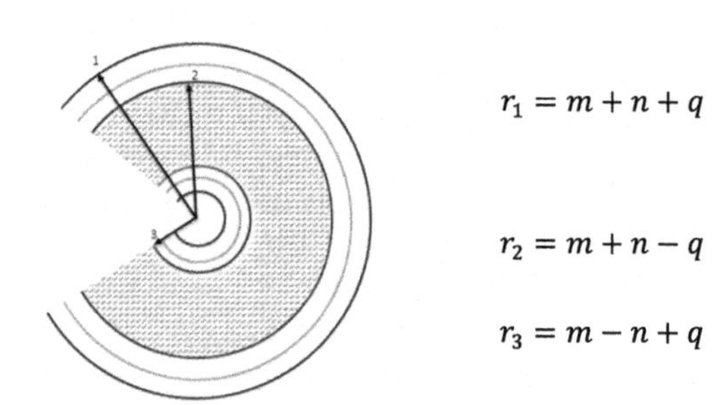

Fig. 6 Optimum structure of the workspace

Table 2 Manipulator D-H table

i	α_i	a_i	d_i	d_i
1	0	0	150 cm	R
2	−90	0	0	R
3	0	65 cm	0	R
4	90	0	50	R
5	−90	0	15	R

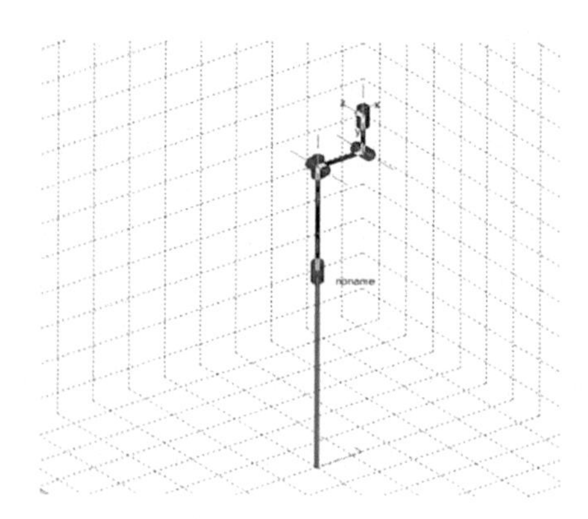

Fig. 7 Manipulator with all joints at the zero position

Earlier in this chapter, the workspace was defined with the base position.

A distance 100 cm is enough for r_2 so $m + n - q = 100\,\text{cm}$

It can be assumed that $q = 15$ cm so $m + n = 115$ cm

If $|m - n| = q = 15$ cm, then $m = 65$ cm and $n = 50$ cm

For L, there is the freedom of choosing the base frame of the structure. From the floor to patient's head is about 130 cm, so 150 cm is a good choice for the first link height. Furthermore, this height can be changed with regard to the base of the robot as mentioned before. The base can also be attached to the sides of the table if desired. So wherever the base is placed, it would not affect the designed structure as long as the height of the first link is taken into consideration, and as long as it is placed in the desired workspace area. Now the whole robot is ready to be put in the D-H table. Table 2 is the complete D-H convention of the robot.

Using the robotic toolbox [24] an illustration of the robot topology was created. In Figs. 7 and 8 two orientations of the robot are displayed.

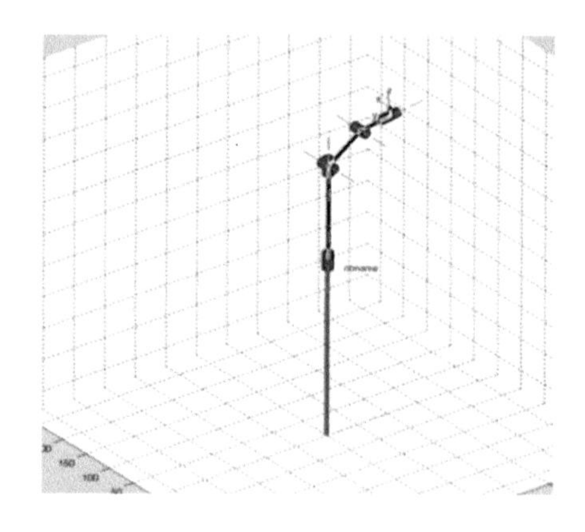

Fig. 8 Manipulator at the desired position

3 Mechanical Simulation

The D-H parameters are not enough for the simulation. The weight of the robot or dynamic specification cannot be obtained from the D-H convention. Using these data and getting help from the capabilities of CAD software, realistic virtual links of the manipulator can be drawn. CAD software provides the ability to draw the parts and specify a material for the part. The software can then calculate the dynamic measurements of the part. The CAD software which is used in this work is Pro Engineer software [25]. Each manipulator link is drawn using this software. A material is chosen for it and after completion it is imported into the simulation environment. Figure 9 shows the parts and the assembled robot.

The robot handles the surgery endoscope as a tool. There are different models of endoscope systems and in this design the aim was to prepare a 3D-drawing close to a real endoscope with the proper weight and dimensions. The CAD drawing of the endoscope is shown in Fig. 10. The camera cable and the light source fiber are connected to the end and middle of the endoscope. These cables are not included in

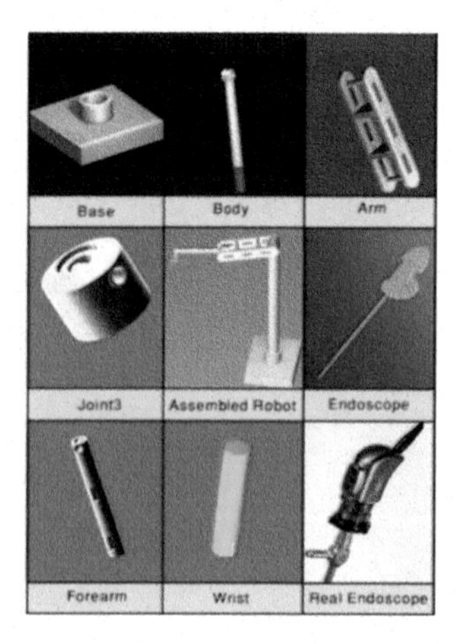

Fig. 9 Parts drawn by CAD

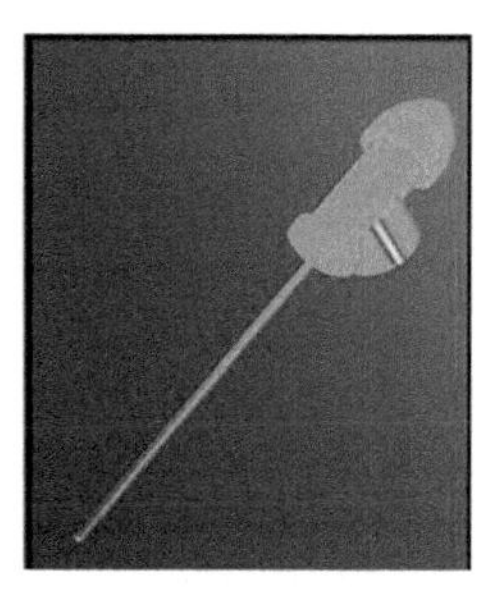
Fig. 10 3D regenerated endoscope

the simulation. In the implementation, these cables should be secured because they may apply unwanted force and torque to the end effector. The best way to handle them is to attach them to the robot body securely.

The simulation environment used here is the Simulink software from Mathworks [26]. The Simscape library can simulate the physical systems in a Simulink environment. Simscape is equipped with different sub-libraries and each is used with a kind of physical system. The sub-library which is used is called SimMechanics. SimMechanics deals with the mechanisms and mechanical systems.

The CAD data of each link is imported into the Simulink environment. Using the assembly mode of the Pro/E software all parts and links are assembled together as shown in Fig. 11.

In practical cases, each joint contains many parts and devices, such as motors, gears, and sensors. These all apply stiffness to each joint. Although the model has none of these parts implemented, each joint can be changed to have a stiffness that is close to reality. This can be done by a special block called a body spring and damper. The stiffness data can be entered into this block, and this block applies the stiffness to the point of a two-body connection (joint). The designed robot has a similar structure and topology to PUMA 560 and as a result the stiffness data from the Unimation PUMA 560 manipulator [27] is used. It should be mentioned that the dynamic parameters of the designed robot are very different from PUMA 560,

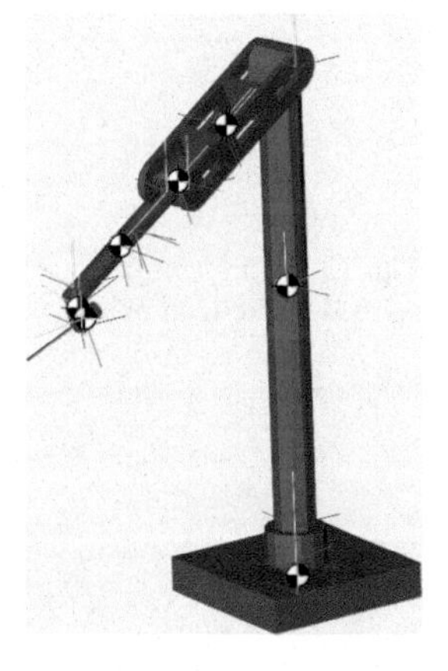
Fig. 11 Assembly of the robot

Table 3 Accepted stiffness data for the robot

Joint	Stiffness (Nm/rad)
1	66,230
2	66,500
3	11,610
4	2020
5	1440

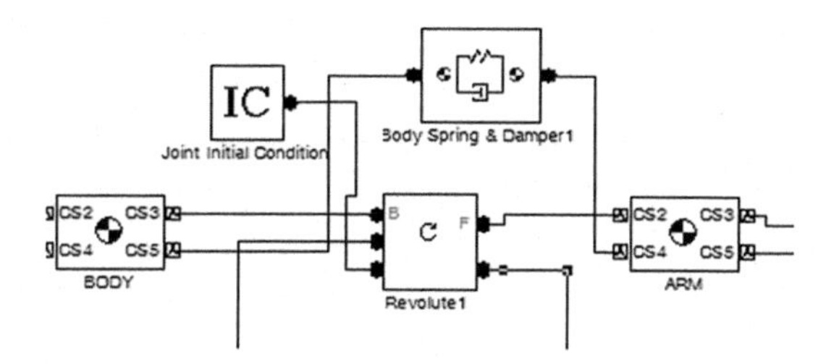

Fig. 12 Joint diagram in Simulink

such as the inertia and mass of the links, but the stiffness is based more on the joint structure data which can be assumed is the same as in the designed robot. Table 3 contains the stiffness data accepted for this manipulator. Moreover, using Initial Condition block for each joint, an initial position is assigned to put the robot in proper orientation before the simulation starts. Figure 12 shows one of joint diagrams in the simulation environment.

The applied controller to the simulation is a PID controller designed for serial robots based on stiffness matrices. This method is introduced in [28]. The block diagram of the position controller is described in Fig. 13. Controller inputs are the position and velocity errors of the joints.

ε is the control parameter in this controller. It should be small enough to provide the closed-loop stability of the whole system. ε has inverse influence on PID controller performance, as mentioned in [28], where "the smaller the value of ε, the larger the region of attraction and the closer the transient performance to the prescribed one."

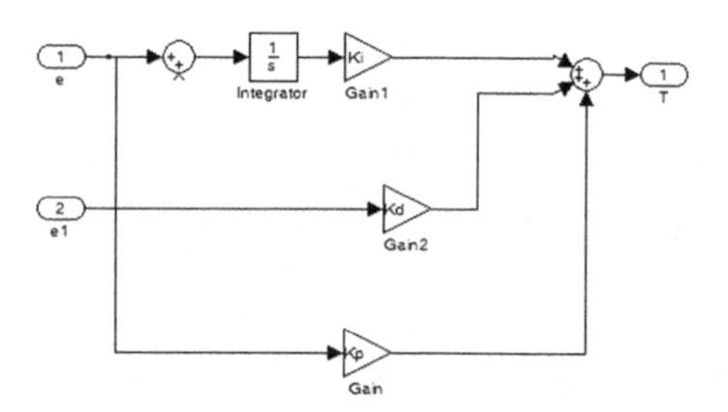

Fig. 13 PID controller applied to the manipulator

4 Results

A five DOF serial manipulator is designed with three joints as translational and two joints as orientation structures. It is designed to manipulate the endoscope in neuroendoscopy surgeries. The simulation results that are given below are based on the manipulator trajectory. Figure 14 shows a set of constant positions applied to the robot joints. This investigates the maintaining situation (gravity compensation) in the robot. This situation can be called a constant trajectory. In Fig. 14 the outputs as well as the applied positions for all five joints are presented. In Fig. 15, $\varepsilon = 0.005$. Delivering the results from all joints is not necessary because the behaviors in the joints are almost the same. So joint5 outputs for different values of ε are discussed. Joint5 is the last joint and it is the smallest one (Figs. 16, 17, 18, 19).

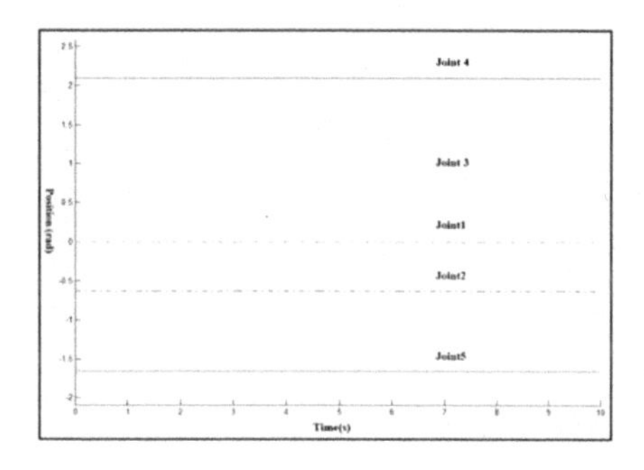

Fig. 14 The generated trajectory, all joints stand still and gravity is compensated

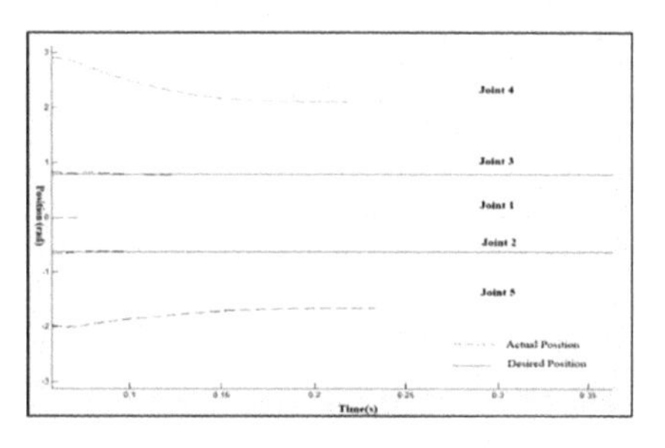

Fig. 15 Joint positions are compared with the trajectory, $\varepsilon = 0.005$

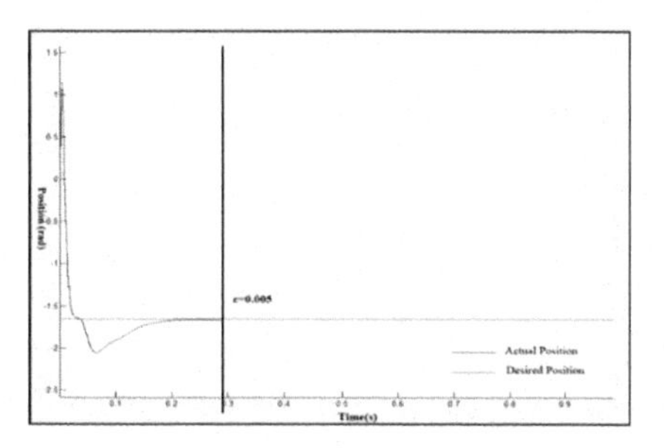

Fig. 16 Joint5 positions are compared with the trajectory (blue line), $\varepsilon = 0.005$

Fig. 17 Joint5 positions are compared with the trajectory (brown line), ε = 0.05

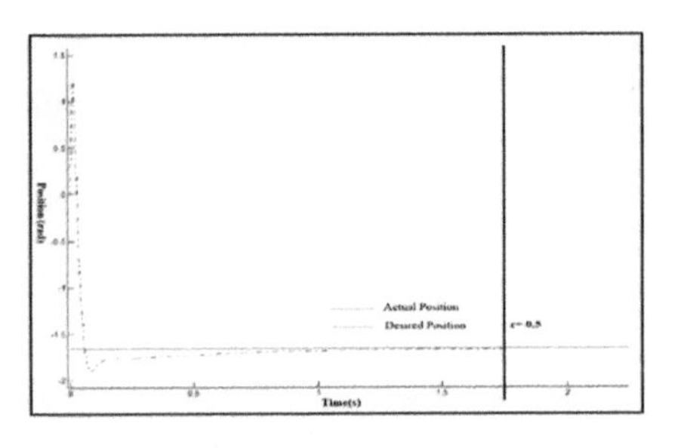

Fig. 18 Joint5 positions are compared with the trajectory (brown line), ε = 0.5

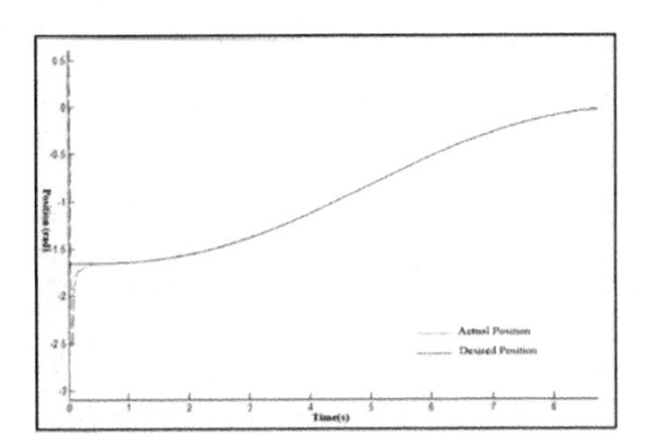

Fig. 19 Joint5 positions are compared with the non-constant trajectory (black line), ε = 0.05

It is obvious that with a smaller ε the system finishes the transaction period sooner. But with a smaller ε the calculation time increases. Table 4 gives the transaction period and calculation time together.

The next test was done with a non-constant applied trajectory while ε = 0.05. Figure 18 shows the system response to a non-constant trajectory.

The reported results are the proof that the design in this paper works properly without making the system unstable. Having the objectives in mind, this

Table 4 Transaction and calculation time for three different values of ε

ε	Transaction period (s)	Calculation time (s)
ε = 0.005	0.29	363
ε = 0.05	0.455	186
ε = 0.5	1.75	160

dissertation was successful in designing the kinematic topology for the endoscope manipulator and a proper low-level controller. The robot and the controller are simulated as per the design.

As stated earlier, in this paper the effort was on designing a robotic arm to handle the endoscope in neuroendoscopy. Going through the introduction sections it is clear that this design should fulfill two main goals. The first goal addresses the need to have a specialized handle for the neuroendoscope. The design approached this issue by adding two DOF more than laparoscopic handles possess. The second issue was about designing the robot to be small enough to save space around the surgeon's hand. Clearly, in the topology of the design the actuators have been moved away from the endoscope to make the mechanical parts as small as possible. Finally, it should be mentioned that all the work done here is based on computer simulation. To precede the project further and shift it to the implementation phase, more simulations are needed. However, it is strongly believed by the writer that this design can be the base design for the implementation of the robotic arm for neuroendoscopic surgery.

Acknowledgements The authors thank the University of Malaya, Kuala Lumpur, for permitting us to carry out the project and providing us with the necessary facilities for successfully completing the project.

References

1. Macarthur D et al (2002) The role of neuroendoscopy in the management of brain tumours. Br J Neurosurg 16:465–470
2. Shahinian HK (2008) Endoscopic skull base surgery. Humana Press
3. Kavoussi LR et al (1995) Comparison of robotic versus human laparoscopic camera control. J urol 154:2134–2136
4. Sekimoto M et al (2009) Development of a compact laparoscope manipulator (P-arm). Surg Endosc 23:2596–2604
5. Gumbs AA et al (2007) Modified robotic lightweight endoscope (ViKY) validation in vivo in a porcine model. Surg Innov 14:261–264
6. Kasalický M et al (2002) AESOP 3000—computer-assisted surgery, personal experience. Rozhledy v Chirurgii: Mesicnik Ceskoslovenske Chirurgicke Spolecnosti 81:346–349
7. Wagner AA et al (2006) Comparison of surgical performance during laparoscopic radical prostatectomy of two robotic camera holders, EndoAssist and AESOP: a pilot study. Urology 68:70–74
8. Hurteau R et al (1994) Laparoscopic surgery assisted by a robotic cameraman: concept and experimental results, vol 3, pp 2286–2289
9. Kommu SS et al (2007) Initial experience with the EndoAssist camera-holding robot in laparoscopic urological surgery. J Robot Surg 1:133–137
10. (2010) Press releases. http://www.freehandsurgeon.com/news/
11. Ma O, Angeles J (1993) Optimum design of manipulators under dynamic isotropy conditions, vol 1, pp 470–475
12. Munro MG (1993) Automated laparoscope positioner: preliminary experience. J Am Assoc Gynecol Laparosc 1:67–70

13. Crouzet S et al (2010) Single-port, single-operator-light endoscopic robot-assisted laparoscopic urology: pilot study in a pig model. BJU Int 105:682–685
14. Richter L et al (2011) Computer assisted neurosurgery. Biomed, Eng
15. Nzue RMA et al (2010) Comparative analysis of the repeatability performance of a serial and parallel robot. In: 2010 IEEE/RSJ international conference on intelligent robots and systems (IROS), pp 63–68
16. Rauf A et al (2004) Complete parameter identification of parallel manipulators with partial pose information using a new measurement device. Robotica 22:689–695
17. Wavering AJ Parallel kinematic machine research at NIST: past, present, and future. In: Parallel kinematic machines, pp 17–31
18. Song J et al (1999) Error modeling and compensation for parallel kinematic machines. In: Parallel kinematic maschines, pp 172–187
19. Balkan T et al (2001) A kinematic structure-based classification and compact kinematic equations for six-dof industrial robotic manipulators. Mech Mach Theory 36:817–832
20. Ottaviano E et al (2006) Identification of the workspace boundary of a general 3-R manipulator. J Mech Des 128:236
21. Vijaykumar R et al (1986) Geometric optimization of serial chain manipulator structures for working volume and dexterity. Int J Rob Res 5:91–103
22. Paul RP, Stevenson CN (1983) Kinematics of robot wrists. Int J Rob Res 2:31–38
23. Craig JJ (1989) Introduction to robotics: mechanics and control, vol 74. Addison-Wesley, New York
24. Corke PI (1996) A robotics toolbox for MATLAB. Rob Autom Mag IEEE 3:24–32
25. Pro P, Pro EW (2003) Toolkit user's guide. Parametric Technology Corporation
26. Simulink M, Natick M (2007) The mathworks. Inc., Natick, MA
27. Kim H, Streit D (1995) Configuration dependent stiffness of the Puma 560 manipulator: analytical and experimental results. Mech Mach Theory 30:1269–1277
28. Alvarez-Ramirez J et al (2000) PID regulation of robot manipulators: stability and performance. Syst Control Lett 41:73–83

Survey on Security in Autonomous Cars

K. V. Harish and B. Amutha

Abstract The improvements made in automotive control systems and sensory technologies have led to the rise in autonomous cars. These cars use a wide range of networking and sensory technologies to control the car and interact with the environment. One recent application which is making headway is in the alliance of the IoT and autonomous cars. Since the IoT focuses on connectivity between different isolated devices found over the web, it is also used to provide services to autonomous cars. It is therefore, imperative to ensure that data privacy and security is maintained by the system. Hence, this paper surveys and discusses the security issues faced in autonomous cars.

Keywords IoT · Autonomous car

1 Introduction

The IoT has begun to incorporate millions of devices around the world. With the increase in speed and coverage of networks, the IoT is has also recently started to integrate itself with autonomous cars [1]. Hence, the car is now no longer just a mechanical vehicle, it has become an interactive machine due to the integration of multiple technologies. Autonomous cars (AC) can deduce their surroundings, locate their position relative to a virtual map, and navigate through the use of a highly specialized control system and sophisticated algorithms. Cars can now talk with each other using vehicle to vehicle (V2V) technology [2] and are now even part of an intelligent grid of cars [3]. Hence, AC have now developed from isolated individual machines to a network of communicating vehicles. However, this level of integration between different physical systems and cybersystems mean that there

K. V. Harish (✉) · B. Amutha
Department of Computer Science and Engineering, SRM University, Chennai, India
e-mail: hari8495@gmail.com

B. Amutha
e-mail: bamutha62@gmail.com

A. Kumar and S. Mozar (eds.), *ICCCE 2018*,
Lecture Notes in Electrical Engineering 500,
https://doi.org/10.1007/978-981-13-0212-1_64

is an increased risk of exploitation by malicious users and hackers. External wireless networks present in the car expose the internal configuration of subsystems to the outside world [4]. Hence, this paper surveys the different network vulnerabilities present in AC and also discusses the how a secure IoT development framework can address all these issues. The paper is organised as follows: Sect. 2 describes the background knowledge required to understand the system. Section 3 talks about the security issues encountered in AC. Section 4 talks about protective strategies that can be undertaken to mitigate the issues. Section 5 describes the IoT secure development framework (ISDF) that can be used to address all of the issues. Section 6 concludes the survey with a discussion of the pending issues.

2 Background

Cars use a fusion of sensors and actuators. This is achieved through the use of ECUs (electronic control unit). With the increase in diverse features and applications, cars today use more ECUs to handle their various subsystems. Moreover, AC today communicate with lot of external devices using wireless interfaces which exposes the car's in-vehicle network, thereby making it more susceptible to cyber attacks from hackers. Let us begin by understanding the background knowledge of ECUs and in-vehicle networks used in AC [4].

1. **Electronic Control Unit** (**ECU**):

Simply put an ECU is an embedded controller that controls one or more of the automobile's subsystems. The controller may be used to control different subsystems such as the engine control unit, body control unit, and transmission control unit, etc. It is observed that there are around 50–70 ECUs to be found in a car. This number will only go up with the increase in functionality and features of the car. In order to ensure cross-compatibility between different car platforms, AC use a standardised development architecture called AUTOSAR [5] (automotive open system architecture) to design software components.

2. **Controller Area Network** (**CAN**):

CAN is the most commonly used bus standard vehicles to different devices to communicate with each other without the use of a computer. It is mainly useful for transmission of measured values and signals. Hence, it is s useful for supporting distributed control systems. It has a data rate of 1 Mbps. Wiring is by a 5 V twisted pair cable. Typical applications are ABS, Power Train and engine control.

3. **LIN** (**Local Interconnect Network**):

LIN is another kind of bus standard used for low cost low-end multiplexed communication in automobile networks. It is used to address high bandwidth and advanced error handling. It can be used with embedded with a UART (universal asynchronous receiver/transmitter). It has a data rate of up to 19.2 Mbps and uses a 12 V single wire cable. It is used mainly for electric seats, mirrors, and tailgates. The choice of usage of LIN over CAN is purely based on usage and the resources allocated [6].

4. **Flex Ray**:

Flex Ray is a deterministic, fault-tolerant, high speed bus. Initially meant to be a successor to CAN, It has a data rate of up to 10 Mbps. It typically uses 2 or 4 wires and is used mainly for steering, traction control and active suspension.

5. **MOST**:

MOST (Media oriented systems transport) is used for multimedia data transmission using optical fiber cables or coaxial cables. It has higher communication speeds compared to CAN, LIN, and Flex Ray. The data rate supported by MOST is up to 23 Mbps. It is used mainly for media players and other infotainment systems present in the car.

External networks used in the car typically include bluetooth, wifi, and 3G/4G telecommunication services. Cars may also receive data from connected flash drives. This next stage in the evolution of ACs include V2V and V2I communication which may in the future automatically stop/slowdown the car based on proximity to nearby cars. This is just an example of how internal actuators may be controlled by external networks.

3 Security Issues

Many studies conducted on the security of vehicle networks show that internal networks such as CAN are extremely susceptible to hackers [7]. This is due to the fact that many of the vehicle's internal subsystems are accessible via the on board diagnostics (OBD) port. Most of the internal details of ECUs are visible from the OBD port. Hence, malicious hackers can cause serious havoc with access to this port. Most in-vehicle networks have little to no security as they possess no authentication and encryption techniques are not employed [8]. Typical scenarios which could compromise security of the system include:

- Bus Vulnerability: Due to the lack of specialized security mechanisms, such as device authentication and cryptographic techniques, the buses are susceptible to unauthenticated access and modifications. Due to this, services on the bus can be stopped with a denial-of-service (DOS) attack and packet integrity on the bus could be lost due to falsification of data or creation of false data.
- Local Attacks: OBD is used in CAN networks to typically identify and retrieve diagnostic data. This communication standard uses OBD plugins of sensors to transfer data. This could be a specific point of entry for hackers. By inserting an additional device into the OBD port, they could gain access to most of the

subsystems. Not only would the hacker gain access to bus traffic but also they could also gain the ability to send frames in the bus. Koscher et al. [9] shows how a hacker could utilize access to one ECU to control other subsytems by sending frames on the bus.

- Remote Attacks: These are made without having physical access to any of the bus systems. They can occur over both short- and long-range wireless networks. The car uses many wireless networks such as wifi, bluetooth, 3G/4G, GPRS, and GSM etc. The implementation of such protocols could be faulty. Such vulnerabilities could be utilised by hackers to compromise the ECU. Technologies like V2V and V2I use a wide range of wireless sensors to control the car. Any malicious signal sent could result in dangerous accidents.

4 Protective Strategies

In light of the above issues, ensuring adequate precautionary mechanisms is a must to prevent misuse and malfunctioning of AC. To ensure secure communication over the various buses, several possibilities are considered. They can be broadly classified into 3 types.

4.1 Authentication and Packet Transmission Using Encryption

CAN typically broadcasts its messages to all connected nodes present in the bus network. Hence, there is a lack of authentication mechanism for the sender. To overcome these flaws, encryption techniques are being implemented to perform authentication of the ECU, integrity checks, and encryption of frames to prevent reading by other nodes. Such implementations can be found in [10] and [11].

4.2 Anomaly Detection Techniques

This method first establishes all the normal operating conditions of the system. Anomalies are detected when the system deviates too much from normal operating condition. Although this provides a reliable model to detect abnormalities, it isn't useful in exceptional cases or whenever the system behaves in an undefined manner.

4.3 Integrity of ECU Software

As mentioned in the above attacks, the ECU is one of the vulnerabilities which could be exploited by hackers through external networks. Hence, ensuring the integrity of ECU software is critical for separating ECU modules from non-ECU modules. Any installations or modifications made to ECUs need to also be handled carefully. Update time is also important as it would make the system more vulnerable during upgrades. Onuma et al. [12] describes one such method to reduce update time by dividing update data.

5 IoT Secure Development Framework

With the increasing complexity of autonomous systems, addressing all these security concerns become even harder. As it is hard to resolve each of the network issues separately, it would make sense to have an architectural model which can provide security and protection in an integrated manner. The IoT secure development framework (ISDF) is one such framework which has incorporated security mechanisms into the design and development phase of the system. The framework is shown in Fig. 1.

End-devices include the sensors, actuators, and controllers used. The communication layer consists of the in-vehicle networks such as CAN, LIN, Flex Ray, and MOST while external networks include wifi, bluetooth, 3G, 4G, GSM, GPRS and other networks. Services provided are usually web-based/cloud services. These services are used by applications to monitor, control, and communicate with the autonomous car. Since each layer describes the attack surface, target functionality and mitigation mechanisms involved. It is easy to deal with all the security issues mentioned in the above sections in a simple and easy manner. Applications on the top layer can communicate with end-devices using any IoT protocols. However, it has to be noted that the choice of IoT protocol needs to be done based on usage, and their ability to provide security [13]. This poses a challenge as most IoT protocols

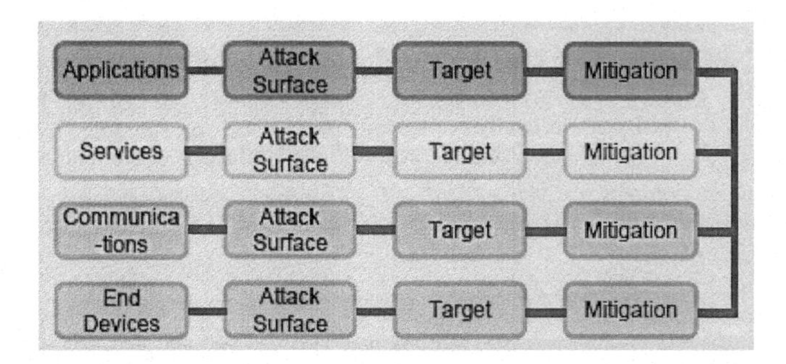

Fig. 1 ISDF framework

provide little to no security as they lack anonymity and authentication features [14]. Hence, required protocols need to be surveyed based on usage and requirement before integration within the framework.

6 Observations

The references discussed, and their advantages, disadvantages, and observations are tabulated as follows (Table 1).

Table 1 Observations

Paper	Concept involved	Advantages	Disadvantages	Future work
[2]	Handling V2V communication using vision based data and GPS signal	Useful when GPS signal is poor or inactive	V2V not reliable in all locations	
[3]	Evolution of intelligent vehicle grids to autonomous, interconnected vehicles in vehicle clouds	Better communication, better storage, better intelligence, better learning capabilities	Prone to vicious security attacks, both internally and externally	Development of intelligent transport systems and development of the vehicular cloud
[4]	Security mechanisms in automotive networks		Internal and external networks prone to hackers and security threats	Future works focus on encrypting the communication, anomaly detection and maintaining integrity of embedded software
[1]	IoT secure development framework	IoT secure development framework	Does not provide security to all layers and functions	Future work is focused on extending security to all layers and functions
[5]	AUTOSAR	Connects different subsystems and provides real-time capabilities and safety	Description of communication protocols at software level	Restructuring of portfolio to meet new demands and requirements
[6]	Hardware efficiency comparison of IP cores for CAN & LIN protocols	Found tradeoffs in area utilization, power consumption, and resource utilization		

(continued)

Table 1 (continued)

Paper	Concept involved	Advantages	Disadvantages	Future work
[15]	Assessing CAN transmission reliability and message scheduling policies using a flexible error-free model	Adaptable to different conditions by tuning parameters	Lacks an error model parameter setting procedure that changes parameters dynamically	Application of this approach to CSMA—CR-based VAN networks
[7]	Cyber-security for CAN protocol	Low communication overhead, no maintenance of global time	Low data rates for increased bus utilization	
[8]	White Box, black box, grey box	ECU messages are secure, interpreting data sent from different ECUs is not possible. Only authorized nodes can talk with ECUs		
[16]	Vulnerabilities of in-vehicle networks & attacking methodologies		Attacks can be easily made on internal networks of the car	Future work includes enhancing the network architecture such as VANETs, cellular networks and cloud
[9]	Security analysis of the automobile		Extent of damage, ease of attack, unenforced control access, increase of attacks	Future work involves securing automobile networks by considering feasibility and practicality of the mechanisms
[10]	ECANDC	Suitable for in-vehicle networks	Limited payload for CAN dataframe	
[11]	CANTrack	Improves CAN bus security, encrypted messages help avoid replay attacks	Payload can't be edited to extend more data bytes, can't change or encrypt message IDs	Use of intelligent algorithms for encrypting, decoding and key generation
[12]	Dividing data to reduce update time	Reduces update time slightly	ECU processing time goes down with addition of new ECU nodes	

(continued)

Table 1 (continued)

Paper	Concept involved	Advantages	Disadvantages	Future work
[13]	CoAP, MQTT, XMPP	Minimal restrictions on security mechanisms	Trade-off between security and lightweight	Future work involves extending secure mechanisms for IoT protocols without causing too much overhead
[14]	CoAP, MQTT	Lightweight protocols	Lack of decentralization and anonymity	Future work is to extend anonymity and decentralization to them

7 Conclusion

This paper describes the various security issues faced in autonomous cars. It can be seen that security issues need to be addressed in all of the internal and external networks to ensure that the overall system is secure. This requires a systematic approach in the addressing all of the mentioned challenges. ISDF is an architectural model capable of covering all of the above-mentioned issues. It is, however, to be noted that IoT protocols to be used in the application layer need to be carefully considered as most of them do not have any form of authentication and anonymity.

References

1. Pacheco J, Satam S, Hariri S, Grijalva C and Berkenbrock H (2016) IoT security development framework for building trustworthy smart car services. In: IEEE conference on intelligence and security informatics, Tucson, AZ, pp 237–242
2. Challita G, Mousset S, Nashashibi F, Bensrhair A (2009) An application of V2V communications: cooperation of vehicles for a better car tracking using GPS and vision systems. In: IEEE vehicular networking conference, Tokyo, pp 1–6
3. Gerla K, Lee EK, Pau G, Lee U (2014) Internet of vehicles: from intelligent grid to autonomous cars and vehicular clouds. In: IEEE world forum on internet of things, Seoul, pp 241–246
4. Studnia I, Nicomette V, Alata E, Deswarte Y, Kaâniche M, Laarouchi Y (2013) Survey on security threats and protection mechanisms in embedded automotive networks. In: 43rd annual IEEE/IFIP conference on dependable systems and networks workshop, Budapest, pp 1–12
5. Fürst S, Bechter M (2016) AUTOSAR for connected and autonomous vehicles: the AUTOSAR adaptive platform. In: 46th annual IEEE/IFIP international conference on dependable systems and networks workshop, Toulouse, pp 215–217
6. Krishnapriya VV, Sikha M, Nandakumar R, Kidav JU (2012) Hardware efficiency comparison of IP cores for CAN & LIN protocols. In: International conference on computing communication & networking technologies (ICCCNT), Coimbatore, pp 1–3

7. Lin CW, Sangiovanni-Vincentelli A (2012) Cyber-security for the controller area network (CAN) communication protocol. In: International conference on cyber-security for the controller area network (CAN) communication protocol, Washington, DC, pp 1–7
8. Khan J (2017) Vehicle network security testing. In: Third international conference on sensing, signal processing and security, Chennai, India, pp 119–123
9. Koscher K et al (2010) Experimental security analysis of a modern automobile. In: IEEE symposium on security and privacy, Oakland, CA, USA, pp 447–462
10. Wu Y, Kim YJ, Piao Z, Chung JG, Kim YE (2016) Security protocol for controller area network using ECANDC compression algorithm. In: IEEE international conference on signal processing, communications and computing (ICSPCC), Hong Kong, pp 1–4
11. Farag WA (2017) CANTrack: enhancing automotive CAN bus security using intuitive encryption algorithms. In: 7th international conference on modeling, simulation, and applied optimization (ICMSAO), Sharjah, pp 1–5
12. Onuma Y, Terashima Y, Kiyohara R (2017) ECU software updating in future vehicle networks. In: 31st international conference on advanced information networking and applications workshops (WAINA), Taipei, pp 35–40
13. Nastase L (2017) Security in the internet of things: a survey on application layer protocols. In: 21st international conference on control systems and computer science (CSCS), Bucharest, pp 659–666
14. Zamfir S, Balan T, Iliescu I, Sandu F (2016) A security analysis on standard IoT protocols. In: international conference on applied and theoretical electricity (ICATE), Craiova, pp 1–6
15. Navet N (1998) Controller area network [automotive applications]. IEEE Potentials 17(4):12–14
16. Liu J, Zhang S, Sun W, Shi Y (2017) In-vehicle network attacks and countermeasures: challenges and future directions. IEEE Netw 31(5):50–58

Identification of Vegetable Plant Species Using Support Vector Machine

K. Deeba and B. Amutha

Abstract This study proposes a method for the identification of vegetable plant species. Each plant leaf has its own features that can be used to identify the species it belongs to. Some of the features of a leaf that enable specific identification of a plant species are its shape, vein pattern, apical and basal features, and color patterns. Those salient features extracted from the leaf image are used along with a data mining algorithm, such as support vector machine, to identify of the species that the leaf belongs to. In this study two vegetable species, namely eggplant and ladies' fingers were considered. Support vector machine is suited to situations where the data need to be classified into two groups.

Keywords Shape information · Vein pattern · Apical and basal features Colour patterns · Support vector machine

1 Introduction

India is a county based on agriculture. It is the second largest country in the production of fruits and vegetables in the world, after China. It is essential to identify vegetable plant species automatically using images. Each plant leaf has its own features that allow it to be classified as different from other plants. Automatic identification of plant species is a type of pattern recognition problem. Images that are taken in real life always tend to show changes in the physical structure. Even in such cases there are few common features that can be used to identify the plant

K. Deeba (✉) · B. Amutha
Department of Computer Science and Engineering, SRM University, Chennai, India
e-mail: deeba.k@ktr.srmuniv.ac.in

B. Amutha
e-mail: amutha.b@ktr.srmuniv.ac.in

A. Kumar and S. Mozar (eds.), *ICCCE 2018*,
Lecture Notes in Electrical Engineering 500,
https://doi.org/10.1007/978-981-13-0212-1_65

species, such as the shape of the leaf, color pattern, and vein structure etc. Some of the salient features considered in this paper are leaf shape, vein pattern, apical and basal features, and color patterns.

Shape descriptors are of two types. One is a region-based descriptor and the other is a counter-based descriptor. A region-based descriptor is used to describe the shape of the object based on boundary and inner pixel information. A counter-based descriptor describes only the outer boundary information of the object and includes conventional representation and structural representation.

Nine different vein patterns are considered widely and for our example we considered two classes; ladies' fingers with palmate venation and eggplant with reticulate venation. After the extraction of the features a support vector machine classifier is built for classification.

This paper is organized as follows: Sect. 2 describes related works, section three discusses proposed work, and Sect. 4 contains the conclusion.

2 Related Work

Leaf recognition based on shape [1, 2] is an old method and when the leaf samples are incomplete this method cannot be used for correct recognition. Crowe and Delwiche [3], developed an algorithm for analyzing apple and peach defects. He obtained an accuracy of 70% by using near-infrared (NIR) images. Nakano [4], applied a neural network for color grading of apples, and achieved an accuracy of 75%.

Pydipati et al. [5], used statistical and NN classification for citrus disease detection using machine vision. He gained an accuracy of 90% using a texture analysis method, depends on lab results. Lei et al. [6], worked on plant species identification based on the neural network algorithm called a self-organizing map (SOM). Liu et al. [7], worked on plant leaf recognition based on a locally linear embedding and moving center hypersphere classifier and achieved an accuracy of 92%. Shilpa et al. [8], used different neural network algorithms for plant species identification and made a comparative analysis.

3 Proposed Approach

This proposed work aims to identify plant species based on the following leaf features: shape information, vein pattern, and apical and basal features. The images used were pre-processed in order to enhance feature extraction.

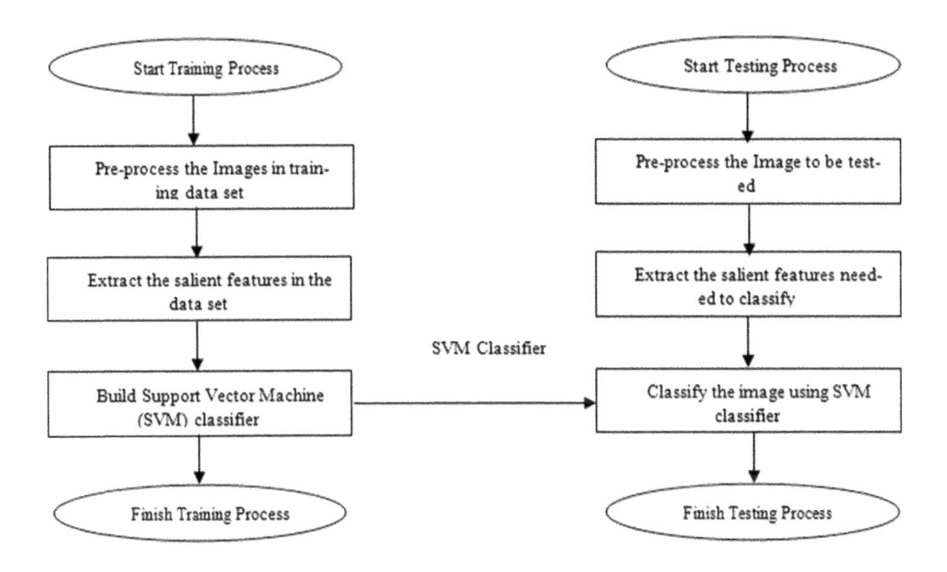

3.1 Pre-processing and Extraction of Shape Information

The major aim of pre-processing is to remove all noise and to discard unwanted descriptions. This step includes converting the image form RDG to a gray-scale image, this gray-scale image will be then converted into a binary image and finally that binary image will be converted into a contour image which only has the external boundary of the image. The following 11 features will be extracted:

1. Area: The area enclosed by the leaf
2. Perimeter: The number of pixels along the margin of the leaf
3. Diameter: The longest distance form a point to another point on the leaf.
4. Length (Major Axis): The distance between base and apex.
5. Width (Minor Axis): The longest line perpendicular major axis connecting two points in leaf.
6. Aspect ratio: The ratio between the major axis (x) and the minor axis (y).

$$\text{Aspect Ratio} = x/y$$

7. Rectangularity: Finding the similarity of the leaf shape to a rectangle.

$$\text{Rect} = (\text{Area/The area of the smallest rectangle that enclose leaf}) * 100$$

8. Circularity: Finding the similarity of the leaf shape to a circle.

$$\text{Circ} = (4 * \pi * \text{Area})/\text{Perimeter}^2$$

9. Eccentricity: Finding the similarity of the leaf shape to a cone.

$$\text{Cone} = \left((x^2 - y^2)/x^2\right)^{1/2}$$

10. Convexity: Number of concaves on the contour.

Apical and basal features:

The two leaf classes that we considered in this study have different apical and basal features. To obtain these features an origin point O is considered for both apex A and base B. Two points were chosen from the origin; one on the left side (L), and one on the right side (R) of the contour. Then the angle is formed form that points to that of the main line connecting apex A and base B point. Theses angles were used as 4 features.

Vein features extraction:

These features are obtained by performing morphological operations on the gray-scale image. The purpose of applying morphology processing to a gray-scale image is to get rid of the gray overlap between the leaf vein and the background. For the proposed plant classification model, the RGB leaf image is converted to a gray-scale image. Then, opening operations are performed on the gray-scale image with a flat disk-shaped structuring element of varying radius (1, 2, 3, 4). The resultant image is then subtracted from the margin of the leaf. Finally, the converted image is binaried. Applying these stages leads to the obtaining of the two leaf vein features.

Dataset:

The images for this work were collected from the fields around Salem district in Tamil Nadu, India, using a Sony Cybershop W810 20.1MP digital camera with 20.1 mega pixels. The images were taken against a white background with 100 images of each species for training and 50 images for testing. Hence, there is a total of 300 images in the dataset.

Leaf classification using support vector machine:

Support vector machine (SVM) is a type of supervised classification algorithm. Here the training samples are associated with the class label and the classifier is built based on the training samples which will be used for classification. With the extracted features of training set with instance label pair (a_i, b_i), i = 1, 2, …, l, where $a_i \in R_n$ and $b_i \in \{-1, 1\}$l, the support vector machine requires the solution of the following optimization problem, $\min_{w,\, b,\, \xi} \frac{1}{2}||w||^2 + C\sum_{i=1}^{l} \xi_i$ showing that $b_i((wT \,.\, \phi(a_i) + b) \geq 1 - \xi_i$, i = 1, 2, … , 1 and $\xi_i \geq 0$, i = 1, 2, … , i.

Here the training sample vectors a_i are mapped into a higher dimensional space by the function ϕ. SVM finds a hyper plane which separates the region in a linear way in the given higher dimensional space. The penalty parameter of the error term $C > 0$ is used and $K(a_i, a_j) = \phi(a_i)^T. \phi(a_i)$ is called a kernel function. With these functions the classifier is built so that the classification of the test samples can be verified. SVM works well for binary classification (classification of samples in two classes).

4 Conclusion and Future Work

Automatic identification of plant species is a type of pattern recognition problem. The images are taken in the real world and hence always tend to have changes present in the physical structure. This study proposes a method for the identification of vegetable plant species. The salient features were extracted from the leaf image and support vector machine (SVM) was used for the identification of the leaf species. Two different varieties of plant species were considered in this study and SVM is well suited to cases where the data need to be classified into two groups. In future work this can be extended to multi-class classification.

References

1. Javed Amin H, Ashraful M (2010) Leaf shape identification based plant biometrics. In: 13th International conference on computer and information technology (ICCIT), pp 458–463
2. Du J-X, Wang XF, Zhang G-J (2007) Leaf shape based plant species recognition. Appl Math Comput 185
3. Crowe TG, Delwiche MJ (1996) Real-time defect detection in fruit: part II. An algorithm and performance of a prototype system. Trans ASAE 39(6):2309–2317
4. Nakano K (1997) Application of neural networks to the color grading of apples. Comput Electron Agr 18(2–3):105–116
5. Pydipati, R, Burks TF, Lee WS (2005) Statistical and neural network classifiers for citrus disease detection using machine vision, by information & electrical technologies division of ASABE in August 2005, Florida Agricultural Experiment Station Journal Series R-10626
6. Zhang L, Kong J, Zeng X, Ren J (2008) Plant species identification based on neural network. In: Fourth international conference on natural computation, 2008. ICNC ‘08. vol 5, pp 90–94
7. Liu J, Zhang S, Liu J (2009) A method of plant leaf recognition based on locally linear embedding and moving center hypersphere classifier. In: International conference on intelligent computing ICIC 2009, LNAI 5755, pp 645–651. Springer

8. Ankalaki S, Majumdar J (2015) Leaf identification based on back propagation neural network and support vector machine. In: 2015 International conference on cognitive computing and information processing (CCIP), pp 1–7
9. Purusothaman G, Krishnakumari P (2015) A survey of data mining techniques on risk prediction: heart disease. Indian J Sci Technol 8(12)
10. Vedanayaki M (2014) A study of data mining and social network analysis. Indian J Sci Technol 7(S7):185–187

Review of Wireless Body Area Networks (WBANs)

B. Manickavasagam, B. Amutha and Priyanka Sudhakara

Abstract This comprehensive study guides the researchers to continue research in Wireless Sensor Networks and understanding of patient monitoring systems, protocold, and communication standards etc. This paper covers general wireless body area network (WBAN) architecture, methodologies, communication standards, and challenges to understanding. We summarize the frequency range, bandwidth, channel capacity, and bit rates of different communication standards and look at how to design sensor nodes and coordinator nodes for WBANs.

Keywords WBAN · Sensor network survey · WBAN standard
Vital parameters · WBAN survey · Patient monitoring system (PMS)

1 Introduction

Over the last one and half decades medical and personal monitoring systems research has grown rapidly. Nowadays, most people (both grownups and children) are affected by chronic diseases, but do not have the patience to continue their treatment in hospital due to the hospital atmosphere being offputting for most people. Researchers have identified these problems and are developing patient monitoring system (PMS) [1], which helps to keep the patients in their favored place (either the home or the workplace) with remote surveillance by a doctor or monitoring equipment. The sensor nodes of WBANs are placed or attached on the remote patient's body following either an on-body or in-body approach [2].

B. Manickavasagam (✉) · B. Amutha · P. Sudhakara
School of Computing, SRM University, Chennai 603203, Tamil Nadu, India
e-mail: bmanickavasagam90@gmail.com

B. Amutha
e-mail: bamutha62@gmail.com

P. Sudhakara
e-mail: miyu_priyanka@yahoo.com

A. Kumar and S. Mozar (eds.), *ICCCE 2018*,
Lecture Notes in Electrical Engineering 500,
https://doi.org/10.1007/978-981-13-0212-1_66

Hence, patient's condition are monitored continuously if there is any change in their schedule based approach observation.

The rest of this paper is structured as follows. Section 2 explain vital medical signals and functionalities. Section 3 explain the WBAN architecture tier format. Section 4 describes the topology standards. Section 5 elucidates the IEEE wireless communication standard for sensor networks (SN). Section 6 contains the mode of WBAN communication, and Sect. 7 explains challenges facing WBAN.

2 Physiological Signals and Functionalities

Most WBAN systems use a combination of 2 to 5 vital signs to predict an emergency. They are temperature, measured in degrees Celsius (°C) or Fahrenheit (°F), blood pressure (BP) measured in millimeters of mercury (mmHg), heart rate (HR) measured in beats per minute (bpm), electrocardiography (ECG) and oxygen saturation levels (S_PO_2). In this section, we will discuss the above-mentioned parameters one by one.

2.1 Temperature

One of the basic vital sign is body temperature, or in other words, the body's capacity to generate heat; this temperature variation helps to find the initial stages of disease, for normal body temperature varies from 36.1 to 37.2 °C [3]. The temperature can be measured either by mouth (oral), rectal, skin or temporal (forehead), and ear (tympanic). It reflects the temperature of the body's core and underarm (axillary area) [4, 5].

2.2 Blood Pressure (BP)

Circulation of blood speed upon the blood vessels walls is termed blood pressure (BP). It is classified as systolic and diastolic. The heart contracts and pushes the blood to the rest of the body through arteries and this is known as systolic pressure, while diastolic pressure is the arterial pressure during the rest between heart beats. Normal human blood pressure is 120/80 mmHg (systolic/diastolic). Another form of calculation for blood pressure is mean arterial pressure (MAP) explained in Eq. (1) [6].

$$MAP = \frac{(2 \times diastolic\ Pressure) + Systolic\ Pressure}{3} \quad (1)$$

2.3 Electrocardiogram (ECG)

The passing of electrical impulse signals from the heart's sinoatrial node (upper right part of the heart or right atrium) called the SA node to rest of the heart. Human heartbeat rhythm produces four stage of signal waves, namely P-Waves, QRS Complex, T-Waves, and U-Waves.

- **P-Wave**: A signal passes from the SA node to the atrioventricular (AV) node. The other form of this process is called depolarization.
- **QRS-Complex**: The AV node passes the electrical impulse to the ventricle through AV bundles and branches.
- **T- Wave**: This represents ventricular repolarization
- **U-Wave**: This represents repolarization of the papillary muscle

2.4 Heart Rate (HR)

Another primary vital sign which helps to diagnose a disease is heart rate (HR). It varies depending on the stage of human, from birth to death. In our body there are four major locations to find the HR or pulse. These are the wrist, the inside of the elbow, the side of the neck, and the top of the foot. While measuring the pulse, some factors are considered and they are air temperature, body position, body size, and use of any medication. In [7] human heart rate variation and its effects are described, Fig. 1 shows the one heart beat's PQRSTU sign waveform and ECG graph and units are explained in Fig. 2.

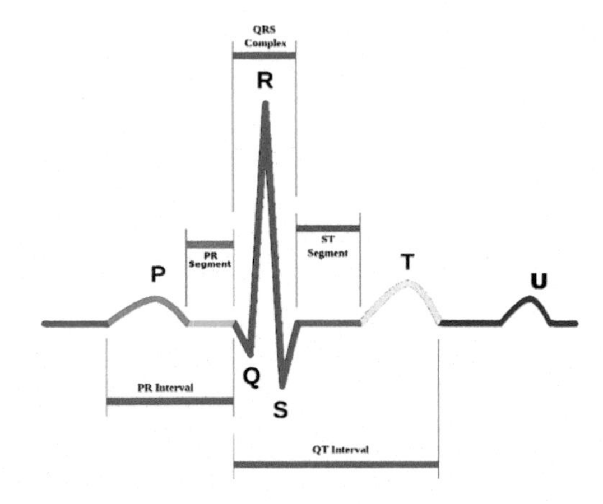

Fig. 1 Electrocardiogram waveform of one heart beat

2.5 Oxygen Saturation (S_PO_2)

This is a method used to find the oxygen level in our body, in other words finding the amount of oxygen-saturated blood hemoglobin in each pumping of blood from each heartbeat. A level of greater than 90% (96–100) of oxygen present in our blood hemoglobin is considered as a normal healthy S_PO_2 level for a human [8, 9].

3 WBAN Architecture

WBANs are categorized into a 3-tier architecture format. These tiers are sensor unit or sensor nodes (SN), gateway or coordinator unit (CU), and doctor's zone, as shown in Fig. 3. Depending upon the application, WBAN may be classified into another three types, namely in-body, off-body, and on-body communication [2, 10, 11].

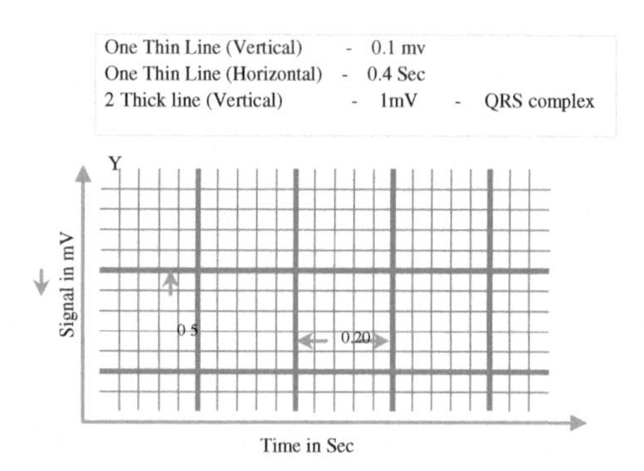

Fig. 2 Sample electrocardiogram graph

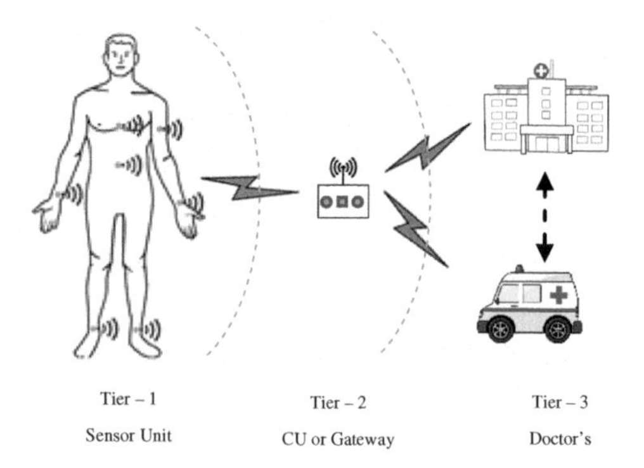

Fig. 3 WBAN architecture

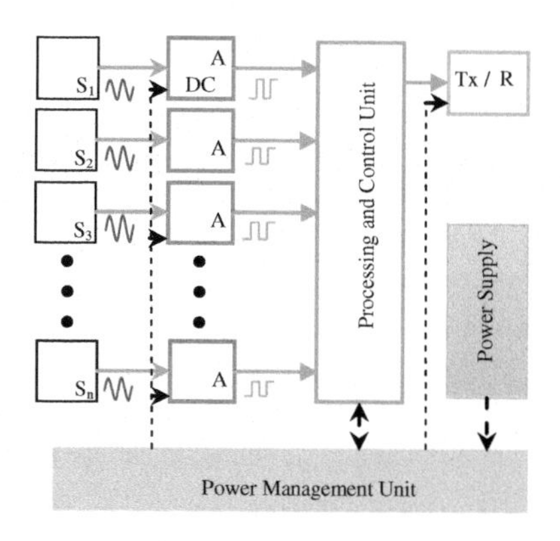

Fig. 4 Sensor unit components

3.1 Sensor Unit

Each sensor unit has different sensor combinations. It acquires or gathers biological signals from the human body. These signals are digitized through an analog-to-digital converter (ADC) and transmitted to a sink node through the processing and transceiver unit. A low voltage battery power supply is required for this entire process and it is managed by a power management unit [11] shown in Fig. 4,

where, S1, 2, 3, … n: sensors, Tx/Rx: transceiver and ADC: analog to digital convertor.

All sensor units transmit data to the sensor unit head present on the same floor or directly transmit to the gateway or coordinator present in tier 2 through near field communication (NFC).

3.2 Coordinator Unit

This unit acquires the all sensor node (SN) raw vital signals [12], processes them and sets the priority level based upon human vital sign thresholds [13]. It transmits the information to tier 3 through wi-fi, mobile communication (GPRS, 3G, 4G), and WiMAX. If a vital sign exceeds the threshold, an alert message is immediately sent to the doctors as well as to an ambulance unit. The coordinator acts as the decision maker for creating priority signals and sending them to the alert notification and immediate diagnosis system (IDS) to avoid unexpected situation.

3.3 Doctor's Zone

This acts as the centralized node, all patient's vital sign information have stored in this location, which helps to manage the patient health details, complete and later diagnosis process, and remote patient's continuous monitor, and it also communicates with the emergency care unit (ambulance service) [11, 14].

4 WBAN Topology

4.1 Star Topology

Sensor nodes communicate to a hub, called the coordinator, or the cluster head of a particular patient's cluster group. In this approach all sensor nodes communicate to the CH within the transmission period without affecting the transmission cycle of other sensor nodes. All sensor nodes are receiving the duty cycle information from Cluster Head. Based on this information cluster group sensor nodes are transferring patient information to CH.

4.2 Mesh Topology

Mesh topology used in both sensor node to coordinator communication, and coordinator to data server communication. In scenario one, sensor nodes communicate to the coordinator through single hop or multihop communication. In scenario two, the cluster head sends the details to a data server or hospital by the shortest and most reliable transmission path in an effective way through multihop communication.

5 WBAN Communication Standard

WBAN utilizes three types of communication technologies, namely short-, medium-, and long-range transceivers. This range of communications utilizes IEEE 802 standard [15], and the following section explains major wireless communication standards and the specifications used for WBAN. Table 1 describes existing research work's communication topology, feature and standards.

Table 1 Existing research work and its methodology and communication standards

Ref. no.	Sensors	Topology	Impact	Standard
[1]	ECG, respiration, temperature, pulse	Point to point	–	IEEE 802.11 b
[3]	Temperature	Point to point	–	IEEE 802.15.4
[11]	Three-Lead ECG, S_PO_2 and heart rate	Cluster		
[16]	Simulation	Multihop	Reduces latency, energy efficient, packet loss rate	CSMA/CA
[17]	Realtime	Point to point	Fall detection, criticality identification	–
[37]	Simulation	Cluster and mesh network	Less time delay, energy consumption	IEEE 802.15.4
[18]	Simulation	–	Energy efficiency, data rate	IEEE 802.15.6
[19]	–	Multihop	Routing, energy efficiency	IEEE 802.15.6
[20]	–	–	Energy efficiency	–
[21]	–	Star topology	–	–
[31]	–	Star topology	–	–
[22, 23]	ECG, S_PO_2 Blood Pressure	Mesh topology	–	–
[24]	Simulation	Star topology	–	–

5.1 IEEE 802.11

IEEE 802.11 is a standard developed for local area networks (LAN), in 1997. The 802.11 physical layer supports frequency hopping spread spectrum (FHSS), direct sequence spread spectrum (DSSS), and infrared (IR). FHSS and DSSS utilize the 2.4 GHz industrial, scientific, and medical (ISM) band [25]. This standard is offered for tier 2 components (see Fig. 3).

5.2 IEEE 802.15.1

From 1989 onward most PMS communicated through Bluetooth technology. Bluetooth technology operates in the 802.15.1 standard within a short range over the ISM band at 2.4 GHz. This communication range varies from 1 m to below 100 meters. The bandwidth is approximately 720 Kbps to more than 20 Mbps and consumes 1 mV for 10-m range communication. In [26], there is a discussion about utilizing Bluetooth communication technology for a patient monitoring system in a home environment in order to detect disease in its early stages.

5.3 IEEE 802.15.4

IEEE 802.15.4 is a standard developed for low-range communication purposes. It helps in industrial monitoring, natural disaster sites, agriculture, and home automation and networking. It also offers a dual physical layer (PHY) that utilizes the DSSS method for low-duty cycles and low-power operation purposes. One PHY operates in the 2.4 GHz ISM band with 250 Kb/s in global, while the other PHY operates in the 868/916 MHz band for specific operation. The 916 MHz band is used in the United States (US), and the 868 MHz band is used by European devices at 20 Kb/s and 40 Kb/s respectively [27, 28].

5.4 IEEE 802.15.6

The final version of IEEE 802.15.6 was released in 2012 to handle the main issues in WBAN, like quality of service (QoS), low energy communication, maximum data rate, high reliability, lower error rate, etc. It supports ultra wide band (UWB), human body communication (HBC), and narrow band (NB) physical frequencies for communication with three different level of security mechanism. These are level 0: communication of unsecured, level 1: only authentication process, and level 2: encryption with authentication. IEEE 15.6 standard properties like, different communication frequencies, channel capacity, data rate and bandwidth capacity, which are all explained in references [29, 30].

6 Communication Scenarios

Sensor nodes communicate to a sink node or hospital server in different circumstances through the sensor network coordinator, and vice versa. The following subsections explaining the different communication scenarios.

6.1 Communication Based on Threshold Values

In this scenario the coordinator converts the analog vital signals into digital data which are acquired from the SN and have been compared with a normal patient's health information. This normal health information is obtained form different patients' health histories by a stochastic method, i.e. patients' gender, age, home environment, job, and any regular activities they carry out. After comparison, if the situation is critical then the coordinator sends the data to the data center, DZ, and the nearest care taker for immediate remedy or any other services required [31]. It is normal for the coordinator to send the details to the data center for future analysis of the patient.

6.2 Communication Based on Scheduling Algorithm

Using this scenario, the SNs' gateway transmits patients' vital information at a particular time interval which has been allotted by one of the controllers. Sender transferred the packets to sink and if it gets back acknowledgement from sink then the particular sender goes to sleep mode till next slot that particular goes to the sleep mode till next scheduling slot. In some critical situations the SNs' coordinator broadcasts the emergency beacon message to the other coordinators to update the scheduling algorithm [31]. Once emergency data have been transferred to the DZ, the entire network resume the previous scheduling approach.

7 Power Consumption

7.1 Energy Efficient Method

While communicating, sensor nodes consume more power in the first three layer levels, namely level 3: network layer (NWK), level 2: medium access control (MAC), and level 1: physical layer (PHY) [32]. To avoid more power consumption in the PHY, sensor networks are capable of transmitting data in the ultra-low power radio frequency band and adopt the periodic scheduled transmission tactic for data transmission instead of continuous transmission except in circumstances of critical and chronic disease monitoring. Radio turn off in no packet transmit and receive time and wake up whenever packet wants to transmit or receive depending up on scheduling or MAC slot allocation and communication scenarios (Sect. 6). Using this tactic we can reduce the energy consumption of sensor networks'. Network layer manages end to end delivery and packet routing, it selects the low cost shortest path for source and destination communication process. Due to this shortest path selection, we can reduce the packet delay, packet loss and more power consumption. In case, network layer selects the same path for n no of nodes communication at a time it leads to packet error. For example, during transmission it is not possible to ignore even one packet failure because in medical applications both reliability and QoS are important factors. Hence, the sensor node retransmits the packet to the sink node. If it's one hop or P2P communication then it affects only a particular sensor, but in the case of multihop communication it will reduce the lifetime of the entire network. In [33–35] authors are explained the energy efficient based data transmission methodologies and channel allocation strategies.

7.2 Self-powering Method

A normal human body generate heats at 20 mW/cm^2 [32]. This will vary depending on the climate, human activity, and clothing material [36]. Using a thermoelectric generator, we can convert human body heat to electrical energy. In [33], a thermoelectric energy harvester attached to a shirt, generated electricity in three situations, which are sitting in the office, outdoor bike parking and outdoors. The generated electrical power was approximately 1 mW @ 24 C, 1mW @ 17 C and 4 MW @ 17–19 °C respectively. Apart from this method, we can generate electricity using air flow, pressure, and vibration.

8 Routing

To increase network lifetime, QoS, reliable communication, and to avoid high traffic congestion, routing plays the main role. Routing is the functionality of network layer it guides to which way the information wants to send from source to destination in the efficient method. Based on the critical situation, sensor network power level, network traffic, communication range, reliability, and network cost, the network layer chooses the best route path for packet transmission [37]. Table 1 summarizes the few existing WBAN-aided PMS and their methodology, standard protocol, and topology.

9 Conclusion

In this review paper, we discussed WBAN architecture framework and its functionalities, how communication occurs between them, and the different communication standards developed and incorporated into PMS and WBAN architecture. Basic vital signs are explained along with their variations that can be used to predict a remote patient's health condition.

References

1. Jiehui J, Zhang J (2007) Remote patient monitoring system for China. In: IEEE Potentials, vol 26, no 3, pp 26–29, May–June 2007. https://doi.org/10.1109/mp.2007.361641
2. IEEE Standard for local and metropolitan area networks—Part 15.6: wireless body area networks. In: IEEE standard 802.15.6–2012, pp 1–271, Feb 29, 2012. https://doi.org/10.1109/ieeestd.2012.6161600
3. Ling THY, Wong LJ, Tan JEH, Kiu KY (2015) Non-intrusive human body temperature acquisition and monitoring system. In: 2015 6th international conference on intelligent

systems, modelling and simulation, Kuala Lumpur, 2015, pp 16–20. https://doi.org/10.1109/isms.2015.17
4. http://www.disabled-world.com/calculators-charts/degrees.php
5. http://baptisthealthblog.com/2015/04/facts-about-fever/#.WD-8MdV97Dd
6. Libii JN (1988) Mean arterial pressure and average blood pressure. In: Proceedings of the 1988 fourteenth annual northeast bioengineering conference, Durham, NH, 1988, pp 208–210. https://doi.org/10.1109/nebc.1988.19386
7. Amutha B, Manickavasagam B, Patnaik A, Karthick N (2015) Erection of comprehensive wellness programme for global healthcare monitoring system using AODV protocol with data clustering schema. Indian J Sci Technol 8(17). https://doi.org/10.17485/ijst/2015/v8i17/65446
8. Lazareck L, Tarassenko L (2006) Detection of apnoeic and breathing activity through pole-zero analysis of the SpO_2 signal. In: 2006 international conference of the IEEE engineering in medicine and biology society, New York, NY, 2006, pp 6573–6576. https://doi.org/10.1109/iembs.2006.260891
9. de Kock JP (1991) Pulse oximetry: theoretical and experimental models. University of Oxford—D.Phil. Thesis, Michaelmas Term
10. Movassaghi S, Abolhasan M, Lipman J, Smith D, Jamalipour A (2014) Wireless body area networks: a survey. In: IEEE Commun Surv Tutor 16(3):1658–1686. https://doi.org/10.1109/surv.2013.121313.00064
11. Lin YH, Jan IC, Ko PI, Chen YY, Wong JM, Jan GJ (2004) A wireless PDA-based physiological monitoring system for patient transport. IEEE Trans Informat Technol Biomed 8(4):439–447. https://doi.org/10.1109/titb.2004.837829
12. Ntouni GD, Lioumpas AS, Nikita KS (2014) Reliable and energy efficient communications for wireless biomedical implant systems. IEEE J Biomed Health Inf 18(6):1848–1856
13. Chowdhury MA, Mciver W, Light J (2012) Data association in remote health monitoring systems. IEEE Commun Mag 50(6):144–149. https://doi.org/10.1109/MCOM.2012.6211499
14. Varady P, Benyo Z, Benyo B (2002) An open architecture patient monitoring system using standard technologies. IEEE Trans Inf Technol Biomed 6(1):95–98. https://doi.org/10.1109/4233.992168
15. https://standards.ieee.org/about/get/802/802.11.html
16. Liu B, Yan Z, Chen CW (2013) MAC protocol in wireless body area networks for E-health: challenges and a context-aware design. IEEE Wirel Commun 20(4):64–72. https://doi.org/10.1109/MWC.2013.6590052
17. Wang J, Zhang Z, Li B, Lee S, Sherratt RS (2014) An enhanced fall detection system for elderly person monitoring using consumer home networks. IEEE Trans Consum Electron 60 (1):23–29. https://doi.org/10.1109/TCE.2014.6780921
18. Thotahewa KMS, Khan JY, Yuce MR (2014) Power efficient ultra wide band based wireless body area networks with narrowband feedback path. IEEE Trans Mob Comput 13(8):1829–1842. https://doi.org/10.1109/TMC.2013.120
19. Wan J, Zou C, Ullah S, Lai CF, Zhou M, Wang X (2013) Cloud-enabled wireless body area networks for pervasive healthcare. IEEE Netw 27(5):56–61. https://doi.org/10.1109/mnet.2013.6616116
20. Liu J, Li M, Yuan B, Liu W (2015) A novel energy efficient MAC protocol for wireless body area network. China Commun 12(2):11–20. https://doi.org/10.1109/CC.2015.7084398
21. Malhi K, Mukhopadhyay SC, Schnepper J, Haefke M, Ewald H (2012) A Zigbee-based wearable physiological parameters monitoring system. IEEE Sens J 12(3):423–430. https://doi.org/10.1109/JSEN.2010.2091719
22. Leonov V (2013) Thermoelectric energy harvesting of human body heat for wearable sensors. IEEE Sensors J 13(6):2284–2291. https://doi.org/10.1109/JSEN.2013.2252526
23. Ivanov S, Foley C, Balasubramaniam S, Botvich D (2012) Virtual groups for patient WBAN monitoring in medical environments. IEEE Trans Biomed Eng 59(11):3238–3246. https://doi.org/10.1109/TBME.2012.2208110

24. Ivanov S, Botvich D, Balasubramaniam S (2012) Cooperative wireless sensor environments supporting body area networks. IEEE Trans Consum Electron 58(2):284–292. https://doi.org/10.1109/TCE.2012.6227425
25. Crow BP, Widjaja I, Kim LG, Sakai PT (1997) IEEE 802.11 wireless local area networks. IEEE Commun Mag 35(9):116–126. https://doi.org/10.1109/35.620533
26. Cheng HT, Zhuang W (2010) Bluetooth-enabled in-home patient monitoring system: early detection of Alzheimer's disease. IEEE Wirel Commun 17(1):74–79. https://doi.org/10.1109/MWC.2010.5416353
27. Callaway E et al (2002) Home networking with IEEE 802.15.4: a developing standard for low-rate wireless personal area networks. IEEE Commun Mag 40(8):70–77. https://doi.org/10.1109/MCOM.2002.1024418
28. Gutierrez JA, Naeve M, Callaway E, Bourgeois M, Mitter V, Heile B (2001) IEEE 802.15.4: a developing standard for low-power low-cost wireless personal area networks. IEEE Netw 15 (5):12–19. https://doi.org/10.1109/65.953229
29. Aranki D, Kurillo G, Yan, P, Liebovitz DM, Bajcsy R (2016) Real-time tele-monitoring of patients with chronic heart-failure using a smartphone: lessons learned. IEEE Trans Affect Comput 7(3) 206–219. https://doi.org/10.1109/taffc.2016.2554118
30. Cavallari R, Martelli F, Rosini R, Buratti C, Verdone R (2014) A survey on wireless body area networks: technologies and design challenges. In: IEEE communications surveys and tutorials, vol 16, no 3, pp 1635–1657, Third Quarter 2014. https://doi.org/10.1109/surv.2014.012214.0000
31. Omeni O, Wong ACW, Burdett AJ, Toumazou C (2008) Energy efficient medium access protocol for wireless medical body area sensor networks. IEEE Trans Biomed Circ Syst 2 (4):251–259. https://doi.org/10.1109/TBCAS.2008.2003431
32. Misra S, Moulik S, Chao HC (2015) A cooperative bargaining solution for priority-based data-rate tuning in a wireless body area network. IEEE Trans Wirel Commun 14(5):2769–2777. https://doi.org/10.1109/TWC.2015.2393303
33. Yi C, Wang L, Li Y (2015) Energy efficient transmission approach for WBAN based on threshold distance. IEEE Sens J 15(9):5133–5141. https://doi.org/10.1109/JSEN.2015.2435814
34. Rezvani S, Ghorashi SA (2013) Context aware and channel-based resource allocation for wireless body area networks. IET Wirel Sensor Syst 3(1):16–25. https://doi.org/10.1049/iet-wss.2012.0100
35. Wang F, Hu F, Wang L, Du Y, Liu X, Guo G (2015) Energy-efficient medium access approach for wireless body area network based on body posture. China Commun 12(12):122–132. https://doi.org/10.1109/CC.2015.7385520
36. Johny B, Anpalagan (2014) A body area sensor networks: requirements, operations, and challenges. IEEE Potent 33(2):21–25. https://doi.org/10.1109/mpot.2013.2286692
37. Manfredi S (2012) Reliable and energy-efficient cooperative routing algorithm for wireless monitoring systems. IET Wirel Sensor Syst 2(2):128–135. https://doi.org/10.1049/iet-wss.2011.010

Association Rule Mining Using an Unsupervised Neural Network with an Optimized Genetic Algorithm

Peddi Kishor and Porika Sammulal

Abstract The best known and most widely utilized pattern finding algorithm in data mining applications is association rule mining (ARM). Extraction of frequent patterns is an indispensable step in ARM. Most studies in the literature have been implemented on the concept of support and confidence framework utilization. Here, we investigated an efficient and robust ARM scheme based on a self-organizing map (SOM) and an optimized genetic algorithm (OGA). A SOM is an unsupervised neural network that efficaciously produces spatially coordinated internal feature representations and detected abstractions in the input space and is the most efficient clustering technique that reveals conventional similarities in the input space by performing a topology maintaining mapping. Hence, a SOM is utilized to generate accurate clustered frequency patterns and an OGA is used to generate positive and negative association rules with multiple consequences by studying all possible patterns. Experimental analysis on various datasets has shown the robustness of our proposed ARM in comparison to traditional rule mining approaches by proving that a greater number of positive and negative association rules is generated by the proposed methodology resulting in a better performance when compared to conventional rule mining schemes.

Keywords Data mining · Association rule mining · Frequent patterns
Positive and negative association rules · Self-organizing maps (SOM)
Optimized genetic algorithm (OGA)

P. Kishor (✉)
R&D Cell, JNTUH, Hyderabad, India
e-mail: kishorpeddi25@gmail.com

P. Sammulal
Department of CSE, JNTUH College of Engineering, Jagtial, India

A. Kumar and S. Mozar (eds.), *ICCCE 2018*,
Lecture Notes in Electrical Engineering 500,
https://doi.org/10.1007/978-981-13-0212-1_67

1 Introduction

Data mining is the name given to the process of mining valuable significant data from Data Warehouse. It is the name given to the way toward mining profitable noteworthy information from vast volumes of normal data. The selected information must be precise as well as intelligible, conceivable, and simple to comprehend. There is a considerable number of data mining tasks, for example, ARs, sequential patterns, classification, clustering, and time series, etc., and there have been many techniques and algorithms for these assignments and numerous types of data in data mining. At the point when the data contain never-ending items, it turns out to be difficult to mine the data and some unique techniques are required. Association mining is used to discover helpful patterns and connection between items found in the database of transactions. For instance, consider the business database of a music CD store, where the records speak to clients and the behavior speak to Music CD. The mined patterns are the arrangement of music CDs mostly every now and again purchased together by the client. It could be that 70% of the general population who purchase old melody cods also purchase chug cods. The store can utilize this data for future deals, self restore of records and so forth. There are numerous application areas for ARM techniques, and these incorporate index configuration, store design, client division, and media transmission etc. (Fig. 1).

ARM has been considered by data mining research groups since the mid-90s, as a method for unsupervised, exploratory data investigation. Association rules were first presented by Agrawal et al. (1993) as a market basket analysis, however, from that point forward they have been connected to numerous other application domains, including science and bioinformatics. An association rule suggests the concurrence of various items in a bit of a transactional database. The objective of this exploratory data examination is to furnish the leader with important information about a specific area demonstrated by a transactional database. The successive presence of at least two items in a similar transaction infers a relationship between them. For instance, the presence of bread and margarine in a similar wicker bin

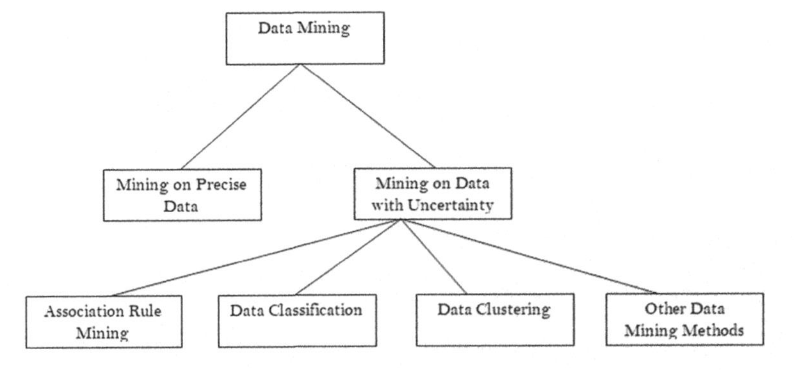

Fig. 1 General structure of data mining processes

infers a conceivable purchasing product design that can be additionally researched so as to enhance the offers of the two items. In addition the conjunction of high articulation estimations of various qualities in similar transactional tests shows a conceivable co-articulation example of these qualities. On the other hand, the uncommon or the completely non-conjunction of two items could likewise infer a negative Association (e.g. a common prohibition) among them. The vast majority of the examination has just call attention to for positive Association rule yet negative Association govern likewise assume critical part in Data mining assignment. Be that as it may, mining negative Association rules is a troublesome assignment, because of the way that there are essential contrasts amongst positive and negative Association rule mining. We will influence consideration on two key issues in negative relationship to rule mining:

(1) How to successfully discovering for negative regular item sets
(2) How to adequately find negative association rules

In spite of the fact that significance of negative Associations, just some of scientists proposed a calculation to mine these sorts of Associations rules. In this research paper, we will go to actualize a novel self-organizing map based frequent pattern generation with positive and negative patterns and furthermore created Association rules utilizing genetic algorithm.

2 Related Works

Numerous algorithms for mining association rules and others that broaden the idea of ARM have been proposed. Agrawal and Srikant [1] proposed Apriori, the main calculation for mining association rules is it continues by distinguishing the regular individual items in the database and extending them to bigger and bigger item sets as long as those item sets show up adequately frequently in the database. It stands out amongst the most prominent data mining algorithms and will be discussed in more detail later. Around the same time Mannila et al. [2] found a variety of Apriori, the OCD calculation. Srikant and Agrawal [3] introduced the issue of mining for summed up Association rules. These guidelines use item scientific classifications (idea progressions) keeping in mind the end goal to find additionally fascinating tenets. Savasere et al. [4] presented this sort of issue. Negative Associations identify with the issue of discovering decides that suggest what items are not liable to show up in an transactional when a specific arrangement of items appears in the transactional. The approach of Savasere et al. requests the presence of a scientific classification and depends on the suspicion that items having a place with a similar parent of scientific categorization are relied upon to have comparative sorts of relationship with different items. The large number of algorithms proposed from that point forward, either enhance productivity, for example, FPGrowth [5] and Eclat [6], or address distinctive issues from different application domains, for example, spatial [7], transient [8] and between value-based association rules [9]. One of the main issues in ARM is the huge number of regularly uninteresting standards extricated. Some methodologies that depend on chains of importance attempt to manage this issue. The sort of negative associations that have been considered concern the mining of fundamentally unrelated items [10, 11, 12]. Kou et al. utilized a particle swarm optimization (PSO) calculation for ARM. The tried their proposed model on a dataset from Microsoft and contrasted it with a genetic calculation. Their outcomes demonstrated that their proposed calculation get better found standards in contrast and Genetic calculation [13]. Djenouri et al. joined Artificial Bee Colony optimization calculation and Tabu Search for mining Association rules. They called their calculation as HBO-TS in which Artificial Bee Colony was utilized for making assorted variety and Tabu Search to look quick. Their outcomes demonstrated that in spite of the fact that this technique had a few challenges in parameters settings, however it can make great guidelines in a worthy time [14]. Bhugra et al. utilized a biogeography-based optimization (BBO) calculation for ARM. They had altered relocation part of this calculation and their outcome demonstrated a decent execution [15]. In another exploration, Ivancevic et al. recognized hazard factors for early youth caries [16]. They made a dataset incorporate of %10 of youngsters matured under 7 years old in Serbia and recognized hazard factors by utilizing Association rules. Kargarfard et al. utilized a mix of classification and association rules in a dataset comprised of 7,000 records, to identify flu infection [17]. They trusted that joining classification algorithms and association rules can prompt improvement of selection assist and master frameworks. Cheng et al. proposed a strategy to find valuable information from previous history and imperfections for development supervisors [18]. Their approach used genetic calculation and after investigation, comes about indicated balanced connection between found factors and existed abandons. Kuo and Shih utilized an ant colony system for finding association rules in a social insurance protection dataset from Taiwan. They put some multi-dimensional requirements on this enormous dataset. Their outcomes demonstrated that their proposed strategy performed better than Apriori calculation. Wang et al. utilized association principles to dissect a dataset of array consumption in compound process [19]. They acquainted another technique to discover association rules. Parkinson et al., use Association rules so as to study and monitor facts in structure records [20]. Their approach comprised of two stages, examination the NTFS to find allowance and using association principle to distinguish unlawful things. In some different method, authors have attempted to utilize metaheuristic algorithms for mining association rules. Martin et al. exhibited another metaheuristic technique using the genetic calculation named NICGAR which alludes to niching genetic algorithm [21]. Their strategy accomplishes more in less time.

3 Proposed Framework

Here, we investigated an efficient and robust ARM scheme based on a self-organizing map (SOM) and an optimized genetic algorithm (OGA). SOM is an unsupervised neural network that efficaciously produces spatially coordinated internal feature representations and detected abstractions in the input space and is the most efficient clustering technique that reveals conventional similarities in the input space by performing a topology maintaining mapping. Hence, a SOM is utilized to generate accurate clustered frequency patterns and an OGA is used to generate positive and negative association rules with multiple consequents by studying all possible patterns. The proposed algorithm is as follows (Fig. 2):

Algorithm: Association Rule Mining (ARM) Step 1: Start
Step 2: Load the sample data set from the memory
Step 3: Initialize the minimum support count and minimum confidence
Step 4: Now, apply the SOM algorithm to clusters of the input data set and thereby get the frequent item sets from the given input dataset
Step 5: Now, generate positive and negative patterns from the obtained frequent patterns in step 3
Step 6: Apply OGA to find the number of positive and negative association rules
Step 7: End

3.1 *Self-organizing Map (SOM)*

A self-organizing map (SOM) is a sort of neural system that uses the rules of focused or unsupervised learning. In unsupervised learning there are no data about a desired output just as in monitored learning. This unsupervised learning approach characterize data by a topology safeguarding mapping of high dimensional input patterns into a lower-dimensional arrangement of output clusters. These clusters compare to every now and again happening patterns of focus among the source data. Because of its straightforward structure and learning system SOM has been effectively utilized as a part of different applications and it has turned out to be one of the successful clustering techniques. The helpful properties of SOM said above have prompted us to explore whether the technique can be connected to the issue of finding frequent patterns in the Association rule method. In particular, it should be resolved whether the idea of finding regular patterns can be reciprocally substituted with the idea of discovering clusters of frequently occurring patterns in the data. There is a connection between the task of isolating continuous data from rare patterns and the responsibility of discovering clusters in data, as a group could speak to a specific regularly happening design in the data. A cluster would, for this

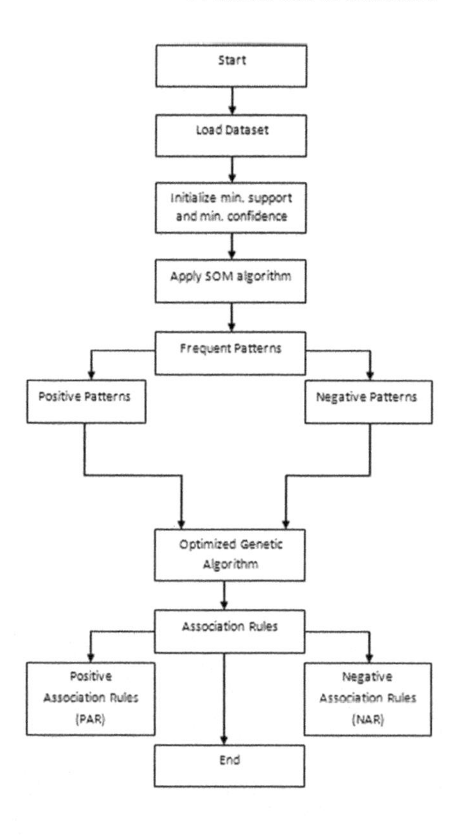

Fig. 2 Flow chart of the proposed system

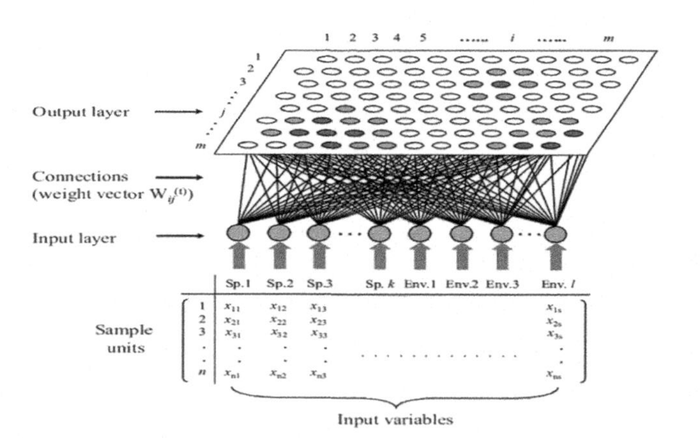

Fig. 3 Basic structure of a self-organizing map

situation, relate to an example, and henceforth it would be intriguing to see whether a group set acquired from a clustering procedure can compare to an arrangement of successive patterns obtained through the SOM approach (Fig. 3).

A cluster would in this case correspond to a pattern, and hence it would be interesting to see whether a cluster set obtained from a clustering technique can correspond to a set of frequent patterns obtained through the SOM approach.

Proposed SOM + FP + OGA Algorithm

1: Set min_con, min_support;
2: nodes are D_j.
3: set decay rate.
4: set alpha
5: set minimum alpha
6: while alpha is > minimum alpha
{
7: for each input vector
{
8: for each node x
{
9: compute: $D_j = \sum_i (w_{ij} - x_i)^2$
10: find index j such that Dj is a minimum. compute Euclidean distance
11: update the weights for the vector at index j and its neighbors:
12: $w_{ij}(new) = w_{ij}(old) + \alpha[x_i - w_{ij}(old)]$
}
}
13: reduce alpha
14: optionally, reduce radius of topological neighborhoods at specific times.
15: generateFrequentPatterns P = I1 $\wedge$ I2 $\wedge$… In (itemset);
16: if(P > = min_support)
17: pattern = "positive"
18: else
19: pattern = "Negative";
20: geneticAssociationRules R = A $\Rightarrow$ B;
21: setBestSolution(R)
22: List_of_solution = FrequentPattern
23: Generate Population
24: R = calculateFitness()//individual fitness
25: if(R best solution with fitness > 0){
26: Support(A $\Rightarrow$ B) = P(A U B)
27: Confidence(A $\Rightarrow$ B) = P(B|A) = posterior probability
30: If(confidence >=min_conf)
31: Rule = "positive";
32: Else
33: Rule = "Negative";
}

3.2 Association Rule

When all is said in done, the association rule is a statement of the form X =>Y, where X is the predecessor and Y is the result. The association rule indicates how often Y has happened if X has just happened relying upon the support and certainty value.

- Support: This is the probability of the item or item sets in the given value-based data base: Support(X) = n(X)/n where n is the aggregate number of transactions in the database and n(X) is the quantity of transactions that contains the item set X.
- Confidence: This is the restrictive probability, for an association rule X => Y and is characterized as confidence(X => Y) = support (X and Y)/support (X).

All the conventional ARM algorithms were created to discover positive relationships between items. Positive associations allude to the relationship between items existing in transactions. Notwithstanding the positive associations, negative associations can provide important data. Underneath there are numerous situations where nullification of items assumes a noteworthy part. For cases: If I examine in SMIT then I don't consider in HIMALAYAN STORE. In the event that I am a male then I am not a female. A negative association rule is a ramification of the form $X \Rightarrow \neg Y$ where X and Y are item sets and $X \cap Y = \Phi$. Mining association guidelines can be separated into the accompanying two sub-issues:

1. Generating all item sets that have support more prominent than, or equivalent to, the client-indicated minimal support. That is, producing all expansive item sets.
2. Generating every one of the guidelines that have least certainty

We can generate the association rule when more than one subsequent item is produced by the accompanying technique:

1. Find the rule in which the number of consequents = 1.
2. For the given tenets $p(x \rightarrow y)$ and $p(x \rightarrow z)$, the rule $p(x \rightarrow yz)$ is created by the crossing point of both the association rules and obtains another rule $p(x \rightarrow yz) = p(xyz)/p(x)$.

3.3 Genetic Algorithm (GA)

Genetic algorithms are a group of computational models inspired by evolution theory of Darwin. As per Darwin the species which are the fittest and can adjust to a changing environment can survive; the others tend to wither away. Darwin likewise expressed that "the survival of a life form can be kept up through the procedure of proliferation, hybridisation and change". GA's essential working instrument is as follows: the calculation begins with an arrangement of arrangements (called as chromosomes) called populace. Arrangements from one populace are taken and used to frame another populace (proliferation). This is driven by positive thinking, that the new populace will be better than the old one. This is the reason that they are frequently named hopeful hunt algorithms. The regenerative prospects are disseminated such that those chromosomes which speak to a superior answer for the objective issue are given a larger number of opportunities to recreate than those which speak to second rate arrangements. They seek through a tremendous blend of parameters to locate the best match. For instance, they can seek through various combination of resources and outlines to locate the ideal mix of both which could bring about a more grounded, lighter and in general, better last item. A genetic calculation is prepared to do viably looking through the issue area and tackling complex issues by re-enacting normal advancement. It perform seek and give close ideal answers for target capacity of an optimization issue. When all is said in done the data formed by Association rule mining algorithms like priori, partition, pincer-seek, incremental, border algorithm and so on, does not consider negation event of the property in them and furthermore these principles have just a single quality in the subsequent part. By utilizing Genetic Algorithm (GAs) the framework can anticipate the guidelines which contain negative qualities in the produced rules alongside more than one attribute in ensuing part. The real preferred standpoint of utilizing GAs in the disclosure of confidence decides is that they perform worldwide pursuit and its versatile nature is less contrasted with different algorithms as the genetic calculation depends on the greedy approach. The main aim of this is to discover all the conceivable streamlined guidelines from the given data set utilizing genetic calculation.

4 Results and Discussion

In this section, we present some of our preliminary experimental results that validate the correctness of the proposed technique and indicate some appropriate ways of using SOM for frequent pattern extraction. The section is split into three parts which are separate from each other with respect to the set of experiments performed. The aim of the first part is to show how a complete set of frequent patterns can be extracted using SOM. Then the set of positive and negative patterns generated is shown in the second part. Finally, the association rules generated by using OGA operators are shown in the third part. Results shows that the proposed algorithm is more useful and faster in discovering association rules in a transactional dataset and it also reaches the answer with less repetition. It obtains better answers compare with other algorithms like Apriori, FP-Growth, genetic algorithms and PNARYCC (Positive and Negative Association Rule mining using Yules Correlation Coefficient method). We have compared our model with FP-Growth, Apriori & BBO meta-heuristic algorithm. We tested our algorithm with various values of minimum support and minimum confidence. Figure 4a demonstrates the analysis of the proposed ARM scheme with constant confidence and variable support values, and we analyzed how the number of association rules (AR) vary

(a)

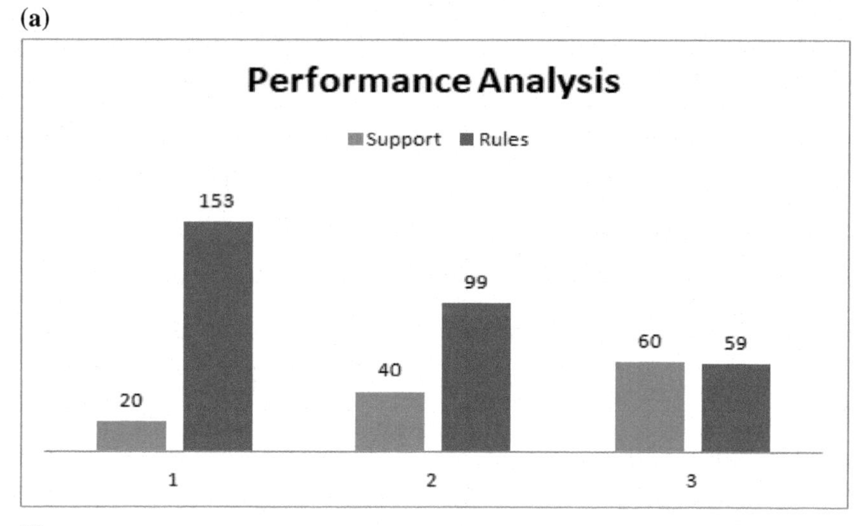

(b)

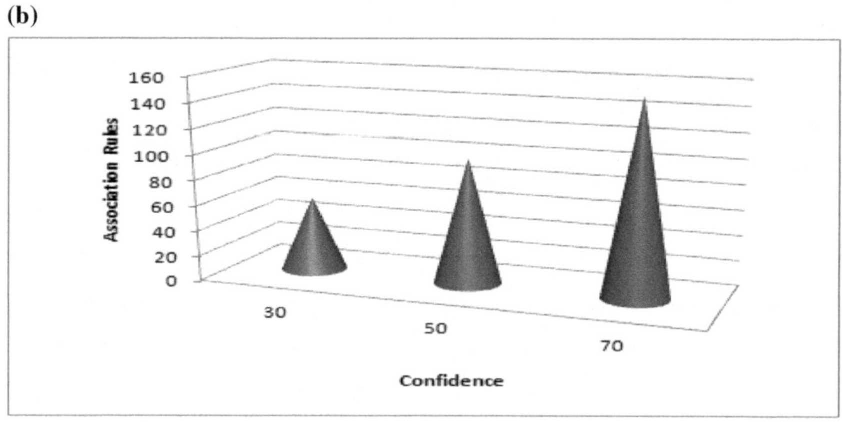

Fig. 4 Performance analysis of the proposed ARM scheme with **a** confidence = 20 with variable support and **b** support = 30 with variable confidence

according to the given inputs. For constant support with variable confidence, the performance of association rules generation is given in Fig. 4b.

Figure 5 show the number of positive association rules (PAR) and negative association rules (NAR) with the generation of association rules for the proposed ARM scheme. As we can see in Fig. 6, the GA created better rules than BBO and FP-Growth, but it was not as good as Apriori. The algorithm proposed in [22] performed well with more rules over conventional BBO, FP-Growth, GA, and even Apriori. Although both FP-Growth and Apriori algorithms are successful in some situations, they use a lot of memory. According to our experimental analysis, our proposed method has performed in a superior way compared with GA, BBO, Apriori, FP-Growth, and even PNARYCC.

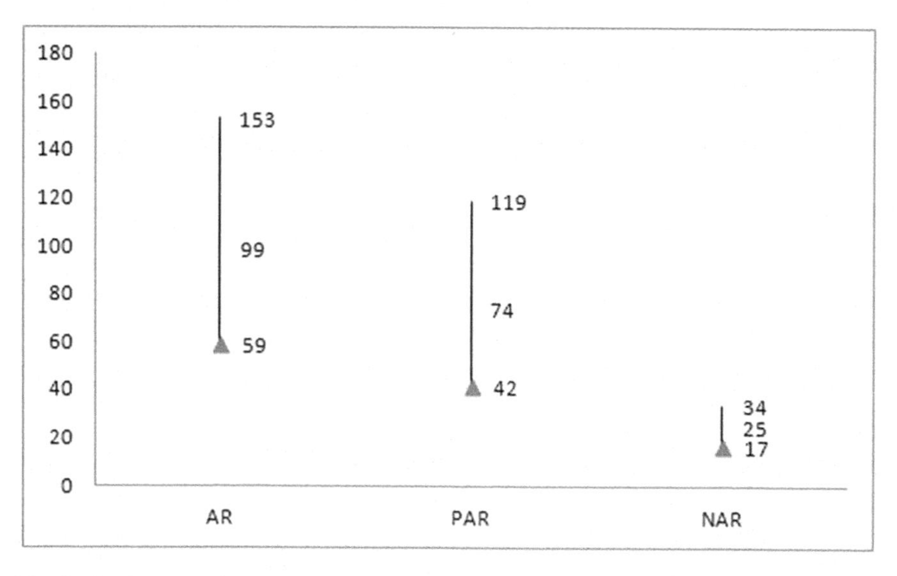

Fig. 5 Obtained association rules with the number of positive and negative association rules for the proposed ARM scheme

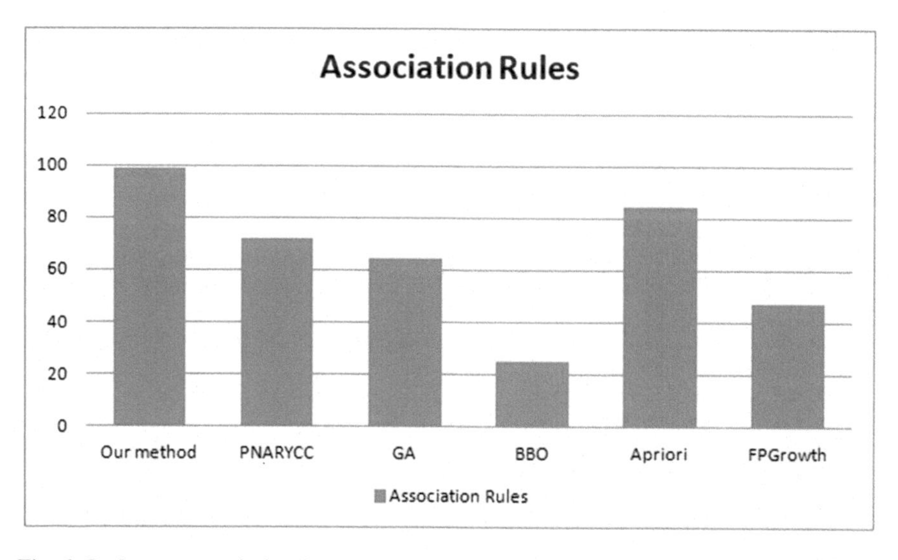

Fig. 6 Performance analysis of the proposed ARM for minimum support (S) = 30 minimum confidence (C) = 20 with conventional ARM algorithms

5 Conclusions

In this study, we utilized an optimized hybrid SOM-based genetic algorithm for the purpose of mining association rules from large transactional datasets. SOM is an unsupervised neural network that efficaciously produces spatially coordinated internal feature representations and detected abstractions in the input space and is the most efficient clustering technique that reveals conventional similarities in the input space by performing a topology maintaining mapping. Hence, a SOM is utilized to generate accurate clustered frequency patterns and OGA is used to generate positive and negative association rules with multiple consequents by studying all possible patterns. Results showed that the proposed algorithm is more useful and faster in discovering association rules from transactional datasets. In addition it reaches the answer with less repetition and it obtains better positive and negative association rules in comparison with other conventional algorithms, such as BBO, GA, Apriori, FP-Growth, and even PNARYCC.

References

1. Agrawal R, Srikant R (1994) Fast algorithms for mining association rules in large databases. In: Proceedings of the 20th international conference on very large databases, pp 478–499
2. Mannila H, Toivonen H, Verkamo AI (1994) Efficient algorithms for discovering association rules. In: Proceedings of AAAI workshop on knowledge discovery in databases, pp 181–192
3. Srikant R, Agrawal R (1995) Mining generalized association rules. In: Proceedings of the 21st VLDB conference, pp 407–419
4. Savasere A, Omiecinski E, Navathe SB (1998) Mining for strong negative associations in a large database of customer transactions. In: Proceedings of the 14th international conference on data engineering, pp 494–502
5. Han J, Pei J, Yin Y (2000) Mining frequent patterns without candidate generation. In: Proceedings of the ACM SIGMOD international conference on management of data, Dallas, Texas, USA, pp 1–12
6. Zaki MJ (2000) Scalable algorithms for association mining. IEEE Trans Knowl Data Eng 12:372–390
7. Koperski K, Han J (1995) Discovery of spatial association rules in geographic information databases. In: Proceedings of the 4th international symposium on large spatial databases, pp 47–66
8. Chen X, Petrounias I (2000) Discovering temporal association rules: algorithms, language and system. In: Proceedings of the 16th international conference on data engineering
9. Tung AKH, Lu H, Han J, Feng L (2003) Efficient mining of intertransaction association rules. IEEE Trans Knowl Data Eng 15(1):43–56
10. Tzanis G, Berberidis C (2007) Mining for mutually exclusive items in transaction databases. Int J Data Warehous Min 3(3), Idea Group Publishing
11. Tzanis G, Berberidis C, Vlahavas I (2009) Machine learning and data mining in bioinformatics. Laura CR, Doorn JH, Ferraggine VE (eds.) Handbook of research on innovations in database technologies and applications: current and future trends, IGI Global
12. Tzanis G, Kavakiotis I, Vlahavas I (2011) PolyA-iEP: A data mining method for the effective prediction of polyadenylation sites. Exper Syst Appl 38(10)

13. Jie CK, Chao CM, Chiu YT (2011) Application of particle swarm optimization to association rule mining. Appl Soft Comput 11.1(2011):326–336
14. Djenouri Y, Habiba D, Chemchem A (2013) A hybrid bees swarm optimization and tabu search algorithm for association rule mining. In: 2013 world congress on nature and biologically inspired computing (NaBIC). IEEE
15. Divya B, Singhania V, Shivani G (2013) Association rule analysis using biogeography based optimization. In: 2013 international conference on computer communication and informatics (ICCCI). IEEE
16. Ivancevic V et al (2015) Using association rule mining to identify risk factors for early childhood caries. Comput Methods Prog Biomed 122(2):175–181
17. Kargarfard F, Sami A, Ebrahimie E (2015) Knowledge discovery and sequence-based prediction of pandemic influenza using an integrated classification and association rule mining (CBA) algorithm. J Biomed Inform
18. Cheng Y, Yu W-D, Li Q (2015) GAbased multi-level association rule mining approach for defect analysis in the construction industry. Autom Construct 51:78–91
19. Wang J et al (2016) Association rules mining based analysis of consequential alarm sequences in chemical processes. J Loss Prev Proc Ind 41:178–185
20. Parkinson S, Somaraki V, Ward R (2016) Auditing file system permissions using association rule mining. Expert Syst Appl 55:274–283
21. Martín D et al NICGAR: a Niching Genetic Algorithm to mine a diverse set of interesting quantitative association rules. Informat Sci 355–356, 208–228
22. Kishor P, Sammulal P (2016) An efficient approach for mining positive and negative association rules from large transactional databases. In: International conference on inventive computational technologies (ICICT)

An Optimal Heuristic for Student Failure Detection and Diagnosis in the Sathvahana Educational Community Using WEKA

P. Vasanth Sena and Porika Sammulal

Abstract The study offered in this paper aims to explore students characteristics and to determine unsuccessful student groups in respective subjects based on their earlier education and the impact of other factors in multiple dimensions. Predictive data mining techniques such as as classification analysis is applied in the analysis process. Datasets used in the investigation were collected from all academic years in the Sathavahana educational community contains different professional disciplines through online. The method adopted is to know the number of students failing in each subject and analyze the reasons for failure using data mining tools like WEKA. This model works effectively with large datasets. It has been tested on WEKA with different algorithms.

Keywords Failure detection · Diagnosis · Educational · Institutions Classification

1 Introduction

Every academic year thousands of students are enter higher educational institutions. As an example, in our Sathavahana educational institutions at least two thousand students enrol every academic year under different disciplines. In the traditional methods, in order to maintain student's details and analysis of pass percentage, precautions to reduce the failure and detentions percentage is difficult, and then the data set exponentially increases.

Existing methods, such as classroom attendance registers for monitoring attendance period-wise, and some computer Excel-based formats that maintain details

P. Vasanth Sena (✉)
R&D Cell, JNTUH, Hyderabad, India
e-mail: vasanthmtechsit521@gmail.com

P. Sammulal
CSE Department, JNTUH-CEJ, Karimnagar, India

A. Kumar and S. Mozar (eds.), *ICCCE 2018*,
Lecture Notes in Electrical Engineering 500,
https://doi.org/10.1007/978-981-13-0212-1_68

are not very well suited to address the problem. Our intention in minimize the pupil detention in education institutions. In this paper we focus on student details relating to subject failure and analyze the reason for that subject not being passed by the student. We are conducting special classes for failure students, without knowing the actual reason to their failure. In this paper we try to find the reasons for failure in multiple dimensions and take special care for each and every student with respect to failure. We attempt to find the reasons for failure of every student from this analysis, and we prepare plans for the students to access counseling and training sessions to gain familiarity in that subject. And to conclude, we look at how student knowledge has improved after the counseling sessions in terms of better results and improvements.

This paper is divided into seven sections. Section 2 briefly explains the background review, Sect. 3 describes the problems we have faced while applying data mining techniques to this kind of data provided by the Sathavahana Educational Institute. Section 4 deals with the preprocessing stage, followed by Sect. 5 which gives an overview of our approach and describes how it is beneficial to the institute to know the results. It also contains the implementation procedures and analysis of results. Section 6 explains the statistical results obtained from the data followed by the conclusion and future progress in Sect. 7.

2 Literature Review

The article named with estimating profile of successful IT student [1], a data mining approach is used to diagnose successful students' track record, which along with research results provide a useful insight into both micro and macro level aspects of the educational process. This can benefit both students and academic institutions using a clustering approach. In this paper [2], the author predicted student performance using data mining tools like R miner and he classified the datasets into three different data mining (DM) goals, namely binary, 5-level classification, and regression, and four data mining methods, namely decision tree [3], random forest (RF) [4], neural networks [5] and support vector machines (SVM) [6], including distinct input selections, i.e. with or without past grades [7], etc.

This case study was based on offline learning [8], while the DM techniques were applied after the data was collected. However, they are not provided online learning environment. Furthermore, they intend to scale to more schools and different years of students [9]. Automatic feature selection methods like filtering or wrapper are explained [3]. The research article entitled "Predictive modeling of the first year evaluation based on demographics data: Case study students of Telkom University, Indonesia" found that gender, selection path, study program and age are the attributes that are most correlated with the probability of passing first year [10].

3 Methodology

The methodology can be split into various categories, such as existing methods, proposed approach and challenges faced in the collection of datasets.

3.1 Existing Methods

In traditional methods, analysis of results takes place with respect to only one dimension, i.e. how many members have cleared the examination with different grades. On the other hand, the failure students details displayed on subject wise. Will be displayed. By this approach, we can know the details of the failure, but are unable to stop the repetition of these issues.

3.2 Proposed Approach

In our approach, we find the details of failure in multiple dimensions. For instance attendance percentage of that student, the aggregate feedback of the faculty member, who deals this subject and student previous histories like backlogs, internal marks in this subject, tenth standard percentage, inter percentage and so many attributes with minimum support count.

3.3 Challenges Faced in the Collection of Datasets

We know that collecting data is a big issue. This problem is addressed by developing enterprise resource planning (ERP) for our institutions, such as students' registrations, attendance maintenance and obtaining results from the university as T sheets. The second one is attendance data set is very complex, we automated the manual process of taking attendance in the register, by providing modern electronic devices such as mobile phones or laptops to our faculty members to enter the attendance in the classroom. This helps to identify absentee students. This helps to identify absentee students, themselves, if there is any wrong entry, it will be corrected immediately by faculty members. The schema of the feedback system was developed following a wide range of conversations among our senior faculty members higher authorities. The attributes of the feedback dataset are (1) presentation skills, (2) subject knowledge, (3) teaching methodology, (4) quality of training, and (5) syllabus covered. The categorical fields for the above attributes are poor, good, very good, and excellent with respective codings of 1, 2, 3, 4, and 5.

4 Preprocessing of Datasets

We have the following datasets: attendance, student registration, and student feedback with respect to subject and faculty. In this section, we explain the procedure of preprocessing each and every dataset used in this implementation. Once the preprocessing is applied, Attendance data set with 11,672 records and schema generated, which was mentioned in Fig. 1. Raw dataset schema of attendance entry table: (194,000 records). After preprocessing the raw dataset of attendance entry table it will generate 11,672 records. Preprocessed is accomplished by applying aggregate functions and the following schema is generated.

Schema of result dataset: (174,446 records): The resulting dataset (T sheets) are collected from the last few years of our students in the following fields. The unique identity of student by hall ticket number, subject code indicates the unique code is given for each subject, name of each subject, some marks are awarded for internal exams and remaining marks are allocated for external examination out of 100 marks, if particular subject is passed, then he/she will get some credits in that subject else zero credits are awarded for that student in that subject only. The prime fields are SSC percentage, inter percentage and degree percentage for predictions of the students in their future academic performance. Remarks 1, Remarks 2, Remarks 3, and Remarks 4 are used to maintain a student's track record in our institution. Admissions Quest is classified into convener, management or spot admission. The Admissions Special Category is further divided into NCC, NSS, and sports categories. We did dimensionality reduction by removing irrelevant attributes, such as identification marks, and remarks fields, such as remark 1, remark 2, remark 3 and remark 4 (Fig. 1).

5 Implementation and Results

The combined dataset is directly imported to a WEKA tool in CSV (comma separated values) format. The file is added successfully, but unable to process the dataset using J48. Which implicitly using C4.5 algorithm, which is based on gain ratio. Since the huge size of the dataset leads to insufficient memory requirement problems.

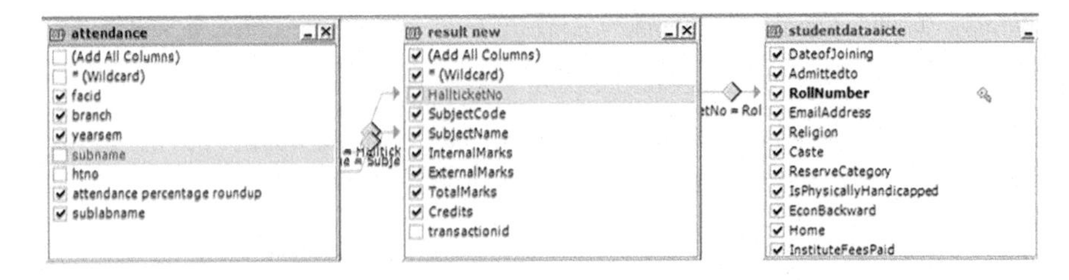

Fig. 1 The equijoin operation of datasets

In order to address the problem, We applied a sprint algorithm to my dataset to generate the decision tree. Based on these algorithms, we classified our entire dataset into class labels of successful and failing students in terms of their academic performance as a training data set. Once classified into the above categories of success and failure, we then targeted the failure sample to try and find out the reasons for failure. The actions required to improve performance are discussed in this paper. Once we apply equijoin operation, we got raw data set. Minimum threshold is the minimum measurements to find the sample of failure students. These were finalized after hours of meetings with the higher authorities of educational institutions (Table 1).

Even though the student failed in that particular subject, it is consider as lack of interest in that subject and consider it has fault, find out that the students, and individual care taken by staff members, and be give the student regular counseling to improve their performance. Once the student begins to receive proper counseling they can improve their performance and pass their subject within the time allowed. This approach is used to increase pass percentages and reduce detention rates, then the management and concerned students both are satisfied with this method. The management try to provide a quality education.

The following bar chart represents the subject code on the x-axis and hall ticket number on the y-axis. The filtering fields are year (from first year to final year), semester (even or odd), and branch-like different disciplines. Credits are zero (the particular student failed in that subject), and subject or lab name filters apply in order. The following bar chart is generated using the above criteria, and hall ticket 11N01AO1224 appeared for the examination but did not pass even through his academic background was good. The other factors was attendance percentage (71), internal marks (22), tenth standard percentage (80), and inter percentage (67) (Fig. 2).

In this way we can extract this information from merging the different fields of heterogeneous datasets. We can also discover the cosine similarity between these numeric fields to watch the consistency of students success in their entire education. (Figs. 3 and 4).

Table 1 Minimum threshold values for attributes

Field	Minimum threshold
Attendance percentage	65
Internal marks	12
SSC	55
INTER	55
Credits	0 (In particular subject)
Feedback for concerned subject faculty	60%

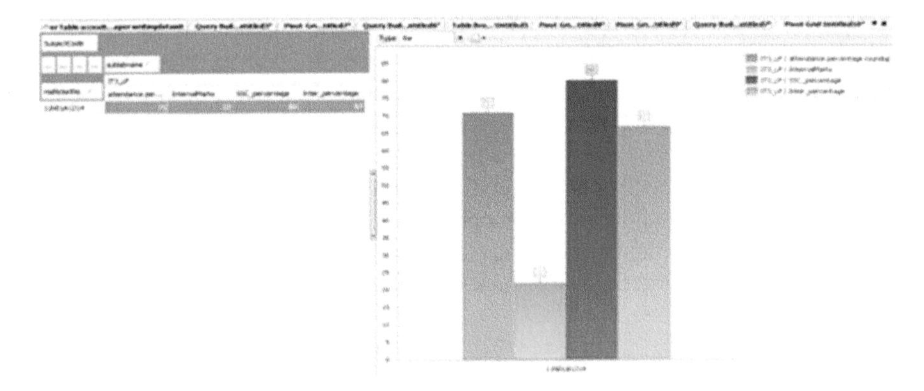

Fig. 2 Bar chart of failing student with different dimensions

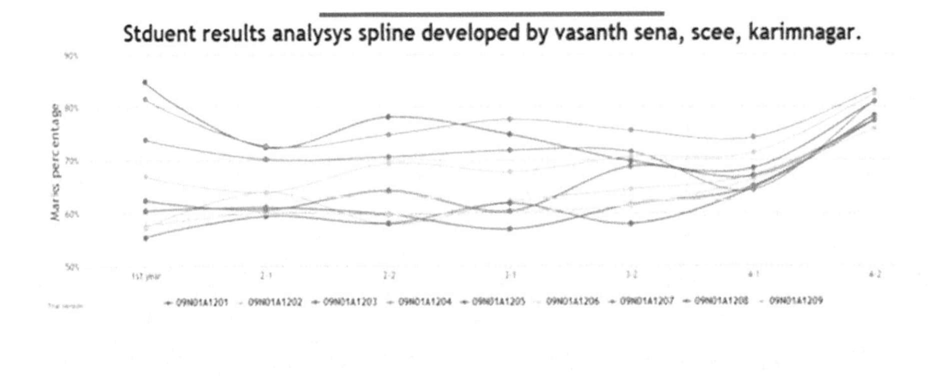

Fig. 3 Student result analysis spline for all semesters in graduation

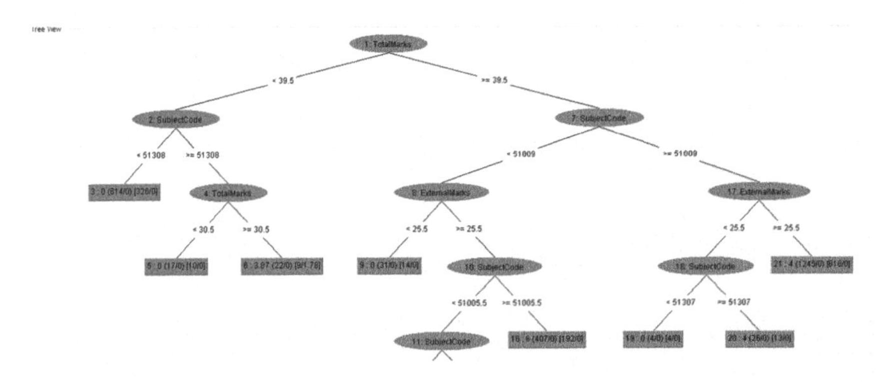

Fig. 4 Student result analysis using WEKA

6 Statistical Analysis

These are the most important statistical findings we discovered. They include:

(1) The age of pupils varies from 17 to 23 years with the mean being 20.346 years and nearly 87% of the students are aged between 17–21 years of age based on their birth data.
(2) The percentage of girls from the 1st year is 80%, 2nd year 78%, 3rd year 89%, and in the final year 88%. Even though our educational institutions are coeducational, most local girls preferred our institutions based on year of studying and gender attributes.
(3) From the religion fields of the dataset, students are classified into four different communities: Hindu, Muslim, Christian, and others. The majority of girls are from the Hindu community (88%), with Muslim (6%), Christian (4%) and others (2%).
(4) Students' pass percentage slowly increases from the 1st year to the final year.
(5) Around 80% of students are dependent on the reimbursement of fees based on field quota as management and convener and their parent income.
(6) 44% of the students were from towns while 50% of them were from a rural village background. The remaining 6% were from other states based on the attribute of permanent address.
(7) Around 65% of the girls' fathers were illiterate and around 32% had a primary education. Most of the students had no plan about their placements (78%) and future goals (90%).

7 Conclusion and Future Developments

As we mentioned in the introduction, an optimal heuristic for student failure detection and diagnosis in the Sathavahana educational community using WEKA is used in a great deal of statistical analysis and supervised learning, especially for the classification of students into two groups based on their academic performance as either successful or unsuccessful students. The groups can be classified further for taking care of individual pupils. The class labels are as follows: First class with distinction students, improving performance class; First class with distinction students, decreasing performance class; First class pupil, improving performance class; First class pupil, decreasing performance class; Second class learners, improving performance class; Second class learners decreasing performance class; Third class learners, improving performance class; Third class learners, decreasing performance class; and Failing students class. For the different classes of students try to reach a higher level, and also from comments on the student feedback form, we plan to introduce one more field with suggestions and improvements as the string.

References

1. Cortez P, Silva A (2008) Using data mining to predict secondary school student performance. In: Brito A, Teixeira J (eds) Proceedings of 5th future business technology conference (FUBUTEC 2008) pp 5–12, Porto, Portugal, April, 2008, EUROSIS, ISBN 978-9077381-39-7
2. Hearst T (1999) Untangling text data mining. In: Proceedings of the 37th annual meeting associative computer. Linguist, 1999, pp 3–10
3. Estimating profile of successful IT student: data mining approach by Dijana Oreški University of Zagreb
4. Predictive modeling of the first year evaluation based on demographics data: case study students of Telkom University, Indonesia, Tora F, Joko LB (eds.) In: 2016 international conference on data and software engineering (ICoDSE)
5. Venkatasubramanian V, Rengaswamy R, Yin K, Kavuri S (2003) A review of process fault detection and diagnosis Part III: process history based methods. Comput Chem Eng 27 (3):327–346
6. Gruber TR (1993) A translation approach to portable ontology specifications. Know Acq 5 (2):199–220
7. Benedittini O, Baines TS, Lightfoot HW, Greenough RM (2009) State-of- the-art in integrated vehicle health management. J Aer Eng 223(2):157–170
8. Vasanth Sena P (2013) Presented a paper on, “A survey on finding frequent items in text by various heuristic data structures” at National conference on advances in computing & networking (NCACN-13) JNTU Manthini, Karimnagar, Telenagana, India on 7 Dec 2013
9. Vasanth Sena P, Ramesh P (2013) Induction decision trees for tentative data. Int J Comp Sci Manag Res 2(10)
10. Vasanth Sena P (2013) Published a paper on 29th July 2013 on “An approach finding frequent items in text or transactional data base by using BST to improve the efficiency of apriori algorithm” at Cornell University Library On arXiv:1307.7513v1 [cs.DB]

Computer Vision Model for Traffic Sign Recognition and Detection—A Survey

O. S. S. V. Sindhu and P. Victer Paul

Abstract Computer vision is an interdisciplinary field which deals with a high level understanding of digital videos or images. The result of computer vision is in the form of a decision or data. This also includes methods such as gaining, processing, analyzing, understanding, and extracting high dimensionality data. Object recognition is used for identifying the objects in any image or video. The appearance of objects may vary due to lighting or colors, viewing direction, and size or shape. The problem we identify here is accuracy at nighttime and in certain weather conditions is less that when compared to daytime and also we enable to detect some signs at the night time. In this paper, we present a detailed study of computer vision, object recognition, and also a study of traffic sign detection and recognition along with its applications, advantages, and disadvantages. The study focuses on several subject, e.g., proposal theme, model, performance evaluation, and advantages and disadvantages of the work. The performance evaluation part is further discussed w. r.t. the experimental setup, different existing techniques, and the various performance assessment factors used to justify the proposed model. This study will be useful for researchers looking to obtain substantial knowledge on the current status of traffic sign detection and recognition, and the various existing problems that need to be resolved.

Keywords Computer vision · Object recognition · Traffic road sign detection Road sign recognition · MSERs

O. S. S. V.Sindhu (✉) · P. Victer Paul
Department of Computer Science and Engineering, Vignan Foundation for Science, Technology and Research, Vadlamudi, India
e-mail: ossvsindu@gmail.com

P. Victer Paul
e-mail: victerpaul@gmail.com

A. Kumar and S. Mozar (eds.), *ICCCE 2018*,
Lecture Notes in Electrical Engineering 500,
https://doi.org/10.1007/978-981-13-0212-1_69

1 Introduction

Computer vision is an interdisciplinary field. It deals with why a computer is suitable for gaining a high-level understanding of the purpose of digital images or videos. This also consists of the methods employed to obtain and interpret digital pictures, and the extraction of high dimensionality of data in order to produce numerical or symbolic data. Computer vision is a subset of image processing and it may use image processing algorithms as a backbone [1, 2]. The result is in the form of a decision or data [3]. In this way the computer vision is difficult. It has many to one mapping and also we could not understand the problem for recognition in computer vision. Some of the system methods for computer vision are image capture, pre-processing, feature extraction, detection/segmentation, recognition, and interpretation [4].

There are various significance in computer vision domain are reliability, simpler and the faster processes, No boundaries like human perception, and picture shooting gadgets are convenient to mount substitute and upgrade. The application of computer vision can be extended to many disciplines, such as automated analysis, species identification, robot-controlled processes, detecting visual surveillance, counting people, human-computer interaction, and object or environment modeling relating to biomedical images. Though it has numerous application possibilities, it has its own challenges like well-founded metric on statistical images and metrics on geometrical shapes, universal shape models, information reduction or "visual attention," input space is an high dimensional for modelling of shape, mapping is nonlinear, generalizing from a few examples, finite memory and computationally time with efficient approximate methods.

2 Object Recognition and Detection

In computer vision, the object recognition process plays a vital role in identifying single or multiple objects in an image or video sequence. The common techniques that are used for object recognition are edge, gradients, histogram of oriented gradients (HOG), Haar-wavelets, and linear binary patterns. Every object looks different under varying conditions, such as changing light or colors, changes in viewing direction or changes in size and shape, E.g. Traffic sign detection and recognition. Some of the applications of object recognition are video stabilization, automated vehicle parking systems, and cell counts in bio-imaging. Object detection is a mechanism that is related to computer vision and the processing of images concerned with identifying objects in the multimedia data like images or videos [4]. It falls in two categories: generative and discriminative. An example is the tracking of a cricket ball in a cricket match and the tracking of the ball in a football match

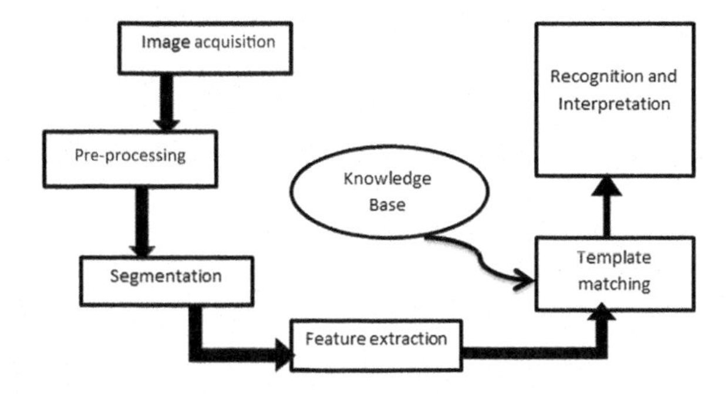

Fig. 1 Fundamental approach to sign detection

3 Road Sign Detection and Recognition

Real-time traffic sign detection and recognition models have gained the attention of many researchers in recent years. This is due to an extensive range of applications like driving assistance systems. It involves a self-learning system that can itself understand and interpret the meaning of new traffic signs [4]. Traffic sign detection is a technology by which the image is pre-handled by different methods, such as color-based, shape-based, and learning-based models. The preliminary model for sign detection and the steps involved are shown in Figs. 1 and 2 respectively. A road sign recognition system has been developed for high resolution roadside video recorded by the PMS video system. Traffic sign recognition is a methodology by which a system is able to identify traffic signs observed on the road, such as speed limits, children ahead, turn ahead, or danger ahead. Recognition deals with classified blobs. Its task is divided into different colors and shapes to improve its speed [3]. Thus, it is possible and easy to track road sign symbols and not cause any accidents involving pedestrians. The importance of road sign detection model is that there are several distinguishing importance of road sign detection model features that are used for recognition and detection of symbols and the performance of the traffic sign recognition system are improved by increasing the number of divided regions.

4 Intelligent Classification Algorithm

This section deals with the study of traffic sign detection and recognition.

A. **Traffic Sign Detection (TSD) for US Roads** [5]

Theme: In [5], it is proposed that TSD can play a crucial role in intelligent vehicles the suitable current state of art traffic, and the american traffic signs works well for colored shape symbols which will be useful, but it may falls under speed limit signs for integral channel features.

Model: The method for detection, which is similar to the Haar-like features for the integration of channels for an input image. ChnFtrs are used for features but

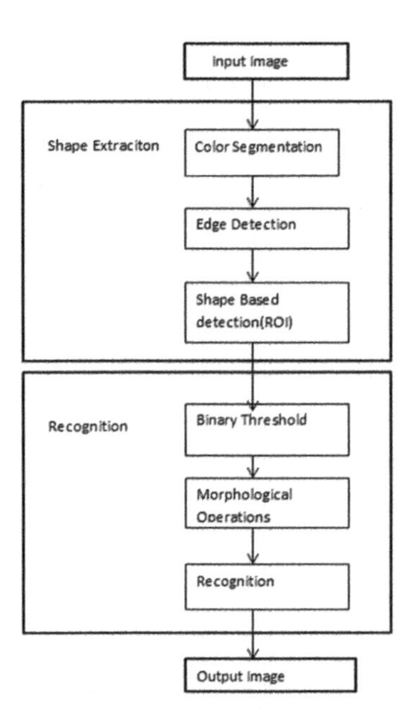

Fig. 2 Steps in the detection of signs

higher-order features are negligible. Tracking is handled by using the Hungarian algorithm for assignments only and Kalman filters for tracking purposes. This tracking is used for minimizing false detections, in which they are not part of a track, but has another property the signs which are seen before can keep it on track.

Performance Evaluation: In the performance evaluation, we report two numbers for test, based on only detection i.e., detection rate which is computed according to the number of correct detections to the actual number of signs and FPPF (false positives per frame) means to analyze the number of frames across all the frames by the number of false positives. In this tracking scenario the system knows which detection belongs to same physically sign, in that manner only we can handle at once handle only at once.

Advantages:

(a) In the detection precision rates for every super class has shown the stop and warning signs.
(b) It is well suited to stop and warning signs.

Disadvantages:

(a) If the signs are missing it is perfectly an speed limit sign.
(b) The speed limit sign is going to get problem in weather conditions.

B. **Traffic Sign Detection and Recognition from Road Scene Pictures** [6]

Theme: According to Malik and Siddiqi [6], automatic detection and recognition of road signs for ADAS has been proposed. In all current research, detection and recognition of road signs remains challenging due to weather and lighting conditions, complex backgrounds, and different viewing angles.

Model: Recognition use three state-of-the-art feature matching techniques. However, every detection is proposed based on color properties of the signs. A Hough transform is used to detect road signs based on their shapes. Weighted mean shift and genetic algorithms have been proposed.

Performance Evaluation: Detection of road sign images include color segmentation which is 0. It is followed by shape analysis and application of geometric constraints. Here, we convert RGB to HSV (hue saturation value). In HSV we have a threshold value to segment red (day time/evening time) and also have saturation value to yellow , blue and green. A binary image is generated with all pixels in the threshold range set to 1, and all others set to 0. Recognition of road signs will be used for SIFT, SURF, BRISK descriptors. For recognition rate, manually segmented road signs are first compared with the signs in the database. Second, the output of the detection system will not be segmented road sign is directly compared to the real world situations.

Advantages:

(a) When the road sign image is converted from RGB to HSV in detection.
(b) In recognition we are using three state-off-the-art SIFT, SURF, BRISK, Brisk speeds up the process when compared to SIFT and SURF.

Disadvantages:

(a) The main challenge arise due to variation in lighting and weather conditions.
(b) The system works on daytime images only. It could be further enhanced to detect and recognize road signs from nighttime images as well.

C. **Robust Traffic Sign Detection in Complex Road Environments** [7]

Theme: Tian et al. [7], discussed the performance and computation cost of traffic sign recognition. In the paper, he proposed a traffic sign detection method based on scoring SVM model, sign color, and color gradient are extracted for the color characteristics shape of traffic sign is computed by voting scheme. The trained SVM model is used to detect traffic signs.

Model: The SVM model is used to detect traffic signs from voting score maps. The method can adapt to various lighting conditions and poor imaging. Two or more methods are used to detect traffic signs, and may achieve a better performance. They use local binary pattern (LBP) to extract local descriptors and discrete cosine transform (DCT) to extract global features from traffic signs. Color invariants and then use the pyramid histogram of the oriented gradient features of shape features. In the deriving color converting model an RGB color space is used in color extraction. In the candidate sign extraction we have a color gradient which will be computed for the gradient magnitude for each pixel orientation and will also remove noise in color conversion. In sign he used shape detector voting of gradient magnitude and N- angle gradient orientation.

Performance Evaluation: In the traffic sign detection method the Chinese traffic sign dataset collected from Chinese roads was used. It contains red circles for prohibitory signs, yellow triangles for danger signs, and blue circles for mandatory signs. In sign detection performance, we use shape voting based on SVM detection to obtain a shape score. A single score value is not strong and may get many false positives, while the scoring SVM model utilizes the context of traffic signs for more precise detection.

Advantages:

(a) Reducing computational cost.
(b) Eliminating falsely detected objects.
(c) Bootstrapping is used in the training process.
(d) We can also adapt to lighting conditions and poor imaging situations.

Disadvantages:

(a) It does not detect traffic signs when processing parallel computing.

D. **Real-Time Traffic Sign Detection and Classification** [8]

Theme: According to Yang et al. [8], TSR plays a major role in assisting the driver. It performs better in real time when compared to the performance of recognition.

They improved efficiency by testing the performance with a state-of-the-art method on German and Chinese roads.

Model: The color probability model used deals with information on colors of traffic signs for building up the specific color traffic signs, i.e. red, blue, and yellow and quelling colors for reducing the search space algorithms. After reducing the detection time, Extracting proposals on the traffic signs detection of slide of window and also like SVM and CNN are used in machine learning. From the original image we get the gray image by HaarHOG and we detect sign color using a probability model.

Performance Evaluation: The dataset of 200 collected images of German and Chinese roads was used to build up the color probability model has shown the effectiveness and the purpose of changing illumination used for strongerness. When the background image and traffic signs are the same color it is not possible to detect the traffic signs using SVM and HST. Thick fog results in a complete failure in the color probability model which also performs superior performance. It reduces the processing time and has a high recall rate and a high detection rate. Classification accuracy is 98.24% and 98.77% for GTSDB and CTSD respectively.

Advantages:

(a) The HOG is a shaping descriptor for TSD due to its superior performance.
(b) Not using its search space size will reduce its recall rate.

Disadvantages:

(a) The performance of the classification model shows less accuracy for danger signs and other signs except stop and prohibitory signs.

(b) The efficiency will improve its computation time, which will be accelerated by the GPU.

E. **Real-time Road Traffic Sign Detection and Recognition** [9]

Theme: According to Greenhalgh and Mirmehdi [9], the authors proposed an automatic detection and recognition of traffic signs system which detects the maximally stable extremal regions (MSERs) of candidate regions, which may changes variations in lighting conditions and recognition is on the support vector machine classifiers are trained using HOG feature. Under the weather conditions the present system is accurate at high vehicle speeds, it will be run at an average speed of 20 frames per second.

Model: Advanced Driver Assistance Systems is important for automatic TSR and TSD. The proposed model consists of two stages. In the first stage detection is performed by novel application of MSERs, and in the second stage recognition occurs using HOG features and is classified using linear SVM. In this work they used an online road sign database for the graphical representation of signs. With large training sets are generated by distortion in the random graphic template like blurred images and capturing of occurrence for real-scene distortion images . Method for detection selected in varying contrast light conditions. Each frame is binarized at each of the different threshold levels. In road sign classification the horizontal and vertical derivatives are found by using the Sobel filter.

Performance Evaluation: For some threshold values, like MSER has compared against the detection of accuracy and execution time in order to enhance working ability of the linear SVM classifier, an appropriate cost for a mis-classification parameter for the color signs. 89.2 and 92.1% is actual70 accuracy of white road signs. Classification results values of white background and color background signs are between the individual road sign classes and the prediction made by the classifier. The overall accuracy for this experiment is 89.2%. This is greater than with the fully synthetic or real datasets.

Advantages:

(a) It gives a high accuracy for vehicle speeds.
(b) It allows large datasets for the training dataset. These are generated from the template's images.

F. **Road Sign Detection on a Smartphone for Traffic Safety** [10]

Theme: According to [10], they developed simple smartphone by using the low cost driver assistance system. By using computer vision techniques and multiple resolution templates for detecting speed limits, it was possible to alert the driver by sending messages whenever the speed was excessive. First, fast filters could restrict the focus on a wide area photo by using its fast normalized cross correlation technique and it detects the speed limit signs. It also compares the actual vehicle speed with the computed vehicle speed in the smartphone and issues a warning message to the driver.

Model: In the paper, one approach is to minimize distracting driving to develop ADAS purpose of the system is to develop alert drivers and performs some functions like monitoring blind spots, detecting lane departure, and emergency braking. In the first step, edge filters are processing by using limited ROI. In the second, detected images are matched with the templates images. Next, there are many numbers of detecting signs to the template's images. This is implemented in the Java programming which will be simple and easy to understand. Focus of attention with the horizontal and vertical edge filters. In the classification, it matches with the number of locations with the speed limit. We can make further improvements by using other algorithms like Sift and AdaBoost.

Performance Evaluation: The range of speed limits was 20–65 on a sunny day. The system processes photos at the rate of approximately one per second with a probability of 0.93 with a 95% confidence level. Failures may occur when the edges are not detected because of sun glare and signs which are blurred during monsoon rain.

Advantages:

(a) It is simple and easy to understand.
(b) It can be applicable for simple android devices.
(c) Speed limit signs are detected and gets the result same.

Disadvantages:

(a) Some signs are not detected because of a combination of glare from the sun and shadows which can cause the edge in its focus of attention.
(b) The Speed limit sign is going to be detected with the low correlation to be proceed for classification.

G. **Traffic Road Sign Detection and Recognition for Automotive Vehicles** [11]

Theme: In [11], the authors implemented and developed a way to extract a road sign from a complex image. The main aim is to detect the detection and extracting image. This plays an important role in specific domains like island, schools, traffic signs, hospitals, etc.

Model: Here we have four different types of signs, i.e. warning signs, prohibitory signs, obligations, and informative signs. Informative signs have an inverted triangular shape and every sign has its own color and shape, and several algorithms are used. Fast gray scale road sign model matching and recognition in this we will be having mobile mapping technique for compiling cartographic information for a vehicle and have the two segments, one is color based allows reducing false positives and other approaches i.e., a genetic algorithm for a grayscale image and for the shape we use Hough transform. Detection by an Adaboost to solve supervised pattern recognition. Here, System evaluates the images later, it pre-processes and will a color threshold and next recognize the object next road sign verification and at last road sign area identifies sign will be shown. Here we use the to find the problem space where the destination object is a road. For edge detection they used a Sobel operator and this can extract our desired road sign. This technique is different from detecting road signs in that it is time effective.

Performance Evaluation: The performance was captured according to the time of day of the images, i.e. morning, noon, evening, and night using 200 images. When evaluated, 80–90% of daytime images are successfully detected but at nighttime only 50%. This means that we have to improve the nighttime performance.

Advantages:

(a) It is easy to detect object recognition using grayscale images.
(b) The input images of our recognition procedure are provided by the weak classifiers cascade detection process.

Disadvantages:

(a) It shows the gray label only for red color captured images.

5 Observations

It has been observed that all the paper included have different techniques and algorithms which all have their own merits and demerits. In general, the algorithms used in most of the works are the Hungarian algorithm and Kalman filters which are used to track and to minimize false detections. A Hough transforms and three state-off-art are used to detect color-based properties. Weighted mean shift and genetic algorithms have also been proposed. A SVM model is used to detect traffic signs from voting score maps and converts RGB to HSV. We can detect shape by using voting of gradient magnitude and N-angle gradient orientation. SVM and CNN models are used to filter out the false positives and also to classify the remaining super classes. We can also detected shape by using edge filters. By using all the mentioned algorithms, it is possible to detect only the daytime images or signs. It is not possible to detect nighttime images or signs and also it is poor under different weather conditions.

6 Future Solutions

In the literature study that we did for this paper, there are many techniques and algorithms for analysis that were applied to traffic sign detection and recognition. This will be useful for users and the researchers who are wishing to do research on this problem. The first step is to get an image, such as warning signs, prohibitory signs, and danger signs. These are taken as an input and later we detect the images using any technique or algorithm. The recognized image is obtained as an output. Here, we improved the performances in the weather conditions and we are enabled to detect signs at night time. By using existing algorithm or technique we will be going to detect the problem in future. Moreover, the performance of object detection can be improved using various bio-inspired algorithms [12–17] which has attracted the attention of image processing researchers.

7 Conclusion

Computer vision is an interdisciplinary field and has many forms, such as sequences of videos, we can view from medical scanner and multiple cameras. In traffic sign detection and recognition we have to improve the overall performance at nighttime and in different weather conditions. With this in mind, a detailed study on the various recent and best-working algorithms for traffic sign recognition has been discussed and the focus was kept on the model, methodology used for performance evaluation, and the factors used. This work will be useful for researchers looking to get a better idea on the current state of research in the domain of traffic sign detection and recognition in computer vision.

References

1. Baskaran R, Victer Paul P, Dhavachelvan P (2012) Algorithm and direction for analysis of global replica management in P2P network. In: IEEE international conference on recent trends in information technology (ICRTIT), May 2012, Chennai, pp 211–216
2. Victer Paul P, Saravanan N, Jayakumar SKV, Dhavachelvan P, Baskaran R (2012) QoS enhancements for global replication management in peer to peer networks. Future Gener Comput Syst 28(3):573–582. Elsevier. ISSN: 0167-739X
3. Lamer J, Cymbalak D, Jakab F (2013) Computer vision based object recognition principles in education, 24–25 Oct. 978-1-4799-2162-1, Stary Mokovec
4. Fu LC (2001) Computer vision based object recognition for vehicle driving, 21–26 May, Seoul, Korea. 0-7803-6475-9
5. Mogelmose A, Liu D, Trivedi MM (2014) Traffic sign detection for U.S roads: remaining challenges and a case for tracking. IEEE 17th international conference on intelligent transportation systems (ITSC), Qingdao, China, 8–11 Oct 2014
6. Malik Z, Siddiqi I (2014) Detection and recognition of traffic signs from road scene images. In: 12th international conference on Frontiers of information technology, Bahria University, Islamabad, Pakistan
7. Tian B, Chen R, Yao Y, Li N (2016) Robust traffic sign detection in complex road environments, China. 978-1-5090-2933-4
8. Yang Y, Luo H, Xu H, Wu F (2016) Towards real-time traffic sign detection and classification. IEEE Trans Intell Transp Syst 17(7):2022–2031, China

9. Greenhalgh J, Mirmehdi M (2012) Real-time detection and recognition of road traffic signs. IEEE Trans Intell Transp Syst 13(4):1498–1506
10. Pritt C (2014) Road sign safety detection on a smartphone for traffic safety. 978-1-4799-5921-1
11. Hossain MS, Hyder Z (2015) Traffic road sign detection and recognition for automotive vehicles. Int J Comput Appl (0975 – 8887) 120(24):10–15
12. Paul PV, Ramalingam A, Baskaran R, Dhavachelvan P, Vivekanandan K, Subramanian R, Venkatachalapathy VSK (2013) Performance analyses on population seeding techniques for genetic algorithms. Int J Eng Technol (IJET) 5(3):2993–3000. ISSN: 0975-4024
13. Victer Paul P, Baskaran R, Dhavachelvan P (2013) A novel population initialization technique for genetic algorithm. In: IEEE international conference on circuit, power and computing technologies (ICCPCT), Mar 2013, India, pp 1235–1238. ISBN: 978-1-4673-4921-5
14. Victer Paul P, Ramalingam A, Baskaran R, Dhavachelvan P, Vivekanandan K, Subramanian R (2014) A new population seeding technique for permutation-coded genetic algorithm: service transfer approach. J Comput Sci (5):277–297. Elsevier. ISSN: 1877-7503
15. Saravanan N, Baskaran R, Shanmugam M, SaleemBasha MS, Victer Paul P (2013) An effective model for QoS assessment in data caching in MANET environments. Int J Wirel Mob Comput 6(5):515–527. Inderscience. ISSN: 1741-1092
16. Victer Paul P, Moganarangan N, Sampath Kumar S, Raju R, Vengattaraman T, Dhavachelvan P (2015) Performance analyses over population seeding techniques of the permutation-coded genetic algorithm: An empirical study based on traveling salesman problems. Appl Soft Comput 32:383–402. Elsevier
17. Dorner J, Kozak S, Dietze F (2015) Object recognition by effective methods and means of computer vision, June 2015, Strbske Pleso, Slovkia. 978-1-4673-6627-4

Color-Texture Image Segmentation in View of Graph Utilizing Student Dispersion

Viswas Kanumuri, T. Srinisha and P. V. Bhaskar Reddy

Abstract The Image segmentation is that for investigation is a noteworthy part of discernment and up to date it is still testing issue for machine recognition. Numerous times of concentrate in PC view demonstrate that dividing a picture into important districts for ensuing preparing (e.g., design acknowledgment) is similarly as troublesome issue as never changing case identification. In this paper work, the proposed one uses the particular sort of frameworks had been taken after to complete shading surface picture division. Division strategies are intended to incorporate more component data, with high exactness and agreeable visual total. The division procedure depends on MSST and understudy's t-conveyance technique.

Keywords Image segmentation · Multi-level graph cuts · Pattern acknowledgment · Multiscale structure tensor · Understudy's t-dissemination strategy

1 Introduction

Image segmentation is key undertaking in PC vision. Shading surface order of Picture is exceptionally fundamental and basic advance in picture handling and example acknowledgment. Specifically the segmentation of normal pictures are testing errand, since these pictures display no noteworthy consistency in shading and surface and furthermore they are frequently demonstrate high level of multi-faceted nature, arbitrariness and anomaly. The principle task of segmentation is to section a picture into significant areas.

Over the years, analysts have contemplated distinctive methodologies for picture segmentation, with the goal of sectioning pictures by utilizing least customer joint effort or acting under supervised way. Progress in science and development and progress in society has staggeringly enriched the division execution as of late. A high degree of precision and unwavering quality is a testing issue for picture

V. Kanumuri (✉) · T. Srinisha · P. V. Bhaskar Reddy
Jawaharlal Nehru Technological University, Hyderabad, India
e-mail: viswasvarmasatya@gmail.com

A. Kumar and S. Mozar (eds.), *ICCCE 2018*,
Lecture Notes in Electrical Engineering 500,
https://doi.org/10.1007/978-981-13-0212-1_70

segmentation techniques. Calculations and strategies proposed in early time are for the most part laced with particular application. For instance to compute text on for platelet investigation segmentation calculation utilizes shading data [1]. So it has limited use and in not really suitable for partitioning typical pictures as they will definitely have rich surface information. Another instance of picture division computation with a pre-defined application is surface examination of stone pictures. This type of calculation utilizes a mix of chromatic data with tiny surface data [1]. To catch applicable auxiliary data by separating surface component utilizing shading grouping and parallel blob pictures together is mind boggling process, additionally it is hard to break down irregular surface shading example. Another delineation is to parcel the bone marrow cell shading picture. Subsequently, we can express that the strategies showed up in early time were normal for particular application and need in exhibiting general thought of surface shading showing. These procedures were not competent enough to depict multi-scale structures and very small structures. So it was essential to discover new strategies able to mastermind different part data for division reason. The image segmentation field in view of shading alongside surface descriptors has been rated broadly finished recent decades distributing gigantic no of computation between 2007 and 2014 years. More or less a thousand papers were produced between 1984 and 2009 [2]. So that, it is crucial to see the way that the surface shading examination is also the one of most investigated zones in the picture getting ready and PC view. The purpose of this paper work is to propose the procedures to generate extremely high quality pictures.

2 Literature Survey

In spite of the fact that there are a few techniques which assess shading highlights or surface components alone, we might want to say that survey in this paper is worried about distributed business related to unsupervised shading surface picture segmentation composed by incorporating different comparing strategy to catch more element data. The techniques of picture segmentation proposed below have been classified into three parts, namely shading surface descriptor, proper assurance of substantial class no, and diminishment strategy for over/mistake segmentation [3].

2.1 The Color-Texture Dissemination Algorithm

Affiliations and Email Addresses. The color-texture descriptor requirements for the change of the element depiction capacity. Just shading construct division calculation deliver district in light of premise of shading contrasts just so it won't be appropriate for pictures containing rich surface data. As result contain little non semantic districts. On the off chance that we consider the surface component a few noteworthy classifications are accessible. The flag preparing strategy has increased wide consideration in PC world as that it could reenact visual observation, and also had a lot of sifting strategies.

Gunjan et al. [4] advanced the Gabor wavelet, along with multi-scale logical capacity; however, the execution of such a technique relies upon a decision of Gabor channels and similarly there are parcels of data excess. A measurement lessening strategy can be utilized to maintain a strategic distance from issue of high measurement. However, this requires a gigantic data memory and a long calculation time due to its complexity and the associated Fourier change. Regardless of the way that various sorts of wavelet and Gabor wavelet computation are available yet wavelet and Gabor wavelet are ordinary and convincing than remaining.

Another strategy given by Deng and Manjunath [5] is a half-managed and 2-mark color-surface picture division strategy, here to portray the surface MSNST technique is connected. This technique treats the shading and surface elements independently and consequently can overlook the connection between changed highlights. The proposed calculation can show both a normal and a concentrated dispersed surface district. Be that as it may, it won't work for a very large-scale surface. This issue can be addressed by add up to variety (television stream) stream calculation.

A more powerful calculation is that derived by Yang et al. [1] which incorporates a smaller MSST surface, RGB color, and a television stream.

2.2 Algorithm to Declare Correct Class Number in Advance

The second part of segmentation is to decide a legitimate class num ahead of time. By and large the legitimate class no cannot be decided ahead of time as various pictures contains distinctive examples of surface shading. So the number is given physically and it is determined to be sufficiently extensive. However, this might result in extremely large computations adding additional time and may not produces a satisfactory result. The answer to the issue is given by D. E Ilen. In this self-sorting out guide (SOM) techniques are utilized to gauge the substantial class number. Restriction of the technique is substantial class number is resolved independently from introductory bunching procedure and that is consistent all through division process. As a result this technique does not give an agreeable arrangement.

Paper [6] utilizes the segment shrewd desire augmentation for multivariate blended understudy's t-circulation (CEM3ST) to demonstrate the color-surface likelihood appropriation. The customary Gaussian blend models (GMM) is not more powerful than multivariate blended understudy's t-appropriation (MMST). Execution of MMST is hearty to separating commotions. This technique likewise conservative the quantity of levels of multilayer diagram.

2.3 Algorithm to Reduce Over/Incorrect Segmentation

The third part of picture division is basically concerned with reducing over/incorrect division. Such a significant number of causes are there for over/blunder division, for example, huge no of little clamor district. Huge numbers of these little clamor locales are futile as a result of which visual whole get bothered. So also division precision and speed of union decreased by finished/mistake division.

The paper by Peel and McLachlan [7] had tended to the issue and recommended repeating the labeling technique to reduce incorrect division. This technique relabelled those locales which are not having spatial association yet exchanges a similar name. This technique redresses these false marking and helps to decrease incorrect division. The restriction of this strategy is that it tends to the main issue of blunder division, no arrangement is accommodated over division issue.

The paper by Chen et al. [8] gives a technique for extra division call as mark consolidating methodology which dependent on MRF show. This kind of technique ascertains the chances rand record (PRI) at pixels level. As this technique delivers low no of division comes about, Gibbs conveyance is very hard to build up.

Technique given by Yang et al. [1] mentioned as local validity consolidating (CRM) incorporates more than 2 strategy. Normal edge in between districts, local nearness relationship, provincial color-surface Divergence and area estimate are coordinated for shaping the CRM which tells us the preferred outcome over the technique specify beforehand. This paper work had composed a powerful emphasis merging paradigm, which guarantees predominant execution of the picture division in acting under a supervision way.

3 Proposed Method

All we talked about there will be numerous calculation existed for unsupervised picture division. Figure 1 demonstrates the stream outline of our derived technique.

3.1 Color-Texture Descriptor

The initial step that is a surface shading descriptor is important for upgrading highlight depiction capacity. This progression for the most part utilizes the multi-scale structure tensor (MSST). This is a non-linear dispersion separating innovation that is utilized for smoothness of the clamor and to upgrade the corners. Multi-scale structure tensor (MSST) is utilized to portray the surface component of a picture. A structure tensor is broadly acknowledged to minimalistically determine this component by the utilization of picture subordinates, which hold the entire introduction data. The MSST is obtained by utilizing a repetitive discrete wavelet

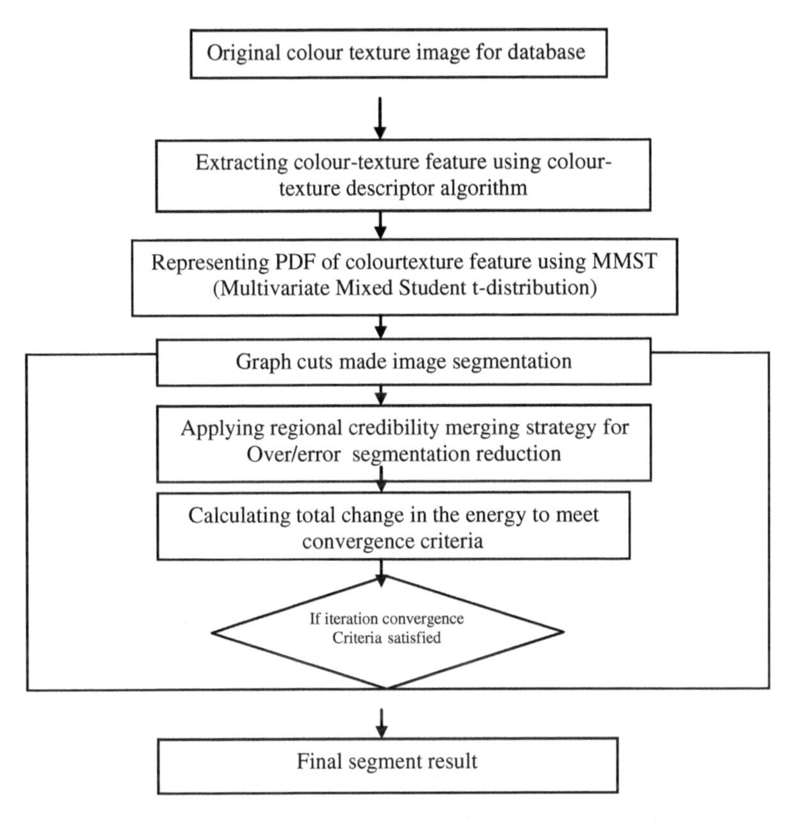

Fig. 1 Flow chart of the proposed work. Probability distribution modelling using multivariate mixed student's t-distribution (MMST)

structure [1, 9]. Ts (the sth scale tensor of MSST) can be developed with the tensor result of inclination,

$$\begin{aligned} \mathrm{Ts} &= \sum_{n=1}^{N} (\nabla(\mathrm{I} * \theta \mathrm{s})\mathrm{n}\ \nabla(\mathrm{I} * \theta \mathrm{s}) \cdot \mathrm{n}^{\mathrm{T}}. \\ &= \sigma^{-2s} \begin{pmatrix} \sum_{n=1}^{N} (D_{ns}^{x})^2 & \sum_{n=1}^{N} (D_{ns}^{x} D_{ns}^{y}) \\ \sum_{n=1}^{N} (D_{ns}^{x} D_{ns}^{y}) & \sum_{n=1}^{N} (D_{ns}^{y})^2 \end{pmatrix} \end{aligned} \tag{1}$$

where, S = 0, 1 … S − 1.

σ The premise of the wavelet, to diminish the cost of calculation and memory stockpiling, this can be settled as 2,

S The aggregate number of scales,

n The nth shading channel of the picture I, and N—the aggregate number of the channels.

MSST network set had diverse element body that than shading vector, and in this way SVD deterioration is utilized [1] for each and every size of MSST in the tensor space. The aggregate variety (TV) stream is utilized to assistant improve the portrayal capacity [1, 10]. The fundamental surface data of Ts is acquire as,

$$\mathrm{Vs} = \eta(\lambda_s \mathrm{V}_s)^{1/2} \tag{2}$$

where, λ+, λ− are the eigen esteems after SVD disintegration, and Vs+, V− are the relating vectors. The above technique is coordinated with RGB shading as RGB, it is high predictable with the shading affect ability of human eyes. At long last nonlinear dissemination separating is embraced to have capable surface shading descriptor with commotion sifting and edge improvement. Separating is embraced to have capable shading surface descriptor with commotion sifting and edge improvement.

The likelihood circulation displaying is necessary to guarantee the precision, also the strength of picture division. As multivariate blended understudy's t-dispersion (MMST) is most vigorous and adaptable, this kind of strategy is utilized for the demonstrating reason. Propel figuring of legitimate class number is difficult, so the strategy recommended in paper [1] called as segment savvy desire augmentation for MMST (CEM3ST) algorithm is executed. This strategy appraises the substantial class number amid the procedure of likelihood dispersion displaying.

Let mean an irregular color-surface component that is situated in the 10th pixel of the color-surface picture. What's more, H is the tallness. At the point when compromise the K parts MMST circulation, the associated PDF can be said as, [1]

$$F(C\Gamma_x^*|\theta) \sum_{k=1}^{K} \omega_k \Gamma_x^*|\theta) \tag{3}$$

where,

K add up to number of blended parts in MMST.
ωk blended weight of the *k*th part
θk {Vk, μk, ∑k} compares to the degrees of opportunity, mean vector and co-variance lattice of the *k*th blended part.

3.2 Multilayer Graph Cut Strategy

Subsequent to knowing the substantial number K-valid the surface shading picture could be portioned in multiple name route into K unique surface shading district.

Here individual class holds special name. The issue of multi mark ideal division could be unraveled by building vitality work along with information terms and smoothness terms and limiting it. The multi-mark vitality capacity can be outlined as [1],

where,

E1 Information
E_2 Smoothness
λ Regularization parameter.

The weighted multilayer diagram [11, 12] is developed for limiting the vitality work E. The objective is for section the primary protest of a picture utilizing division technique in light of chart. The picture with K classes takes the K − 1 layers. The estimation of K is introduced utilizing K valid number that is now computed. In between any two neighboring layers of chart there exist similar n-connections and t-links. Let G = (U, V) be the multilayer diagram, where U and V are the arrangements of corners and vertices. The arrangement of hubs is equivalent to the arrangement of pixels in the picture. An individual pixel is associated with its d-neighborhood (d = 4, 8).

The component similitude between any two pixels is demonstrated by the weight. In the wake of developing the weighted multilayer chart, a maximum-stream/minimum-slice hypothesis is utilized to comprehend vitality minimization by augmenting the stream across the network.

3.3 Algorithm to Reduce Over/Incorrect Segmentation

A natural color-texture image will have alternativeness, decent variety and many-sided quality. In condition (4) arranging regularization parameters is difficult.

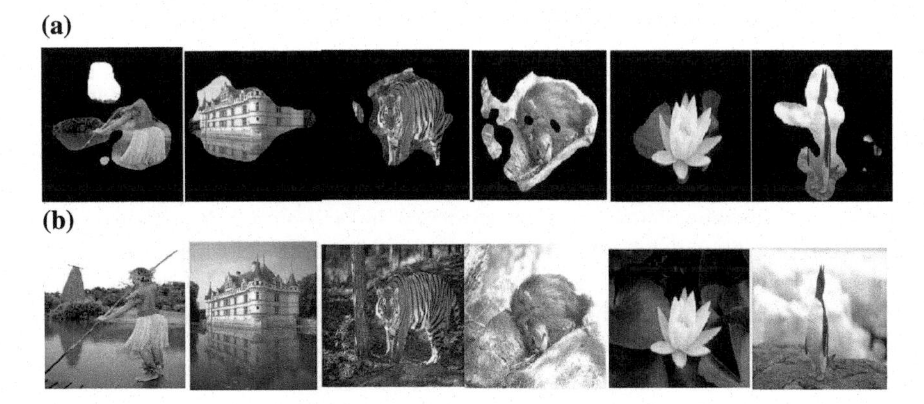

Fig. 2 Preparing of local believability combining. **a** Unique surface shading picture. **b** Result of over/incorrect division compared to multilayer diagram cuts. **c** Result of division after local validity consolidating

For the most part, a small λ is favored. The procedure of Regional validity consolidating (RCM) [1] enhances the visual sum/consistence of the real sectioned districts by limiting across/mistake division phenomenon [1] (Fig. 2).

$$RCMD_{ij} = \{-\varphi(R_i, R_j)\} \cdot \{|R_i||R_j|/|R_i||R_j|\} \cdot \{\mathrm{J}(R_i, R_j)\} \cdot \{\exp\left[-1/|E_{ij}| \sum(x_{ij}, y_{ij})\right]^{-1}\} \tag{4}$$

$|R_i|$ and $|R_j|$ denote the region sizes of R_i and R_j respectively.

4 Results

4.1 Parameter Settings

The purpose of clearness and respectability, a few parameters of derived technique likewise should be fittingly defined before. The parameters will be considered from the [1]. Hence the primary surface data is for the most part moved in the initial 3 scales, the scale No S in recipe (1) is given as 3. The improvement factor η in Eq. (2) are settled as 5.0. The smoother weight λ (4) is picked as 10.

4.2 Software

These calculations are executed utilizing MATLAB 2010a and tested on the storage BSD300.

4.3 Qualitative Evaluation

The execution of a given technique is shown in Fig. 3. A Berkeley division database BSD300 [13] is utilized. Here we have chosen some intricate pictures from BSD500 for doubting as demonstrated as follows. All things considered, these chosen pictures are illustrative and they have a distinctive measure of surface data. Initially, we incorporated some of the pictures with rich surface data, for example, the tiger and the winged creature. Second, there are likewise some pictures with direct surface and shading data, for example, the structures and the stream. Third, the flower and the man are the surface shading pictures with little surface data but a lot of shading data.

The fragmented yield picture is compared to the accessible ground truth [13]. By contrasting the images, genuine positive (TP) and false opposite (FN) values are computed.

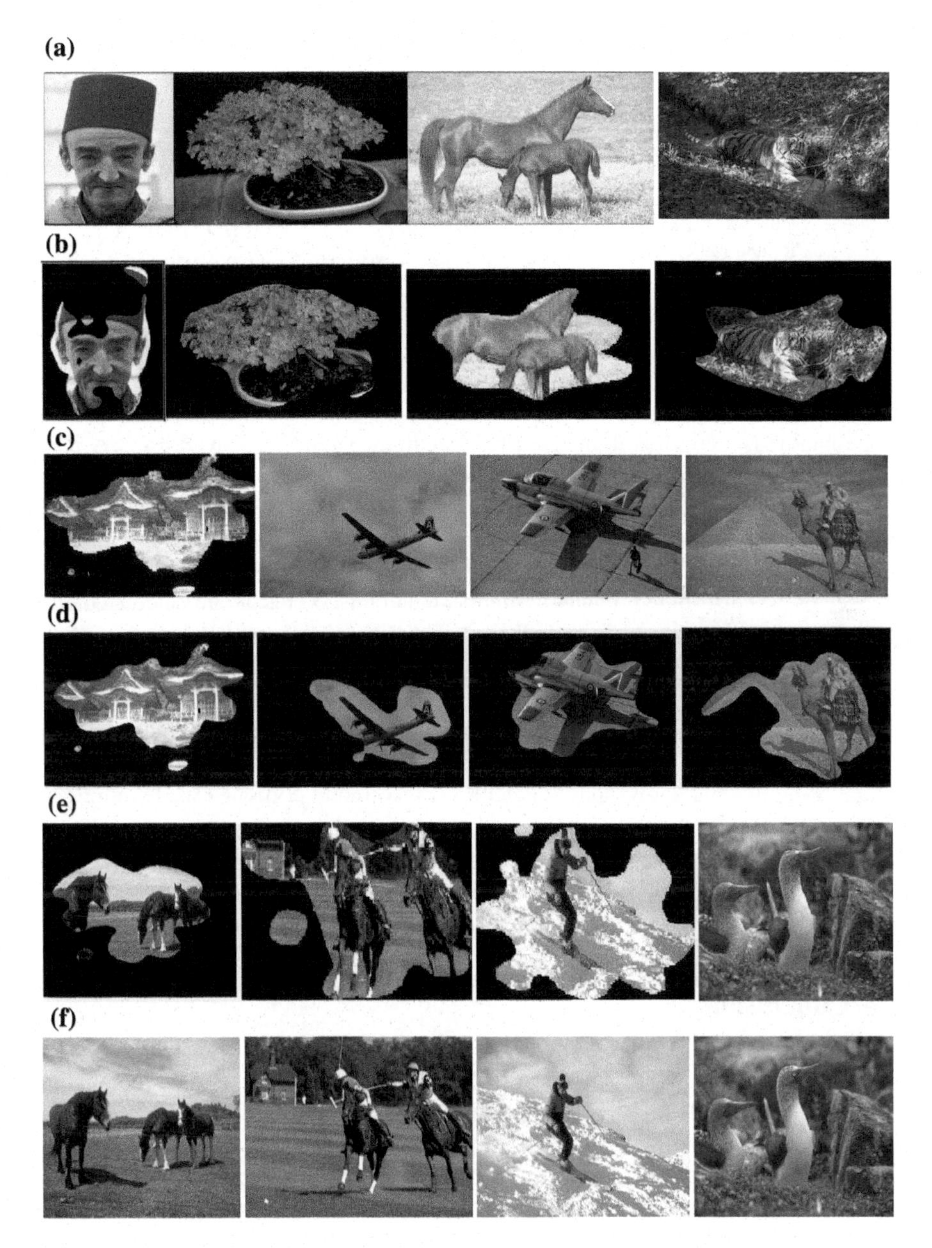

Fig. 3 Original images from BSD300 and corresponding segmented results

Table 1 Performance in terms of PRI and NPR values

Images	Fig. 3a	Fig. 3b	Fig. 3c	Fig. 3d	Fig. 3e	Fig. 3f
PRI value	0.9979	0.9976	0.9991	0.9959	0.8898	0.8788
NPR value	0.8750	0.9874	0.9643	0.9354	0.8675	0.8775

Table 2 Mean and variance values of PRI and NPR

Mean value (PRI/NPR)	1.01874/1.0123
Variance value (PRI/NPR)	0.0023/0.1222

4.4 *Quantitative Evaluation*

For quantitative consideration the PRI (occurrence rand list) qualities and NPR (institutionalized probabilistic rand) values are processed along with the mean and the change values and these are shown in Table 1, thinks about to the divided pictures showed up in Fig. 3. Table 2 demonstrates the quantitative outcomes of PRI and NPR on the Berkeley division database BSD300 (using delayed consequence of 300 distributed images). The mean and the co-change are calculated, and it can be seen that the mean is high and the distinction regard is at a minimum. Quantitative evaluation displays the prevalence of our proposed work as far as exactness, vigor and shading-surface segregating capacity.

Genuine positive esteem demonstrates how close the division yield is to the ground truth picture. The estimation of TP shifts from 0.7733 to 1.0595. The value of TP being under 1 shows that the picture is incorrectly portioned while an esteem of more than 1 demonstrates the over division.

A desktop PC was used in every one of these trials. It was fitted with a 1.7 GHz Intel(R) Pentium(R) 3558u CPU and 4 GB RAM.

5 Conclusion

The significant goal of this paper is to section a picture on the theory of reconciliation of shading surface identifying. After looking at previous papers it was discovered that earlier strategies used for division were particular to applications. Such techniques claimed the restricted usefulness to fragment regular pictures and do not have the general thought of shading surface displaying. This made it critical to propose new and encouraging techniques for the division of pictures with rich surface shading.

This paper determines the procedure of acting under supervised color–surface division. The calculations are executed for shading surface descriptor, to decide substantial class no adaptively, and for lessening of blunder/across division, which tells us the agreeable visual sum with their palatable outcomes. The calculations talked about here are outflanking and get the shading surface division comes about, with high correctness's.

References

1. Yang Y, Han S, Wang T, Tao W, Tai X-C (2013) Multilayer graph cuts based unsupervised color-texture image segmentation using multivariate mixed student's t-distribution and regional credibility merging. Pattern Recognit 46:1101–1124
2. Ilea DE, Whelan PF (2011) Image segmentation based on the integration of colour-texture descriptors—a review. Pattern Recognit 44:2479–2501
3. Warudkar S, Kolte M Colour-texture based image segmentation using effective algorithms: Review. IJARCCE 5(6), June 2016
4. Gunjan VK, Shaik F, Kashyap A, Kumar A An interactive computer aided system for detection and analysis of pulmonary TB. Helix J 7(5):2129–2132, Sept 2017 (ISSN 2319–5592)
5. Deng Y, Manjunath BS (2001) Unsupervised segmentation of color-texture regions in images and video. IEEE Trans Pattern Anal Mach Intell 23:800–810
6. Ilea DE, Whelan PF (2008) CTex-an adaptive unsupervised segmentation algorithm based on colour-texture coherence. IEEE Trans Image Process 17:1926–1939
7. Peel D, McLachlan GJ (2000) Robust mixture modelling using the t distribution. Stat Comput 10
8. Chen SF, Cao LL, Wang YM, Liu JZ (2010) Image segmentation by MAP-ML estimations. IEEE Trans Image Process 19:2254–2264
9. Ma WY, Manjunath BS (1997) Edge flow: a framework of boundary detection and image segmentation. In: IEEE computer society conference on computer vision and pattern recognition, pp 744–749
10. Han SD, Tao WB, Wang DS, Tai XC, Wu XL (2009) Image segmentation based on GrabCut framework integrating multiscale nonlinear structure tensor. IEEE Trans Image Process 18:2289–2302
11. Brox T, Weickert J (2004) A TV flow based local scale measure for texture discrimination. In: Computer vision—ECCV 2004. 8th European conference on computer vision 2, pp 578–590
12. Tao WB, Chang F, Liu LM, Jin H, Wang TJ (2010) Interactively multi-label image segmentation based on variational formulation and graph cuts. Pattern Recogn 43:3208–3218
13. Liu L, Tao WB (2011) Image segmentation by iteratively optimization of multilabel multiple piecewise constant model and Four-Color relabeling. Pattern Recogn 44:2819
14. Han SD, Tao WB, Wu XL Texture segmentation using independent-scale component-wise Riemannian-covariance Gaussian mixture model in kl measure based multi scale nonlinear structure tensor space, Mar 2011
15. Mignotte M (2010) A label field fusion bayesian model and its penalized maximum rand estimator for image segmentation. IEEE Trans Image Process 19:1610–1624

A Novel Approach for Digital Online Payment System

M. Laxmaiah and T. Neha

Abstract Nowadays digital transaction security is seen as essential in an online payment system. Earlier, cryptographic authentication techniques were used to make transactions very secure with third-party verification. However, in recent times, digital transactions have developed to allow online payments to be made directly from one party to another, without the intervention of the third party. This kind of P2P network transaction is achieved by utilizing the blockchain innovation. A blockchain uses the idea of Bitcoin. It isn't observed by the central authority, but its clients direct and approve exchanges when one individual pays another for merchandise or administrations, dispensing with the requirement for outside confirmation. All finished exchanges are freely recorded as a block and each block is added one after other to form as a blockchain. In this paper, we present the workflow of a digital online payment system using the blockchain technique. We explain the secure sign-in procedure of the blockchain strategy. We elaborate on the effects of blockchain technology on the online transaction management system in terms of security and usability.

Keywords Blockchain · Bitcoin · Cryptocurrency · Distributed ledger

1 Introduction

In recent years, the advanced digital market has heard the important buzzwords Bitcoin, blockchain, and cryptocurrency on an increasingly frequent basis. Blockchain is an innovation of online cash transactions using digital currency. This digital currency is widely known as Bitcoin, for which the blockchain innovation was developed. A cryptocurrency is a medium of trade, for example, the Russian

M. Laxmaiah (✉) · T. Neha
Department of Computer Science and Engineering, CMR Engineering College, Kandlakoya (V), Medchal (D), Hyderabad, Telangana, India
e-mail: laxmanmettu.cse@gmail.com

T. Neha
e-mail: thotapalli.neha30@gmail.com

A. Kumar and S. Mozar (eds.), *ICCCE 2018*,
Lecture Notes in Electrical Engineering 500,
https://doi.org/10.1007/978-981-13-0212-1_71

ruble or Indian rupees. In advanced exchanges encryption systems are used to control the production of financial units and to confirm the exchange of funds. In such situations, a blockchain is a digitized and decentralized public ledger of all cryptocurrency transactions. The latest exchanges are recorded and added to it in a consecutive manner as blocks of chain. The blockchain helps to keep track of all records of the transactions occurred in the network. In the network, each computer is associated with a correspondence computer system to get a duplicate of the blockchain which can be later downloaded by each user. Initially, this innovation was used with the end goal of enabling online record keeping of virtual money.

The Bitcoin is a virtual currency [1] developed by Santhoshi Nagamoto in 2009. It can be exchanged from one individual to another without a bank or central person. No exchange expenses are gathered for any transaction. This kind of money can be enabled clients to buy or offer bitcoins by using different types of monetary standards occurring in their geographies. The client can trade Bitcoin monetary standards [2] using their neighborhood monetary standards. Presently, in many business applications, the Bitcoin innovation is being used. The digital records are a database which can be used to share and synchronize over network system hubs, various sites, and geographies. This innovative currency is used to check exchanges through computerized currencies. It is made feasible for digitizing the money for any online transactions and the essential records of transactions can be archived for future references. In addition, the record's credibility will be confirmed by the entire group utilizing blockchain rather than central computing authority or a person.

Blockchain technology (BCT) offers a fabulous opportunity for the decentralization of transactions [3]. It has the ability to create businesses and operations that are both flexible and secure. The success of this technology can create products and service consumers but whether they trust and adapt to it remains to be seen. However, this is emphatically a space for investors to watch. The demand for blockchain-based services is growing at a rapid pace and the technology is also maturing and advancing. Possible applications for of blockchain technology are still either in the development stage or in beta testing. If more money is poured into the market to promote blockchain-based start-ups, consumers should not be surprised to see distributed ledger technology services and products in the near future.

Blockchain is an apparently a data structure, and suites for related virtual money conventions that have often taken the world of financial technology. It is becoming more famous though it's problematic advancements and historical applications in the advanced cryptographic Bitcoin currency. Today, the blockchain innovation has been used in more number of new applications, for example, smart contracts, is the pieces of the software that provides facility of blockchain to transfer the financial transactions information automatically to multiple parties according their agreement.

Blockchain technology depends heavily on the principles of cryptography and data security, particularly in terms of message verification, focused on unalter proof and unalter strength. In its most dynamic shape, a blockchain might be portrayed as an alter clear record shared inside a system of entities, where the record holds a record of exchanges between the entities. To accomplish alter confirm in the record; blockchain makes use of cryptographic hash functions.

In this paper, Sect. 2 describes the related research work carried out by various researchers. Section 3 discusses the working mechanism of BCT and Sect. 4 briefs the sign-in process involved in the blockchain technique. Sections 5 and 6 explore the various applications of BCT in industrial sectors, such as banking, IoT, music industries etc. and look at research areas for the near future. Section 7 concludes the paper.

2 Related Work

The blockchain is best known as the innovation running the Bitcoin digital currency. It is an open record framework for maintaining the trustworthiness of exchange information [4]. It was first used when the Bitcoin cryptographic currency was introduced. Bitcoin the most commonly used application that employs the blockchain innovation [5]. It is a decentralized distributed advanced cash installment Framework that comprises an open exchange record called blockchain [1]. The basic element of Bitcoin is transfer of the money from one place to another without participation of a person, bank or organization in control [6]. There are some familiar currencies, such as USD, EUR and Indian rupees that are always in cash trade markets [7]. Bitcoin has attracted the attention of different groups and is presently the most advanced currency utilizing the blockchain innovation [8].

Bitcoin utilizes principles of the public key cryptography (PKC) [9]. In PKC, the client has one set of public and private keys. A person with a public key is utilized as a part of the address of the client's Bitcoin wallet, and the private key is for the confirmation of the client. The exchange of Bitcoin involves people in the public key of the sender and private key of receiver. Within around 10 minutes, the exchange will be composed in a block. This new piece of block is then connected to a previously composed block. All blocks including data about each exchange made are then stored in hubs or network nodes.

The blockchain is the decentralized system of Bitcoin, and is used for issuing transaction cash for Bitcoin clients. This system can boost people in wide-ranging record of all Bitcoin transactions that have ever been executed, without an outsider control [10]. The benefit of blockchain is that the overall grouping blocks cannot be changed or erased after the information has been declared by all the hubs. So, blockchain is outstanding in terms of its information correctness and security attributes [11]. This innovation can also be connected to different types of employment. It can, for instance, create a situation for computerized contracts and distributed information sharing in a cloud computing. The motivation behind the blockchain technology is to use it for different processes and applications.

With the increasing practice of Bitcoin as an approach to drive installments and exchanges, security methods and their effect on the financial wealth of Bitcoin clients have expanded. In literature some of the recognized papers exhibited security breaches that had happened to the Bitcoin currency [12]. The researcher, Lim et al. [13] describes the pattern of a security breach in Bitcoin and their measures. As indicated by the authors, it had experienced every single possible form of security breach. The blockchain system is composed with the view that legitimate hubs control the system [1].

Luu et al. [14] present a security attack named the verifier's problem, which drives normal miners to skip verification wherever the verifying transactions require major computational resources in Bitcoin and particularly in a Bitcoin named Ethereum. The authors formalize a consensus model to give incentives to miners by limiting the amount of work required to verify a block.

Armknecht et al. in their work explained about how to support security and privacy in the Ripple system, which is one of the consensus-based circulated payment protocols. The paper discusses the difference between the protocol of Ripple and Bitcoin-focused Blockchain fork. A fork can occur if two conflicting ledgers get a clear majority of votes, and might lead to double spending attacks. The researchers, Decker and Wattenhofer [15], found that the propagation delay in the Bitcoin network is the main reason for Blockchain forks and discrepancies among duplications. This is done by analyzing Blockchain synchronization mechanism.

In Bitcoin, the private key is the real verification component. Verification in a digital currency controls self-certification. There have been a few examples involving validation. For instance, there is the outstanding case in Mt. Gox, where a Bitcoin wallet organization was attempted. In the attack, Mt. Gox's stored private keys of their client was stolen [16]. Many researcher are focused on security aspects of the Bitcoin and in some papers addressing about the issues facing in validation process of the Bitcoin.

3 The Workflow of Blockchain Technology

Cryptocurrency is digital money created from code. It is a medium of exchange between users. It is created and stored electronically in the blockchain using encryption techniques to control the creation of money units and verify the transfer of funds. Cryptocurrency is a new electronic cash system that uses a peer-to-peer network to prevent double spending [17]. It is completely decentralized with no server or central authority. Every peer (node/computer) has a record of the complete history of all transactions and thus the balance of every account. Cryptocurrency has no intrinsic value. It is not redeemable against another commodity. It has no physical form and exists only in the network. Its supply is not determined by a central bank and the network is a completely decentralized process. The workflow of blockchain technology is depicted in Fig. 1. We can also understand the working principle of blockchain technology using the following sequence of steps.

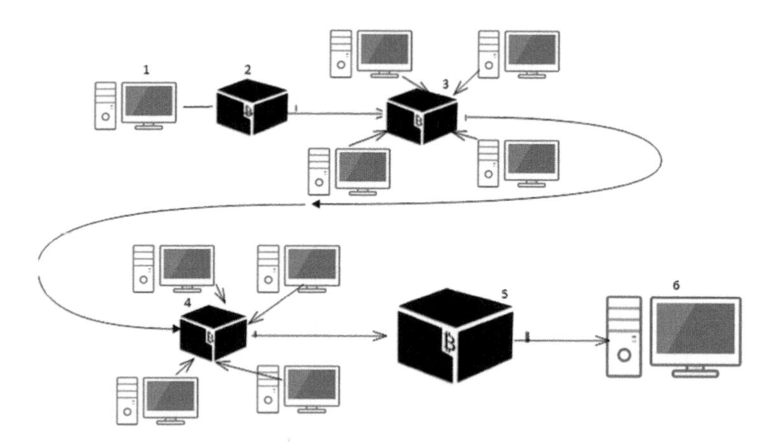

Fig. 1 The working mechanism of blockchain technology

The steps in blockchain technology are:

1. One requests the transaction.
2. The requested transaction is broadcast to a P2P network consisting of computers, known as nodes.
3. The network of nodes validates the transaction and the user's status using algorithms.
4. A verified transaction can involve cryptocurrency, contacts, records or other information.
5. Once verified, the transaction is combined with other transactions to create a new block of data for the ledger.
6. The new block is added to the existing block chain in a way that is permanent and unalterable.
7. The transaction is completed.

The above process is continued for all transactions without involvement of third parties or a bank. Currency exchanges, such as Okcoin, Shapeshift, and Poloniex, facilitate the trade of hundreds of cryptocurrencies such as Bitcoin, Litecoin, Ethereum, and Dogecoin etc. These all have their own advantages and disadvantages. The daily trade volume of these technologies exceeds that of major European stock exchanges. As a result, many organizations are looking to adopt their own cryptocurrencies to run their business effectively.

4 The Secure Sign-In Procedure in Blockchain Technology

BCT is used in the most well-known cryptocurrency, named Bitcoin. It is a very important technology which can be applied in many ways. It focuses on challenges that have arisen in digital transactions, such as double spending and currency reproduction. It reduces the cost of the transaction and also provides greater security and authenticity for users' accounts. It eliminates the need for third-party verification, custodians, and payment devices. It is being used to protect sensitive records and to authenticate the identity of the users. It has a simple keyless cryptography infrastructure which stores data hashes. It runs hashing algorithms for transaction verification. Public key cryptography is more vulnerable. The simple login processes in the network result in the generation of vulnerabilities. Weak passwords are giving opportunities to attackers to access network infrastructures.

In this method, the decentralization of the network helps in providing an understanding between the two different parties for verification of transactions through Blockchain-based security certificates. It verifies the integrity of transactions and also helps to check account balances, making the attacks on the transaction mathematically impossible.

The login process in the blockchain network is explained in Fig. 2. The transactions and how they are signed and linked together can be seen below. Let us consider the middle transaction, which is transferring bitcoins from source transaction B to destination transaction C. The contents of the transaction are hashed together in such a way that the hash of the previous transaction along with B's private key. However, B's public key is also included in the transaction.

By performing several steps, anyone who is in this network can verify that the transaction is authorized by B. First, B's public key must correspond to B's address

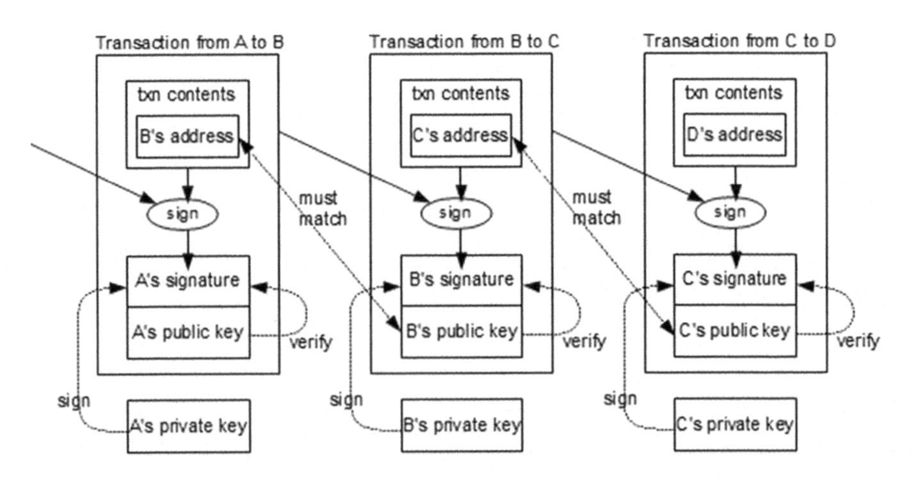

Fig. 2 Secure transaction using blockchain technology

in the previous transaction, proving the public key is valid. The address can easily be derived from the public key as explained above.

Next, B's signature of the transaction can be verified using B's public key in the transaction. These steps ensure that the transaction is valid and authorized by B. One unexpected aspect of Bitcoin is that B's public key isn't made public until it is used in a transaction. With this system, bitcoins are passed from one address to another address through a chain of transactions. Each step in the chain can be verified to ensure that the bitcoins are spent validly.

5 Applications of Blockchain Technology

BCT is becoming more powerful by the day in online transaction management systems. It applications are spreading to many industrial domains. The potential applications areas of blockchain technology can be broadly divided into two categories.

5.1 Financial Applications

The banking sector [18] is looking at various use-cases of BCT that are occurring across the world and is trying to find suitable cases to be implemented in their organization.

- *Digital Currency*: Electronic money, which is an early form of advanced cash, is formally characterized as value put away electronically in a gadget, for example, a chip card or a hard drive in a PC. The value put away and exchanged should be designated in a sovereign cash to be thought of as e-cash; while, much of the time computerized monetary standards are not named in or even fixed to a sovereign money, yet rather are named in their own units of significant worth. Digital money is a type of advanced cash intended to function as a medium of trade utilizing cryptography to secure exchanges and to control the formation of extra units of the money. Cryptographic forms of money have their own particular focal points and drawbacks as detailed below:
- *Control and Security*: In the blockcain, the users are in control of their transactions, without foregoing their privacy while overcome identity stealing. Due to the fact that blockchain transactions cannot be reversed, do not carry with them personal information, and are protected, merchants are secured from potential sufferers that might occur from fraud.
- *Transparency*: All completed transactions in the blockchain are accessible for anyone to view and check the transactions conveniently. The transactions are openly available for all the users who are the network. The transactions occurred in the blockchain method cannot be controlled by single central authority or a

bank or an organization. It is easy to send and get the cash from anywhere and anytime across the globe.

- *Very Low Transaction Cost*: Recently, the fees of blockchain have been very low. On a priority basis users also include fees to process the transactions. Digital currency exchanges help merchants to process transactions by converting them into fiat currency by charging lower fees than credit cards and PayPal.
- *Risk and Volatility*: Digital currencies are very volatile mainly due to the fact that there is a limited amount of coins and the demand for them is increasing each passing day.
- *Risk and Volatility*: Digital currency has a limited amount of coins due to its volatility and the demand for them is increasing day by day.
- *Trade Finance*: One of the very important areas of business is trade finance.
- *BCT in Capital Markets*: BCT is a distributed ledger technology which is used in many global financial markets, involving central counter parties, brokers, custodians, investment managers, and central securities depositories.
- *Pre-Trade*: The BCT system is used to store and facilitate KYC (Know Your Customers) information which helps in reducing the cost and removes a require number of KYC checks. It helps to maintain transparency and present credit ratings and reduced credits ratings of the customers.

5.2 Non-financial Applications

Business intelligence [19] expects companies to flesh out their blockchain IoT solutions. It is a promising tool that will transform parts of the IoT and enable solutions that provide greater insight into assets, operations, and supply chains. It will also transform how health records and connected medical devices store and transmit data. Blockchain won't be usable everywhere, but in many cases it will be a part of the solution that makes the best use of the tools in the field of the IoT. It can help to address particular problems, improve workflows, and reduce costs, which are the ultimate goals of any IoT.

6 Research Directions

In recent times, Bitcoin has gained more popularity in the market and has attracted more number of people into the world of cryptocurrency (digital money). Therefore; the people are buying and trading bitcoins everyday. So, Bitcoin has become a prominent future research topic for the researchers due to its significant applications. The industry and academia also may conduct more research on bitcoins from technical and Industrial perspectives.

In digital world, the Bitcoin uses blockchain technology even though there are many numbers of other cryptocurrencies in the market. It is may become a more attracted area for the researchers. In our broader study of the blockchain technology, we found some of the papers are discussing about the licensing, IOT, and smart contracts in a blockchain environment. Furthermore, the security and privacy of blockchain will also be a future area of research for interesting researchers.

7 Conclusions

The blockchain innovation runs Bitcoin digital money for online transactions. It is a decentralized domainfor exchanges, where every one of the exchanges is recorded to an open record and made available to every user. It provides security, usability, and clarity to all its users. However, these traits set up a considerable measure of specialized difficulties and impediments that should be addressed for it to become a proven technology in the near future.

Acknowledgements We would like to thank our *CMREC* management and principal who encouraged us at every stage while writing this paper.

References

1. Nakamoto S (2012) Bitcoin: a peer-to-peer electronic cash system. Consulted 2008
2. Yli-Huumo J, Ko D, Choi S, Park S, Smolander K (2016) Where is current research on blockchain technology?—a systematic review
3. Nakamoto S (2009) Bitcoin: a peer-to-peer electronic cash system, Bitcoin project registered at SourceForge.net
4. Swan M (2015) Blockchain: blueprint for a new economy. O'Reilly Media, Inc.
5. Coinmarketcap, Crypto-Currency Market Capitalization (2016). https://coinmarketcap.com/. Accessed 24 Mar 2016
6. Kondor D, PoÂsfai M, Csabai I, Vattay G (2014) Do the rich get richer? an empirical analysis of the Bitcoin transaction network. PLoS ONE 9(2):e86197. https://doi.org/10.1371/journal.pone.0086197 PMID: 24505257
7. Herrera-Joancomart J (2015) Research and challenges on Bitcoin anonymity. In: Garcia-Alfaro J, Herrera-Joancomart J, Lupu E, Posegga J, Aldini A, Martinelli F et al (eds) Data privacy management, autonomous spontaneous security, and security assurance. Lecture notes in computer science, vol 8872. Springer International Publishing, pp 3–16. http://dx.doi.org/10.1007/978-3-319-17016-9_1
8. Bitcoincharts (2016). https://bitcoincharts.com. Accessed 24 Mar 2016
9. Housley R (2004) In: public key infrastructure (PKI). Wiley. http://dx.doi.org/10.1002/047148296X.tie149
10. Kitchenham B, Charters S (2007) Guidelines for performing systematic literature reviews in software engineering
11. Massias H, Avila XS, Quisquater JJ (1999) Design of a secure time stamping service with minimal trust requirements. In 20th symposium on information theory in the Benelux, May 1999

12. Vasek M, Moore T (2015) There's no free lunch, even using bitcoin: tracking the popularity and profits of virtual currency scams. In: Bhme R, Okamoto T (eds) Financial cryptography and data security. Lecture notes in computer science, vol 8975. Springer Berlin Heidelberg, pp 44–61. http://dx.doi.org/10.1007/978-3-662-47854-7_4
13. Lim IK, Kim YH, Lee JG, Lee JP, Nam-Gung H, Lee JK (2014) The analysis and countermeasures on security breach of bitcoin. In: Murgante B, Misra S, Rocha AC, Torre C, Rocha J, Falco M (eds). Computational science and its applications ICCSA 2014. Lecture notes in computer science. vol 8582. Springer International Publishing. pp 720–732. http://dx.doi.org/10.1007/978-3-319-09147-1_52
14. Luu L, Teutsch J, Kulkarni R, Saxena P (2015) Demystifying incentives in the consensus computer. In: Proceedings of the 22nd ACM SIGSAC conference on computer and communications security CCS'15. ACM, New York, NY, USA, pp 706–719. http://doi.acm.org/10.1145/2810103.2813659
15. Decker C, Wattenhofer R (2014) Bitcoin transaction malleability and MtGox. In: Kutyowski M, Vaidya J (eds) Computer security ĐESORICS 2014. Lecture notes in computer science, vol. 8713. Springer International Publishing. pp 313–326. http://dx.doi.org/10.1007/978-3-319-11212-1_18
16. Bos JW, Halderman JA, Heninger N, Moore J, Naehrig M, Wustrow E (2014) Elliptic curve cryptography in practice. In: 18th international conference on financial cryptography and data securityĐ, FC 2014, Christ Church, Barbados, pp 157–175, 3–7 Mar 2014, Revised Selected Papers. http://dx.doi.org/10.1007/978-3-662-45472-5_11
17. Double-spending (2016). https://en.bitcoin.it
18. Applications of blockchain technology to banking and financial sector in India. In: IDRBT, Jan 2017
19. Lansiti M, Lakhani KR (2017) The truth about blockchain. Harv Bus Rev. Harvard University

Ensemble-Based Hybrid Approach for Breast Cancer Data

G. Naga RamaDevi, K. Usha Rani and D. Lavanya

Abstract Classification of datasets with characteristics such as high dimensionality and class imbalance is a major challenge in the field of data mining. Hence to restructure data, a synthetic minority over sampling technique (SMOTE) was chosen to balance the dataset. To solve the problem of high dimensionality feature extraction, principal component analysis (PCA) was adopted. Usually a single classifier is biased. To reduce the variance and bias of a single classifier an ensemble approach, i.e. the learning of multiple classifiers was tested. In this study, the experimental results of a hybrid approach, i.e. PCA with SMOTE and an ensemble approach of the best classifiers obtained from PCA with SMOTE was analyzed by choosing five diverse classifiers of breast cancer datasets.

Keywords Classification · PCA · SMOTE · Ensemble approach
Breast cancer data

1 Introduction

The main issues in the field of classifying datasets are how to handle the class imbalance and high dimensionality and how to reduce the variance and bias of classifiers. To avoid the problem of unbalanced data, instances should be sampled with sampling techniques such as oversampling and undersampling.

G. Naga RamaDevi (✉)
Department of CSE, CMRIT, Kandlakoya, Medchal Road, Hyderabad, India
e-mail: ramadeviabap@yahoo.co.in

G. Naga RamaDevi · K. Usha Rani
Department of Computer Science, SPMVV, Tirupati, India
e-mail: usharani.kuruba@gmail.com

D. Lavanya
Department of CSE, SEAGI, Tirupati, India
e-mail: lav_dlr@yahoo.com

A. Kumar and S. Mozar (eds.), *ICCCE 2018*,
Lecture Notes in Electrical Engineering 500,
https://doi.org/10.1007/978-981-13-0212-1_72

The SMOTE oversampling technique [1] creates new synthetic instances to the minority class instances rather than adding the existing minority instances. Feature extraction techniques are very helpful in overcoming the problem of high dimensionality. The popular feature extraction technique of principal component analysis (PCA) was chosen for our study.

In the literature it was proved that an ensemble approach is often more accurate than any of the single classifiers in the ensemble [2]. Generally, people will seek a second, third, or sometimes even a fourth opinion before making a decision because the risk factors are highly influential in areas such as finance, medicine, and the social sphere, etc. By aggregating the individual views of several experts in a particular field the most informed final decision may be obtained. This procedure is called an ensemble-based system and produces more favorable results than single-expert systems in a variety of scenarios for a broad range of applications.

An ensemble of classifiers is a combination of multiple classification techniques and has a great advantage over using a single classifier. An ensemble reduces variance and bias. Many women across the world fall victim to breast cancer and this is a major cause of death in women. Hence breast cancer datasets are considered suitable for experimentation and analysis. Our previous papers [3, 4] related to the comparison of classifiers with and without feature extraction technique, PCA and PCA versus SMOTE results.

This paper focuses on the performance of five popular and diverse classifiers using the hybrid approach, i.e. PCA with the combination of SMOTE and an ensemble method. The classifiers which were considered in the study are: k-nearest neighbor (k-NN), support vector machine (SVM), logistic regression (LR), C4.5 decision tree algorithm, and random forest (RF).

In Sect. 2 related work is presented. A description of the datasets is given in Sect. 3. The ensemble method and hybrid approach is presented in Sect. 4. Experimental results are presented in Sects. 5 and 6 contains the conclusion.

2 Related Work

Naseriparsa et al. [5] conducted experiments on lung cancer datasets along with PCA and SMOTE resampling to boost the prediction rate in the datasets. A naive Bayes classifier was used to classify the data.

Bidgoli et al. [6] proposed a hybrid feature selection method which is the combination of SMOTE and a consistency subset evaluation method. Experiments were conducted using five medical datasets of the UCI Repository.

Mustafa et al. [7] proposed a new algorithm using PCA and a farther distance-based synthetic minority oversampling technique (FD_SMOTE) to classify the medical datasets. This study revealed that the new algorithm increases the performance of area under the curve (AUC) metrics and accuracy metrics on different biomedical datasets.

Sanchez et al. [8] performed an analysis of SMOTE, RUS (random undersampling) with a PCA method using the classifiers C4.5 and neural network on hyper spectral images. Results proved that the order of SMOTE + PCA gives a good level of accuracy.

Sato et al. [9] presented a local PCA approach along with SMOTE for classifying a telecommunication dataset.

Derrick et al. [10] conducted experiments on biomedical datasets to address the problem of imbalanced data and dimensionality. Learning vector quantization-SMOTE (LVQ-SMOTE) along with PCA was recommended to classify biomedical data.

Seema et al. [2] experimented on micro array gene expression cancer datasets by applying an ensemble of classifiers with a 10-fold cross-validation technique. Experimental results outperformed single classifiers.

Gouda et al. [11] experimented on breast cancer datasets with various combinations of multi-classifiers. Results showed that multi-classifiers lead to better performance than individual classifiers.

3 Description of Datasets

In this study to evaluate the performance of classifiers, only breast cancer datasets were considered as breast cancer is a prime cause of death in women across the world as well as in our own country. The different breast cancer datasets considered were: Breast cancer (BC), Wisconsin diagnostic breast cancer (WDBC), Wisconsin breast cancer (WBC) and Wisconsin prognostic breast cancer (WPBC). These datasets are publicly available at the UCI Machine learning Repository [12]. A description of the datasets is given in Table 1. For the analysis of the performance of classifiers on the breast cancer datasets experiments are conducted with 10-fold cross-validation using the open source data mining tool Weka.

Table 1 Description of breast cancer datasets

Dataset	No. of instances	No. of attributes	No. of instances under each class	
			Major class (−ve)	Minor class (+ve)
BC	286	10	NR—201	R—85
WBC	699	10	B—458	M—241
WDBC	569	32	B—357	M—212
WPDC	198	34	NR—151	R—47

NR No Recurrence, *B* Benign, *R* Recurrence, *M* Malignant

4 Ensemble Method

Ensemble methods have been found to be very useful in a range of diverse tasks relating to computer-aided medical diagnosis, particularly in terms of increasing the reliability of diagnosis [2, 13]. Ensemble methods train multiple learners to solve the same problem.

In contrast to ordinary learning approaches which try to construct one learner from training data, ensemble methods try to construct a set of learners and combine them.

For an ensemble of classifiers there are various approaches: forward selection, backward elimination, and best model, etc. An ensemble of classifiers is nothing but the average of the performance of all the classifiers considered for the ensemble. For an ensemble all the classifiers are combined and the performance compared. However, this study is concentrated on the classifiers with the best and second best accuracy rates out of a group of classifiers for the generation of an ensemble model with better performance because to get a good ensemble, base learners should be as accurate as possible.

4.1 Hybrid Approach: A Hybrid Approach of Classifiers with PCA and SMOTE

In this study a hybrid procedure for the construction of an ensemble model is proposed. We considered a best model approach for an ensemble of classifiers.

The hybrid procedure is as follows:

1. Train the base classifiers with PCA and SMOTE with various percentages on datasets.
2. Out of a set of classifiers select the best and second best classifiers with PCA and SMOTE on a particular dataset based on their accuracies.
3. Construct the best ensemble model.
4. Test the ensemble model.

5 Experimental Results

To assess the performance of the classifiers on the breast cancer datasets, experiments are conducted with 10-fold cross-validation using an open source data mining tool, WEKA (3.6.0). After preprocessing the data experiments are conducted with five diverse classifiers: k-nearest neighbor (k-NN), support vector machine (SVM), logistic regression (LR), C4.5 decision tree algorithm, and random forest (RF). The performance of the classifiers on the datasets was calculated along

Table 2 Performance of classifiers (accuracy) on datasets with PCA and various percentages of SMOTE

Dataset	PCA + SMOTE (%)	k-NN	SVM	LR	C4.5	RF
		Accuracy (%)				
BC	**50**	74.39	65.29	66.46	71.03	73.17
	75	**78.51**	69.05	68.48	64.46	**71.63**
	100	78.44	67.38	67.92	64.95	70.08
WBC	**50**	97.31	96.34	96.09	97.06	97.19
	75	97.84	96.70	96.58	97.49	97.38
	90	**97.93**	96.94	96.51	97.27	**97.82**
WDBC	**50**	95.85	96.29	97.03	93.33	96.15
	68	**96.92**	96.64	**97.34**	94.25	95.52
WPBC	**50**	73.75	75.11	76.47	72.85	79.18
	75	75.11	75.54	75.54	73.39	77.68
	100	**79.59**	74.69	73.87	72.65	**78.36**

with PCA and with various percentages of SMOTE. The results are presented in Table 2.

For each dataset the best two classifiers based on the accuracy are considered for the ensemble of classifiers model. By observing the above values for the BC dataset the best classifier (based on accuracy) is KNN with 78.51% with PCA and SMOTE 75%. Hence, the second best is also selected from the same combination, i.e. RF with 71.63%. Similarly for the WBC dataset KNN and RF (SMOTE 90%), for the WDBC dataset LR and KNN (SMOTE 68%) and for the WPBC dataset KNN and RF (SMOTE 100%) are selected. A best ensemble model is generated with the selected classifiers and the experiments are conducted on the datasets. The results are tabulated in Table 3.

From the results it can be seen that the accuracies of the ensemble models is high, and the Kappa values are also increased. We know that Kappa values with a value of 1 or lose to 1 will be in good agreement. We can also see that the RMSE values are also larger than the MAE values.

The confusion matrices of all the datasets with PCA and corresponding SMOTE is specified in Table 4:

Table 3 Performance of ensemble models on datasets

Dataset	PCA + SMOTE (%)	Ensemble model	Accuracy (%)	Kappa	MAE	RMSE
BC	75	KNN + RF	97.99	0.9589	0.1327	0.1694
WBC	90	KNN + RF	100	1.0	0.0121	0.0427
WDBC	68	LR + KNN	98.32	0.9664	0.0283	0.1159
WPBC	100	KNN + RF	100	1.0	0.0984	0.1249

From the results we can see that the ensemble models performed better when compared with the performance of the single classifiers. That comparison is presented in Table 5.

Table 4 Confusion matrices of ensemble models

Dataset		True prediction	False prediction
BC	No recurrence	198	3
	Recurrence	4	144
WBC	Benign	457	0
	Malignant	0	457
WDBC	Benign	353	5
	Malignant	8	350
WPBC	No recurrence	151	0
	Recurrence	0	94

Table 5 Comparison of accuracies (%) of various classifiers with the ensemble model

Classifiers	Without PCA	With PCA	PCA + SMOTE	Ensemble model
BC dataset				
K-NN	71.67	64.68	**78.51**	KNN + RF 97.99
SVM	69.58	71.32	69.05	
LR	68.88	69.93	68.48	
C4.5	**75.52**	**72.37**	64.46	
RF	68.18	69.58	71.63	
WBC dataset				
K-NN	94.99	95.85	**97.93**	KNN + RF 100
SVM	**96.85**	**96.56**	96.94	
LR	96.56	**96.56**	96.51	
C4.5	94.99	95.56	97.27	
RF	96.14	**96.56**	97.82	
WDBC dataset				
K-NN	96.13	93.67	96.92	LR + KNN 98.32
SVM	**97.89**	96.84	96.64	
LR	93.32	**97.54**	**97.34**	
C4.5	92.97	93.84	94.25	
RF	96.66	96.31	95.52	
WPBC dataset				
K-NN	68.68	66.66	**79.59**	KNN + RF 100
SVM	76.26	76.26	74.69	
LR	79.79	79.29	73.87	
C4.5	75.25	72.73	72.65	
RF	79.29	80.30	78.36	

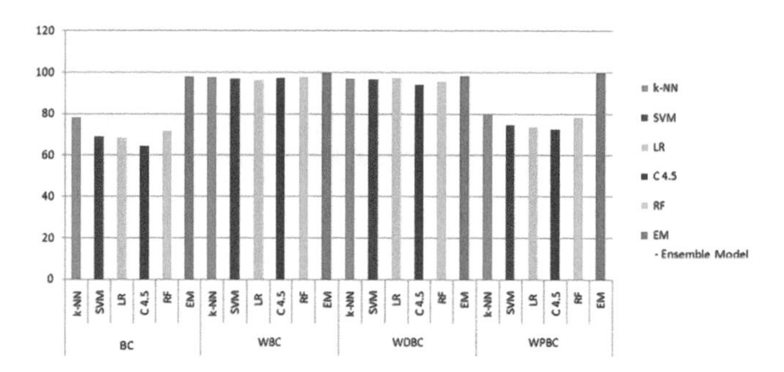

Fig. 1 Comparison of classifier and ensemble model accuracies

From the results it can be seen that the ensemble model (k-NN + RF) for the BC dataset has a higher accuracy, i.e. 97.99% than the performance of single classifiers with various approaches. Similarly for the WBC and WPBC datasets the ensemble model (k-NN + RF) performed with a 100% accuracy. The ensemble model (LR + k-NN) for the WDBC dataset has a better accuracy (at 98.32%) than the performance of single classifiers with various combinations. This is clearly represented in Fig. 1.

6 Conclusion

To overcome the problems of high dimensionality and imbalance related to classifying datasets a hybrid approach is proposed in this study with a combination of PCA, SMOTE and an ensemble of classifiers. The proposed method was tested with five classifiers on breast cancer datasets and the results are presented with an analysis. It is concluded that the proposed hybrid approach has proved to be the best method.

References

1. Chawla NV, Bowyer KW, Hall LO, Kegelmeyer WP (2002) SMOTE: synthetic minority oversampling technique. J Artific Intelligen Res 16:321–357
2. Seema S, Srinivasa KG, Supriya J, Bhavana C (2012) Ensemble classification with stepwise feature selection for classification of cancer data. Int J Pharmaceut Sci Health Care 6(2). ISSN 2249-5738
3. Naga Ramadevi G, Usha Rani K, Lavanya D (2015) Importance of feature extraction for classification of breast cancer datasets—a study. Int J Sci Innov Math Res (IJSIMR) 3(2):763–768, ISSN 2347-307X (Print)
4. Usha Rani K, Naga Ramadevi G, Lavanya D (2016) Performance of synthetic minority oversampling technique on imbalanced breast cancer data. In: Proceedings of the 10th INDIACom; INDIACom-2016; IEEE conference ID: 37465 3rd 2016 international

conference on "computing for sustainable global development", 16th–18th March, 2016 BharatiVidyapeeth's Institute of Computer Applications and Management (BVICAM), New Delhi, India
5. Naseriparsa M, Riahi Kashani MM (2013) Combination of PCA with SMOTE resampling to boost the prediction rate in lung cancer dataset. Int J Comput Appl 77(3)
6. Bidgoli A-M, Naseri Parsa M (2012) A hybrid feature selection by resampling, chi squared and consistency evaluation techniques. World Acad Sci Eng Technol Int J Comput Informat Eng 6(8)
7. Mustafa N, Memon RA, Li J-P, Omer MZ (2017) A classification model for imbalanced medical data based on PCA and farther distance based synthetic minority oversampling technique. Int J Advanc Comput Sci Appl 8(1)
8. Sanchez JS, Garcia V, Mollineda RA (2011) Exploring synergetic effects of dimensionality reduction and resampling tools on hyperspectral imagery data classification. In: International workshop on machine learning and data mining in pattern recognition MLDM 2011: machine learning and data mining in pattern recognition pp 511–523
9. Sato T, Huang BQ, Huang Y, Kechadi MT, Rim P (2010) Local PCA regression for missing data estimation in telecommunication dataset. In: International conference on artificial intelligence PRICAI 2010: PRICAI 2010: trends in artificial intelligence, pp 668–673
10. Derrick KR, Varayini P (2015) A one-dimensional PCA Approach for classifying imbalanced data. J Comput Sci Syst Biol ISSN: 0974-7230
11. Gouda IS, Abdelhalim MB, Magdy A-Z (2012) Breast cancer diagnosis on three different datasets using multi- classifiers. Int J Comput Informat Technol 01(01):2277–2764
12. http://archive.ics.uci.edu/ml/datasets.html
13. Lavanya D, Usha Rani K (2012) Ensemble decision tree classifiers for breast cancer data. Int J Informat Technol Convergen Serv 2(1):17–24

Probabilistic-Based Rate Allocation Flow Control Technique for Traffic Governance in Wireless Sensor Networks

Sudha Arvind, V. D. Mytri and Siddapuram Arvind

Abstract The proposed control mechanism delivers higher data traffic flow over a wireless sensor network. The proposed work is evaluated based on the probability of rate flow control method where the queue length is controlled by the traffic model based on the given traffic flow. The approach defines traffic governances based on the node mobility approach, where nodes are dynamically moved from short range to a more distant range. The link overhead and end-to-end delay are minimized in this technique, in comparison to the conventional controlling technique. Due to an early evaluation of congestion probability in the buffer unit, the blockage probability has been controlled.

Keywords Quality of service (QoS) · Traffic blockage control
Congestion

1 Introduction

Message updates can be obtained easily through a wireless sensor network (WSN) [1–4]. The routing protocol pertaining to the topology plays a crucial role while developing applications. Energy resources, bandwidth constraints, instability of low-power wireless links, and data redundancy are a few issues to consider while designing an efficient routing protocol which saves energy and provides better throughput [5]. The widespread nodes lead to multipath routing.

S. Arvind (✉)
CMR Technical Campus, Hyderabad, India
e-mail: sudha_chandrika@rediffmail.com

V. D. Mytri
Appa Institute of Technology, Gulbarga, India
e-mail: vdmytri.2008@rediffmail.com

S. Arvind
CMR Institute of Technology, Hyderabad, India
e-mail: scarvi@rediffmail.com

A. Kumar and S. Mozar (eds.), *ICCCE 2018*,
Lecture Notes in Electrical Engineering 500,
https://doi.org/10.1007/978-981-13-0212-1_73

Having nodes spread widely over a network leads to a realization of multipath routing from source to destination, extending lifetime and balancing consumption of energy by message shunting [6–8]. The congestion control mechanism [9] which cannot be ignored during multipath transmission has become a key research area.

2 Rate-Controlled Approach

During communication, the registered user sends packets to the registered network which works as a router to forward data to the next node. In this process, under random user entry or elimination causes congestion at the network level. To overcome the observed congestion, a rate allocation procedure is defined in [10]. In this rate-controlled congestion controlling technique, the congestion control is derived by referring to the back pressure at a given node defined by a CON-signal as,

$$\{(if\ congestion\ observed\ then \\ CON = 1\ else\ CON = 0 \tag{1}$$

based on the CON status, the allocated data rate is then set as [10],

$$R_i(t) = \begin{cases} R_i(t) + \Delta r & if\ R_i(t) < R_u, \quad CON = 0 \\ R_i(t) & if\ R_i(t) = R_u, \quad CON = 0 \\ \frac{R_i(t)}{2} & if \quad\quad\quad\quad\quad\ CON = 1 \end{cases} \tag{2}$$

In this technique CON is set as 1 if congestion is observed. Congestion is observed when the buffer queue length has reached the higher limit value, i.e. $L_{current} \cong L_{\max}$.

When the stated condition is satisfied congestion is detected in accordance to the set CON value. However, the status signal is set high at the upper bounding limit of the congestion, on the congestion level reaching to $L_{\max.}$ This technique gives a higher probability of total node blockage in the network.

3 Proposed Probabilistic Control Approach

To avoid such node failure probability, an adaptive rate-controlling technique based on the computed probability of congestion at the buffer, rather than direct queue length correlation is proposed. To develop the proposed probabilistic rate allocation, a probability of congestion is computed by setting two limiting values of L_{min} and L_{max} as two tolerance values, as shown in Fig. 1.

The technique of a dual tolerance limit reduces the congestion probability, and provides an initialization of congestion controlling at a lower stage of data buffering

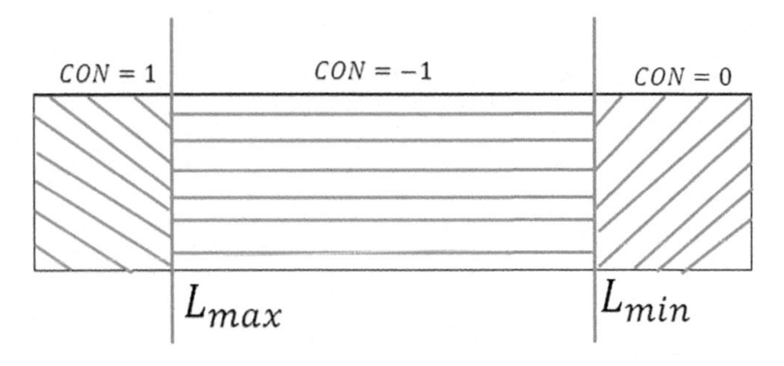

Fig. 1 Proposed dual threshold buffer logic

rather than an upper limit as in the conventional case. In such a coding technique the data rate for the flow of data from the buffer is R_u. The data are buffered into the buffer till the lower limit L_{min} is reached. Once the lower limit L_{min} is reached the data rate is then controlled by the probability of congestion at the buffer logic.

The probability of congestion at the buffer level is then defined as,

$$P_{cong} = \frac{P_{Blk}}{1 - P_{pkt}} \tag{3}$$

where,

P_{pkt} – *no of packets transfered*
P_{Blk} – *blockage rate*

The blockage rate for the buffer logic is defined as,

$$P_{Blk} = B_{current} - \left(\frac{R_{alloc}}{R_{max} - R_{alloc}}\right) \tag{4}$$

where,

$B_{current}$ – *current blockage rate*
R_{alloc} – *allocated data rate*

The current blockage at the buffer level is computed as,

$$B_{current} = \frac{L - L_{current}}{L} = 1 - \left(\frac{L_{current}}{L}\right) \tag{5}$$

where,

L – *total buffer length* and,
$L_{current}$ – *current queue length measured*

For such a buffer control operation, the congestion control signal CON is split into three logical values rather than two values as stated in [10]. The CON signal is assigned with CON = (−1, 0, 1), and Eq. (1) then becomes,

CON = 0 indicates no congestion.
CON = 1 indicates Max limit congestion
CON = −1 indicates a congestion with a probability inbetween Min and Max limits.

In this case the allocated rate updated by the probability of congestion and Eq. (2) is then updated as:

$$R_i(t) = \begin{cases} R_i(t) + \Delta r & \text{if } R_i(t) < R_u, \quad CON = 0 \\ R_i(t) + (\Delta r - P_{cong}) & \text{if } R_i(t) < R_u, \quad CON = -1 \\ R_i(t) & \text{if } R_i(t) = R_u, \quad CON = 0 \\ R_i(t) - \frac{R_i(t)}{P_{cong}} & \text{if } R_i(t) = R_u, \quad CON = -1 \\ \frac{R_i(t)}{P_{cong}} & \text{if } R_i(t) = R_u, \quad CON = 1 \end{cases} \tag{6}$$

The data rate allocated for congestion control based on probability of congestion, leads to minimization of node overhead. Due to early initiation of congestion controlling, buffer queuing is reduced, which results in a higher traffic flow through each node. This improves the overall network throughput, and this in turn improves network performance. An evaluation of network performance based on the suggested control mechanism is carried out and the observations obtained are outlined in the following section.

4 Observations

To evaluate the operational efficiency of the proposed technique, a randomly distributed wireless network is simulated with the control mechanism of rate control technique [10] and the proposed probabilistic control technique. A network is defined with the following network parameters (Table 1):

For the analysis of the proposed probabilistic-based coding technique compared to the conventional rate control technique, different networks with varying node densities are used. A node density of 50 nodes is considered for evaluating the impact and observations are obtained. The observations for such a network are illustrated in Fig. 2.

Under high density node condition with a mobility scenario:

The node variation, coverage area, neighbor links and selected path for communication are shown in Figs. 2, 3, 4 and 5 above.

Table 1 Parameters of the simulated network

Network parameter	Values
Node placement	Random
Mobility	Random
MAC protocol	IEEE 802.11
Power allocation	Random
Transmission range	40 units
Network area	200 × 200
Number of nodes	50–1000
Memory size/node (M)	3 M
L_{min}	0.15 × M
L_{max}	0.75 × M
Initial blockage probability	0.1

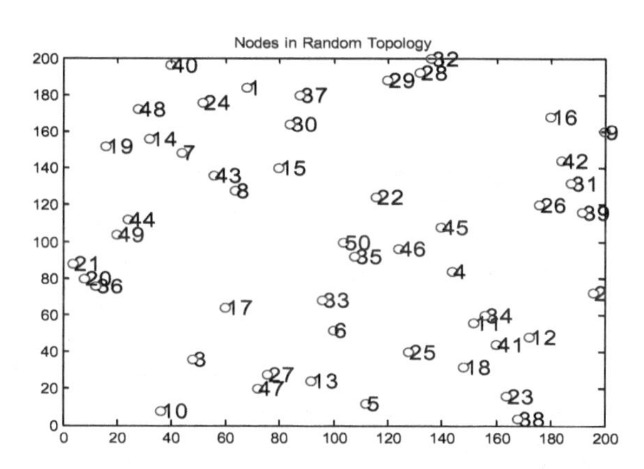

Fig. 2 Network topology for 50 nodes

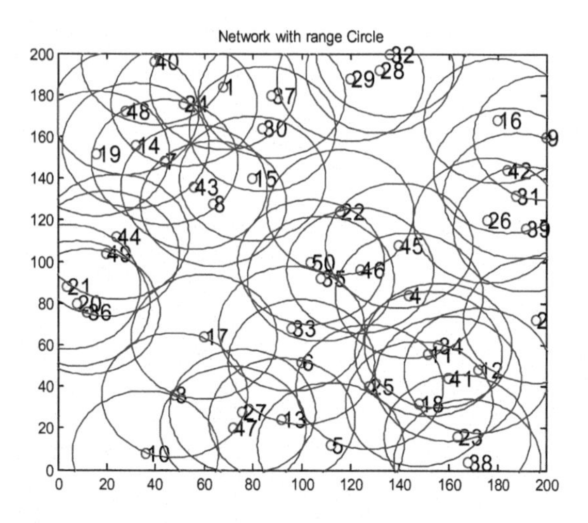

Fig. 3 Coverage area for each node

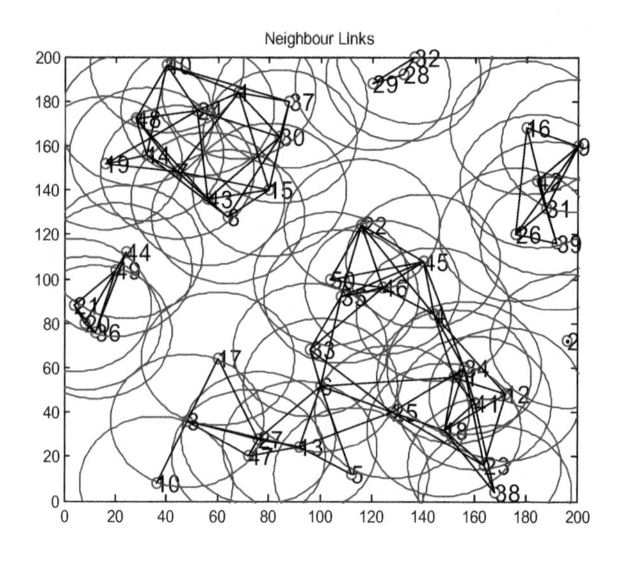

Fig. 4 Network with 1-hop neighbor links

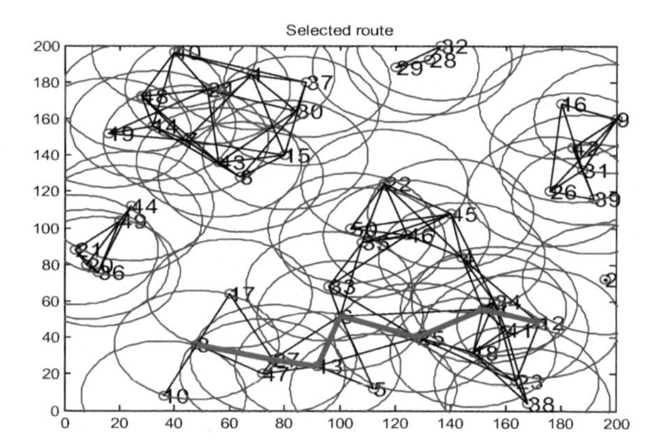

Fig. 5 Selected route communication network

5 Simulation Results

The network overhead is reduced due to the existence of nodes in large numbers for data exchange as shown in Fig. 6, resulting in a quick exchange of data. Due to the non-optimal rate controlling technique, the overhead was a little higher in the rate control method compared to the probabilistic method.

Network throughput has improved in the case of the probabilistic method and reduced in the rate control method, due to an increase in data buffering per node as shown in Fig. 7. A higher data rate for packet forwarding leads to a greater number of packets received, resulting in higher throughput in the network. The throughput resulted over received packets for an observing communication over a period of time is similar to that of 1-hop forwarding and buffering is reduced, hence improved allocated data rate. However, the increase in forward packets leads to a rise in

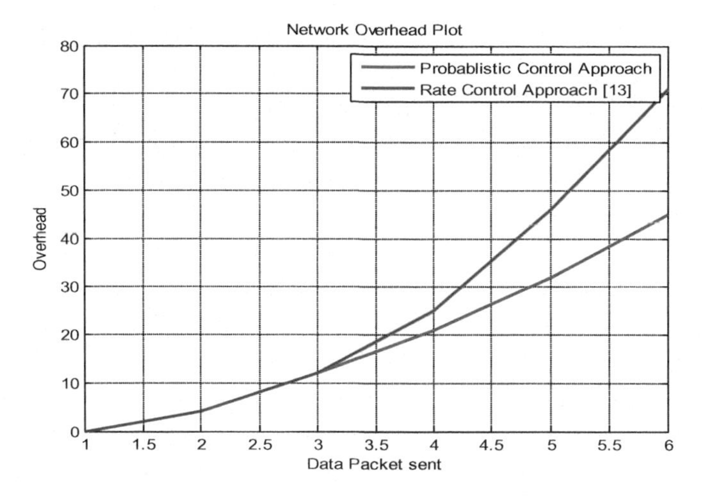

Fig. 6 Network overhead

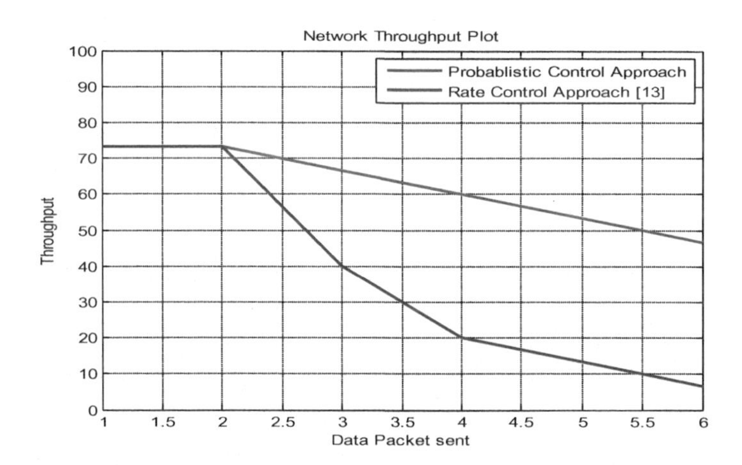

Fig. 7 Network throughput for a node density of 50

congestion level resulting in a reduction of throughput, which is maintained to be higher and linear in case of probabilistic coding. Figure 8 depicts the end-to-end delay factor for the developed system. The delay in the rate control method is found to be higher than in probabilistic coding. The congestion level developed in the probabilistic method, wherein buffered packets based on the rate control technique build the forwarding delay for each node.

The network overhead reduced from 70 to 45 due to the existence of a greater number of nodes, resulting in a quick exchange of data. Due to the non-optimal rate-controlling technique, the overhead is found to be higher in the rate-controlling technique compared to the probabilistic technique.

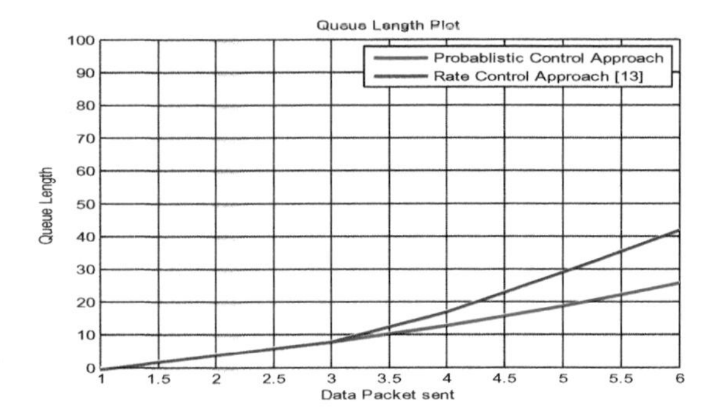

Fig. 8 End-to-end delay in the developed network

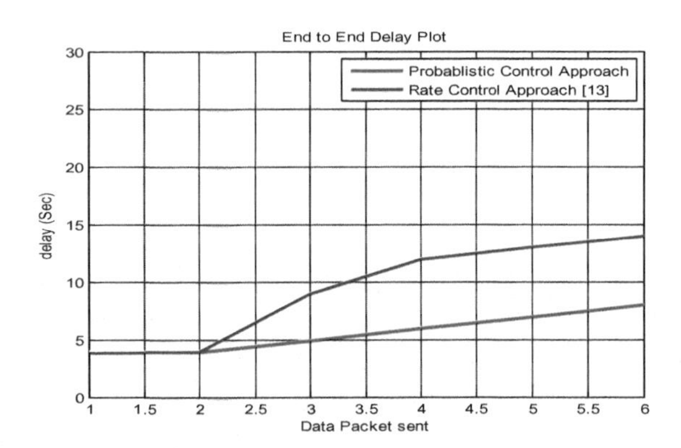

Fig. 9 Q-Length at the nodes

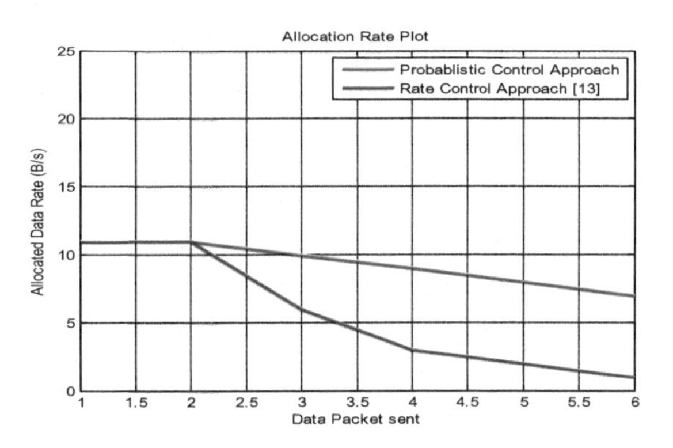

Fig. 10 Data rate allocation

The network throughput has improved from 8 to 48 in the case of the probabilistic-based control method and was found to be lowest in the rate-controlling method due to higher per node buffering of data (Figs. 9 and 10).

For high density nodes, the delay factor is reduced to 8 from 14, the Q length from 42 to 25, while the allocable data rate is increased from 2 to 7 in this probabilistic method.

6 Conclusion

A probability-based control technique to achieve higher throughput is proposed. This improvement results in faster data transfer, which in turn results in improvement to data quality at the receiver point. In such a network, measured parameters are transferred at a faster rate with a maximum level of precision. The qualitative parameters measured for this proposed technique illustrate the improvement of performance in such a network. The link overhead and end-to-end delay are observed to be minimized in this technique, in comparison to the conventional controlling technique. Due to an early evaluation of congestion probability in the buffer unit, the blockage probability is controlled. This control technique will lead to an improvement by providing better performance with respect to reliability and operational ability in distributed monitoring and controlling units. The quality of service (QoS) offered is observed in terms of service quality level, measured as a parametric value in heterogeneous network.

References

1. Wang S, Yan Y, Hu F (2007) Theory and application of wireless sensor networks
2. Xiaoyang L, Enqing D, Fulong Q, Bo C, Vertex coloring based distributed link scheduling for wireless sensor networks. In: The 18th Asia-Pacific conference on communications. pp 156–161, Oct 15–17 2012
3. Dong E, Qiao F, Zou Z, Wang J, Zhang, D, Li L (2014) Energy efficient distributed link scheduling protocol. Opt Precis Eng 22(2):474–480
4. Dong E, Zou Z, Zhang, D, Song J, Li L (2013) Time synchronization protocol based on dynamic route list for wireless sensor network. Opt Precis Eng 21(11):2951–2959
5. Couto DSJD, Aguayo D, Bicket J, Morris RA (2005) High throughput path metric for multi-hop wireless routing. wireless networks. pp 419–434 Nov 2005
6. Lou W, Liu W, Zhang Y (2006) Performance optimization using multipath routing in mobile Adhoc & WSN. In: Combinatorial optimism communication network
7. Key P, Massoulie L, Towsley D (2007) Path selection and multipath congestion control. In: Proceedings of the 26th IEEE international conference on computer communications. pp 143–151, Anchorage, AK, USA, 6–12 May 2007
8. Popa L, Raiciu C, Stoica I, Rosenblum D (2006) Reducing congestion effects in wireless networks by multipath routing. In: Proceedings of the 2006 IEEE international conference on network protocols. pp 96–105, Santa Barbara, USA, 12–15 Nov 2006
9. He T, Ren F, Lin C, Das S (2008) Alleviating congestion using traffic aware dynamic routing in wireless sensor networks. In: Proceedings of the 5th annual IEEE communications society conference on sensor, mesh and Ad hoc communications and networks (SECON 08). pp 233–241, San Francisco, USA, 16–20 June 2008
10. Peng Q, Enqing D, Juan X, Xing L, Wei L, Wentao C (2014) Multipath routing protocol based on congestion control mechanism implemented by cross-layer design concept for WSN. In: IEEE 17th international conference on computational science and engineering

Amended Probabilistic Roadmaps (A-PRM) for Planning the Trajectory of Robotic Surgery

Priyanka Sudhakara, Velappa Ganapathy, B. Manickavasagam and Karthika Sundaran

Abstract Trajectory planning is an essential aspect of research into the use of pliable needles for surgical processes. Sampling-based algorithms can generate trajectories and reach the target avoiding obstacles. However, the trajectories cannot match the physical constraints of injecting the pliable needle into human flesh, as the trajectories are not continuous. Aimed at solving this problem, an enhanced probabilistic roadmap (PRM) is used in this work. A PRM generates trajectories for surgeries that are minimally invasive and simultaneously guarantees the effectiveness and continuity of the trajectory. In this research work, the classical PRM method is enhanced by using a shape preserving piecewise cubic hermite interpolation (PCHIP) technique, used to generate smooth trajectories, which are important for navigating the curved path of the pliable needle in surgery. Trajectories that have been generated using the PRM satisfy direction constraints approach in terms of both source and target positions. As a result, the trajectories produced by the pliable needle are dynamically and geometrically feasible. Results of simulations performed show the validity of the algorithm implying that it can be efficiently used in trajectory planning of pliable needles in real-time surgical operations.

Keywords Shortest path · Trajectory planning · Probabilistic roadmaps
Pliable needle robot · Interpolation technique

P. Sudhakara (✉) · V. Ganapathy · B. Manickavasagam · K. Sundaran
School of Computing, SRM University, Kancheepuram 603203, Tamil Nadu, India
e-mail: priyanka.k@ktr.srmuniv.ac.in

V. Ganapathy
e-mail: ganapathy.v@ktr.srmuniv.ac.in

B. Manickavasagam
e-mail: bmanickavasagam90@gmail.com

K. Sundaran
e-mail: karthika.su@ktr.srmuniv.ac.in

A. Kumar and S. Mozar (eds.), *ICCCE 2018*,
Lecture Notes in Electrical Engineering 500,
https://doi.org/10.1007/978-981-13-0212-1_74

1 Introduction

Transdermic insertion of a needle into human tissue is one of the most widely used techniques in surgery. In order to make the surgical process easier, minimally invasive interventional treatment was developed. The development of this treatment has seen an inevitable shift in the surgical world. The application of this treatment is used extensively in the delivery of local drugs, cancer therapy, and pathological examination of living tissues. The pliable needle, which is traditionally different from a linear motion rigid needle, has enough flexibility to be inserted into the tissue in curved-flexible trajectories. By using the deformation of pliable needles, it is possible to circumvent obstacles like the skeleton and vital organs like blood and nerve vessels. This helps the pliable needle to reach its destination precisely, which is not always possible with classical linear motion rigid needles.

Trajectory planning is one of the research areas in the surgical world in order to gain prime motion control of pliable needle insertion. Techniques like rapidly-exploring random tree (RRT), probabilistic roadmaps (PRMs), search method, inverse kinematic method and object function method have all been proposed by various researchers for trajectory planning of the insertion of pliable needles. The insertion of a robotic pliable needle into soft human tissue has been frequently discussed after the proposal of the unicycle model [21]. It was proposed in order to describe the motion of slant-tip pliable needles. Trajectory planning and simulation results for the trajectory with which the pliable needle enters human flesh were obtained in [1]. Later, a trajectory planning algorithm based on the Markov decision process for uncertainty of the pliable needle [2] was proposed to circumvent obstacles in a 2D environment. Offline trajectory planning using a probability density function [16] was proposed to plan the trajectory of the pliable needle by calculating the probability density value of the needle's sharp tip. Yet, the applicability of this technique was not favorable for real-time applications as the calculation of the traversing time to reach the target was relatively long. To enhance the speed of trajectory planning, the heuristic RRT technique was embraced in [15]. Using feedback from two-dimensional ultrasound images, the control and planning of the pliable needle's motion took place online. Many other methods of online trajectory planning are proposed in [4, 5, 7, 19, 20].

Aimed at considering the features of pliable needle motion and solving the issues of trajectory planning of the insertions of pliable needles, the unicycle robotic model is adopted in the present study. An amended PRM method is then proposed to plan the trajectory of the insertion of the pliable needle. In this proposed work, instead of generating the geometry-arbitrary curve using the classical PRM method, executable curves trajectories have been obtained by using a shape preserving (PCHIP) interpolation technique along with the PRM method. This ensures the continuity and effectiveness of the trajectory that is generated. In addition, an evaluation function is established in order to choose the optimal trajectory that carries the minimum number of arcs and has the shortest superior collision-free trajectory in the presence of obstacles. When compared to the classical PRM method, an amended PRM is more appropriate for planning the trajectory of insertion of a pliable needle.

2 Design of Trajectory Planning for the Insertion of a Pliable Needle

2.1 Design of the Trajectory for the Insertion of a Pliable Needle

The body of the pliable needle has sufficient flexibility for the needle to enter human tissues and has a slant-tip. Consequently, the body of the pliable needle will flex when it is inserted into human tissue because of the side force generated by the human flesh. With an inconsistent radius, the needle can also move in a circular motion which is evidently different from classical linear motion rigid needles. The extrinsic pliable needle's twisting and insertion [18] controls the motion of the slant-tip needle when inserted into soft human flesh. Based on the unicycle model, the pliable needle's trajectory is contemplated as variable and continuous radii curves

The inserting input controls the needle when the twisting input is set to zero and the trajectory of the needle's motion will be a curve with minimal radius r_{min}. The geometric and material characteristics of both the needle and human tissue decide the value of r_{min} [22]. The trajectory of the needle's motion will be a curve with radius r, when the inserting input is aggregated with a duty-cycled twisting input, where $r_{min} \leq r \leq +$. These styles of controls and properties of motion have often been used in needle steering control and planning research [14, 18, 22].

The trajectory of the needle's motion is described as a curved pattern as shown in Eq. (1):

$$Trajectory = \sum_{i=1}^{n} Curve_i(a_i, u_i.r_i, h_i) \tag{1}$$

where h_i denotes the length of the $Curve_i$, u_i denotes the unit tangent vector of $Curve_i$ at a_i and is along the forward direction of the path, ai denotes the starting point of $Curve_i$, b_i is the end point of $Curve_i$, r_i denotes the radius of $Curve_i$, and C_i is the center of circle of $Curve_i$. The design of the trajectory considered by the pliable needle is illustrated in Fig. 1.

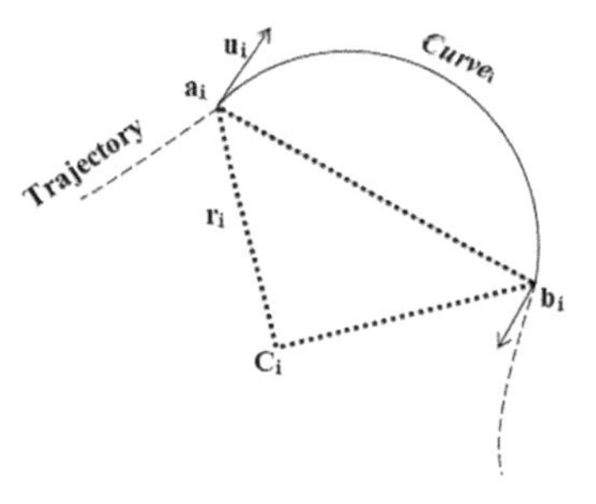

Fig. 1 The design of the pliable needle's trajectory

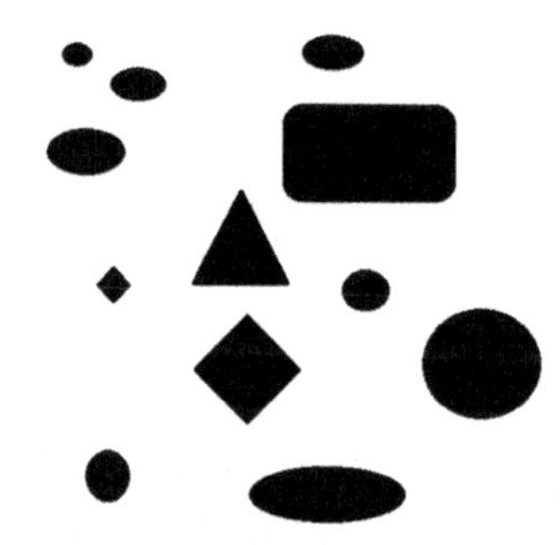

Fig. 2 The 2D scenario map for the insertion of a pliable needle

2.2 The Framework of the Scenario with Static Obstacles Where the Pliable Needle Has to Be Inserted

Figure 2 illustrates the 2D framework scenario for the insertion of a pliable needle. The 2D scenario is comprised of two spaces, an obstacle-free space (Cfree) and an obstacle space (Cobstacles). The static obstacles, which are randomly dispersed in the scenario, manifest as shapes filled with black. The free spaces are represented by the white areas and the obstacle spaces are represented by the black-filled shapes. The starting and destination positions will be assigned after creating the 2D framework scenario for the planning of the trajectory of the pliable needle.

3 Trajectory Planning for the Insertion of a Pliable Needle

3.1 Enhanced Probabilistic Roadmaps

A probabilistic roadmap (PRM) [10] is one of the methods used with sampling-based algorithms. Global trajectory planning is provided by other sampling-based trajectory planning algorithms [6, 12, 13, 17, 23], such as RRT and D* by using sampling of the grid space to capture the connectivity of the free configuration space. PRM in particular constructs trajectories that resemble a web of roads called roadmaps. They are acquired by sampling a huge set of connecting neighboring configurations called waypoints through obstacle avoided collision-free trajectories. Then various tasks such as shepherding, homing, or target finding [3, 8] are performed by the roadmaps. Combinational methods of PRM with shape preserving piecewise cubic hermite interpolation (PCHIP) are used as a technique for the non-holonomic robots to reach the target and also at the same time to preserve the exact pattern [11]. The shape preserving PCHIP technique is applied using a curve fitting tool in MATLAB. Unlike any other spline techniques with highly deviating curves, this PCHIP technique maintains the shape of the trajectory that has been traced. This is because in the resulting fitted curved trajectory, each and every pair of consecutive waypoints is linked by different cubic polynomial functions which are described using four coefficients.

Threshold distance connections between waypoints and the number of waypoints generated on the map are the two prime properties of the PRM technique. Waypoints that are generated from the technique are positioned on the map and each and every waypoint is connected to all the other waypoints positioned on the map with threshold connecting distances between them. The higher the number of waypoints and their connections, the higher the possibility of choosing the efficient-optimal trajectory is. However, the computational time needed to select the feasible trajectory gets maximized as the amount of waypoints in the PRM technique increases. Hence, a sufficient amount of waypoints to occupy the positions on the map may yield a better outcome. This is because the connection threshold distances between the waypoints have a direct influence on the amount of linkages connected between the waypoints. The numbers of accessible trajectories are minimized by setting a lower threshold distance between the waypoint connections. Therefore, PRM is used to select an ultimate trajectory which is an obstacle-free path from the accessible trajectories. An appropriate combination of number of waypoints in the PRM is needed for the framework scenario map with an appropriate suitable threshold distance connection. For a map in a basic scenario with a higher number of distance connections, fewer waypoints and a fewer obstacles positioned, efficient outcomes can be obtained. Generally, the possibility of finding an optimal trajectory in a complex unknown environment can be enhanced by positioning a larger number of waypoints with a smaller number of threshold distance connections. It is observed that the effect of positioning different combinations of number of waypoints can lead to an optimal result. In our case, Figs. 3, 4, 5, 6 and 7 illustrate five divergent occurrences of PRM obtained when the numbers of waypoints are varied.

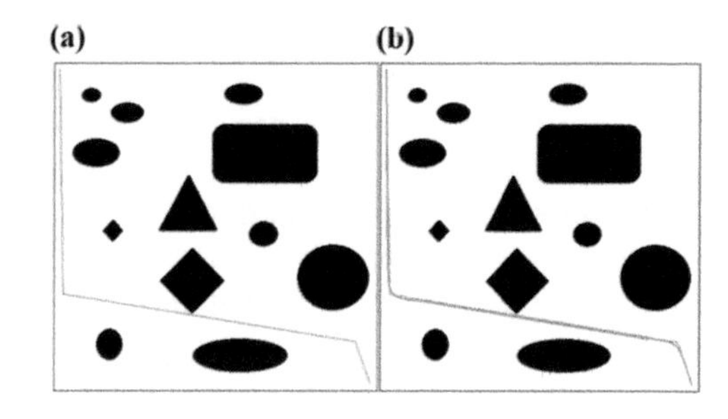

Fig. 3 Case 1: PRM with 10 waypoints

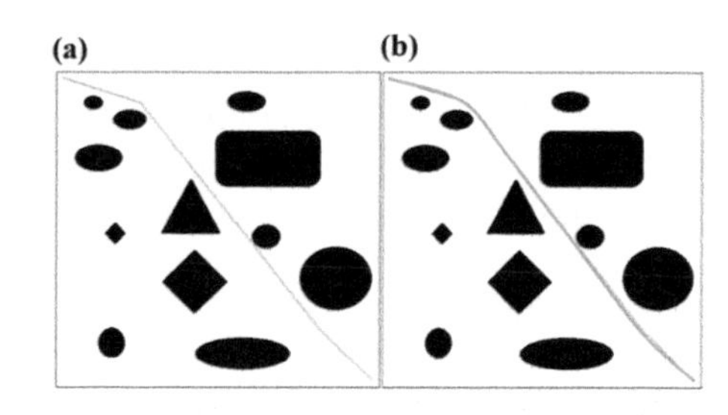

Fig. 4 Case 2: PRM with 50 waypoints

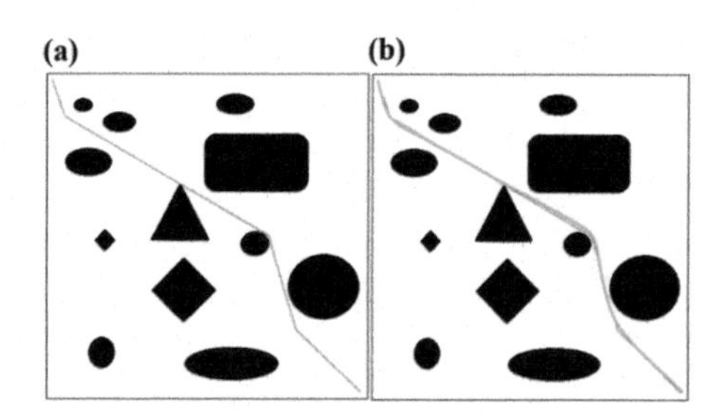

Fig. 5 Case 3: PRM with 100 waypoints

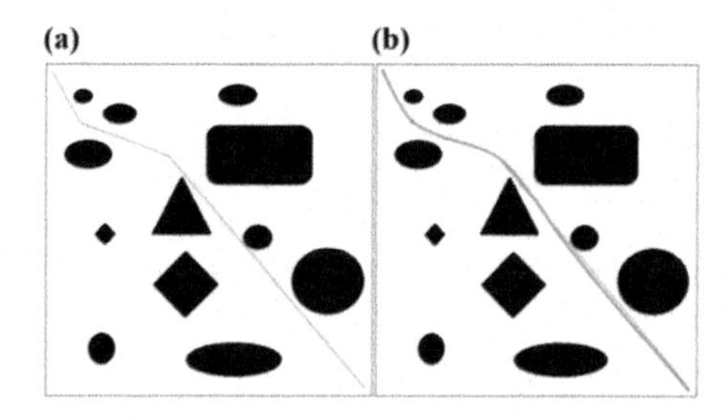

Fig. 6 Case 4: PRM with 150 waypoints

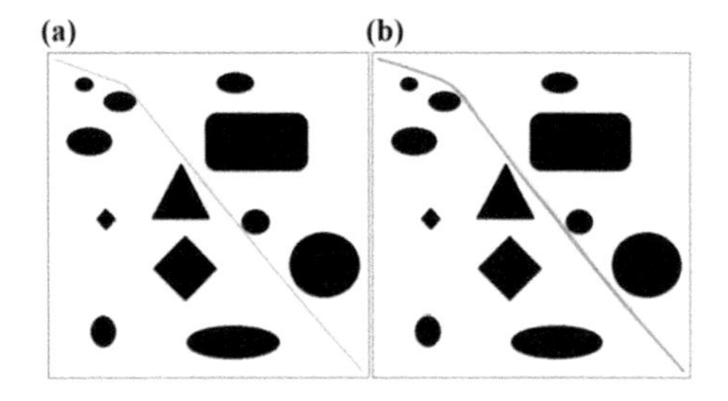

Fig. 7 Case 5: PRM with 200 waypoints

4 Simulation Results

To validate the performance of pliable needle trajectory planning using PRM, we carried out several simulation experiments in a static obstacle environment for five separate waypoint cases. The trajectory planning scenario utilized in this paper comprises an original map resolution of 500 × 500, fixed source and target points at [10, 10] and [490, 490] respectively for the five varied scenarios. Figures 3, 4, 5, 6 and 7 illustrate the simulation results for same scenario framework with five different sets of waypoints. Simulation results (a) and (b) in each scenario show the trajectory traced by the robot using a classical PRM method, and the trajectory traced after applying a PCHIP technique to the PRM method respectively. Based on the work done in [9], we have enhanced the PRM using a PCHIP interpolation technique.

From the simulation results shown in Figs. 3, 4, 5, 6 and 7, we have found that as the number of waypoints increases, the trajectory traced is very much smoothed. As the pliable needle used for the path planning is curved, it will be easy for the surgical robot to turn and move forward. This will reduce the time taken by the pliable needle robot to reach the goal as it need not stop while changing direction. It is evident that there is a poor possibility of tracing an optimal trajectory by setting a smaller number of waypoints, 10 in this case, with the same threshold distance connections. As the number of waypoints is increased, the PRM technique has a higher probability of tracing an optimal trajectory. Therefore, it is better to harmonize the scenario with a greater number of waypoints with the same threshold distance connections. Figures 3, 4, 5, 6 and 7 present occurrences of trajectories produced by PRM and amended PRM methods. Waypoints in the range of 10 to 200 in steps of 10 have been fixed and the results have been generated. For convenience, only 5 sets of waypoints are illustrated in the figures given below. A model with a 500 × 500 area has been created where the waypoints have been varied from 10–200 in steps according to the scenario.

From Table 1 and Fig. 8, it is evident that selecting waypoints from 40 to 70 for the chosen scenario would produce the minimal trajectory length with considerably reduced processing time. This proposed method can be applied to any predefined specific environment in which there can be any number of obstacles present. The

Table 1 Length of the trajectory traced and processing time respective to the waypoints

Number of waypoints	Trajectory length (cms)	Processing time (s)
10	869	22.3
20	795	45.4
30	750	43.7
40	701	50.1
50	697	43.8
60	699	47.2
70	700	49.5
80	709	55.8
90	709	56.5
100	716	57.3
110	696	60.1
120	695	61.6
130	699	67
140	695	71.6
150	697	75.9
160	691	75.6
170	690	86.5
180	697	90.6
190	694	111.9
200	691	104.2

proposed method can be used to select the shortest and smoothest trajectory with minimal processing time depending on the real-time scenario.

5 Conclusion

In this research paper, a trajectory planning approach for surgical robots is proposed. The proposed amended PRM method uses a shape preserving PCHIP technique in order to generate trajectories with enhanced smoothness and continuity without any additional computation. To validate the performance of the proposed work, simulations were carried out and the simulation results show that this method is feasible in both fixed and dynamic unknown environments. It is also optimally asymptotic and can satisfy a restricted directional movement approach on both source and target locations. From the graph shown in Fig. 8, it can be inferred that for a fixed environment, when we select waypoints between 40 and 70, the trajectory length is minimal with considerably reduced processing time. This proposed method will be utilized to select the shortest traversing path with a smooth trajectory and a lower processing time for any type of predefined environment. The application of this proposed method optimizes the cost function and can be utilized

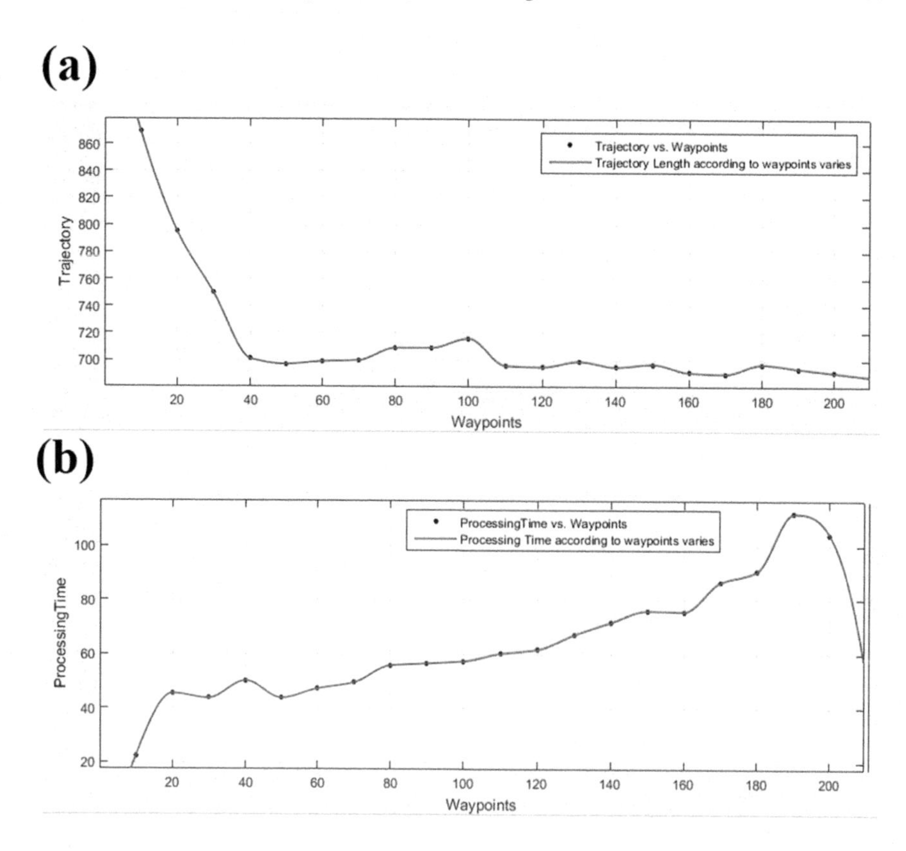

Fig. 8 **a** Graph illustrating trajectory length versus number of waypoints, and **b** Graph illustrating processing time versus number of waypoints

in various other similar fields. In future work, this method can be used in scenarios with different environments and with different dynamic obstacles positioned. In addition, when this method is considered for dynamic obstacles, each and every instance of the scenario is mapped in order to recalculate and reorient all the trajectories that have been traced.

References

1. Alterovitz R, Goldberg K, Okamura A, (2005) Planning for steerable bevel-tip needle insertion through 2D soft tissue with obstacles. In: IEEE proceedings of the IEEE international conference on robotics and automation 1640–1645
2. Alterovitz R, Branicky M, Goldberg K (2008) Motion planning under uncertainty for image-guided medical needle steering. Int J Robot Res 27(11–12):1361–1374

3. Bayazıt OB, Lien J-M, Amato NM (2005) Swarming behavior using probabilistic roadmap techniques. In: Swarm robotics. Springer, pp 112–125
4. Bernardes MC, Adorno BV, Poignet P et al (2013) Robot-assisted automatic insertion of teerable needles with closed-loop imaging feedback and intra-operative trajectory replanning. Mechatronics 23(6):630–645
5. Bernardes MC, Adorno BV, Borges GA et al (2014) 3D robust online motion planning for steerable needles in dynamic workspaces using duty-cycled rotation. J Control Autom Electr Syst 1–12
6. Denny J, Amato NM (2013) Toggle PRM: a coordinated mapping of C-free and C-obstacle in arbitrary dimension. Springer Tracts Adv Robot 86:297–312
7. Ganapathy V, Sudhakara P, Jie TTJ et al (2016) Mobile Robot Navigation using Amended Ant Colony Optimization Algorithm. Indian J Sci Technol 9(45):1–10
8. Harrison JF, Vo C, Lien J-M (2010) Scalable and robust shepherding via deformable shapes. In: Motion in games. Springer, pp 218–229
9. Kala R (2014) Code for robot path planning using probabilistic roadmap. Indian Institute of Information Technology Allahabad. Available at: http://rkala.in/codes.html
10. Kavraki LE, Svestka P, Latombe JC, Overmars MH (1996) Probabilistic roadmaps for path planning in high-dimensional configuration spaces. IEEE Trans Robot Autom 12(4):566–580
11. Krontiris A, Louis S, Bekris KE (2012) Multi-level formation roadmaps for collision-free dynamic shape changes with nonholonomic teams. In: IEEE International Conference on Robotics and Automation
12. LaValle SM, Kuffner JJ (2001) Randomized kinodynamic planning. Int J Robot Res 20 (5):378–400
13. LaValle SM (2011) Motion planning: The essentials. IEEE Robot Autom Mag 18(1):79–89
14. Minhas DS, Engh, MM, Fenske et al (2007) Modeling of needle steering via duty-cycled spinning. In: IEEE proceedings 29th annual international conference of the IEEE engineering in medicine and biology society, pp 2756–2759
15. Patil S, Alterovitz R (2010) Interactive motion planning for steerable needles in 3D environments with obstacles. In: IEEE Proceedings of 3rd IEEE RAS and EMBS international conference on biomedical robotics and biomechatronics, pp 893–899
16. Park W, Kim JS, Zhou Y et al (2005) Diffusion-based motion planning for a nonholonomic flexible needle model. In: IEEE Proceedings of the IEEE 2005 international conference on robotics and automation. Piscataway, NJ, USA, pp 4600–4605
17. Plaku E, Kavraki LE, Vardi MY (2010) Motion planning with dynamics by a synergistic combination of layers of planning. IEEE Trans Robot 26(3):469–482
18. Reed KB, Majewicz A, Kallem V et al (2011) Robot-assisted needle steering. IEEE Robot Autom Magazine 18(4):35–46
19. Sudhakara P, Ganapathy V (2016) Trajectory Planning of a Mobile Robot using Enhanced A-Star Algorithm. Indian J Sci Technol 9(41):1–10
20. Sudhakara P, Ganapathy V, Sundaran K, Balasubramanian Manickavasagam (2016) Optimal trajectory planning based on improved ant colony optimization with ferguson's spline technique for mobile robot. Int J Control Theory Appl 9(40):623–634
21. Webster III RJ, Cowan NJ, Chirikjian GS et al (2004) Non-holonomic modeling of needle steering. In: Proceedings of the 9th international symposium on experimental robotics. Singapore, pp 35–44
22. Webster RJ III, Kim JS, Cowan NJ et al (2006) Nonholonomic modeling of needle steering. Int J Robot Res 25(5–6):509–525
23. Yeh HY, Thomas S., Eppstein D, Amato NM (2012) UOBPRM: a uniformly distributed obstacle-based PRM. In: IEEE/RSJ international conference on intelligent robots and systems, pp 2655–2662

Region-Based Semantic Image Clustering Using Positive and Negative Examples

Morarjee Kolla and T. Venu Gopal

Abstract Discovering various interest of users from massive image databases is a strenuous and rapid impel expedition region. Understanding the needs of users and representing them meaningfully is a challenging task. Region-based image retrieval (RBIR) is a method that incorporates the meaningful description of objects and an intuitive specification of spatial relationships. Our proposed model introduces a novel technique of semantic clustering in two stages. Initial semantic clusters are constructed in the first stage from the database log file by focusing on user interested query regions. These clusters are further refined by relevance feedback in the second stage based on probabilistic feature weight using positive and negative examples. Our results show that the proposed system enhances the performance of semantic clusters.

Keywords RBIR · Negative example (NE) · Positive example (PE) Relevance feedback (RF) · Semantic image clustering

1 Introduction

Content-based image retrieval (CBIR) still faces difficulties when searching content from the large amount of image databases. Another challenge in this area is reducing the semantic gap. CBIR mainly consists of feature extraction and similarity matching. Current research focuses on CBIR systems that fetch an exact cluster of meaningful images. The process of a typical CBIR is as follows.

M. Kolla (✉)
Department of Computer Science and Engineering, Jawaharlal Nehru Technological University, Anantapur, Ananthapuramu, Andhra Pradesh, India
e-mail: morarjeek@gmail.com

T. Venu Gopal
Department of Computer Science and Engineering, Jawaharlal Nehru Technological University Hyderabad College of Engineering Jagtial, Nachupally, Telangana, India
e-mail: t_vgopal@rediffmail.com

A. Kumar and S. Mozar (eds.), *ICCCE 2018*,
Lecture Notes in Electrical Engineering 500,
https://doi.org/10.1007/978-981-13-0212-1_75

An image in a collection is indexed by its category tag and/or image descriptor. For example, given an image of a dog, a label "dog" can be assigned, and/or an image feature vector can be extracted. In the query phase, the user can search for images by specifying a tag or a query image. A tag-based search can be performed by simply looking for images with the specified keyword, while an image-based search can be accomplished by performing a nearest-neighbor search on feature vectors. Despite the success of current CBIR systems, there are three fundamental problems that narrow the scope of image searches. (1) Handling multiple objects: Typically, an image is indexed by a global feature that represents a whole image. It hard to search by making multiple queries with a relationship such as "a human is next to a dog" because the global feature does not contain spatial information. (2) Specifying spatial relationship: Even if multiple objects can be handled by some other means, specifying a spatial relationship of objects is not straightforward. Several studies have tried to tackle this problem by using graph-based queries [1, 2] that represent relationships of objects; however, such queries are not suitable for end-user applications because their specification and refinement are difficult. Several interactive systems [3] can specify a simple spatial query intuitively, but cannot specify complex relationships between objects. (3) Searching visual concepts: Tag-based searches with keyword queries are a simple way to search for images with specific visual concepts, such as object category ("dog") and attribute ("running'"). However, managing tags is not so easy if we have to consider multiple objects. For example, consider an image showing a human standing and a dog running; it is not clear whether we should assign a tag "running" to this image. Moreover, the query of a tag-based search is limited to be within the closed vocabulary of assigned tags, and annotating images involves huge amounts of manual labor. In the region-based approach, the image is divided into a pool of segments, where each segment is defined by image features [4]. We created an interactive RBIR system (Fig. 1) that solves the above three problems as follows: (1) Our system provides an interactive spatial querying interface by which users can easily locate multiple objects [5]. (2) Our system provides a region of interest function that make it easy for users to concentrate on part of the image instead of the entire image.

Clustering is an unsupervised learning technique that can partition data into groups of similar objects. Image clustering is an important technique used to organize a large amount of data into clusters, such that intra-cluster images should have a similar meaning and inter-cluster images should have a dissimilar meaning. Image clustering is used to solve the problems of image segmentation, compact representation, search space reduction, and semantic gap reduction. Visual level abstractions are used to find objects and their relations. Visual features are mapped to semantic concepts in order to form concept description that narrow the semantic gap. Research into solving the semantic gap and extracting relevant images is moving towards semantic clustering and it is focusing on direct mapping of visual features to semantic concepts.

Semantic image clustering (SIC) is the concept of combining unstructured images based on fragment of implication. In an image retrieval system, semantic

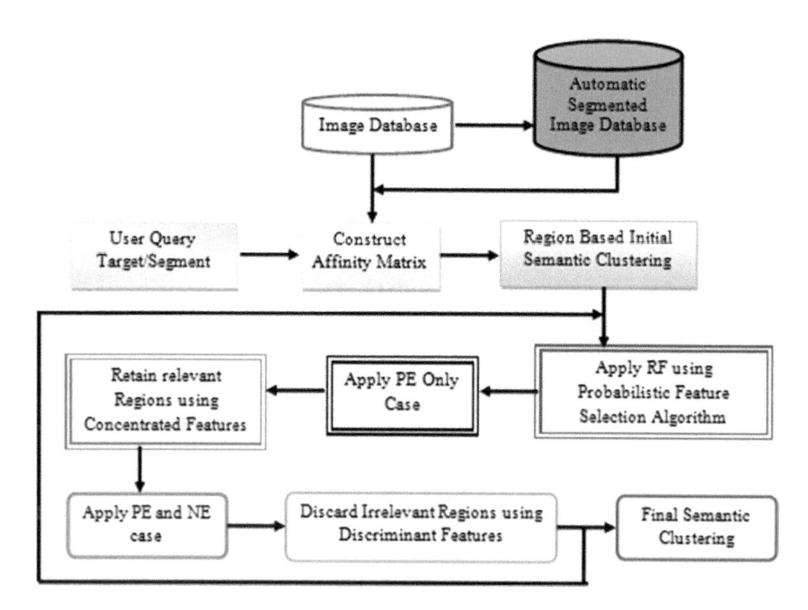

Fig. 1 The block diagram of our proposed system

clustering plays a vital role by retrieving user interested meaningful images. The main objective of semantic clustering is to reduce the search space and semantic gap. Searching images relevant to a user query as a whole will lead to an increase in the search space. A user is searching to retrieve relevant images from the database through the query. Image similarity is subjective and task dependent as per human perception. In general, image segments are represented according to their low-level features. One image segment may have several meanings [1]. For example, a "Blue Car" can be assigned to both the "Blue" cluster and the "Car" cluster. Our proposed model performs region-based clustering before image retrieval with semantic regions of user interest, which can reduce the search space.

Relevance feedback is used to define the best similarity measures to identify relevant images [6]. The main issue in RF is effective utilization of the information provided from user feedback to increase retrieval accuracy. This means that a RF mechanism must maximize the relevance of the results as per the user query. Query refining using relevance feedback has gained much attention in CBIR. Three different challenges facing RF are its inability to extract high-level semantics, lack of feedback samples, and real-time processing. Measuring the similarity between images requires feature weighting, because similarity is defined in terms of features. Relevant features helps to group similar images into cluster and distinguishes between the images belonging to different classes. In this paper, we describe a RF model that learns the importance of each feature based on a feature weighting algorithm [7].

2 Related Work

Numerous studies have been carried on semantic image clustering. Some of focuses have been on indexing images, annotations, relevance feedback, and ontology. Retrieval systems are developed by incorporating large visual codebooks and spatial information [8–10], and include query-based [11, 12] and representative approaches [6, 13], among others [14, 15].

Semantic image clustering was developed based on RF [16–18] and ontology-based approaches [2]. User control in image clustering is incorporated first, which can adjust the clustering results according to user interest. Semantic image clustering using relevance feedback [19] and association rule hyper graph partitioning algorithms [20] are based on an entire image. Rather than viewing each image as a whole, it is viewed as a pool of segments [1]. A number of clusters can be predicted based on prior knowledge. It is very hard to acquire prior knowledge. However, our method can construct the semantic clusters without prior knowledge.

Relevance feedback was earlier built on the vector model [21, 22], classification problem [7], and learning problem [7]. Ranking and assigning a given image to a class are difficulties in the classification problem. Most of the existing methods do not consider negative examples (NE), which can improve retrieval accuracy. Another drawback is the lack of data. The basic idea is to increase the features of the example images that are relevant to the user query.

2.1 Our Approach

In this paper, we consider RF as a probability optimization problem [6], which makes it possible to take a region-based query into account. Our semantic clustering algorithm is an extension of our earlier work [23]. This method has advantages, such as multiclass query support, better feature selection, and better understanding of the data. The main aim of taking positive examples (PE) is to give more weight to the features that the user is interested in. Relevant features are identified by applying characteristic rules from PE images. Relevance can be measured with respect to a user query and its features. By taking negative examples, we can easily identify the outliers by reducing the feature weight and then discard them to reduce noise [14]. NE is used to refine the results of PE. The images retained in the PE case participate in the PE and NE case. In this case, we enhance discrimination features and rank the candidate images with respect to the similarity and dissimilarity of PE and NE respectively. The NE case only neglects important common features from the PE case. To avoid this we are taking both PE and NE cases. In the first step, PE only reduces the search space while in the second step the PE and NE case discards undesired images. Region-based semantic clusters initially formed by using log information require some refinement [12]. These initial clusters are unable to reflect user interest. Semantic meanings are often ambiguous, subjective, and different in all aspects. We need to focus more on a user's interest, not only in terms of user interested regions, but also in terms of user interested features to perform semantic clustering. Our main intention is to give more importance to user interested regions as well as features at the time of retrieval.

2.2 *Outline of the Paper*

In Sect. 3, we explain our proposed method with the help of a block diagram. In Sect. 4, experimental results are discussed. In Sect. 5, we conclude our approach.

3 The Proposed Method

Our proposed system performs two-stage semantic clustering. In the first stage, initial semantic clustering is performed based on an affinity matrix constructed from a user's log file. In the second stage, refinement of initial semantic clusters based on the RF model occurs.

The block diagram is shown in Fig. 1. The log query of the system will collect the user feedback and stores it in the form of an affinity matrix [1]. In this matrix rows are query segments and columns are images. Depending on user query segments, entries in the matrix are positive integers, negative integers, or zeros. The image databases and their automatic segmented image regions from the Blobworld [24] are considered for user feedback. All the entries in this matrix are initially set to zero. Positive integers in this matrix indicate corresponding query region matches with the image region. Negative integers indicate that there is no match between the query region and the image regions. Zero indicates no relevance [1].

The proposed system algorithm is as follows:

Input: User Query Segment

Output: Semantic Image Cluster as per user query.

Procedure:

1. Select a query image segment similar to the user query segment to retrieve from image database.
2. Automatically segment the image database and store them with indices.
3. Construct the affinity matrix by query segments as rows and images as columns.
4. Initial semantic clusters are formed based on an affinity matrix.
5. Refine the initial semantic results by using relevance feedback.
6. Select the relevant images using the PE only case.
7. Discard the irrelevant images using the PE and NE case.
8. If user satisfied.

Display the final semantic image clusters
Else
Go to step 5
End

3.1 Initial Semantic Image Clustering

After constructing the affinity matrix by using a query log with user feedback, initial clustering is performed by grouping positive regions into a respective query segment. Initially we are assuming that query regions are the cluster centers. For positive regions, the regions of the image which has the shortest Euclidean distance will be added to their corresponding query regions. For negative regions, the nearest query center is assigned. In this way we obtain initial semantic clusters from the first stage.

3.2 Refinement of Initial Semantic Clusters Using Relevance Feedback

In the second stage, for each semantic cluster we apply relevance feedback using a probabilistic feature selection algorithm [7]. A positive example (PE) only case is used to retain relevant image regions using user focused features. We use an EM-type algorithm to estimate the parameters of our model [25, 26]. The PE and NE case is used to discard irrelevant regions using discriminate features. Our objective is to give more importance to discriminate features. The output semantic cluster is compared with the threshold value of the performance of the cluster. Several iterations of RF are executed to overcome the threshold as shown in Fig. 1. Initial semantic clusters are refined by removing outliers at this stage.

4 Experimental Results

A Flickr 25 k image dataset with an aggregate of 2862 images from 15 categories labeled as, Building, Beach, Bird, Food, Car, Dog, Flowers, Girl, Lake, Night, Sky, Snow, Sun, Tree and Water is used to conduct our experiment. This comes to 7442 segments as per Blobworld [24] and each segment is represented by a 32-dimensional feature vector. Semantic clusters are obtained from the initial semantic clustering that is refined in the successive stage. Precision and recall of experiments are evaluated with different input queries in Fig. 2. Table 1 shows the experimental results of different input queries. For some categories like Food, Girl, and Night our results are low because RF is unable to meet user perception. For remaining categories, our results prove the efficacy of our system.

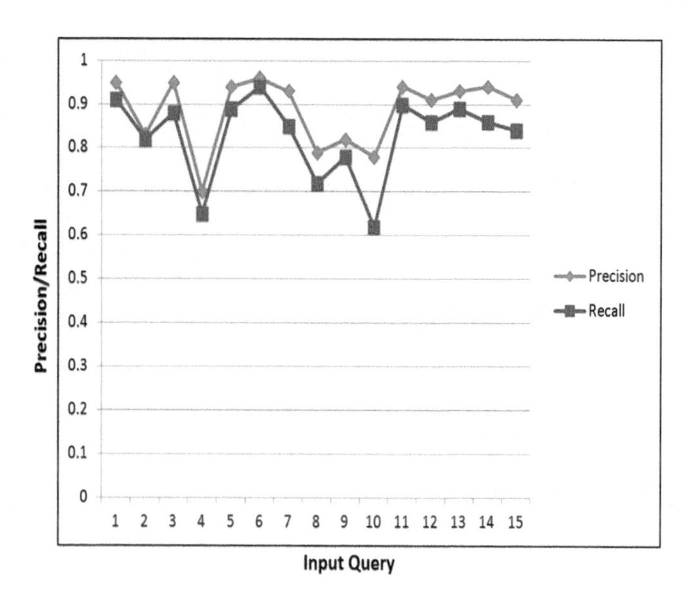

Fig. 2 Precision and recall of input queries

Table 1 Precision and recall values for different input queries

S.No	Input query	Precision	Recall
1	Building	0.95	0.91
2	Beach	0.83	0.82
3	Bird	0.95	0.88
4	Food	0.70	0.65
5	Car	0.94	0.89
6	Dog	0.96	0.94
7	Flowers	0.93	0.85
8	Girl	0.79	0.72
9	Lake	0.82	0.78
10	Night	0.78	0.62
11	Sky	0.94	0.90
12	Snow	0.91	0.86
13	Sun	0.93	0.89
14	Tree	0.94	0.86
15	Water	0.91	0.84

5 Conclusion

This paper proposes a region-based approach for clustering using positive and negative examples. Each cluster is considered as a meaningful unit. Semantic clustering is performed in two stages. Initial semantic clusters are constructed in the first stage and refined by RF in the second stage. The key advantage of our model is to reduce the search space and the semantic gap. These are the two main challenges faced by CBIR. Instead of searching entire images, our model searches only user interested regions from semantic clusters, which can reduce the search space. Focusing on user interested features through RF reduces the semantic gap. Our model gives more importance to user interested regions as well as features. With the improvement in retrieval accuracy, our results proved that the proposed system produces semantic clusters efficiently.

References

1. Liu Y, Chen X, Zhang C, Sprague A (2009) Semantic clustering for region-based image retrieval. J. Vis Comun. Image Represent
2. Chen N, Prasanna V (2012) Semantic image clustering using object relation network. In: Proceedings of the first international conference on computational visual media
3. Smith JR, Chang SF (1996) Visualseek: a fully automated content-based image query system. In ACMMM
4. Yang J, Price B, Cohen S, Yang MH (2014) Context driven scene parsing with attention to rare classes. In: CVPR (2014)
5. Carson C, Tomas M, Belongie S, Hellerstein JM, Malik J (1999) Blobworld: a system for region-based image indexing and retrieval. Inf Syst J 1614(1999):509–516
6. Jegou H, Perronnin F, Douze M, Sanchez J, Perez P, Schmid C (2012) Aggregating local descriptors into compact codes. In: PAMI
7. Kherfi ML, Ziou D (2006) Relevance feedback for cbir: a new approach based on probabilistic feature weighting with positive and negative examples. IEEE Trans Image Process 15(4):1017–1030
8. Philbin J, Chum O, Isard M, Sivic J, Zisserman A (2007) Object retrieval with large vocabularies and fast spatial matching. In: CVPR
9. Avrithis Y, Kalantidis Y (2012) Approximate gaussian mixtures for large scale vocabularies. In: ECCV
10. Hinami R, Matsui Y, Satoh S (2017) Region-based image retrieval revisited. https://doi.org/10.1145/3123266.3123312
11. Chum O, Mikulik A, Perdoch M, Matas J (2011) Total recall II: query expansion revisited. In: CVPR
12. Tolias G, Jegou H (2014) Visual query expansion with or without geometry: refining local descriptors by feature aggregation. Pattern Recognit
13. Perronnin F, Liu Y, Sanchez J, Poirier H (2010) Large-scale image retrieval with compressed fisher vectors. In: CVPR
14. Radenovi F, Jegou H, Chum O (2015) Multiple measurements and joint dimensionality reduction for large scale search with short vectors. In: ICMR
15. Tolias G, Furon T, Jegou H (2014) Orientation covariant aggregation of local descriptors with embeddings. In: ECCV

16. Morrison D, Marchand Mailet S, Bruno E (2014) Semantic clusters of images using patterns of relevance feedback. Comput Vis Multimed Lab
17. Gong Z, Hou L, Cheang CW (2005) Web image semantic clustering, Springer, pp 1416–1431
18. Gunjan VK, Shaik F, Kashyap A, Kumar A (2017) An interactive computer aided system for detection and analysis of pulmonary TB. Helix J 7(5):2129–2132. ISSN 2319-5592
19. Patino-Escarcina RE, Ferreira Costa JA (2008) The semantic clustering of images and its relation with low level color features. In: IEEE international conference on semantic computing, pp 74–79
20. Yin X, Li M, Zhang L, Zhang HJ (2003) Semantic image clustering using relevance feedback. In: Proceedings of the international symposium on circuits and systems (ISCAS)
21. Duan L, Chen Y, Gao W (2003) Learning semantic cluster for image retrieval using association rule hypergraph partitioning. In: Proceedings of the 2003 joint conference of the fourth international conference on information, communications and signal processing and the fourth pacific rim conference on multimedia, pp 1581–1585
22. Ishikawa Y, Subramanya R, Faloutsos C (1998) Mind reader: querying databases through multiple examples. In: Proceedings of 24th intinternational conference on very large data bases. New York, pp 433–438
23. Vasconcelos N, Lippman A (1999) Learning from user feedback in image retrieval systems. Neur Inf Process Syst
24. Kolla M, Gopal TV (2015) Semantic image clustering using region based on positive and negative examples. In: ICICC. pp 261–264 Feb 2015. ISBN 978-93-82163-59-6
25. Carson C, Belongie S, Greenspan H, Malik J (2002) Blobworld: image segmentation using expectation—maximization and its application to image querying. IEEE Trans Pattern Anal Mach Intell 24(8):1026–1038
26. Dempster AP, Laird NM, Rubin DB (1977) Maximum likelihood from incomplete data via the EM algorithm. J Roy Statist Soc B 39(l):l–38

A Cost Effective Hybrid Circuit Breaker Topology for Moderate Voltage Applications

D. S. Sanjeev, R. Anand, A. V. Ramana Reddy and T. Sudhakar Reddy

Abstract Compared to mechanical circuit breakers, with respect to speed and life, solid state circuit breakers based on modern high power semiconductors offers considerable advantages. During a short circuit the voltage profile of the power grid can be improved since the fault current is reduced. The distortion in voltage caused by a three-phase short circuit can be limited to fewer than 100 µs. In this paper, a theoretically approached active thyristor circuit based on a new hybrid topology of connecting the semiconductor devices in series and parallel is proposed. This permits an increase in supply voltage and fault clearance without arcing. In turn, and to get benefit from a current limitation, the circuit breaker is realized with the extinction of an electrical arc when the breaker is opened. Hence the proposed topology has led to a wider integration of solid state circuit breakers (SSCB) in existing power grids because of their cost effective nature when compared to turn-off semiconductor devices.

Keywords Circuit breaker · Fault current · Voltage profile · Electrical arc Active thyristor circuit · Hybrid circuit breaker · SSCB

D. S. Sanjeev (✉) · R. Anand · T. Sudhakar Reddy
Department of EEE, CMR College of Engineering & Technology, Kandlakoya Medchal, Secunderabad, Telangana, India
e-mail: dandolesanjeev@gmail.com

R. Anand
e-mail: anandrajkumar108@gmail.com

T. Sudhakar Reddy
e-mail: sudhakar943@gmail.com

A. V. Ramana Reddy
Department of EEE, BV Raju Institute of Technology, Rajiv Gandhi Nagar Colony, Hyderabad, Telangana, India
e-mail: venkatanchoori@gmail.com

A. Kumar and S. Mozar (eds.), *ICCCE 2018*,
Lecture Notes in Electrical Engineering 500,
https://doi.org/10.1007/978-981-13-0212-1_76

1 Introduction

1.1 Circuit Breaker

An electrical switch that is operated automatically to protect electrical circuits from overload or short circuit.

Circuit breaker importance: As the power system network is very complex, it requires protection switchgear in order to operate the system safely and efficiently in both normal and abnormal situations. The circuit breaker acts like a switch along with the fuse, having added advantages with complex features.

1.2 Mechanical Circuit Breaker

The circuit breaker consists of two contacts which are placed in an insulating medium. The insulating medium facilitates two basic functions:

(1) Quenching the arc established between two contacts when the circuit breaker is operated.
(2) Providing insulation between contacts and also with the earth.

Even though mechanical circuit breakers play a vital role in power system protection, their major drawbacks are listed as i. Requirement of frequent reconditioning of the quenching medium after every operation. ii. Complexity in maintenance of compressor plants. iii. Chance of explosion when the hydrogen and oil combine with air during arcing. iv. The dielectric strength of the oil is reduced due to arcing. v. A breaker like the SF6 is costly due to the insulating gas (SF6).

2 Technological Advancement of Circuit Breakers

2.1 Hybrid Circuit Breakers

A theoretically approached active thyristors circuit based on a new hybrid topology by connecting the semiconductor devices in series and parallel is proposed. This permits an increase in supply voltage and fault clearance without arcing. In this study the advantages and disadvantages of mechanical switch-based hybrid circuit breakers and semiconductor-based circuit breakers are considered so as to obtain an improved version of a SSCB owing to less conduction loss. To achieve fast and longer life systems, modern power electronic devices are finding the best fit instead of conventional mechanical part of the circuit breakers. The thermal loss created due to voltages drops is the major drawback for static circuit breakers. The fault current gets interrupted by deflected branch current in the semiconductor device,

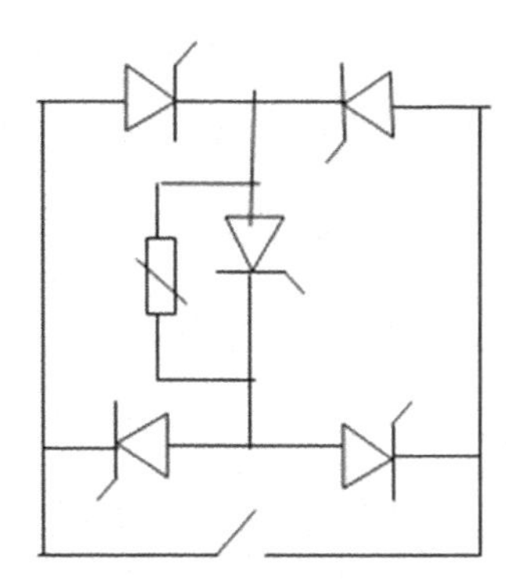

Fig. 1 Hybrid circuit breaker

created due to a small arc voltage drop when the mechanical switch gets opened. IGCTs (integrated gate turn-off thyristors) are being successfully used for moderate voltage applications (Fig. 1).

2.2 DC Solid State Circuit Breaker

A DC SSCB has two terminals connected to the dc source. The voltage generated by the circuit breaker is proportional to the circuit current which is utilized to control a solid state switch which in turn interrupts the circuit. The enhancement of turn-off time is achieved by the feedback of the regenerative voltage.

2.3 Solid State Circuit Breaker

The inclusion of sensible loads in the electrical system leads to power quality issues. Moreover, handling of short circuits in moderate voltage grids is an important issue in improving the power quality.

The use of power semiconductor devices can reduce the short circuit current along with voltage distortions in the event of short circuit failure.

Merits of SSCB:

1. Operating Speed and life of a SSCB is superior to mechanical circuit breakers.
2. Voltage profile will be improved during a short circuit because of the reduction in fault current.
3. No arcing and restriking is present.
4. Eco friendly since there is no chance of ionization of gases or oils.
5. Highly reliable since no moving parts and low on maintenance.
6. Small in size, weight, running cost, and response is fast.

7. For a better power quality profile in distribution systems, solid state technology is used. The reaction time of solid state switches is less when compared to mechanical switches.

3 Topologies

3.1 Topology 1

Topology 1 uses modern high power semiconductors like gate commutated turn-off thyristors (GCT) instead of mechanical circuit breakers. In self turn-off converters, thyristors like GTOs, GCTs, and IGBTs were used using forced turn off. These topologies are utilized for an adequate amount of time despite their complexity. The limitation of a forced commutation circuit is its switching frequency and also the switching sequence. Application related to solid circuit breakers the behavior of the switching is on smaller side because of the absence of switch at high frequency. During the short circuit period the solid circuit breaker must be turned off actively and should be ready for the next turn-off period after various line periods. Therefore the limitation with respect to forced commutation circuits will be on par least important in solid circuit breakers applications. Initially, for forced commutation circuits a standard topology is proposed. The addition of a varistor is the only difference when compared to inverter classical solutions. These varistors are used to reduce the size of the capacitor and also to keep the breakdown voltage of the device under permissible limits (Fig. 2).

The two main thyristors are in the ON state under normal conditions of operation and these two thyristors also offer greatly reduced on-state losses. Precharged

Fig. 2 Topology 1

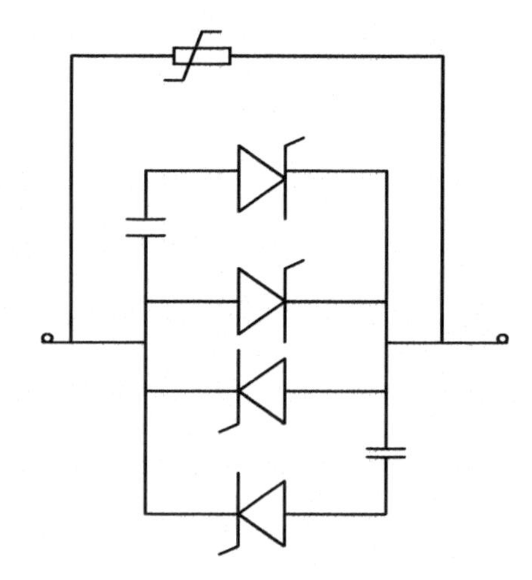

capacitors are used in this topology. In the case of a failure the auxiliary thyristor is fired, which is associated with the main thyristors. With this operation the current is commutated instantly followed by the capacitor discharge. Therefore the SCR current is forced to go to zero. This gives rise to condition where the thyristor is turned off at the zero crossing. To turn off the short circuit current successfully, the commutation capacitor with stored energy becomes more critical. With this design of the commutation circuit, precharged voltage can be selected which proves to be one of the added advantages.

3.2 *Topology 2*

Various design topologies are developed to reduce the capacitor number so as find an economical solution (Fig. 3).

In order to avoid a second capacitor in this case the two additional thyristors are used. Switching of the capacitor in the required direction is achieved by firing the two thyristors. This topology behavior is the same as a standard forced commutation circuit.

3.3 *Topology 3*

A transformer with very low inductance is used in this topology and also the thyristors have a maximum voltage of 1.5 kV along with the capacitor as shown in Fig. 4. A second path is associated with a semiconductor switch. In this paper GCTs are used. In this topology we may find that the requirement for thyristors and the transformer is not necessary but these find their importance when the losses in

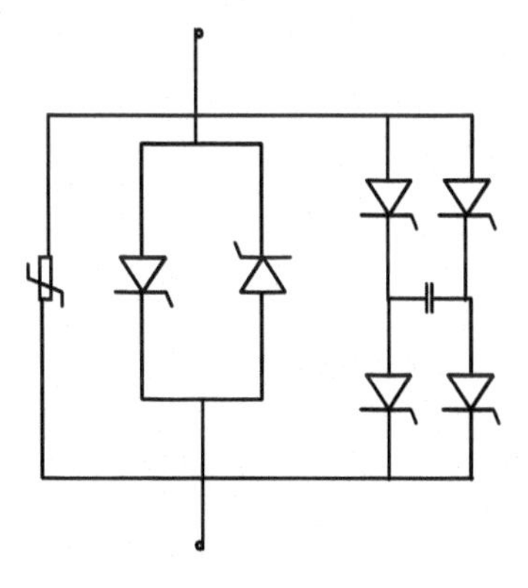

Fig. 3 Topology 2

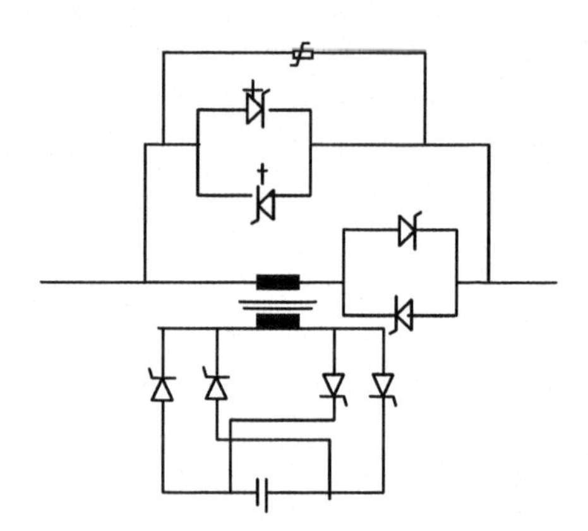

Fig. 4 Topology 3

the GCTs are considered as these losses are very high in power grids. A second advantage of this solution is that much smaller, i.e. cheaper, and symmetric GCTs can be used in this application, because the on-state behavior of the GCTs is of no relevance during normal operation. Other benefits of this topology are the smaller size, availability, cost, symmetric GCTs are used in this application due to on-state behavior of GCTs is at a low importance under normal operating conditions.

The thyristors and inductance carry current during normal operation. In the event of a failure the auxiliary circuits are turned on. A capacitor with stored energy is able to demagnetize the inductance during a failure and the current in the active circuitry is forced to come down to zero. The short circuit current in the GCTs is turned off once the main thyristor crosses the hold-off interval.

Due to the small leakage inductance of the transformer, the capacitor required for the current commutation process is very small when compared to convectional forced commutation circuits. To demagnetize the leakage inductance and to prevent positive voltage caused by the GCT, sufficient energy is stored inside the capacitor.

The voltage at the primary winding is responsible for the commutation time of the current to the auxiliary path. Current commutation time in the auxiliary path is affected by the primary voltage of the transformer. Since the hold-off time of a thyristor strongly depends on the current fall time or slew rate, an optimal operation point exists where the turn-off time is minimal. Since thyristor hold-off time depends on slew rate an optimal operation will exist with minimal turn-off time.

4 Technical Comparison Among Topologies

Topologies 1 and 2 offer exactly the same performance. However, selection based only on the technical performance of the two topologies is not reasonable.

To demagnetize the current topology 3 does not adopt large capacitors when compared to topology 1 and 2. Interruption to the line current is provided directly in topology 3 with active a device which does not increase the di/dt in the hold-off period. Compared to convectional commutation circuits the GCTs commutate the current more rapidly leading to reduced current peaks.

The on-state losses in topology 3 due to transformer resistance are much less and can be neglected. Hence by taking the technical aspects into consideration, topology 3 provides the best performance when compared to the solutions presented earlier.

5 Economical Comparison of Topologies

A predominate role is played by economic factors in the design and adaptability of new technologies in the industry. Therefore an unavoidable economic comparison of the different topologies is presented (Fig. 5).

In this comparison chart the cost of GCT is related with all per unit costs. While investing, the cost for mechanical built-up, cooling, control is taken into consideration. Since the investment cost of varistors protection systems is very low, the variation in cost of the different topologies can be neglected.

The cost is calculated for a transmission power of 25 MVA with a blocking voltage of 32 kV and a hold-off time interval of 500 μs.

Therefore it is very much true that conventional circuits provide the lowest cost of investment compared to other systems. The step of using a thyristor bridge to reduce the commutating capacitor in topology 2 was taken to reduce the cost and turned out to be more costly than topology 1.

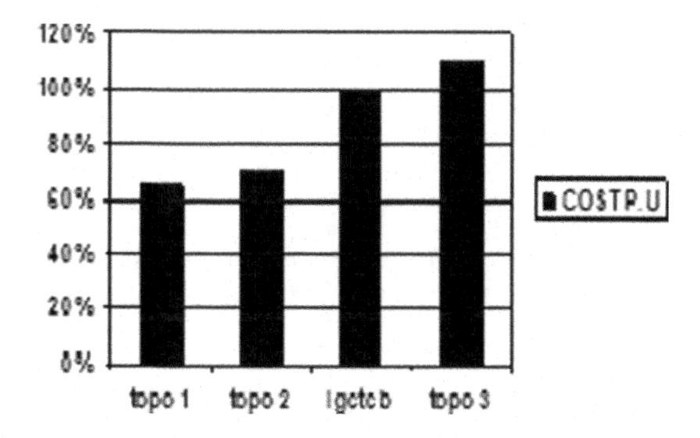

Fig. 5 Cost analysis

6 Conclusions

In this paper, different solutions for a cost effective hybrid circuit breaker topology for moderate voltage applications are presented. Even though forced commutation circuits are not the latest technology, the application in moderate voltage SCB requires improvements. Compared to the latest turn-off devices, thyristors provide more advantages, such as extremely low on-state losses and low-cost trigger circuits. Active turn-off devices offer fewer advantages when compared to an active current limiter.

Since the varistors are cheap the overall impact due to the charging circuit is negligible and also the on-state losses are very low compared to a GCT circuit breaker. Due to the existence of natural redundancy reliability of SCR circuits is very high, in the event of commutation failure the thyristors will commutate at the preceding zero crossing. Therefore circuit breaker failure is quiet unlikely.

To fulfill the required necessary functions of the latest breakers, a varistor coupling based is the best fit. Aiming for rapid reclosing a topology with constant charging capacitor is presented.

A requirement for additional diodes in the phases of the grid is much costlier when we go topology using varistor coupling for all phases. However, to compensate the leakage current of the capacitor there is no need for any voltage source.

References

1. Mehta VK Principles of power system
2. Wadhwa CL Electrical power systems
3. Klingbeil L, Kalkner W, Heinrich C (2001) Fast acting solid-state circuit breaker using state-of-the-art power- electronic devices. In: Proceedings of European conference on power electronics application (EPE), Graz, Austria
4. Smith RK et al (1993) Solid state distribution current limiter and circuit breaker: application requirements and control strategies. IEEE Trans Power Deliv 8(3):1155–1164
5. Ekström A, Bennich P, De Oliveira M, Wilkström A (2001) Design and control of a current-controlled current limiting device. In: Proceedings of EPE'01 conference, Graz, Austria
6. Sugimoto S, Neo S, Arita H, Kida J, Matsui Y, Yamagiwa T (1996) Thyristor controlled ground fault current limiting system for ungrounded power distribution systems. IEEE Trans Power Deliv 11(2):940–945
8. Schwartzenberg JW, De Doncker RW 15 kV medium voltage static transfer switch. In: Proceedings of IEEE 30th industry applications society annual meeting, vol 3, pp 2515–2520, 8–12 Oct 1995
9. Meyer M (1974) Selbstgeführte thyristor-stromrichter. Technical Report, Siemens AG
10. Meyer C, Schröder S, De Doncker RW (2004) Solid-State circuit breakers and current limiters for medium-voltage systems having distributed power systems. IEEE Trans Power Electron 19(5):1333–1340
11. Eupec (2001) Additional technical information for high power b phase control thyristors release 2.2. Technical Report (Online). www.eupec.de

12. Blume D, Schlabbach J, Stephanblome T (1999) Spannungsqualität in elektrischen netzen. VDE-Verlag, Munich, Germany, pp 49–49
13. Kunde K, Kleimaier M, Klingbeil L, Hermann HJ, Neumann C, Pätzhold J (2003) Integration of fast acting electronic fault current limiters (EFCL) in medium-voltage systems. In: Proceedings of 17th international conference electricity distribution (CIRED), Barcelona, Spain, pp 148–152
14. Meyer C, Schröder S, De Doncker RW (2003) Integration of solidstate switches into medium-voltage grids. In: Proceedings of European conference power electronics applications (EPE'03), Toulouse, France
15. Celli G, Pilo F, Sannais R, Tosi M A custom power protection device controlled by a neural network relay. In: Proceedings of IEEE power electronics society Summer meeting, Seattle, WA, 16–20 Jul 2000, pp 1384–1389
16. Cannas B, Celli G, Fanni A, Pilo F (2001) Automated recurrent neural network design of a neural controller in a custom power device. J Intell Robot Syst 31:229–251

Multi-criteria Decision Analysis for Identifying Potential Sites for Future Urban Development in Haridwar, India

Anuj Tiwari, Deepak Tyagi, Surendra Kumar Sharma, Merugu Suresh and Kamal Jain

Abstract Decadal population growth and increasing demand for land have led to the present study to identify potential sites for future development in Haridwar City, Uttarakhand. This study is conducted using remote sensing (RS) and geographical information system (GIS) using various thematic layers, such as slope, elevation, land use land cover (LULC), a digital elevation model (DEM), normalized difference vegetation index (NDVI), urban landscape dynamics (ULD), and other physical parameters which can affect the growth of urban expansion. GIS provides an opportunity to integrate various parameters with population and other relevant data associated with features which will help to determine potential sites for expansion. The appropriate weights are assigned to each layer using an analytical hierarchical process (AHP), with a multi-criteria decision analysis (MCDA) technique. The weights are assigned using expert opinion on the factors which are most suitable to least suitable for urban expansion according to their importance and are used in the study to extract the best result from the given data.

Keywords Geographic information system (GIS) · Remote sensing
Urban · Analytical hierarchical process (AHP) · Multi-criteria decision analysis (MCDA)

A. Tiwari (✉) · S. K. Sharma · K. Jain
Geomatics Engineering, Indian Institute of Technology Roorkee, Roorkee, India
e-mail: anujtiwari.iitr@gmail.com

D. Tyagi
Department of Civil Engineering, College of Engineering Roorkee, Roorkee, India

M. Suresh
R&D Centre, CMR College of Engineering & Technology, Hyderabad, India

A. Kumar and S. Mozar (eds.), *ICCCE 2018*,
Lecture Notes in Electrical Engineering 500,
https://doi.org/10.1007/978-981-13-0212-1_77

1 Introduction

Urban development and the migration of population from rural to urban areas is a global phenomenon. Many small and isolated population centers are rapidly changing into large metropolitan cities, and hence the conversion of natural land to urban use is quite obvious [1]. According to a United Nation's Population Division report published in 1975, about 38% of the earth's population was living in urban areas and by 2025 this proportion is expected to rise to 61%. This implies that about 5 billion people out of a total world population of 8 billion will be living in urban areas.

The expansion of urban areas is a great problem for many countries, especially in the developing world, where the rate of urban growth is much higher than in developed countries. Many urban areas situated in developing countries have already crossed the permissible limit of growth and only limited access to urban services is available for large parts of the city population.

Therefore, there has been worldwide concern about urban growth over the last two decades. As everything is changing and developing in the world, the needs of humans also keep changing. There is a continuous pressure towards urbanization. It is important that urban planning should be carried out in such a way that it satisfies human requirements [4]. Urban planning is a complex task and it requires a huge volume of data to support any decision. Urbanization is a dynamic phenomenon, therefore, for proper urban planning, accurate and timely data are required. Planners need a wide variety of data for analysis of an ever changing landscape [3]. Since 1972, the developments in remotely sensed imagery and related data products have enhanced planner's ability to map urban spatial structures. With the emergence of fast and efficient computer-based GIS during the last few years, it has been possible to analyze a complex relationship that was difficult to handle manually. GIS and remote sensing provide a broad range of tools for urban area mapping, monitoring, and management to achieve optimization in their utilization and conservation [5]. Remote sensing data with its unique characteristics of a synoptic view, repetitive coverage, and reliability have opened immense possibilities for urban area mapping and change detection. Spatial data stored in a digital database of GIS, such as DEM, along with the capability of GIS to integrate different datasets, can be used to evaluate the suitability of land for urbanization. Satellite images that are acquired by high-resolution satellites can be used within GIS, which will enable new and flexible forms of output, tailored to meet particular needs.

2 Urbanization in an Indian Context

The trends of urbanization indicate that the share of urban population compared to rural population in India has increased from 238.4 million in 1951 to 2011 million in 2011 as shown in Fig. 1. Similarly, the number of towns/urban areas has grown

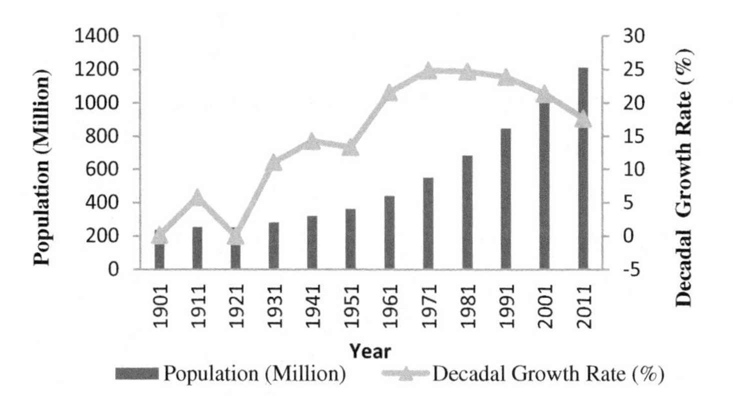

Fig. 1 Trends in population size and growth

from 1827 in 1901 to 7935 in 2011, and markedly so in the last ten years as shown in Fig. 2.

Thus, the patterns of urban population growth and concomitant industrial development have resulted in the growth of non-compatible and non-conforming land uses. This has created various problems of traffic congestion, solid waste disposal, and an unequal distribution of infrastructural facilities and services in almost all big cities. Thus, planning of urban areas on suitable sites is required in order to avoid these problems.

Migration from rural areas to small towns and from small towns to cities creates problems for urban planners in the formulation of urban development plans. Planners are unable to predict how many people would migrate to the city from outside. Nowadays, the problem of population growth is faced by all developing countries. Especially in urban areas, the natural growth of population in addition to migration from rural areas has resulted in the overcrowding of cities. The rapid increase of urban population and the transforming urban economy has resulted in an ever increasing load on the urban environment in terms of urban structure of cities, creating many problems, such as unplanned sprawl, inadequate housing facilities, traffic congestion, insufficient drainage, sewerage facilities, and a lack of other amenities [2].

Haridwar, a district in the State of Uttarakhand, has been experiencing rapid urbanization and other developmental activities over the last couple of years. To overcome undesirable growth, it is important to plan and monitor the urbanization process in a systematic manner. It is therefore necessary that an urban land use suitability analysis should be carried out to evaluate the suitability of land for urbanization.

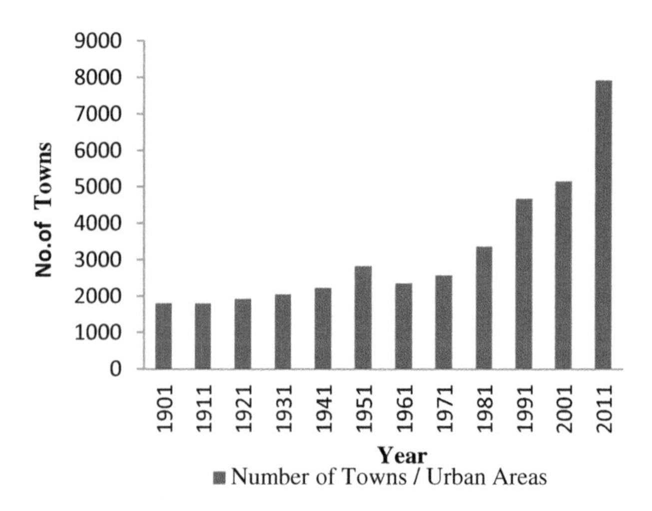

Fig. 2 Growth in urban areas

3 Study Area

Haridwar is regarded as a religious city by Hindus and is the second largest city of Uttarakhand. It is situated in the south-western part of Uttarakhand and is chosen as the study area. Haridwar is situated on the bank of River Ganga and many other seasonal streams flow through it. Most of the region of this city is covered with forest. Haridwar is referred to as the main religious, commercial, and financial center of the state. Having such a high importance and such a densely populated area this city offers a plethora of education and business opportunities, providing shelter to many people from nearby villages and towns. Haridwar is facing urbanization at a rapid rate which means that its resources should be properly classified and sustainably used.

Haridwar, a district of Uttarakhand is situated at cardinal points at latitude 30.080481 and longitude 78.143989 to latitude 29.848912 and longitude 78.089107. It has an altitude of 1030 ft. above mean sea level and covers an area of 34301.28 ha. The study area is focused on metropolitan as well as forest regions of Haridwar which starts from undulating terrain of northern region and lasts up to the plainer area of the southern region as shown in Fig. 3. Haridwar has a subtropical climate which on average lies from 29 to 39 °C. It generally has high humidity and three distinctive seasons, i.e. winter, summer, and the rainy season.

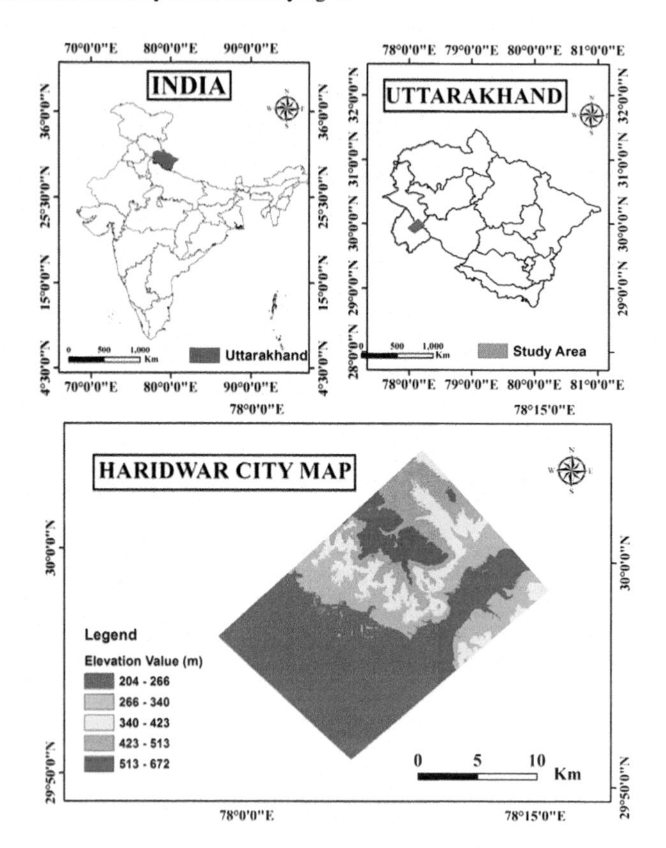

Fig. 3 Map of the study area

4 Data Used

One satellite imagery has been taken from Landast-8 OLI data acquired on 19 October 2016 from the US Geological Survey's (USGS) Global Visualization Viewer (Table 1). The obtained Landsat data was georeferenced to a UTM zone 43 North projection using WGS-84 datum with the help of topographic maps of the area. Other satellite data have been acquired from BHUVAN (an Indian geographic satellite portal) on the elevation model of the region (Table 2).

Table 1 Specification of satellite data used

Satellite	Sensor	Path	Row	Acquisition date	Time (GMT)	Resolution (m)
Landsat 8	OLI/TIRS	145	44	19 October 2016	20:26:34	30
Cartosat I	DEM	–	–	25 May 2012	–	30

Table 2 Description of the Landsat 8 satellite and its spectral bands used in the study

Satellite	Sensor	Band	Resolution (m)	Spectral Band (μm)
Landsat 8	OLI/TIRS	Coastal/Aerosol	30	0.435–0.451
		Blue	30	0.452–0.512
		Green	30	0.533–0.590
		Red	30	0.623–0.673
		NIR	30	0.851–0.879
		SWIR-I	30	1.566–1.651
		SWIR-2	30	2.107–2.294
		Cirrus	30	1.363–1.384
		Pan	15	0.503–0.676
		TIR-I	100	10.60–11.19
		TIR-2		11.50–12.51

5 Methodology

To perform site suitability analysis we compare Earth's physical parameter, i.e. elevation, slope, and other socioeconomic factors, i.e. medical facilities, schools, railway stations, and other transport networks. Study of these parameters in preparation of the site suitability map is very important. There are several methods used by researchers to study site suitability analysis and mapping. They can be classified into three groups: expert evaluation, statistical method, and deterministic method. The successful use of one method over the other strongly depends on many factors such as the scale of the area, accuracy of the expected results, availability of data, parameters considered, etc.

In the present study, AHP is used to determine the site suitability. AHP is a multi-criteria decision-making approach and a technique introduced by Saaty [7]. AHP is a decision support tool that helps to solve complex decision-making problems. It uses weights of different classes to generate result. AHP is implemented in three simple steps [7]

- Computing the vertex of criteria weights
- Computing the matrix of option scores
- Ranking the options

The first step includes identification of all parameters, criteria, sub-criteria and alternatives which are responsible for the problem. The second step is to set datasets in the hierarchic structure with the help of expert human decision. Experts can rate the comparison as either good, medium, moderate, or poor. The third step is pair wise comparisons of the multi-criteria generated in step 2 are these are organized in square matrix formats (m × n). This comparison matrix gives the relative importance of the multi-criteria being compared. Comparisons made by this method are subjective and the AHP tolerates inconsistency through the amount of redundancy present in the approach. If this consistency index has failed to reach a required level then answers to comparisons may be re-examined.

$$CI = (\lambda_{max} - n)/(n - 1)$$

where

- λ_{max} is the maximum Eigen value of the judgment matrix.
- CI can be compared with the random matrix, RI.

The ratio derived CI/RI, is termed as the consistency ratio, CR. Saaty suggests the value of CR should be less than 0.1. The rating of each alternative is to be multiplied by the weights assigned to the sub-criteria and then aggregated to get local ratings with respect to each criterion. The local ratings are then multiplied by the weights of the criteria and aggregated to get global ratings. The AHP produces weight values for each alternative based on the importance of one alternative over another with respect to common criteria.

To generate a site suitability map, multilayer datasets are used in the present study, such as satellite images, Google Earth data, and other data. Figure 4 shows the detailed methodology adopted in the study. A Landsat 8 image has been acquired from the USGS's Glovis online portal, and has been used in the preparation of several thematic layers. Other data from Cartosat has been acquired from the BHUVAN online portal.

In order to generate a LULC map, a supervised classification method is adopted using a maximum likelihood classifier with ERDAS software. An NDVI layer is developed using a band ratio method (NIR and RED band). DEM data have been captured from the CARTOSAT satellite. Slope, aspect and elevation are developed using DEM data in Arc GIS software. The soil layer is developed by georeferencing a soil map of India obtained from the National Bureau of Soil Survey and Land Use Planning (NBSS & LUP). Software such as ArcGIS and ERDAS have been used for editing, digitization, and topology criteria. Then, by combining all these raster layers, the final model is prepared as shown in Fig. 5 using ArcGIS software.

The next step is to assign weight values to each raster layer (based on expert judgment) as shown in Table 3. Then these steps are applied in AHP for pair wise comparison matrix and normalized matrix etc. AHP is a structured technique for organizing and analyzing complex decision-making problems based on a psychological and mathematical method which was developed by Saaty [6]. Figure 5 shows the present AHP model used in study which was made in ArcGIS software. Finally, the site suitability map is generated using these parameters.

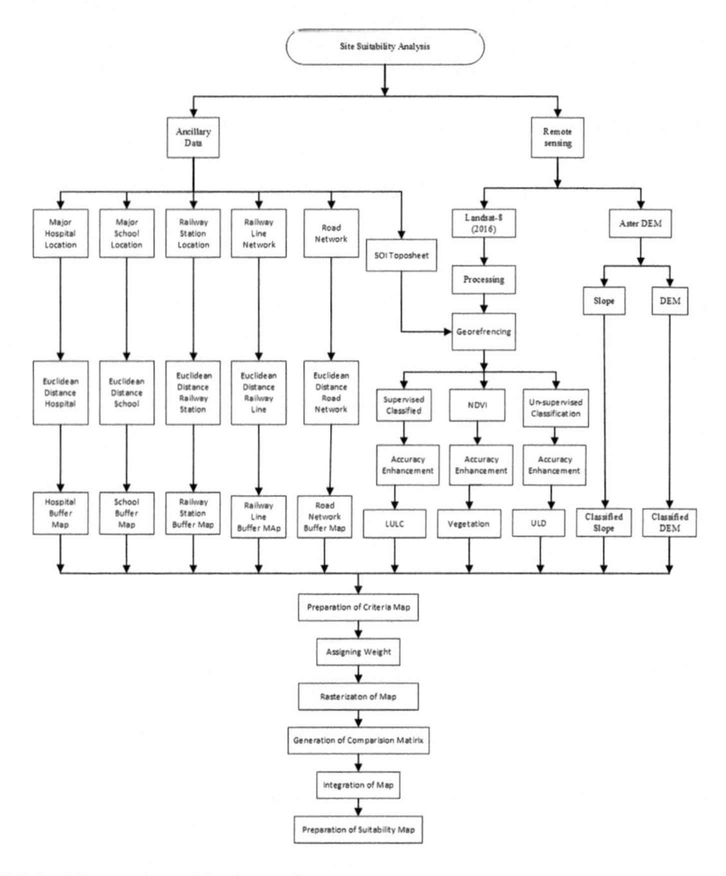

Fig. 4 Methodology adopted in the study

6 Results and Discussion

The results are developed using various thematic map layers, and these layers are defined in the following section.

6.1 Digital Elevation Model (DEM)

DEM is derived from a Cartosat satellite whose data is present on BHUVAN. The study area has an almost flat topography with small variations in the northern region due to the presence of mountains. DEM is a raster layer and each pixel contains a fixed elevation value. DEMs are well suited to calculations, manipulation, and analysis of areas based on their elevation profile. ArcGIS software has many built-in features which can convert elevation maps into derivative maps.

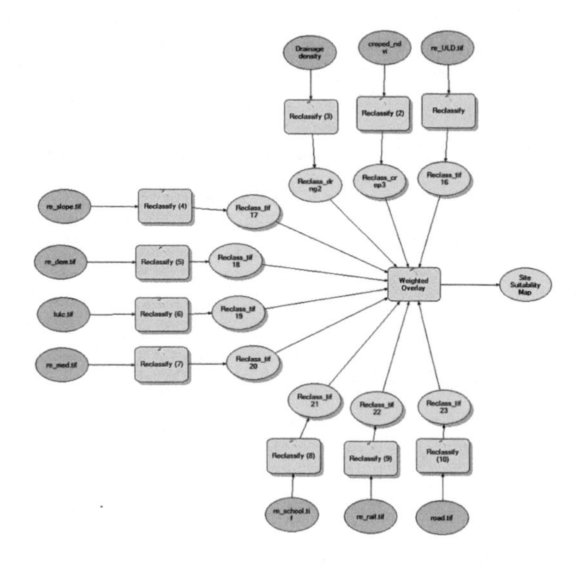

Fig. 5 Model used in the present study

Figure 6a represents the elevation map of the study area. The elevation map is divided into 10 classes as depicted in the figure.

6.2 Slope Map

Site suitability of an area is highly dependent on the evenness of slope, bedding of rocks, and extent of faulting and folding of rocks. As the evenness of a slope remains constant, the probability of site suitability increases. Figure 6b represents the slope map with 10 classes for the study area. The slope of the study area lies between 0 and 61.06, and this area is medially ranked as an area that has flat terrain in some regions while it also includes some mountainous regions. The class having a higher value is categorized with a relatively lower rank due to high run-off and low recharge.

6.3 Normalized Difference Vegetation Index (NDVI)

NDVI is a mathematical method by which we can analyze the vegetation cover of a target area. NDVI is a band ratio method which uses the amount of reflection of the near infrared region (NIR) and visible (red) bands of the electromagnetic spectrum. Areas of healthy vegetation cover absorb most of the red light and reflect most of the NIR band while the area with less vegetation or dead vegetation comparatively reflects more of red band and less of the NIR band. In order to calculate NDVI we

Table 3 Ranking for different parameters in site suitability zone mapping

Compare Criteria	Weight	Sub-criteria
Good	6	ULD
Medium		
Moderate		
Poor		
Good	3	NDVI
Medium		
Moderate		
Poor		
Good	11	Drainage density
Medium		
Moderate		
Poor		
Good	30	Slope
Medium		
Moderate		
Poor		
Good	16	Elevation
Medium		
Moderate		
Poor		
Good	9	LULC
Medium		
Moderate		
Poor		
Good	4	Hospital
Medium		
Moderate		
Poor		
Good	3	School
Medium		
Moderate		
Poor		
Good	2	Railway
Medium		
Moderate		
Poor		
Good	16	Road
Medium		
Moderate		
Poor		

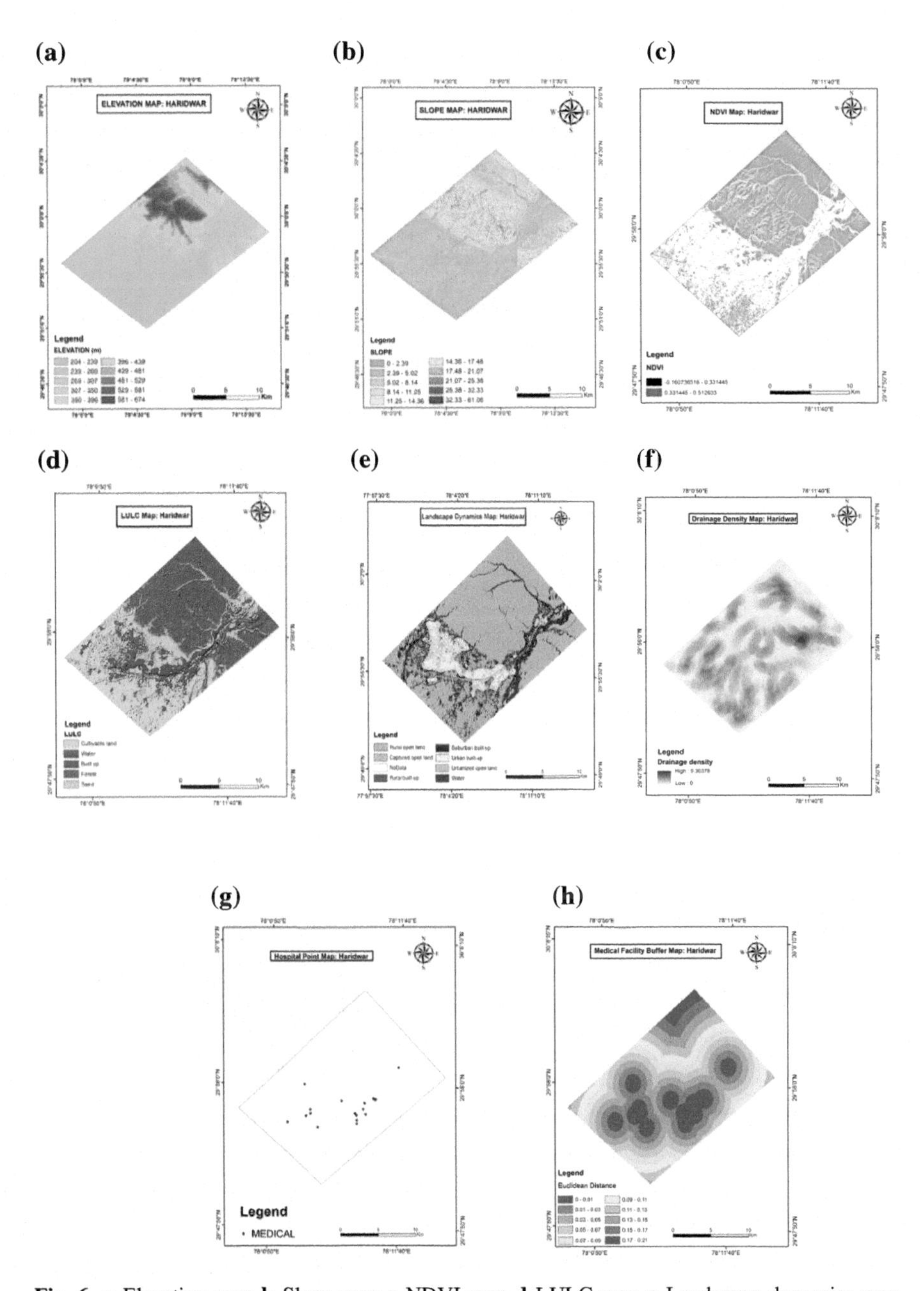

Fig. 6 **a** Elevation map **b** Slope map **c** NDVI map **d** LULC map **e** Landscape dynamics map **f** Drainage density map **g** Hospital location map **h** Hospital buffer map **i** School location map **j** School buffer map **k** Road map **i** Road buffer map **m** Railway line map **n** Railway line buffer map

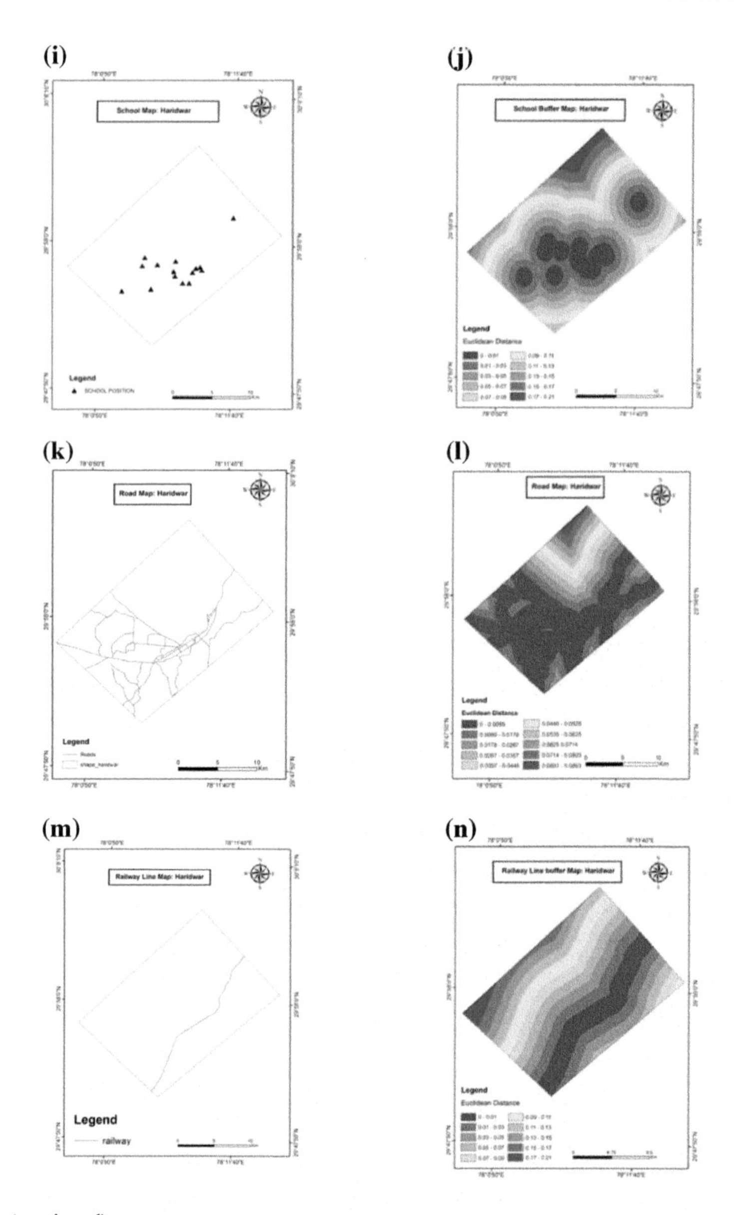

Fig. 6 (continued)

have to subtract the red band from the NIR band in the numerator and add the NIR and red bands in the denominator.

$$\mathrm{NDVI} = (\mathrm{NIR} - \mathrm{RED})/(\mathrm{NIR} + \mathrm{RED})$$

The values are calculated for each pixel giving a value range from -1 to $+1$. The pixels showing a value of more than 0.6 are considered to be in a healthy vegetation cover area. Haridwar has well-vegetated areas due to the forest range in the upper hilly region, and these areas are permanently unsuitable for expansion. While the area in the southern part of Haridwar has well-covered vegetation areas which can be considered for future expansion under site suitability analyses. Figure 6c shows the NDVI map of the Haridwar region.

6.4 Land Use Land Cover (LULC)

The land use land cover layer is generated by supervised classification with a maximum likelihood algorithm which is applied in ERDAS Imagine software. A maximum likelihood algorithm is one of the most widely used algorithms of supervised classification, and is used with remote sensing image data. Land use land cover is the most significant method for depicting the urban development of a particular study area.

In most cases it is used to find areas of open land, vegetated land, urban areas, and water. Plants and vegetation play an important role in cleaning the environment by removing harmful gasses. Image classification resulted in four land use land cover classes, namely open land, vegetation, water, and urban as shown in Fig. 6d.

6.5 Urban Landscape Dynamics (ULD)

The basic relationship between the physical landscape and past societies is a recurring theme in archeological research. Here, analysis of geological composition, landscape analysis based on GIS, and 3D modeling of fieldwork data are valuable tools. GIS-based landscape analysis typically utilizes modern and historic maps, digital altitude modeling, geological data, environmental data, airplane photogrammetry, and other sources. The method can identify changes in geographical objects through time and space. The detailed data strata sequencing includes:

- Origin of strata
- Continuous depositing or rebidding
- Presence of a past growing zone in strata
- Water impact of strata

- Evidence of human activity on strata
- Possible movement of lines on the landscape

Thus this raster layer has been divided into eight classes, i.e. rural open land, captured open land, rural built-up, suburban built-up, urban built-up, urbanized open land, water, and some areas where no were data captured as shown in Fig. 6e.

6.6 Drainage Density Map

The drainage network of the area is calculated using ArcGIS software. The drainage networks of an area are important as they decide the runoff and groundwater recharge level of the area. Drainage density of this area is calculated using ArcGIS software, so we can easily locate areas with high density. The places nearer to these sites have a high potential for future development as this factor is of high importance with regard to settlement in any area. Figure 6f shows the drainage density map.

6.7 Hospital Buffer Map

ArcGIS software can be used to create buffer zones around the selected features. The user can indicate the size of buffer and join together this information with incidences data to establish different cases fall within the same buffer. Buffer or proximity analysis can be used to map the impact zones of vector breeding sites, where the concern is required to be most strengthened. Hospital buffer gives the areas which are closer to the hospital and can be more useful for people living nearby as shown in Fig. 6g, h.

6.8 School Buffer Map

School mapping is the art and science of building geospatial databases with relational databases of education, demographic, social, and economic information for school and educational directorates to support educational planners and decision makers. Decision makers play a major role in training and guiding and can be considered as an important link between geospatial databases and stakeholders who are managing the data for the study.

The major role is to ensure the efficient use of available data, models, developed software, and basic theoretical knowledge. By applying these we can ensure educational decision support systems are useable and applicable. Figure 6i, j show a school position map and a school buffer map respectively.

6.9 Road Buffer Map

Road networks are several interconnected lines and points joining a system of roads in a given area. A buffer is a region used to temporarily hold output and input data. In the case of geographical information systems, units of buffering are lines, points and polygons. The term buffer refers in these cases to the creation of zones of specified width around the specified feature. There are two types of buffer:

- Constant width buffer
- Variable width buffer

Both buffers are used according to their need in the study and are generated by covering specified feature using attribute values. These zones are used to insert queries on entities to determine whether a feature lies within or without the buffer zone. Figure 6k, l show a road map and a road buffer map.

6.10 Railway Line Buffer

These are the sites most suitable for industrial purposes as most industries prefer to be situated near a railway line so they can easily transport goods. However, on the other hand, these sites are worst for residential purposes as the continuous noise from locomotives disturbs the people residing nearby. Figure 6m, n shows a railway map and a railway buffer map.

6.11 Preparation of Land Suitability Map

All the criteria in the map are converted in raster format, so that each pixel can be calculated and a result can be determined. All the criteria are integrated and overlaid to generate a site suitability map with a specific weightage according to their role in future development.

$$\text{Suitability map} = \sum (\text{Criteria map} \times \text{weight})$$

$$\begin{aligned}\text{Suitability index} = \{ & [\text{ULD} \times (0.06)] + [\text{NDVI} \times (0.03)] \\ & + [\text{Drainage density} \times (0.11)] + [\text{Slope} \times (0.30)] \\ & + [\text{Elevation} \times (0.16)] + [\text{LULC} \times (0.09)] + [\text{Hospital} \times (0.04)] \\ & + (\text{School} \times (0.03)) + [\text{Railway} \times (0.02)] + [\text{Road} \times (0.16)]\}\end{aligned}$$

Table 4 Area covered by each resultant class

Suitable Area	Area in sq. Meter	Percentage
Most suitable	82543695.85	0.24
Moderately suitable	1,1847,8426	0.34
Less suitable	133567659.5	0.39
Permanently not suitable	6628085.53	0.019

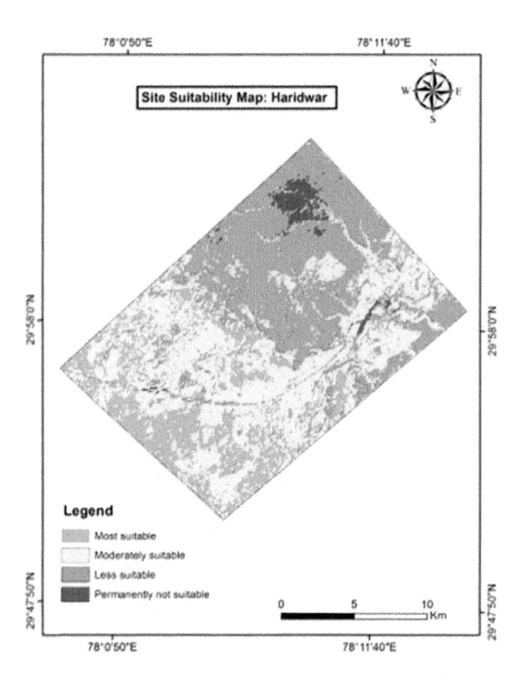

Fig. 7 Site suitability map

After applying weights a final raster layer map of a potential site for future urban development is developed as shown in Fig. 7 and the total area is divided into four classes as displayed in Table 4.

7 Conclusion

This study is mainly focused on the selection of sites as most suitable, moderately suitable, less suitable, and permanently unsuitable land for urban expansion. An AHP matrix along with GIS is used in the analysis of the different criteria considered for site suitability. AHP along with GIS was found to be very useful for site suitability identification. The result is adopted for the decision-making process of site suitability analysis of an area. The study includes different physical, social and economic parameters to determine the expansion opportunities. A decision support system has been preferred to identify appropriate sites for the best location for people to live in Haridwar City. Key steps of site suitability analysis were as follows:

- To determine best suitable site in city to establish site suitability of area.
- Survey of city in study area.
- Analytical hierarchy process (AHP).
- Site evaluation for future expansion.

These factors help to find the most suitable sites in the city which are the most likely to be included in future urban expansion. Two major tools are used; the first tool is GIS and the second is AHP. The result of the proposed methodology is based on assigning of weights to each class which is done according to their role in urban expansion. As a change in one class can lead to a huge difference in future expansion, the assigning of weights is of great importance.

References

1. Hauser PM, Gardner RW, Laquian AA, Shakhs SE (1982) Population and the urban future. State University of New York Press, Albany, p 187
2. Liu Y (1998) Visualizing the urban development of Sydney (1971–1996) in GIS. In: Proceedings of the spatial information research centre's 10th colloquium. University of Otago, New Zealand, pp 189–198
3. Malczewski J (2011) A gis-based multicriteria decision analysis approach for mapping accessibility patterns of housing development sites: a case study in Canmore, Alberta. J Geogr Inf Syst 3:50–61
4. Myagmartseren P, Buyandelger M, Brandt SA (2017) Implications of a spatial multicriteria decision analysis for urban development in ulaanbaatar, mongolia. Hindawi Math Probl Eng 2017
5. Nyeko M (2012) GIS and multi-criteria decision analysis for land use resource planning. J Geogr Inf Syst 2012(4):341–348
6. Saaty T (1977) A scaling method for priorities in hierarchical structures. J Math Psychol 15 (3):234–281
7. Saaty T (1980) The analytic hierarchy process, McGraw-Hill International Book Company, USA

Configurable Mapper and Demapper for the Physical Layer of a SDR-Based Wireless Transceiver

Zuber M. Patel

Abstract In this paper, field programmable gate array (FPGA) implementation of an adaptive constellation mapper and demapper is discussed for use in a SDR-based wireless transceiver. The adaptive functions at the physical layer of wireless technologies play a key role in achieving optimum performance. These functions include forward error correction (FEC) coding, puncturing, orthogonal frequency division multiplexing (OFDM), constellation mapping, sub-channelization, and frame assembly etc. We present here constellation mapping and demapping that supports gray-coded BPSK, QPSK, 16QAM, and 64QAM modulation schemes. This block can be configured at run time so that you can use them in multi-user systems where each user may be operating with a different modulation scheme. The proposed mapper/demapper block is implemented on a Xilinx Virtex-II pro FPGA and it is characterized in terms of functional correctness, area, power, and speed. Experimental results show that our design takes 1360 gates and can run at a maximum frequency of 320 MHz.

Keywords Physical layer · Software-defined radio (SDR) · Mapper and demapper · FPGA

1 Introduction

The future of wireless technologies will heavily depend on configurable/adaptive processing to deliver optimized and flexible services. One possible proposal towards achieving this goal is software-defined radio (SDR) which has been widely researched. For a reconfigurable hardware design of a wireless system, we first explore SDR technology.

Z. M. Patel (✉)
Department of Electronics Engineering, S. V. National Institute of Technology, Surat, India
e-mail: zmp@eced.svnit.ac.in

A. Kumar and S. Mozar (eds.), *ICCCE 2018*,
Lecture Notes in Electrical Engineering 500,
https://doi.org/10.1007/978-981-13-0212-1_78

1.1 SDR System

The demand of wireless services and applications will continue to grow in the future. Wireless devices are becoming more common and users are demanding the convergence of multiple services and technologies into single piece of equipment. To cater for this, wireless communication technologies need to advance and evolve. Conventional methods of fixed functions will become unsuitable whereas configurable radio systems are a better solution for meeting the needs of users. Hardware accelerated SDR [1] is one such technology that offers all these features. It is defined as a radio implementation technique in which the hardware components of physical (PHY) layer functions of a wireless communication system are realized in configurable hardware.

One of the key objectives of the vision of the 5G wireless network is to support efficient and flexible usage of the wireless spectrum and radio access technologies along with the easy adoption of innovative mechanisms as they appear. Industry and standardization bodies are in the early phase of the definition of 5G technologies, and major demands that are shaping design are already known, such as heterogeneous connectivity and 1000-times more capacity. To meet such challenging demands, two approaches to enabling the implementation of air interfaces for 5G networks are proposed. The first approach utilizes massive MIMO techniques to achieve higher spatial gains and increase spectral efficiency. The second approach leverages hyper-dense deployments of heterogeneous and small cell networks (HetNets) [2] located close to wireless users to cope with the increasing traffic demands in indoor environments. HetNets could support the offloading of outdoor traffic in cases where mobile users are close to small cells. Radio resource management for HetNets plays a crucial role in achieving the benefits of this advanced architecture.

SDR also enables rapid prototyping of wireless radio transceiver systems using reconfigurable hardware platforms such as FPGA and has significant advantages over conventional fixed digital hardware centric designs. In particular, time and cost can be saved by design reuse, flexibility, and reconfigurability. The intention of our work is to develop a reusable mapper/demapper intellectual property (IP) core for use in SDR applications using hardware description language (HDL) coding and synthesis. We test the design using a simulator and verify it using a FPGA platform.

1.2 FPGA and Reconfigurable Computing

Traditionally wireless network technologies have favored the implementation of hardwired logic circuits on an application specific integrated circuit (ASIC) for performance and power efficiency so that the battery life of portable wireless devices can be extended. Unfortunately, ASICs are hardwired chips and therefore are not reconfigurable. On the other hand, general purpose processors (GPP) are completely flexible and suited to SDR systems. But sequential execution of GPP degrades computing throughput and real time response. Configurable hardware platforms, such as FPGA strike a good balance between performance and flexibility. They offer more processing power and GPP-like flexibility at the expense of somewhat higher power consumption.

FPGAs are made of highly reconfigurable logic blocks and memory cells together with a programmable switch matrix to route signals between them [3]. Their flexibility and speed have made them popular and they are preferred as reconfigurable hardware platform for SDR [1]. The flexibility and parallelism characteristics of FPGAs enable computationally intensive and complex tasks to be processed in real time and with better performance. For these reasons, they are promising contenders for use in the reconfigurable hardware platforms to be used in the next generation of wireless systems.

An important attribute of FPGA implementation is the design reuse. A library of IP cores for FPGA [4] programming makes reuse effective. A designer can achieve a better trade-off amongst delay, area, and power with proper selection of IP cores that best suit the requirements of the target application. This will lead to improved productivity and enhanced system level SDR performance.

2 Configurable Blocks of the PHY Layer of Wireless Systems

Today's wireless technologies use adaptive mechanisms at medium access control (MAC) and PHY layers for efficient operation. While most of the MAC layer functions are realized through software, realization of adaptive/configurable MAC operations is relatively easy and straightforward. On the other hand, the physical baseband layer processing [5, 6] is hardware based and realization of adaption in hardware within constraints is slightly complex. So we discuss the physical layer functions and explore various adaptive parameters.

The PHY layer consists of baseband processing blocks in the transmit path and complementary baseband processing blocks in the receive path as shown in Fig. 1a, b respectively. Therefore, the physical layer is also called a transceiver. In the transmit path, a MAC protocol data unit (PDU) is taken from the MAC layer and is then passed through various blocks which involves scrambling, FEC encoding, interleaving, mapping, piloting, IFFT, cyclic prefix (CP) insert, and packet assembly. On the reception path we have synchronization, CP remove, FFT, depiloting, demapping, deinterleaving, FEC decoding, and descrambling. Table 1 lists some PHY layer parameters which are configurable. The PHY layer is the most intensively researched stage of SDR since it has room for algorithm evolution to improve throughput and performance leading to better performance within constraints.

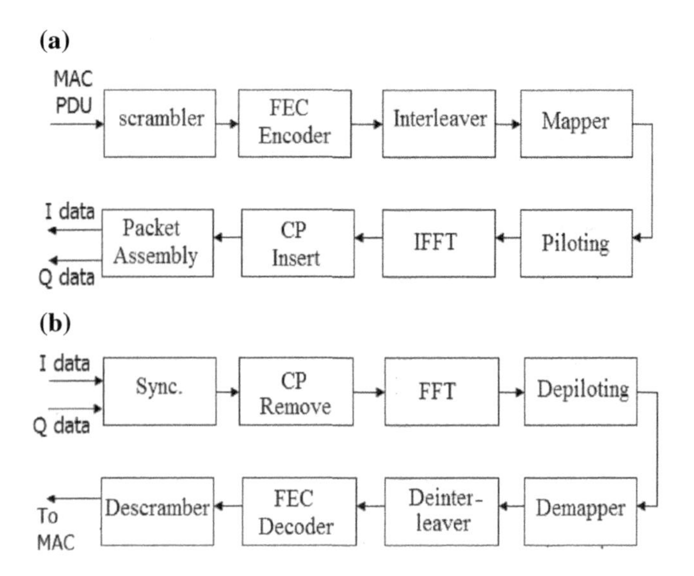

Fig. 1 PHY layer baseband processing blocks: **a** Transmission path **b** Receiving path

Table 1 Configurable parameters supported by wireless PHY layer

Baseband block	Configurable parameters
FEC encoder	Polynomial, code rate
Interleaver	Inter-symbol and intra-symbol interleaving
Modulation	Modulation scheme, data mapping value
Piloting	Pilot position, pilot value, size
FFT	Symbol size, guard prefix

FEC Encoder/Decoder:

This error correcting block uses convolutional or block code to generate coded stream at the transmitter and decodes at the receiver. Configurable parameters here are polynomials and code rate.

Interleaver/Deinterleaver:

The interleaver permutes data across a block to improve the correction of burst errors and average power across the symbol. Interleaving can be done either via inter-symbol or intra-symbol permutation of data that is either between data within a single frame or between data in adjacent frames. Similarly, the deinterleaver permutes the received data back in order, based on the vector used for interleaving at the transmission site.

Modulation/Demodulation:

This performs an adaptive constellation mapping and demapping operation using BPSK, QPSK, 16QAM or 64QAM modulation schemes.

IFFT/FFT:

These operations implement a scalable OFDM [7] modem where transmitter maps input data to subcarriers through IFFT (frequency to time domain conversion) and the receiver recovers the data from subcarriers using FFT (time to frequency domain conversion). The OFDM symbol size varies depending on the standard being implemented and so does the FFT/IFFT size.

Piloting/Depiloting:

This operation has configurable parameters corresponding to the pilot position, the pilot value, and the symbol size. The flexibility developed into the piloting controller is used to allow the implementation of non-contiguous OFDM by allowing specific subcarriers to be nulled during processing.

Frame Assembly:

This performs assembly of OFDM symbols. In this last stage of the transmitter, OFDM data symbols are appended one after another and provided with a preamble and header to create a complete physical layer PDU frame, which is now ready for transmission over a channel.

3 Design of Adaptive Mapper and Demapper

Constellation mapping is performed in a digital modulator before OFDM. A constellation mapper takes a serial bit stream as its input and segments the stream into *k*-bit symbols, which are mapped to coordinates in the signal constellation. It is worth noting that input bits in all modulation types are gray coded where adjacent entries in columns differ only by one bit. The benefit of this is that a single out-of-position by one error results in only a single bit error in received data. The coordinates of each point in a two-dimensional signal constellation represents the baseband in-phase (I) and quadrature-phase (Q) components that modulate the orthogonal IF carrier signals. Because the constellation mapper defines the shape and dimension of the signal constellation, it defines the modulation scheme that is implemented. 2-D schemes require two coordinates to specify the position of a signal point $\boldsymbol{S_k}$ [8]

$$S_k = I_k + jQ_k \tag{1}$$

The components $\boldsymbol{I_k}$ and $\boldsymbol{Q_k}$ can each be modulated onto two orthogonal sinusoidal carriers. The resulting waveform spectra are centered on the same frequency and can occupy the same bandwidth. However, the components are uniquely separable at the demodulator because there is a 90-degree phase shift between the carrier signals. Therefore, it is possible to send twice the amount of data for a given signal bandwidth over quadrature carriers than over a single carrier.

In the PHY layer, the OFDM sub-carrier is modulated by using BPSK, QPSK, 16QAM or 64QAM modulation depending on the PHY mode selected for data transmission. The interleaved binary serial input data is divided into groups of NBPSC (1, 2, 4 or 6) bits and converted into complex numbers representing BPSK, QPSK, 16QAM or 64QAM constellation points. The conversion is performed according to Gray coded constellation mappings, illustrated in Figs. 2, 3, and 4 with input bit *b1* being earliest in the stream. The symbols are normalized so that each constellation has an equal average power, as described by the specification. The mapping process multiplies each constellation point by a scaling factor.

Constellation demapping is performed in a digital communications receiver and involves the translation of the demodulated I-Q coordinates back to the *k*-bit symbols they represent. This is achieved through either soft-decision [9] or hard-decision. In our design we selected a hard-decision demapper to reduce complexity. Demapping is complicated by the fact that the transmitted symbols can be corrupted by a noise-inducing channel. The demodulated I-Q coordinates may no longer correspond to the exact location of a signal point in the original constellation. The demapper must detect which symbol was most likely transmitted by finding the *smallest distance* between the received point and the location of a valid

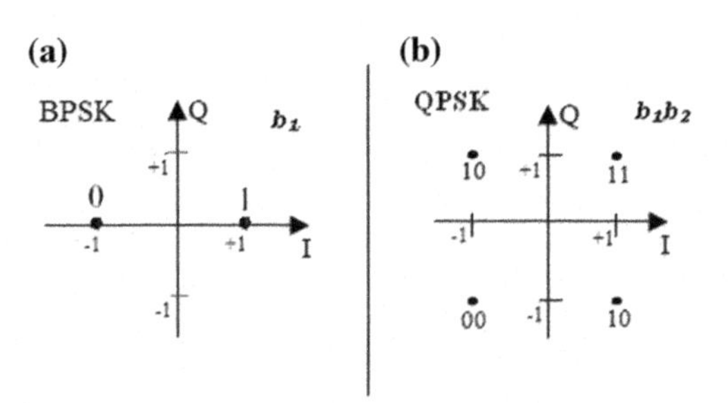

Fig. 2 Constellation for **a** BPSK **b** QPSK

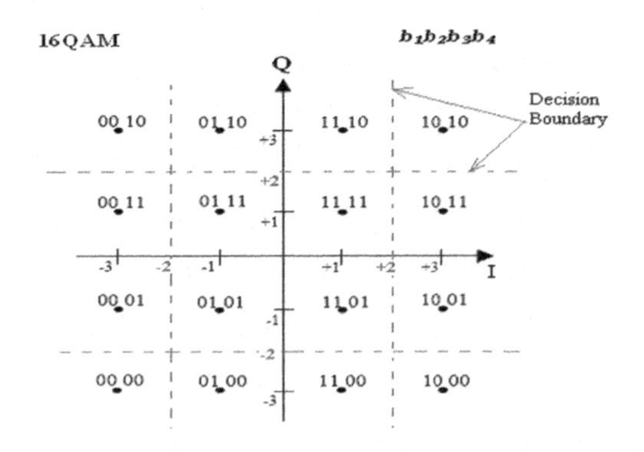

Fig. 3 Gray coded constellation for 16 QAM

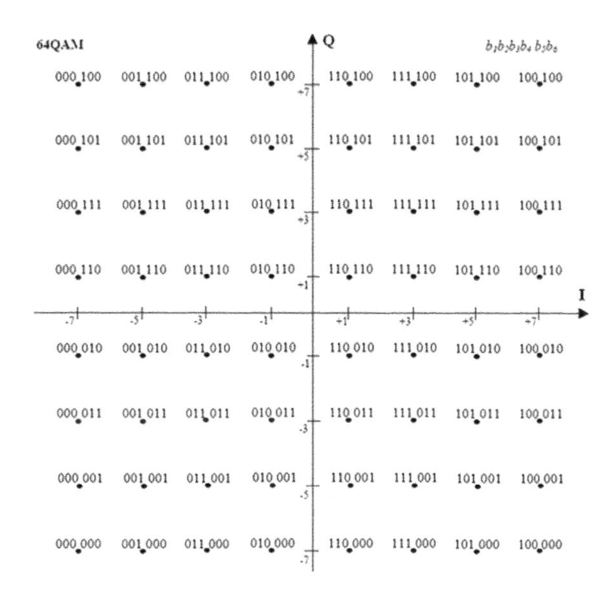

Fig. 4 Gray coded constellation for 64 QAM

symbol in the signal constellation. To decide on the combination of bits the design divides the complex plane into equally sized regions that correspond to each constellation point and outputs the bit combination of the region that the received signal appears in, known as hard decision decoding. Having detected the most likely coordinates, the constellation demapper then performs the decoding procedure to output the correct k-bit symbol represented by those coordinates.

4 FPGA Implementation and Results

The configurable constellation mapper and demapper modules are developed using very high speed integrated circuit hardware description language (VHDL) in a Xilinx ISE tool. The design is synthesized for a target FPGA device, a Virtex-II Pro xcvp30 (1.5 V, 130 nm) [10]. Post-synthesis simulation waveforms shown in Fig. 5 verify the correct working of our design with dynamic mode change. The FPGA resource utilization is shown in Table 2 in terms of flip-flops, LUTs, slices and gate count. Our design can be operated at a maximum frequency of 320 MHz and dissipates 63 mW at maximum frequency.

Timing Summary:

Minimum period: 3.123 ns (maximum frequency: 320.169 MHz)
Minimum input arrival time before clock: 3.013 ns
Maximum output required time after clock: 3.340 ns

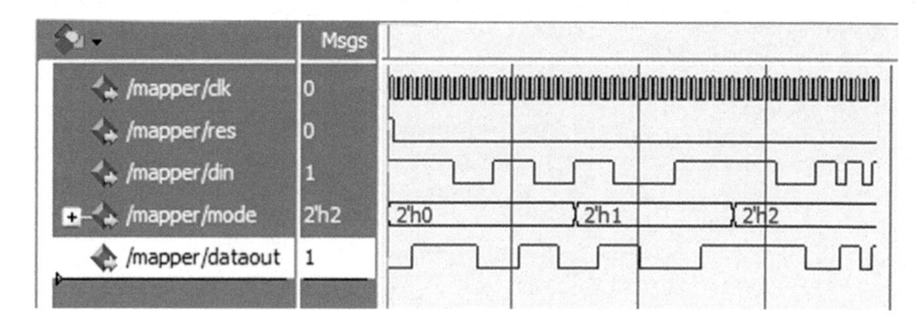

Fig. 5 Post-synthesis HDL simulation

Table 2 FPGA synthesis results for the mapper/demapper block

Baseband block	FFs/Latches	LUTs	Slices	Gate count
Mapper/Demapper	73	120	73	1360

Power:

Dynamic Power consumption = 63 mW@320 MHz.

5 Conclusion

The intention of our work is to realize functions of SDR on configurable hardware to meet the requirement of future adaptive wireless technologies. We investigated PHY layer reconfigurable modules which can be realized in programmable hardware. We implemented mapper and demapper blocks in FPGA to realize adaptive modulation which supports BPSK, QPSK, 16 AQAM, and 64 QAM schemes. On Virtex-II pro FPGA, it consumes 1360 gates, dissipates 21 mW@100 MHz and operates at a maximum frequency of 320 MHz.

References

1. Kazaz T, Praet CV, Kulin M, Willemen P, Moerman I (2016) Hardware accelerated SDR platform for adaptive air interfaces. In: ETSI workshop on future radio technologies: air interfaces. Sophia Antipolis, France, pp 1–10
2. Hwang I, Song B, Soliman SS (2013) A holistic view on hyper-dense heterogeneous and small cell networks. IEEE Commun Mag 51(6):20–27
3. Olivieri SJ, Aarestad J, Pollard LH, Wyglinski AM, Kief C, Erwin RS (2012) Modular FPGA-based software defined radio for CubeSats. In: Proceedings of IEEE international conference on communications (ICC'12). pp 3229–3233 June 2012

4. Tsoeunyane L, Winberg S, Inggs M (2017) Software defined radio FPGA cores: building towards a domain-specific language. Int J Reconfigurable Comput 2017, Article ID 3925961, 28 pp
5. Al-Ghazu N (2013) A study of the next WLAN standard IEEE 802.11ac physical layer. Master of science thesis, Signal Processing Group, KTH Royal Institute of Technology, Sweden, January 2013
6. Gu C, Li X (2011) Design and implementation of IEEE 802.16 baseband system on FPGA. In: Proceedings IET international conference on communication technology and applications (ICCTA). https://doi.org/10.1049/cp.2011.0794
7. Chacko J, et al (2014) FPGA-based latency-insensitive OFDM pipeline for wireless research. In: Proceedings of IEEE international conference on performance extreme computing conference (HPEC). https://doi.org/10.1109/hpec.2014.7040969
8. Constellation Mapper and Demapper for WiMAX, Application Note 439, Altera, 2007
9. Aung NY (2016) Digital radio baseband and testbed for next generation wireless system. Technical report no UCB/EECS-2016-87, Department of Electrical and Computer Science, University of California at Berkeley
10. Xilnix (2007) Virtex-II Pro and Virtex-II Pro X FPGA User Guide. Version 4.2. https://www.xilinx.com/support/documentation/user_guides/ug012.pdf

Experimental Investigation to Analyze Cognitive Impairment in Diabetes Mellitus

Vinit Kumar Gunjan, Puja S. Prasad, S. Fahimuddin and Sunitha Devi Bigul

Abstract Metabolic disorders and cognitive impairment are very common age-associated disorders. Alzheimer's disease and diabetes are two common diseases and their possibility of occurrence increases as the age of a person increases. This paper proposes the percentage of type 2 diabetes patients that have a chance of developing Alzheimer's disease. Pathophysiology and clinical patterns share common features in both the diseases as shown by epidemiological studies. There are a number of studies that show evidence of a connection between type 2 diabetes and Alzheimer's disease but there is no proof of any biochemical mechanisms present yet. We focused our study and experiments on whether the alteration of factors like glycogen synthase kinase-3β activity, olfactory function, and ApoE genotypes can identify early cognitive impairment, which leads to Alzheimer's disease.

Keywords Diabetes mellitus · Mild cognitive impairment · Olfactory scores Genotypes · Alzheimer

1 Introduction

In the past few years, several studies have shown that there is a sharp connection between people having diabetes that leads them towards developing Alzheimer's disease (AD) if their sugar is uncontrolled for a long time. In the journal *Neurology* it was published that the brain "tangles" commonly seen in a person having AD are more likely develop in a person having type 2 diabetes (T2DM). A study from Albany University in New York shows that the extra insulin secreted in a person having diabetes damages and disrupts the chemistry of the brain leading to a

V. K. Gunjan (✉) · P. S. Prasad · S. D. Bigul
CMR Institute of Technology, Hyderabad, Telangana, India
e-mail: vinitkumargunjan@gmail.com

S. Fahimuddin
AITS Rajampet, Rajampet, AP, India

A. Kumar and S. Mozar (eds.), *ICCCE 2018*,
Lecture Notes in Electrical Engineering 500,
https://doi.org/10.1007/978-981-13-0212-1_79

generation of toxic protein, which damages brain cells. This toxic protein is found in people with AD. AD is a neurodegenerative disease in which progressive brain cell damage affects different people in different ways, such as facing difficulty in remembering things, difficulty in thinking of common words while speaking, being hesitant while speaking, spelling, and writing errors, speech as well as changes in personality, behavior also seen etc.

Both AD and T2DM are common age-related diseases with effects that increase with age [1, 2]. Janson et al. [3] used both clinical and pathological studies to observe the existence of a common risk for AD and T2DM. When looking at both studies together, the clinical and pathological studies show a possible relationship between the neurodegenerative disorder that is one of the reasons for cortical brain cell damage in AD and also the reduction in T2DM of β-cells. Biessels et al. [4, 5] found that using neuroimaging or brain imaging cognitive dysfunction and abnormalities are linked with diabetes mellitus (DM). A number of studies also focus on why certain people with DM are more inclined to progress to the neurocognitive disorder while some are unaffected. They worked on finding the various clinical changes that lead toward to the neurocognitive disorder. They found that the cognitive disorder that occurs due to DM occurs crucially in two periods of life; when the brain is growing in childhood and when the brain undergoes neurodegeneration due to aging. Knowing the period of risk and conditions for the development if the cognitive disorder are the focuses of the studies in this scenario.

Huang et al. [6] revealed that DM that was newly diagnosed was much more related with an increased risk of future AD. Pathogenesis of AD was found to be linked with T2DM. There are some mechanism that show the relationship between AD and DM. this mean that Furthermore, neither monotherapy nor combination therapy with oral anti-diabetic medications were found to be associated with a risk of AD occurrence. However, combination therapy with insulin was found to be associated with a risk of AD occurrence. For example, hyperglycemia, which results, increased oxidative stress in brain leading to functional and structural abnormalities of the brain [7]. One of the reason for decreased apoE mRNA may be the processing of the primary transcript of the apoE gene or due to a defect in transcription suggested by Mc Khann et al. [1].

Stumvoll et al. [8] suggest that T2DM, i.e. DM is turning into an epidemic, and virtually all doctors have patients who suffer from the disease. Whereas insensitivity towards insulin is an early phenomenon that is related to obesity, the decline of pancreatic beta cells occurs gradually with time [9]. Furthermore, the disease has a genetic component to it, but only a few of genes have been recognized so far as being involved in diabetes.

Exalto et al. [10] gives an update about the risk of dementia in a person having T2DM. This paper reviews the relationship of T2DM with dementia subtypes, i.e. AD and vascular dementia. Vascular lesions in the brain also increase the risk of dementia in people with T2DM person. There is also evidence that atherosclerosis, microvascular complications, and a severe hypoglycemic condition increase the risk of dementia [11]. Lehrner et al. [12] worked on olfactory dysfunction. Olfactory dysfunction is early sign of AD. MCI or mild cognitive impairment patient also and olfactory dysfunction has been found in mild cognitive impairment (MCI). They work on the ability of odor identification and self-reported olfactory function having patients with different types of MCI [2]. Tan et al. review is that they provide an indication of biomarkers for preclinical Alzheimer's disease, with an importance on neurochemical and neuroimaging biomarkers. They conclude with a debate of the future direction of research into biomarkers of Alzheimer's disease [3].

2 Methods

A Folstein test (FT) was used for the evaluation of cognitive ability. This test is also called a minimal mental state examination. Petersen's criteria are used to diagnose mild cognitive impairment and provide a clinical dementia rating. A Western or dot blot test was used for testing GSK-3beta activity in platelets. A Connecticut (CCCRC) olfactory test was used to detect olfactory dysfunction.

2.1 Study Design and Participants

We recruited T2DM patients from five hospitals in India between January 2013 and May 2016, and the T2DM patients were diagnosed by the World Health Organization (WHO). The classification of different subjects are as follow:

T2DM patients aged over 55 years who are literate and understand the different questions of the test or the neuropsychological test and are interested in this study.

We have not included subjects that have a history of any type of trauma, such as epilepsy, stroke, brain tumor, coma, transient ischemic attack, the presence of dementia before T2DM, previous thyroid disease, depression, schizophrenia, and other psychological disorders.

Petersen et al. [13] give different criteria called Petersen's mild cognitive impairment criteria based on which T2DM patients were divided into two groups [8]:

(a) T2DM with MCI (T2DM-MCI) which met objectives and subjective cognitive difficulty
(b) T2DM without MCI Alzheimer's disease diagnosis is standard on the National Institute of Neurological and Communicative Disorders and Stroke and the Alzheimer's Disease and Related Disorders Association (NINCD-SADRDA) criteria [4].

2.2 Procedures

All subjects underwent a complete assessment as follows: a comprehensive assessment of cognitive and behavioral functions called neuropsychological evaluation, smell identification to test the function of an individual's olfactory system called an olfactory score test, platelet GSK-3β activity measurement and determination of the specific Apo lipoprotein E (APOE) genotypes in patients called ApoE genotyping. The neuropsychological evaluations were based on the clinical dementia rating (CDR), and minimum mental state examination (MMSE) and were performed by six examiners who already had neurology training and wide experience with neurophysiologic techniques. There is one score ranging from 0 to 30 used for commonly measuring cognitive impairment, where having lower scores points to greater cognitive impairment. The severity of symptoms of dementia is classified into six areas: community affairs, memory, orientation, home and hobbies, judgment and problem solving ability, and personal care. In Petersen's MCI criteria, MMSE and CDR scores for MCI diagnosis were used respectively (Fig. 1).

To measure the olfactory score a Connecticut chemosensory clinical research center (CCCRC) test was used, which was done blindly to different measures by the investigators. A reserve solution with 4% butanol was diluted with distilled water as a one-third ratio to twelve dilutions in twelve flasks. Subjects were then exposed to those flasks as well as blank flasks with low to high concentrations of butanol to identify the strength of the solution. Erroneous choices start another blank paired having next higher concentration, whereas right choices direct to another arrangement of the same concentration in another bottle and an empty flask. Six correct choices in a row led to the end of the test, and the number of this concentration marked the score. Higher scores indicated greater olfactory impairment.

To test biochemical indicators, all subjects were asked to compute fasting blood glucose, serum insulin, serum magnesium, hemoglobin A1C (HbA1c), postprandial

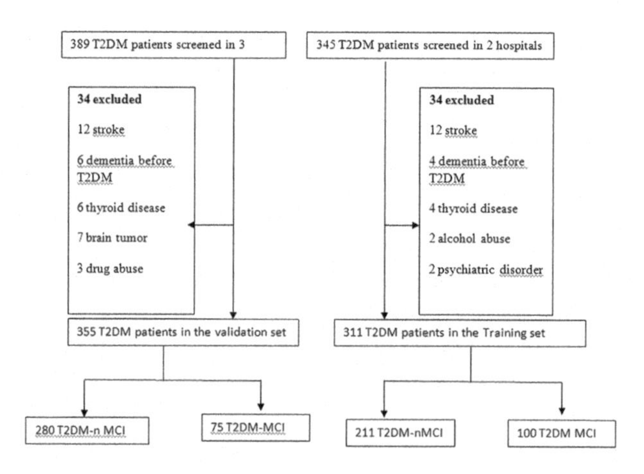

Fig. 1 Study Profile

Table 1 Baseline characteristics of T2DM patients in the training set and validation set

Characteristic	Validation set		Training set	
	T2Dm-nMCI (n = 280)	T2DM-MCI (n = 75)	T2DM-n MCI (n = 211)	T2DM-MCI (n = 100)
Age (year)	61.31 ± 7.34	66.34 ± 9.37*	64.24 ± 9.00	68.91 ± 9.68
Male (%)	43.31	38.75	46.77	35.00
T2DM years	7.29 ± 5.27	8.49 ± 6.86	8.45 ± 6.89	8.80 ± 7.85
Hypertension (%)	48.40	39.75	54.33	64.00
Hyperlipidemia (%)	6.95	4.81	9.00	10.75
Coronary disease (%)	8.95	4.53	8.96	16.00
Complication (%)	65.23	59.83	48.26	32.00
Insulin treatment (%)	48.69	45.71	39.79	35.00
MMSE	30.05 ± 0.95	27.00 ± 1.75*	30.10 ± 0.90	25.94 ± 2.99*

Data are mean ± SD or n (%). T2DM = type 2 diabetes mellitus. MCI = mild cognitive impairment. T2DM-nMCI = T2DM without MCI group. T2DM-MCI = T2DM with MCI group. MMSE = the minimum mental state examination

blood glucose, and next day. On receiving by centrifugation leukocytes, serum and platelet samples were separated from EDTA and stored at −85 °C until analysis. ELISA and dot blotting were used to measure total GSK-3β (tGSK-3β) and serine-9 phosphorylated GSK-3β. The ratio of tGSK-3β/pS9GSK-3β was used as a measure of GSK-3β activity. In dot blotting, the levels of tGSK-3β and pS9GSK-3β were normalized using the same control people (non-T2DM). By using an enzyme activity assay kit, the biochemical activity of GSK-3β in platelets was measured.

Gerard's method with modifications in the multiplex amplification refractory mutation system PCR (ARMS) was applied for ApoE genotyping [4, 14]. Using a DNA extraction kit, genomic DNA was extracted from peripheral blood leukocytes (Aidlab Biotechnologies Co., Ltd., China). The primers were designed and synthesized by Invitrogen Life Technologies (Shanghai, China). The PCR protocol was performed with 13.5 μl Taq MasterMix, 60 ng genomic DNA, 0.5 μM Cys112/Arg112 primers or 0.8 μM Cys158/Arg158 primers, 0.8 μM common reverse primers, and 2 μl DMSO in a total of 25-μl reaction. PCR was initiated with denaturation at 96 °C for 2 min, amplification by 36 cycles of 96°C for 44 s, 63 °C for 46 s and 73 °C for 46 s, with a final extension at 72 °C for 6 min (Table 1).

3 Results

We analyzed 734 T2DM patients between Jan 2013 and June 2015 and the number of subjects, which comes under all the inclusion criteria are 666. Based on hospitals 355 patients from three hospitals were assigned to the validation set and 311 from two hospitals to the training set. In the next step, of those 355 patients, two sets were prepared: (a) T2DM-MCI including 75 patients and (b) T2DM-nMCI including 280 patients of the validation test. Similarly, in the training test out of 311 patients, two sets were prepared: (a) T2DM-MCI including 100 patients and (b) T2DM-nMCI including 211 patients. Peterson's MCI criteria were used for this division.

3.1 Demographics and Clinical Characteristics of the Participants

The demographic and clinical data of all the subjects in the validation set and the Training set. The T2DM-MCI group had an older age (68.91 ± 9.68 vs. 64.34 ± 9.00 and 66.34 ± 9.37 vs. 61.31 ± 7.34) and lower MMSE score (25.94 ± 2.99 vs. 30.10 ± 0.90 and 27.00 ± 1.75 vs. 30.05 ± 0.95) than the T2DM-nMCI group in the training set and the validation set, respectively. There were no significant differences for T2DM with respect to years, sex, hypertension, coronary disease, hyperlipidemia, insulin treatment, and complications.

3.2 Comparison of Olfactory Score and ApoE Genotypes Between T2DM-MCI and T2DM-NMCI Patients

All 666 subjects were tested for HbA1c level, ApoE genotypes, and olfactory score. The T2DM-MCI group had an olfactory score higher than the T2DM-nMCI group (Table 2). Negative correlation was marked between MMSE score and olfactory

Table 2 Potential markers of T2DM patients in the training set and validation set

Characteristic	Validation set		Training set	
	T2DM-nMCI (n = 100)	T2DM-MCI (n = 280)	T2DM-n MCI (n = 75)	T2DM-MCI (n = 211)
Olfactory	7.75 ± 1.40	8.45 ± 2.05*	7.85 ± 1.55	9.55 ± 1.97*
HbA1c	8.00 ± 2	8.27 ± 2.35	7.82 ± 1.89	8.20 ± 1.99
ApoE €2 (%)	12.92	12.76	24.88	23.00
ApoE €3 (%)	97.00	94.00	93.23	87.05
ApoE €4 (%)	13.08	27.06	10.45	25.00
tGSK3β	1.38 ± 1.93	1.97 ± 3.16	1.86 ± 3.62	3.74 ± 3.00
pS9GSK3β	2.17 ± 4.05	2.94 ± 6.53	3.12 ± 3.98	3.74 ± 4.55
rGSK3β	0.78 ± 0.28	1.20 ± 0.67	0.75 ± 0.46	1.62 ± 1.03*

Data are mean ± SD or n (%). T2DM = type 2 diabetes mellitus. MCI = mild cognitive impairment. T2DM-nMCI = T2DM without MCI group. T2DM-MCI = T2DM with MCI group. HbA1c = hemoglobin A1C. ApoE = Apo lipoprotein E. GSK-3β = glycogen synthase kinase-3β. GSK-3β ratio = total GSK-3β/Ser9-GSK-3β. Pb 0.05 versus the T2DM-nMCI group

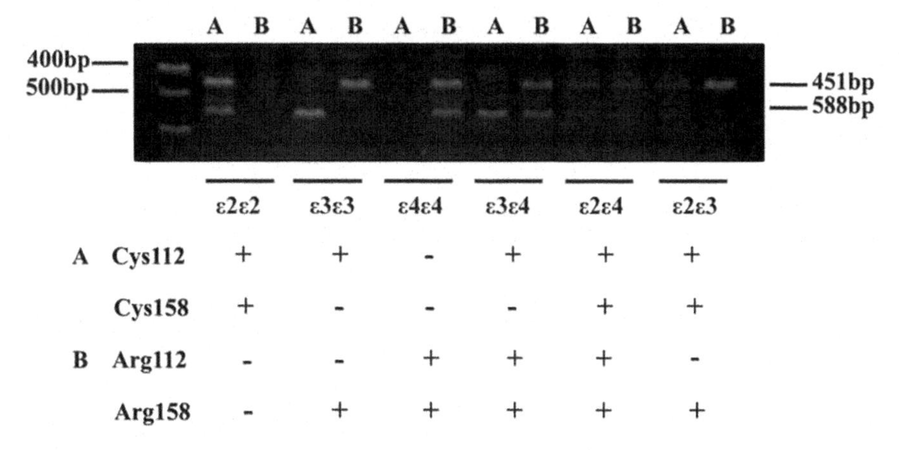

Fig. 2 Six ApoE genotypes found in the study

score in T2DM patients by ARMS-PCR and six ApoE genotypes, including ε2ε2, ε3ε3, ε4ε4, ε2ε3, ε2ε4, and ε3ε4, were detectable in T2DM patients (Fig. 2). From the analysis, it was found that genotype ε3ε4 was higher and ε3ε3 was lower in T2DM-MCI patients than T2DM-nMCI patients. No changes was found in the other genotypes.

3.3 Comparison of Platelet GSK-3β Activity Between T2DM-MCI and T2DM-NMCI Patients

Using ELISA, it was found that GSK-3β serum level was lower in T2DM-nMCI (n = 7) subjects compared to T2DM-MCI (n = 10) (Fig. 3).

In another test performed using an enzyme activity assay kit, we found that Platelet GSK-3β serum level had a higher value so that it became a biomarker for the identification of MCI in T2DM patient dot blotting is used for the analyses of pS9GSK-3β tGSK-3β proteins of all 666 participants. The platelet GSK-3β activity was represented using a ratio of tGSK-3β/pS9GSK-3β with higher ratios indicating higher GSK-3β activity. We found that levels of pS9GSK-3β (the inactive form of the kinase) and tGSK-3β were not significantly changed, while the level of rGSK-3β was significantly increased in T2DM-MCI patients compared to T2DM-nMCI patients.

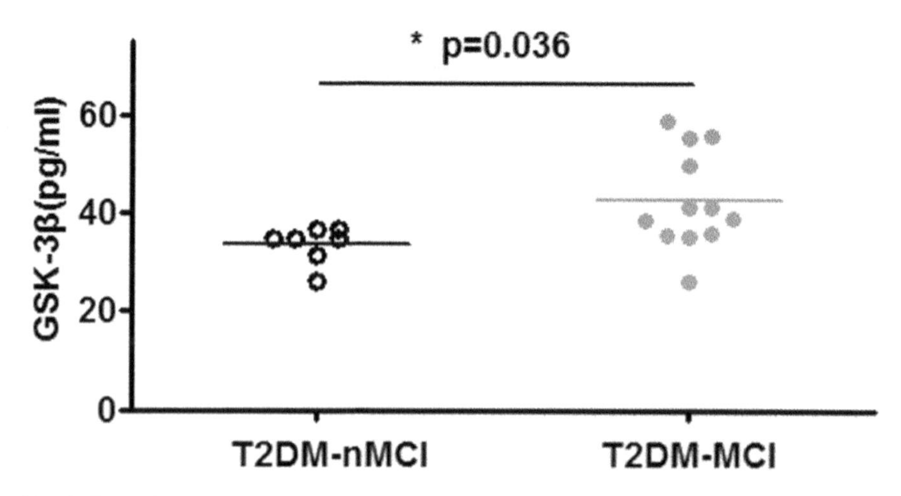

Fig. 3 This diagram shows the difference in GSK-3β expressions in serum between the T2DM-nMCI group (n = 7) and the T2DM-MCI group (n = 10). The expressions of GSK-3β in serum of the T2DM-MCI group were higher than those in the T2DM-nMCI group

4 Discussions

Millions of people all over the globe have been affected by AD and T2DM as they are age-associated problems. There is incidence of 15.2% in diabetic patient AD increases and 82% have irregular or abnormal blood sugar levels. In these scenarios biomarkers for the diagnosis of AD in T2DM patients is essential. In our study, we found that what would be biomarkers of cognitive impairments in T2DM patients has validated the diagnostic of the biomarkers we studied. We divided the patients into two groups T2DM-MCI and T2DM-n MCI and as age is a strong factor for cognitive impairment in T2DM patients, we discuss on four parameters age, GSK-3β, ApoE and Olfactory for diagnosis of MCI in diabetic patient. In this research, it was observed that activation of GSK-3β in peripheral is one of the main causes of cognitive impairment in T2DM. Patients' levels of GSK-3β, protein, and messenger RNA in leucocytes were also seen to increase in clinical reports. People having T2DM-MCI have a lot of GSK-3β activity compared to non-MCI patients. Another finding of the above research was that the expression of the ApoE ε4 gene is strongly associated with mild cognitive impairment in T2DM patients.

References

1. McKhann G, Drachman D, Folstein M et al (1984) Clinical diagnosis of Alzheimer's disease report of the NINCDS-ADRDA work group* under the auspices of department of health and human services task force on Alzheimer's disease. Neurology 34(7):939

2. Lehrner J, Pusswald G, Gleiss A, Auff E, Dal-Bianco P (2009) Odor identification and self-reported olfactory functioning in patients with subtypes of mild cognitive impairment. Clin Neuropsychol 23(5):818–830
3. Arvanitakis Z, Wilson RS, Bienias JL et al (2004) Diabetes mellitus and risk of Alzheimer disease and decline in cognitive function. Arch Neurol 61(5):661–666
4. Janson J, Laedtke T, Parisi JE et al (2004) Increased risk of type 2 diabetes in Alzheimer disease. Diabetes 53(2):474–481
5. Biessels GJ, Deary IJ, Ryan CM (2008) Cognition and diabetes: a lifespan perspective. Lancet Neurol 7(2):184–190
6. Biessels GJ, van der Heide LP, Kamal A, Bleys RL, Gispen WH (2002) Ageing and diabetes: implications for brain function. Eur J Pharmacol 441:1–14
7. Huang CC, Chung CM, Leu HB et al (2014) Diabetes mellitus and the risk of Alzheimer's disease: a nationwide population-based study. PLoS One 9(1):e87095
8. Medina M, Avil J (2011) The role of glycogen synthase kinase-3 (GSK-3) in Alzheimer's disease. In: Alzheimer's disease pathogenesis-core concepts, shifting paradigms and therapeutic targets, Sept 2011
9. Stumvoll M, Goldstein BJ, van Haeften TW (2005) Type 2 diabetes: principles of pathogenesis and therapy. Lancet 365(9467):1333–1346
10. Donohoe GG, Salomaki A, Lehtimaki T, Pulkki K, Kairisto V (1999) Rapid identification of apolipoprotein E genotypes by multiplex amplification refractory mutation system PCR and capillary gel electrophoresis. Clin Chem 45(1):143–146
11. Exalto LG, Whitmer RA, Kappele LJ, Biessels GJ (2012) An update on type 2 diabetes, vascular dementia and Alzheimer's disease. Exp Gerontol 47(11):858–864
12. Kim DS, Lee HS, Choi SI, Suh SP (2000) Modified and improved ARMS PCR method for apolipoprotein E genotyping. Korean J. Clin. Pathol. 20(2):150–156
13. Petersen RC, Smith GE, Waring SC et al (1999) Mild cognitive impairment: clinical characterization and outcome. Arch Neurol 56(3):303–308
14. Strupp M (2009) Dementia: Predicting its risk, APOE effects and disclosure of APOE genotype. J Neurol 256(10):1784–1786
15. Anchors JM, Gregg RE, Law SW, Brewer HB (1986) ApoE deficiency: Markedly decreased levels of cellular ApoE mRNA. Biochem Biophys Res Commun 134(2):937–943
16. Corzo L, Fernández-Novoa L, Zas R, Beyer K, Lao JI, Alvarez XA, Cacabelos R (1998) Influence of the apoE genoptype on serum apoe levels in Alzheimer's disease patients. In: Progress in Alzheimer's and Parkinson's diseases, pp. 765–771
17. Type 2 diabetes in adults. Staged diabetes management, pp 77–137, Nov 2011
18. Type 2 diabetes: general introduction. Practical diabetes care, pp 83–101, Feb 2011
19. Giacomazza D Di Carlo M (2013) Insulin resistance: a bridge between T2DM and alzheimer's disease. J Diabetes Metab 4(4)
20. Umegaki H (2014) Type 2 diabetes as a risk factor for cognitive impairment: current insights. Clin Interv Aging 9:1011–1019
21. Tan CC, Yu JT, Tan L (2014) Biomarkers for preclinical Alzheimer's disease. J Alzheimers Dis 42(4):1051–1069
22. Lo H-J (2016) The comparison of different health education tools and their effects in glucose management for type II diabetes patients. Diabetes Res Clin Pract 120:S160
23. Introduction to alzheimer's disease, alzheimer's disease decoded, pp 73–98, Oct 2016
24. Galligan A, Greenaway TM (2016) Novel approaches to the treatment of hyperglycaemia in type 2 diabetes mellitus. Internal Med J 46(5):540–549
25. The history of alzheimer's disease," alzheimer's disease decoded, pp 121–138, Oct 2016

Author Index

A. Kumar and S. Mozar (eds.), *ICCCE 2018*,
Lecture Notes in Electrical Engineering 500,
https://doi.org/10.1007/978-981-13-0212-1

Printed by Printforce, the Netherlands